这里全部是快乐的记忆……

20110912

思纳史密斯(集团)中国
SMITHGROUP cna
Architecture · Urban Planning · Landscape Design
Interiors Design · Art Consultant · Lighting Design
Since 1853
www.cna-group.com

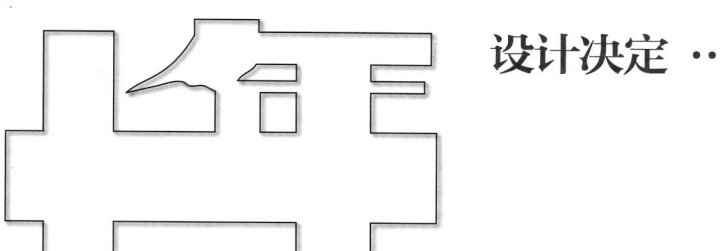

设计决定 ……

TenYears
SmithGroup CNA

Design Determines …

天津大学出版社
TIANJIN UNIVERSITY PRESS

十年······设计决定······
Ten Years·SmithGroup CNA......Design Determines…
by SmithGroup CNA
Copyright © 2012
Originally published in China by SmithGroup CNA, Publishers, Shanghai, 2012.

Executive Editor：Zhu Yijun
Author：SmithGroup CNA

Editors：Zhu Yijun, Zhang Zhentao, An Qingdong, Li Chao, Cao Jie, Wang Mei, Fu Wei, Yang Dongdong
Essays Contributors：Zhu Yijun, Zhang Zhentao, Wang Mei, Fu Wei, An Qingdong, Li Chao, Cao Jie

Translators: Li Chao, Qin Lin, Zheng Xianjiao
Photographer: Zhu Yijun, Cao Jie, Tan Weiqi etc.
Designer: C_Lucida Art

Format: 240mmX330mm
Printed Sheet: 46.75
Revision: First published in July 2012
Impression: 3,000
colour Separations: ShangHai Rui Shi Printing Co., Ltd
Printed by: ShangHai Rui Shi Printing Co., Ltd

Published in July 2012
Printed in China

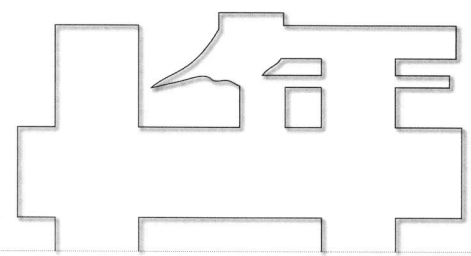

Ten Years
SmithGroup CNA
Design Determines …

Contents

Design Determines ...

设计决定……

SmithGroup CNA

As an architectural design practice and management
organization, registered in the United States,
Canada and China, SmithGroup CNA is one of the
acknowledged leading architectural design firms in the
world. The company has 3 regional offices and more than
20 subsidiary offices around the world, with more than
3,000 staff. The firm has outstanding designers in each
specialized field including architecture, engineering,
interior design, city planning, pre-phase planning,
budget estimating, art-layout architectural engineering
technology, environment analysis, building energy
analysis, project management and facility management,
etc.

SmithGroup CNA is committed to meeting different
customers' market requirements by using art and
technique method. The mixed-uses of urban planning,
residence, retail, recreation and exhibition buildings
are integrated as a creation and advocacy of a Happy
Spending design strategy that is put into practice in
various retail and planning developments. Groups
of diverse people from different countries with
considerable international project experience working in
collaboration have brought success to large mixed-use
projects.

Established in 2,000, SmithGroup CNA (China) is a
design and business supervisor and decision-maker in
China, available to provide full services in China and is
devoted to making our dreams come true.

前言 |08
Foreword |08

建筑设计部分 |019
Architecture Design |019

规划设计部分 |497
Urban Planning Design |497

景观设计部分 |669
Landscape Design |669

Foreword

Design Determines . . .
设计决定……

1	2	3	4	5	6	7	8	9	10 · · · · · ·

这里全部是快乐的记忆……

即使在以往的十年中经历那么多无助、悲伤、懊恼，也丝毫不能影响凝聚在这里的快乐、喜悦和激情。

您看到的我们就是有理想、有抱负、天天想着向上的一群人。

这里目力所及，能触及的不过是十年来的沧海一粟，如同管中窥豹，虽仅一鳞半爪，我们仍希望能由点及面，和您一起回忆过去这些年中那些有意义的项目和它们的设计过程，既以从容的心态重温往事，铭记那份宝贵的心境与体验，更用客观的目光再次审视这三千六百五十天的成长，以期获得不同的视角，然后，迎接更加精彩的又一个十年。

于安龙小站

All here is our happy reminiscence…

Even though we have experienced so much helplessness, sorrow, and chagrin in the past ten years, that can not affect our happiness, joy and passion cohered here in the least. We, as you see, are just a generation who have dreams, aspirations and want to make progress day by day. Here as far as our eye can reach, we can touch only a drop in the ocean over 10 years, like a peek at a leopard through a tube. Although it is only a small fragment, we still hope we can grasp the whole through points, and remember those meaningful projects and their design processes in those years together with you, review past events with an easy mood, enshrine that valuable mental state and experience, and still survey objectively the growth of these three thousand six hundred and fifty days once again, to try to acquire different angles of view, then to welcome another ten years of more magnificence.

朱轶俊 ~ 思纳史密斯（集团）中国 (SmithGroup CNA) 中国区执行总裁

两个问题

建筑设计在今天的中国有着相当重要的地位，从政府到公众都给予了这个行业超过历史上任何时期的重视和关注。虽然多数设计师依然延续着"匠人"的命运，但是必须承认我们赶上了一个好的时代！然而，我们还面临着两个更重要的问题，即使现在还没有找到答案，看到问题也总比对问题视而不见要好。

一、无论大家认为建筑设计该从哪里出发，最终都必须实现建筑的基本功能和使用需求，也必须去解决技术的可行性问题。当然，得承认建筑设计的变革可能会给生活甚至社会带来革命性影响，但是作为一个设计师在决定"变革"之前，应充分地去研究功能、需求和技术，因为建筑设计不是"私人"作品，它和绘画、雕塑不一样，建筑的公共属性决定了建筑无法成为"纯"艺术。但是，目前整个设计界都有越来越视觉化的倾向，有的设计甚至演化成了"行为艺术"，这不仅仅和决策者及公众的价值取向有关，也清楚地反映出整个社会的深层情绪。

二、从技术角度来看，"绿色""环保""节能""科技""智能"等名词是最近这段时间被提及最多的概念，但是我们应该要知道"永动机"是不存在的，所有的先进技术都是需要付出代价的，更何况很多的新技术其实是"伪技术"甚至是技术的倒退。因此在设计当中也不应该陷入新概念和所谓新技术的陷阱，而忘记了最传统最有效的技术手段和经验。

2010 年，公司走过了整整十年，而本人也入行快二十年了，有很多的感慨和领悟，也有很多的困惑和迷茫。但是无论如何，作为一个设计者是需要有信念的，即使在"消费至上""娱乐到死"甚至"礼崩乐坏"的年代也要保持一份真诚、一份坦率、一份独立的思考和一颗"求道"的心。

Two Questions

Architectural design occupies a quite important position in today's China, and both the government and the public lay great store on this industry, as has never been seen before in history. Although most designers are still continuing the lot of a "craftsman", yet it must be recognized that we have met with a good era! But we are still facing two important questions, and even if at present no answers to them has been found yet, setting eyes on the questions is always better than turning a blind eye to them.

No matter whence architectural design should be starting as people may think, at last the basic functionality and utility demand, as well as the technical feasibility must be achieved. Of course, it has to be acknowledged that a reformation of architectural design may bring about a revolutionary influence on life and even on society, and yet a designer, before he decides to "reform", should adequately study the functionality, demand and technology, since architectural design is not "personal" work and it is also different from painting and sculpture, and the public attribute of a building determines that it can not become "pure" art. But at present, the whole design industry has the tendency of becoming more and more visualized, and some designs even become an "action art", which not only is related to the value orientation of policy-makers and the public, but also clearly reflects the whole society's deep sentiments.

In terms of technology, such terms as "green", "environmental protection", "energy saving", "technology", "intelligence", and so on, are the concepts most frequently mentioned in the recent period of time, but we should know that no "perpetual motion machine" exists, and all advanced technologies will require costs, not to mention that many new technologies in fact are "false technologies" and even a retrogression of technology. Therefore in design, we should not run into the pitfall of new concepts and so-called new technologies, or forget the most traditional and effective technical means and experience.

In 2010, the Company has journeyed for ten years, and I have also engaged in the industry for nearly 20 years, and have many sentiments and comprehensions, as well as many puzzles and bewilderments. But in any case, a designer needs to have faith, and even in an era of "consumption paramount", "recreating until death", and even "ritual collapse and music decline", and should keep sincerity, guilelessness, independent thinking, and a "pursuing-the-truth" heart.

张振涛～思纳史密斯（集团）中国 (SmithGroup CNA) 上海区总经理，施工图总监

"一年之计，莫如树谷；十年之计，莫如树木；终身之计，莫如树人。一树一获者，谷也；一树十获者，木也；一树百获者，人也。"——《管子 权修》

历经十年，思纳已经成长为枝繁叶茂的参天之木了，现在，在中国东西南北的各个大城市都能看到思纳的作品矗立起来，还有更多的标志性建筑正在一幢幢崛起，这真是一件让所有思纳人非常骄傲和自豪的事情！

历经十年，思纳已经从史密斯集团的中国代表处，发展成四百多人的大型集团设计公司，规模堪比大型国营设计院、业务范围覆盖整个中国，思纳的核心竞争力引发的是一场令人震撼的"核聚变"！

历经十年，思纳成就了一大批建筑精英，当初而立之年的年轻人如今已是沉稳睿智的资深建筑师，初出茅庐的学生如今已是独当一面的中坚力量。思纳所聚集的是一个朝气蓬勃、充满激情的设计师群体！

开放的平台是思纳发展的基础！

无限成长的空间是思纳吸引人才的秘诀！

对建筑艺术的不懈追求是思纳作品成功的动力！

思纳一定会不断进步，让我们期待思纳未来的辉煌！

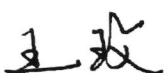

"If you want one year of prosperity, then grow grain; if you want ten years of prosperity, then grow trees; if you want one hundred years of prosperity, then you grow people. What you can harvest after one year of cultivation is the crop; what you can harvest after ten years of cultivation is the tree; and what you can harvest after one hundred years of cultivation is the talented person."

Through ten years, CNA has already grown into a mighty oak with thriving branches and profuse leaves, now, CNA's works can be seen in all major cities all around China, also more iconic buildings are still towering one after another, and this really is a great thing for all CNA people's great pride and glory.

Through ten years, CNA has already grown from the Chinese representative office of Smith Group into a large group design company with more than 400 staff, with its scale equivalent to large state-run design institute and its business scope covering all China. CNA's core competitiveness has generated an amazing "nuclear fusion".

Through ten years, CNA has brought forth a great number of architectural elites, the young people have already become composed and sagacious senior architects now, and the young and inexperienced students have already been the core strength working on their own now. What CNA has congregated is a design group full of vigor and enthusiasm!

An open platform is the foundation of CNA's development!

A space of boundless growth is CNA's secret of attracting talents!

The unremitting pursuit of architectural art is the motivation for the success of CNA's works!

CNA will keep moving, and let's expect CNA's glories future!

王玫 ~ 思纳史密斯（集团）中国 (SmithGroup CNA) 成都创作中心执行总裁

成都，一座表面平静的城市。2004年，集团将大中华西南地区的分部设在这个来了就不想离开的地方，对于一个已在成都生活并成长了二十多年的"外乡人"来说，从这一刻起我开始真正按照自己对建筑的理解来构建成都的未来，还原建筑的真正思想，筑就人文的有生命的建筑。

我们的思想就像宗教一样有它们自己的纪念碑，人类没有任何一种重要的思想不被建筑艺术写在石头上，建筑与人类有着更为密切与深刻的关系，建筑的巨大的艺术容量和强烈的艺术表现力，建筑与人类心灵直接相通的抽象性所赋予的巨大囊括力，都使建筑具有不同寻常的文化意义。

无论从建筑概念的内涵还是外延看，我们都不能仅仅把对它的理解停留在普通房屋的层次上。从其内涵来看，建筑与房屋的概念有着巨大的差异，建筑的内涵不仅包含了房屋所指代的所有对象，也显示了比"房屋"这一概念要丰富得多的内容，因为它在满足人的生存需要的同时，更包含了人对文化的追求、人的心理倾向和人的审美趣味等精神内容。从外延来看，"建筑"所指的范围要大于"房屋"所包含的范围，建筑不仅囊括了房屋所指定的各种对象，还包括大量的构筑物体范畴。

当下的中国建筑界，表面上轰轰烈烈，实际上仍然处于一种被"遮蔽"的状态，这主要是因为我们把本来是外在的、作为手段与方法的东西当作了目的，遗忘了建筑设计的真正、基本的目标和方向，以及内在的人文价值与尺度，从而导致了目前建筑活动中普遍的本末倒置的追求。在这种追求下，当代建筑日益远离人的日常生活世界，日益放弃其人文价值取向而步入了物化和异化的境地。

一切建筑设计活动，无论历史的、现在的还是未来的都应当被视为一种与人的生存和生命活动直接关联的活动，我们不能将那些"显赫""重大"的建筑和建筑事件排斥于视野和关心之外，相反，它们恰恰是要赋予这些建筑和建筑活动以人的品格和属性，而不是像我们所习惯的那样，仅仅把它们当作一个个孤立的、"超凡脱俗"的、与我们和生活不相干的"建筑"现象。它的目的就是为了在建筑与人类生活之间建立起根本、内在的联系，而这一点，对于整个当代建筑文化的建构和发展是至关重要的。

要树立建筑的人文目标，要让建筑设计活动回归人的生活世界。我认为当代建筑设计要回归自己真实的人文目标，避免无意义的思想和行为，必须回归生活世界，立足生活世界。我们把建筑作为人类的一种以物质形式存在的创造物和文化现象，是我们赋予它独特的个性和品格特征，是我们使它具有了自己独特的文化内容。我们要知道建筑不是纯粹物质上的东西，建筑的延续往往是生命的延续，历史的延续和文化的延续。

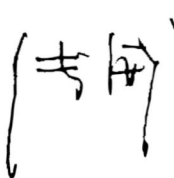

建筑设计类
Architecture Design

....

TenYears
SmithGroup CNA
Design Determines …

TenYears
SmithGroup CNA
Design Determines …

SMITHGROUP cna

Architecture · Urban Planning · Landscape Design
Interiors Design · Art Consultant · Lighting Design
Since 1853
www.cna-group.com
Design Determines . . .

设计决定······

01 公共建筑
Public Building

扬州国际会议中心暨运河论坛永久会址

Yangzhou International Convention Center & The Permanent Venue Of World Canal Forum

扬州国际会议中心暨运河论坛永久会址
Yangzhou International Convention Center & The Permanent Venue Of World Canal Forum

项目名称：扬州国际会议中心暨运河论坛永久会址
用地规模：2.36 万平方米
建筑面积：2.01 万平方米
容积率：0.85
项目地点：中国 / 江苏 / 扬州
编制时间：2009 / 02

古城是扬州凝固的历史，运河是扬州流动的文化。作为中国大运河申遗的牵头城市，扬州着手建造一个"物化"的永久性会址，这里将成为与世界运河文明对话的永久性平台。唐风辉映大运河，整个建筑群落将成为"运河名城、精致扬州"的新地标。

扬州京杭之心项目位于扬州 CBD 广陵新城一期启动区，西临京杭大运河，东临运河风情带。其中国际会议中心将成为世界运河论坛永久会址。

中外建筑师团队在漫长的设计过程中，围绕"什么才能代表扬州？"这一核心议题进行了大量科学论证，最终确立以"唐""永恒""现代""简洁"四大关键词作为设计的战略定位。在设计理念上，围绕关键词"唐"，通过解析唐文化建筑风格，科学运用传统细节和元素，力图表达建筑的历史延续之美；围绕关键词"永恒"，建筑空间及造型追求唯一性、庄重感和礼仪性，力图表达建筑的文化传承之美；围绕关键词"现代"，建筑外观运用现代手法，力图表达建筑的先进文明之美；围绕关键词"简洁"，凸显强烈、辉煌的地标建筑形象和气质，力图表达建筑的雄浑壮阔之美。

Ancient city is the solid history of Yangzhou, while canal is the flowing culture of Yangzhou. As the leading application city of Grand Canal World Heritage, Yangzhou is going to construct a "materialized" permanent venue. Here is a permanent platform for dialogue with canal civilization of the world. Tang style influences Grand Canal. The main target of the building community will be "Famous Canal City, Delicate Yangzhou ".

The Yangzhou Heart of Beijing-Hangzhou project is located in the initiating zone of Yangzhou CBD Guangling New City, Phase I, adjacent to Beijing-Hangzhou Grand Canal in the west and canal landscape belt in the east. The International Convention Center will become the permanent venue of World Canal Forum.

In the design process, Chinese and foreign architects made a lot of researches on the core issue of "What can represent Yangzhou?" and finally established four key words of "Tang", "eternal", "modern", "simple" as the design strategy. For the design concept, surrounding the key word "Dang", analyzing the architectural style of Tang culture, scientifically using the traditional details and elements, the design is trying to express the beauty of historical continuity; with the key word "eternal", in pursuit of uniqueness, solemness, and courtesy of architectural space and shape, the design is trying to express the beauty of cultural transimission; with the key word "modern", the exterior elevations use modern techniques to express the beauty of advanced civilization; with the key word "simple", the design is trying to highlight the landmark's image and temperament, and convey the magnificent and powerful feeling of the buildings.

镂空方格网
成球花图案(糖)

石材表面雕花.

在唐风、现代、后现代、抽象之间游走，你何从选择？

总体沿湖效果图

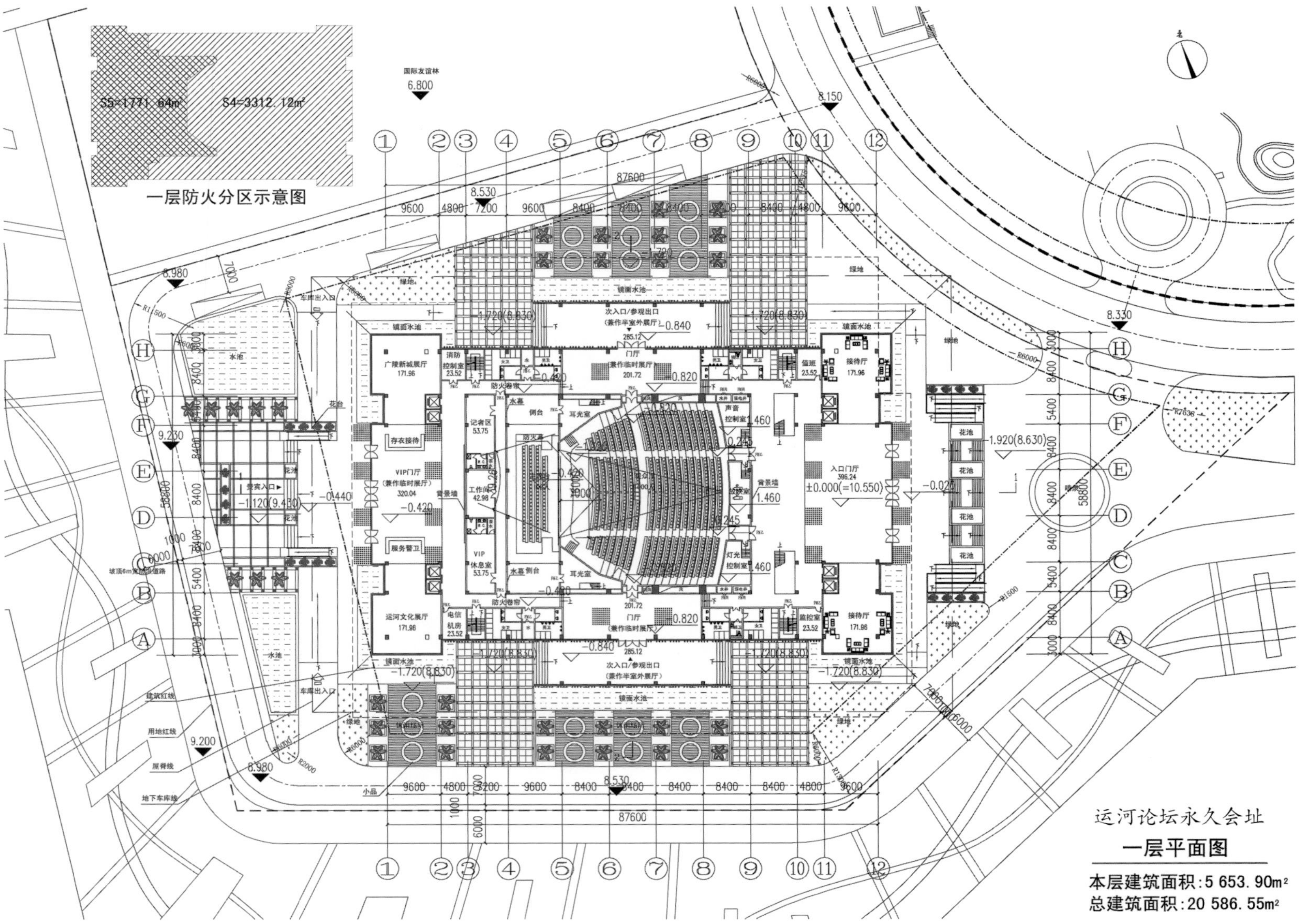

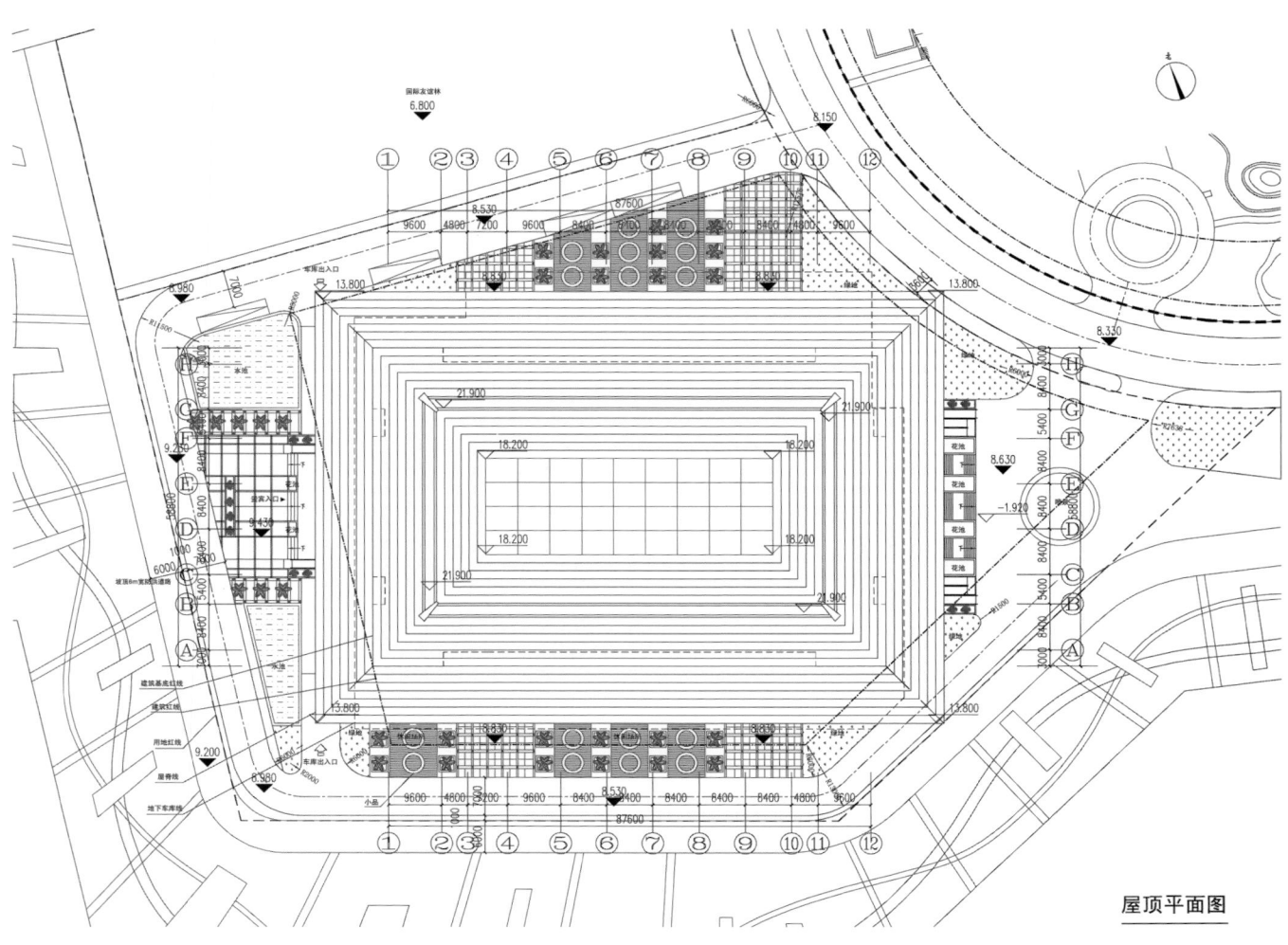

一层防火分区示意图

S5=1771.64㎡　　S4=3312.12㎡

运河论坛永久会址
一层平面图
本层建筑面积：5 653.90㎡
总建筑面积：20 586.55㎡

屋顶平面图

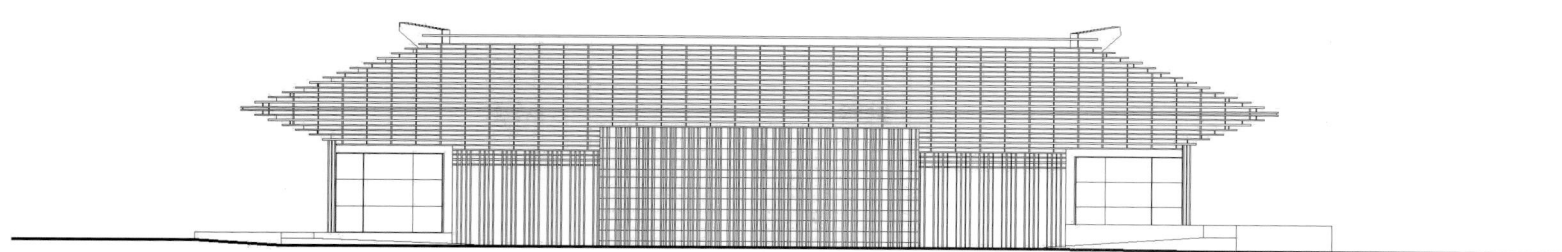

南立面图

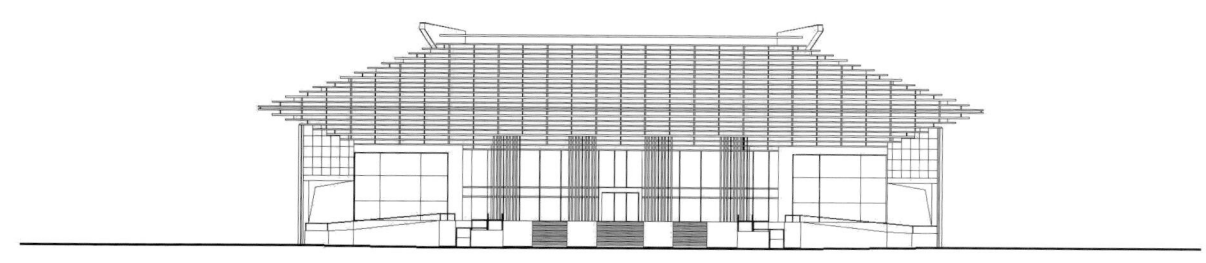

东立面图

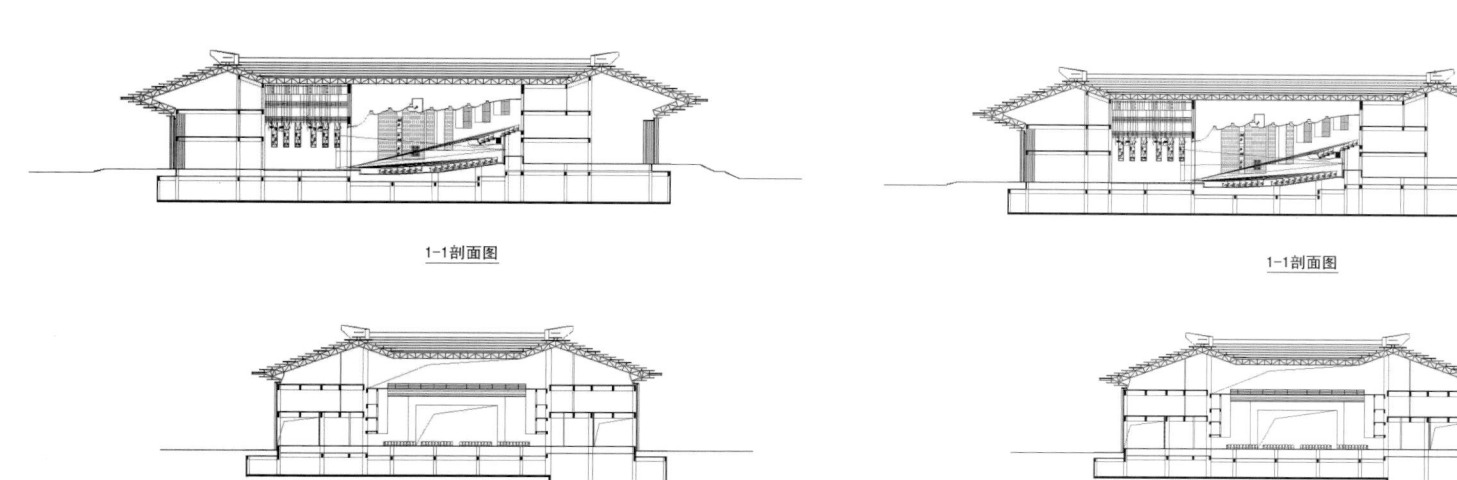

1-1剖面图　　　　　　　　　　　　　　1-1剖面图

2-2剖面图　　　　　　　　　　　　　　2-2剖面图

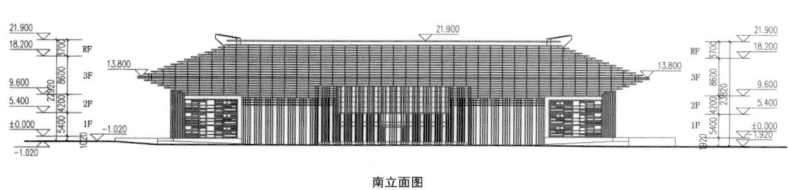

南立面图

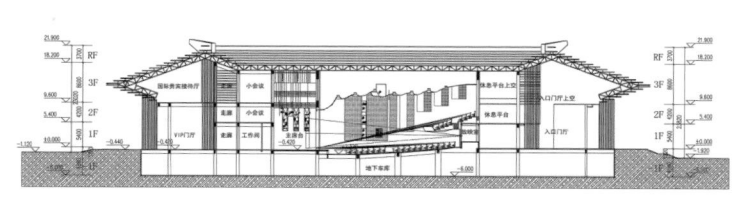

1-1剖面图

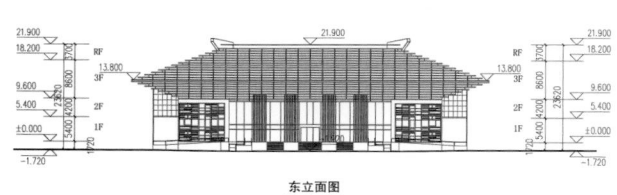

东立面图

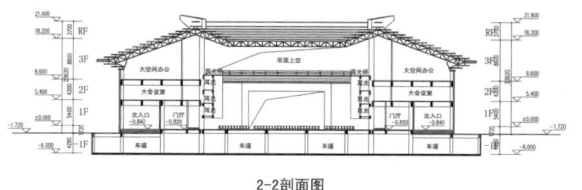

2-2剖面图

上海京剧院迁建工程和朵云轩艺术中心

Shanghai Peking Opera Theater Relocation
Project and Duoyunxuan Art Center

海京剧院迁建工程和朵云轩艺术中心
Shanghai Peking Opera Theater Relocation Project and Duoyunxuan Art Center

项目名称：上海京剧院迁建工程和朵云轩艺术中心
用地规模：1.5 万平方米
建筑面积：3 万平方米（地上）
容积率：2.0
项目地点：中国 / 上海
编制时间：2009 / 10

项目地处徐汇区龙华西路与天钥桥路交汇处，毗邻徐家汇商圈及龙华寺商圈。项目由朵云轩艺术中心和上海京剧院两部分组成，拟建成为一处集观演、博览、购物、餐饮等多种功能为一体的城市综合体。其中上海京剧院工程已经被列为 2010 年上海市重大工程。项目不仅将对新型文化商业的探索做出贡献，更重要的是将搭建一处弘扬及传播民族文化的平台。

在总体规划的创意上，方案设计用龙和凤两个中华传统图腾来比拟朵云轩和京剧院两个设计主体，又用龙凤呈祥来描绘两者的互动关系。此种动静结合、阴阳拓扑的规划理念源于对中华民族传统文化审美精髓的深刻理解。

在建筑形态塑造方面，方案利用大实大虚的对比手法，凸显了文化建筑的体量感与纪念性。实体部分采用横向条纹的材质肌理和开窗方式，隐喻中华文化积淀之深厚，而虚体量采用异型切割的手法，使得转折的玻璃幕墙面犹如宝石般晶莹剔透。整体立面造型通过此种构成手法，寓意两大国粹犹如两块艺术瑰宝从古老的文化地层中破土而出。强烈的虚实对比，动静对比，新旧对比让建筑更新锐、灵动，能够很好地反映海派文化的内在精神。

This project is located in the intersection of West Longhua Road and Tianyaoqiao Road, Xuhui District, adjacent to the Longhua Temple and Xujiahui bussiness circles. It consists of Duoyunxuan Art Center and Shanghai Peking Opera Theater, trying to build an urban complex with the functions of performance, exhibition, shopping, restaurant, etc. Shanghai Peking Opera Theater project has been listed as 2010 Shanghai major project. The project not only contributes to business exploration, but also builds a platform for the promotion and dissemination of national culture.

For overall planning idea, the design uses the traditional Chinese totems of dragon and phoenix to describe the interaction of the two buildings. Such planning philosophy containing the combination of motion and stillness, and the theory of yin and yang comes from the deep understanding of the essence of Chinese traditional culture.

For building's shape, the project uses a large amount of structural and decorative elements to highlight the volume and memorability of the buildings. For structural part, using horizontal stripes texture structure and open windows to metaphorically describe the deep Chinese culture, while for the decorative part, the abnormity cutting through the virtual volume makes the turning point of the glass curtain wall surface like a jewel-like crystal. By such means, the design indicates two national cultural heritages as two pieces of art treasures just breaking out of cultural soil. Strong contrasts between the virtual and the real, motion and stillness, the old and the new make the architecture more brilliant, which can reflect the inner spirit of Shanghai culture.

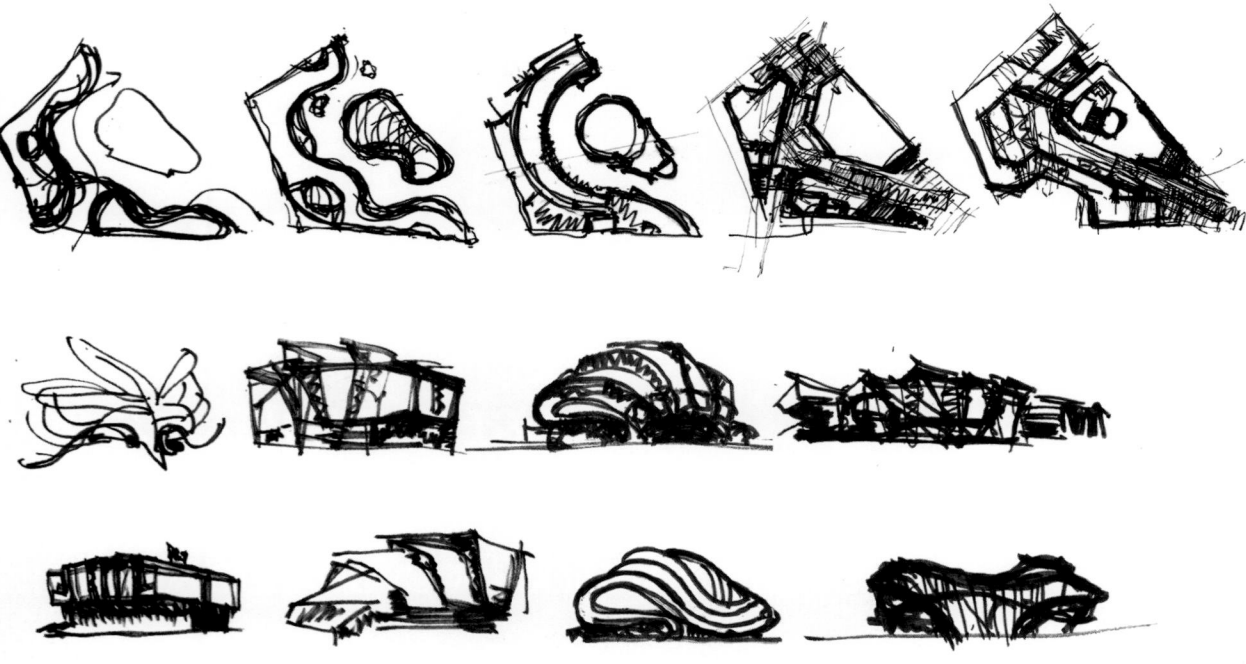

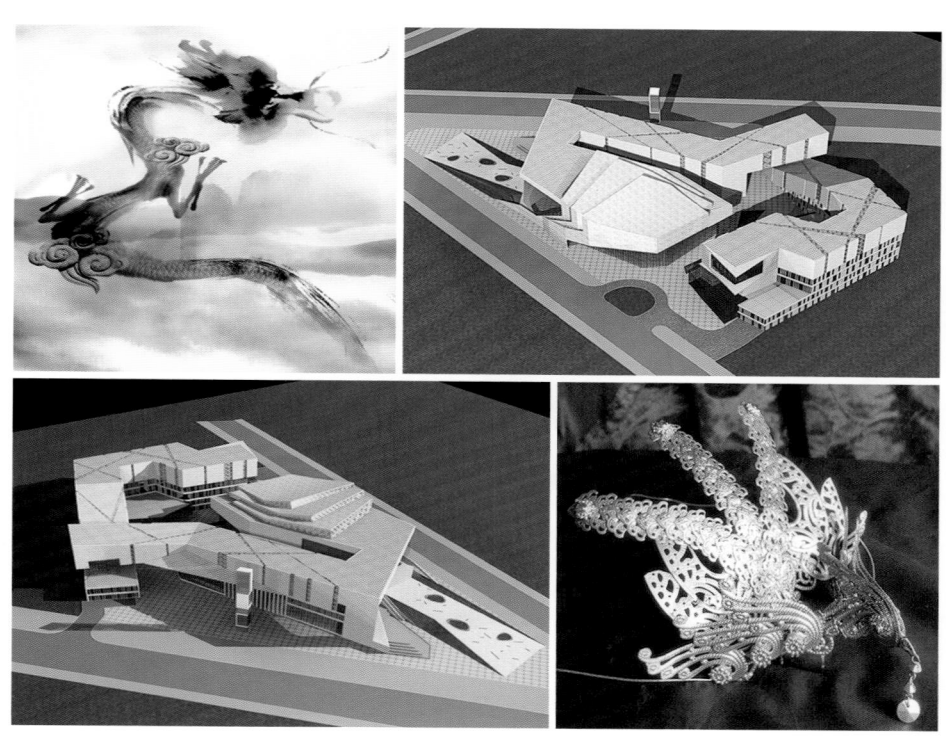

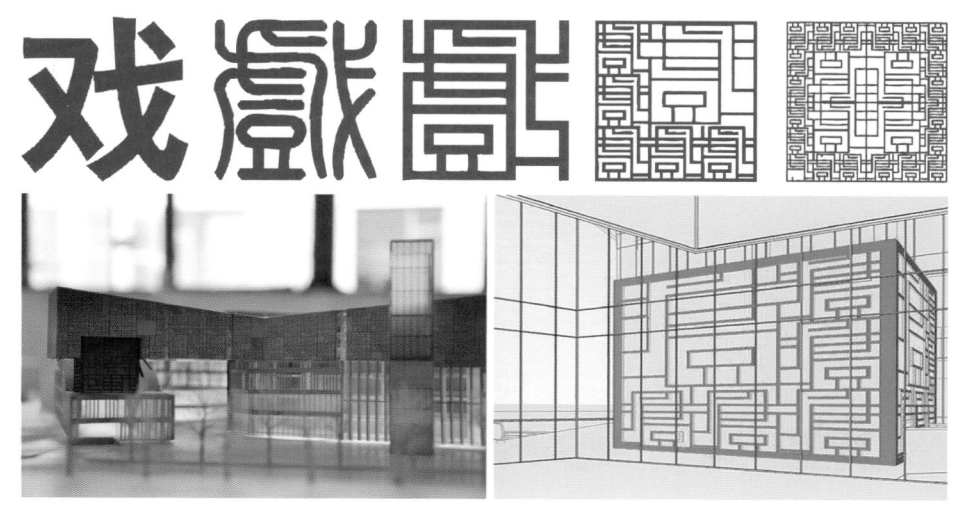

总平面图

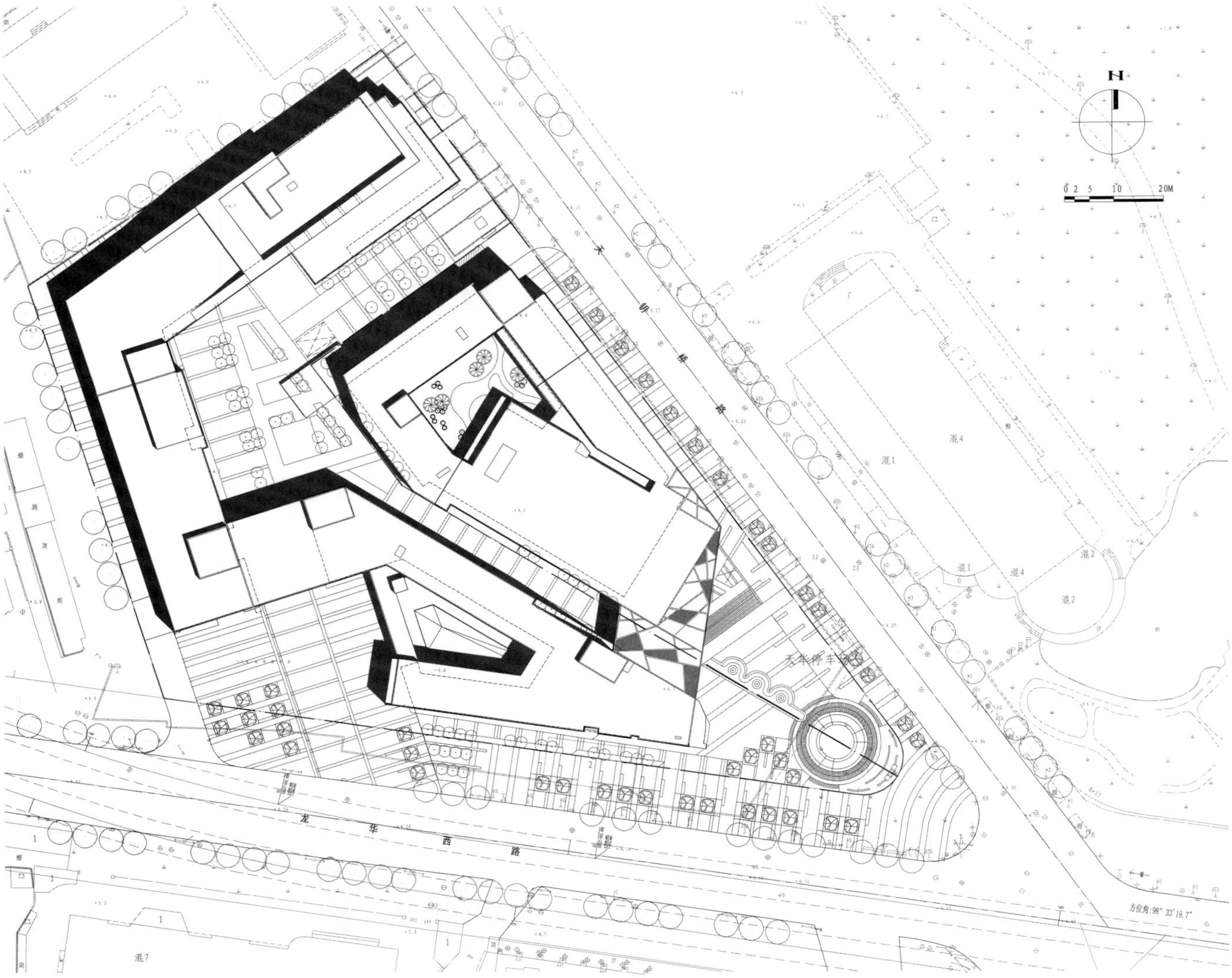

一层平面图

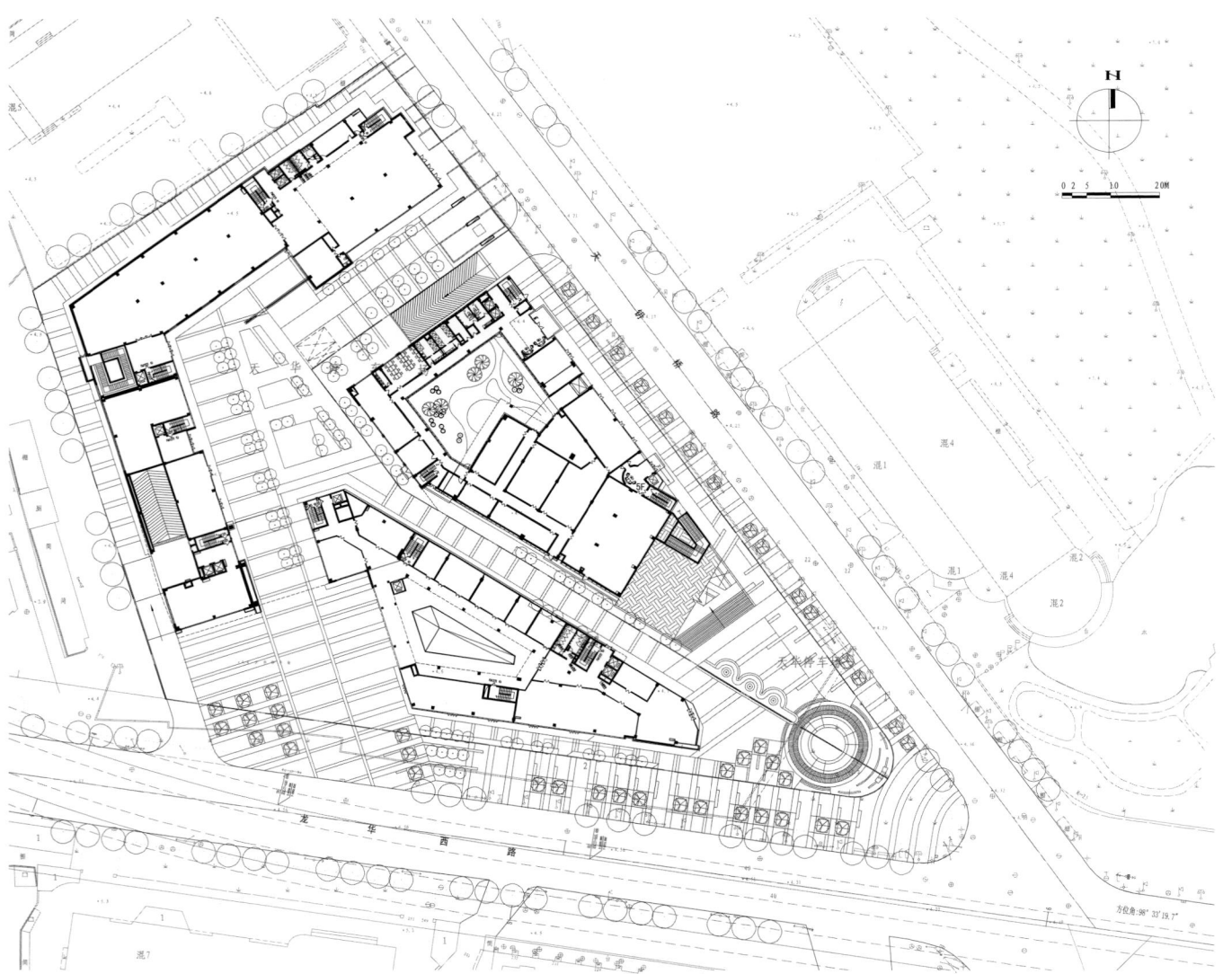

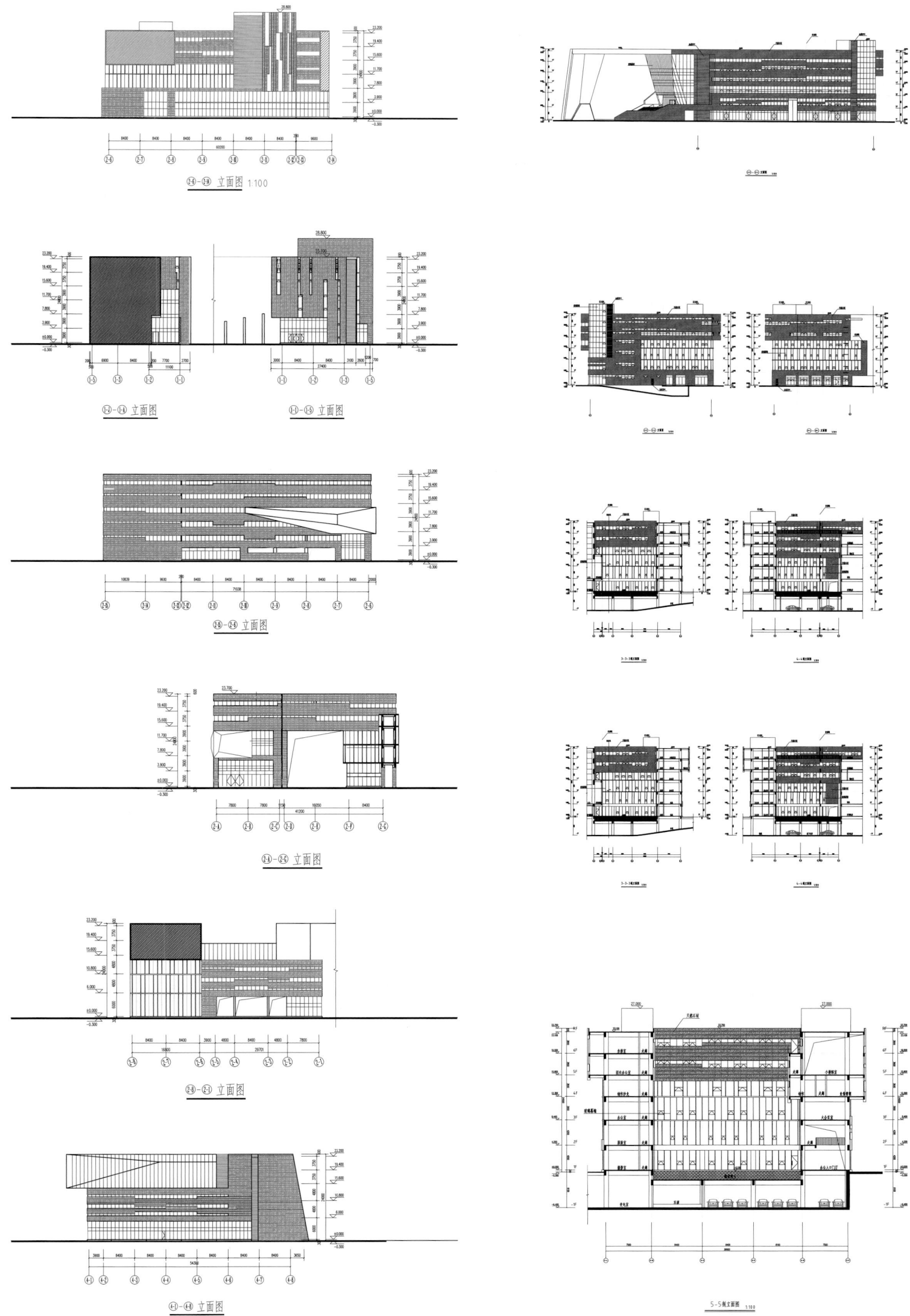

哈尔滨医科大学附属第一医院群力分院
The No.1 Affiliated Hospital (Qunli Branch) of Harbin Medical University

项目名称：哈尔滨医科大学附属第一医院群力分院
用地规模：3.44 万平方米
建筑面积：8.06 万平方米
容积率：2.34
项目地点：中国 / 黑龙江 / 哈尔滨
编制时间：2009/05

项目位于哈尔滨市群力开发区的中心位置，是集医疗救治、康复疗养、健康体检及医学教育于一体的综合性甲级医院，项目分门（急）诊和住院部两部分。

设计理念

1. 体现天人合一的生态设计原则，营造园林化的休闲医疗环境

通过建筑布局的错动与围合，形成大小不同的院落、庭院空间，建筑、环境、人三者相互协调，和谐共生；通过一条医疗轴线串联不同部分，达到设计构思与哲学思想在空间序列上的交融互动，结合外部滨河景观，营造丰富、多元、安静、舒适的疗养空间。

2. 体现传统与现代的结合

通过对哈尔滨城市建设历史与新区建设目标的理解，运用现代建筑的语言与设计理念，表现新区的建筑特色，实现现代建筑与传统建筑的结合，作为群力医院标志性建筑形态的体现，创造具有特色的城市空间。

3. 以人为本的设计思想

全面考虑医院管理者、医生、病人、家属等各类人群的需求，既充分体现和努力创造一个有利于医生治疗病人身体疾病的理想场所，也能提供给病人安抚内心的舒展空间。

4. 和谐的城市空间序列

医院门诊楼以弧形的展开面沿着城市道路交叉口展开，以"虚"的姿态迎合城市界面，综合楼与住院部、门（急）诊楼围合出一个中心庭院，面向城市，将开敞的景观向城市空间渗透，创造一个崭新的医疗形象和人文环境。

The project is located in Harbin, center of Qunli Development District. It is a comprehensive hospital with medical treatment, convalescence, physical examination and medical education and it is divided into clinic (urgent) and in-patient departments.

Design Concept

1. It Reflects Eco-design Principles, Creating leisurely medical Environment

Through closure of the layout, forming courtyards of different sizes, three elements of building, environment, and people are harmoniously coordinated, reaching symbiosis; through a medical axis, different parts have been connected, combining design idea and philosophy with external riverside landscape, creating a quiet and comfortable recuperation space.

2. The Combination of Tradition and Modernity

Through the understanding of the history of Harbin city and construction objectives of the new district, we use modern architecture languages and design concept to represent the new architectural features, combining modern architecture and traditional architecture, creating a distinctive urban space.

3. Humanist Design Idea

Considering the needs of hospital administrators, doctors, patients, family members, the design tries to create an ideal environment for treatment, as well as a comfortable space for people to relax.

4. Harmonious City Space

The hospital out-patient service building is of arc shape. In-patient department building and urgent clinic building enclose a new courtyard which faces the city. The open landscape creates a brand-new medical image and human environment.

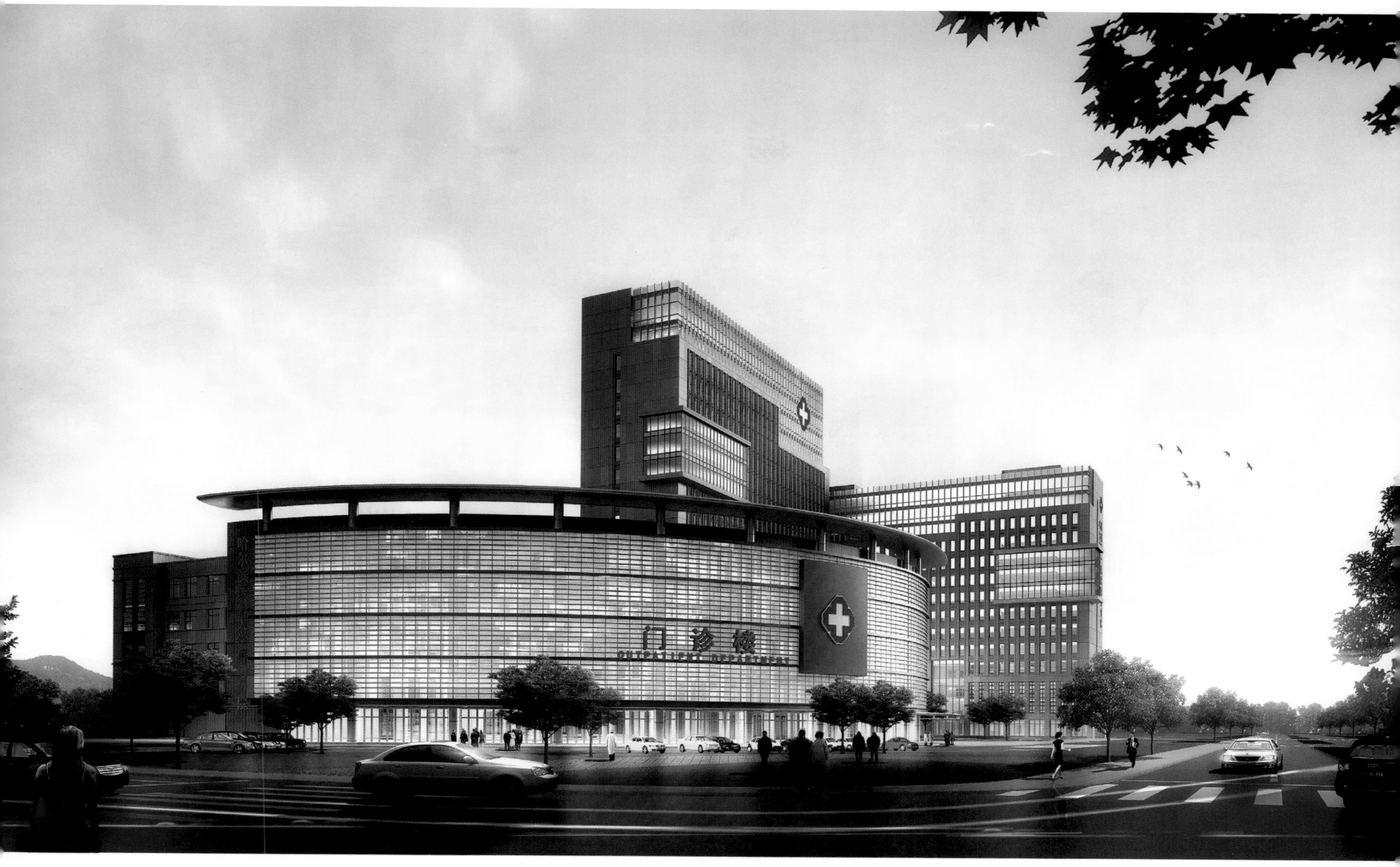
门诊楼
OUTPATIENT DEPARTMENT

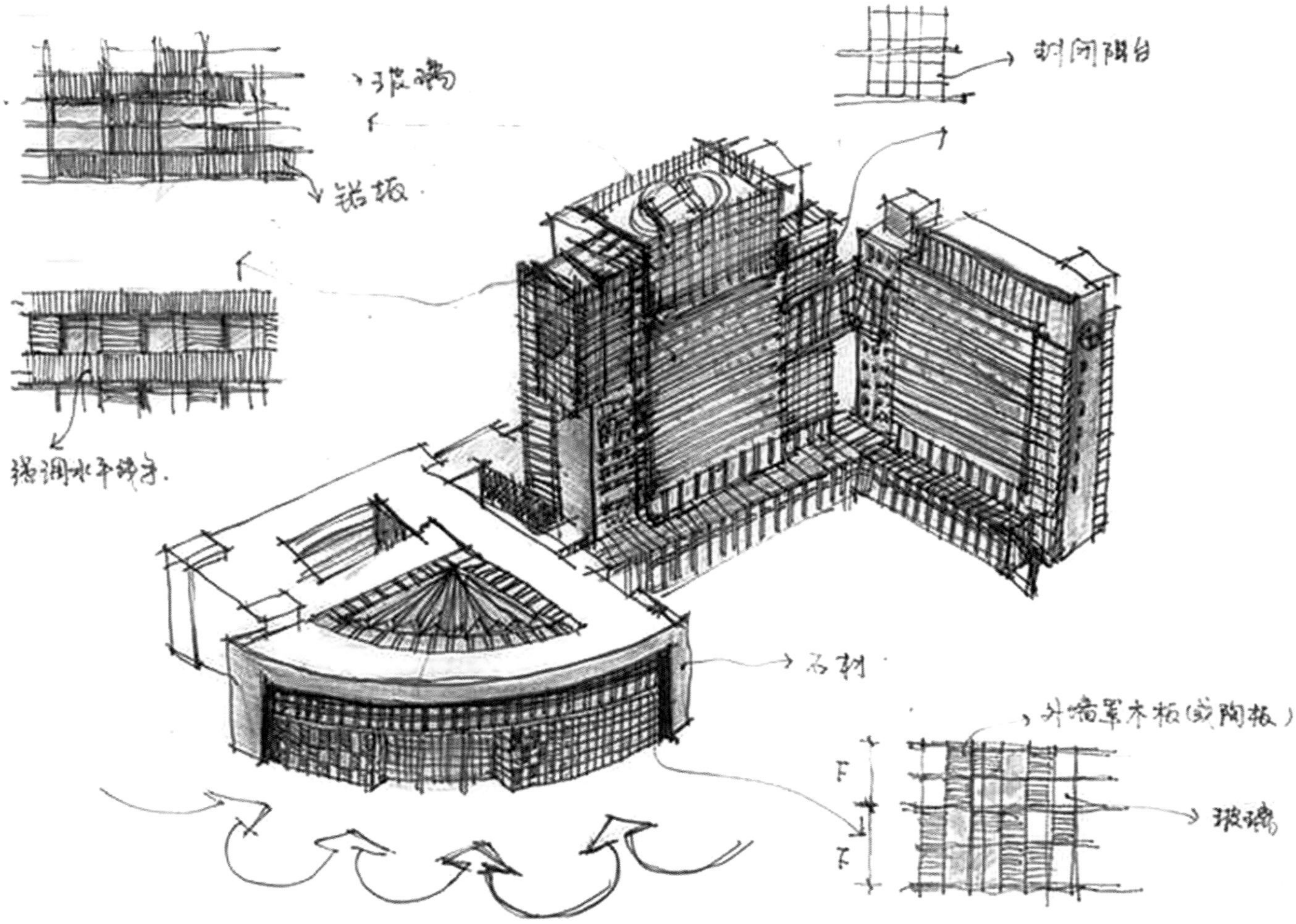
玻璃
铝板
封闭阳台
石材
玻璃

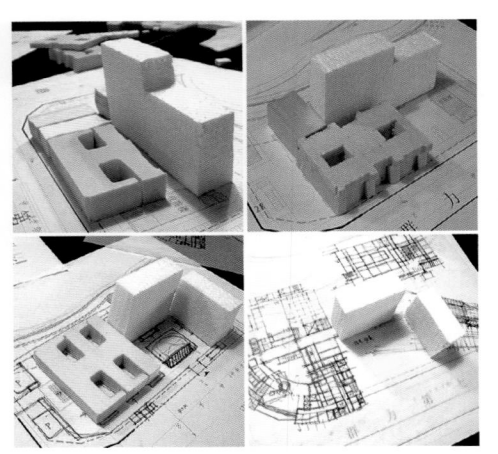

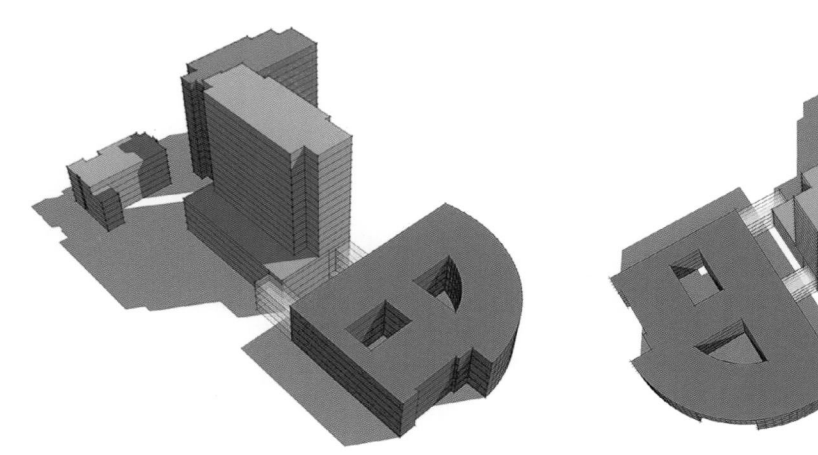

a. 规划演变

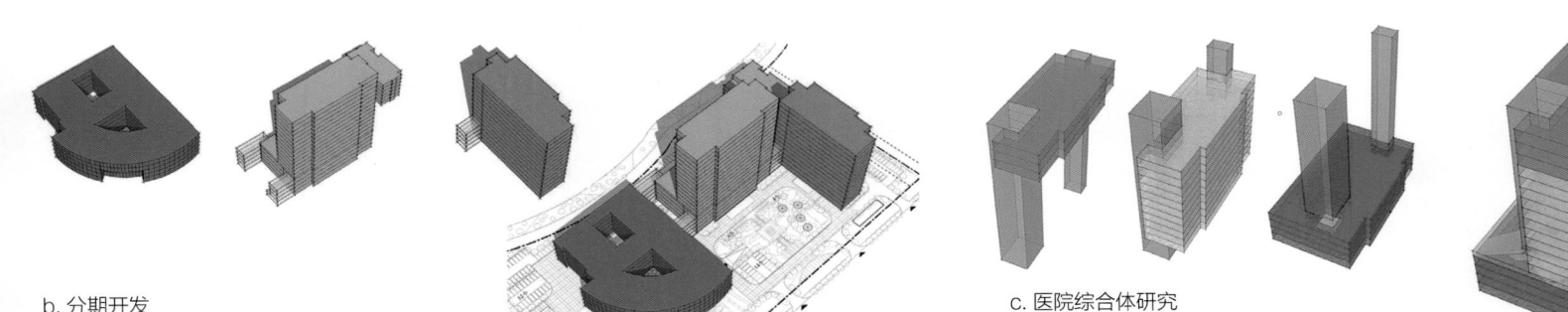

b. 分期开发
● 一期工程 ● 二期工程 ● 三期工程

c. 医院综合体研究
● 行政办公 ● 护理单元 ● 医技中心 ● 垂直交通

c. 功能分析
● 体检中心 ● 门诊 ● 急诊 ● 医技部 ● 住院部
● 行政办公 ● 专家公寓 ● 垂直交通

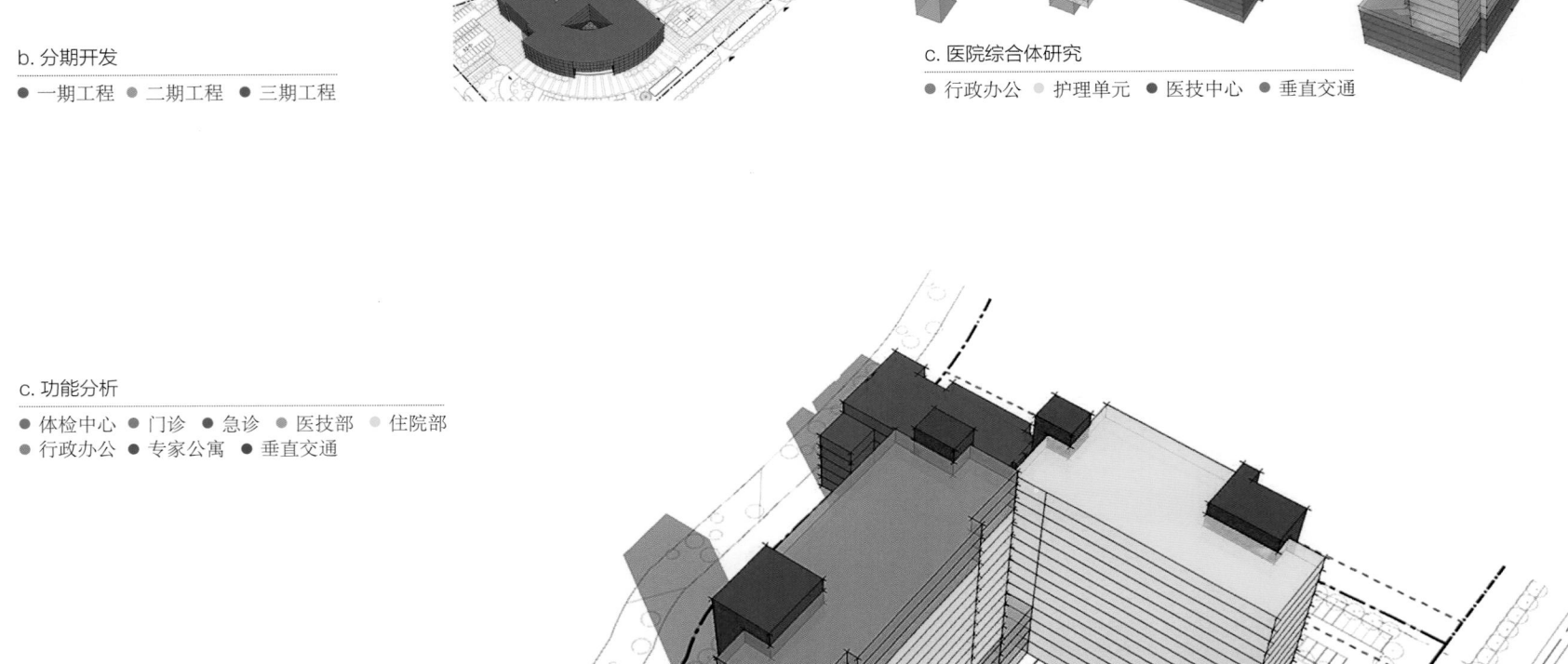

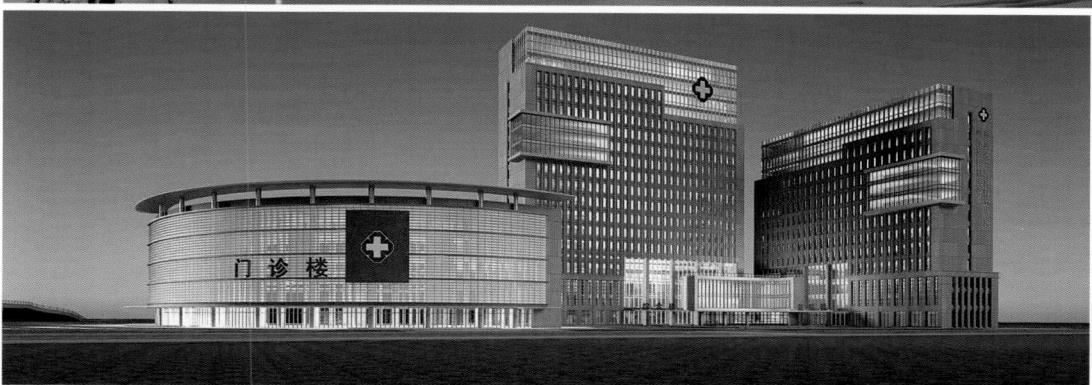

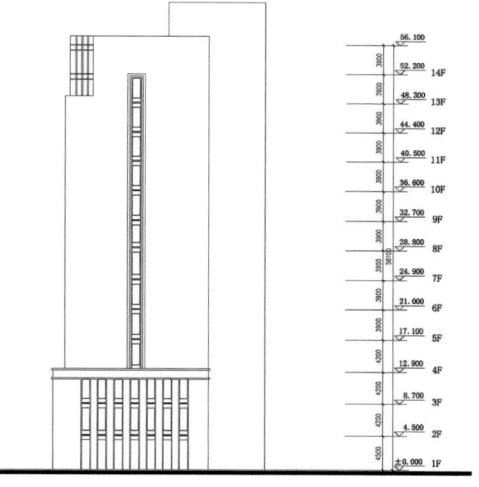

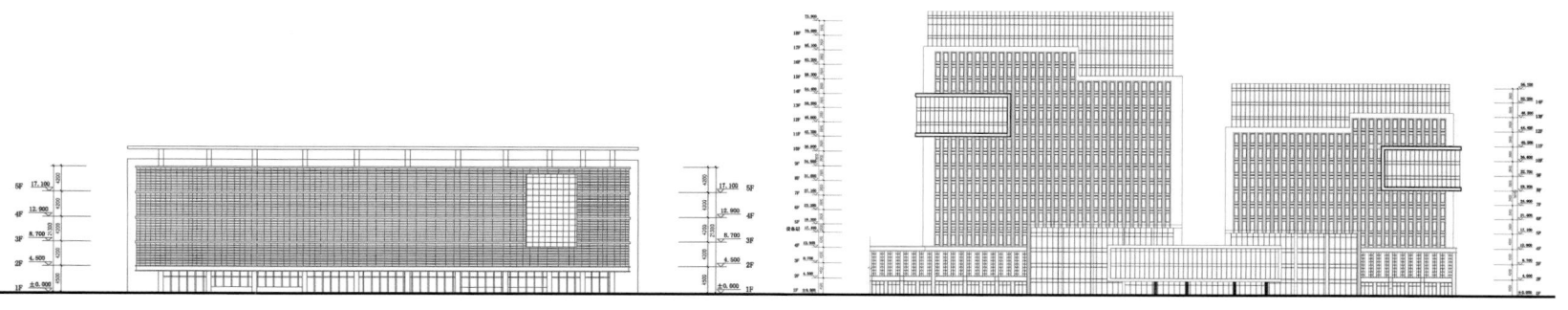

吉林大学第一附属医院扩建工程
Extension Project of No.1 Affiliated Hospital of JiLin University

项目名称：吉林大学第一附属医院扩建工程
用地规模：7.88 万平方米
建筑面积：13.59 万平方米
容积率：1.72
项目地点：中国 / 吉林 / 长春
编制时间：2006 / 09

本项目所在地位于长春市中心。由于基地处于城市中央的特殊位置，清晰地体现如何尊重和平衡新建筑与周围现有历史性建筑和景观条件之间的关系是至关重要的。基地的独特位置还表现在东北面与城市主要的中央广场相邻，南面与步行公园相接，为其增加了活泼而生机勃勃的双重性格。我们的设计旨在与环境相呼应并保持广场和公园在构成中的主导元素。此处人们感受到的是一种风格和特质的过渡，从广场上严谨的建筑所体现出的传统、庄重和对称的平衡渐而过渡到公园中交错而亲切的树荫空间所散发的随意而又休闲的有机和谐。

项目的美在这里被赋予了双重定义。这种双重性源于独一无二的基地条件。一方面我们拥有的是广场和历史建筑的庄重，另一方面是公园的活泼性和随意性。在设计中这种双重性得到了充分的反映。于是我们考虑到庄重的建筑和广场的线条、比例、规模和材料，与此同时在面向公园的一侧采用了活泼而又自然的形式。形式、材料、细节变化、功能上的结合，与自然的和谐、光与影的交映最终创造出这个能与美形成共鸣的设计。

从拟人化的角度，建筑融合了阳刚和阴柔之美。建筑面向广场的结构（外科和门诊）具有的是阳刚、坚毅等父亲的特征；而面向公园的病房闪烁的是阴柔而又温柔的母性光环。消除隔离感的同时，希望病人能体会到那份温馨，就仿佛住在浮于苍翠之上的空中楼阁里俯瞰公园。

（空中景观从公园到广场再到街道）对项目整体构成的考虑将不仅仅局限于基地红线之内，而是密切联系城市文脉和整体环境。设计对如何将项目构成融于环境作了着重的考虑，进而从广场、街道和公园等不同的角度进行了周密的思考。本设计的精髓来自换位思考，即当一个人真实地穿越和移步于设计建筑中所持的观点。

以呼应独一无二的基地条件为前提，设计的意图在于通过各项目功能之间的和谐而营造出与环境的互动，并与相邻的基地相照应，从而与整个城市融为一体。

This project is located in the center of Changchun City. Due to its special position, how to respect and balance landscape relationships between new buildings and existing historical buildings is essential. Unique position also reflects its connection with the main central square and the park, which increases the lively and vibrant character. Our design aims to connect its rhythm with the park and the square. Here you get a transition of style and character, from the tradition, solemness, and symmetry reflected by buildings to the organic harmony expressed by the shades in the park.

The beauty of the project is endowed with double definitions. This duality comes from a unique base condition. On the one hand, what we have is a solemn feeling of square and the historical buildings; on the other hand it is vividness and randomness of the park. This duality is adequately reflected in the design. So we take into account the stately buildings and square lines, proportion, size and material, and at the same time on the other side take the lively and natural form. Forms, materials, detailed changes and functional combination, accompanied by natural harmony and reflection of light and shadow create this nice design.

From the perspective of personification, the building combines masculine and feminine beauty. The structure of the building faczing the square has fatherly features, while the hospital wards are feminine. Eliminating the sense of isolation, we hope the patient can feel the warmth and overlook the park from the attic in the sky above greenery.

(Aerial view from the park to the square and then to the street) The overall consideration of the project is not just limited by the base limited line but closely related to urban context and the overall environment. Focusing on the correspondence between the project construction and the environment, the design makes close consideration of the square, street, park and others. The essence of the design is empathy; that means it delivers the true opinions of one personally going through and traveling among these buildings.

Based on the correspondence with the unique base conditions, the design intends to achieve harmonious functions, interaction with the environment and the correspondence with adjacent base, thus to integrate with the entire city.

建筑的二个面表现出完全不同的特点，
朝向公园的面带有很强的侵略性，
希望更多地获得对自然景观的吸取，
而临街的一面则表现出绅士的矜持……

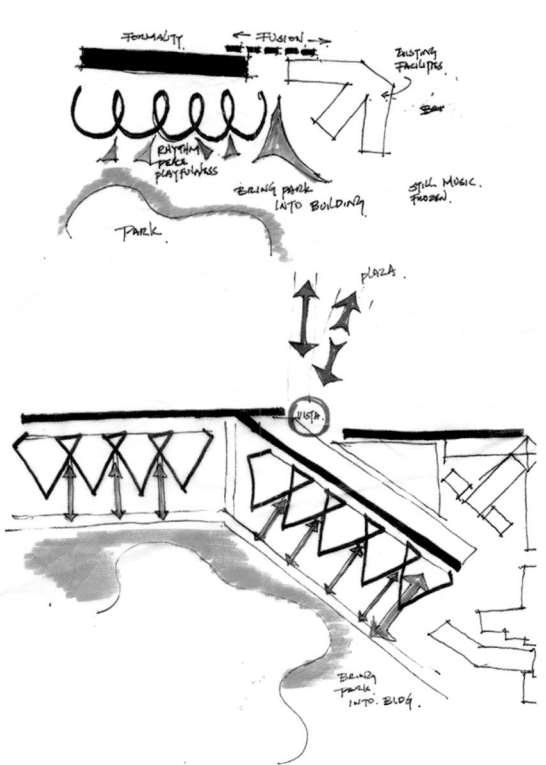

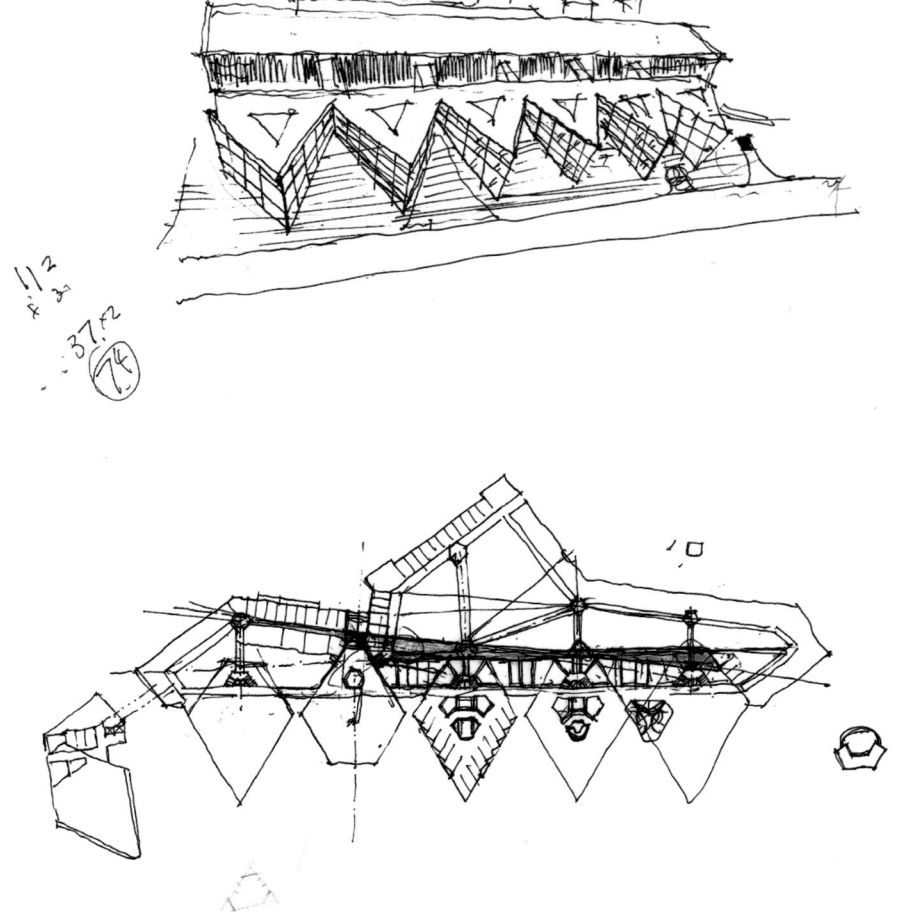

总体鸟瞰图

江苏省人民医院
Jiangsu Provincial People's Hospital

项目名称：江苏省人民医院
用地规模：12.39 万平方米
建筑面积：24.83 万平方米
容积率：2.0
项目地点：中国 / 江苏 / 南京
编制时间：2005 / 03

项目位于南京市鼓楼区西南部地域，规划在原有基础上进行增建，成为集临床、教学、科研、预防于一体的大型现代化综合医院。由于基地位于南京城中的一处上下落差极大的坡地上，总体布局采用分散的"集中式"布局，通过设计师提出的"绿色建筑""医疗大街"的核心理念，采用一条经过设计的医院时尚长廊将整个医院的新老建筑群完美地连接起来，以达到功能紧密联系、资源有效使用和节约能源的目的。

This project is located in Gulou District, Nanjing City. The plan is to build a new inpatient building, outpatient center, and medical center, as well as increase relevant ancillary facilities to build a modern comprehensive hospital with functions of clinic, teaching, research, and prevention.
Due to the huge height difference of the site, overall layout uses scattered "concentrated layout". Under the core concept of "green building" and "medical street", the design connects the old buildings and the new buildings through a fashionable hospital corridor, reaching the aim of functions in close contact, effective use of resources and energy saving.

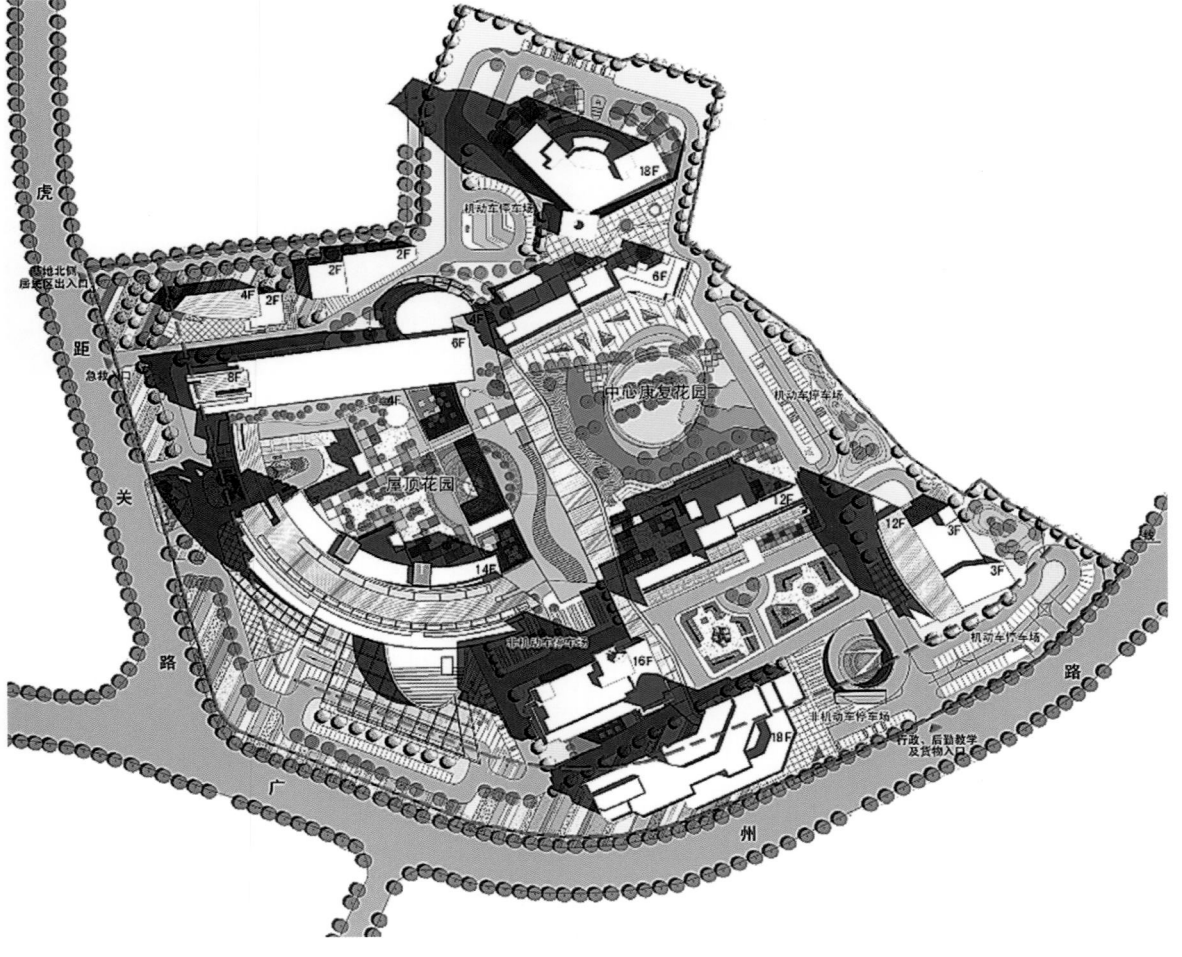

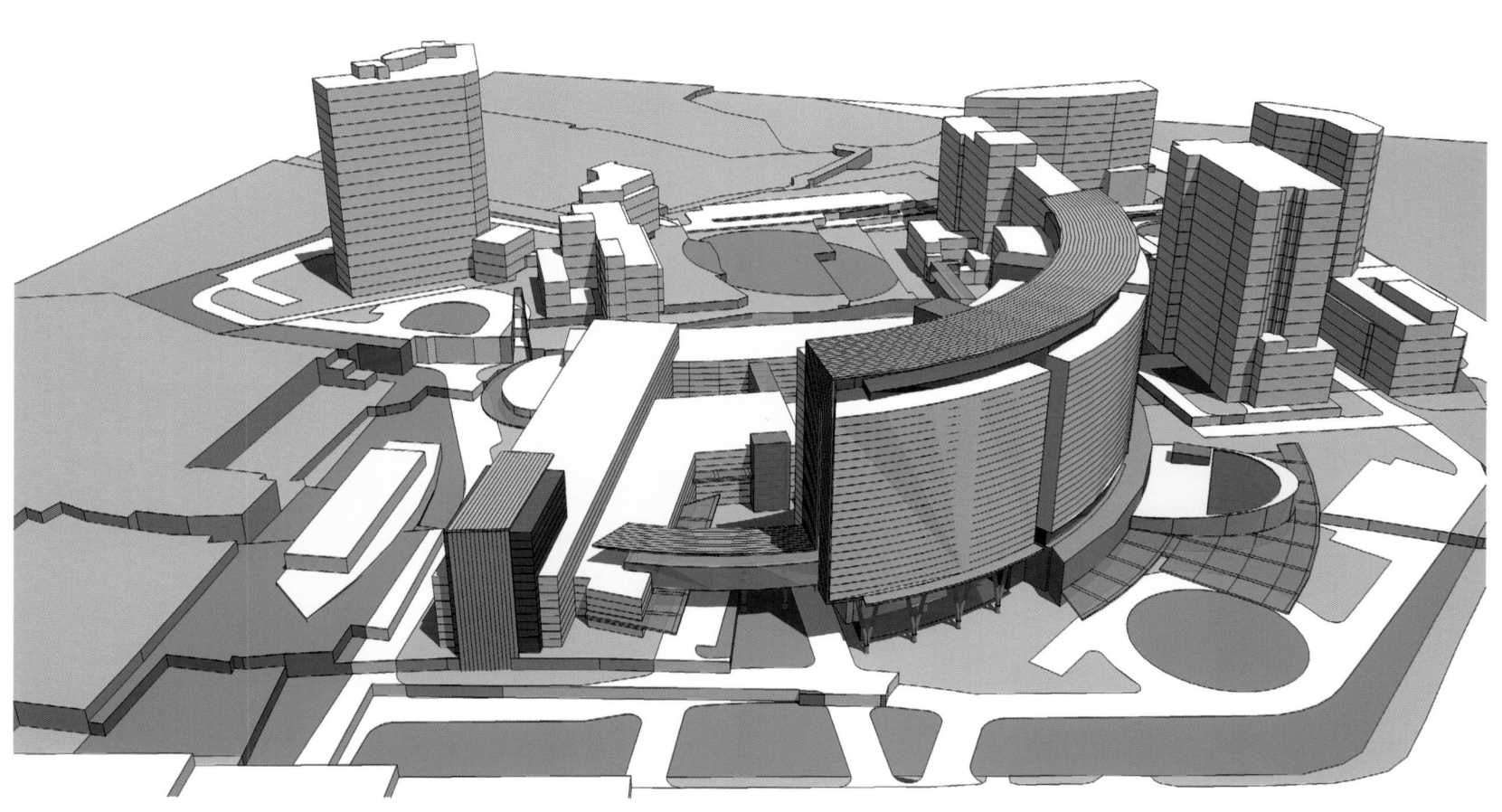

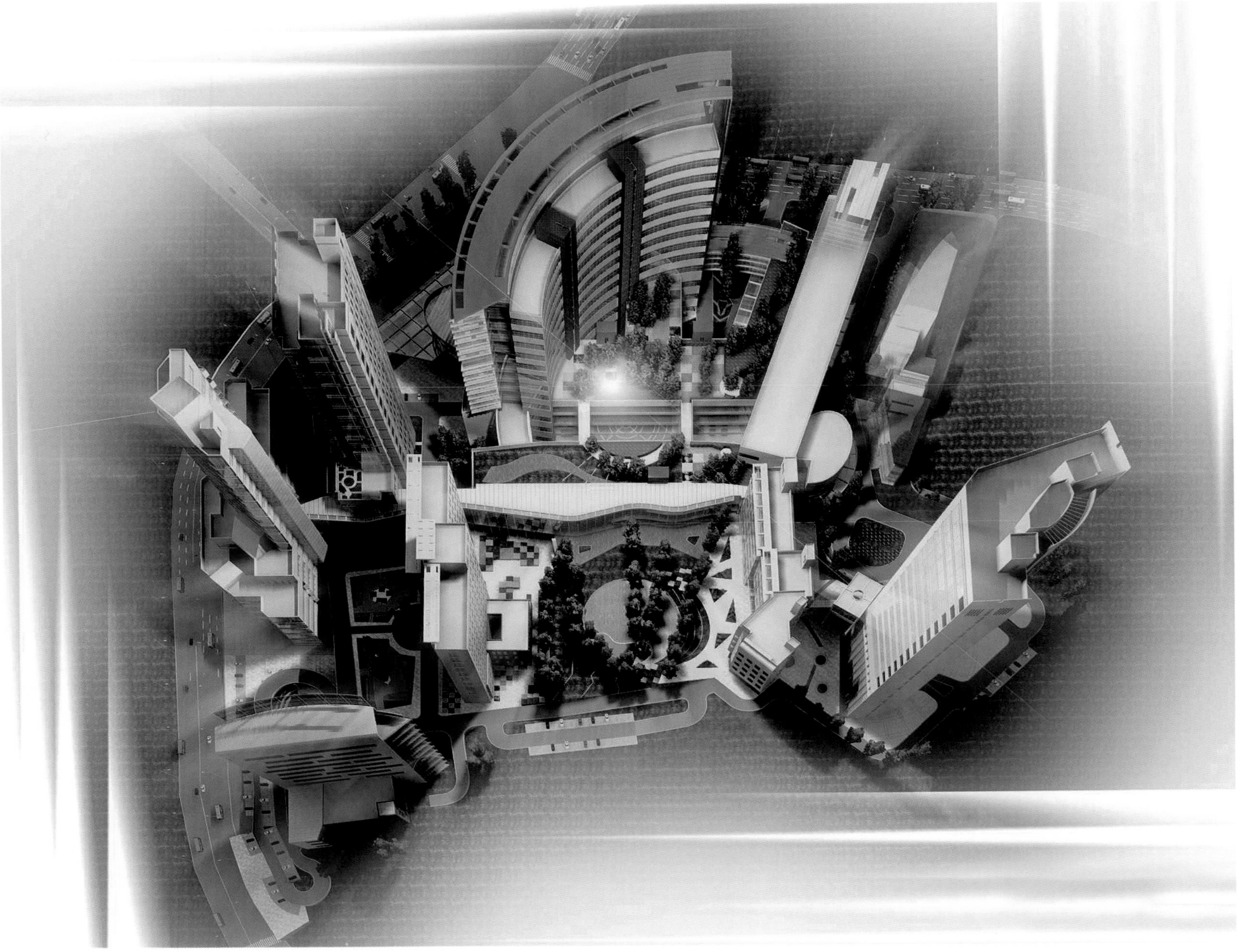

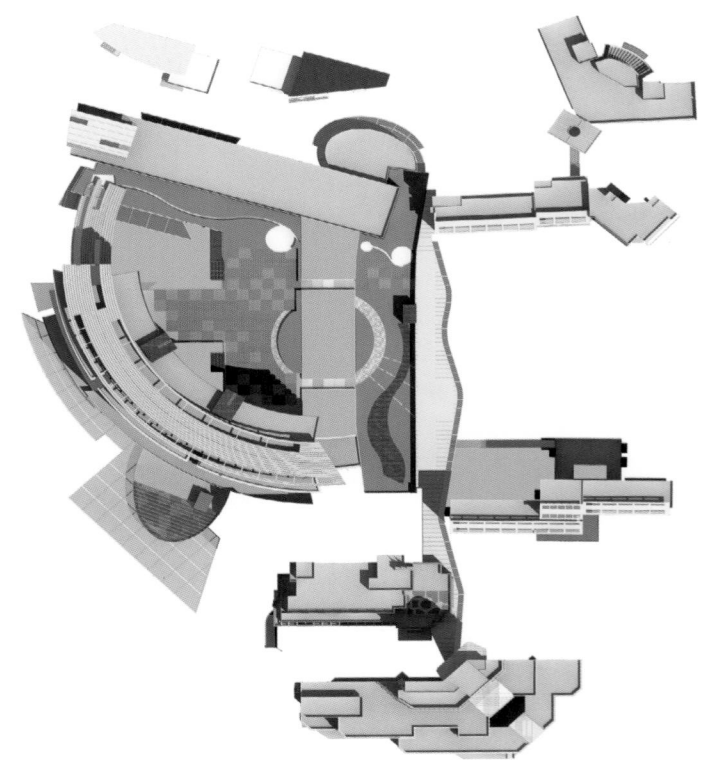

大型建筑本身和群体建筑之间都需要更多的连接来加强医院
的各科室和医技部门之间的联系……

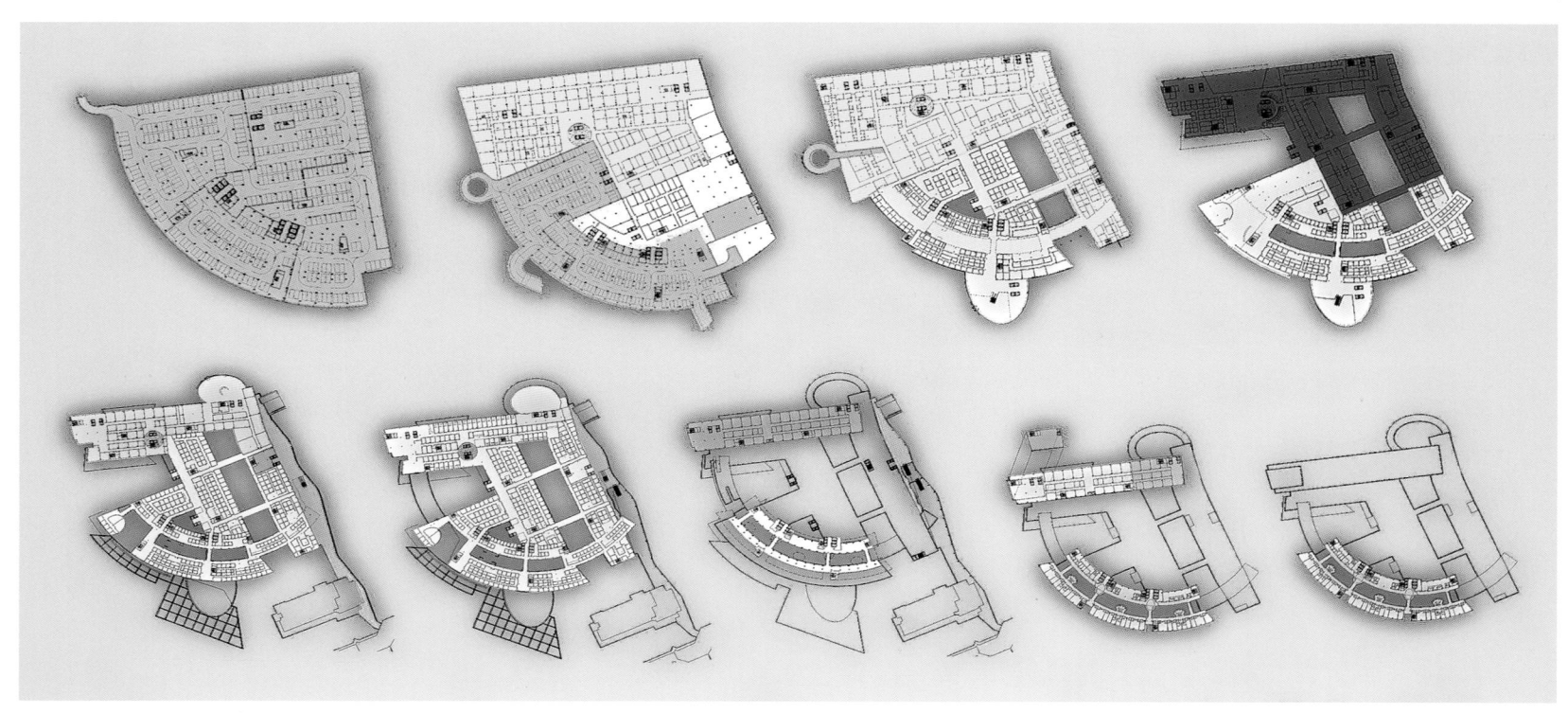

上海市东方医院改扩建项目
Reorganization and Expansion Project of Shanghai East Hospital
项目名称：上海市东方医院改扩建项目
用地规模：2.29 万平方米
建筑面积：6.07 万平方米
容积率：2.65
项目地点：中国 / 上海
编制时间：2005 / 06

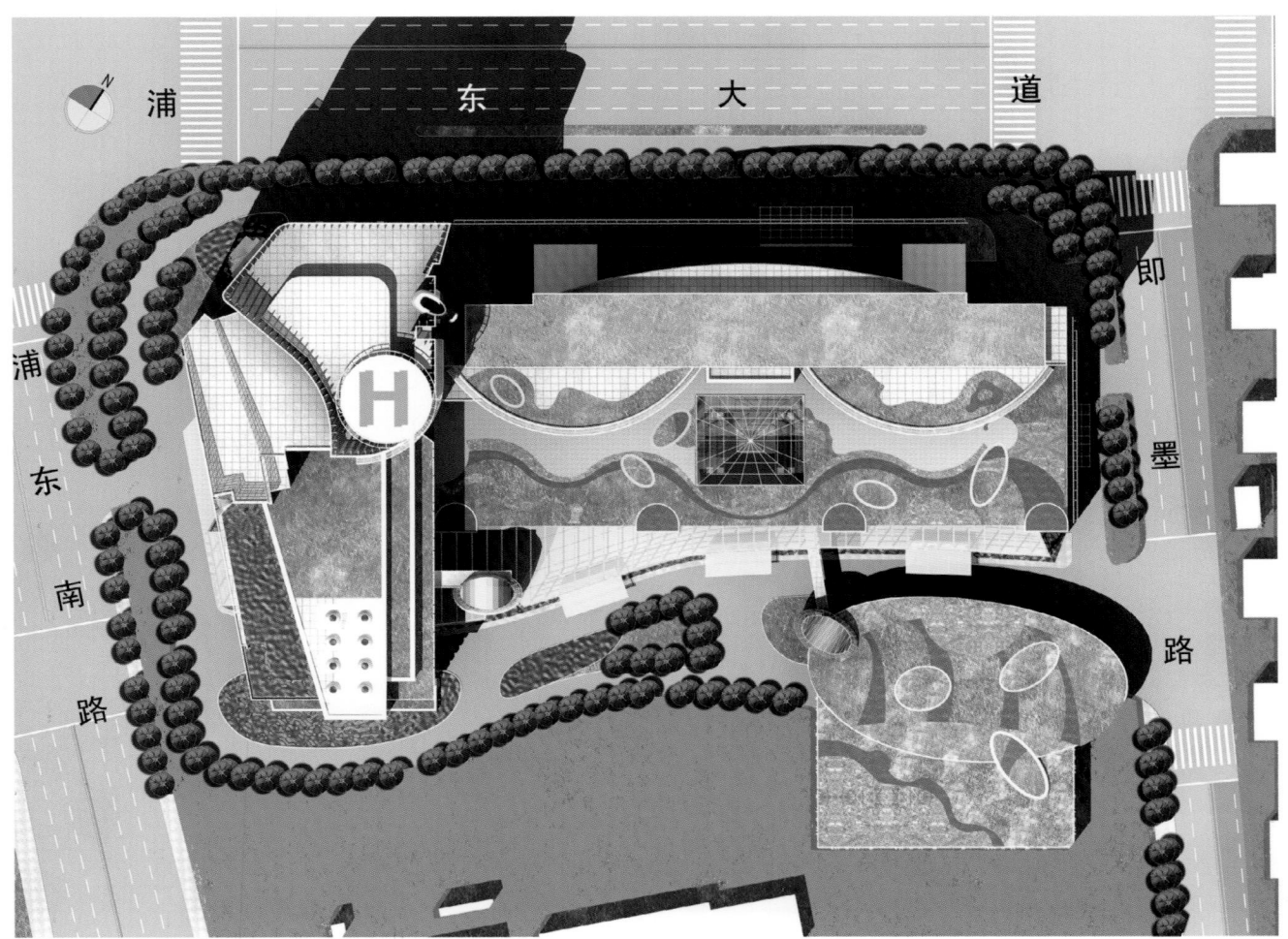

上海市东方医院是一所集医疗、教学、科研于一体的三级综合医院，医院位于上海浦东即墨路 150 号。北临浦东大道，东至即墨路，西靠浦东南路。设计上考虑并力求满足日新月异的现代医疗技术对医院建筑提出的要求，平面设计上采用比较规则的布局，尽最大可能地追求可容性、可变性；在总体规划上采用明确的功能分区，将医院的综合性科目集中起来，并且与原建筑各层的功能相衔接，以利于今后统一发展的需要，可满足医院的长期发展战略。

Shanghai East Hospital is a third-class comprehensive hospital with medical, treatment teaching and scientific research. The hospital is located on the NO.150 Jimo Road in Shanghai Pudong New District. Seeking to meet requirements of modern medical technology, the plane design uses regular layout to achieve maximum capacity and variability. The overall function partition is clear, and comprehensive subjects of the hospital have been concentrated up, and connected with the functions of the original building. This is advantageous for the unified development in the future which can meet the long-term development strategy of the hospital.

主要路口效果图

医院总体鸟瞰图

病房楼效果图

上海儿童医院普陀新院
Putuo New Hospital of Shanghai Children's Hospital

项目名称：上海儿童医院普陀新院
用地规模：2.6 万平方米
建筑面积：5.2 万平方米
容积率：2.0
项目地点：中国 / 上海
编制时间：2007 / 09

设计理念

1. 现代科技与人本主义的完美结合

新建成的儿童医院和妇婴保健院将是一个现代化的设施齐备的大型医疗机构，它将成为妇幼保健和儿童治疗、康复的场所，同时，它也是一个儿童交流、游戏、学习的场所。通过适宜的建筑尺度，富有情趣、充满活力的建筑元素以及多层次的绿化景观系统，提供一个都市里的"休疗养花园"。

2. 符合儿童心理的建筑语言

① 折纸

以儿童手工课上的折纸为主线，构建整个地块的建筑布局和立面元素，通过建筑布局的错动与围合，形成大小不同的院落、庭院空间，建筑、环境、人三者相互协调，和谐共生；随着一条"纸"带的折叠，串联起不同的部分；建筑元素上也采用折纸的直线与弧形的曲线相结合，形成丰富的具有变化的立面效果，同时也体现了儿童世界的纯真与自由和女性世界的干练与柔美。

② 万花筒

每个儿童都会被万花筒里奇妙的世界所吸引，设计灵感来源于儿时的玩具——万花筒，将万花筒里丰富的色彩抽象处理后，作为建筑语言，应用于建筑立面，形成色彩丰富的建筑效果，创造更加人性化的医疗空间。

3. 以人为本的设计思想

全面考虑医院管理者、医生、病人、家属等各类人群的需求，充分体现和努力创造一个有利于医生治疗病人身体疾病的理想场所，也能提供给病人安抚内心的舒展空间。

4. 生态长廊——空中花园与儿童娱乐场

设计采用多层次景观绿化设计，形成一个自然、生态的生活、学习、康复空间，一条长长的空中花园，作为儿童娱乐活动场所，与一层庭院分开，在减少与其他人流交叉、干扰的同时，更加体现了对患者人性化的关怀。

Design Concept

1. Perfect combination of modern technology and humanism

The newly built children's hospital will be a fully equipped large modern medical institution. It will not only be a place for feminine and children's health, children's treatment and rehabilitation, but also be an ideal place for children's communication, gaming and learning. With appropriate building scale, interesting and vibrant building elements and multi-level greenery landscape system, it is designed to be a "recreation and rehabilitation garden" in the city.

2. Architectural Language in Line with Child Psychology

① Paper folding

Inspired by paper folding on the handwork class of children, the design forms courtyards of different sizes through the scattered and enclosed layout, adopts lines and curves in paper folding to form a rich and changing facade effect, which at the same time reflects children's innocence and freedom and women's capacity and charm.

② Kaleidoscope

No child can withstand the temptation of the wonderful kaleidoscope. The design is inspired by a childhood toy-kaleidoscope. The design makes abstract treatment of the rich colors in kaleidoscope to be architectural language and applies it to the building facade which can represent colorful architectural effect and create a more humane health care space.

3. People- oriented Design Idea

The design takes into full account of demands of the hospital administrators, doctors, patients, families and other people, fully embodies and strives to create an ideal place for the work of doctors and for the inner calm of patients.

4. Ecological Corridor- Sky Garden and Children's Amusement Park

Multi-level landscaping is designed to form a natural and ecological space for living, study and rehabilitation. A long sky garden separated from the courtyard on the first floor is designed as children's entertainment venue that can not only reduce the interference by others but also reflect the humane care for patients.

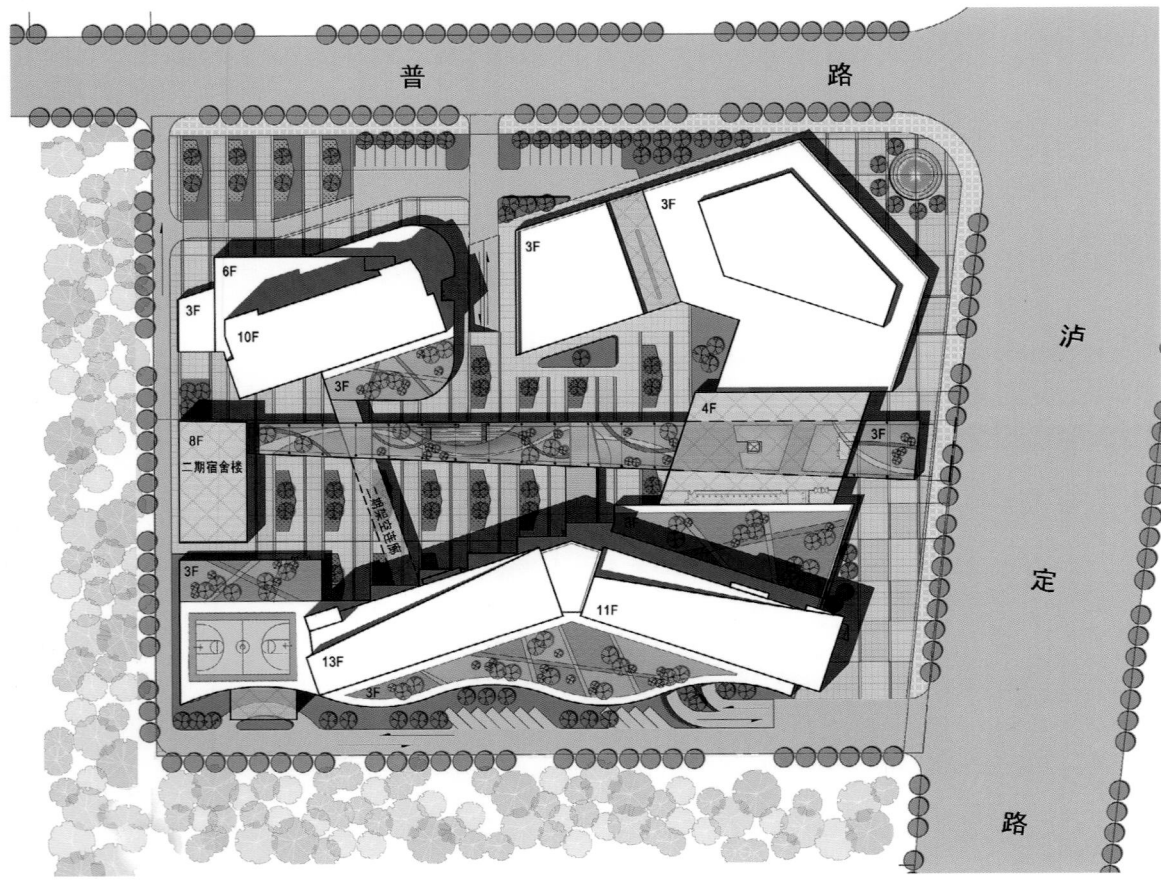

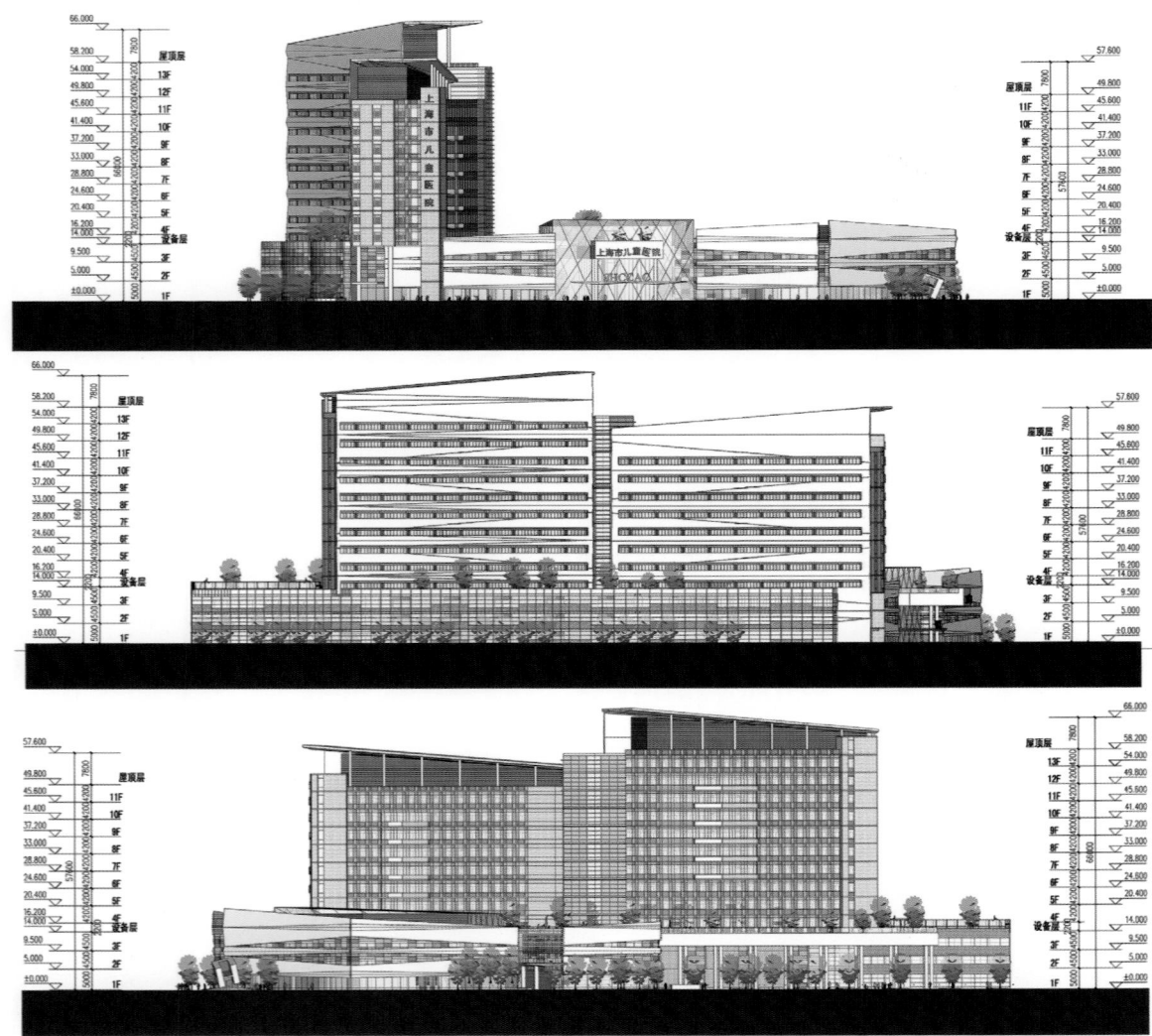

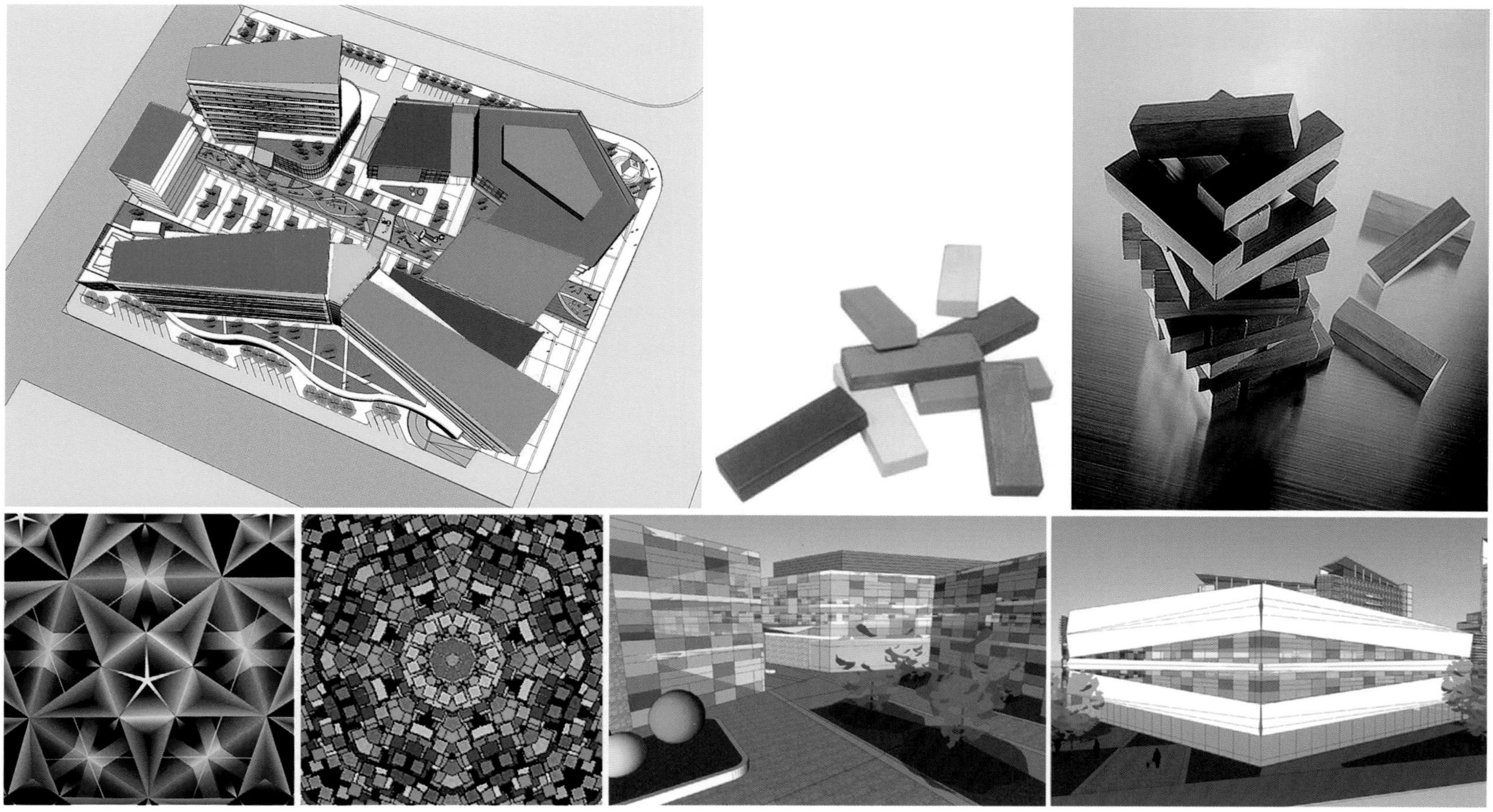

设计思想中色彩、积木和折纸等元素的运用

上海曙光医院
Shanghai Shuguang Hospital

项目名称：上海曙光医院
用地规模：10.85 万平方米
建筑面积：8.23 万平方米
容积率：0.76
项目地点：中国 / 上海
编制时间：2006 / 03

曙光医院为上海市唯一的三级甲等中医院和全国示范中医院，也是上海唯一被卫生部授予全国"百佳"医院称号的中医院，其综合实力在上海乃至全国的中医学术界处于领先地位。

设计理念

1. 体现中医哲学的思想

易曰："一阴一阳之谓道，继之者，善也。成之者，性也。"

通过对中医哲学，阴阳运动变幻思辩的研究，取基地天然之灵气，分阴阳二翼布置，而由阴、阳互继其中，达到设计构图与哲学思想在空间序列上的交融互惠的作用。

2. 体现传统与现代的结合

通过运用现代建筑材料和设计理念表现传统中国建筑特色元素的方法，实现现代建筑与传统建筑的结合，作为曙光医院中医特色在建筑形体上的重要表现。

3. 处处以人为本的考虑

全面考虑医院管理者、医生、病人、家属等各类人群的需求，既充分体现和努力创造一个有利于医生治疗病人身体疾病的理想场所，也能提供给病人安抚内心的舒展空间。

4. 和谐的城市空间秩序

医院的主楼呈弦月状，垂直于圆弧方向轴线延伸的是规划中的中医医科大学及上海中医药谷。这个城市空间序列的相互呼应隐寓着中医医科大学和药谷与医院不可分割的组带关系。

总体布局

中医药科学以传统中国哲学——易学为哲学基础，从宏观整体出发，以运动变幻思辩，取天然药食精粹，治疾患于未然。我们在设计中充分利用基地内河道通过的地理现状，将医院建筑整体与园林水景大致沿河道两侧分开布局，同时又将园林水景引入医院整体区域，将特需医疗保健中心放置在园林水景区域，中间用连廊相接，以此来达到构图和空间序列上的阴、阳交合的效果。

Shuguang Hospital is a famous traditional Chinese medicine hospital in Shanghai and it is the academic community leader in China.

Design Concept

1.Reflect the Chinese Medicine Philosophy

Through the research on Chinese medicine philosophy, and yin and yang movement changes, the design takes the natural aura of the base split by yin and yang, each following one side, combining the design philosophy and the design composition.

2.Reflect a Combination of Tradition and Modernity

Through the use of modern building materials and design concep to show traditional Chinese architectural features, it achieves the combination of modern and traditional architecture which is an important architectural reflection of the traditional Chinese medicine.

3. People-oriented Considerations Everywhere

Fully considering the needs of hospital administrators, doctors, patients, families and other people, the design tries to embody and create an environment where doctors treat patients ideally, providing comfortable space for the patient.

4.Harmonious Order of Urban Space

The shape of the main building looks like a crescent moon. The extension perpendicular to the arc axis is the planned Chinese Medical University and Shanghai Traditional Chinese Medicine Valley. This coherence of urban space indicates the close relationship between the hospital and Traditional Chinese Medicine as well as University Traditional Chinese Medicine Valley.

Overall Layout

Chinese medicine is based on traditional Chinese philosophy - the philosophical basis of Yi. Based on the geographical situation of river, the hospital building and garden are separated along the two sides of the river, introducing landscape water features into the hospital area. Special medical service centers will be placed in the water garden area, connecting through the corridor in the middle part in order to achieve the composition and spatial sequence of yin and yang effect.

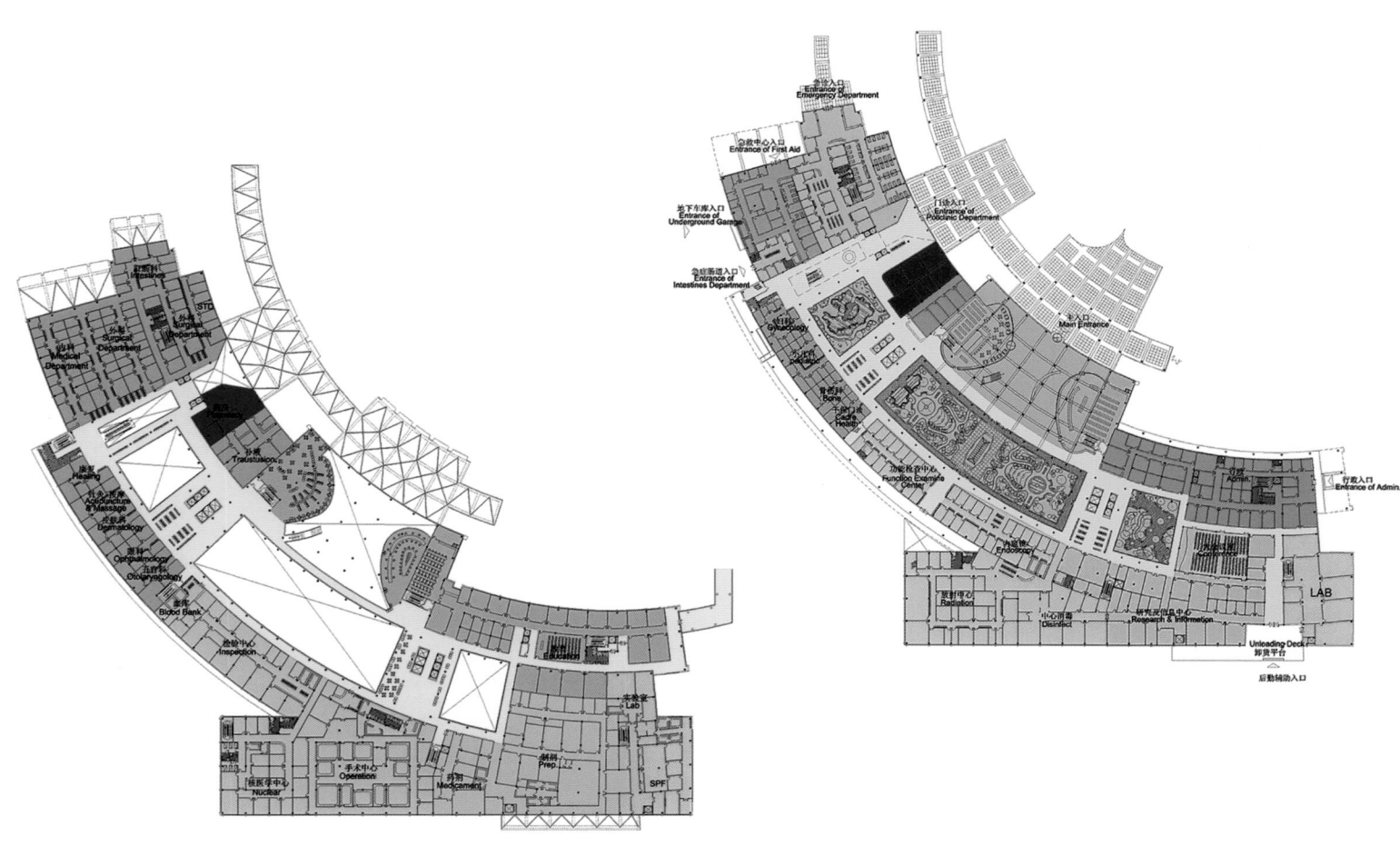

主要平面图

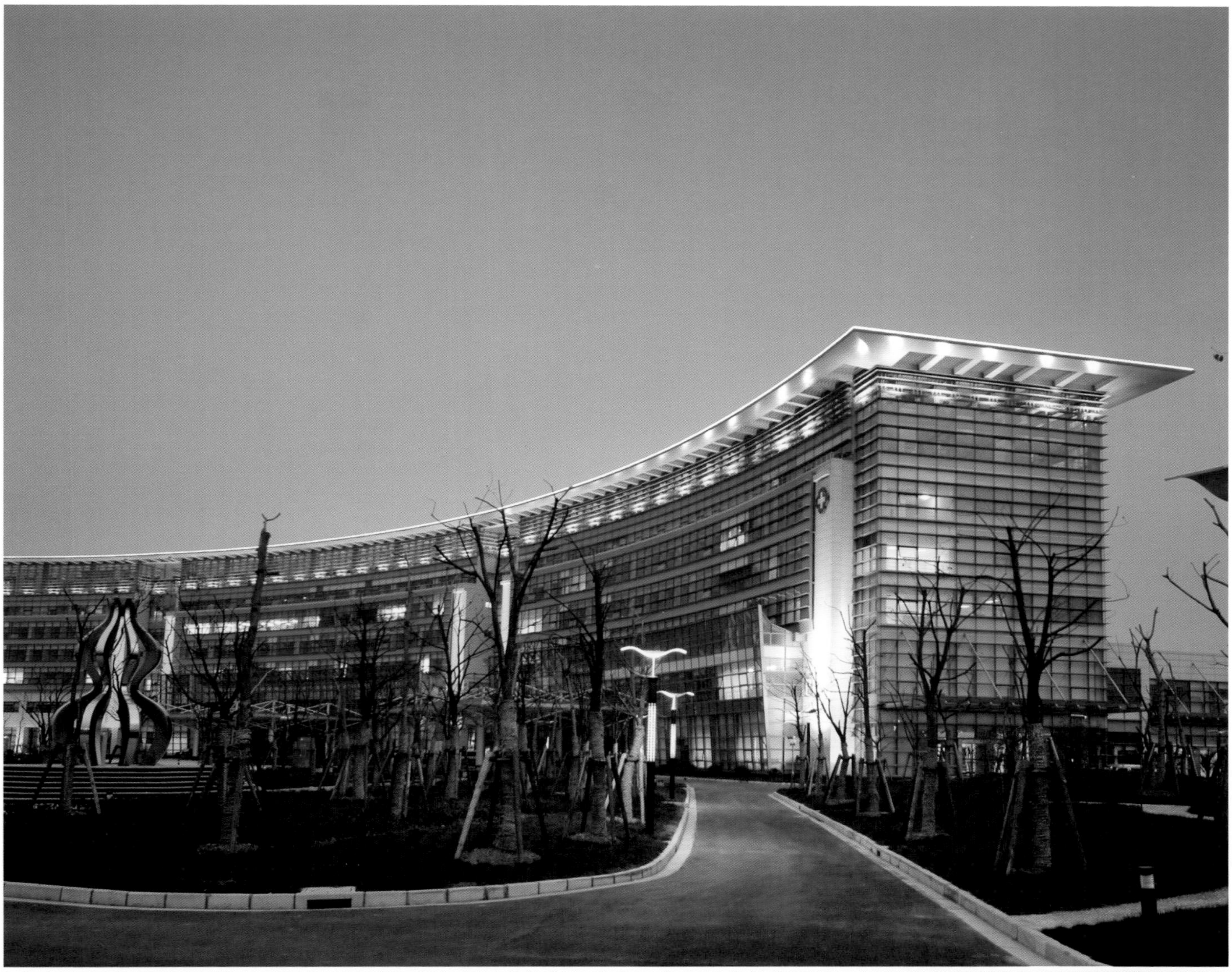

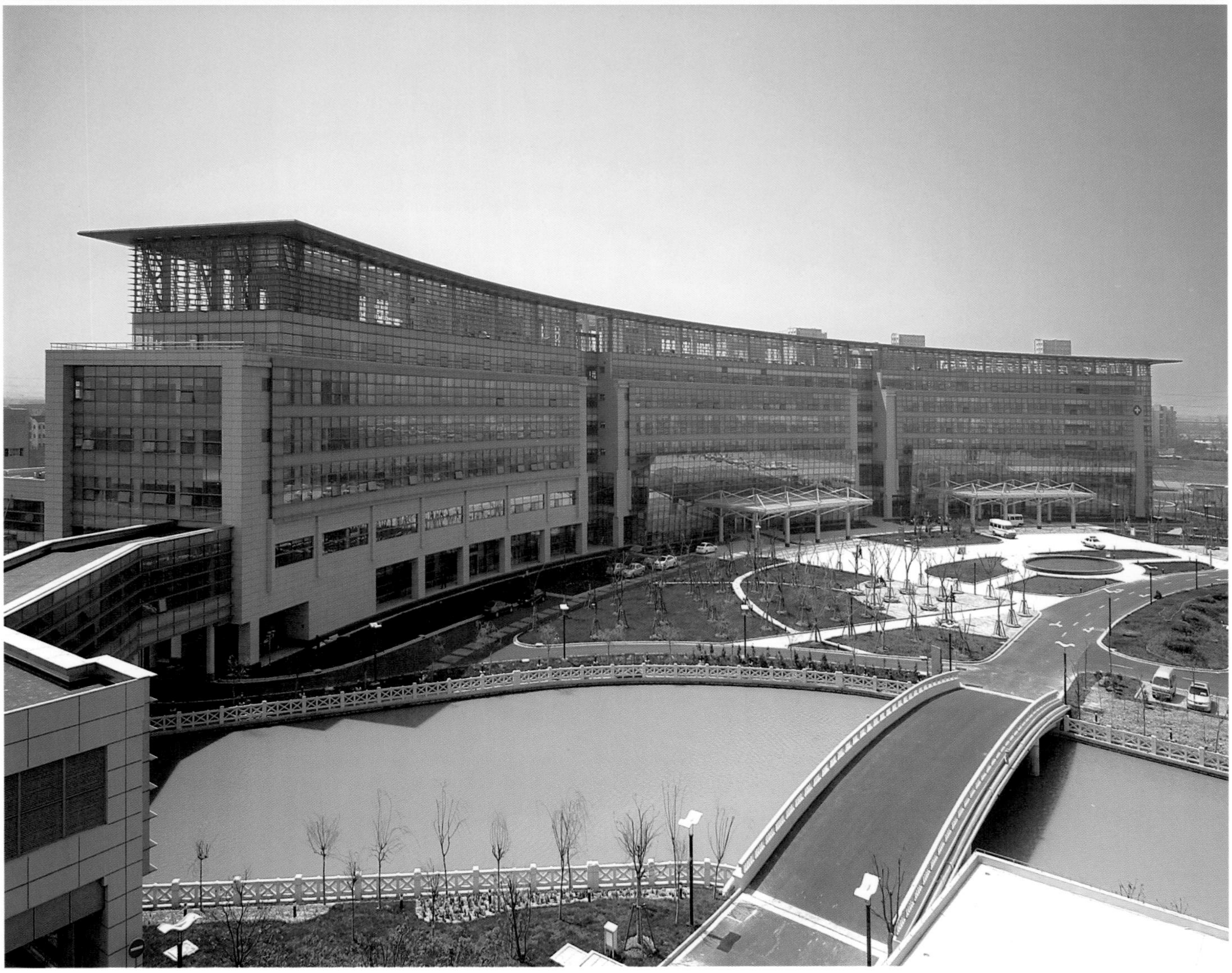

扬州新区医院
Yangzhou New District Hospital

项目名称: 扬州新区医院
用地规模: 6.89 万平方米
项目地点: 中国 / 江苏 / 扬州
编制时间: 2003 / 07

依靠现代医院建筑设计理念，结合医疗规划，达到布局紧凑，形成使用方便、高效能的有机体，实现布局设计超前性、内部空间独特性、建筑造型标志性。

将"以人为本"作为设计的指导思想，合理组织医疗空间，尽可能缩短病人就医流程，为病人创造方便、优美的就医环境，为医护人员创造便捷、高效、舒适的工作条件。正确、合理地安排各部分面积，尽可能既使各部门集中设置方便联系和使用，又使各类房间能有直接的通风采光条件以节省能源。

With the concept of modern hospital building, combined with the health care plan to achieve convenient, high-efficient organism, the layout should be ahead of time, unique in space, outstanding in architecture.

Using "people-oriented" concept as a design guideline, the design rationally organizes medical space, shortening patient treatment process, making convenience for the patient, creating a beautiful hospital environment. At the same time it creates convenient, efficient, and comfortable working conditions for the medical staff. With correct, reasonable arrangements for each part of hospital, it tries to make easy contact for each department, as well as using direct ventilation-lighting conditions in order to save energy.

郑州颐和医院
Zhengzhou Yihe Hospital

项目名称：郑州颐和医院
用地规模：18.07 万平方米
建筑面积：24.03 万平方米
容积率：1.33
项目地点：中国 / 河南 / 郑州
编制时间：2003 / 07

1. 以方便医患为主要原则。设计需要合理组织医疗空间，尽可能缩短病人就医流程，为病人细致考虑，并且为医护工作人员创造便捷、高效、舒适的工作条件。

2. 正确合理安排各部分面积，相关科室的布置相对集中，以满足中心化的需要，各科室既能联系方便，又使各类房间能有直接的通风采光条件以节省能源。

3. 保持生态平衡，实现人与自然相融的可持续性发展理念，营造医院的绿化景观、小品，给病人和医护人员以最大享受。

4. 确立以医疗大街为医院空间主轴线，采用了多翼端步开放的布局，有利于今后的改扩建，各医疗功能分区均留有充分发展余地。同时医疗大街融入了交通组织、绿化休闲，二层增加辅助设施，真正为病人提供了一个可停可行、导向明确、环境幽雅的综合性空间。

1. "People-orientation" has been used as a design guideline. The design rationally organizes medical space, shortening patient treatment process, carefully considering the patients. It creates convenient efficient and comfortable work conditions for the medical staff.

2. Concentration of relevant departments is achieved so that departments are given convenient connection and rooms are given direct ventilation-lighting conditions in order to save energy.

3. Keep the balance between man and nature, with sustainable development concept to provide patients and healthcare workers with the greatest pleasure.

4. Medical street space has been established as the main axis, and medical functional zoning has sufficient development room in order to benefit expansion and reconstruction in the future. At the same time, the medical street is combined with street traffic organization, the green lounge and facilities that have been added to the second floor, giving patients a clearly oriented and elegant environment.

复旦中学体育馆
Fudan Middle School Gymnasium

项目名称：复旦中学体育馆
建筑面积：2 517 平方米
项目地点：中国 / 上海
编制时间：2002 / 01

本方案位于上海市长宁区复旦中学内，东侧面临华山路，周围环境充满了文化气息。拥有百年历史的复旦中学，沉淀下来的历史，足以使它具有无限的魅力。

在设计之前，我们既充分考虑了学校所体现出的文化内涵，又考虑到校内各建筑对它的作用。于是在形体上，运用了简洁、大方、规整的矩形体块作为建筑主体，而建筑立面上开设窗户的大小和位置，都充满了建筑的节奏韵律感，这种开窗的方式正与周边现有的教学楼开窗方式类似，体现了整个建筑对周边环境的尊重，是对学校百年历史文化的一种传承。建筑的实墙面则运用了清水砖，与周边学校内建筑墙面运用的材料一致，充满了校园气息。在墙面材料，开窗方式等地方都与周边建筑相呼应，使其融入其中。而体育馆的入口部分面向学校主体楼，使之产生建筑与建筑之间的对话，交相辉映。建筑的顶部则设计成一片具有动感的弧形屋顶，既突出了体育场馆的特殊性，又延续了整个建筑简洁、大方的整体风格。面向华山路的立面则设计得现代、时尚，入夜华灯初上，漂亮的立面将为百年华山路增添一道绚烂的风景。

The project is located in Changning District, Shanghai, inside Fudan Middle School. Facing Huashan Road in the east. The surrounding environment is filled with cultural atmosphere. Fudan Middle School's one-century-old history is enough to make it charming.

We take into full account the cultural connotations of the school. So in the shape, we use rectangle block as the main element, which is simple, natural, and structured. The window size and position on a building facade fully reflect rhythm of architecture. This fenestration is similar to the surrounding buildings, and embodies the respect for the school history and cultural heritage. Plain-brick is used on solid walls of the building, which is similar to the wall materials of the surrounding school buildings and reflects the school atmosphere. The main entrance of the stadium is oriented to the main building so that a dialogue between the two buildings has been established. The top of the building is designed as a curved roof, highlighting the particularities of the stadium and continuing the overall style of simpleness and excellence. The facade facing Huashan Road is designed to be modern, trendy. When evening arrives, beautiful facades create a gorgeous scenery for Huashan Road.

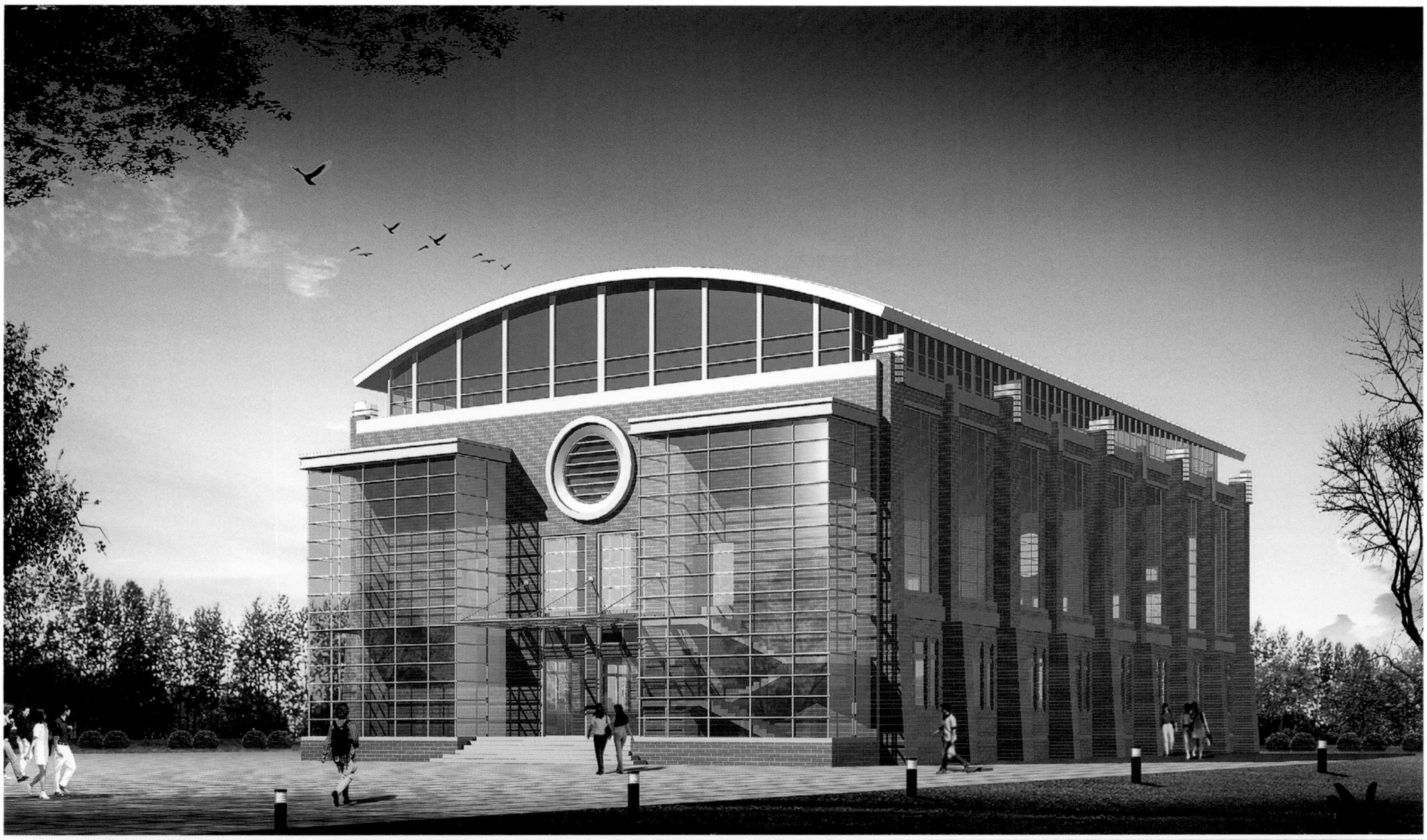

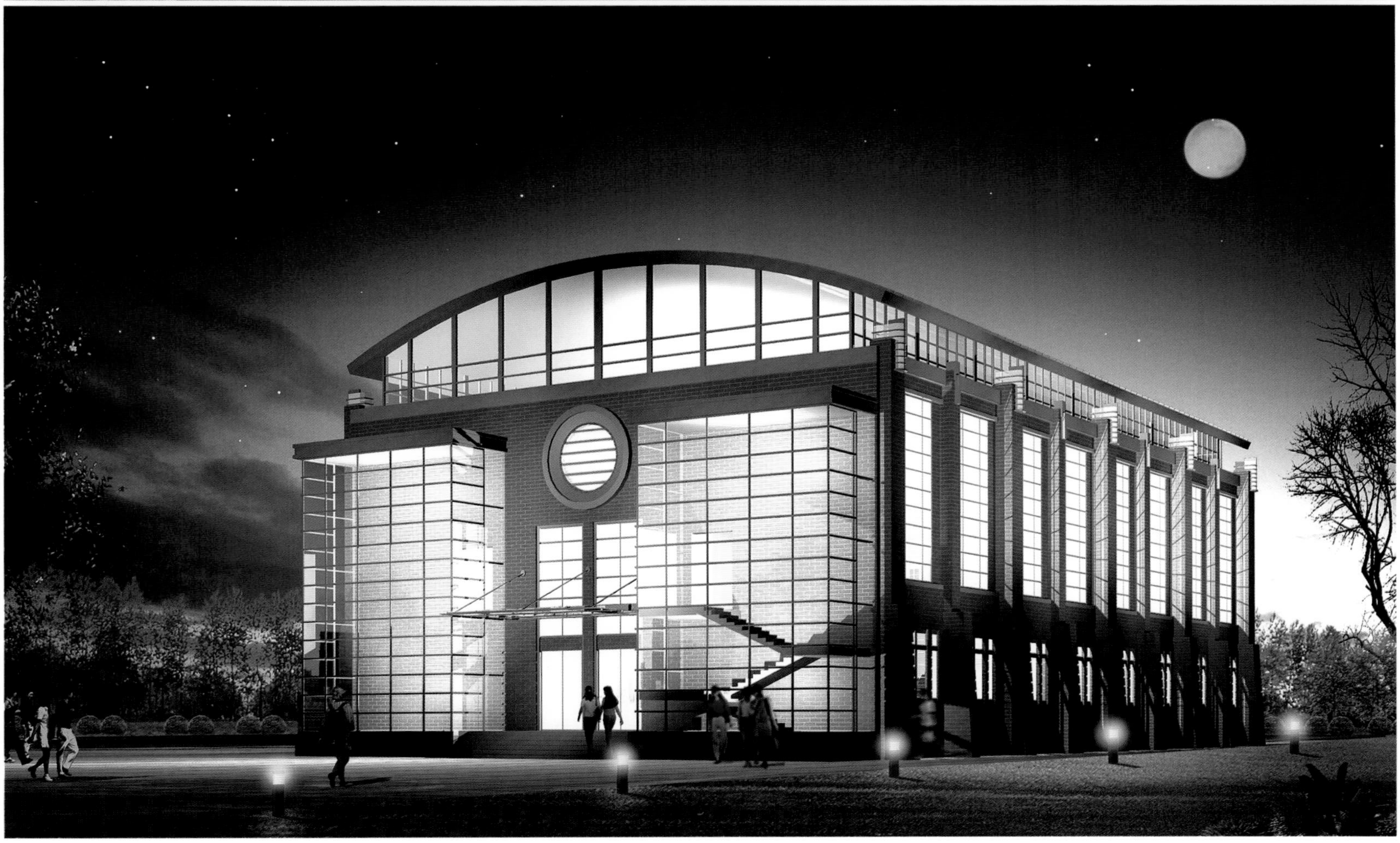

杭州和家园 K 组团学校项目
School Project of Hangzhou Hejiayuan Garden K-Group

项目名称：杭州和家园 K 组团学校项目
用地规模：2.55 万平方米
建筑面积：1.75 万平方米
容积率：0.69
项目地点：中国 / 浙江 / 杭州
编制时间：2008 / 12

本工程为杭州和家园 K 组团，内容为社区配套小学，基地西邻杨梅山路，北接规划支路与达利公司相望，东靠和家园 A 组团，南面隔规划支路为农居点改造区域。整个基地南面高，北面低，基地周围城市交通便捷，市政基础设施完善。

规划在基地西边的中部和南端设置了两组由综合楼与教学楼组合而成的建筑群，综合楼层数为 5 层，教学楼层数为 4 层，内部以连廊进行联系；中部综合楼与教学楼沿杨梅山路内凹形成人行主入口广场，利用基地高于杨梅山路的地形高差，设置主入口景观大台阶；在综合楼一层设置了配套用房，主要作为仓库、器材室及教材室，在教学楼下方设置了设备用房，使地块的利用更具有合理性和经济性，同时又为整个项目沿道路提供了良好的形象和主入口景观。

基地内部有一条贯穿南北的车行道，与位于北侧的规划支路上的车行次入口相接，车行主入口位于东侧的规划道路，由一条 6 米的车行道与内部道路相接。

基地西北侧为食堂、会议室等配套辅助用房，为整个学校提供后勤保障，一层为食堂及厨房部分，二层为 350 人的会议室，会议室向南有连廊与中部综合楼相连，方便学校师生使用，东侧与体育馆入口平台也以连廊相连，既加强了与体育馆的联系，也在北面入口处产生围合感。

在基地东北侧设置了体育馆，在其南面设置了风雨操场和篮、排球场等室外活动场所，整个基地功能分区明确、闹静分离，也为校园场地向公众开放提供了管理上的方便。

创造理念是打造中国式伊顿公学——让成功者的后代青出于蓝胜于蓝。

The project is an auxiliary elementary school project of Hangzhou Hejiayuan K Group. The whole base is higher in the south and lower in the north. There is convenient transportation around the base with complete municipal infrastructure.

The project is planned to set two architectural complexes of comprehensive building and classroom building; they are located in central and southern area of the site's west part. There are 5 floors for the comprehensive building and 4 floors for the classroom building. The two buildings are connected through an internal corridor; and the main entrance plaza is just in front of these two buildings, and we set big step landscape by using the height difference between the base and the Yangmeishan Road. Supporting houses are set on the 1st floor of the comprehensive building, mainly used for warehouse, equipment room and teaching material rooms. Equipment room has been set at the bottom of the classroom building, so that the site has been used more rationally and economically, at the same time providing a good image and the main entrance landscape for the entire project.

There is a north-south driveway, connecting vehicle sub-entrances. The main car entrance is located on the planning road east of the site, connecting internal roads with a 6 m carriageway.

The northwest side of the site is refectory and conference rooms and other ancillary support rooms, providing logistical support for the whole school. The first floor is refectory and kitchen. The second floor is a conference room of 350 people, connected with the central comprehensive building through a corridor. The east side also connects the stadium through a corridor , which not only strengthens connection with the stadium, but also produces a sense of enclosure at the north entrance.

The stadium is located at the northeast of the base. There is a playground, basketball court and volleyball court to the south of the stadium. These facilities are open to the public and convenient for management.

The design concept is to build Chinese-style Eton College-to make the new generations much better.

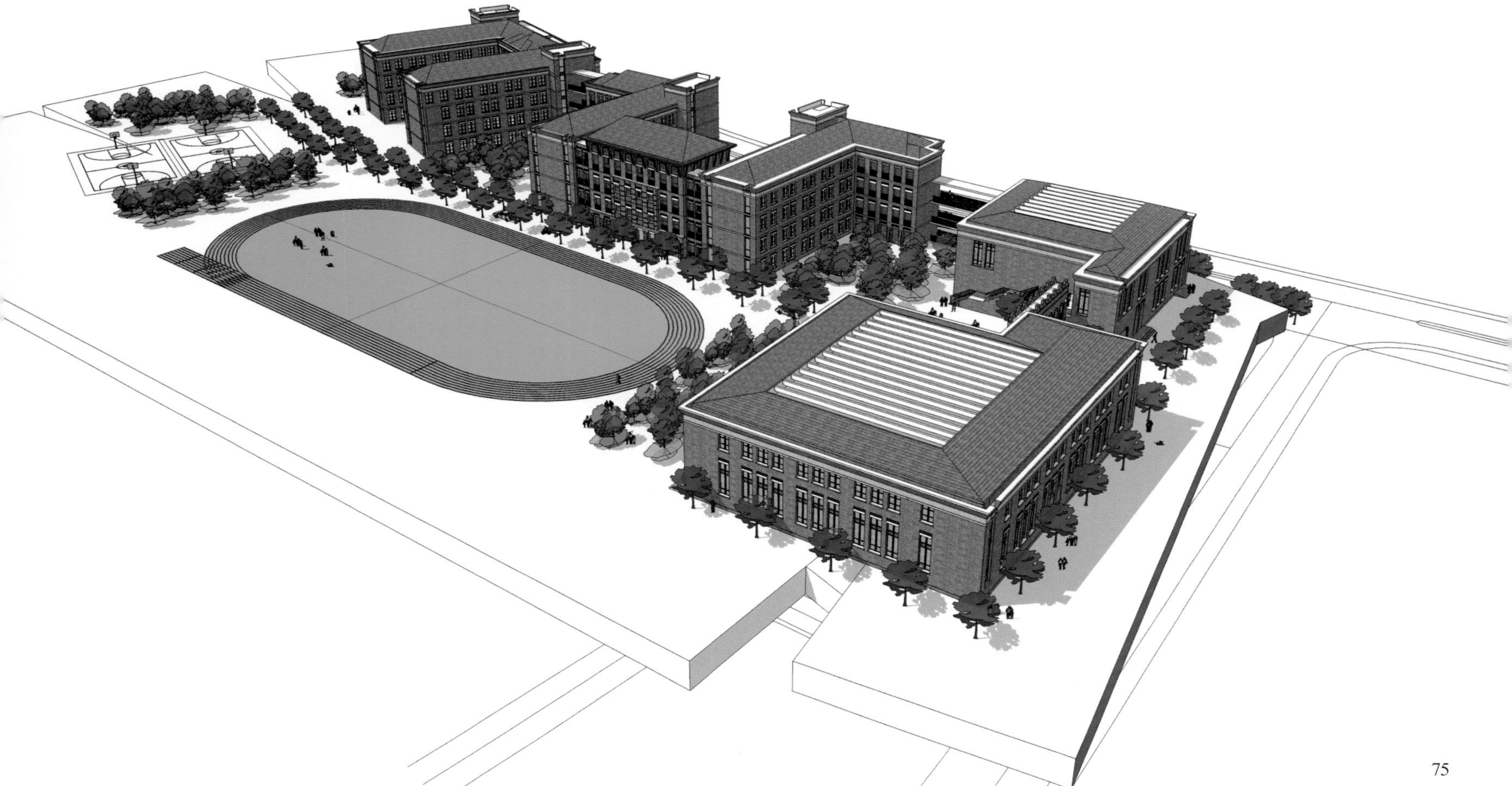

上海天山中学
Shanghai Tianshan Middle School

项目名称：上海天山中学
用地规模：3.63 万平方米
建筑面积：2.31 万平方米
容积率：0.64
项目地点：中国 / 上海
编制时间：2003 / 04

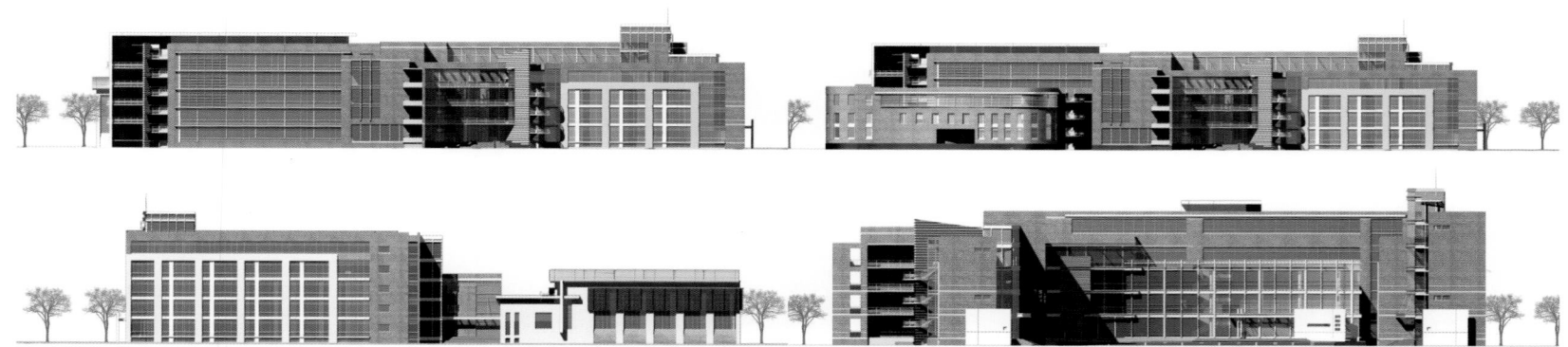

我们在总体布局中充分考虑了天山中学绿化环境的优势，基本保留了学校原有的两块大型绿地并加以重新规划利用，使其在新的天山中学里发挥更大的作用，为师生提供更理想的景观生态环境。

在设计中不仅注意了绿地的"量"，还注重绿化的"质"，特别引入了"庭院"的概念，使绿化能渗透到建筑内部来，改善建筑内部环境，使绿化存在于每一个老师及学生的周围，最大限度发挥绿地的作用，体现以人为本的设计思想。

主体墙面部分采用砖红色面砖和白色涂料，体现出传统教育建筑沉稳、严肃的教学氛围，使其与传统的红色外墙产生对比，具有较强的现代气息，与虹桥开发区的建筑风格相协调。

We fully take into account the green environmental advantages of Tianshan Middle School. Basic retention and redesign of the original two greenbelts will make them play a greater role in the new middle school. Students and teachers will be provided with a better green environment.

Paying attention not only to the "quantity" of green space, but also to the "quality" of the greening , we introduce the "courtyard" concept so that green can penetrate into the interior building, improving interior building environment, making green space around the teachers and students, maximizing use of green space to embody the "people-oriented" design concept.

The designers use red brick and white paint for part of the main walls, reflecting stable and solemn atmosphere of traditional education in contrast with the traditional red walls. Its strong modern feeling is consistent with Hongqiao Development zone.

江苏省美术馆
Jiangsu Provincial Art Museum

项目名称：江苏省美术馆
用地规模：1.06 万平方米
建筑面积：2.21 万平方米
容积率：2.1
项目地点：中国 / 江苏
编制时间：2005 / 09

在本项目中设计师是在为业主创造一种记忆的符号。该符号由象征着钟摆的云锦轴线和颇具城市回忆色彩的墨色砖墙实体穿插的饱和手法组合而成，象征城市文化的传承和艺术圣殿的地位。建筑既隐喻着过去和未来的传承关系，同时也隐喻着一种空间的力量在一座城市、一个组群、一所校园内的穿叉和流动。设计师立意将它建成汇江苏省文化之大乘，传承南京文史的文化隐喻建筑和一处体验园区。在这所体验园内，不仅有造型优美的建筑形体，众多的建筑元素如大块城砖、屋顶草坪、室内花园、水池、雕塑等也都被和谐地运用，而这些元素也通过色彩、材质、光线、透明度把单纯展示的原型变形成为体验园区。

美术馆主要由四部分组成，即作品展示区、管理培训区、典藏保护区、交流共享区。典藏保护区根据业主的意见分为地下和顶楼二部分，中间有一部专门的货运电梯。

In this project, designers create a symbol of memory for the owners. The symbol is composed by the pendulum axis and urban black brick wall of memories, symbol of urban cultural heritage and art sanctuary status. The building metaphorizes the inheritance relationship between the past and the future, as well as the power flowing in a city, a group and a campus. Designers are willing to build it as a building with cultural heritage of Nanjing, and an experiment park. In this park, not only beautiful architectural bodies are included but also a large number of architectural elements like big city tile roof, roof gardens, lawns, pond, sculpture are used harmoniously. Through color, texture, lighting, transparency, the originally displayed prototype has been converted to the experiment park.

The art museum mainly consists of four components: works display area, management training area, protection area of the collections, exchanging area. Protection area of the collections is divided into ground floor and top floor, connected through a freight elevator in the middle.

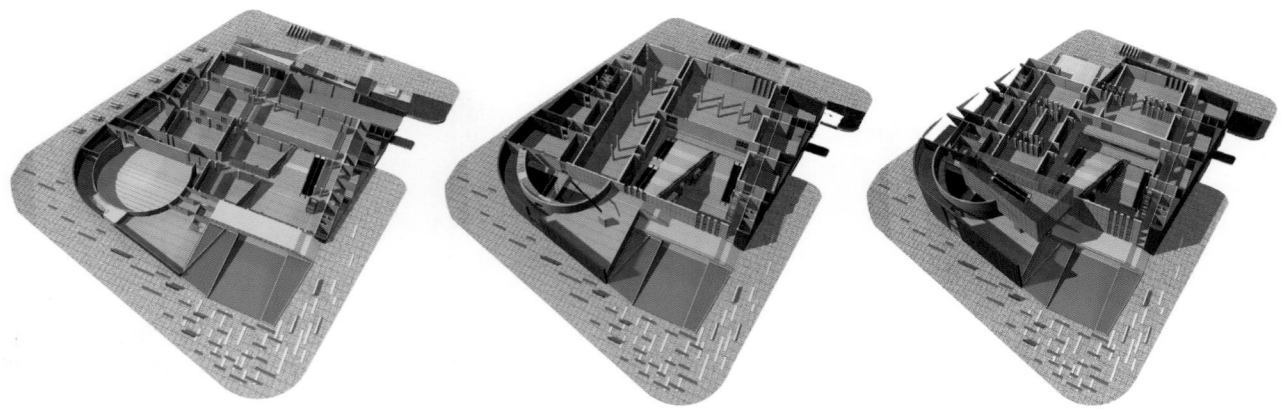

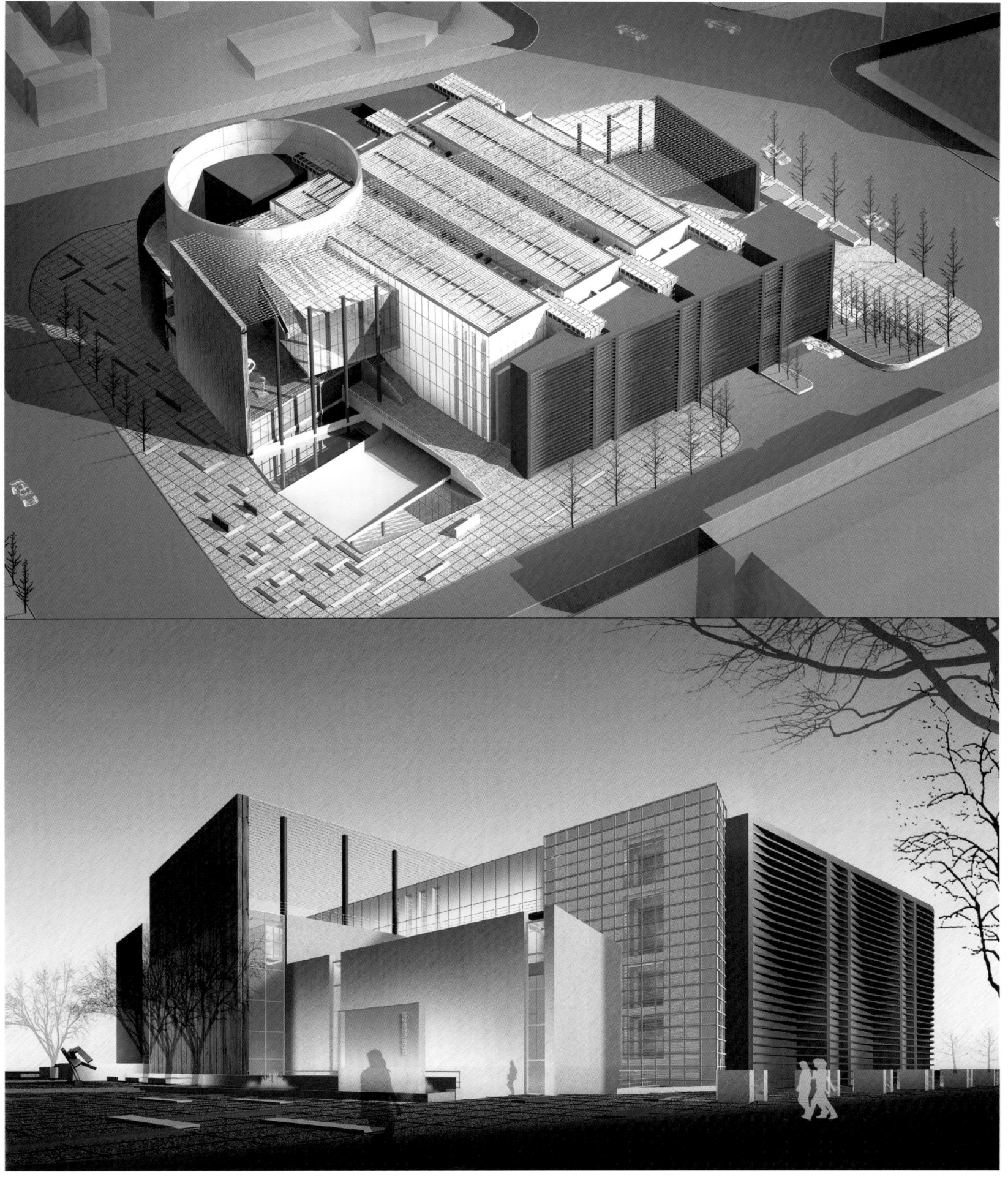

梅陇新中心
Meilong New Center

项目名称：梅陇新中心
建筑面积：1.60 万平方米
项目地点：中国 / 上海
编制时间：2004 / 03

规划的造型设计采用传统的对称设计手法，以达到建筑内容的直接体现，用材料对比、光线穿透的方式避免建筑造型上单一生硬的可能。总体上，力求大实大虚、虚实相连，形成强烈有力的形体效果，体现出办公建筑端庄、大气的形态。

本设计由四部分组成，第一部分是政府办公用房，共9层，主楼用对称的手法进行功能上的布置和设计，得以体现政府大楼的庄严、肃穆。同时，运用高档石材与大面玻璃幕墙的结合对比做法，摆脱了传统对称建筑过于笨重、呆板的缺点。主楼前的两层裙房主要具有接待来访的用途。第二部分是与主楼连接紧密的会议中心，会议中心在基地的北端设有独立的出入口，方便政府人员以外的人士进行会议活动。小型会议室、休息室围绕主会议室布置，使得流线不互相干扰。第三部分是在基地东部的休闲体育运动中心，设有健身房、室外网球场等体育娱乐设备，使政府人员在繁忙的工作之余能得到最好的运动锻炼，保持高水平的工作效率。第四部分是与会议中心衔接的餐厅，餐厅独立的设置为会议人员、政府人员的就餐活动提供了方便，倒锥台体的造型、临水的大玻璃面，使得餐厅的布置十分自由，适当地调节紧张的工作气氛。

Using the traditional symmetrical approach, the design directly embodies building content, using solutions of materials contrast, light penetrating method to avoid single architectural style. The design uses real-virtual combination to form a strong powerful body effect that reflects the elegant and grand shape of the office building.

This design is made up of four parts. The first part is the government office building with a total of 9 floors. Symmetrical approach is used for main building functional layout and design to embody solemn feeling of government building. At the same time, using the contrast approach of combination of top-grade stone and large glass curtain wall, we can get rid of the bulky feeling. The two-story podium in front of the main building is used to host the visitors. The second part is the conference center connected to the main building. Conference Center has an independent entrance at the north for outside people to conduct activities. Small conference room and lounge are set around the main meeting room so that flow lines do not interfere with each other. The third part is the eastern recreational sports center equipped with gym and outdoor tennis courts, where staff can do some exercise after busy work. The fourth part is a canteen that connects the Conference Center. Independent canteen provides easy access for staff and government personnel.Inverted cone shape with large glass surface near the water makes the layout of the restaurant very flexible, which properly regulates intense work atmosphere.

85

南京金陵图书馆
Nanjing Jinling Library

项目名称：南京金陵图书馆
用地规模：3.49 万平方米
建筑面积：2.25 万平方米
容积率：0.64
项目地点：中国 / 江苏 / 南京
编制时间：2005 / 02

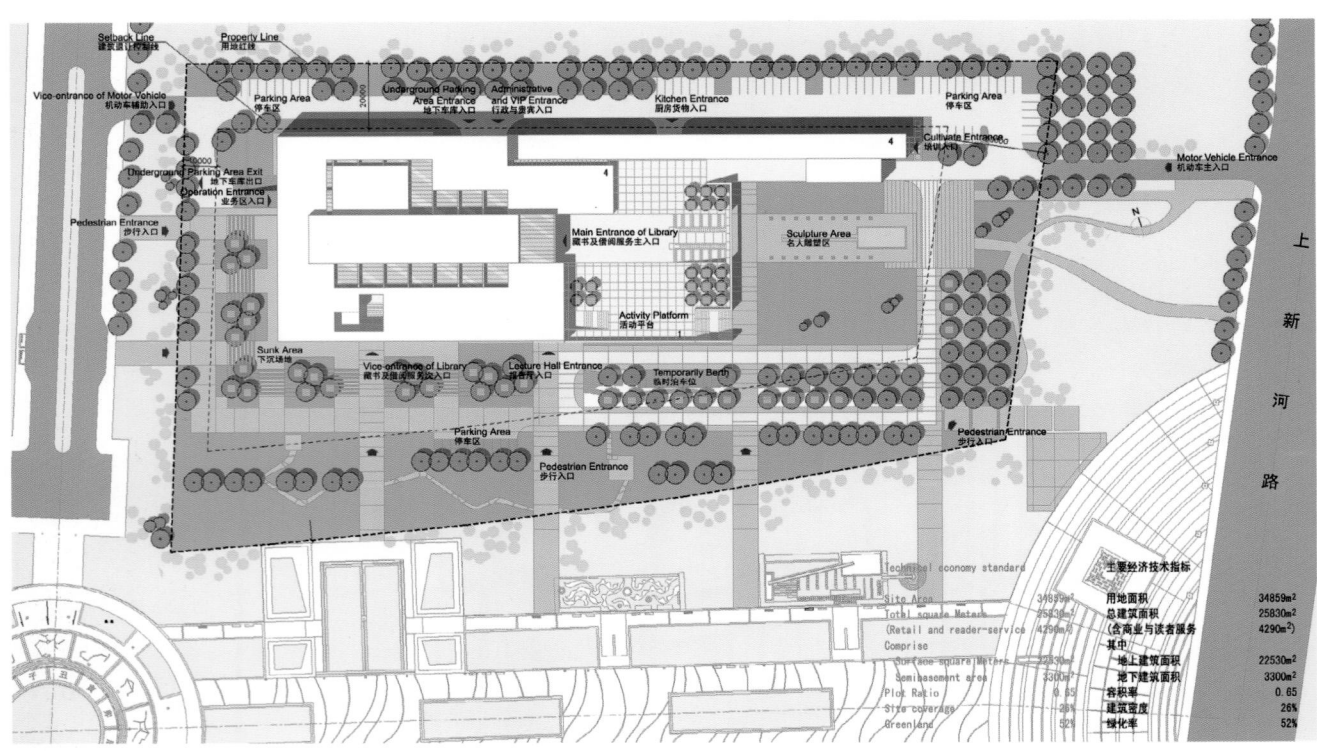

设计将集中储存的书放入一个巨大的玻璃盒子中，作为全馆的核心，并展现在中庭里，书库中的书不再是惰性物体，而成为生动的图书馆主题，它使我们铭记知识是文化的基本元素。非但如此，我们还用一块特别的地板将书库衬托出来，当人们沿着这条地板指引的道路一步步登堂入室时，会有一种传统的仪式感。

建筑的内外一致是内部功能逻辑的外部反映。

实墙与大面积玻璃的区分体现了开架书库与读者阅览部分的区别，其间的分层与分区通过楼板在外立面上表现出来。

在总体布局上充分考虑世纪轴线景观带与艺兰斋美术馆的影响，形成呼应与场所限定。商业与读者服务设于半地下室与底层周边，人流可方便到达。阅览室外墙采用双层玻璃幕墙，双层玻璃中间使用可调式遮阳百叶，通过调节双幕墙上下两个位置的通气口优化能源效率，以达到节能目的。局部外窗也可以通过平开或上旋的方式直接进行自然通风。中庭采光天窗则采用可调节角度的遮光板来避免强烈的日光。

Storde books have been placed in a big glass box as the core of the library and shown in the courtyard. Stacks of books are no longer inert objects of the library but lively themes, which gives us an idea that knowledge is the fundamental element of culture. Furthermore, a special board is designed to highlight the books; when people walk along this road step by step to the hall, they will have a sense of a traditional ceremony.

Consistence-the surface can reflect the internal functions and logic.

Solid walls and large areas of glass distinguish the open-shelf library shelves from the reading area. Detailed floors and zoning are reflected by the envelope.

The overall layout takes into full account the influence of century axis landscape zone and the Yilanzhai Art Gallery to create the feelings of correspondence and location restriction. Commercial and reader services are provided in the semi-basement and around the basement, easy to access. Double-layer glass curtain wall is adopted for the facade of the reading room with adjustable shading shutter in the middle. People can optimize energy efficiency by adjusting top and bottom vents on the glass curtain wall to achieve energy-saving purposes. Natural ventilation can also be achieved through flat open or upward open of some external windows. Courtyard skylight uses shading plates that can adjust the angle to avoid strong sunlight.

上海意大利中心
Shanghai Italy Center

项目名称：上海意大利中心
建筑面积：0.71 万平方米
项目地点：中国 / 上海
编制时间：2010 / 09

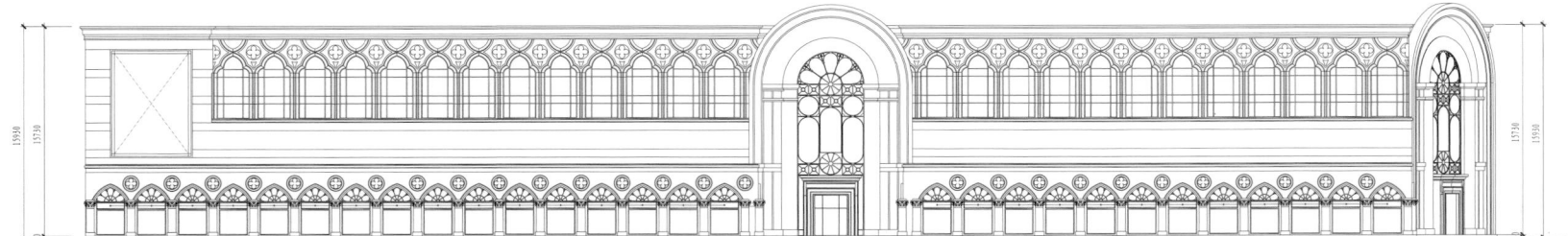

设计理念

1. 精心研究、设计、构筑清晰的建筑外界面，营造既与上海展览中心的建筑风格相协调，又能体现意大利风情的建筑形象，彰显该区域地段的独特景致。

2. 营造具有历史场所感的商业空间氛围，演绎古典情节，传承永恒的美学要义。

3. 除了对以小资女性为主的目标消费群体行为心理方面的考虑，还应体现现代商业购物空间对建筑空间舒适度的要求，调适古典与现代的契合临界点。

4. 线状建筑的景观及地理位置虽然较好，但处于群楼环绕之中，应在考虑城市空间规划的基础上，合理地扩建，尽可能地使建筑空间得到最大范围的使用，最大限度地发挥该建筑的商业价值，并尽可能地使建筑内部的视野扩大。

5. 本案虽然受到现状建筑的结构限制，但屋顶空间可以自由发展，充分利用这一空间将有助于商业空间功能完善，并形成本案的独特之处。

6. 建筑平面狭长，所以常立面需要避免单调，从交通及景观方面考虑，建筑的西侧沿铜仁路侧的一面适合设置建筑主要出入口，应为建筑造型处理的重点，而西南角临近城市干道，为铜仁路与延安路的交叉口，也应该成为建筑造型处理的重点。

7. 现有建筑屋顶上采用流线型造型设计的广告牌，既可以为屋顶上的活动遮挡高架路的车流，又能体现现代高科技的商业氛围和此广告位巨大的商业价值。

Design Concept

1. Carefully study and design a clear building facade that can not only be harmonious with the architectural style of Shanghai Exhibition Center but also show the Italian style, highlighting the unique view of this area.

2. Create the commercial space with historical sense, interpreting classic plot and continuing the eternal elements of aesthetics.

3. Take white-collar women as target consumers and consider their psychological appeals; ensure the special comfort of the modern commercial and shopping center; find a balance between classical and modern aesthetics.

4. In spite of the favorable landscape and location conditions of the linear architecture, it is surrounded by buildings; consider appropriate expansion based on the urban space planning to maximize the utilization of the building space; create the biggest commercial value of the building and expand the field of vision inside the building as far as possible.

5. In spite of the restriction of the existing building structure, free development can be made for the roof; taking full advantage of this space will help improve the unique commercial space features.

6. Due to the narrow building plan, it is necessary to avoid dull outlook of envelopes; from the traffic and landscape points of view, the west side of the building along Tongren Rd. is suitable to present the main entrance so that it should be the key of this project; besides, the southwest side near the trunk line of the city faces the intersection between Tongren Rd. and Yan'an Rd. and should be another focus of the building design.

7. The billboard on the roof uses streamlined design, which can not only be a screen for activities on the roof from the busy elevated traffic on Yan'an Rd. But also embody the modern high-tech commercial atmosphere and the great advertising value.

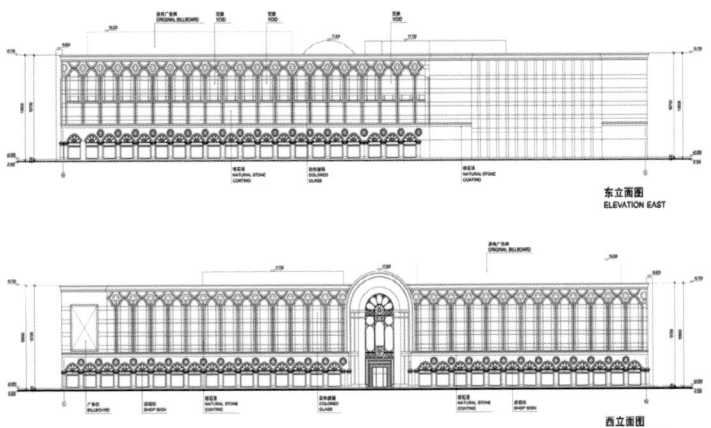

东立面图
ELEVATION EAST

西立面图
ELEVATION WEST

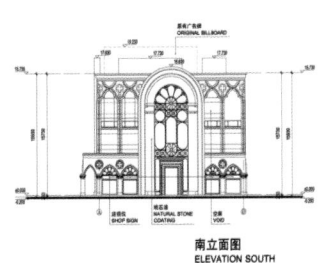

南立面图
ELEVATION SOUTH

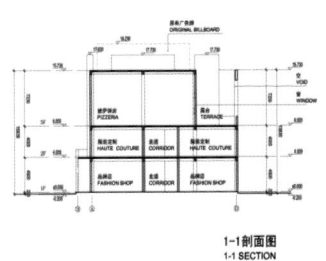

1-1剖面图
1-1 SECTION

东立面效果图

沈阳音乐厅
Shenyang Concert Hall

项目名称：沈阳音乐厅
用地规模：4.5 万平方米
建筑面积：2.19 万平方米
容积率：0.49
项目地点：中国 / 辽宁 / 沈阳
编制时间：2006 / 10

音乐厅是位于沈阳市浑河以北的地标性建筑，是充满浑河的精神和力度并十分优雅的时代产物，它将成为这座城市的大门，设计师将沈阳的历史文化和人民意志以及浑河本身作为这个项目的信息和资源，在平静流淌中创造纪念的力量和推动的力量，并以此启发人们去探究其自身曾经存在的历史真相，建筑带着灵动的曲线骄傲地耸立，雕塑般的形态承载了令人印象深刻的交响乐中不同的音调，向人们展现其本身所代表的浪漫和雄浑的篇章。

Shenyang Concert Hall is a landmark building located to the north of Hun River. It is full of Hun River's spirit and intensity. It will become the gate of this city. Designers use the history and culture of Shenyang, the will of people and Hun River itself as information and resources of this project, to inspire people to seek the historical truth. Sculpture-shape is like the impressive different tones in the symphony, showing people the romantic chapters it represents.

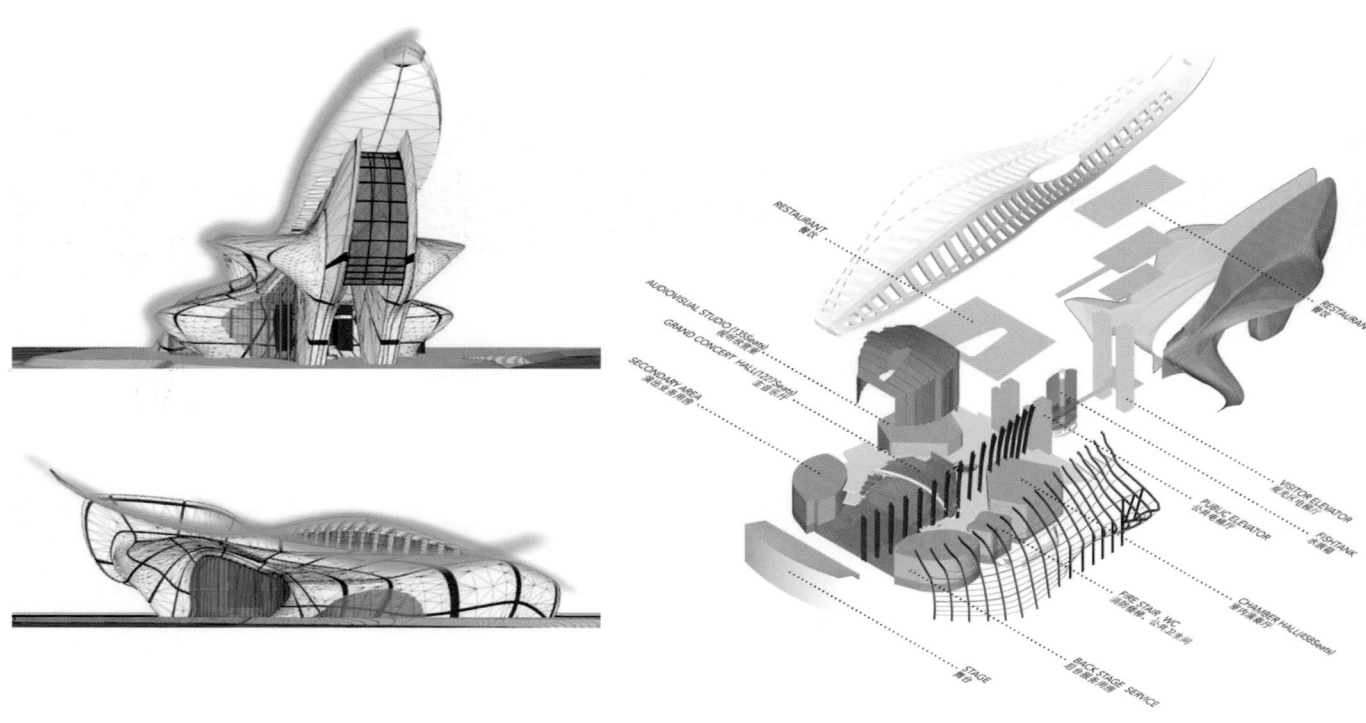

浑河视角效果图

它山博物馆
Tashan Museum

项目名称：它山博物馆
用地规模：18.10 万平方米
建筑面积：31.61 万平方米
容积率：1.75
项目地点：中国 / 浙江 / 宁波
编制时间：2007 / 09

本项目是以石雕为主题的私人博物馆群落，基地位于宁波市郊，项目分为 4 个区域组团，分别是展示和配套组团、古村落改造及它山文化会馆组团、水休闲中心及水岸文化院落组团、博物馆院落组团。各组团之间相互独立，根据地势以最小地修改地形为基本原则，让建筑如自然生长出来一般散落其间。

The project is located in the suburbs of Ningbo. The private Museum communities are divided into 4 regional groups: Presentation and Supporting Groups, Reconstruction of Ancient Villages and Tuoshan Mountain Culture Hall Group, Waterfront Recreation Center and Cultural Complex Groups, Museum Courtyard Group. Various groups are independent of each other. According to the principle of minimal modification of terrain, buildings scatter like a natural growth.

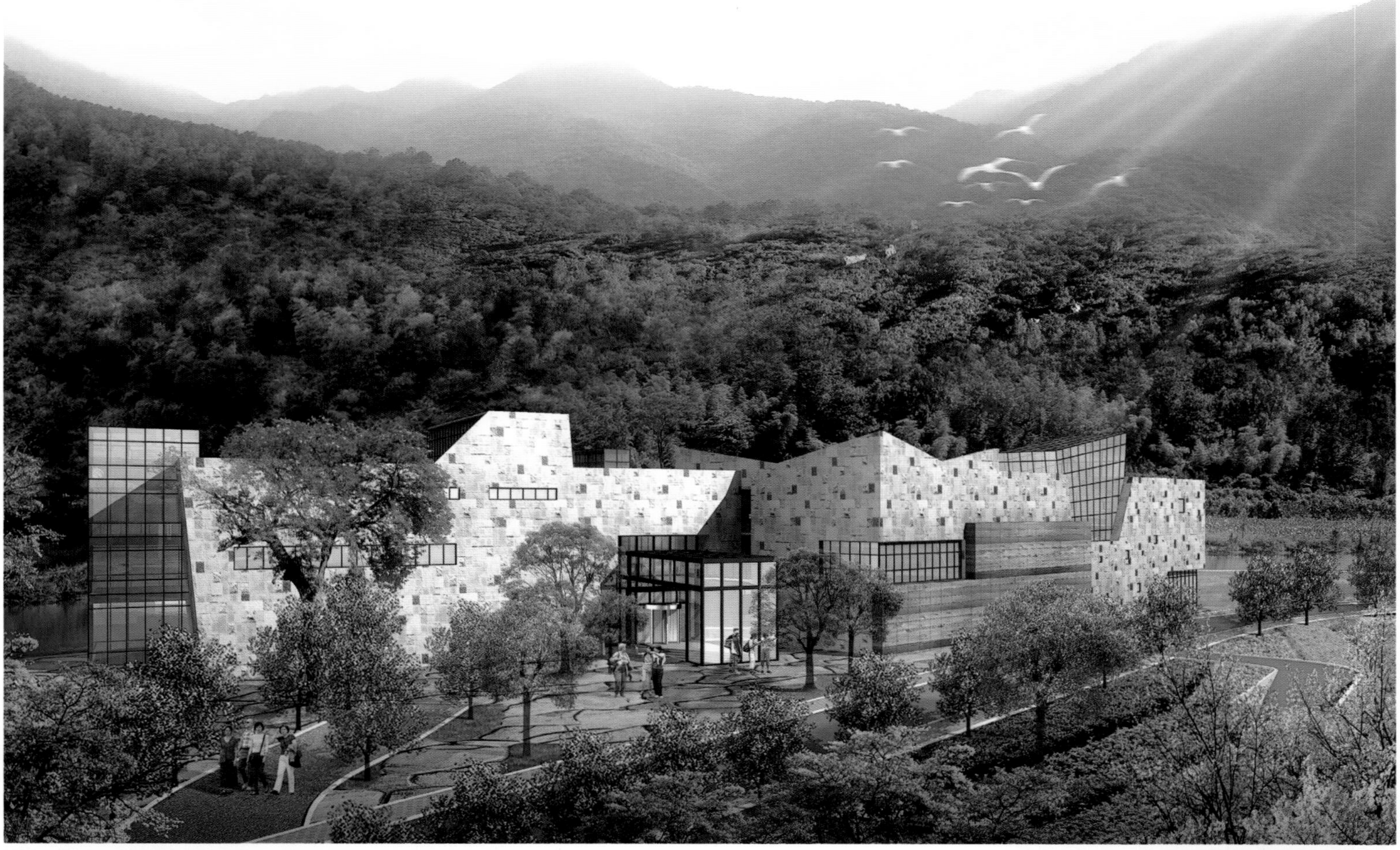

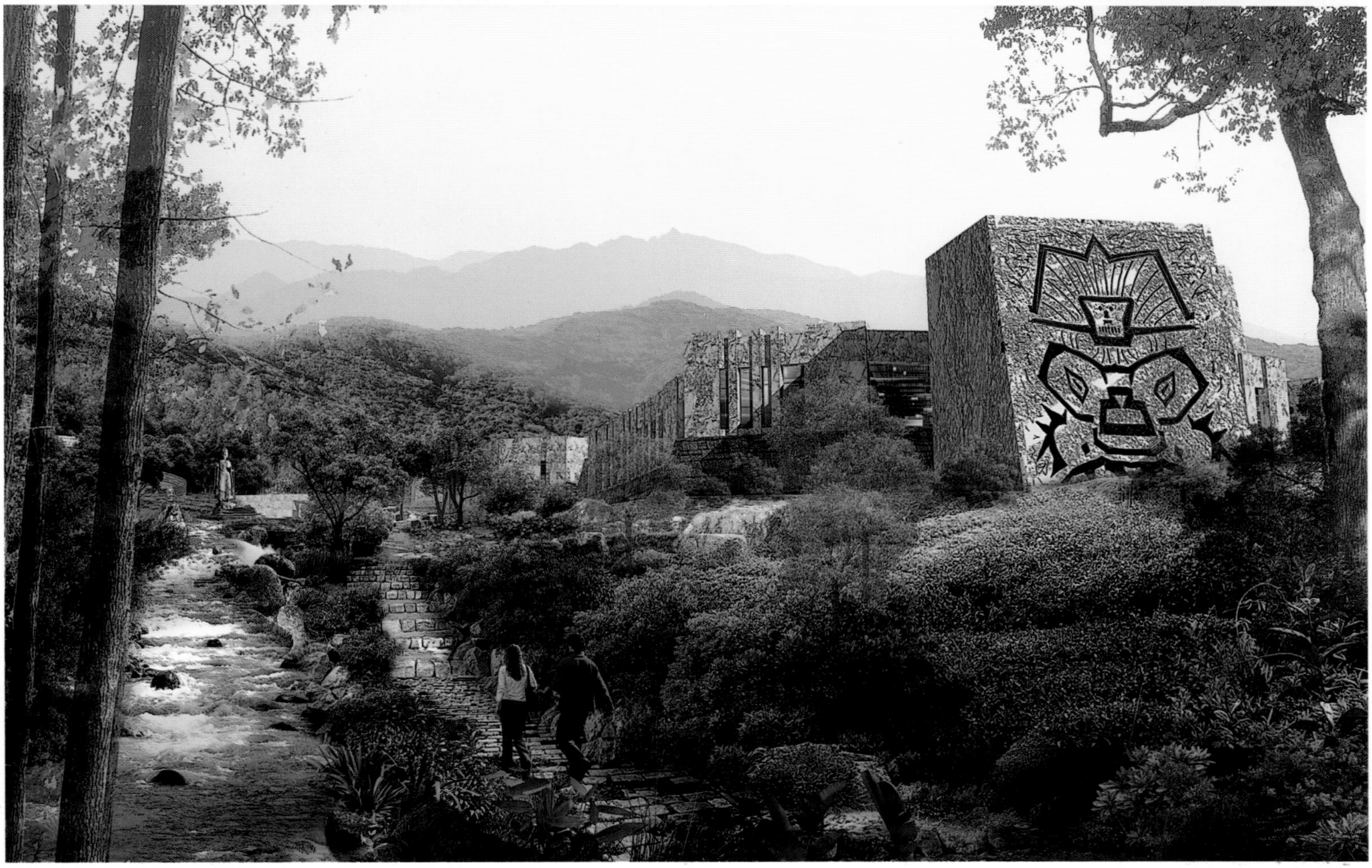

山涧效果图

天津城市规划建设展览馆
Tianjin Urban Planning Exhibition Hall

项目名称：天津城市规划展览馆
用地规模：2.87 万平方米
建筑面积：2.08 万平方米
容积率：0.72
项目地点：中国 / 天津
编制时间：2007 / 03

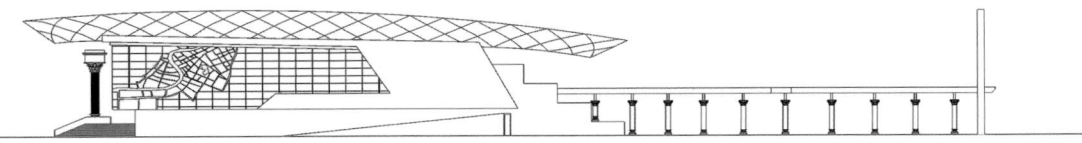

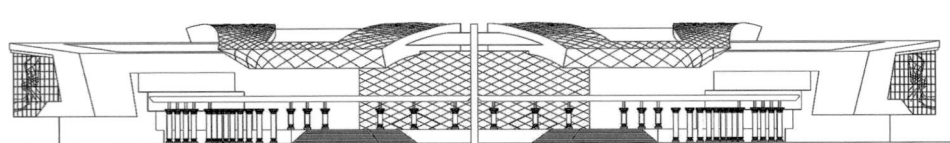

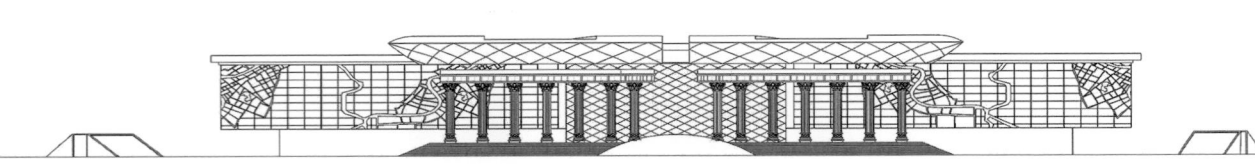

本项目坐落在海河中心广场上，是天津市中心城区的几何中心，项目主要由展览馆本身和海河中心广场两个部分组成，是展示天津未来发展和回忆城市历史的重要舞台，设计通过对天津本土文化和城市发展的图像表达，渲染一种无声的空间语言，使参观者从走入基地的那一刻起就完全融入于建筑所表达的意向中。

This project is located in Haihe Central Plaza and it is the geometric center area of Tianjin City. The project is consisted of two parts, which are Exhibition Hall and Haihe Central Plaza; it is an important platform to display the history and future trends of Tianjin. Through the image expression of Tianjin's native culture and urban development, the design attracts people to fully get into certain design intention as soon as they move into the base.

总平面图：

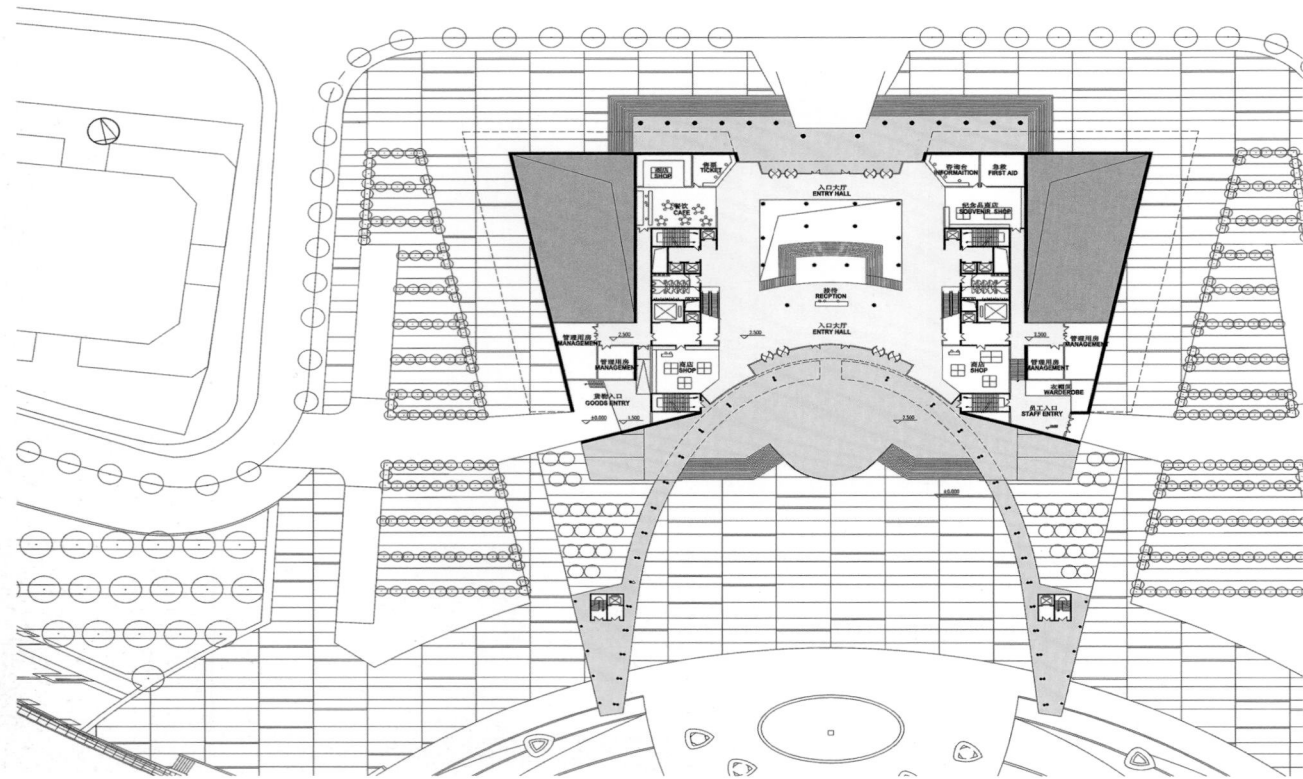

G EXHIBIT HALL

黄昏总体鸟瞰图

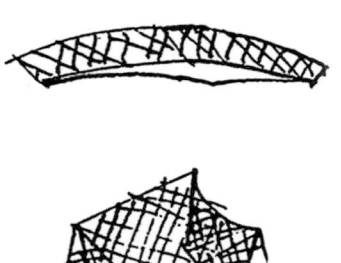

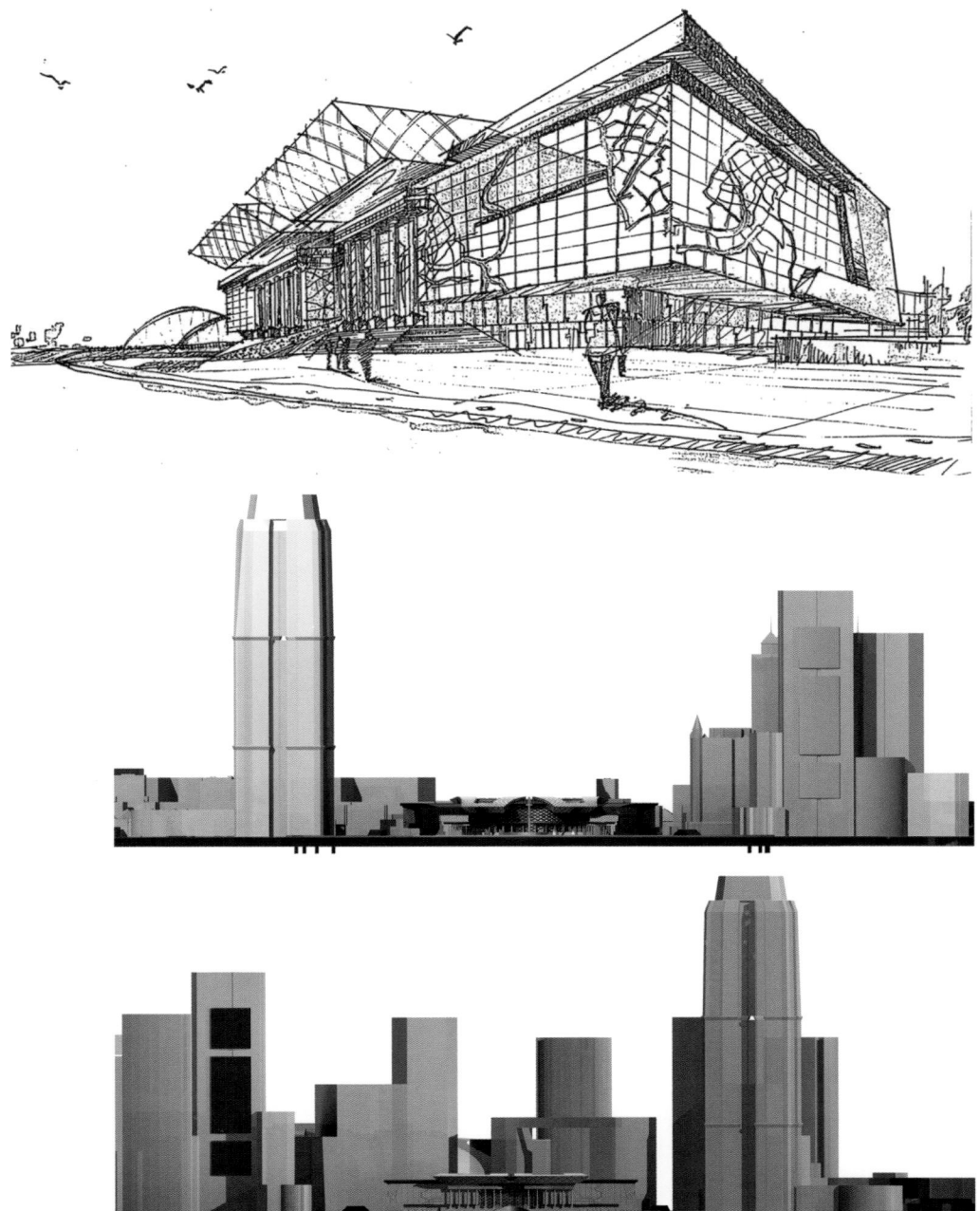

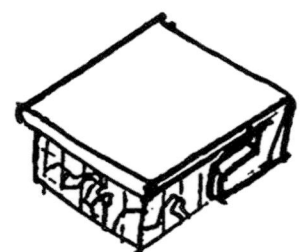

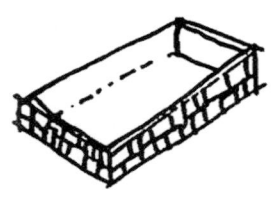

107

天津柳林桥
Tianjin Liulin Bridge

项目名称：天津柳林桥
项目地点：中国 / 天津
编制时间：2005 / 06

主跨椭圆形封闭钢梁体两端设置强大的环形封闭式横梁，桥面以上横梁部分设冷铸锚锚固斜拉锁，桥面以下椭圆形钢梁在与内部混凝土主梁连接部位预设剪力钉；边跨采用钢——混凝土叠合梁，钢梁为纵——横梁体系，钢梁顶面设置焊钉，与桥面湿接头部分有效连接成整体；塔柱采用拱形，塔顶向两岸倾斜，以自重平衡拉锁向河心方向产生拉力。

Powerful ring shape-closed beams are set near the steel. Above the beam part, cold cast is set anchoring oblique zip, deck following oval steel beams in and internal concrete main beams received parts preset shear force nail; edge across using steel-concrete stack hop beam, steel beams for longitudinal-beam system, top surface set welding nail, and deck wet connector effectively connects the overall; tower column uses arch; tower tilts to Strnit; weight balance zips to direction of rally.

项目名称：天津柳林桥
项目地点：中国 / 天津
编制时间：2005 / 06

当你可能被一次次的吸入这座有着彩色屋顶的大桥后，出来的心情一定大好！

长春国际会展中心综合馆
Changchun International Conference and Exhibition Center

项目名称：长春国际会展中心综合馆
用地规模：34.45 万平方米
建筑面积：10.23 万平方米
容积率：0.3
项目地点：中国 / 吉林 / 长春
编制时间：2007 / 01

长春国际会展中心综合馆布置于基地东侧，临北海路及世纪大街，主轴线指向城市中心的世纪广场。综合分析基地周边条件和基地内部各建筑布局特征，综合馆采用中心对称的圆形为主体、主入口面向东侧世纪广场的布局方案。

Changchun International Conference and Exhibition Center is located in the east of the site, next to Beihai Road and Century Avenue. The exhibition center uses centro-symmetric circle layout, the main entrance facing the east side of square, to meet the needs of social image. Analyzing the surrounding condition and architectural layout of the site, the Center uses layout of symmetric circle as its main shape, with main entrance facing the East Century Plaza.

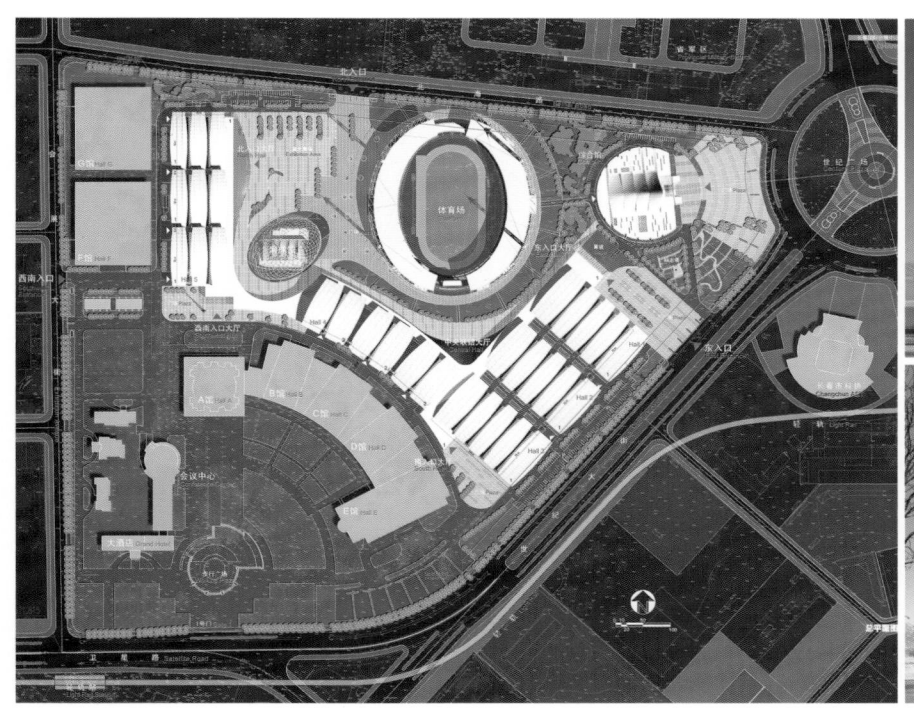

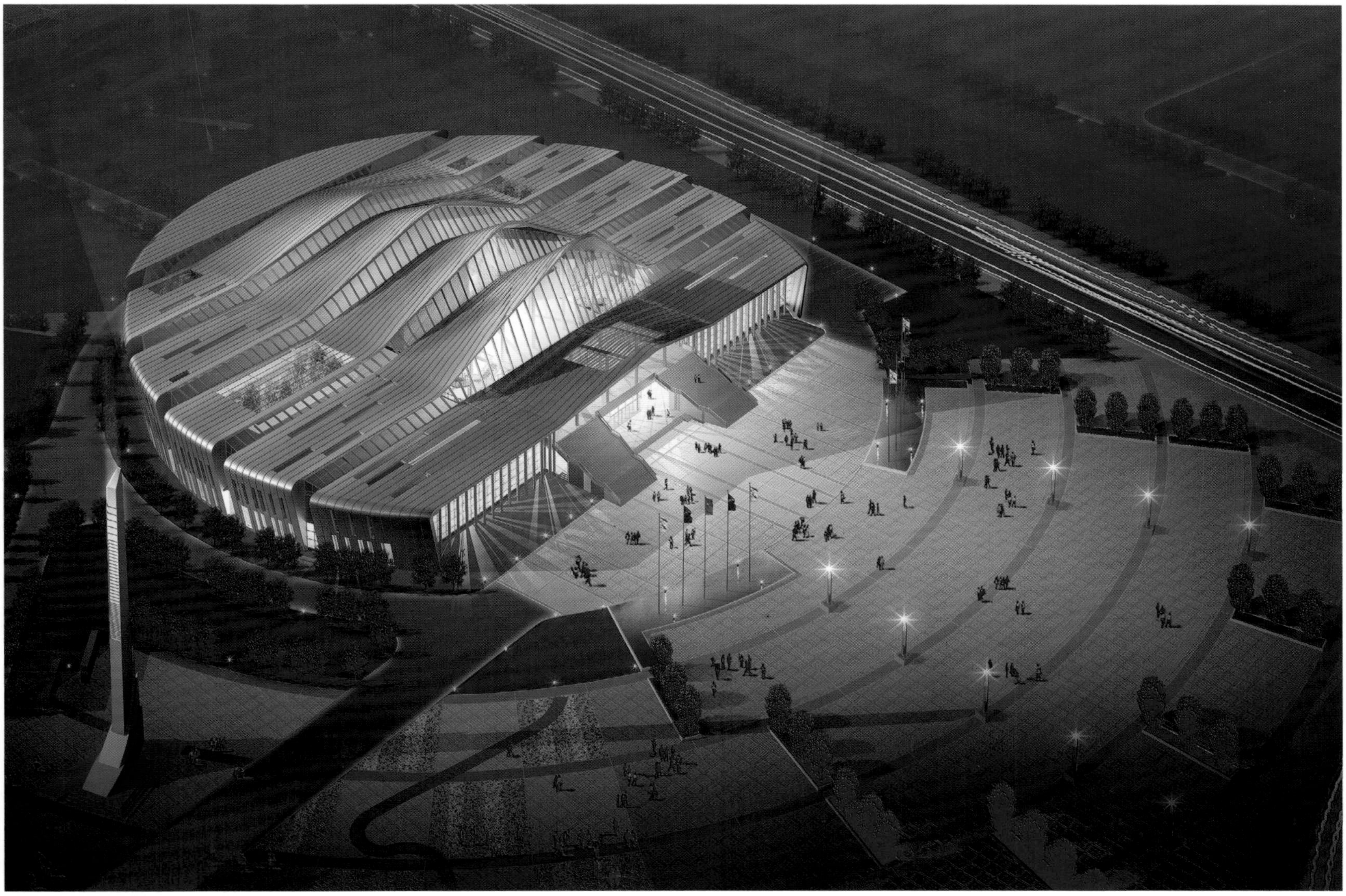

长春科技文化中心
Changchun Science and Technology Center

项目名称：长春科技文化中心
用地规模：34.45 万平方米
建筑面积：10.23 万平方米
容积率：0.3
项目地点：中国 / 吉林 / 长春
编制时间：2007 / 01

长春科技文化中心是一组低能耗、生态化的新型建筑，其形态将长白山黑、白、水的意境融入其中，威严地耸立在长春经济开发区内，既现代又不出格，自然地融入周边国际会展建筑群中，以其独特的姿态体现出与时俱进的时代精神，引领新一代的城市建筑潮流。

Changchun Science and Technology Center is a sustainable, ecological new building, which is a bold innovation to perfectly connect the architectural language with mountain and lake, as well as combine the space language with traditional architectural culture. The whole building straightly stands at Changchun Economic Development District. The configuration is modern while elegant. The unique spirit leads the fashion of the city.

113

总体鸟瞰图

长春科技文化综合馆
Changchun Science and Culture Pavilion

项目名称： 长春科技文化综合馆
用地规模： 6.06 万平方米
建筑面积： 2.18 万平方米
容积率： 0.36
项目地点： 中国 / 吉林 / 长春
编制时间： 2007 / 01

长春科技文化综合馆是长春国际会展中心的主要组成部分，以其独特的形态和人性化合理的功能布局，像长春的市花君子兰一般坐落在长春经济开发区内，成为长春的标志性建筑和东北地区的展示中心。设计将城市功能有机融合起来，在带来整体文化繁荣的同时带动周边地块价值的提升与振兴。

Changchun Science and Culture Pavilion, with its unique shape and reasonable layout, like a clivia in the Changchun Economic Development Zone, has become the landmark of Changchun City and the exhibition center of the northeastern China. The design combines with urban functions, bringing prosperity while enhancing the value of the surrounding plot.

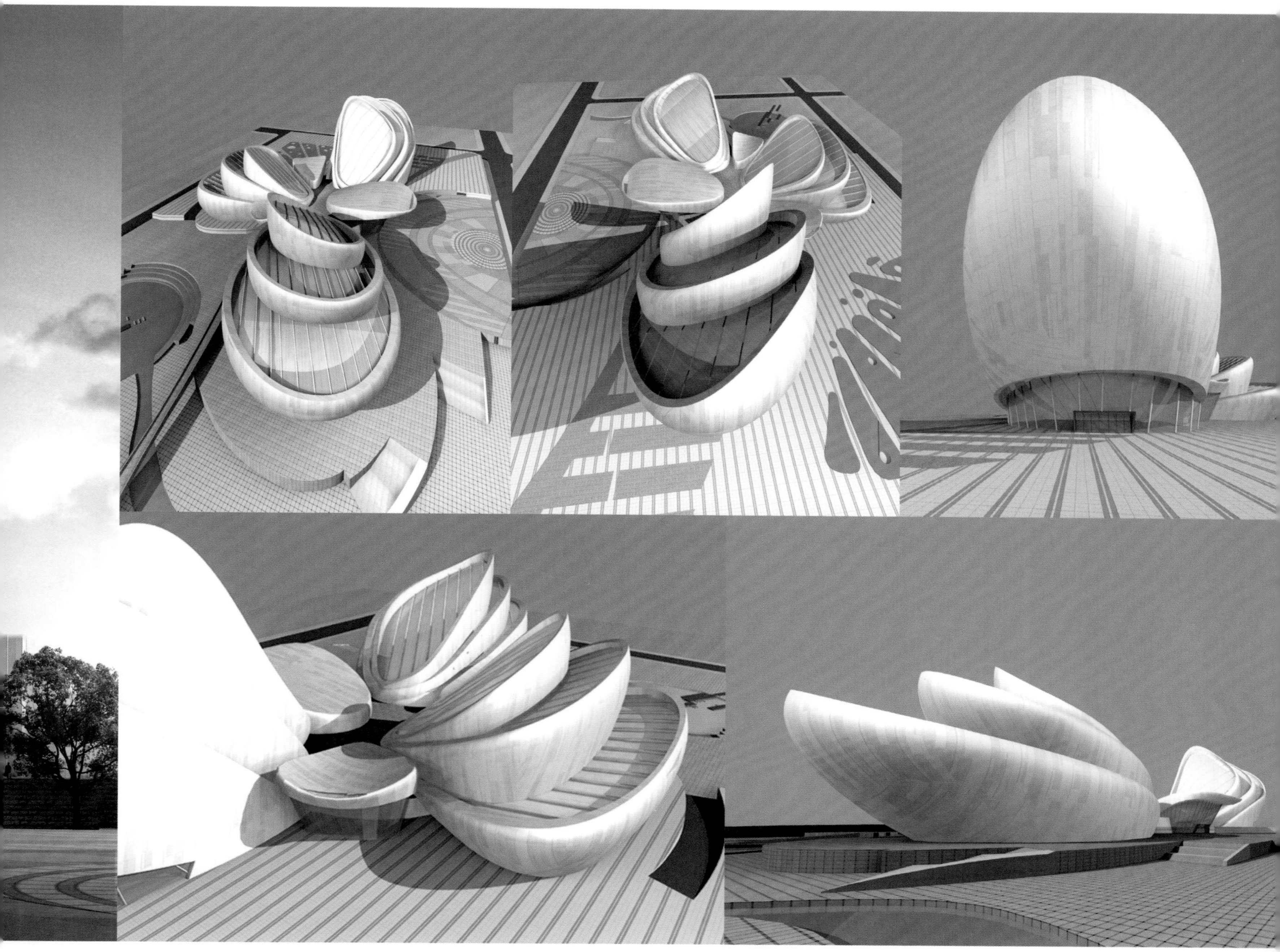

青城山道教博物馆
Mount Qingcheng Taoism Museum

项目名称：青城山道教博物馆
用地规模：3.67 万平方米
建筑面积：1.96 万平方米
容积率：0.53
项目地点：中国 / 四川 / 青城山
编制时间：2007 / 06

道教是"活"的本土宗教：既有"博大精深"的文化内涵，又有"喜闻乐见"的群众基础；既有"体悟修炼"的深远哲学思维，又有"身体力行"的寻常处世原则；所以，得"道"应该依循两个线索——分别是精神层面的"悟"和实践层面的"行"。

作为青城山脚下的道教博物馆，它的社会意义不应仅仅停留在文物"展示"上，而是通过营造一种特殊的场所精神，使参观者在游历过程中深入思索、开启心智、提升修养，从而让传统的道教文化对现代生活产生积极意义。这是一个心灵的"体验"历程，而并非简单的"参观"过程，由此导出我们的设计理念——体验式博物馆。

Taoism is the "living" native religion: it has both a "broad and profound" cultural content and a "lovable to see and hear" mass basis, both the far-reaching philosophical thought of "experience, comprehension and exercise" and the common behavioral principle of "earnestly practicing what one advocates"; therefore, to achieve "Nirvana", one should follow the two clues-the "comprehension" at the spiritual level and the "action" at the practical level respectively.

As a Taoist Museum at the foot of Mount Qingcheng, it should not only have its social meaning on the mere "exhibition" of cultural relics, but also make visitors think in depth, open the mental intellect, and enhance the cultivation during their touring process, so as to allow the traditional Taoist culture to generate a positive bearing on modern life, through the creation of a special locale spirit. This is an "experience" of the soul, rather than a simple "visit" process. From there we derive our design concept-Experience-type Museum.

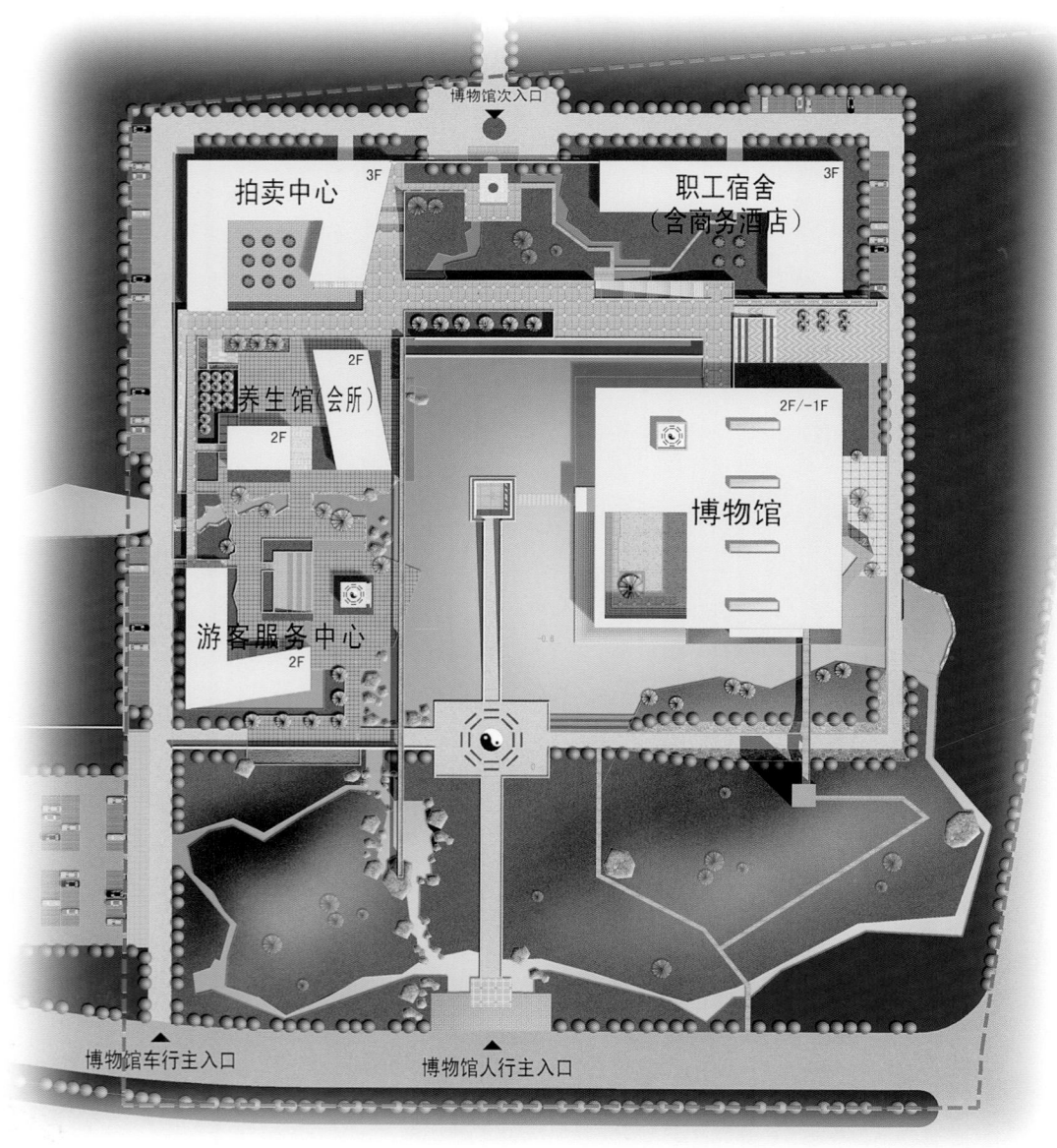

企僧束立跨偕朿竹自見偕
不明自是者不彰自伐偕飛
功自矜偕不處箕在道也日
命食赘行物或蕆此故劳道
偕不處

企僧束立跨偕朿竹自見偕
不明自是者不彰自伐偕飛
功自矜偕不處箕在道也日
命食赘行物或蕆此故劳道
偕不處

企者不立 跨者不行 自見者不明 自是者不彰 自伐者無功 自矜者不長 其在道也曰 餘食贅行 物或惡之 故有道者不處

SMITHGROUP cna

Architecture · Urban Planning · Landscape Design
Interiors Design · Art Consultant · Lighting Design
Since 1853
www.cna-group.com

Design Determines . . .

设计决定……

绿地花桥企业总部
Greenland Huaqiao Headquarter

Architecture Design | 2007·06

绿地花桥企业总部
Greenland Huaqiao Headquarter

项目名称：绿地花桥企业总部
用地规模：13 万平方米
建筑面积：9.13 万平方米
容积率：0.7
项目地点：中国 / 江苏 / 昆山
编制时间：2007 / 06

1. 人本——充分考虑现代都市的生活休闲方式，形成一种绿意盎然，自然和谐，经典高尚的居住与商务休闲环境。

2. 自然——贯彻"尊重自然"与"可持续发展"的思想，贯彻生态原则、文化原则与效益原则，力求塑造一个具有优雅环境、丰富文化内涵、显著经济效益和鲜明个性的花园式经典高尚居住空间和办公商务区。

3. 文化——体现绿地 21 城特有的文化底蕴，使之在城市中独树一帜，充分体现区域特色。

4. 融合——讲求人与环境的融合、建筑与整体规划布局的融合、居住行为与购物行为的融合、建筑与绿化环境的融合。

5. 科技——强调高科技在建设、管理和生活中的运用，以科技为向导，加强市政基础设施的合理规划，注重基本配套，强调节能建筑。

6. 安全——适度的人车分流，既消除人车混行的不安全因素，又避免由此造成的时段性消极空间，确保车行与步行系统的利用率。针对不同出行设置道路及出入口，避免交叉干扰。合理组织不同功能区的客、货流，使其合理化、便捷化。

建筑形象设计

对于绿地大道充分考虑了其天际线，使建筑群的轮廓起伏优美，外形简洁富有力度。以玻璃与实体墙面组成简洁的外墙，玻璃的通透和建筑组群的错落形成一种进行式的动态空间。同时，沿绿地大道方向考虑了整个区域的整体性和连续性，通过相近的设计手法与材质、细部的处理，使商业、塔楼、总部办公在绿地大道方向上既有自己的个性特征，又相互和谐统一。

商业区的形体以穿插、灵活多变为特色。流动变幻的商业空间本身作为"人的活动"景观为全区提供了生机与活力。滨河区建筑作为一组连续的景观建筑，利用建筑的退台、极简的体块、水中的倒影形成"景"与"景"的互动，与二期滨河商业区共同形成徐公河上一组"漂浮"的、亮丽的风景线。

1. Humanism-we take full account of modern urban life and leisure to form a harmonious and natural residential and business leisure environment.

2. Naturalism-with implementation of "respect for nature" and "sustainable development", we strive to create an elegant environment with rich cultural connotations, economic efficiency, individual noble garden-style and classic residential area and office district.

3. Culture-reflect unique cultural heritage of Green 21 City to show its uniqueness and regional feature.

4. Integration-we attach great importance to integration between man and the environment, architecture and overall layout, residential behavior and consuming behavior, architecture and green environment.

5. Technology-we emphasize on application of high-tech in life, construction and management, to enhance municipal infrastructure planning.

6. Safety-appropriate division between passenger and vehicle flows can not only eliminate safety risk but also avoid period of negative space caused therefrom, thus ensuring the full utilization of vehicle and passenger systems. Design different roads and entrances for different travels to avoid cross-interference. Ensure rational organization of passenger and cargo flow in different functional areas to be easy and convenient.

Architecture Image Design

We take full account of the skylines of Green Road. The building has a simple and rich appearance with ups and downs. Glass walls and entities make up the simple exterior walls. Permeability and randomly strewn cluster of glass form a dynamic space. At the same time, along the Green Road we should consider the integrity of the entire region and continuity. Through similar design techniques and materials with the treatment of details, Headquarters Office and commercial buildings on the Green Road have their own personality traits, and are harmonious with each other.

Downtown's body has its insertion and flexible features. Business flow changes of space itself as "human events" landscapes for the whole region offer vigor and vitality. Riverfront area of architecture as a contiguous set of landscape architecture, makes use of the leveraging building retreat, a very simple body block and the reflections in the water to form the interaction between sceneries, formed jointly with phase II Riverside Commercial on the Xugong River, and to present a set of "floating", and beautiful scenery.

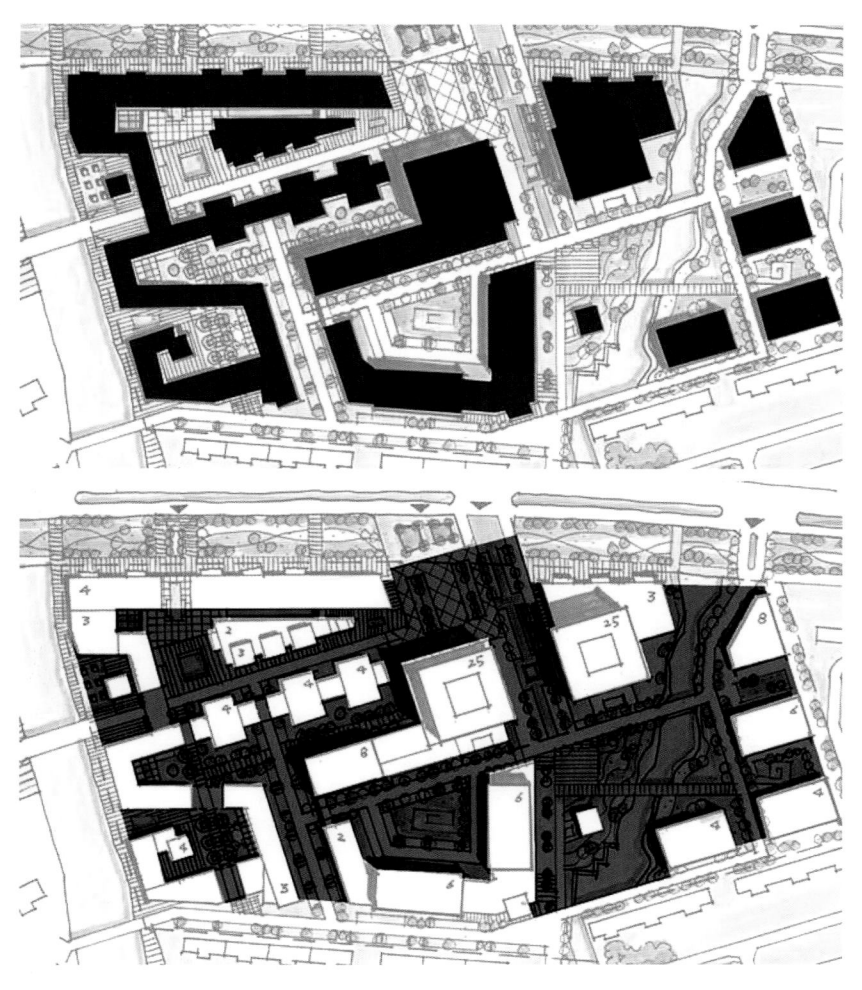

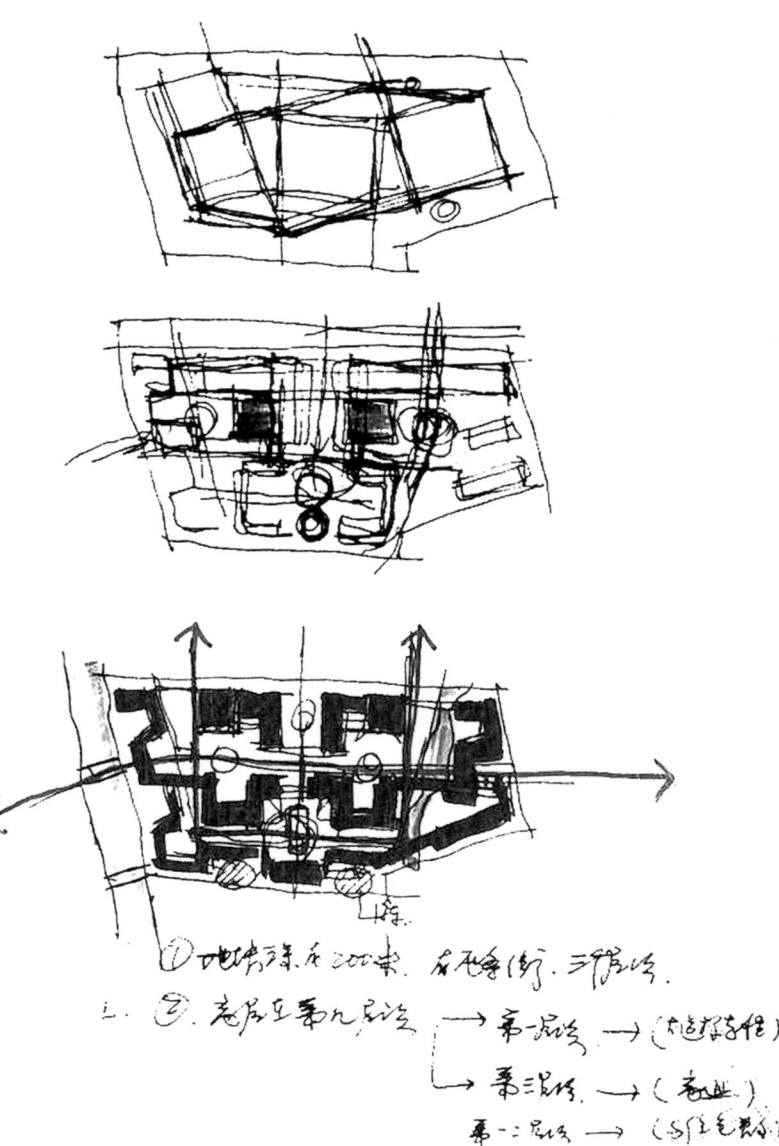

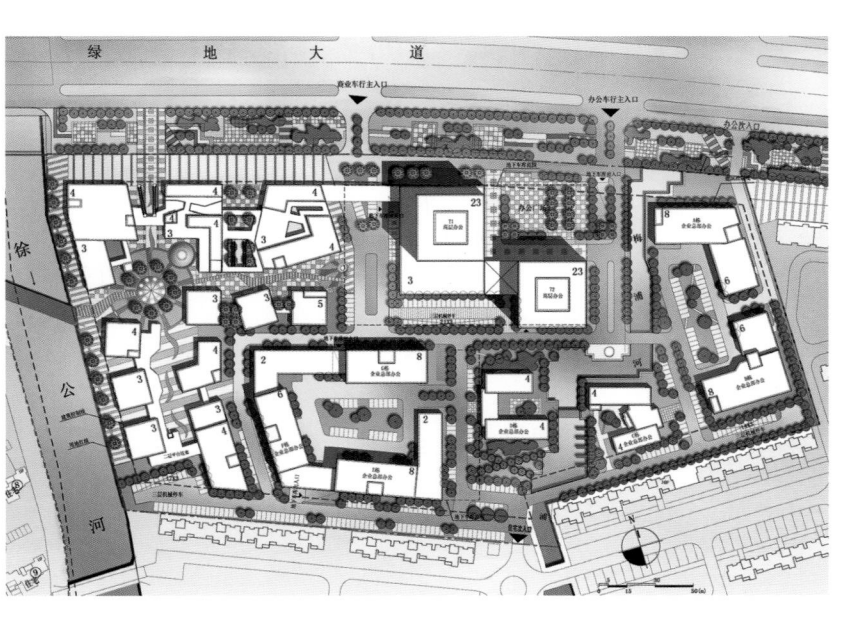

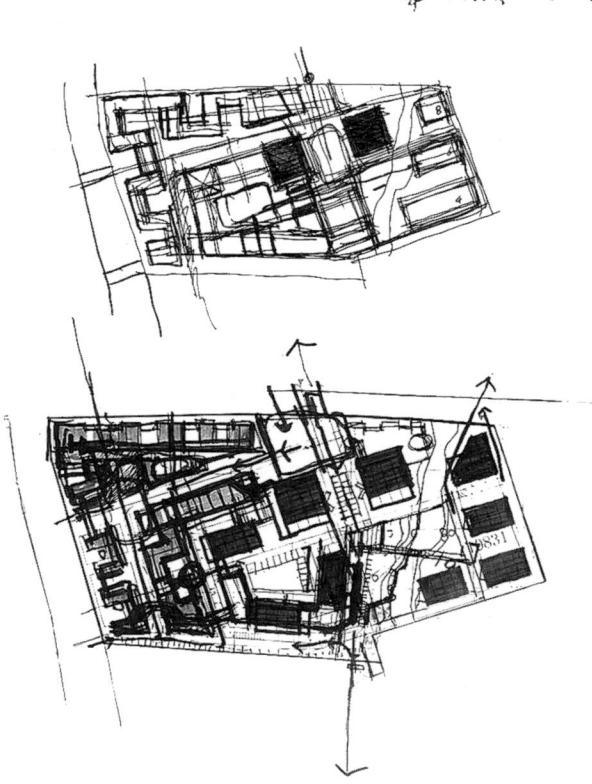

现房抢购
8800元/m²
021-39217001

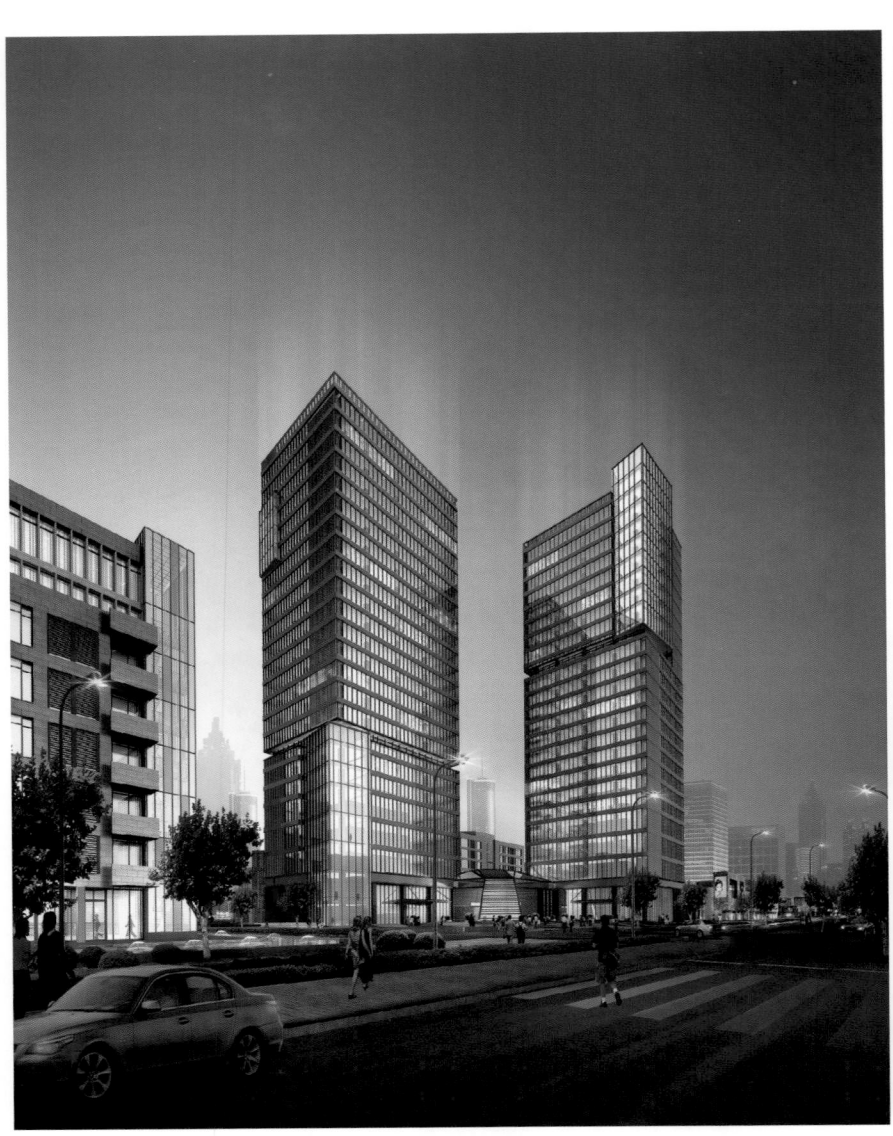

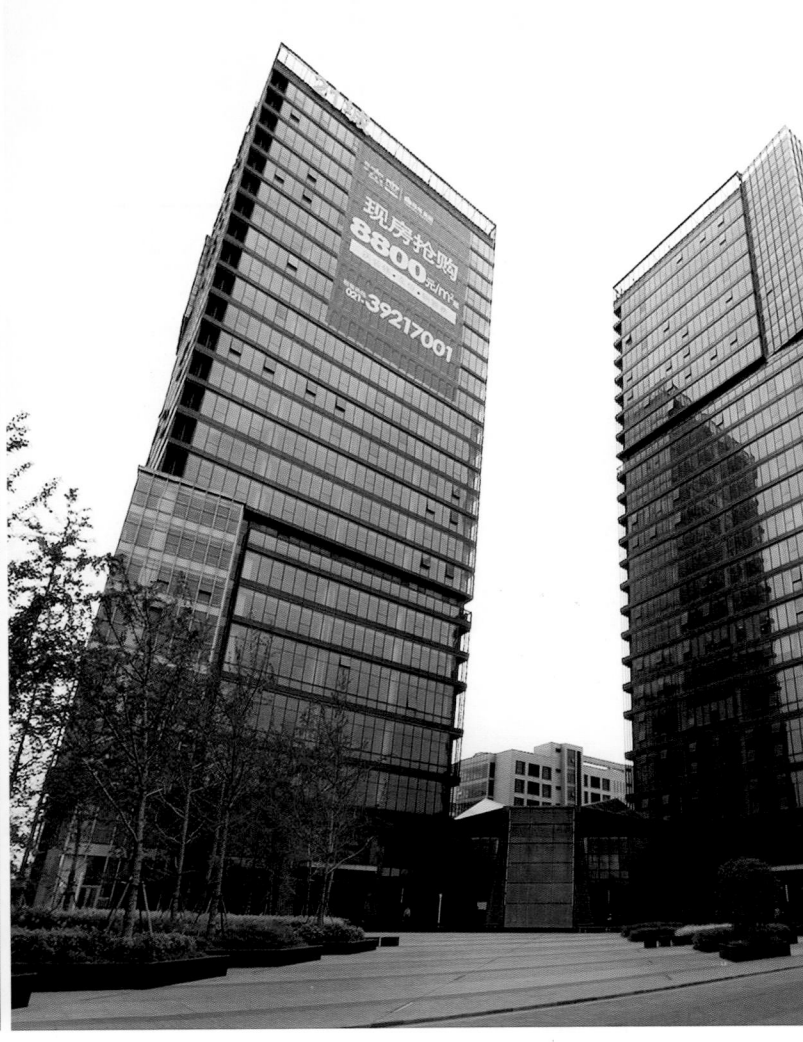

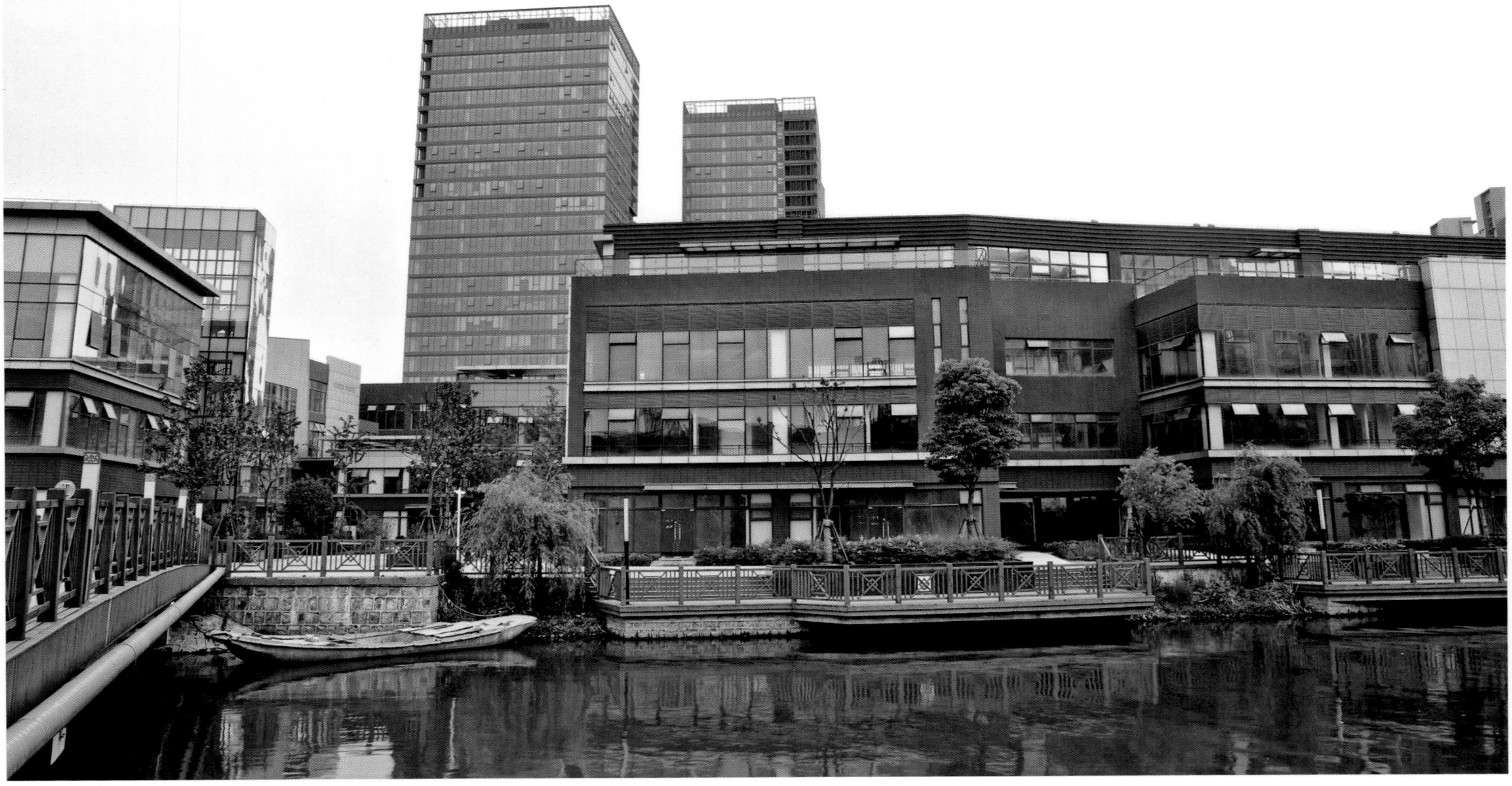

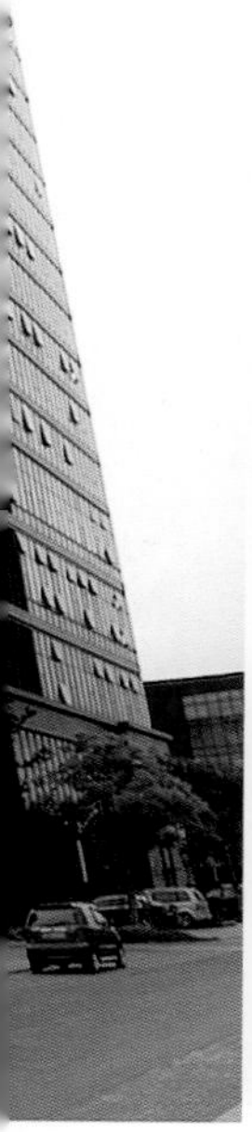
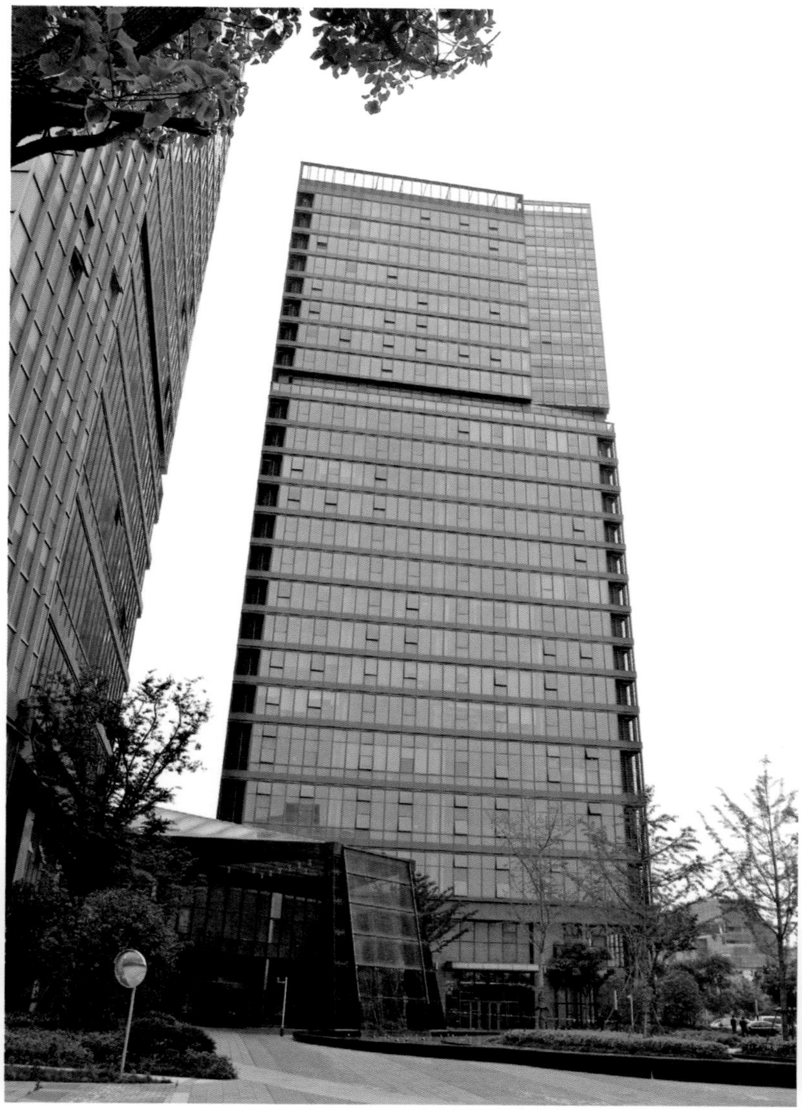
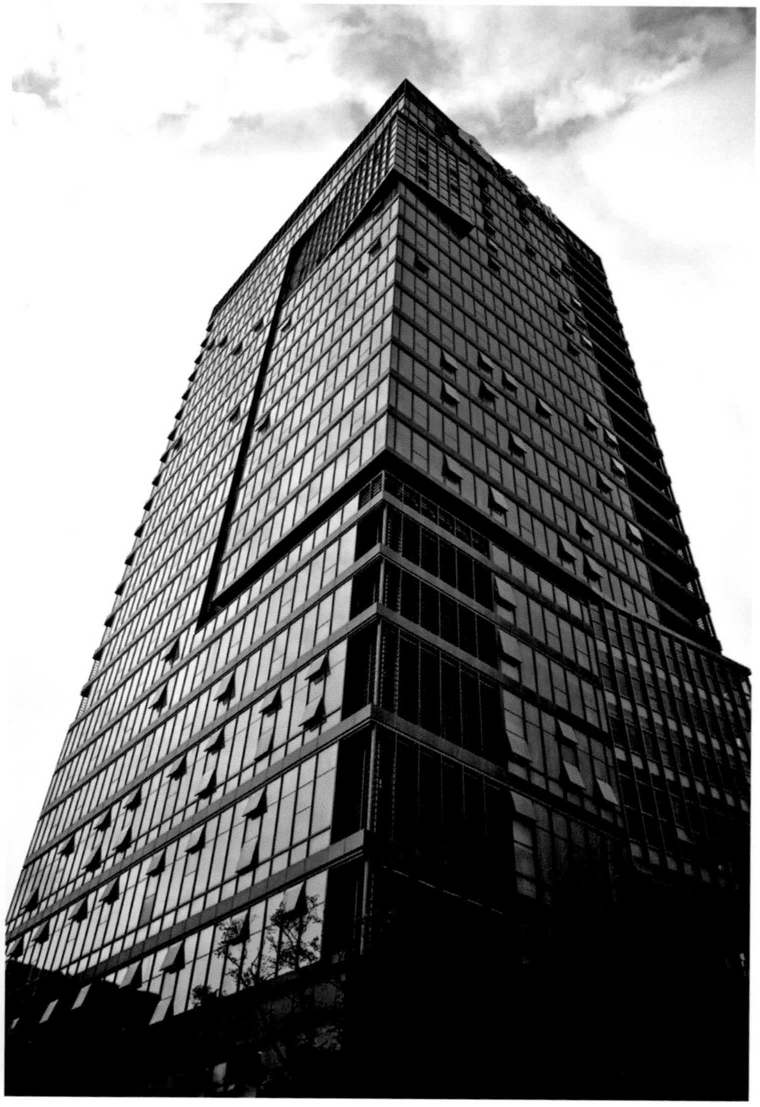
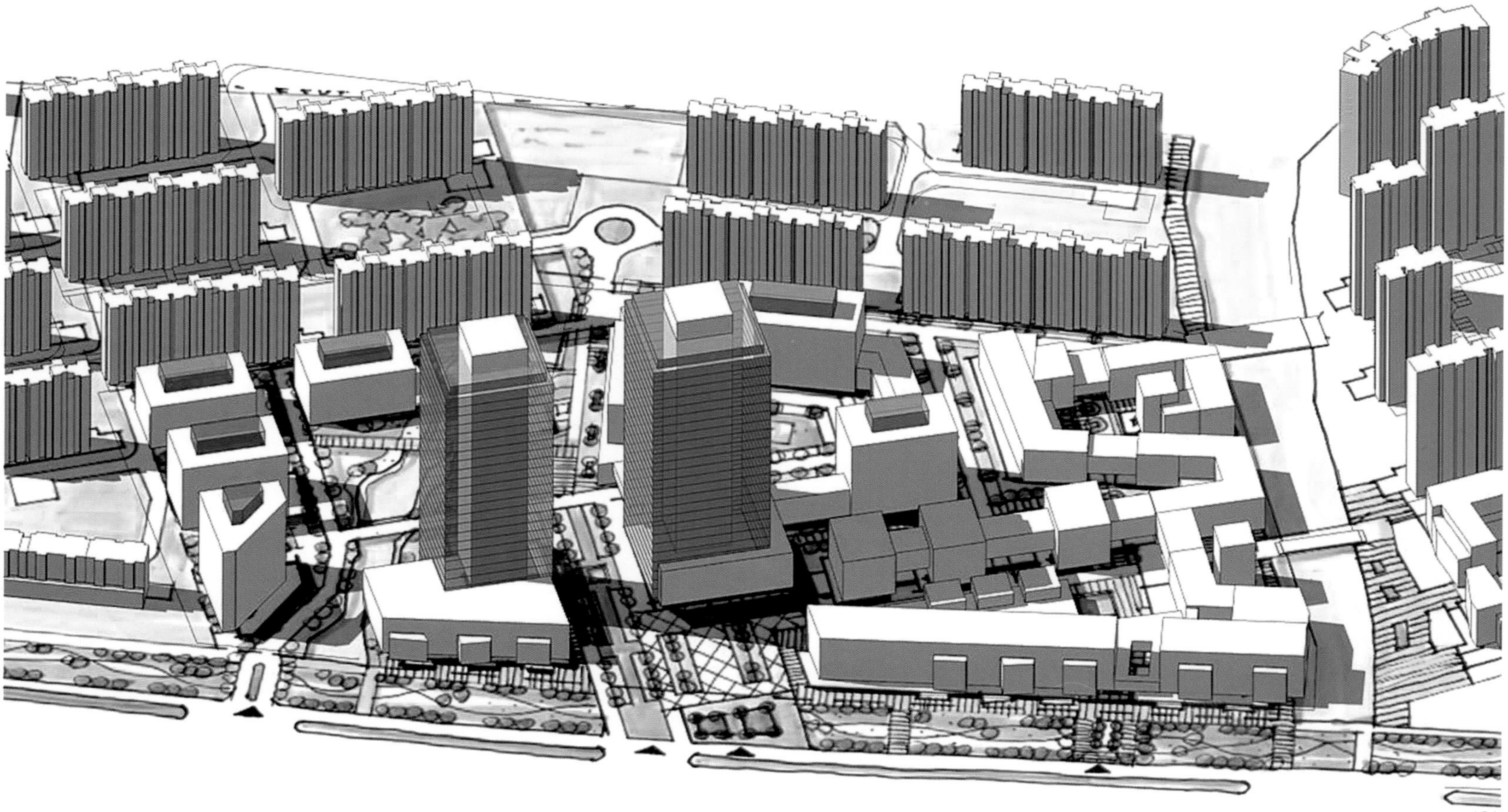

总体鸟瞰图

长春环球贸易中心
Changchun World Trade Center

Architecture Design | 2010·09

长春环球贸易中心
Changchun World Trade Center

项目名称：长春环球贸易中心
用地规模：4.49 万平方米
建筑面积：21.53 万平方米
容积率：4.8
项目地点：中国 / 吉林 / 长春
编制时间：2010 / 09

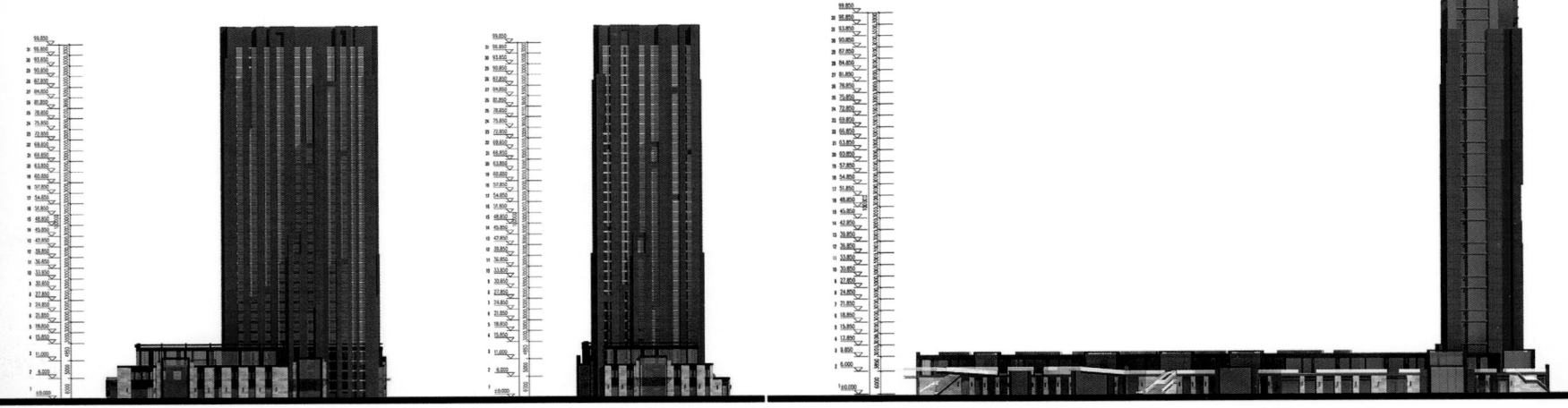

基地位于长春绕城高速东南向内侧，老环城路外侧，卫星路南侧，其中彩宇大街、102 国道穿越其中，交通条件优越，属于城市地标性区位。北面有老城商业中心，南面有净月 EBD，西面有新城 CBD。本项目为一大型综合体，包括了五星级酒店，办公，商业，住宅与公寓等功能。以玻璃、混凝土与陶板的结合将现代建筑的时尚、活跃和周围环境的距离拉近，并使得中间的商业步行街充满活力。

本项目整个规划围绕着周边环境严谨地展开，在面对广场的一面我们设置了商业、酒店与办公场所，在南面设置的是住宅与公寓。为了更好地组织人流与降低高层建筑给行人近距离的压迫感，设计在地块的中心部位设置了逐级下沉的景观商业广场，并且把住宅与公寓的入口设在了南面，这巧妙地避免了居住人流与商业人流的相互干扰。酒店、办公与商业的集中设置也使得其功能被更加高效地利用。

The site has a good city landmark location. North of the site is the old city commercial center; south of it is Jingyue EBD; west of the site is New City CBD. This project is a large complex, including a five-star hotel, office buildings, commercial area, residential apartments and other functions. Building materials are mainly glass, concrete and ceramic panel.

Based on the surrounding environment, we put commercial area, hotel and office on the face side of the square, residence and apartments in the south. In order to better organize the traffic flow and reduce the plot ratio, the sunken plaza is set in the central part of the site. The entrance of the residential building is located in the south which avoids the interfering between the residential flow and the business flow. The centralized organization of hotel, office and commercial area also makes their function be more efficiently utilized.

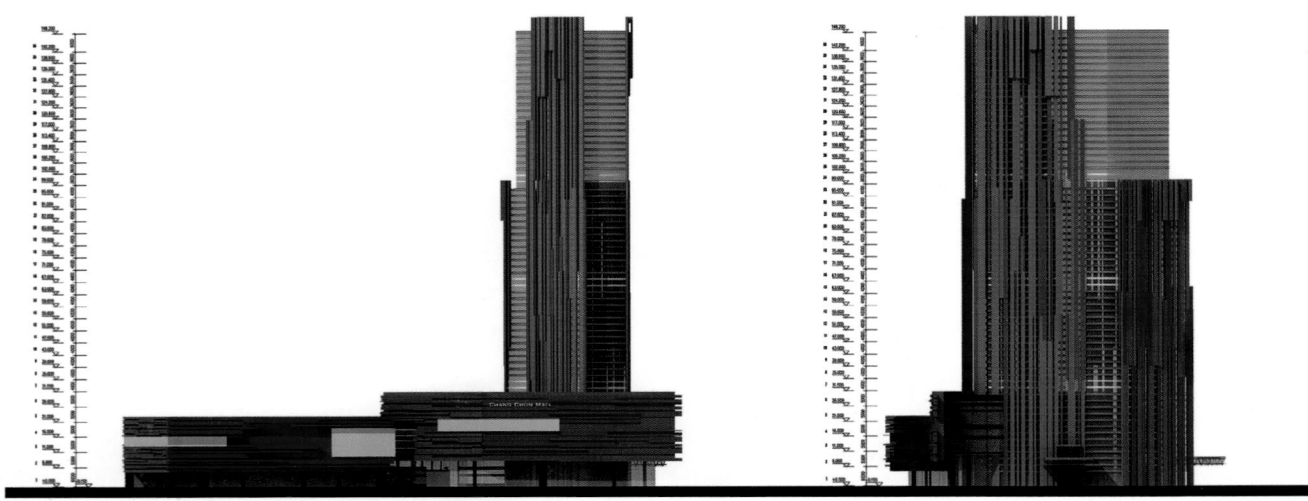

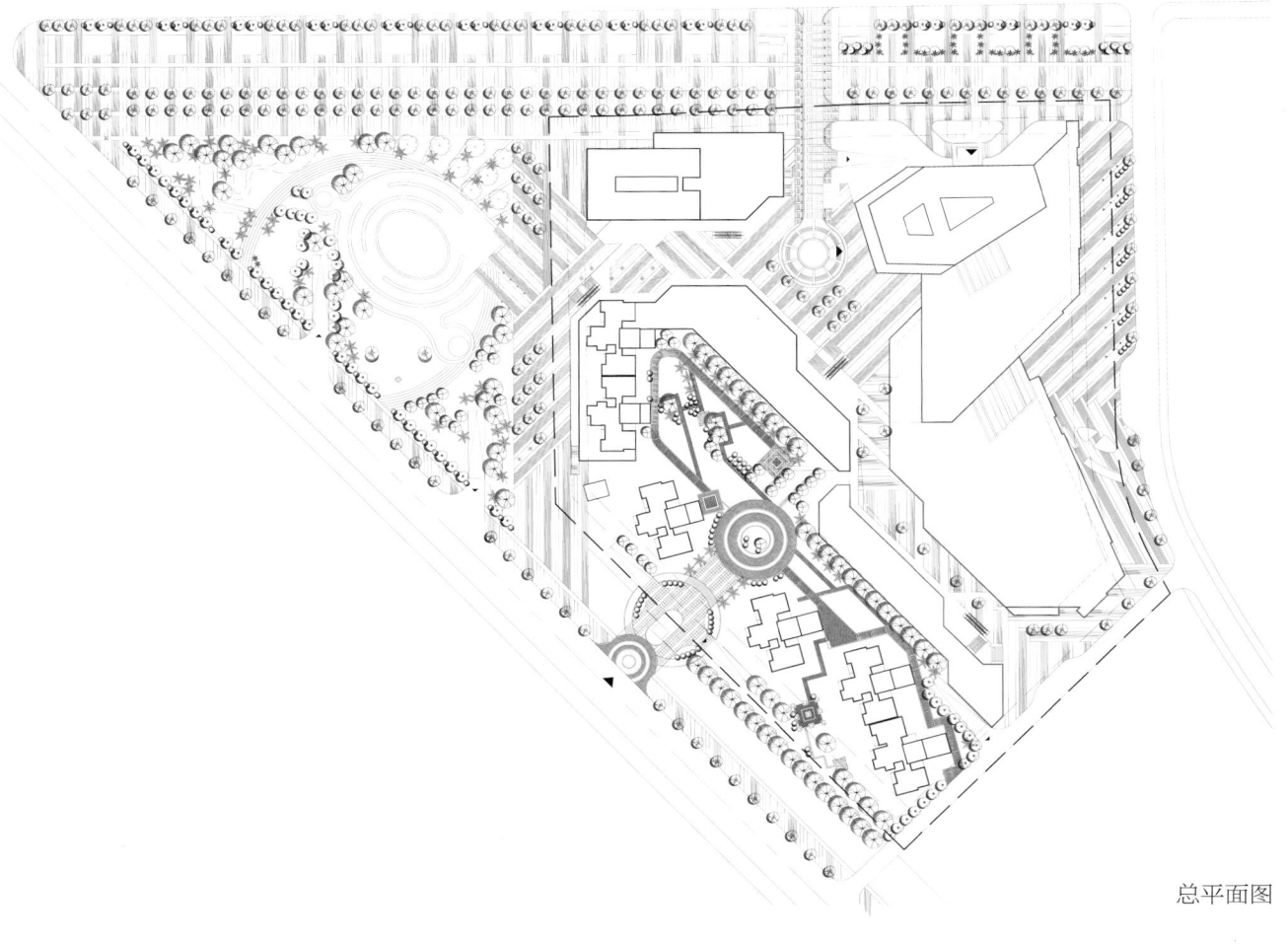

总平面图

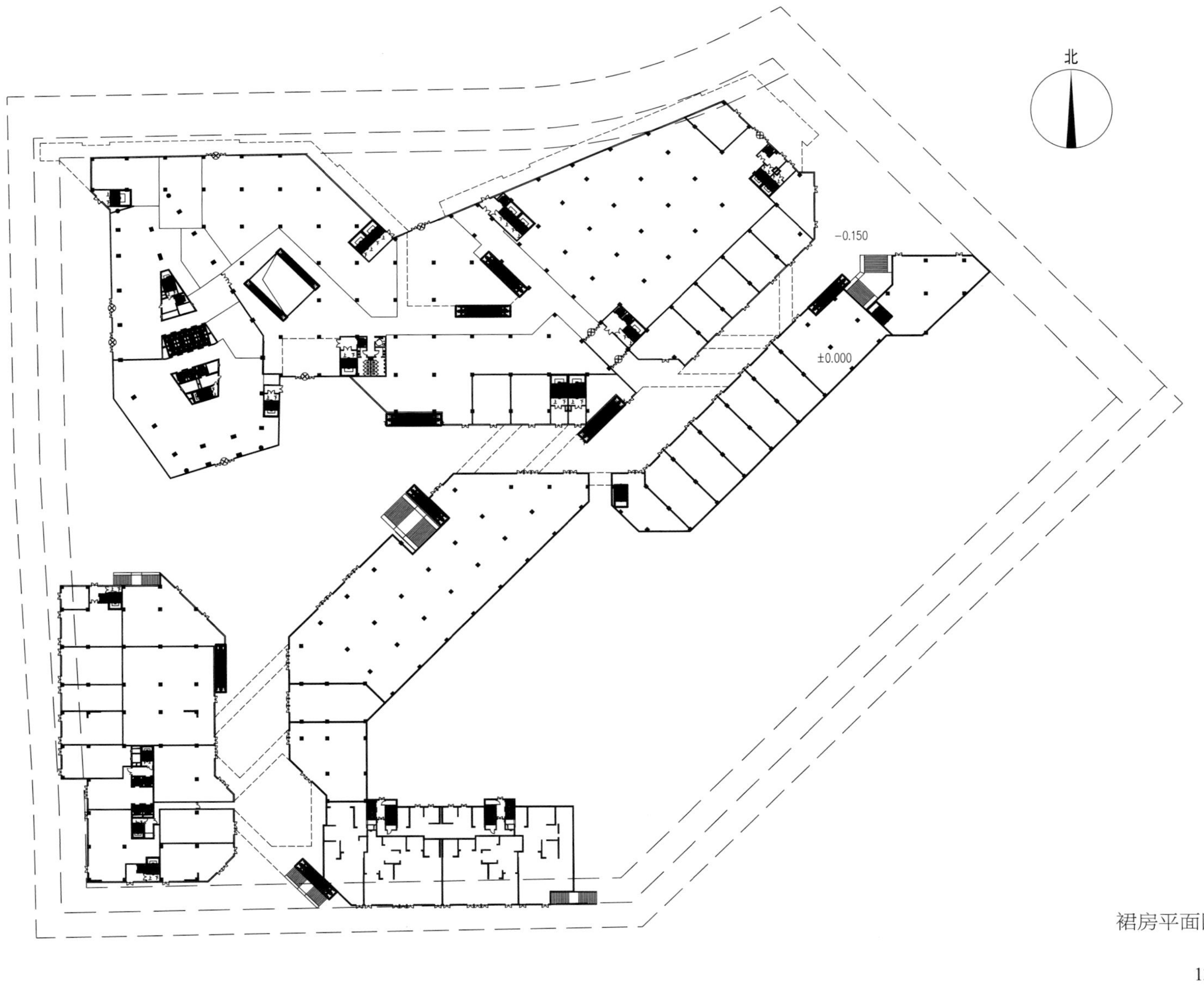

北

−0.150

±0.000

裙房平面图

各行车视角透视图

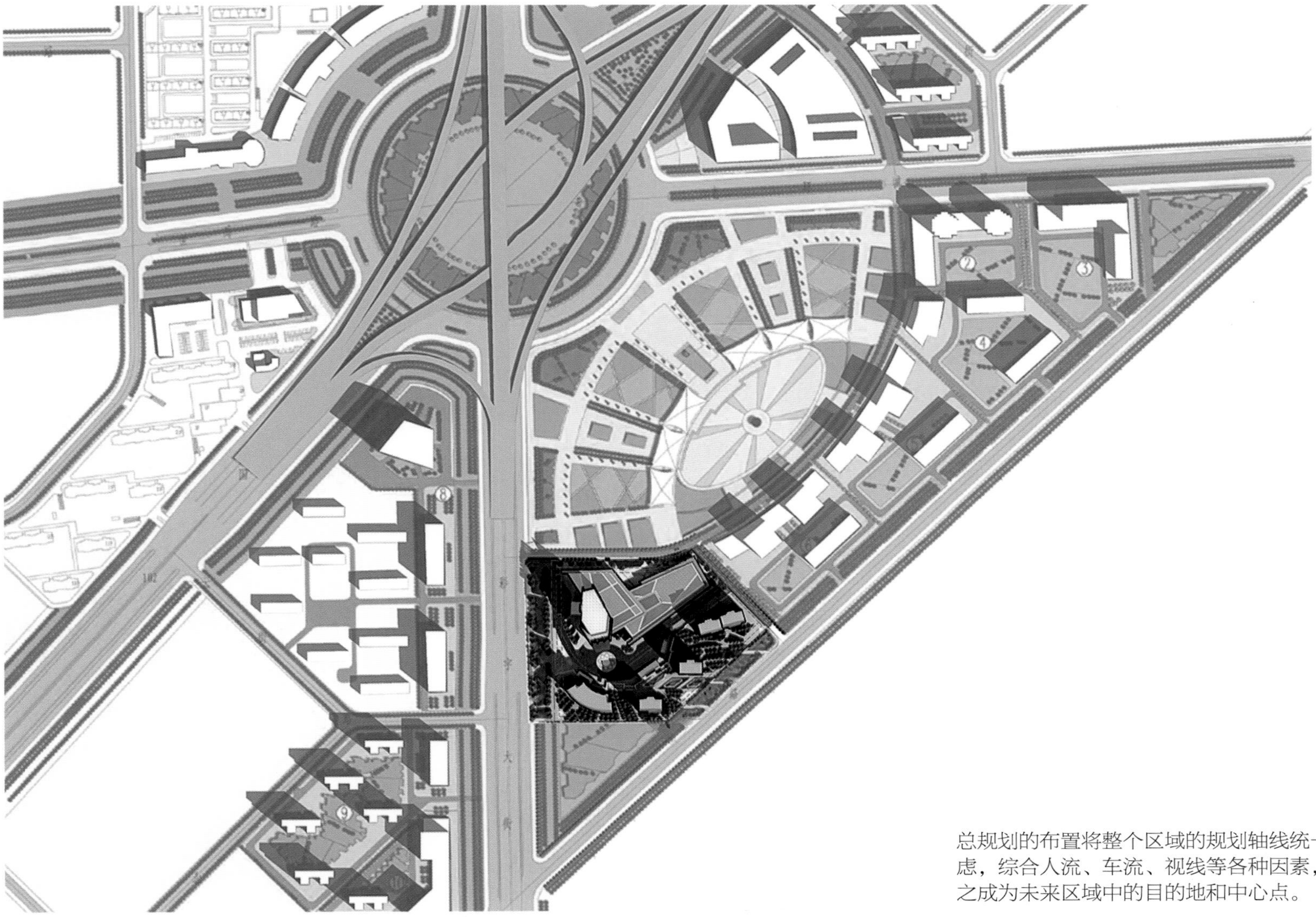

总规划的布置将整个区域的规划轴线统一考虑，综合人流、车流、视线等各种因素，使之成为未来区域中的目的地和中心点。

舞动着的红色芭蕾舞鞋
旋转的图腾

贸易中心酒店式公寓效果图

商业内街入口效果图

蚌埠中华第一街
Bengbu The First Street of China

项目名称: 蚌埠中华第一街
用地规模: 36.18 万平方米
建筑面积: 62.45 万平方米
容积率: 1.73
项目地点: 中国 / 安徽 / 蚌埠
编制时间: 2007/ 03

安徽省蚌埠市地理位置良好，交通发达，承东启西，是中国南北交通的重要枢纽，也是华东地区安徽省的最大城市之一，中华第一街项目地块位于正在建设中的蚌埠中心城区的南侧，项目总用地面积约 36.2 万平方米，我们将其设计为未来蚌埠市新的金融、贸易、生活居住中心，使之成为蚌埠市民的"心中之城"。

Bengbu, Anhui enjoys a good location with favorable traffic connecting east and west. It is an important north-south transport hub and one of the largest cities in Anhui Province. The First Street of China project is located in the south of the Bengbu downtown area under constructing. The total floor area of the project is about 362 thousand square meters. It is a core part in the "downtown area" identified by the general planning of Bengbu City. Through the comprehensive development this time, The First Street of China will become a new financial, trading and living center of Bengbu in the future and the "city in heart" of Bengbu people.

上海金山新城综合社区
Shanghai Jinshan New City Comprehensive Community

项目名称：上海金山新城综合社区
用地规模：46.0 万平方米
建筑面积：141.0 万平方米
容积率：3.06
项目地点：中国 / 上海 / 金山
编制时间：2007 / 11

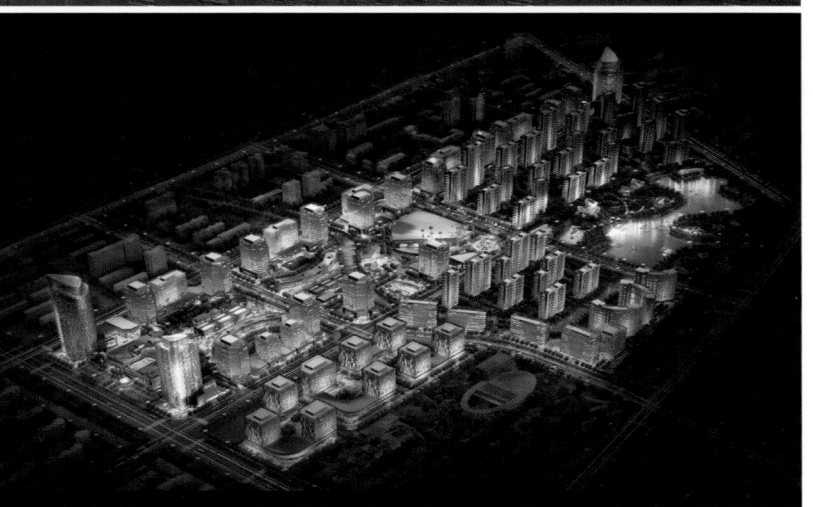

项目沿龙翔路以南区域形成综合性的公共设施，计划设置金山新城的商业金融和商务办公中心，成为该地区的一处全新的综合性区域，并展示金山新城的现代化形象风貌，设计同时结合了娱乐、餐饮、旅游等功能，也是金山新城对外发展和辐射长三角南翼的功能核心。

The project is a mixed-use project in south of LongXiang Road. It aims to set a commercial finance and business center for JinShan New City, making it a brand new mixed-use area which shows its modern image. The design carries the function of entertainment , restaurant and tourism, also the function core for JinShan New City's development and south Yangtze River Delta.

富阳世界贸易中心
Fu Yang Global Trade Center

项目名称：富阳世界贸易中心
用地规模：1.47 万平方米
建筑面积：5.52 万平方米
容积率：3.76
项目地点：中国 / 浙江 / 富阳
编制时间：2003 / 09

设计宗旨

一个在新世纪、新富春江时空条件下，国际一流的集金融、旅游、商业、宾馆、办公、会展、会议为一体的综合性建筑是本设计的宗旨。该大厦将成为富阳市最重要、最有特色的标志性建筑。

设计理念

经过对地域环境、城市发展、功能设施、用户要求和业主利益的全面、详细和深入的分析后，定出"二个定位"和"三个关系"作为本设计的理念，贯彻到设计细节。二个定位为建筑定位——富阳市一流的宾馆、办公、商业为一体的综合性建筑总部；设计定位——再创时期的精品。三个关系即与富春江的关系，与区域建筑的关系和与相邻建筑的关系。

Design Purpose

The design aims to create a complex with unified functions of finance, tourism, commerce, hotel, office, exhibition and conference. This complex will be the most important and distinctive signature building in Fuyang City.

Design Concept

After detailed and thorough analysis on the regional environment, urban development, functional facilities, residents' demands and the owner's interests, the design concept for this project has been set as "two orientations" and "three relations", which should be reflected throughout the detailed design process. Two orientations are architectural orientations to be first class complex of hotel, office and commerce in Fuyang City, and design orientation, to create a masterpiece. Three relations mean the relations, with Fu Chun River, regional architecture and neighboring architecture.

沈阳皇姑车辆厂工业产业园
Shenyang Huanggu Vehicle Factory Industrial Park

项目名称：沈阳皇姑车辆厂工业产业园
用地规模：85.00 万平方米
建筑面积：280.50 万平方米
容积率：3.3
项目地点：中国 / 辽宁 / 沈阳
编制时间：2007 / 08

基地位于沈阳市的老工业核心区，新的规划将原沈阳最大的工业企业皇姑沈阳车辆厂定义为在东北地区具有示范意义的工业产业与时尚产业相结合的基地，是一处生态景观型、数字化、人性化的"沈阳曼哈顿"，也是一处工业设计极具发达的设计型产业基地，大规模高科技工业产品荟萃，白领机械迷云集，是人们休闲、交易的"中国工业艺术世界"。这次规划不仅为基地空间上的整合，同时还为整个项目今后的运营进行了全面的业态安排，使项目具有实际的操作意义。

The site is located at the core district of old industry park. Our new planning redefined the vehicle factory as the integration of demonstration and fashion industry site. It would become a "Manhattan of Shenyang" for the natural environment digitization and humanizing space design. It is also an "industrial art world of china" for the converge of various high-tech industry design, leisure and trade functions. The planning has also made the arrangement for the entire operation in the future, making the project pretty practical.

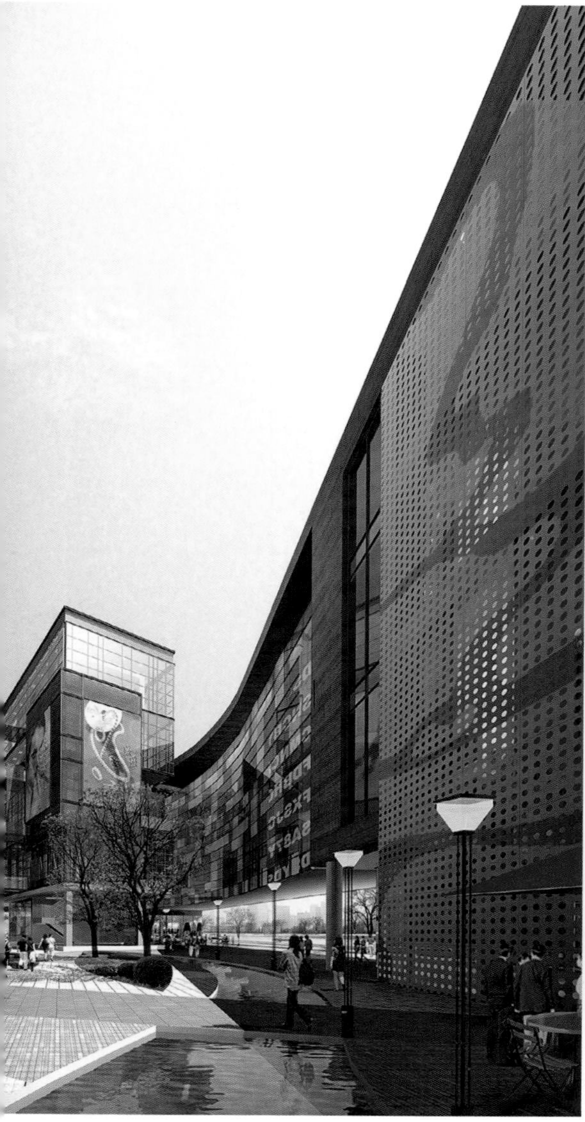

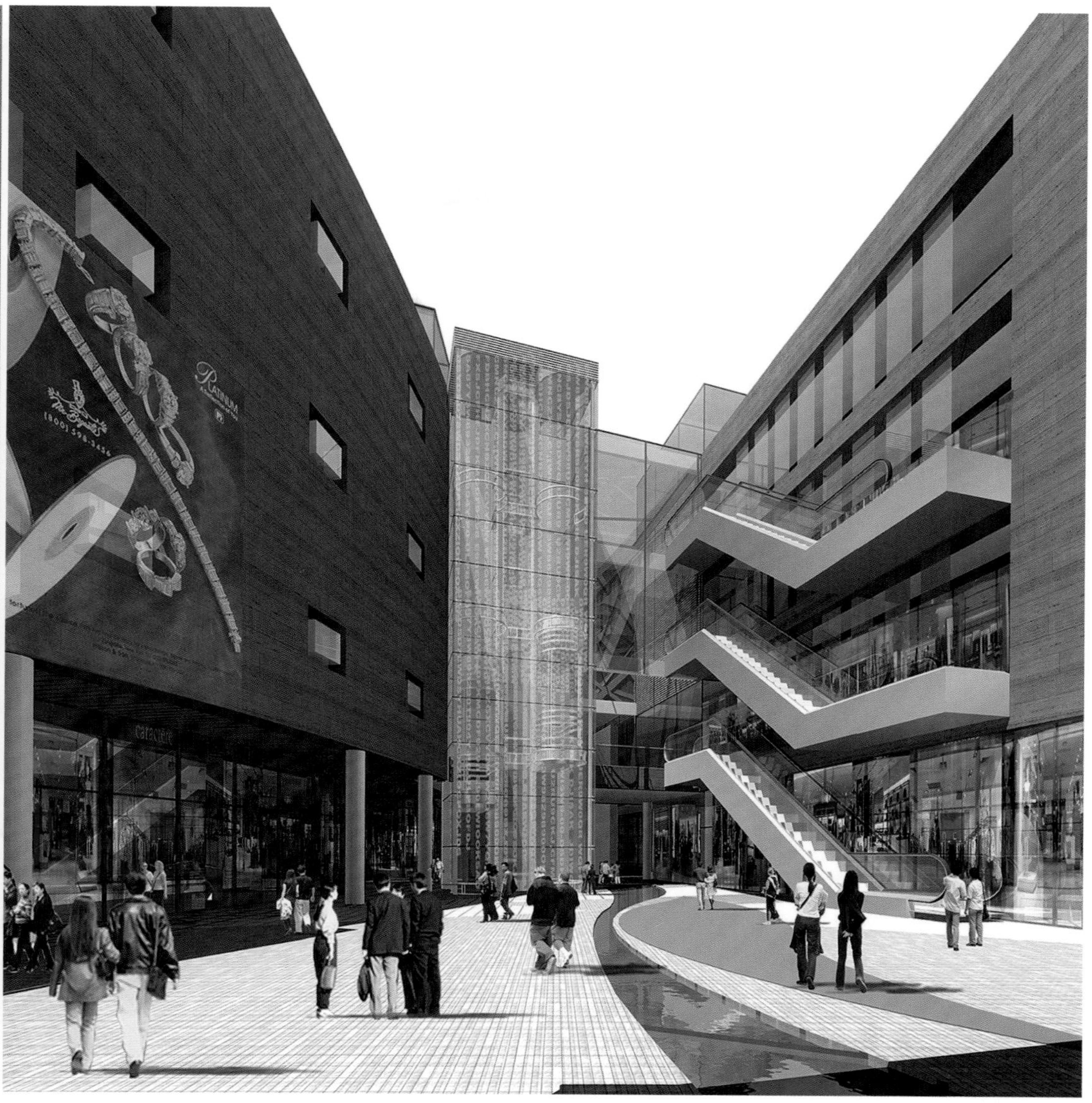

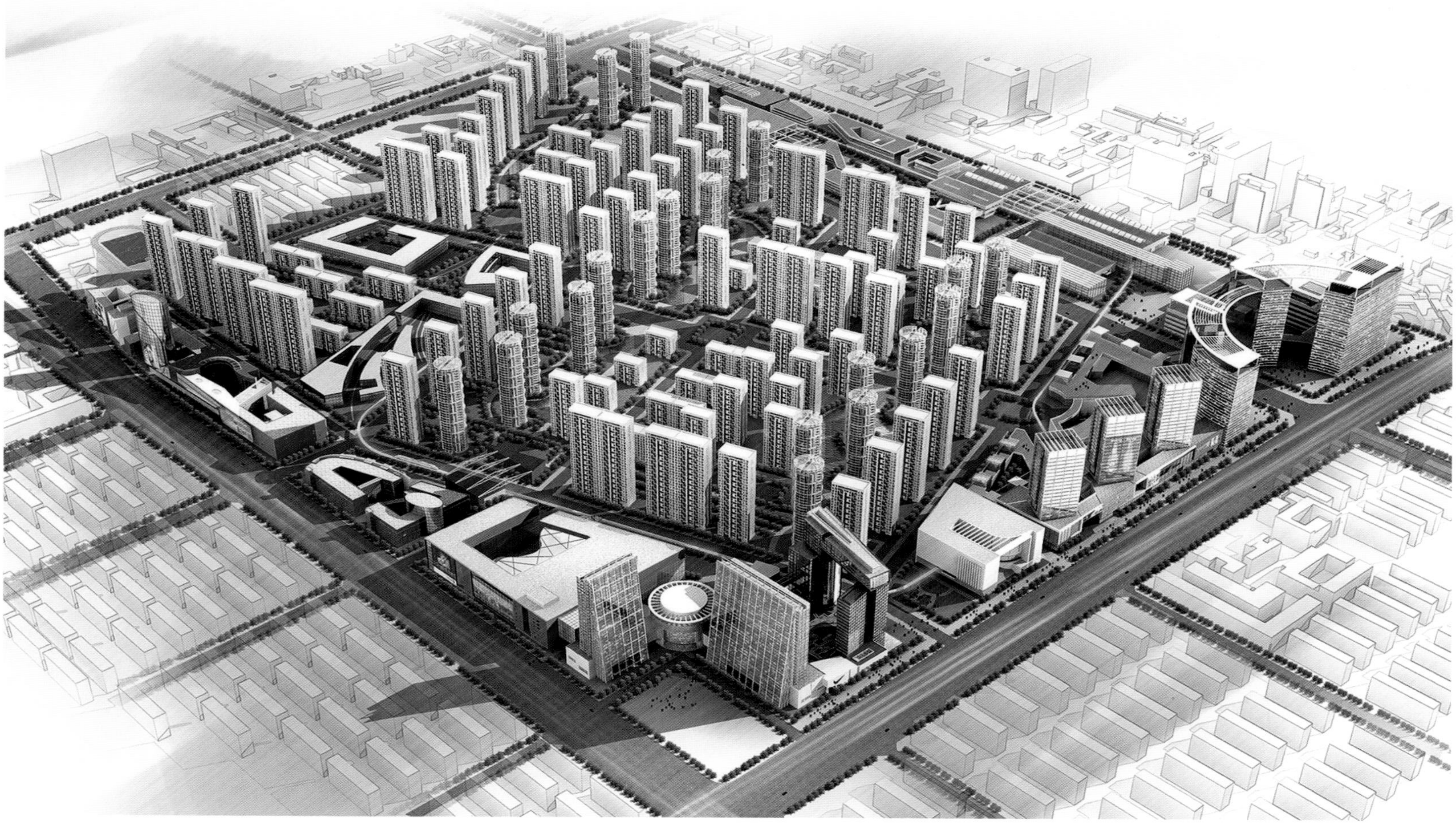

绿地牡丹江景福新都会
MudanJiang Greenland JingFu New Metropolis

项目名称：绿地牡丹江景福新都会
用地规模：1.45 万平方米
建筑面积：8.65 万平方米
容积率：5.97
项目地点：中国 / 黑龙江 / 牡丹江
编制时间：2007 / 09

景福新都会位于牡丹江市旧城的核心价值地段，地理位置十分优越。用地北侧临近牡丹江市火车站，南抵景福街，东临牡丹江市唯一的步行商业街东一条路，西临汇集了大量传统商业的太平路。

景福新都会是一座大型建筑群，集购物中心、餐饮、电影院、超市、办公楼、酒店式公寓于一体。商业购物中心布置在地块东区，该区处于"T"字形路口，沿街面长、商业价值高的部分为大型商业的旗舰店和主要入口，电影院布置在购物中心的第六层。考虑到办公楼和酒店式公寓需要相对安静的环境和人流，两者则布置在地块西区，紧挨太平路的狭长地块布置了 3 层高的餐饮楼，功能上相对主体较为独立，造型上和主体有所呼应，超市则布置在地下一层，在购物中心的南入口广场设有一人行坡道直通地下超市，并与规划中的东一条路地下商业街相连。

The project is located in the core area of Mudanjiang old city with superior geographical location. It is north to Mudanjiang train station, south to Jingfu Street, east to Dongyitiao Road, the only pedestrian street in Mudanjiang City, and west to Taiping Road.

It is a large complex with shopping centers, restaurants, cinemas, supermarkets, office buildings and apartments. Commercial shopping center is in the east, and located in "T" cross road with long street side and high commercial value. The flagships and main entrance of large scale commercials are arranged here. The cinema has been arranged on the 6th floor of the Mall. Office buildings and apartments are arranged in the west, the relatively quiet environment that the office buildings and apartments require. There is a 3-storey restaurant building in the narrow site next to Taiping Road. The restaurant is independent in function, consistent in style with the main building. The supermarket is in B1, with one ramp going down through the south entrance of the shopping mall, connecting to the Dongyitiao Road planning underground commercial street.

上海惠南镇南奉绿地项目
Shanghai Huinan Town Nanfeng Greenland Schematic Design

项目名称：上海惠南镇南奉绿地项目
用地规模：25.6 万平方米
建筑面积：82.8 万平方米
容积率：3.2
项目地点：中国／上海／南汇
时间：2008／01

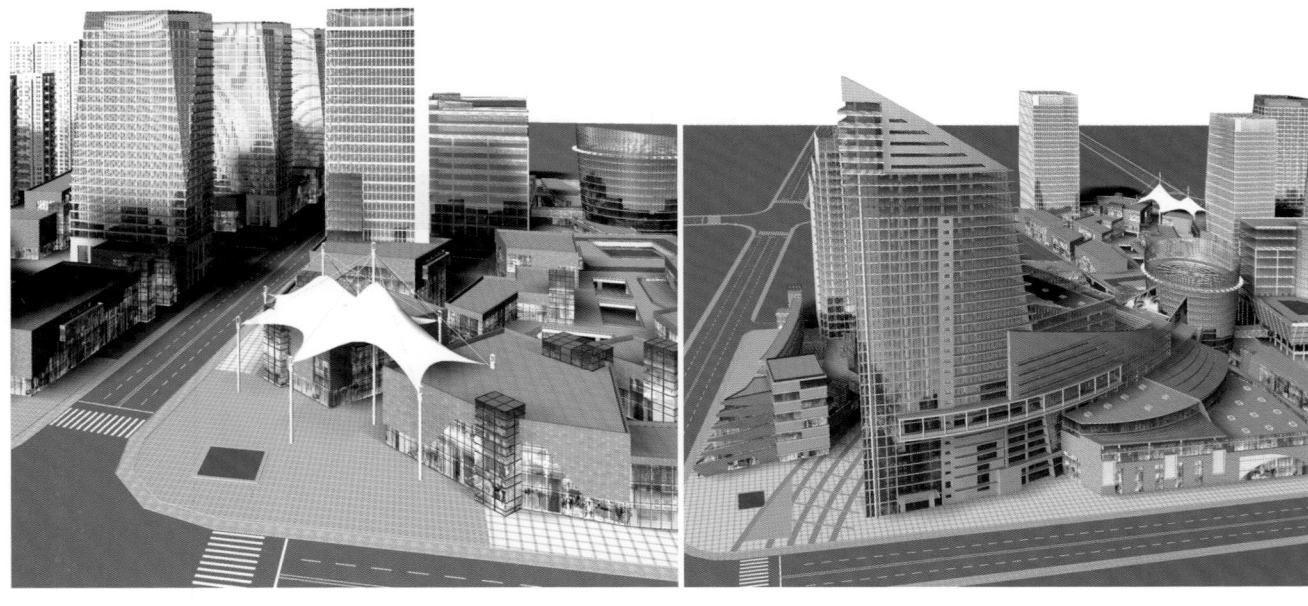

本项目致力于打造一个符合都市生活功能的综合体来提升
该区域的档次，构建一个都市型商业中心，设计师通过将
商务、休闲娱乐、购物、餐饮等消费多业态产品在建筑空
间上的集合，形成片区的 24 小时人气链；以高星级酒店、
甲级标准办公楼、商业中心在街区品质、形象、人气上进
行的全面提升，作为撬动项目整体的经济杠杆的组合支点；
再以 SOHO、LOFT 产品、酒店式服务公寓、住宅导入并
沉淀人气，与主题商铺共同形成业态互补、功能高度复合
的城市综合体。

The project aims to produce a complex urban life facility to
improve the quality of core district. The designer builds a
converge of commerce, entertainment, restaurant functions
on the architecture space and forms a 24 hours hot
commercial destination. By involving SOHO, LOFT and
serve apartments as support facilities, the community turns
into a convenient complex place combined with theme
shops.

南通中央商务区
Nantong CBD

项目名称：南通中央商务区
用地规模：58.12 万平方米
建筑面积：38.29 万平方米
容积率：0.66
项目地点：中国 / 江苏 / 南通
编制时间：2004 / 01

景观水环——逐水而居，沿水而聚，延续南通传统护城河的空间模式，借鉴传统老城肌理，以内外两条景观水环组织整体规划空间，区分并联系各功能组团，成为规划结构中主导性的景观构架，极大地提升了 CBD 地区的景观品质和休闲品位。

绿色服务环——在两条景观水环中间，架构一条以绿化步行系统为基底的休闲服务环，补充景观水环以外的商业、休闲服务功能，布局以点状服务设施和带状绿化结合，串联办公、居住等各功能组团，为市民提供全面、优质、便捷和高质量的服务。

中央广场轴——作为城市新中心区中央轴线的组成部分，该轴线在规划范围内的部分将被规划为中央广场轴，形成由绿化广场和硬质广场组成的三段式景观结构。两端绿化广场强调自然山水，局部造景，注重与周边地块轴线设计的延续性；中心硬质广场强调和周边建筑群的基底关系，加强对广场中心地位的空间感知，力求形成空间品质优异的 CBD 公共活动中心。轴线整体上利用南北连续性的建筑小品，突出轴线的连续性和中心感，为整个轴线寻求出一种整体性的空间结构。

Water landscape ring-continue the spatial mode of Nantong's traditional moats to be located in waterside; draw lessons from the traditional old city texture; organize the overall planning space by two landscape rivers inside and outside to achieve the distinction and connection among various functional groups. Thus, water landscape will be the major landscape in the planning; highly enhance the landscape quality and recreational tastes of the CBD.

Green service ring-design a leisure service ring based on the green pedestrian system between the two water landscape rings to add commercial and leisure services for the exterior places. Point-lick service facilities and belt-like greening are combined together to connect office buildings, residence and other functional groups to provide comprehensive, convenient and high quality services for citizens.

Central square axis-as a part of the new downtown area central axis, it will be planned to be the central square axis to form a three-section landscape structure by green square and hard quality square. Greening squares on both ends emphasize on natural landscape, local landscaping and the continuity with the axis design of surrounding plots; the central hard quality square emphasizes the harmony with surrounding buildings, strengthens the spatial perception of the central position of the square, striving to form a CBD public activity center with high spatial quality. The axis uses the continuous exquisite buildings in south-north direction to highlight the continuity and the central position, thus seeking the integral spatial structure of the entire axis.

上海漕溪路商业广场
Shanghai Caoxi Road Commercial Plaza

项目名称：上海漕溪路商业广场
用地规模：0.98 万平方米
建筑面积：3.98 万平方米
容积率：4.06
项目地点：中国 / 上海
编制时间：2007 / 08

在一处相对狭窄的基地上，设计师通过对平面元素的有机组合丰富了地块给人的空间感受，大量动感的曲线放大了项目的商业展示面，相互交合在一起的几大功能区块清晰而且明确地将交通和零售单元划分清楚，围合而成的内部活动广场开合自如，有很强的人流引入作用，使得项目产生了小中见大的整体环境效果，增强了商业的煽动效果。

By logically organizing many plane elements in a comparatively narrow site, the designer puts lots of curved lines which zoom out the commercial edge. The major functional blocks getting together clearly separate the traffic and retail units, and the inner activity square has the strong attractiveness for people flow, all of which manifests the entire environment effect.

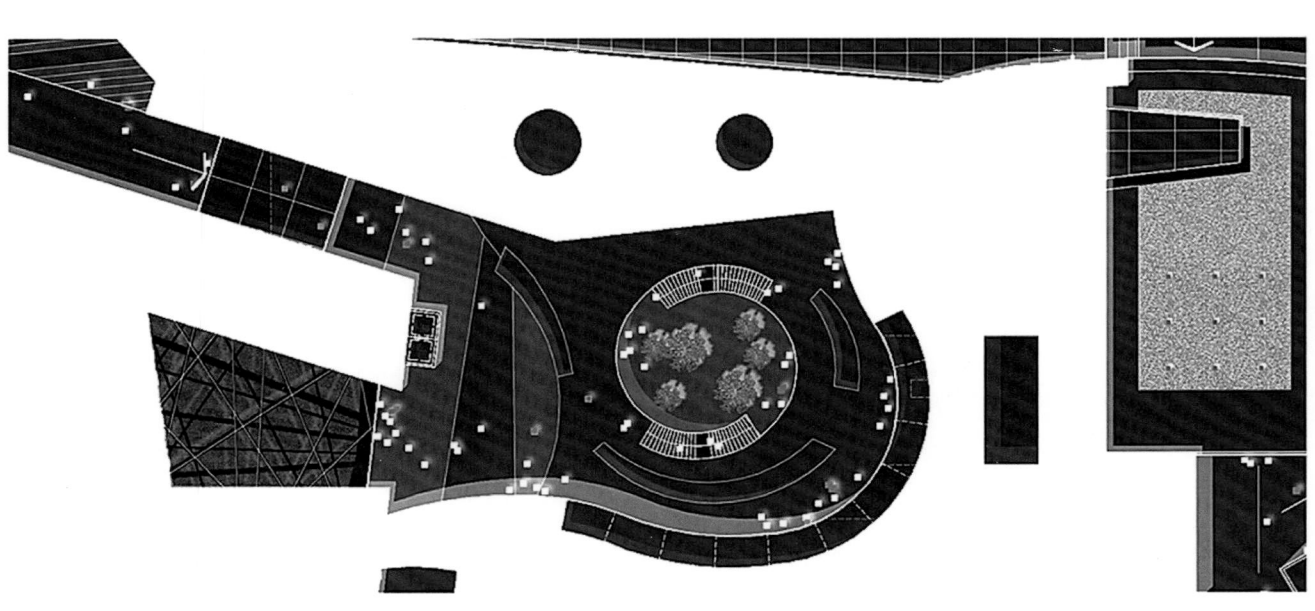

上虞财富广场
Shangyu Fortune Plaza Schematic Design

项目名称：上虞财富广场
用地规模：5.94 万平方米
建筑面积：14.85 万平方米
容积率：2.5
项目地点：中国 / 浙江 / 上虞
编制时间：2007 / 06

项目位于浙江省上虞市城北新区，是上虞新区的综合性核心办公商务中心区。设计注重空间景观的人性化和可用性，将办公塔楼面向中心绿地，围合形成中央景观区，在保持项目整体庄重感和价值性的同时，创造亲水办公环境，形成一种轻松自由的人流组织布局。

The project is located in Chengbei New District, Shangyu City, Zhejiang Province the CBD of New District Shangyu City. Design emphasizes the humanity and usability of landscape space, and the administration building faces the central greenland. While preserving the entire sobriety and value of project, it creates a waterfront relaxing environment.

157

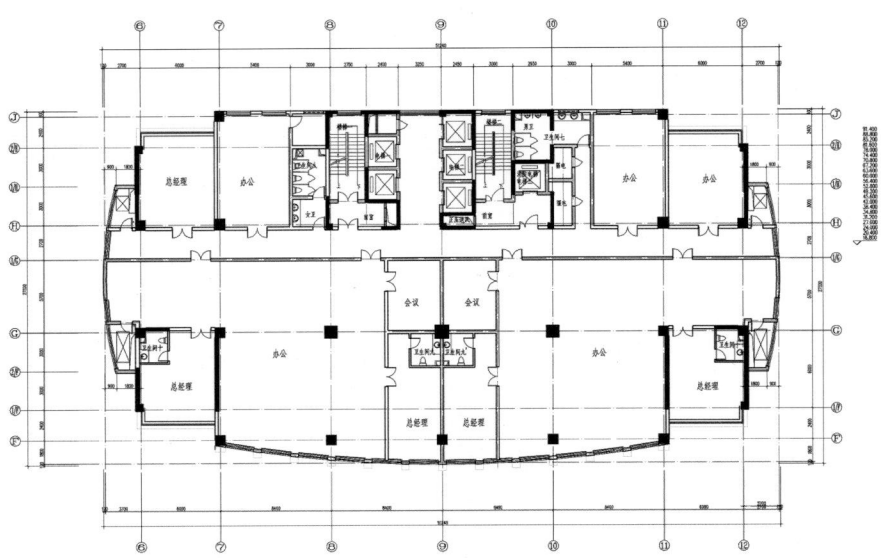

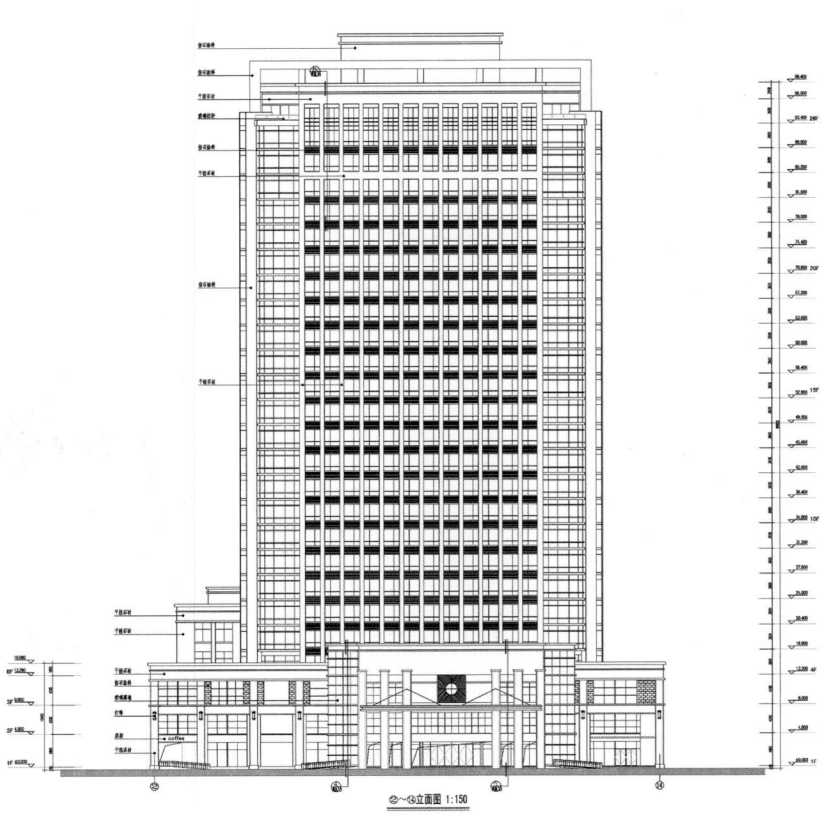

②～⑬立面图 1:150

上虞国际时代广场
Shangyu International Times Square Schematic Design

项目名称：上虞国际时代广场
用地规模：10.48 万平方米
建筑面积：20.95 万平方米
容积率：2.0
项目地点：中国 / 浙江 / 上虞
编制时间：2006 / 06

上虞国际时代广场地块位于上虞市城北新区市民大道北侧，江杨路东侧，凤鸣路西侧，周边规划是以金融商贸、休闲娱乐、生活居住、科技教育等功能为主的公共设施，同时又濒临西横河和百沥河两条主要河流，是未来城市的核心商业区。

上虞时代广场是集贸易、购物、餐饮、休闲、娱乐、公寓为一体的大型综合体。这个项目的目标是创建一个流动的城市，它包括大规模的步行商业街、标志性的办公楼、精致典雅的酒店式公寓、一条由花园和树木设置于两旁的生动的步行道路、精致的花园广场。砖、石头、玻璃、钢的利用增强了现代与传统的结合，细微之处如钢铁结构和金属镶板的使用突出了与上虞工业城市联系的紧密。

Shangyu International Times Square is located in the north of the Citizen Avenue, Chengbei New District, Shangyu City, in the east of Jiangyang Road and west of Fengming Road. The surrounded area is planned with public facilities with finance and commerce, entertainment, residence, education and other functions. Meanwhile, it is also adjacent to two major rivers, Xiheng River and Baili River. Thus, there will be the core central business district of the city in the future.

Shangyu International Times Square is a large complex integrating trade, shopping, dining, leisure, entertainment and apartment. This project aims to create a mobile city which includes a large-scale commercial pedestrian street, landmark office buildings, grand hotel-style apartments, a lively pedestrian road with gardens and trees lined both sides and an exquisite garden square. The combination between tradition and modern is presented by the use of brick, stone, glass and steel. The use of steel structure and metal panels at detailed place highlights the close contact with Shangyu as an industrial city.

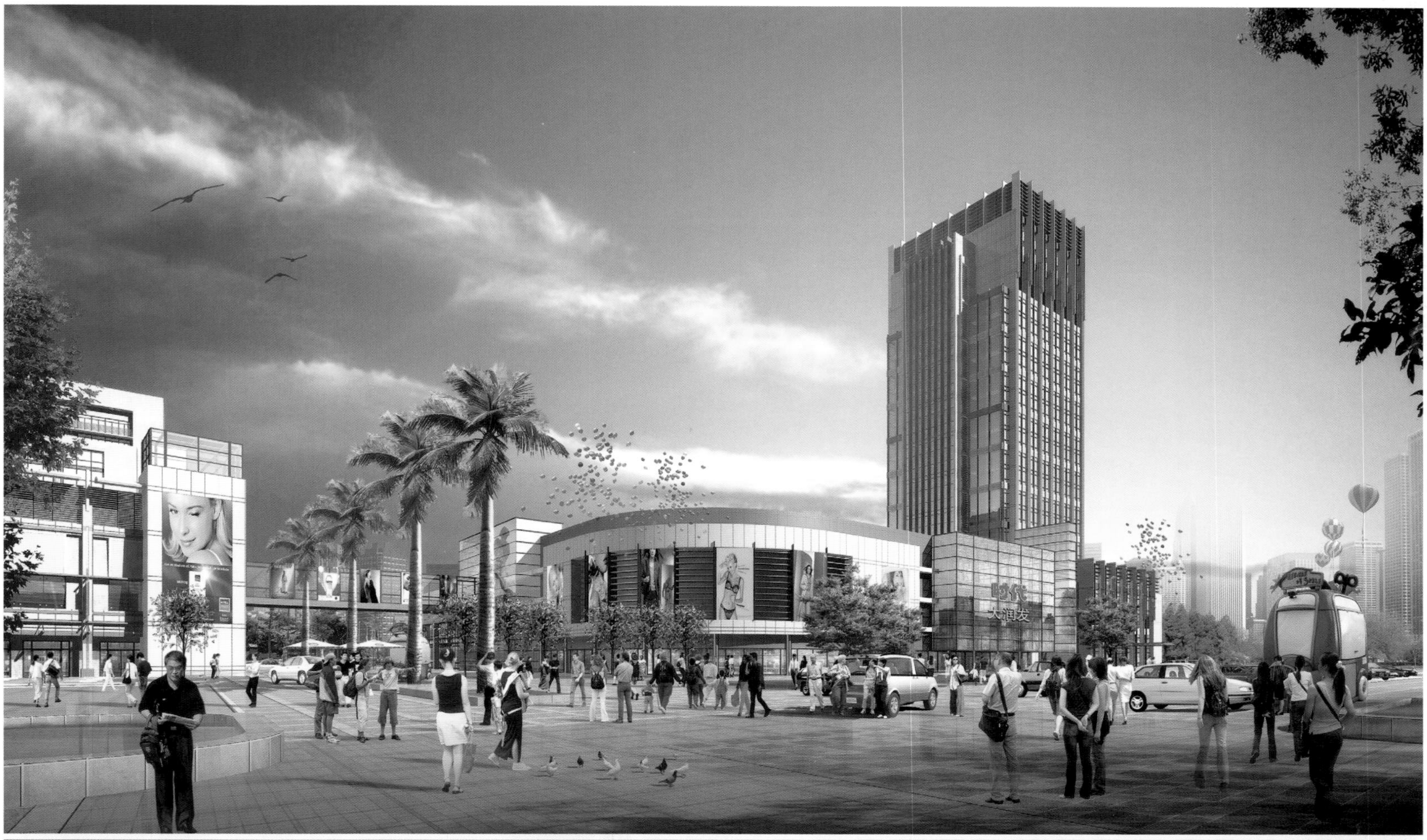

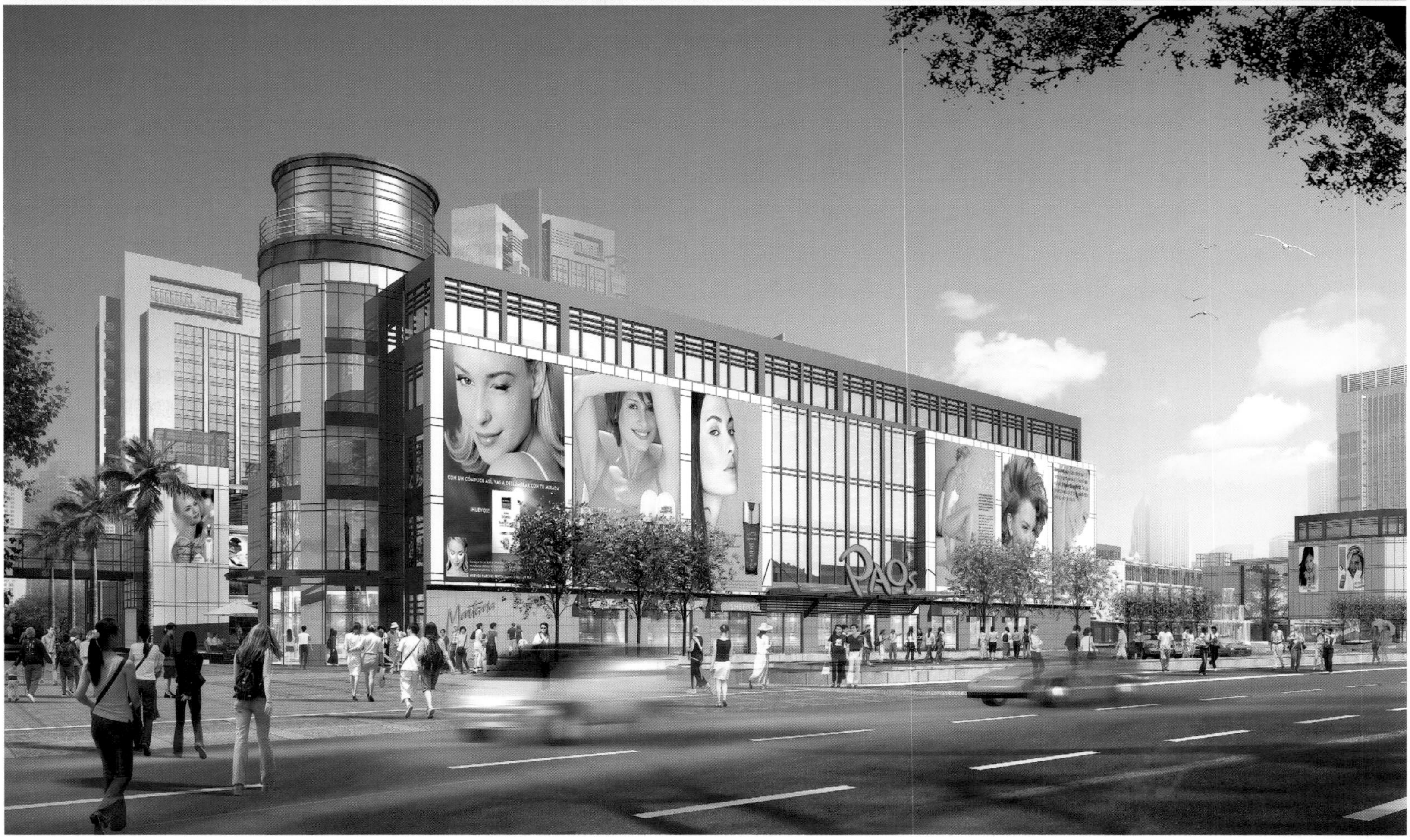
PAO's

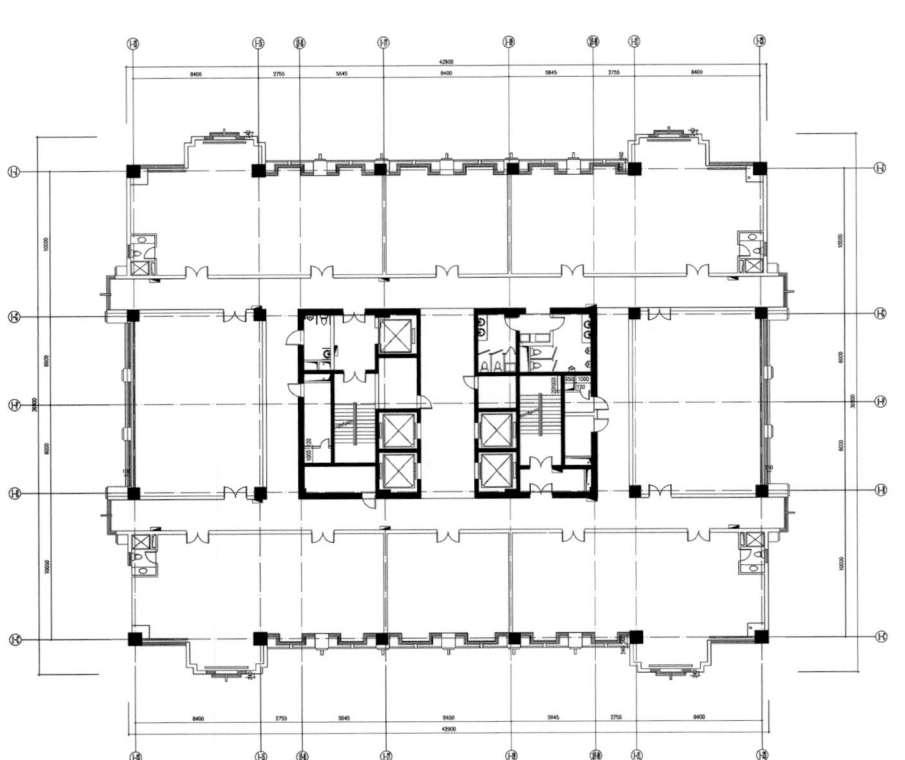

尝试高效的空间使用和销售的结合，
诀窍由我们掌握……

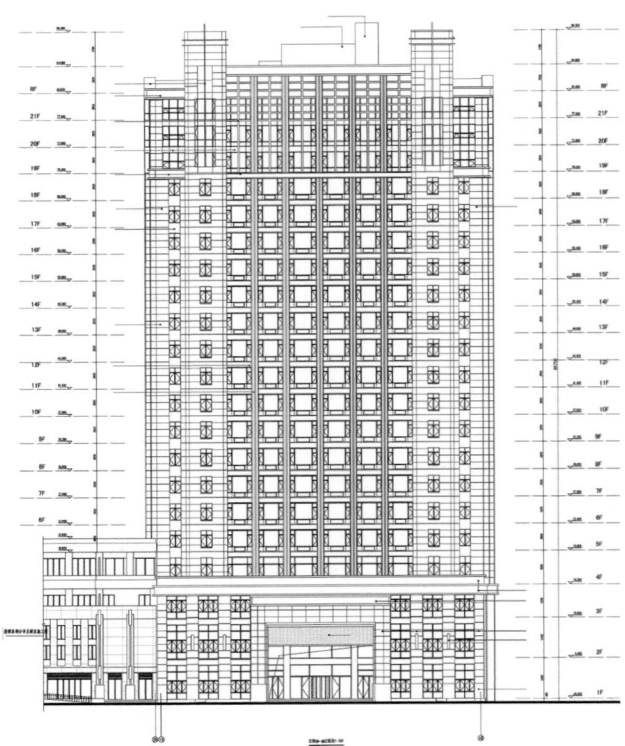

沈阳金茂中心
Shanyang Jinmao Center

项目名称：沈阳金茂中心
用地规模：7.53 万平方米
建筑面积：53.55 万平方米
容积率：7.0
项目地点：中国 / 辽宁 / 沈阳
编制时间：2005 / 08

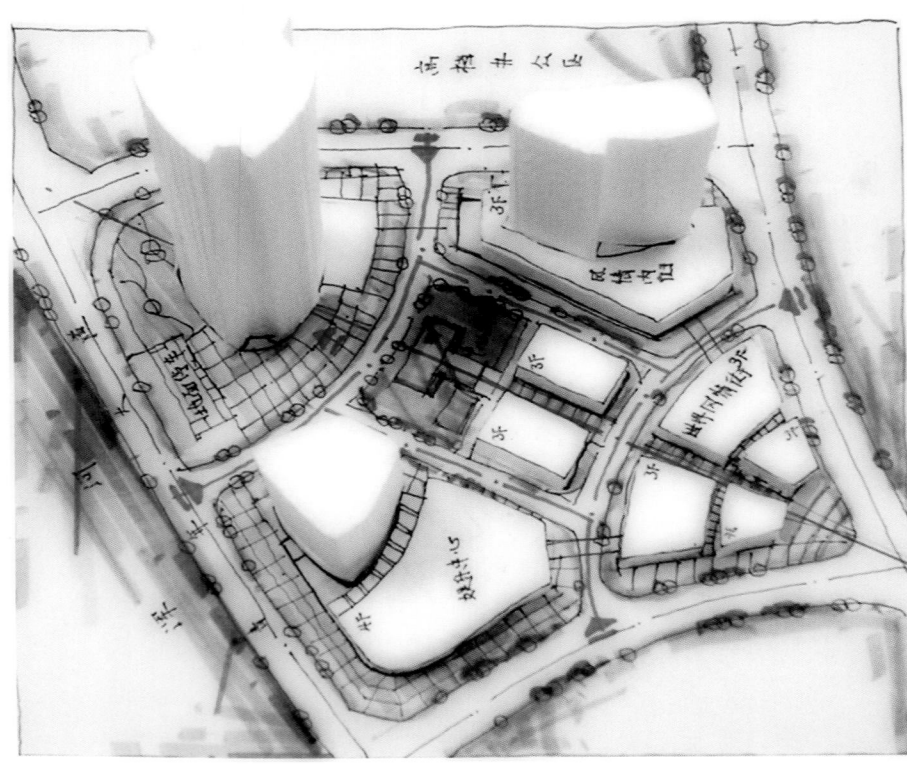

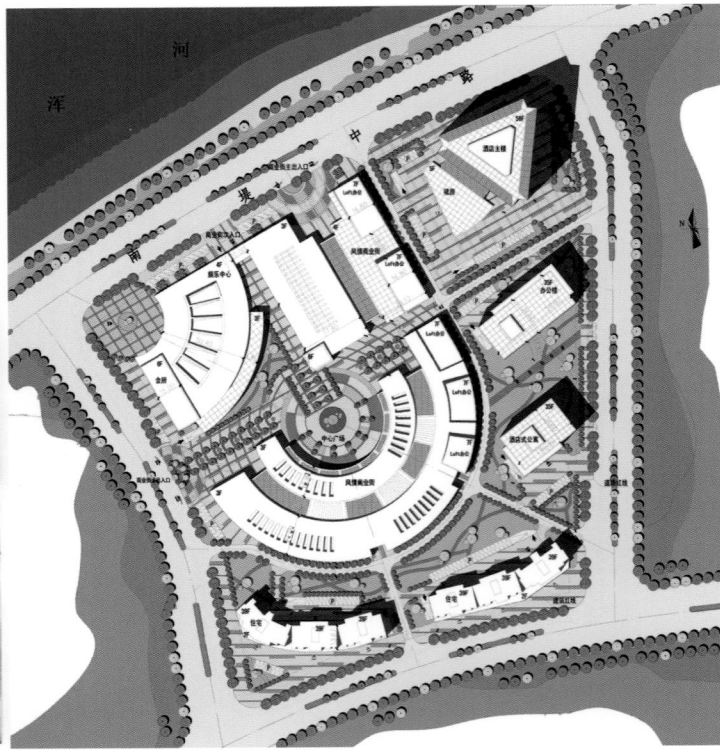

本方案将沈阳金贸主塔楼布置在临主干道南堤中路一侧，占据东北拐角，保证沿南堤大街一侧的视线不受遮挡，奠定了其核心地位，也使之迎向主要人车流；将商业区及KTV 夜总会也布置在临南堤路一侧，迎向主导人车流，扇形的夜总会占据西北拐角，丰富了城市街景效果，将住宅、办公楼、酒店式公寓布置在较安静的临鑫园路和东侧的城市支路地段。这样布局的好处是各区位置合理，"闹""静"分明，各区之间连接顺畅，互不干扰。最重要的是使所有建筑都能朝向浑河与五里河公园，最大限度引入景观而不让高层建筑挡住低层商业建筑。这样的布局给外空间设计注入了活力与激情。

In this case, Shenyang Jinmao main tower building is designed on the Nanti Central Road side near the trunk line occupying the northeast corner so as to ensure unobstructed view along Nanti Street. This not only contributes to its core status, but also attracts more passenger and vehicle flows; commercial area and KTV night clubs are also arranged on the Nanti Road side, facing main passenger and vehicle flows; the fan-shaped night club is arranged on the northwest corner so as to enrich the city's streetscape effect; residence, office buildings and hotel apartments are arranged in the quiet Linxinyuan Road and the urban trunk section in the east. Such layout is quite rational with clear distinction between quiet and busy areas. The various sectors are well connected without interference to each other. The most important thing is that all buildings can face Hunhe River and Wulihe Park. Landscape can be introduced to the maximum and the high-rise buildings will not block up the low-rise commercial buildings. In this layout, the entire space is full of vitality and passion.

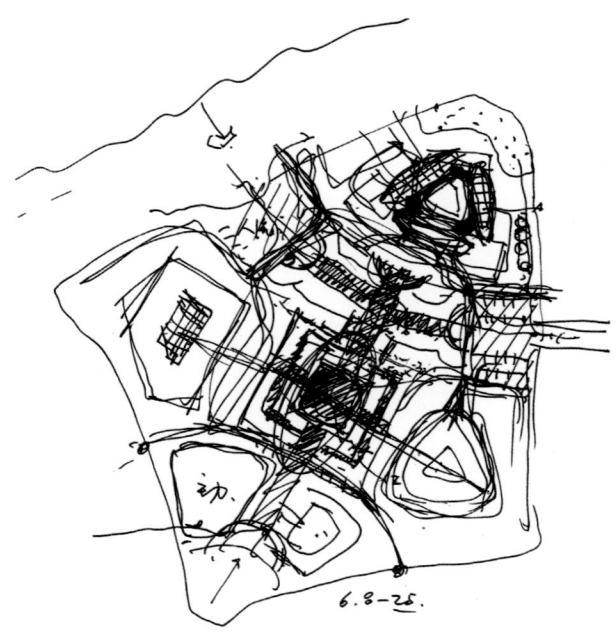

天津金奥国际中心
Tianjin Jinao International Center

项目名称：天津金奥国际中心
用地规模：7.13 万平方米
建筑面积：15.07 万平方米
容积率：2.1
项目地点：中国 / 天津
编制时间：2006 / 03

项目位于天津南开区宾水西道与凌宾路交汇处，整个地块以水为主题，呈现绿色奥运和回归自然的理念。280 米的高度衬托出奥体中心的壮丽与气魄。设计延续奥体"水"的主题以及"水滴"的概念，创造柔美的建筑体形和流动的界面，以达到对整个地块空间的承载。

The project uses water as its design theme and expresses the emotion of Olympic "Green" spirit and concept of returning the nature. The 280 m building emerges the magnificent feature and the water drop configuration creates a beautiful and smooth edge.

苏州石路金座广场商务楼
Suzhou Shilu Golden Tower Schematic Design

项目名称：苏州石路金座广场商务楼
用地规模：0.5 万平方米
建筑面积：2.49 万平方米
容积率：4.9
项目地点：中国 / 江苏 / 苏州
编制时间：2008 / 01

项目地块地处苏州古城西侧的石路商业区，紧邻外护城河，是仅次于古城观前商业区的重要地段。项目为集商业、办公为一体的综合性建筑，力图打造一个富有活力，现代时尚的，集购物、休闲、娱乐、办公于一体的标志性建筑。

设计一方面要使建筑融于城市空间和自然环境中，另一方面又要追求别致、新颖、大方的艺术品味，以时尚、现代的建筑形体，打造地区地标建筑，从而提升该区域的城市空间景观质量，提升地缘价值。

通过现代、简洁的设计手法，塑造一个典雅、清晰的建筑形象。建筑造型以适合基地形状的体块为基础，立面以窗户自身的韵律、凹凸、排比构成肌理，来创造优雅大方的艺术格调。外立面结合不同的使用功能采用不同的建筑元素，重点体现竖向线条，使建筑形象干净挺拔。

The project is located in the Suzhou Shilu Business District, adjacent to the outer moat, which is the second important place other than the Ancient City Guanqian Business District. Our objective is to build a vibrant, modern and fashionable landmark building with shopping, leisure, entertainment and office in one.

On one hand, the buildings should be integrated into the urban space and the natural environment; on the other hand, with the pursuit of unique, innovative, elegant and artistic taste, it has a sleek, modern architectural form, creating landmarks, so as to enhance the quality of urban space and landscape of the region, and promote geo-value.

Through modern and simple design, the design creates an elegant and clear image of the building. Architectural modeling based on the body block fits the shape of the base, with the composition textures combined by rhythm, bump and parallelism of the facades to the window itself, to create a graceful art style. Built with different architectural elements combined with different architectural features, with a focus on embodiment of vertical lines, the architectural image is clean and upright.

天津弘泽洞庭路项目
Tianjin Hongze Dongting Road Project

项目名称：天津弘泽洞庭路项目
用地规模：27.10 万平方米
建筑面积：69.55 万平方米
容积率：2.57
项目地点：中国 / 天津
编制时间：2006 / 05

本基地项目位于天津塘沽区，紧邻天津市经济技术开发区核心区。基地东临洞庭路，南至规划杭州道，西侧为规划城市支路，北至规划广州道，是基地与区域联系的主要交通轴。基地沿洞庭路向南即与新港四号路相接，具备与天津城区快捷、高速的交通联系。

项目定位为开发区大型的复合高端社区，具有多元化的居住空间、宜人的商业设施，并保证在未来三到四年内在地区市场的持续领先。户型产品具有完整的序列和多样化的设计，主要面对塘沽和开发区的"银领"阶层，营造的是恒久文化氛围下超越时尚的生活。

The project is located in Tanggu District, Tianjin, close to the core area of Tianjin Economic and Technological Development Zone. The base is adjacent to Dongting Road in the east, the planned Hangzhou Avenue in the south, the planned Urban Minor Road in the west and the planned Guangzhou Avenue in the north. These are major roads connecting with other regions. Going to the south along Dongting Road is the Line 4 of Xinggang Road so that high-speed transport is enjoyed by the base with Tianjin City.

The project is positioned as a large complex high-end community in the Development Zone with diversified residential space and pleasant commercial facilities and the ensured leading place in the regional market in the next three to four years. House type and products have complete sequence and diversified design. The target customers are "silver collars" in Tanggu and the Development Zone and the theme is the fashionable style in the permanent cultural context.

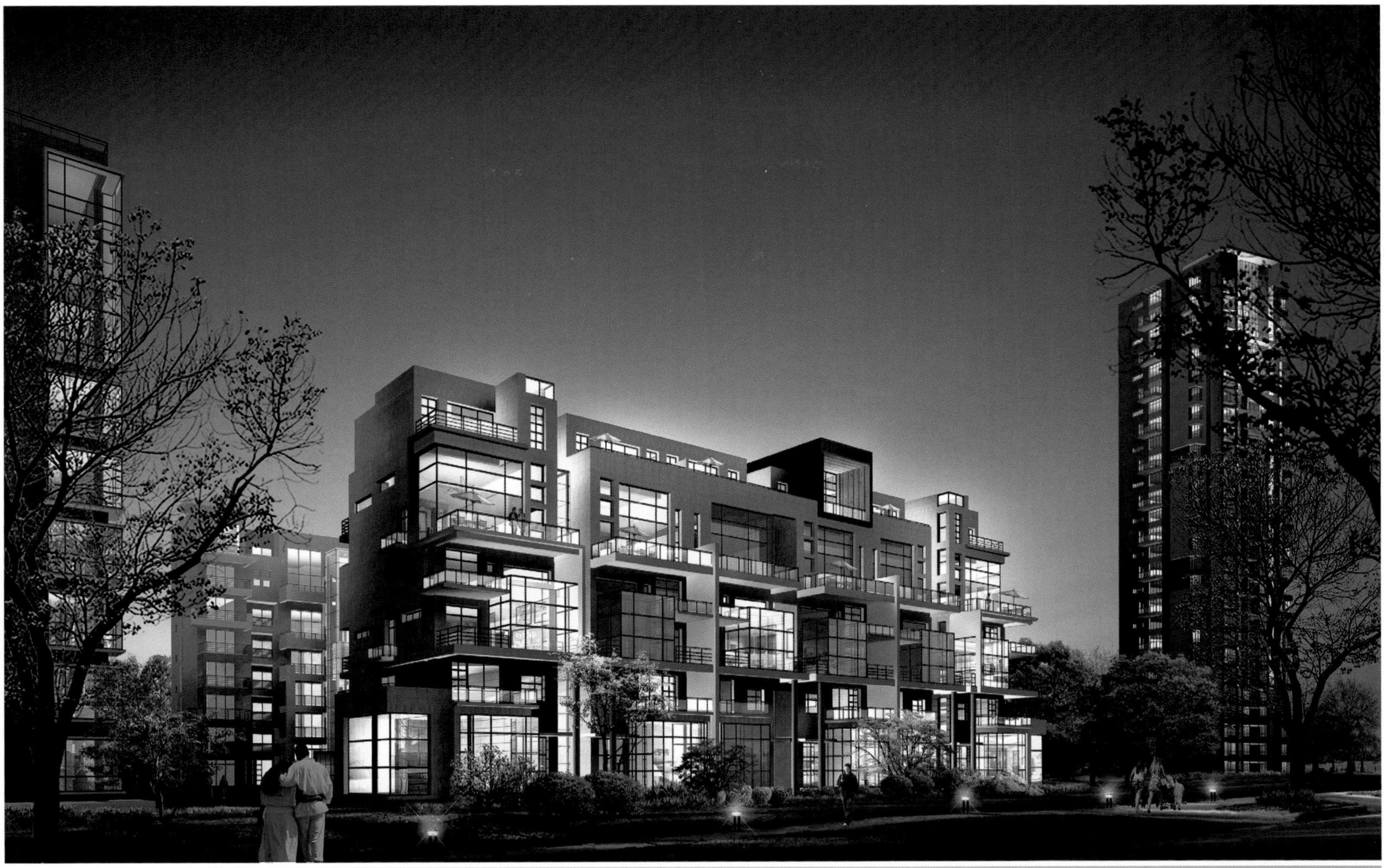

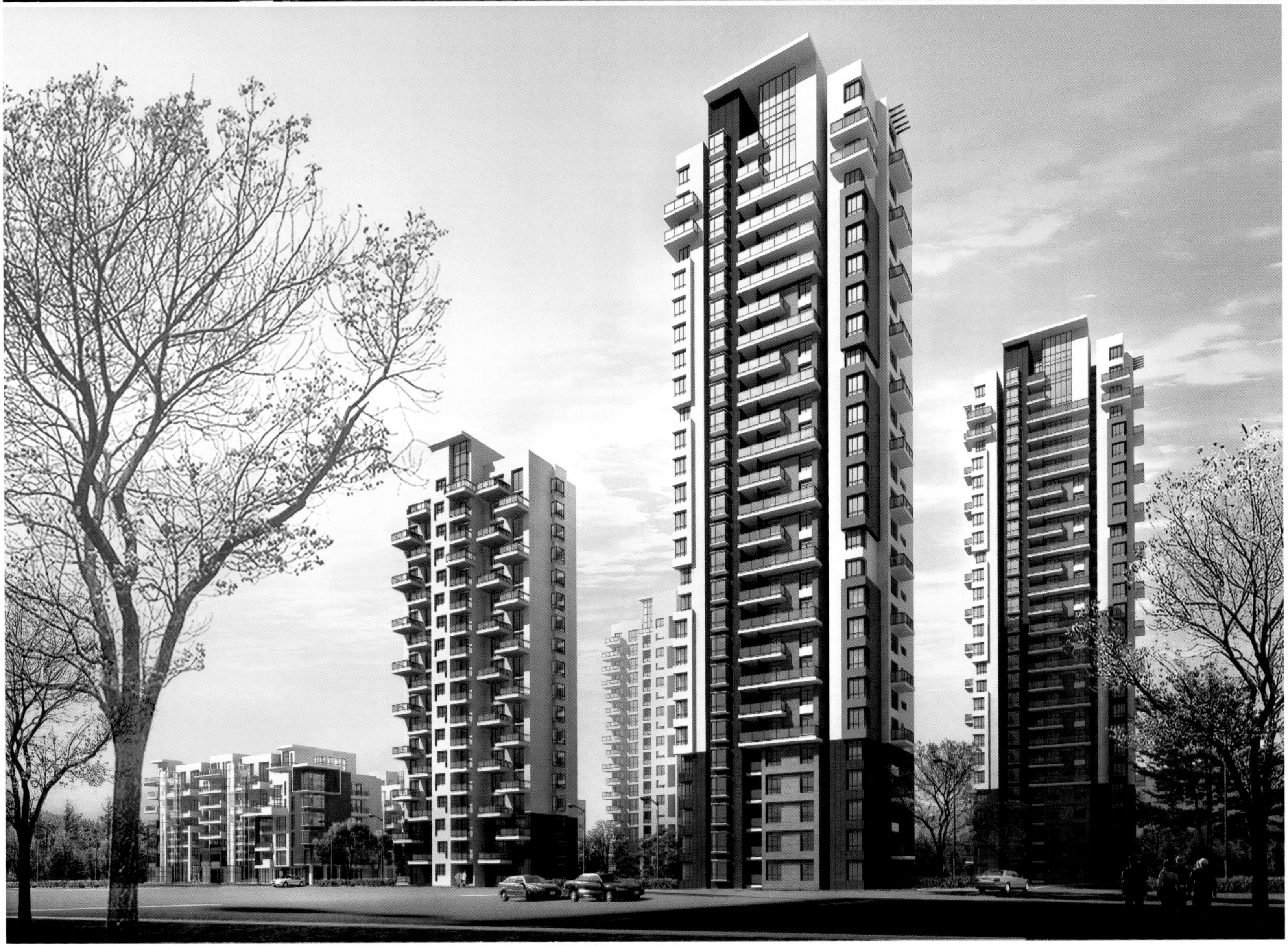

天津世贸中心
Tianjin World Trade Center

项目名称：天津世贸中心
用地规模：8.90 万平方米
建筑面积：48.15 万平方米
容积率：5.4
项目地点：中国 / 天津
编制时间：2004 / 06

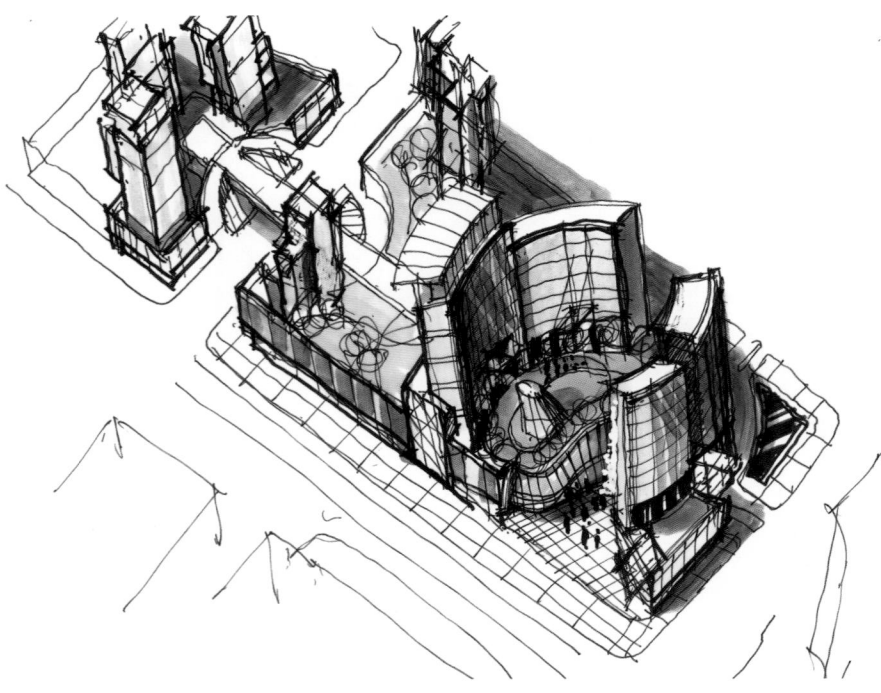

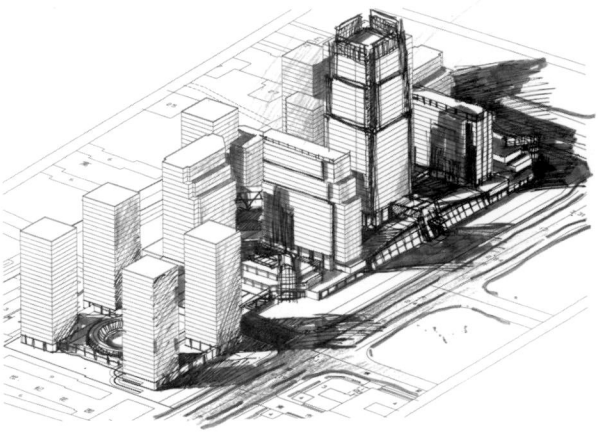

设计理念

它将是一个复杂的综合体用来满足复杂的人类活动需要，使人们能够进行完全的商务活动，享受购物与生活共同的乐趣；成为日常生活的剧场和都市精神的熔炉。成为一个充满活力的、富有现代城市精神的、地标性的新城市空间；最终形成一处具有功能互促性、空间丰富性、景观互动性、形态多元性的城市目的地。

总体布局

整个基地被服装街分成了 B、C 两地块，设计将商业、办公、酒店及公寓等几大功能区进行功能整合，通过轴线控制，形成脉络清晰的架构。在 B 地块，100 000 平方米的写字楼由一栋 60 000 平方米的主塔楼和两栋 20 000 平方米的副楼组成，临近南马路一侧，靠近地铁，便于未来写字楼员工上下班，写字楼的地下同商业区连通起来，构成一个相对融合的整体，同时保证写字楼的高效、安全的环境。在写字楼主塔楼的南面，是由酒店和酒店式公寓共同组成的群体，酒店的总建筑面积为 50 000 平方米，包含一座酒店和其相应的配套设施，平台层设置酒店大堂，酒店裙房部分在 2~3 层分层设置中餐厅、西餐厅、自助餐厅与风味餐厅，4 层设置商务会议中心，含一个大型多功能厅，休闲健身功能围绕游泳池设在 5 层，并在地下二层设置娱乐区。6 层到 22 层为酒店客房，客房按高标准高起点设计，采用 4.5 米开间，每层有 27 个标准间，顶部的三层为豪华套房，拥有区内开阔视野。酒店房间的数量可以达到 495 间。酒店远离地铁人流较多的通道和出入口，保证了酒店始终处于相对安静的环境中。

Design Concept

It will become a complicated complex that will be used to satisfy sophisticated requirements of activities; become the theatre of daily life and the smelting furnace of urban spirit. It will become a new urban space full of energy and modern city spirit; and finally achieve the mutual promotion between different functions, richness of space, interactive feature of landscape, ecological feature of environment, and diversity of forms.

General Layout

All the base is separated into B and C lots. The commerce, office and hotel and apartment functions are integrated by the axis , therefore form a clear structure. In Lot B, there is an office building which is 100,000 m² including a major 60,000 m² tower and two 20,000 m² supporting towers. It is near the NanMa Road and the metro. Therefore it's more convenient for the people working in the office building and it connects the office building and the commerce center to form a relatively integrated one. At the same time, it ensues the effectiveness and security of the office building. On the south of the office building, it's a combination of hotel and service apartment. The total architecture is 50,000 m² , which includes a hotel and back-up facilities. On the platform, there is a hotel lounge. The swimming pool is on the fifth floor. And the entertainment section is on the second floor below the ground. From the sixth to the swenty-seventh floors are the suites of the hotel. They are designed to aim at high-end market. It has 27 units and the top three floors are the luxury suites. with a wide view. The total number of the hotel rooms can reach 495. The hotel is far away from the corridor, entry and exit of the metro, which ensures the hotel to have a relatively peaceful environment.

徐州绿地城市广场
Xuzhou Greenland City Plaza

项目名称: 徐州绿地城市广场
用地规模: 5.50 万平方米
建筑面积: 19.80 万平方米
容积率: 3.60
项目地点: 中国 / 江苏 / 徐州
编制时间: 2007 / 09

设计师对几大功能区进行功能整合, 形成脉络清晰的构架, 提供从混合到清晰的演进以创造白昼黑夜循环不息的活力之城。规划认为未来的商业业态及相互关系必须不同于旧城区的商业, 建筑群的空间及景观的创造必须有助于市民形成关于这种差异的很强的识别力, 以成为独特的无法被廉价复制的和有别于传统市中心的新体验经济示范场, 并让人在消费中休闲, 获得精神的放松和满足。

The integration of different function areas is applied in order to form a clear structure to give a progress from medley to clear, which creates a sleepless city with spirited day and night. The future business states and their own relations will be different from the old town mode. The building space and landscape design have to make the residents easy to distinguish the difference from the old town in order to form a model place, which is impossible to be cheaply copied. Our aim is to make the residents be entertained and gain relaxation and satisfaction when they are consuming.

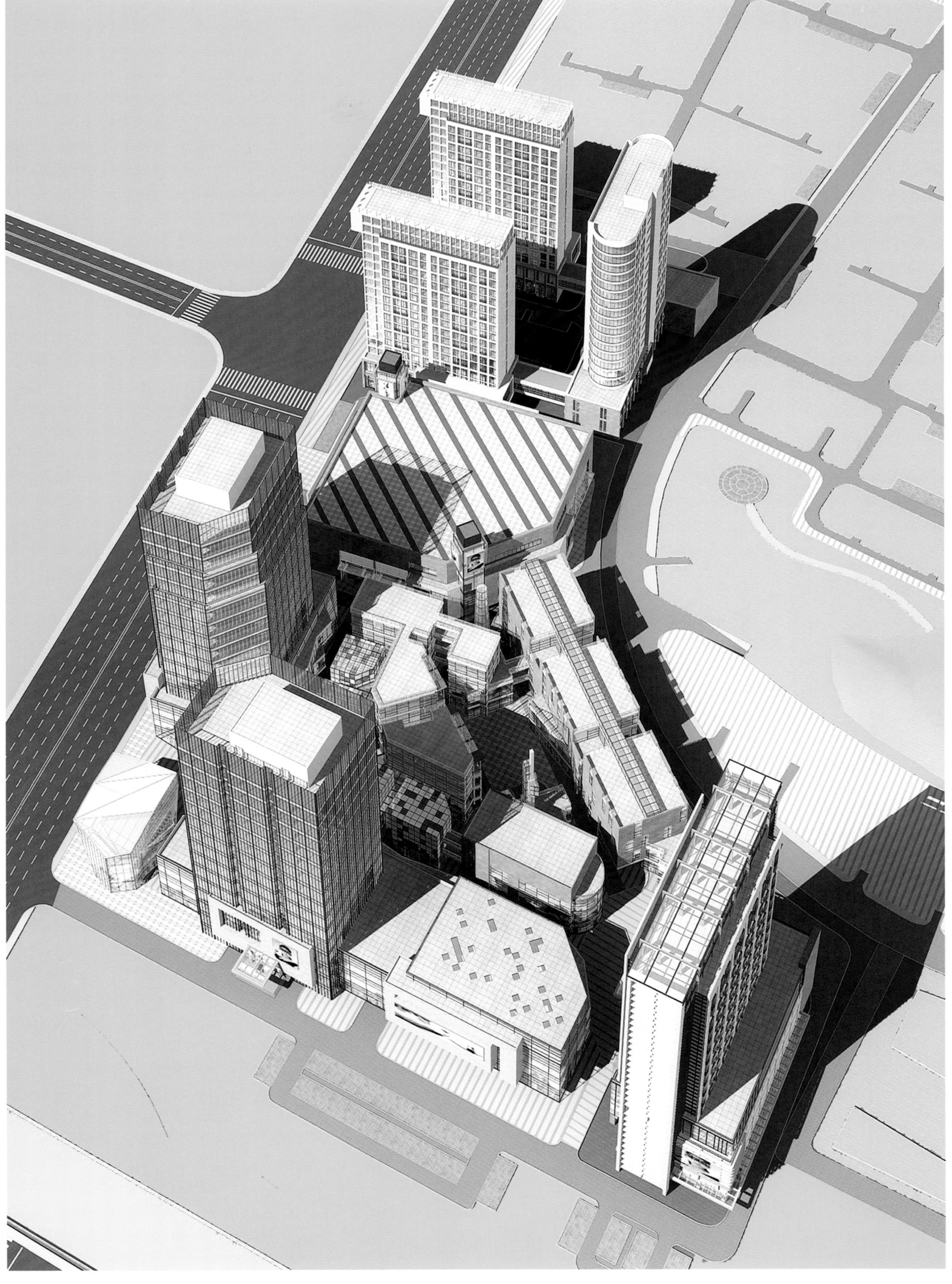

175

余姚国际广场
Yuyao International Plaza

项目名称：余姚国际广场
用地规模：8.16 万平方米
建筑面积：38.86 万平方米
容积率：4.76
项目地点：中国 / 浙江 / 余姚
编制时间：2005 /10

就现代人的需求而言，人们在物质、精神上的需求与满足感，正在逐步多样化，物质商品意义以外人类其他的需求和欲望，正由体验和交流等活动过程来呈现。当人们的工作、日常生活逐渐成为格式化的形式后；我们开始寻求种种不寻常的体验与交流活动，来丰富自己的生活。

余姚国际广场是集商务、购物、高档住宅区于一体，新型多元化的现代楼盘。它拥有地标式高层办公、公寓、商务楼群，以提供给人们休闲、娱乐、工作等不同序列空间组成的大型购物商业街、环境优美的高档住宅区、体现余姚历史文化的大型沿河景观绿地。

我们试图通过余姚国际广场突破常规的单纯购物概念，将复合场所、空间演化为一种体验交流的空间序列，为购物、休闲娱乐、消费过程提供多样丰富的空间环境和内容线索，融入人们的体验，留下轻松而难忘的深刻记忆。

Modern people's material and spiritual demands are becoming diversified. However, the desire beyond material is embodied by experiencing and communicating. When the daily life has become uniformed, people began to seek other extraordinary experience and communication to enrich their lives.

The Yuyao International Plaza is a modern architecture combing business, shopping and senior residential areas, including landmark high office building and apartment. The large-scale commercial street provides ample space for recreation, entertainment and work. The residential area with nice environment is in line with the landscape greenbelt along the river embodying the culture and history of Yuyao.

In the project of Yuyao International Plaza, we try to break the limit of sheer shopping, and develop the complex area into a sequence of space for experiencing and communicating to provide abundant space for shopping, recreation and entertainment. This space will be integrated into people's experience and thus leave them a pleasant and deep impression.

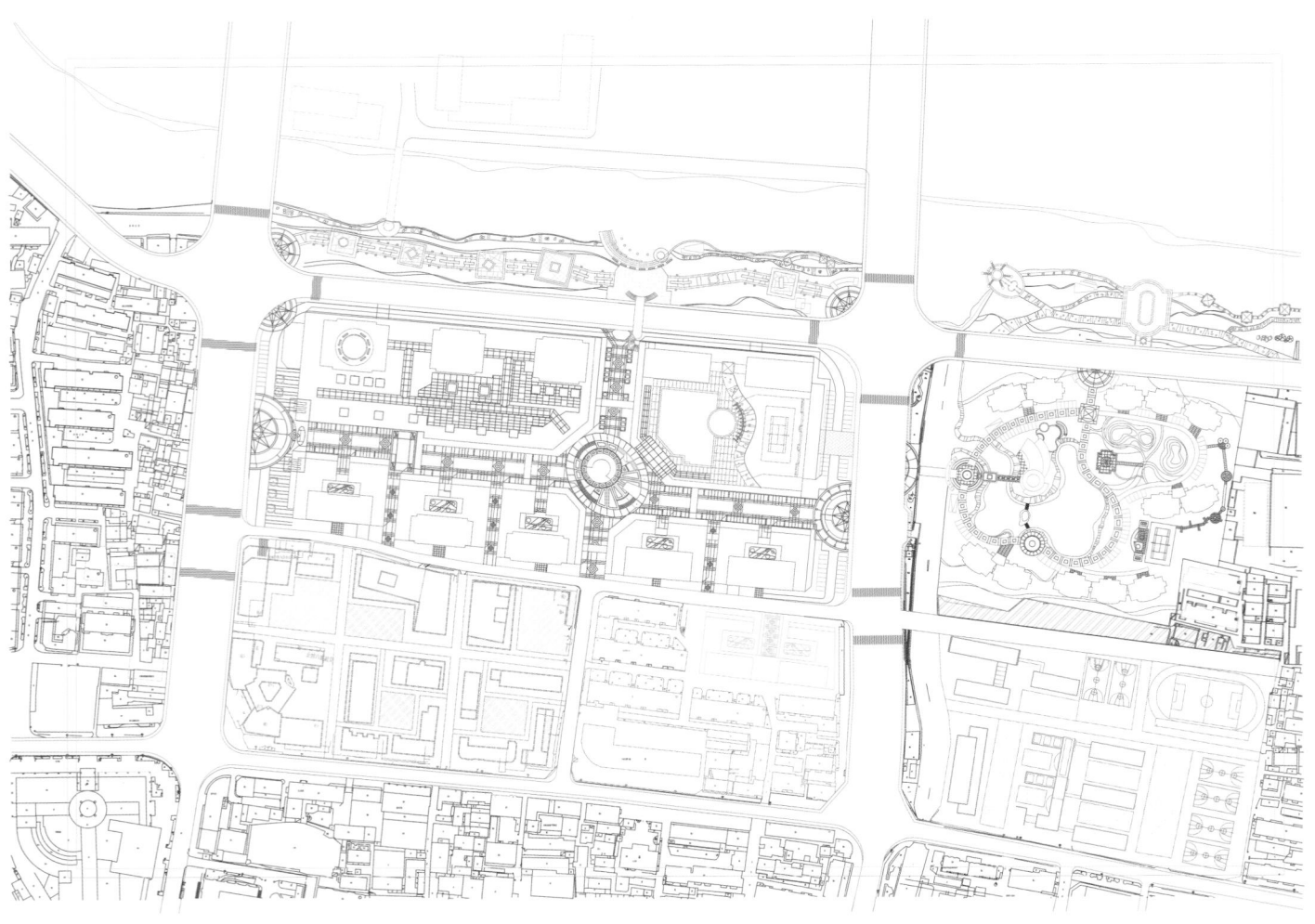

上普财富 · 又一城
Shangpu Fortune • Another City

项目名称：上普财富 · 又一城
用地规模：20 万平方米
建筑面积：34 万平方米
容积率：1.7
项目地点：中国 / 四川 / 成都
编制时间：2006 / 11

在城市主要路口形成集中的大规模城市商业形态，体现出整个楼盘的价值感。商业与住宅分区的规划结构有利于提升住宅区的品质，独树一帜的城市坡地、台地景观成为项目的特色。贯穿的商业中庭空间成为都市生活活力的展示舞台，丰富的建筑群轮廓线、独特的立面风格使本项目成为区域性城市生活中心。

A centralized large-scale urban business modality is formed in the main crossroads of the city, so that the value sense of the whole housing project is embodied. The planned structure of business and residence being zoned helps to enhance the quality of the residential area, and the unique urban sloping-land and table-land landscape turns into a feature of the project. The commercial atrium space becomes the display stage of the vitality of urban life. The rich outlines of building clusters and the unique facade style turn this project into a center of the regional urban life .

成都明宇金融广场
Chengdu Mingyu Financial Plaza

项目名称：成都明宇金融广场
用地规模：1.6 万平方米
建筑面积：11.7 万平方米
容积率：7.3
项目地点：中国 / 四川 / 成都
编制时间：2008 / 11

作为 21 世纪成都市新建的一座高度达 200 米的超高层大楼，如何在经济性及合理性的前提下，创造一个具有视觉冲击力的地标性建筑，是本项目研究的主要课题。在整体建筑意向上，其灵感来自于三星堆出土文物玉璋，结合玉璋代表权力和尊贵的文化延伸及形体暗示，在平面和立面造型上大量运用弧线，结合 200 米的高度形成舒展圆润的体态，一座富含象征意义的地标性建筑从方正的建筑群中脱颖而出。

As a skyscraper up to 200 meters high newly built in Chengdu in the 21st century, how to create a visually-impressive landmark building under the premises of economy and rationality is the main task for this project research. In the total intent of the building, its inspiration is derived from the cultural relic jade tablet excavated from Sanxingdui. With the cultural extension and physical implication of the jade tablet representing power and dignity considered, curves are plentifully applied in plane and elevation modeling, and an extended round posture is formed with the height of 200 meters. A landmark building full of symbolic meaning stands eminently out from the clusters of square buildings.

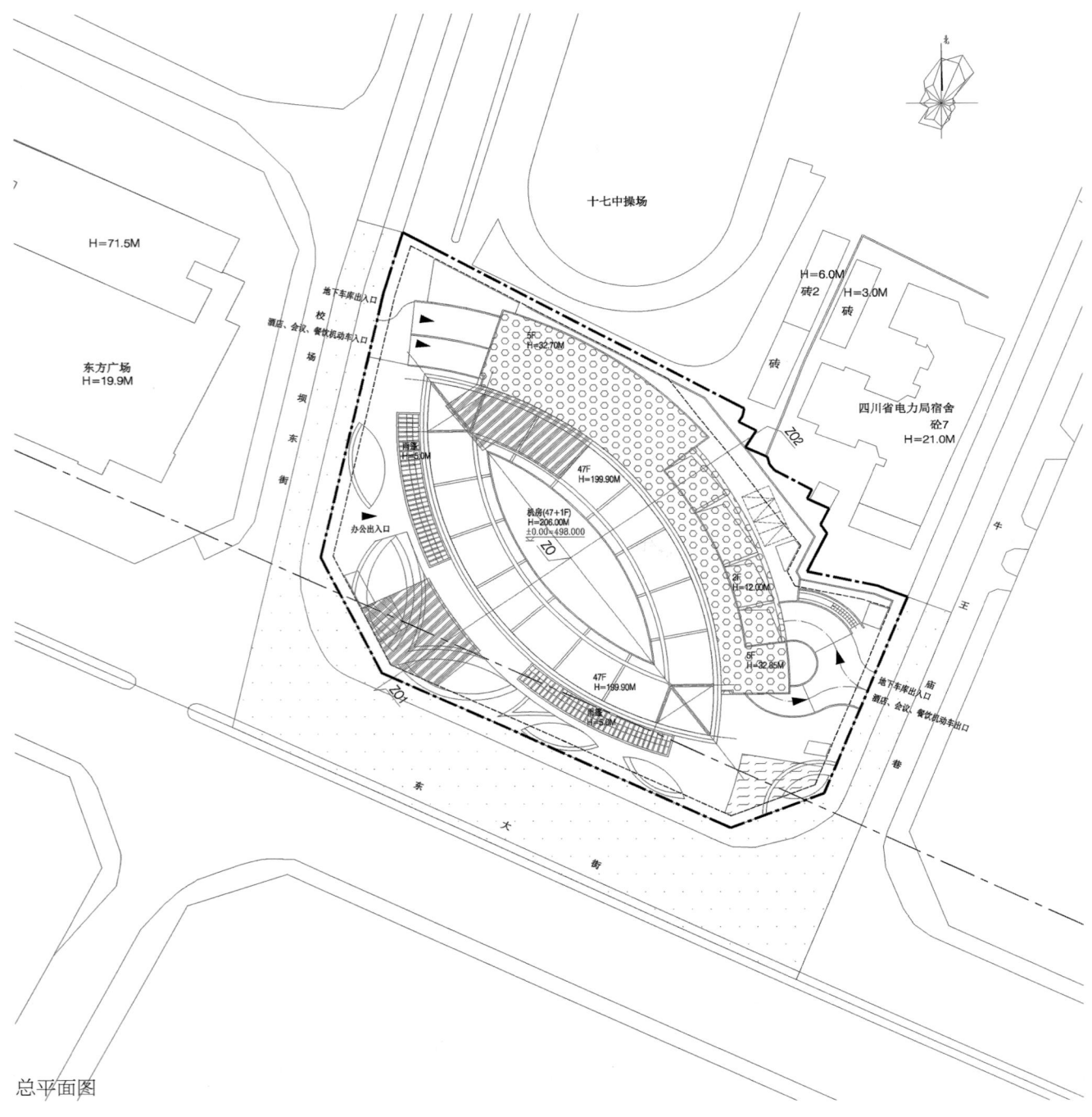

总平面图

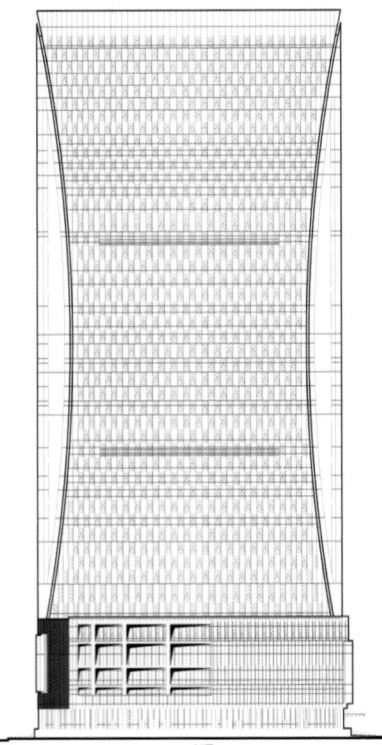

立面图 1:150

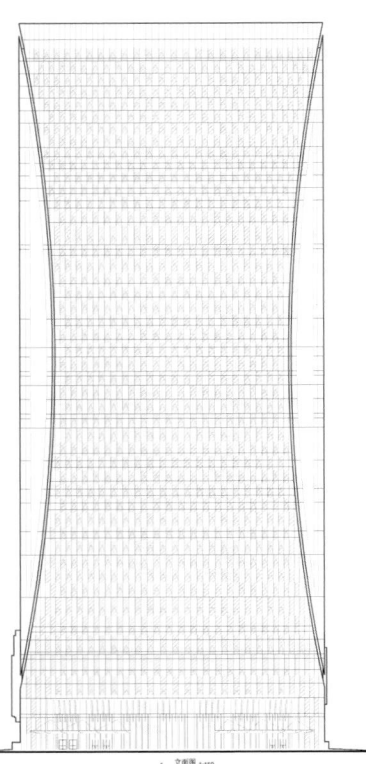

立面图 1:150

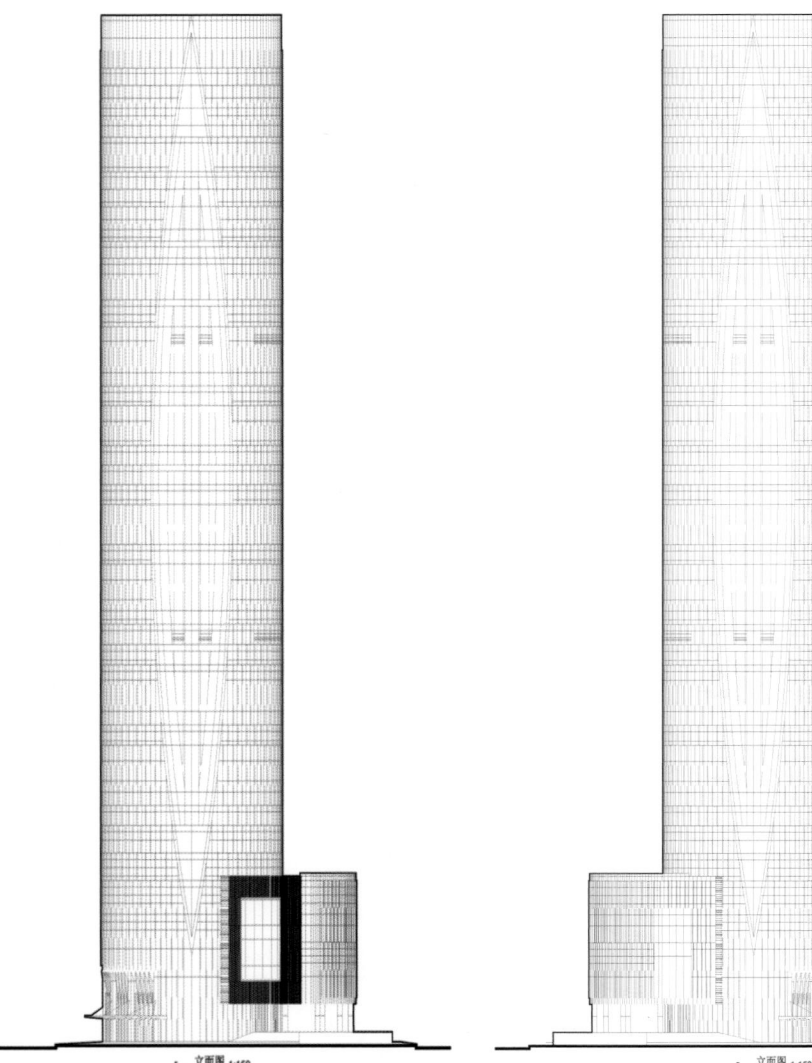

立面图 1:150

立面图 1:150

蜀都中心
Shudu Center

项目名称：蜀都中心
用地规模：3.25 万平方米
建筑面积：26.1 万平方米
容积率：8.03
项目地点：中国 / 四川 / 成都
编制时间：2009 / 11

成都是源远流长的历史名城，经历了 4 500 年以上的文明发展历程，既是"水绿天青不起尘、风光和暖胜三秦"的生态城市，又是"诗人自古列到蜀、文宗自古出巴蜀"的文化城市。本案基地位居成都城南副中心的心脏区位，属于天府新城 CBD 区域，是新世纪成都发展的"标杆"，是一座集科技、商务、国际、时尚、宜居于一体的现代化科技商务新城。

作为天府新城 CBD 先期开发的超高层商业综合体项目，"蜀都国际广场"追求的是在经济性和科学性的前提下，成为具有视觉冲击力的城市标志性建筑。以兼具中国书法艺术和雕刻工艺之精华的传统篆刻艺术为文化基底、以现代建筑艺术和技术为文化载体、"蜀都国际广场"所表达的是一种更具生命活力的"新文化"艺术：它积淀历史又开拓未来、既为人熟知又新颖独特，最终以"文化"的魅力赋予建筑鲜明的个性，形成具有文化内涵的现代商业综合体。

南向的两栋办公楼和商务楼共同围合成一个超过 3 600 平方米的广场，阳光充沛、绿树成荫，将是商务人士休憩的最佳选择，它给城市街道空间一个谦虚的退让，优雅的环境"软化"了高层建筑对街道和行人的压迫感，形成舒缓的城市空间节奏，有利于城市 CBD 区域的商务氛围培养；中央商务楼由于南北开敞，具有良好的日照和视觉效果，特别是北向可以直接俯瞰 D4 地块的城市公园；如此布局使得这三栋 150 米高的超高层建筑日照通风环境良好而且保证了相互之间没有视线干扰。

住宅楼邻近城市公园呈东西向布局，与商务楼形成面向北侧的"凹"形空间，避免了大片建筑阴影投向城市公园，为街区公共环境品质作出了贡献。商务楼北向裙房层层跌落的屋顶花园呼应了城市公园的绿色景观，为建筑群体空间增添了弥足珍贵的绿色生机。

Chengdu is a famous historical city. Having experienced a civilization development journey of more than 4,500 years it is both an eco-city where "the water is green, the sky blue, and no dust arises, the mild scenery excels that of San-Qin", and a cultural city of "since ancient times poets would come to Shu, and master literati used to emerge from Shu". The base of the project lies in the heart location of the city-south subcenter of Chengdu and belongs to the Tianfu New City CBD. It is the "marker post" of Chengdu's development in the new century, and is a modern technology business new city integrating technology, business, internationalization, fashion, and habitability into a whole.

As the super-high-rise business complex project developed early in the Tianfu New City CBD, the Shudu International Square is intended to achieve a city landmark of visual impact under the premises of economy and science. Based culturally on the traditional seal carving art with the piths of both Chinese calligraphy and artistic carving, taking the modern architectural art and technology as the cultural carrier, Shudu International Square is to express a "new culture" art which possesses more vitality and vigor: it both accumulates the history and opens up the future, and is both well-known and novel. At last the architecture is endued with a brilliant individuality by the charm of "culture", and a modern business complex with cultural contents is formed.

The two office buildings facing the south and the business buildings together close round to form a square of more than 3,600 m², where the sunshine is flush and the green trees make pleasant shades. There will be business people's best choice for their rest and leisure. It makes a humble back-off for the city's street space. The elegant environment "softens" high-rise buildings' constriction on the streets and pedestrians, and forms a relaxed city space rhythm, beneficial to the cultivation of the city CBD's business atmosphere; the central business building has excellent sunlight and visual effect for its south-north opening, and especially in the north direction the urban park in the D4 plot can be directly overlooked; this overall arrangement makes the sunlight and ventilation environment of these three 150m super-high-rise buildings excellent, and ensures that there is no interference of sight lines between them.

The residential buildings, adjacent to the urban park, are laid out in the east-west direction. Together with business buildings they form a "concave"-shaped space facing the north side, prevents a large expanse of shadow from buildings from casting onto the urban park, and contributes to the public environmental quality of the block. The roof gardens layer upon layer on the northward annexes to the business buildings correspond with the green landscape of the urban park, and add a highly precious green vitality to the cluster of buildings.

武昌府商业广场
Wuchangfu Commercial Plaza

项目名称：武昌府商业广场
用地规模：8.25 万平方米
建筑面积：37 万平方米
容积率：4.48
项目地点：中国 / 湖北 / 武昌
编制时间：2009 / 12

长沙华盛世纪城
Changsha Huasheng Century City

项目名称：长沙华盛世纪城
用地规模：29 万平方米
建筑面积：50.75 万平方米
容积率：1.75
项目地点：中国 / 湖南 / 长沙
编制时间：2010 / 08

项目名称：长沙华盛世纪城
用地规模：29 万平方米
建筑面积：50.75 万平方米
容积率：1.75
项目地点：中国 / 湖南 / 长沙
编制时间：2010 / 08

长春万通四季城
Changchun Wantong Four Season Flower City

项目名称：长春万通四季城
用地规模：32.10 万平方米
建筑面积：54 万平方米
容积率：1.68
项目地点：中国 / 吉林 / 长春
编制时间：2006 / 11

设计理念

1. 多功能复合的充满朝气和活力的城市商业居住空间

本设计力求从土地利用和土地使用方面综合考虑，在商业的主导功能之外，引入居住、休闲、餐饮、批零兼营等多种城市功能，丰富补充地区的功能类型，注重整体性，使万通四季城成为长春东部地区一个充满朝气和活力、富有"现代、独特、科技"精神的新城市空间。

2. 人性化的四季景观空间

万通四季城在塑造区域景观空间的过程中，不但充分强调空间景观的视觉感受，而且注重空间景观的主题性、人性化和可用性，实现穿越四季的流动商业景观节点。

3. 与城市充分融合和独特商业文化共存

地块本身作为区域中心，与经济开发区乃至二道区的发展有密切的联系，并且作为一个重要的区域融入新的城市建设中。

4. 实现 365 天的商业环境价值

打破北方无四季商业购物理念，回避地域性气候与现代化商业运营的冲突性特征，通过精心的布置和适宜的流线组织安排，构建四季商业的规划轮廓，在增加规划设计中文化内涵、创造精彩节点的同时，充分满足今后地区性商业中心无限制购物的需求。

Design Concept

1. Multi- functional Commercial Living Space Full of Energy and Vitality

Having taken land utility and use into consideration, the design is willing to introduce multi-functions like living, entertainment, dining, which makes Wantong Four Season Flower City become a new city space of "modern, unique and technological spirit" in the eastern area of Changchun.

2. People- friendly Seasonal Landscape Space

Not only does the center fully highlight visual inspiration of the space landscape, but it also concerns about its theme, people-orientation and accessibility with floating commercial landscape.

3. Full Integration with the City and Unique Mutual Existence of Commerce and Culture

As the regional center, the block is closely associated with development of the economic development area and Erdao Area and has been included in new urban construction.

4. Achieve 365- day Commercial Environment Value

It breaks the concept that there is no four- season commercial shopping in north China and avoids conflicts between regional climates and modern commerce. Its planning outline is established through subtle layout and proper line-shaped arrangement, which creates great cultural details and fully meets the demand of non-restricted shopping in regional commercial centers.

SMITHGROUP cna

Architecture · Urban Planning · Landscape Design
Interiors Design · Art Consultant · Lighting Design

Since 1853

www.cna-group.com

Design Determines . . .

设计决定······

1	2	3	4	5	6	7	8	9	10	······

03 复合社区
Composite Community

成都青城山泰达研发社区
Chengdu Qingcheng Mountain Taida R&D Community

Architecture Design | 2008 · 05

成都青城山泰达研发社区
Chengdu Qingcheng Mountain Taida R&D Community

项目名称：成都青城山泰达研发社区
用地规模：36.30 万平方米
建筑面积：26.50 万平方米
容积率：0.73
项目地点：中国 / 四川 / 都江堰
编制时间：2008 / 05

项目位于成都市青城山镇的发展区域带上，是一个中央有序、充满趣味的办公、商业和居住复合多元化研发社区，设计师将传统的建筑元素肌理进行有机的整理，并完成向时尚品质延续和过渡的过程。整个项目希望在记忆中获得足够的力量来突破一般认为的传统带来的束缚，使社区具有禅悦的快感。

The project is locatesd on a developing zone of Qingcheng Mountain. It's an interesting community with administrative, commercial and residential part. The designer organized the traditional architecture elements and made a transition to the fashion quality. The whole process was a powerful break of the cultural chains and made a pleasant improvement for the community.

总平面图

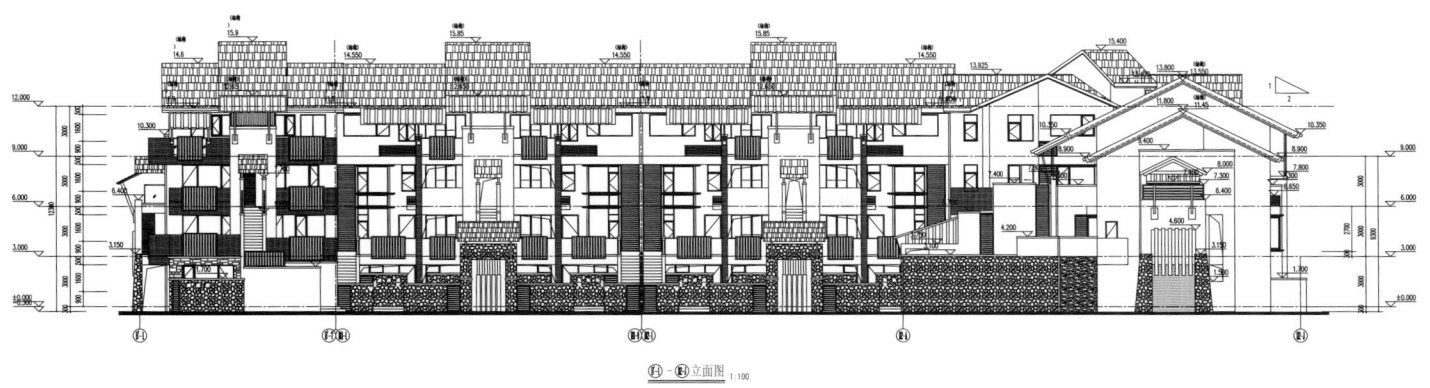

Ⓐ−Ⓗ立面图 1:100

Ⓗ−Ⓐ立面图 1:100

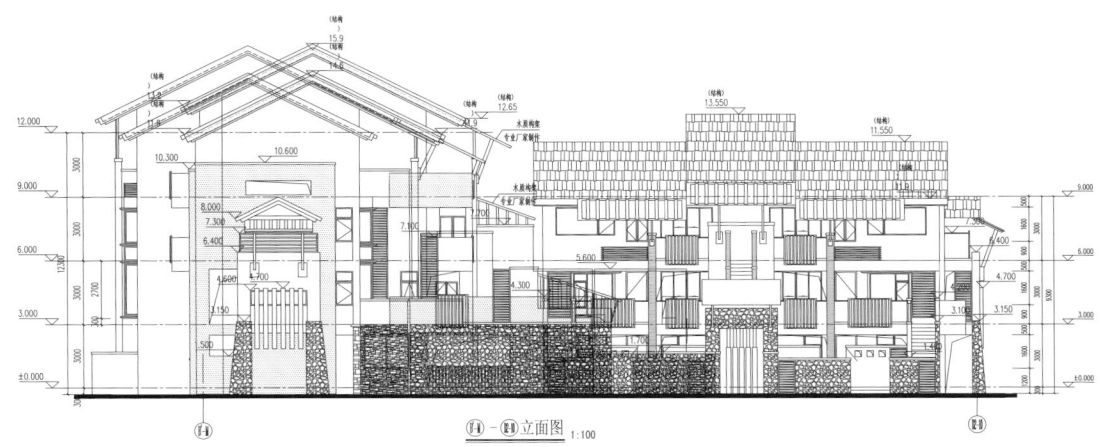

Ⓐ−Ⓗ立面图 1:100

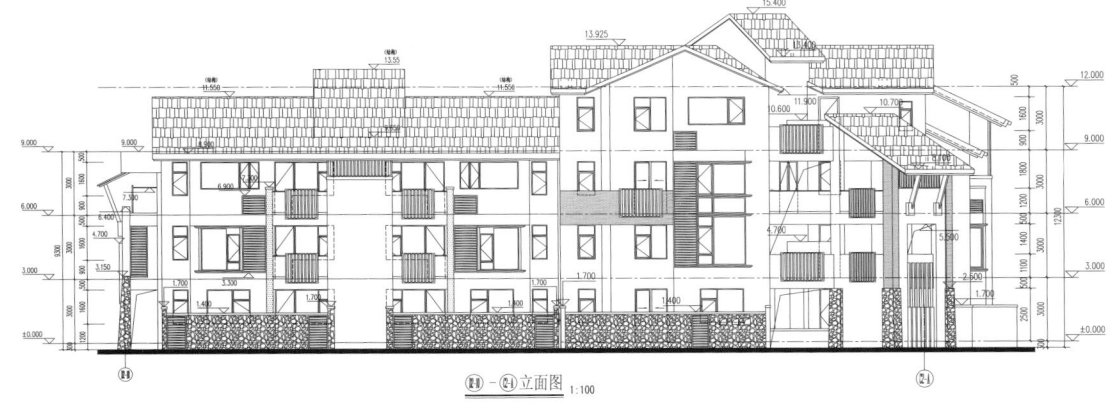

Ⓗ−Ⓐ立面图 1:100

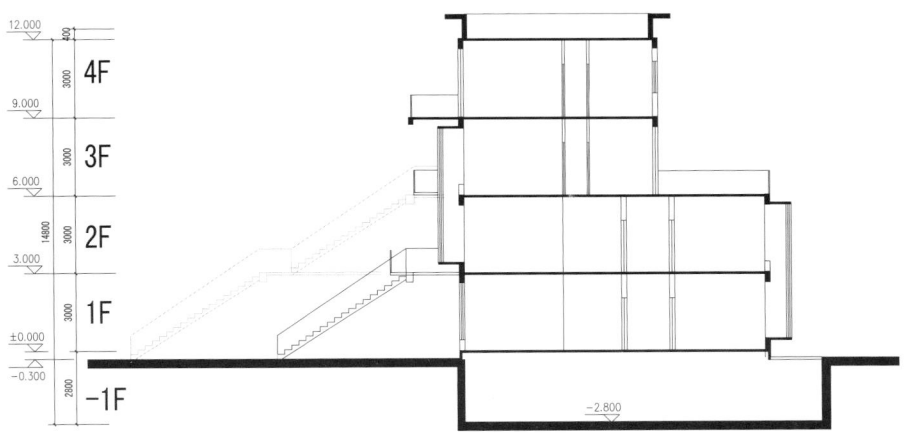

12.000	
	4F
9.000	
	3F
6.000	
	2F
3.000	
	1F
±0.000	
−0.300	
	−1F
−2.800	

3 FLOORS

4 FLOORS

OPTION 2
ELEVATION

3 FLOORS

4 FLOORS

OPTION 1
ELEVATION

OPTION

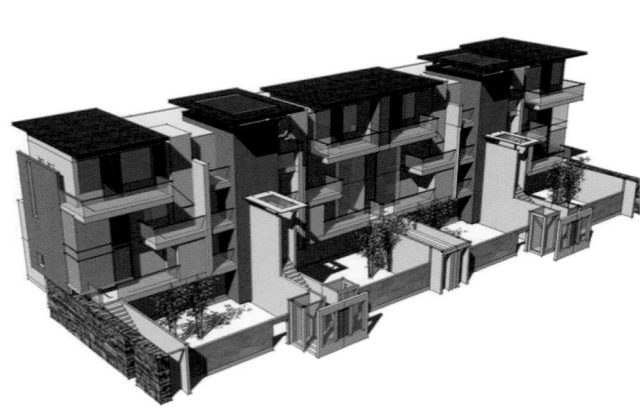

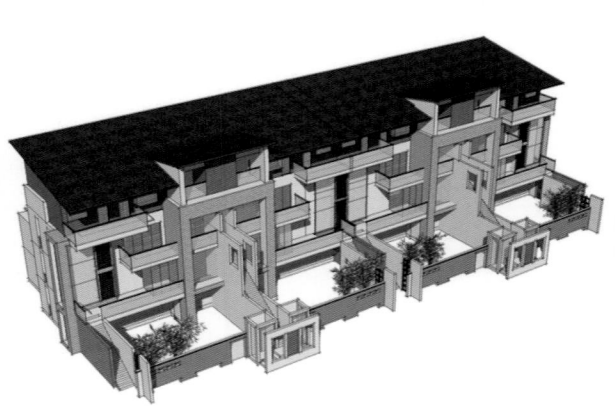

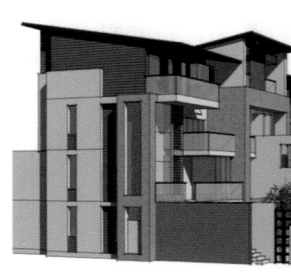

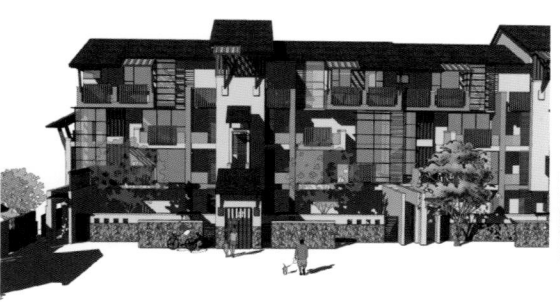

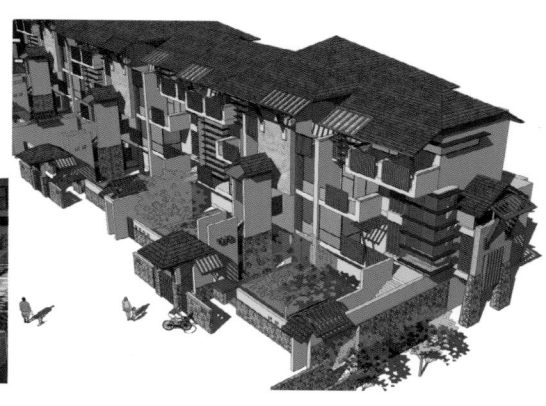

过程的力量

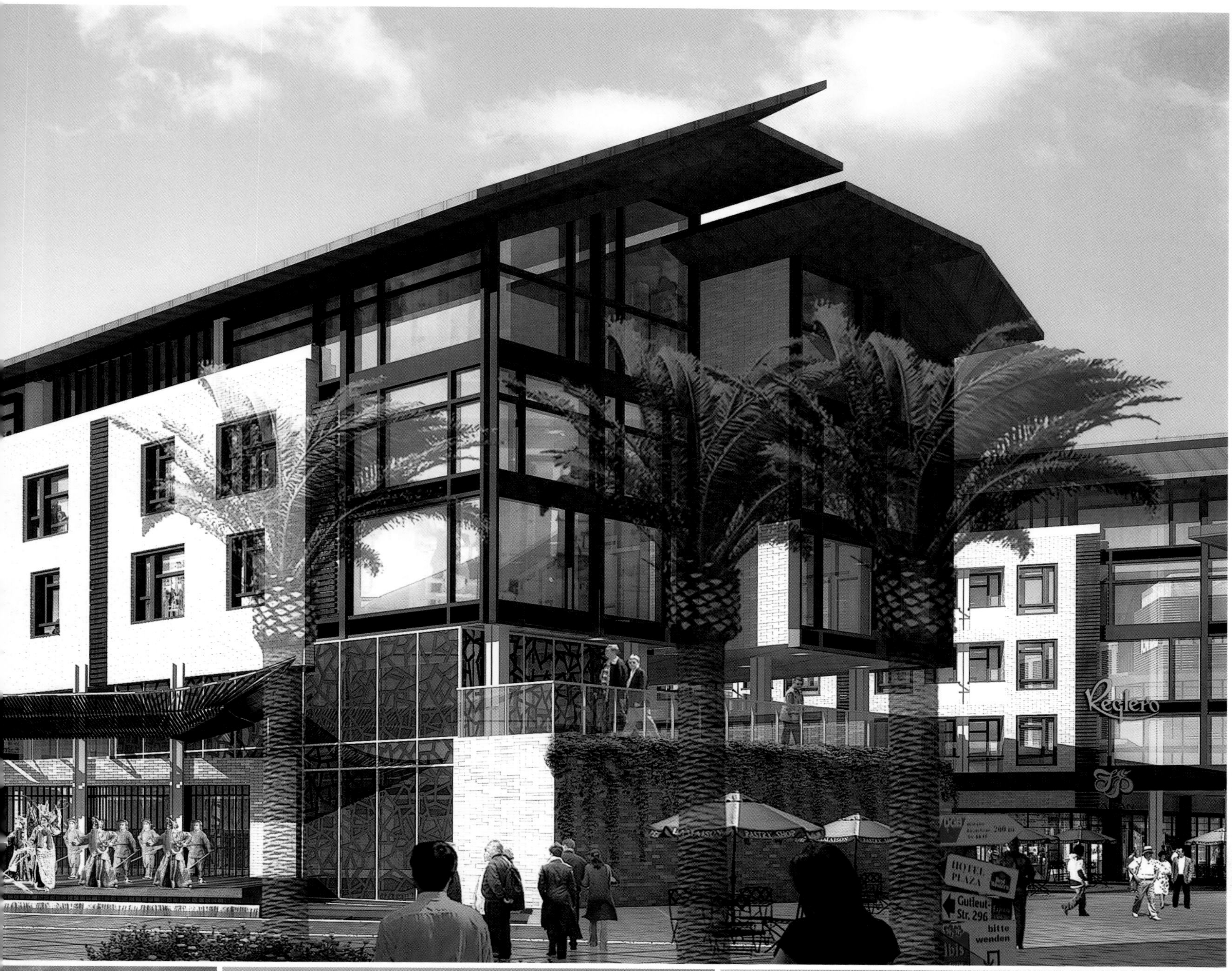
Reglero
HOTEL PLAZA
Gutleut-
Str. 296
bitte
wenden

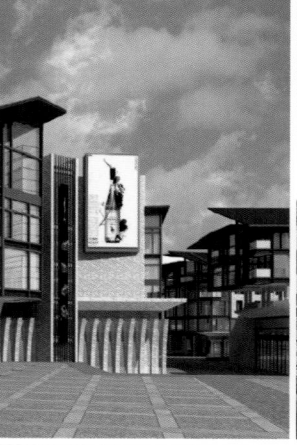

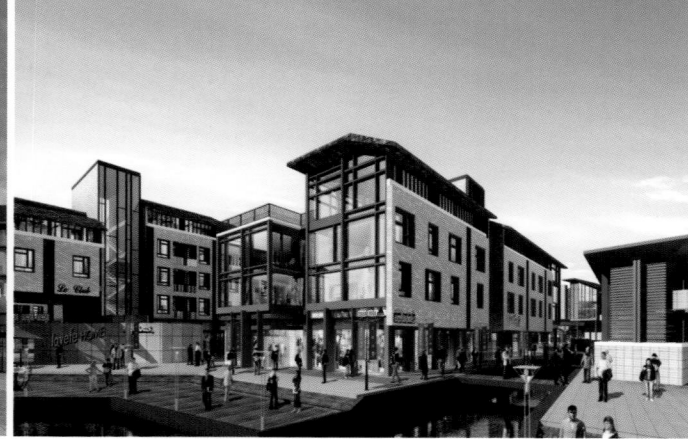

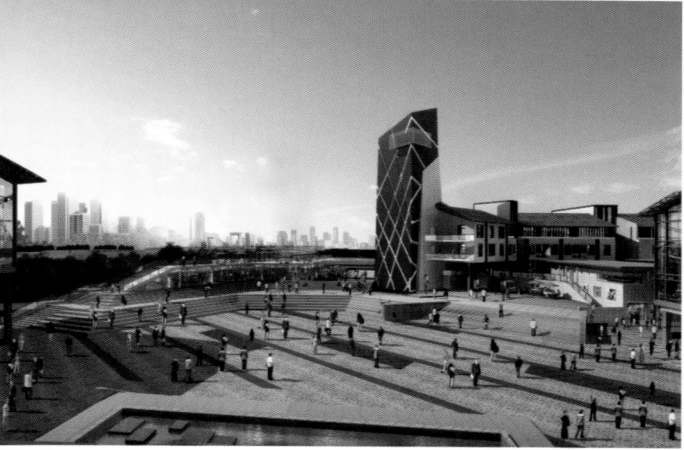

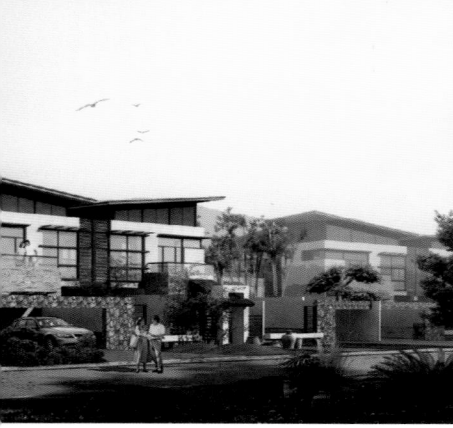

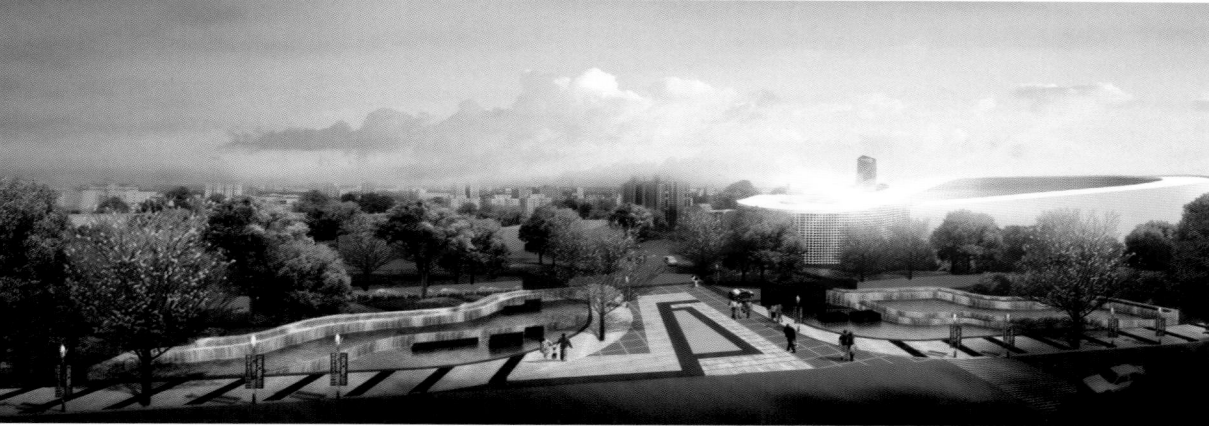

给予一个园区生活活力和商业的自由度及兴奋度是十分重要的。在此我们结合园林团队和
商业团队的力量，将景观的丰富性和材料以及业态布局等各关键要素组合进去，形成建筑
丰富、环境丰富、商业业态布局丰富等状态下的新区商业和居住规划模式。

上海松江科技绿洲总部企业园区

Shanghai Songjiang Science And Techonology Oasis Headquarters Enterprise Park

Architecture Design | 2010·07

方案【A&B】

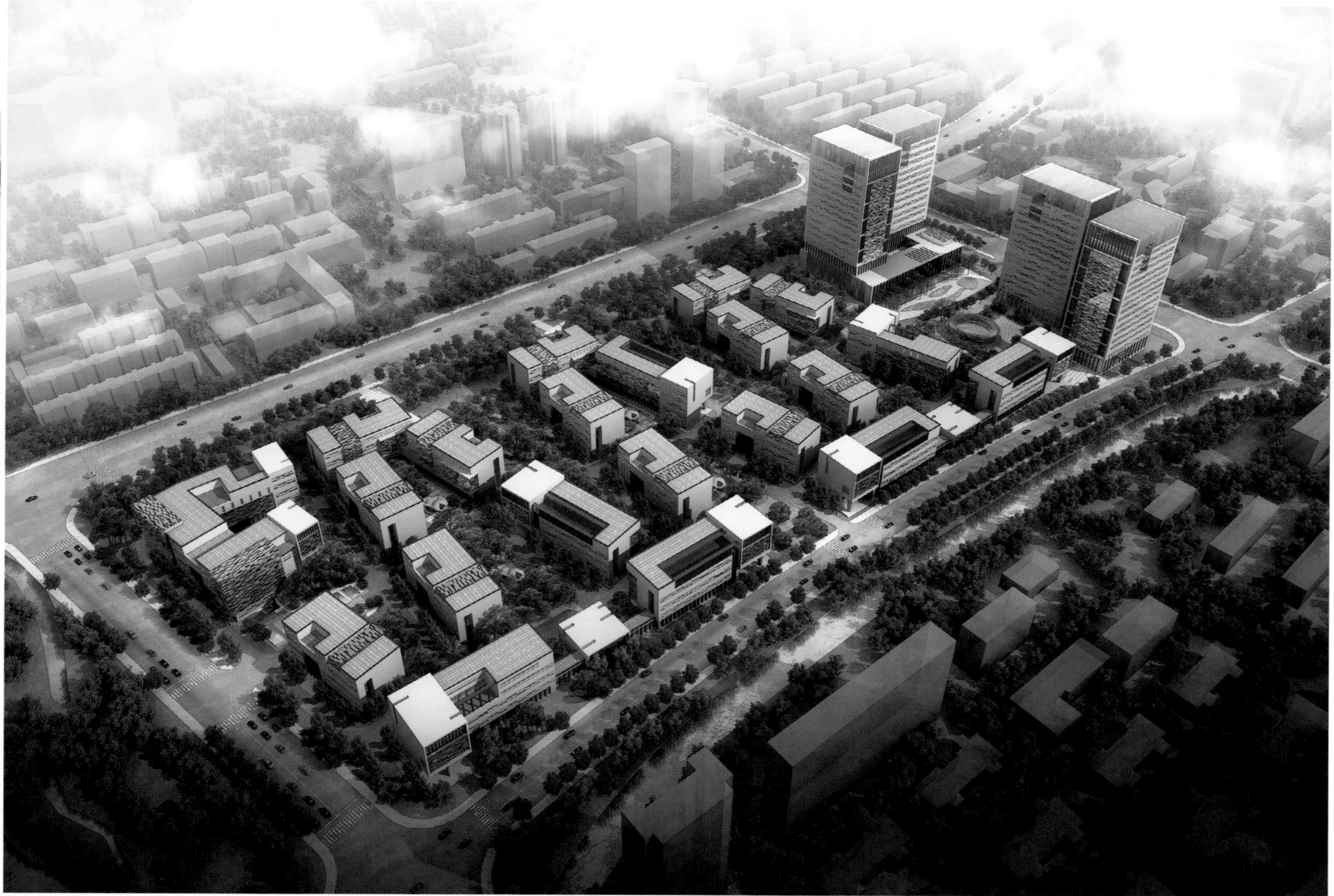

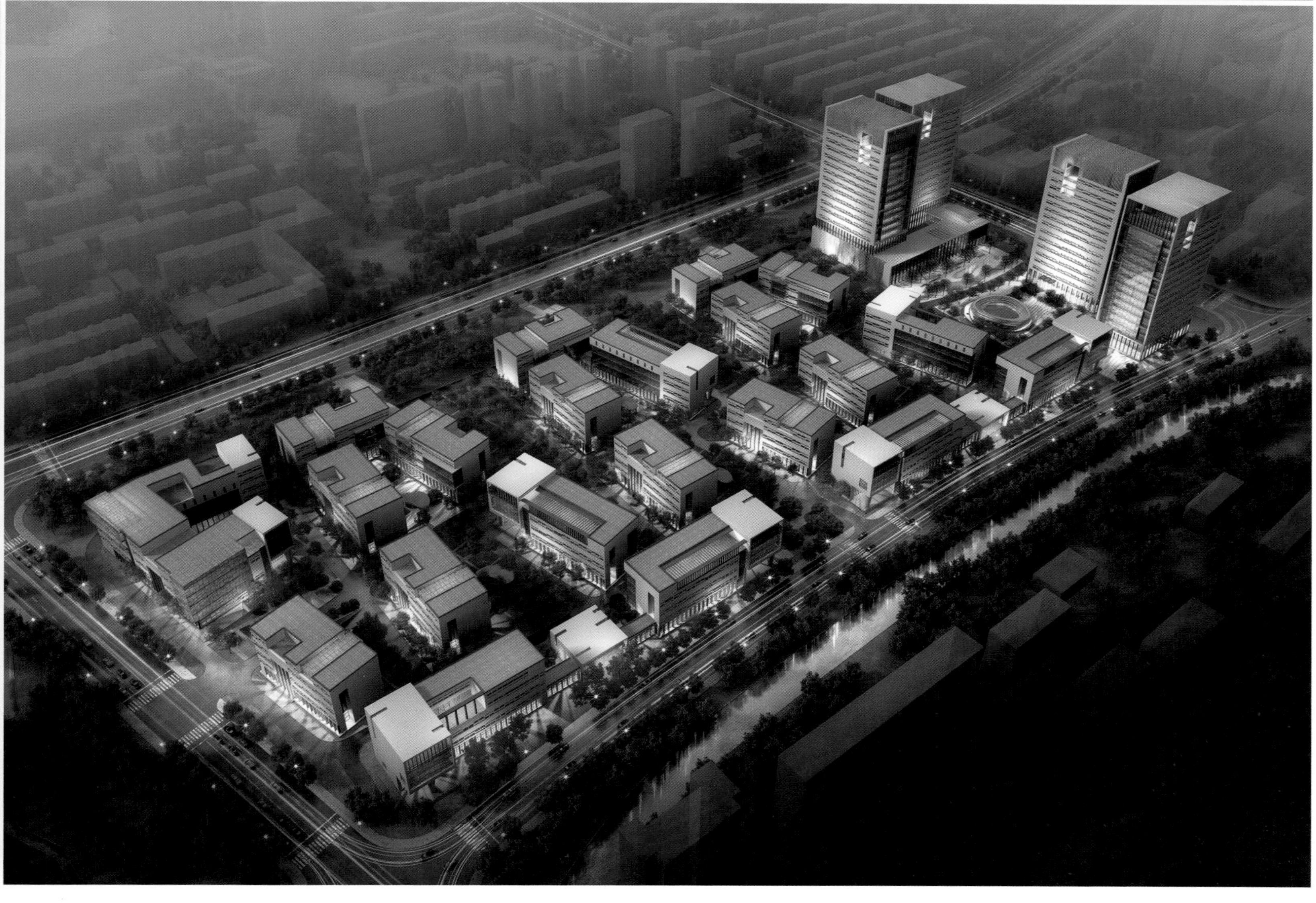

上海松江科技绿洲总部企业园区
Shanghai Songjiang Science and Technology Oasis Headquarters Enterprise Park

项目名称：上海松江科技绿洲总部企业园区
用地规模：13.33 万平方米
建筑面积：21.5 万平方米
容积率：1.6
项目地点：中国 / 上海
编制时间：2010 / 07

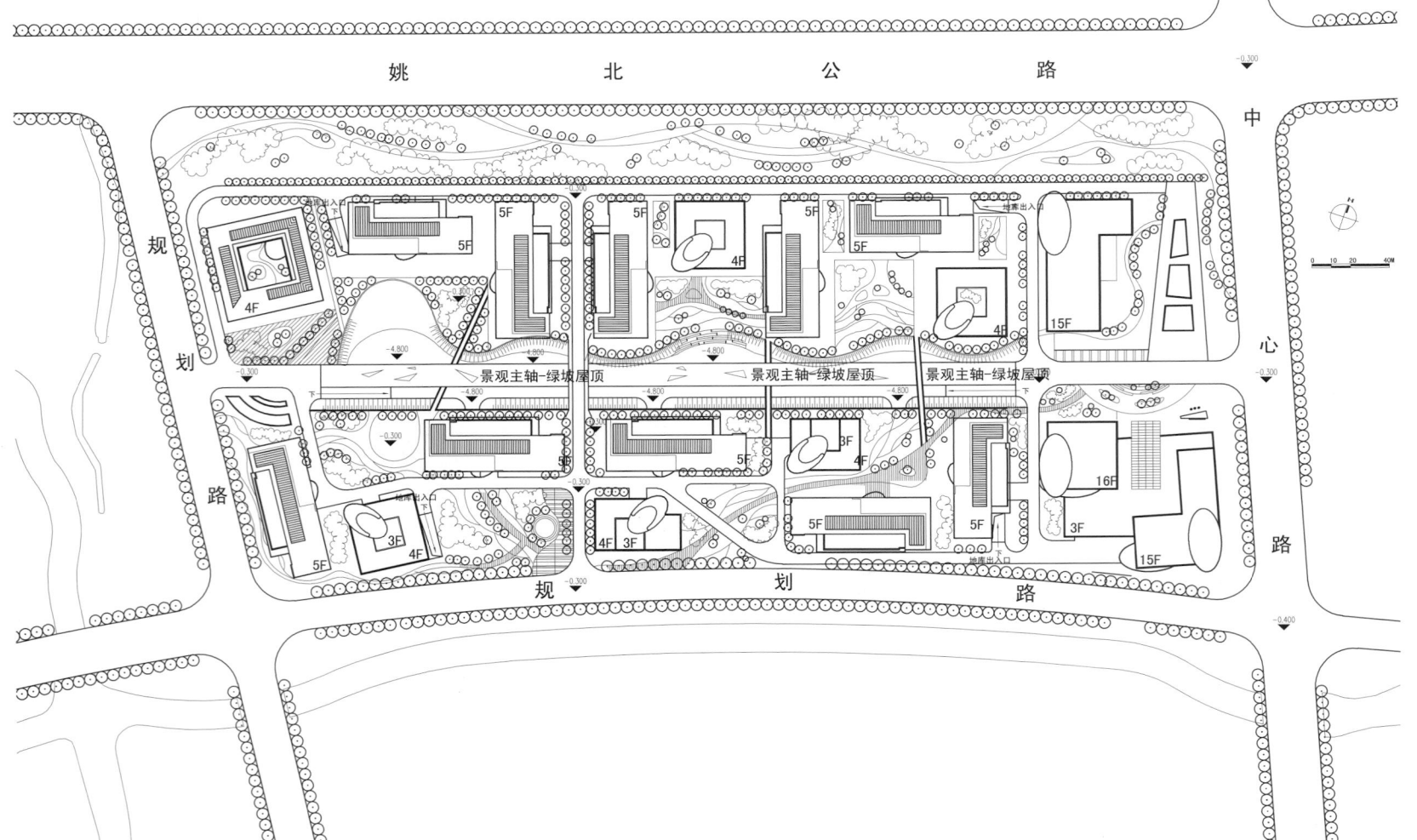

设计师在建立景观中的建筑的同时也在意图建立建筑中的景观，设计用消解建筑的第五立面的手法将景观系统与建筑系统相叠加，使室内外空间互相渗透，从而获得多变、亲切、新奇的工作场所。

项目位于上海漕河泾开发区松江园区内，占地 200 亩，综合容积率 1.6，建筑面积 21.5 万平方米，其中总部基地办公园区为 12.5 万平方米，配套及研发区域为 9 万平方米，是为新一代高科技企业提供区域总部办公、产品研发的区域。

The buildings in landscape and the landscape in buildings - the landscape system and the building system are superposed over each other by the technique of dissolving the fifth facade, so that the indoor and outdoor spaces permeate each other, thus changeful, genial and novel work locations are provided.

The project is located in the Shanghai Caohejing Songjiang Hi-Tech Park, covering an area of 13.33 hectares (200mu), with the FAR of 1.6, and the building area of 215,000 square meters, therein the area of the headquarters base office zone is 125,000 square meters, and the supporting and R&D area is 90, 000 square meters.

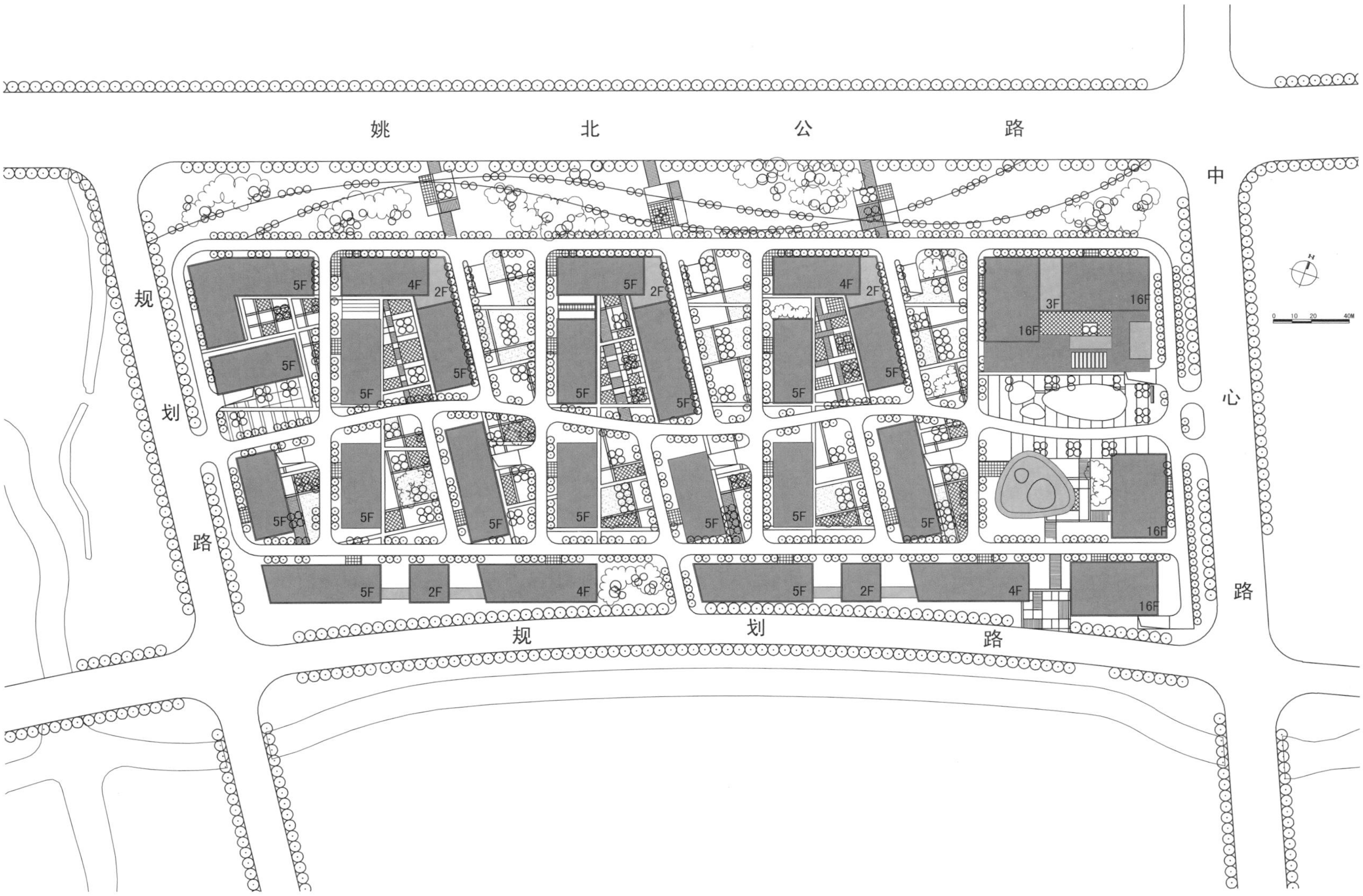

姚　北　公　路

规

划

路

规　　　划　　　路

中

心

路

5F　　　4F　　2F　　　5F　　2F　　　4F　　3F　16F

5F　5F　　5F　　5F　5F　　5F　　5F　16F

5F　5F　　5F　　5F　5F　　5F　　5F

5F　5F　　5F　　5F　5F　　5F　　16F

5F　2F　　4F　　2F　　4F　　16F

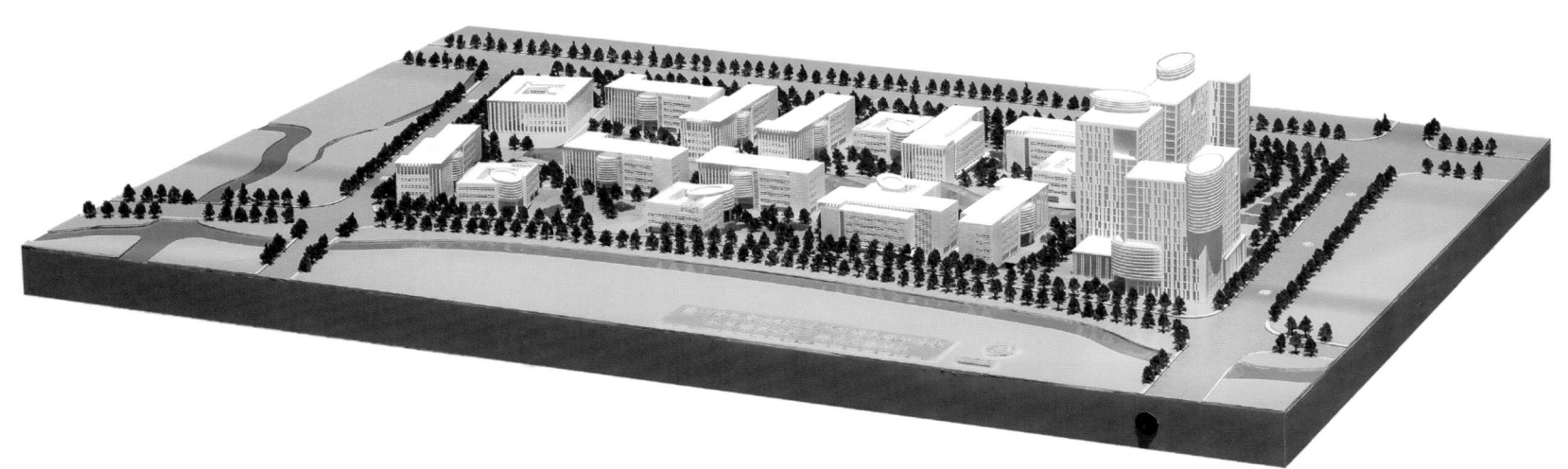

没有被建成的不一定就是不好的……

215

大会议室
空中花园
休闲娱乐
小会议
办公
办公入口大堂
电梯间

5F
4F
3F
2F
1F
21.300
17.100
12.900
8.700
4.500
±0.000
−0.150

70800
3600
8400
办公
办公入口大堂
水景

屋顶层
5F
4F
3F
2F
1F
−1F
21.300
17.100
12.900
8.700
4.500
±0.000
−0.150
−4.500
办公
电梯厅
办公大厅
车库

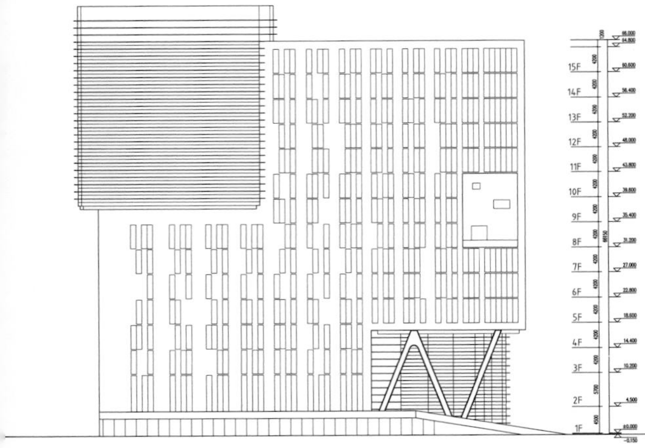

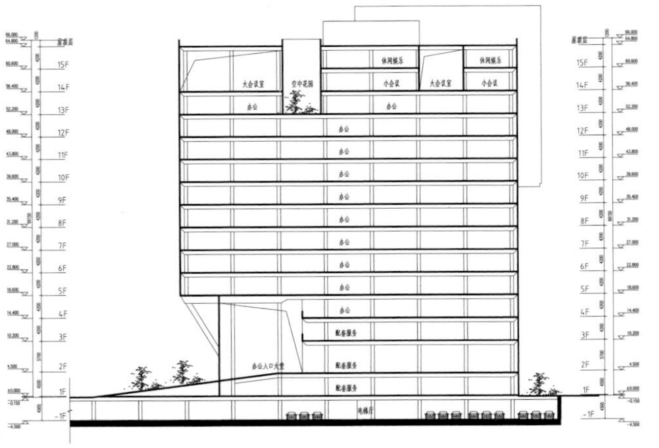

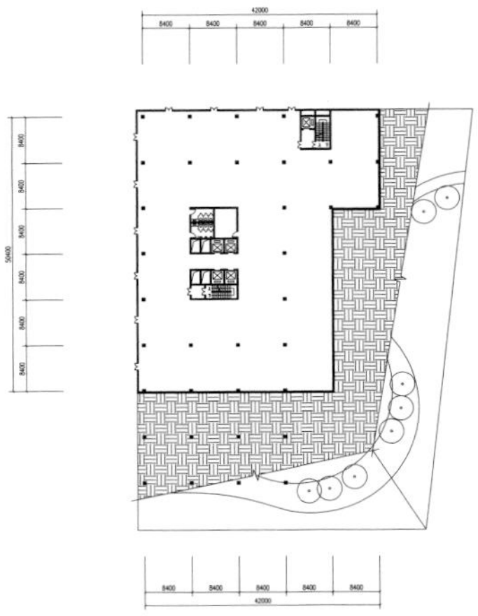

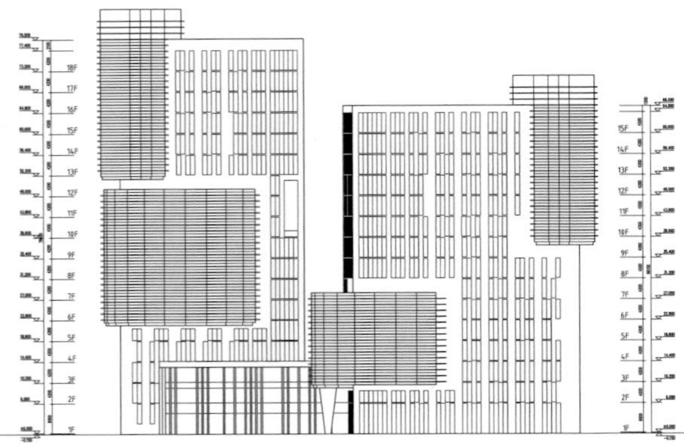

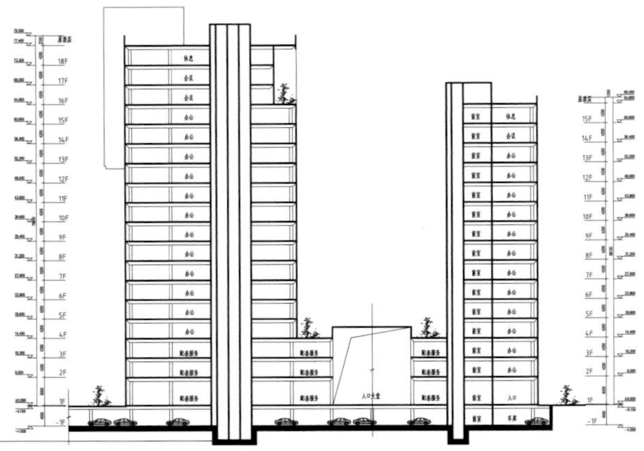

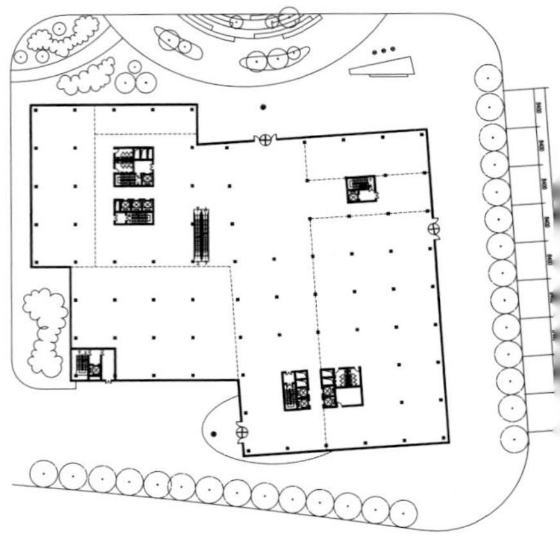

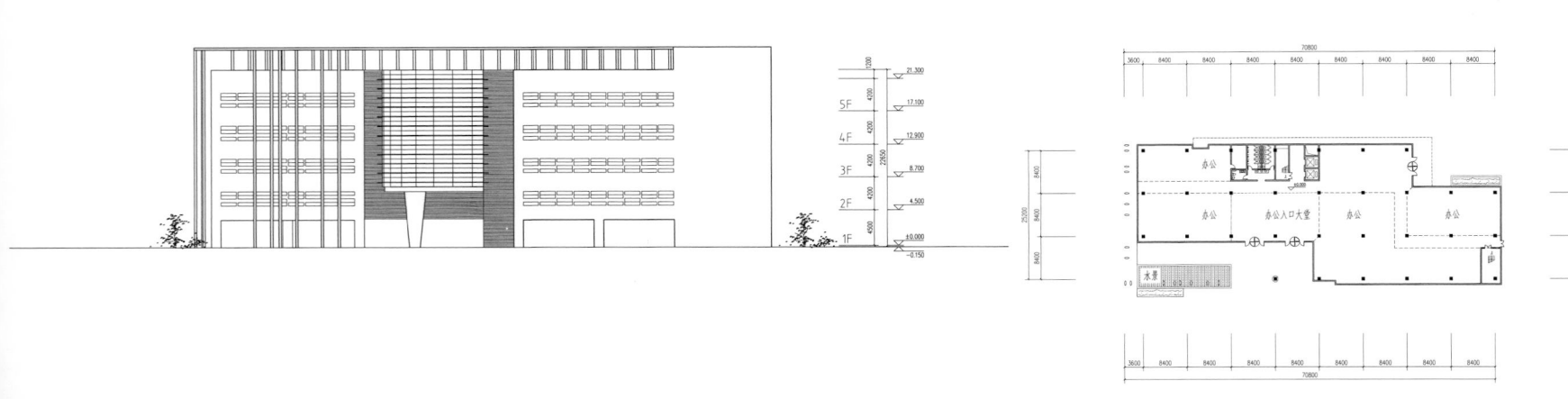

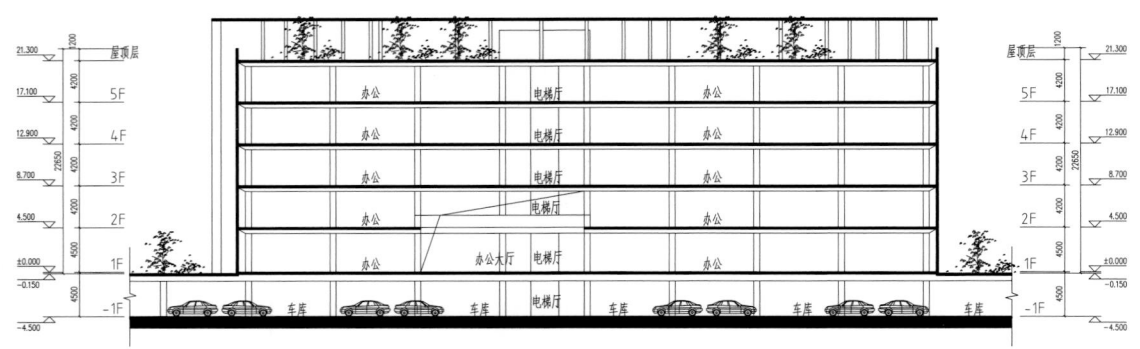

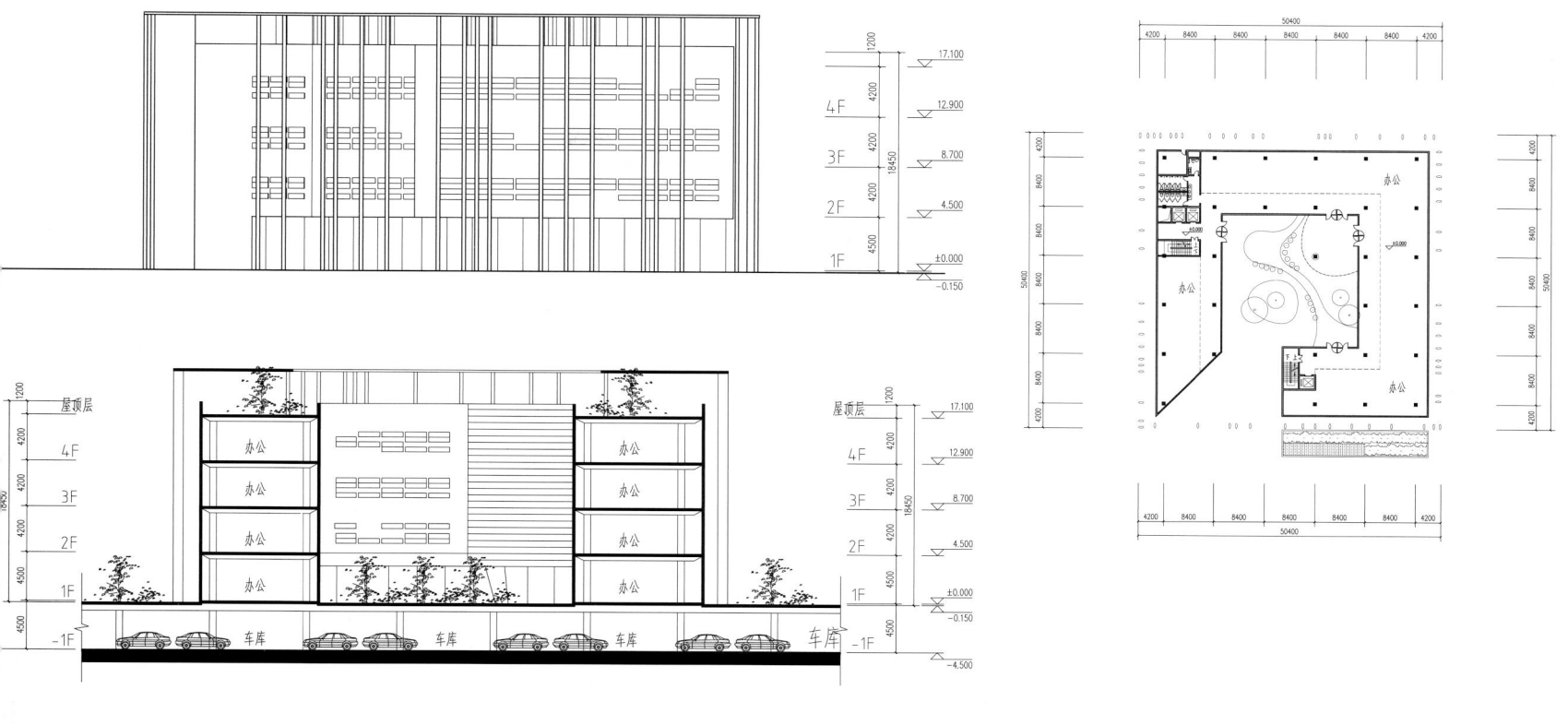

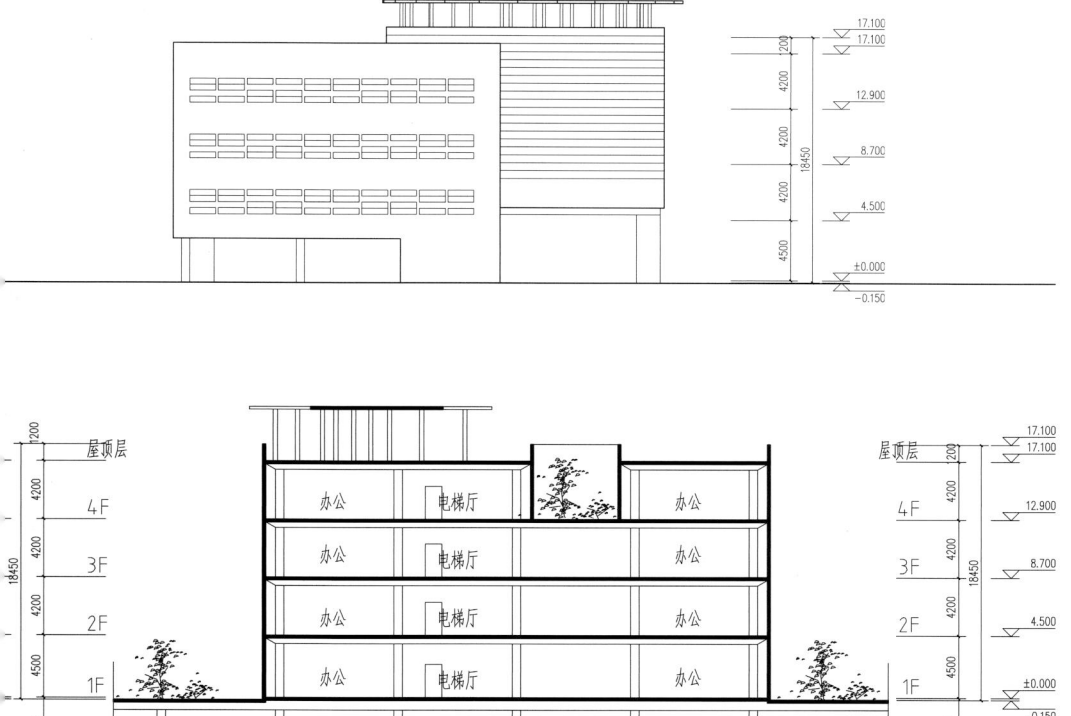

常州华润国际社区会所
Changzhou Huarun International Community Club

项目名称：常州华润国际社区会所
用地规模：7 200 平方米
建筑面积：2 963 平方米
容积率：0.41
项目地点：中国 / 江苏 / 常州
编制时间：2010 / 06

项目为常州华润国际社区的会所兼销售中心。在我们设计师进行改造设计时，该建筑的主体和室外工程已全部竣工，正进行内部装修。

在设计之初对原有建筑进行了充分的理解和深刻的分析之后，在不对结构作变动的情况下，针对原有建筑设计所存在的一些问题，给予目标明确的改善，争取以最小的改动获得最大的价值感的改观，新的建筑外观以获得与整体社区同样的价值感和对时尚的期待为目标。

This is a facade transformation project for the club and sales center of Changzhou Huarun International Community located in Changzhou City, Jiangsu Province. When we undertok the transformation design project, the main body and outdoor works of the building had been completed and the internal decoration was under construction.

Through full understanding and deep analysis of the original building, we made targeted improvement for some defects of the original building design without structural change, striving to obtain the greatest appearance change through the least change.

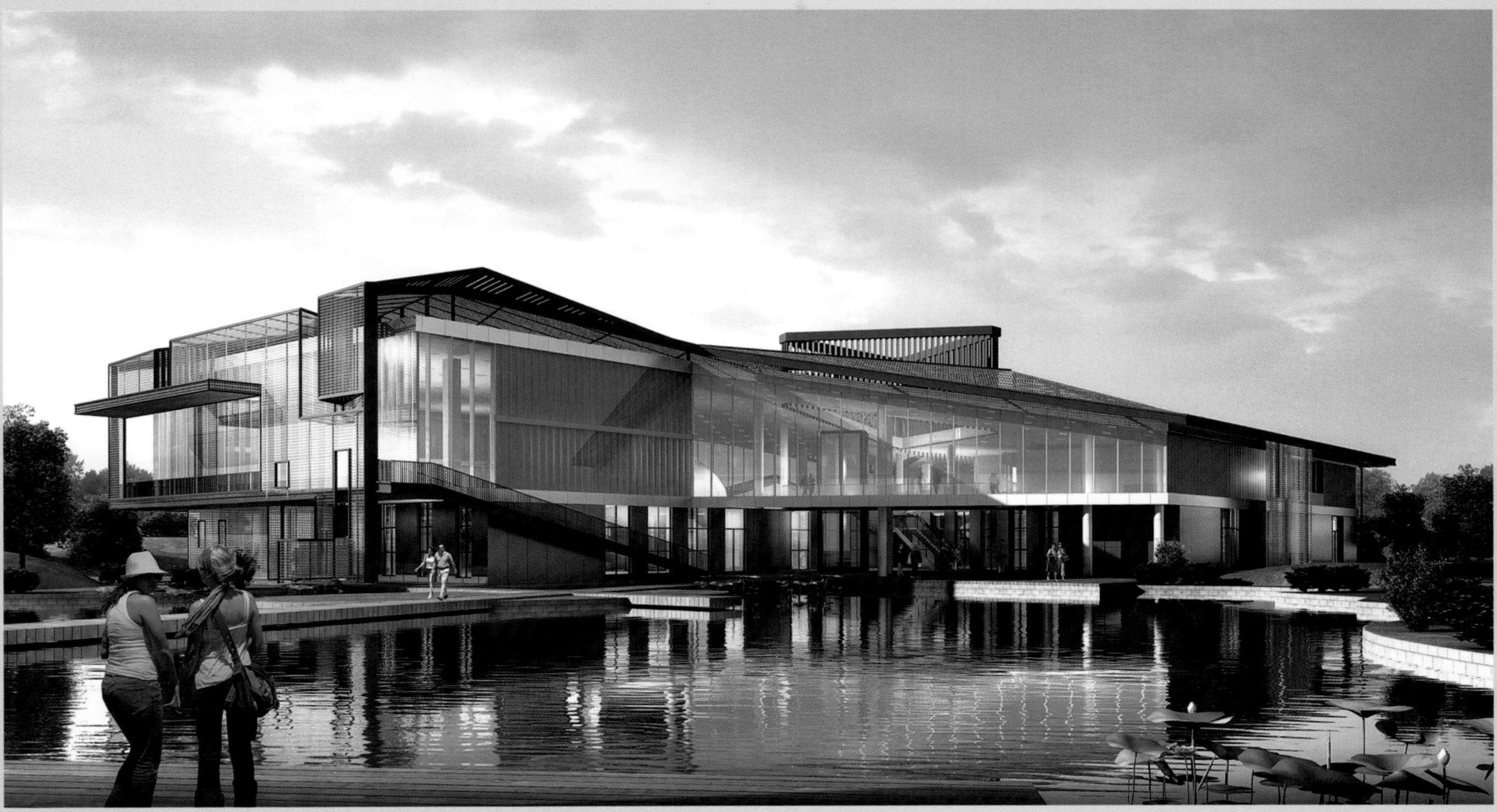

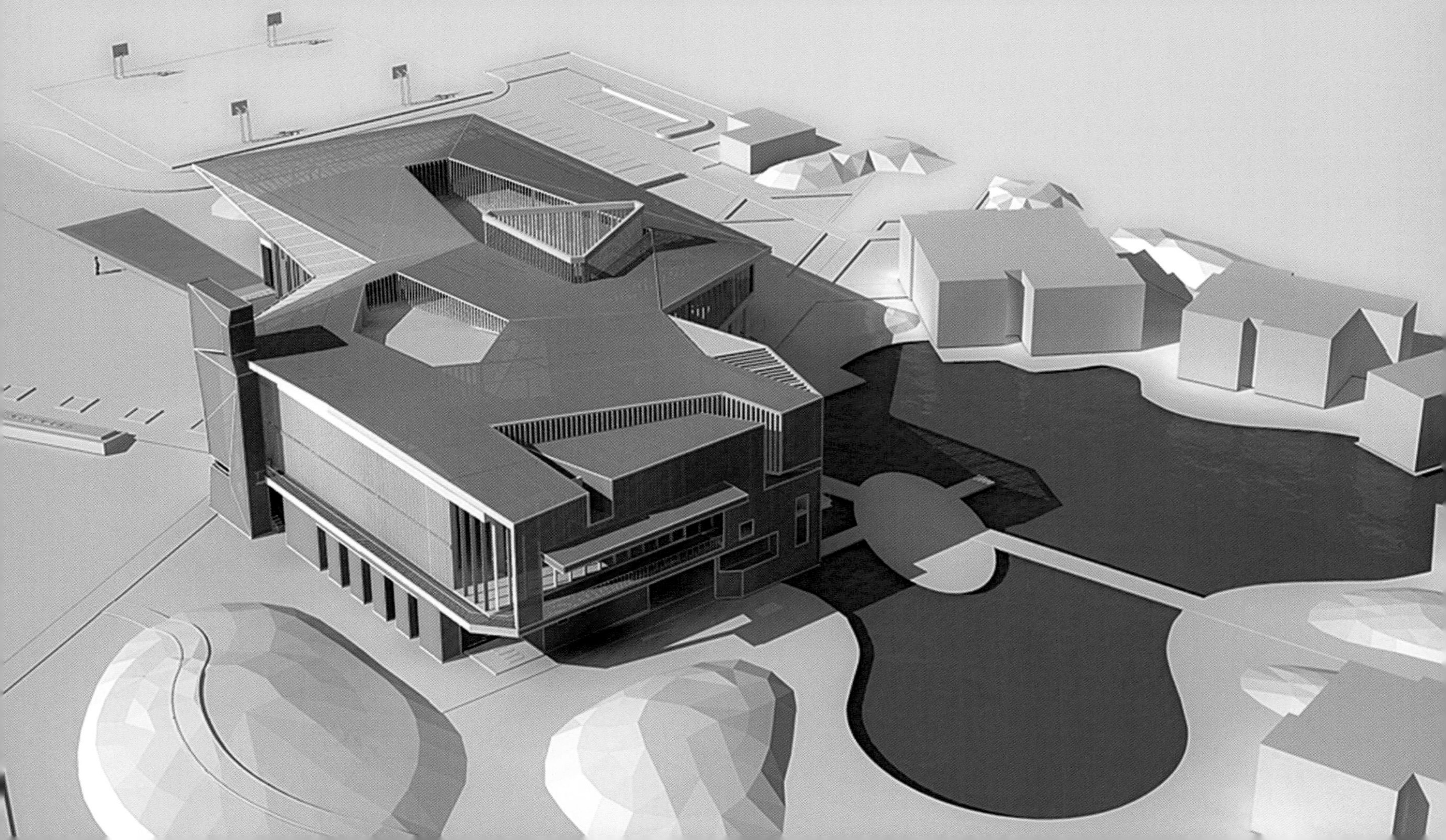

成都鹤鸣山道教文化旅游区
Chengdu Heming Mountain Taoism Culture and Tourism Centre

项目名称：成都鹤鸣山道教文化旅游区
用地规模：24.0 万平方米
建筑面积：70.0 万平方米
容积率：2.9
项目地点：中国 / 四川 / 成都
编制时间：2007 / 11

成都鹤鸣山道教文化旅游区是以中国道教发源地遗址、古村落和自然景观相结合的现代宗教主题综合旅游产业发展规划，项目包括道教殿堂建筑，研究机构，会议论坛，休闲疗养酒店，高尔夫马球场，古商业街特色居住等大量的综合内容。

Chengdu Heming Taoism Culture and Tourism centre is located at the Heming Mountain. It's the origin place of Taoism culture. The project creates a comprehensive community which integrates ancient heritage and natural landscape. The project includes Taoism Hall, Research Facility, and various places for leisure and vacation.

观
音
河
桥
街
边
半
戏台
茶楼
特色餐饮
特色商品
客栈
养身SPA
茶文化
咖啡
酒家
N

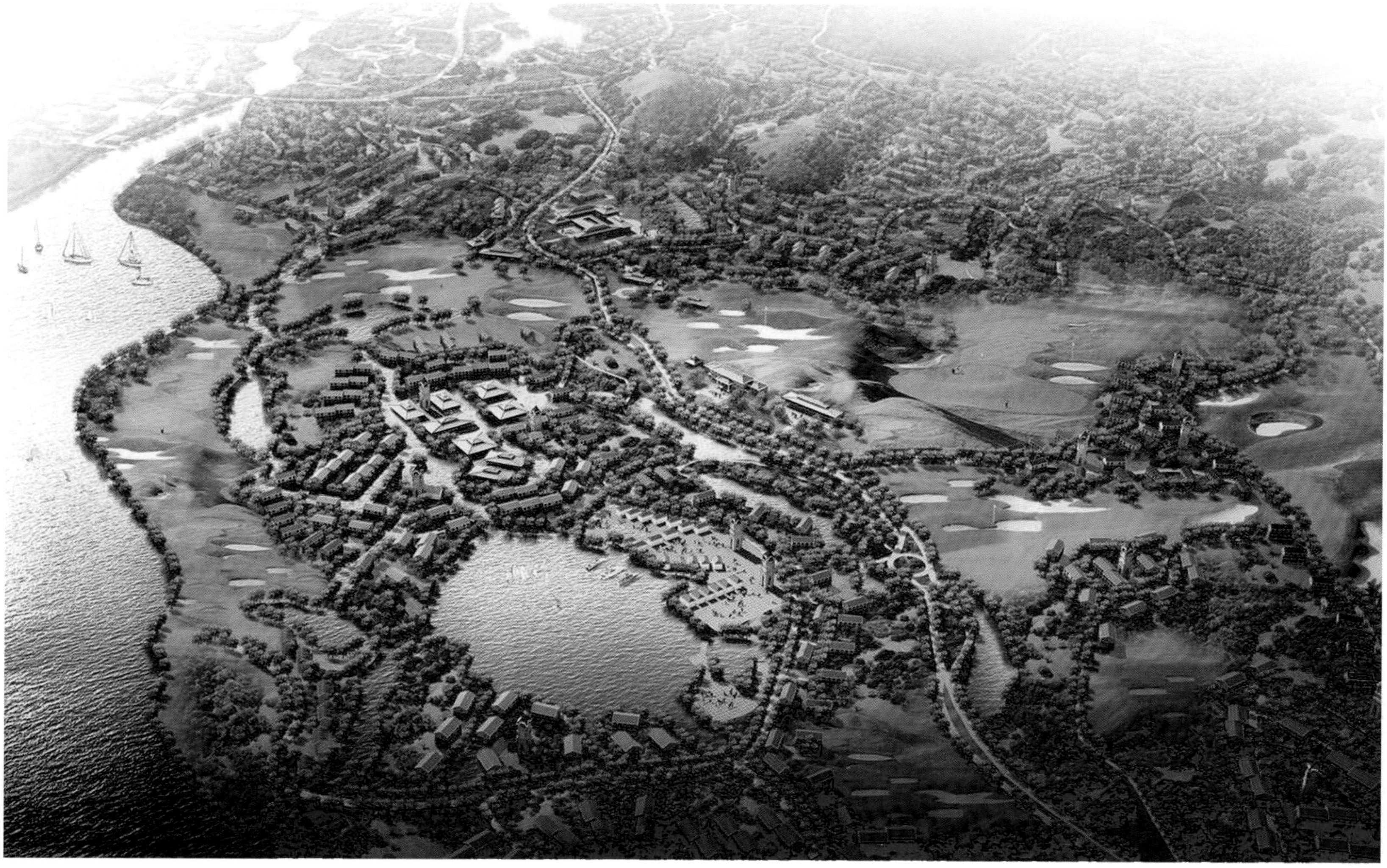

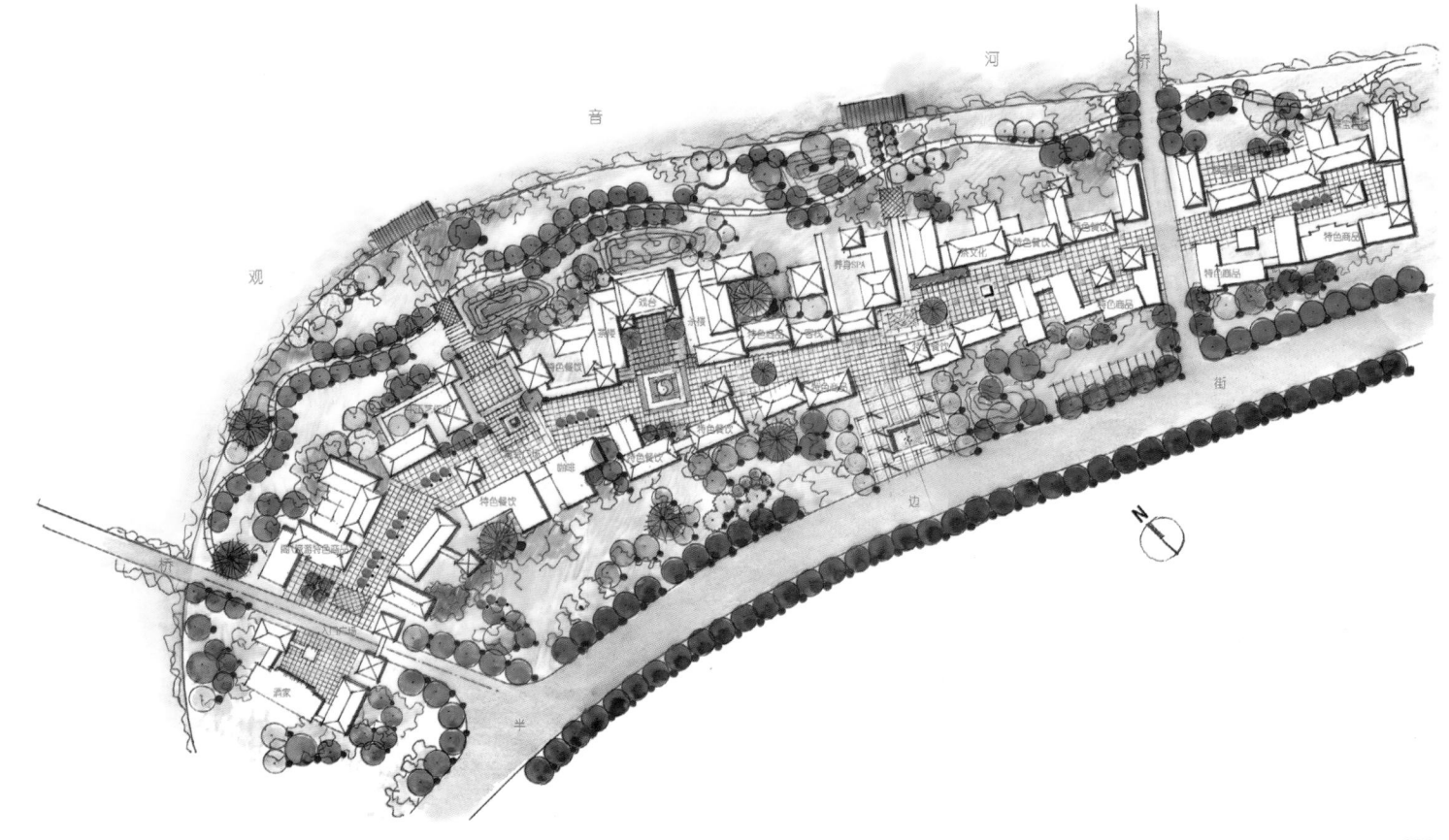

成都青羊工业科技创新园 C、D 区
Chengdu Qingyang Industrial Science Park C&D Area

项目名称：成都青羊工业科技创新园 C、D 区
用地规模：13.25 万平方米
建筑面积：6.89 万平方米
容积率：0.52
项目地点：中国 / 四川 / 成都
编制时间：2006 / 05

项目位于成都市青羊区科技创新园区内，周边具有完善的配套设施和优良环境，建筑风格现代简洁，色彩明快，小区内部作下沉处理，形成了一个有趣的景观峡谷，创造了另一种亲切、幽静的空间感受，使得整个区域的空间层次更为丰富，充满趣味性和惊喜。

Locatd at the Qingyang Industrial Science Park, the project enjoys well supported service and excellent environment. The architectures are of simple, colorful modern style. Parts of the community buildings are arranged o the semi-underground square to form an interesting landscape valley. The hospitable, quiet space creates a zone with rich spacial layers and enjoyment.

南通世纪海港乡村俱乐部
Nantong Century Harbor Country Club

项目名称：南通世纪海港乡村俱乐部
用地规模：1.80 万平方米
建筑面积：2.44 万平方米
容积率： 1.36
项目地点：中国 / 江苏 / 南通
编制时间：2010 / 06

本基地位于南通市如东县小洋口镇经济开发区，洋口镇跨港大桥西侧，用地周边路网已初步形成，道路交通条件良好，整个用地地形规整平坦，区内小市政给水（区内深井）、广电等条件一应俱全。基地南北两端为球场内 1 号洞，自然水系围绕，西侧为球场内 19 到 25 号洞区域，东侧为 20 米宽规划道路，与二期建设的高档住宅小区相邻。基地东西向较长，南北向较窄，东侧为球场主干道，因此考虑设置两个机动车出入口连接基地。

借助基地原有的纯天然资源，运用"以人为本，人无我有，人有我优"的思想观，"融和共生"的生态设计手法，力图打造一个坐拥自然壮阔、清新优雅的迷人景致，远离尘嚣跟纷扰，天堂般的全景高尔夫水榭风情旅游度假休闲酒店区。使其成为具有优美的居住环境、高品质住宅商务配套设施、丰富的文化氛围的休闲度假区。和谐的空间尺度和生态的景观环境完全融入整个酒店区域。

The project is located in the Economic Development Zone in Xiaoyangkou Town, Rudong County, Nantong City, in the west side of the Kuagang Bridge in the town. Surrounding roads have taken shape and the traffic conditions are good. The entire plot is regular and flat and enjoys complete electricity, water supply, and broadcasting and TV facilities. The north and south ends of the base are surrounded by natural water in No. 1 hole; the west is for No. 19-25 holes; in the east, a 20m-wide road will be planned, adjacent to the high-grade residential area that will be built in Phase II. The base is long in east-west direction and narrow in south-north direction. Trunk road of the golf course is located in the east, therefore, two vehicle entrances are designed here to provide access to the base.

Relying on the original natural resources of the base, the design follows the concept of "people-oriented, unique and outstanding" and the "harmonious coexistence" eco-design practice and strives to create a fantastic golf leisure resort community with fresh and elegant natural landscape free from trouble and distress. There are beautiful living environment, high quality residential and business facilities and rich cultural atmosphere with harmonious spatial scale and ecological landscape environment.

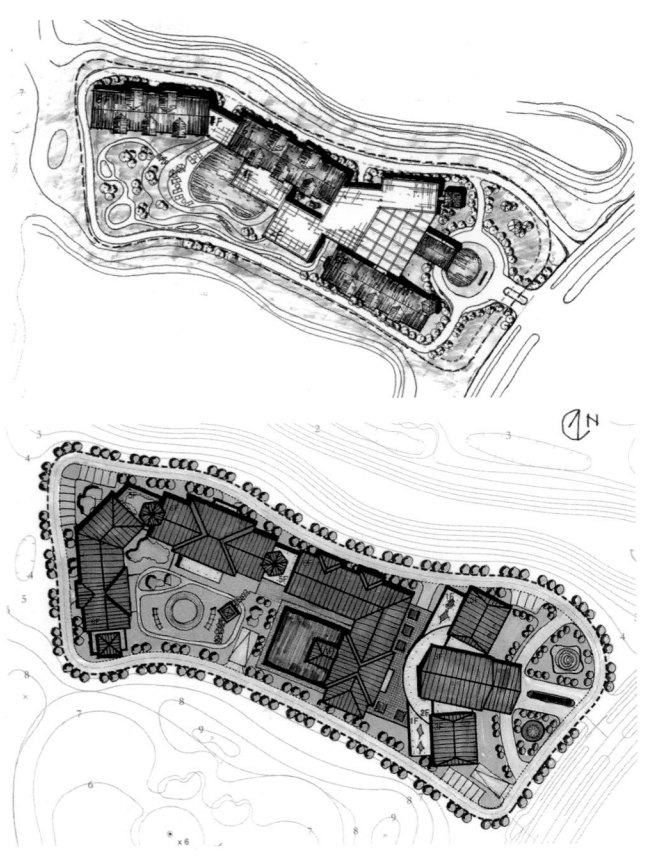

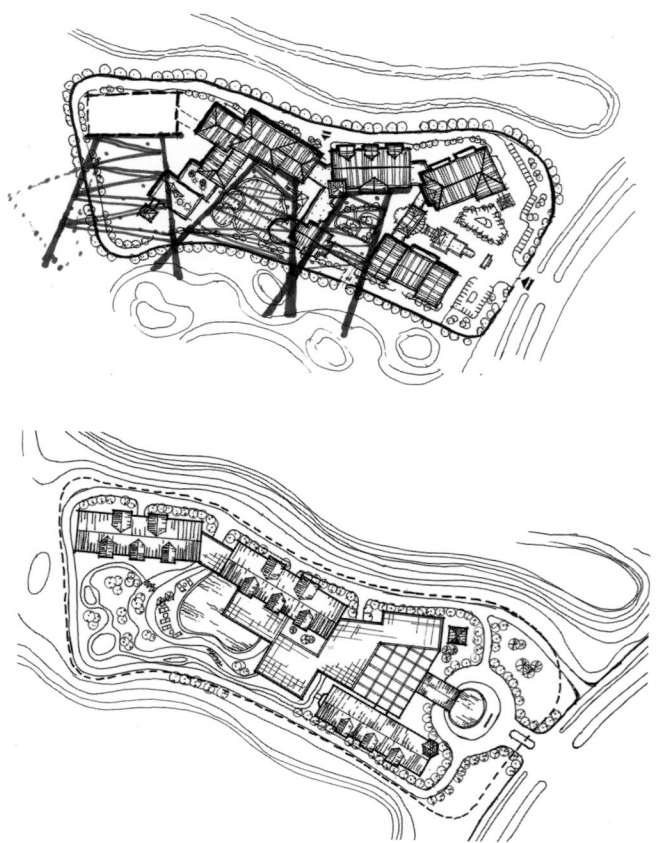

宁波东钱湖东方花博园
Ningbo Dongqian Lake Oriental Flower Exhibition

项目名称：宁波东钱湖东方花博园
用地规模：22.0 万平方米
建筑面积：4.02 万平方米
容积率：0.18
项目地点：中国 / 浙江 / 宁波
编制时间：2007 / 05

项目位于宁波市东钱湖东北侧，背山面湖，地形为东西向，是两边青山环绕的狭长地带。依托东钱湖旅游资源，是一座以"花博会"为主题，以花的海洋为中心，休闲养身为重点的特色花卉度假博览园，是集观光、都市休闲、度假旅游的复合功能于一体的旅游地产。

The project is located on the north-eastern side of Dongqian Lake, Ningbo. Based on the great natural resource of traveling and the complex traveling project which integrates the function of sightseeing, leisure and holiday is subjected in "Flowers Park".

上海崇明鹏欣会所
Shanghai ChongMing Pengxin Club Project

项目名称：上海崇明鹏欣会所
用地规模：400 万平方米
建筑面积：0.07 万平方米
项目地点：中国 / 上海
编制时间：2008 / 06

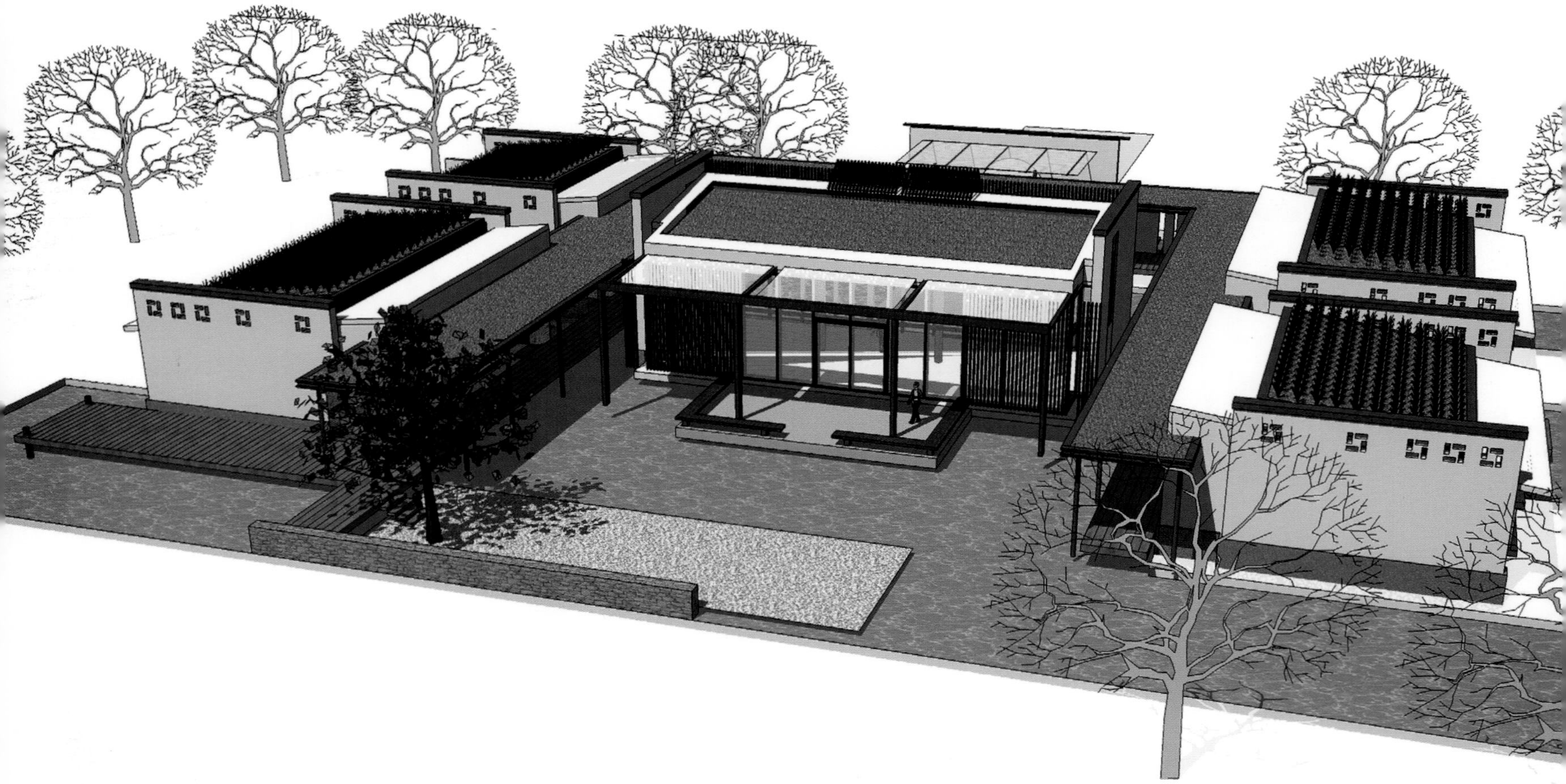

建立一个新中式庄园会所，在崇明岛体验田园牧歌般的生活。

本项目的开发设计基于以上理念，在长三角地区最接近大自然的良好生态环境中，在 6 000 亩恢宏土地上，依托政府对生态观光农业的大力推进和发展，力图打造一个现代的休闲农庄会所。

New Chinese manor club-idyllic life experience

According to the planning of Chongming Government, it is positioned to be a forest garden island, ecological living island, holiday island, green food island, marine equipment island and science and technology research and innovation island.

Based on the above background and the government's great promotion and development of eco-tourism agriculture, the project intends to design a modern farm on the grand land amounting to 6,000 acres in the Yangtze River Delta region with best ecological environment.

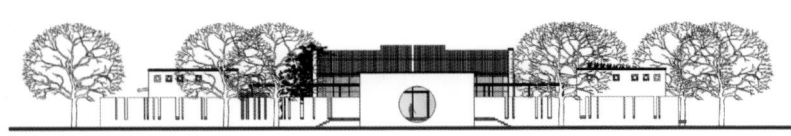

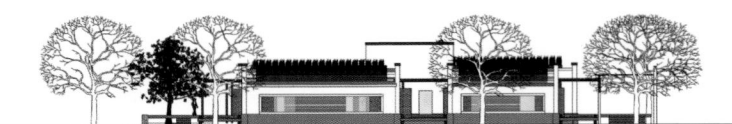

上海金山生产性服务业集聚区
Shanghai Jinshan Productive Service Integrated Area

项目名称：上海金山生产性服务业集聚区
用地规模：9.07 万平方米
建筑面积：9.0 万平方米
容积率：0.99
项目地点：中国 / 上海 / 金山
编制时间：2006 / 03

项目位于上海市金山区沿海仓储服务区内，作为该区的生产性服务业聚集区，主要为附近生产人员和厂家提供生活服务、办公服务和小型的生产服务设施。建筑单体整体上强调简约的几何体块，并采用体量分级和模块化设计，充分考虑将来使用的灵活性和项目在市场上的接受度。

The project is located on the storage district of Jinshan. The main function is to provide the living service, administrative service and production service for the factory workers and enterprises. The architecture design involves the geometry concept to achieve the flexibility of building.

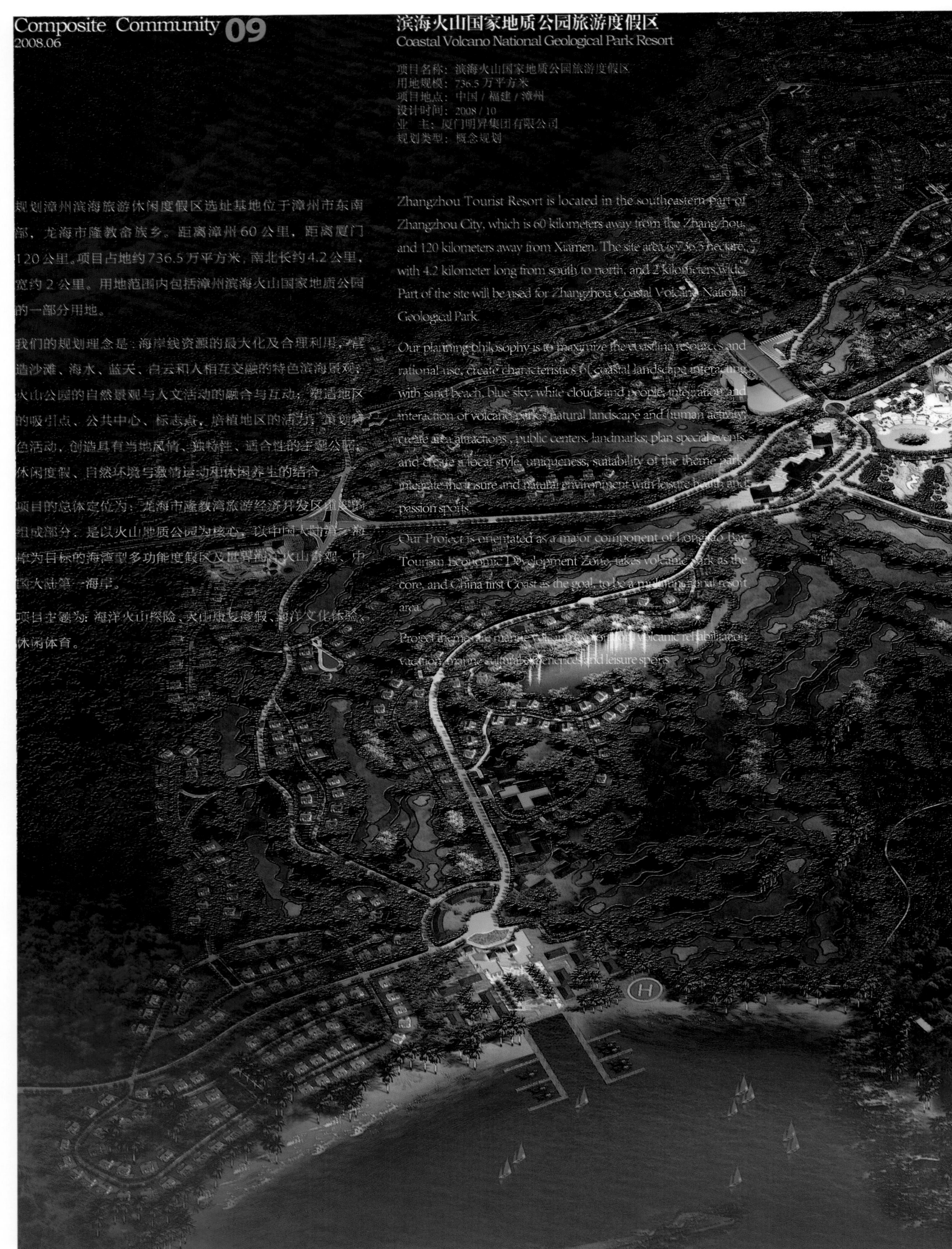

滨海火山国家地质公园旅游度假区
Coastal Volcano National Geological Park Resort

项目名称：滨海火山国家地质公园旅游度假区
用地规模：736.5万平方米
项目地点：中国 / 福建 / 漳州
设计时间：2008 / 10
业　主：厦门明昇集团有限公司
规划类型：概念规划

规划漳州滨海旅游休闲度假区选址基地位于漳州市东南部、龙海市隆教畲族乡。距离漳州60公里，距离厦门120公里。项目占地约736.5万平方米，南北长约4.2公里，宽约2公里。用地范围内包括漳州滨海火山国家地质公园的一部分用地。

我们的规划理念是：海岸线资源的最大化及合理利用，营造沙滩、海水、蓝天、白云和人相互交融的特色滨海景观；火山公园的自然景观与人文活动的融合与互动；塑造辖区的吸引点、公共中心、标志点，培植地区的活动，策划特色活动，创造具有当地风情、独特性、适合性的主题公园，休闲度假、自然环境与激情运动和休闲养生的结合。

项目的总体定位为：龙海市隆教湾旅游经济开发区组成部分，是以火山地质公园为核心，锁中国大陆第一海岸为目标的海湾型多功能度假区及世界级的火山奇观、中国大陆第一海岸。

项目主题为：海洋火山探险、火山康复度假、海洋文化体验、休闲体育。

Zhangzhou Tourist Resort is located in the southeastern part of Zhangzhou City, which is 60 kilometers away from the Zhangzhou, and 120 kilometers away from Xiamen. The site area is 736.3 hectare, with 4.2 kilometer long from south to north, and 2 kilometers wide. Part of the site will be used for Zhangzhou Coastal Volcano National Geological Park.

Our planning philosophy is to maximize the coastline resources and rational use, create characteristics of coastal landscape interacting with sand beach, blue sky, white clouds and people, integration and interaction of volcano parks natural landscape and human activity, create area attractions, public centers, landmarks; plan special events and create a local style, uniqueness, suitability of the theme park, integrate the leisure and natural environment with leisure health and passion sports.

Our Project is orientated as a major component of Longjiao Bay Tourism Economic Development Zone, takes volcanic park as the core, and China first Coast as the goal, to be a multifunctional resort area.

Project theme are marine volcano exploration, volcanic rehabilitation vacation, marine cultural experiences and leisure sports.

上海大宁 313 地块
Shanghai Daning 313 Plot

项目名称：上海大宁 313 地块
项目规模：3.88 万平方米
建筑面积：8.35 万平方米
容积率：2.15
项目地点：中国 / 上海
编制时间：2010 / 05

项目位于规划的上海多媒体谷园区内，一期多媒体大厦和清华科技园已经投入使用，且基地南侧两个地块也以商务办公为主，区域内商务办公环境相当成熟。

本项目定位为，为科技企业提供甲级办公楼及相应配套商业服务设施，以完善整个区域内商务办公的服务功能，提高区域内的商务价值。

The project is located in the Multi-Media Valley Park under planning. Currently, the multi-media building and Tsinghua Science Park built in Phase I have been put into use. Two plots in the south of the base are also dedicated to business office purposes, so that the business office environment in the region is becoming mature gradually.

According to the project positioning, grade A office will be the main feature. Supporting commercial service facilities will be built to improve the region's business office functions and improve the business value of the region.

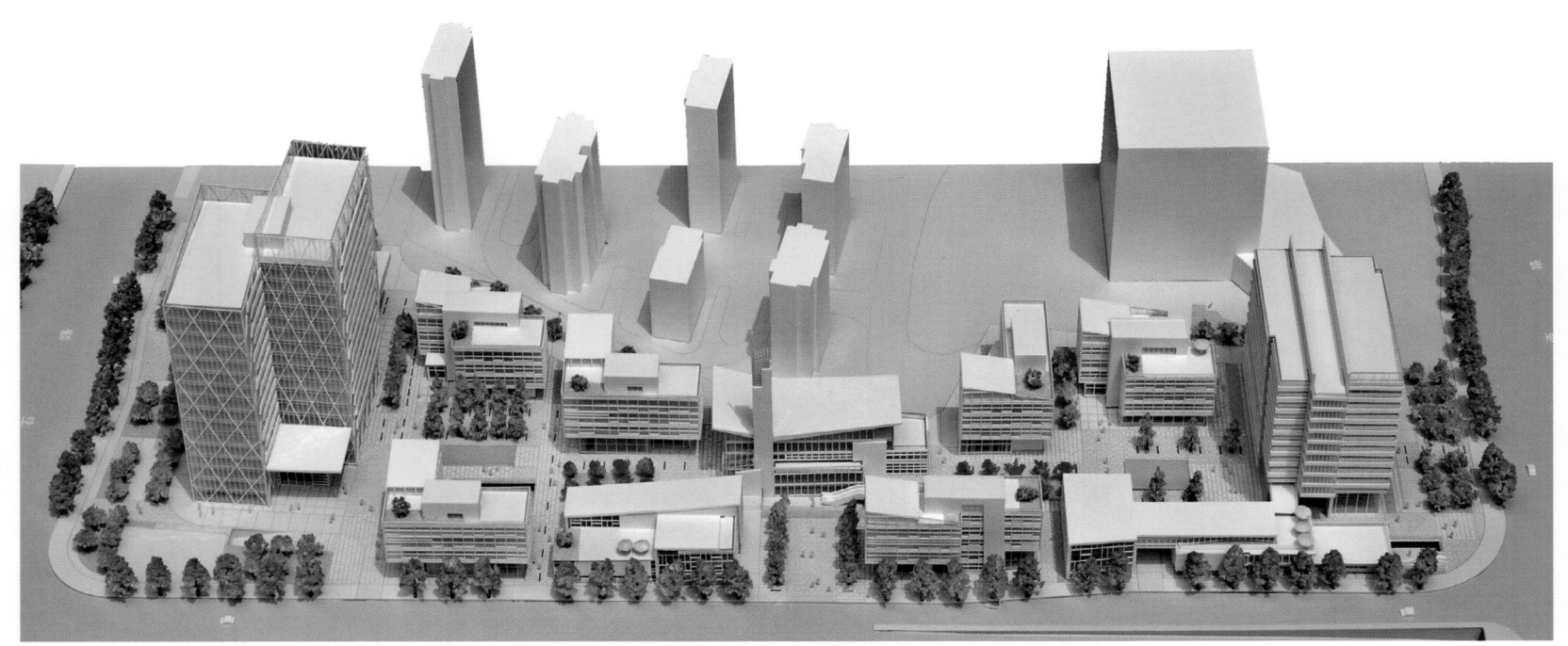

Composite Community
2007.08

上海江桥绿地创想城
Shanghai Jiangqiao Greenland Creative City

项目名称：上海江桥绿地创想城
建筑面积：5.03 万平方米
用地规模：16.78 万平方米
容积率：3.0
项目地点：中国／上海／江桥
编制时间：2007／08

项目对几大功能区进行功能整合，用商业外街将办公与商
业进行分隔和联系，创造新的活力空间，通过空间、色彩、
立面材质的组合，使巨大的建筑体量得以花瓣，成为人们
愿意亲近的场所。

The project integrates plenty of functions. The commercial
street divides the administration part and the commercial
part. The arrangement creates an active space. Great volume
is separated by combination of space, color and elevation
materials.

上海国际医学园
Shanghai International Medical Park

项目名称：上海国际医学园
用地规模：10.33 万平方米
建筑面积：8.86 万平方米
容积率：0.86
项目地点：中国 / 上海 / 南汇
编制时间：2006 / 06

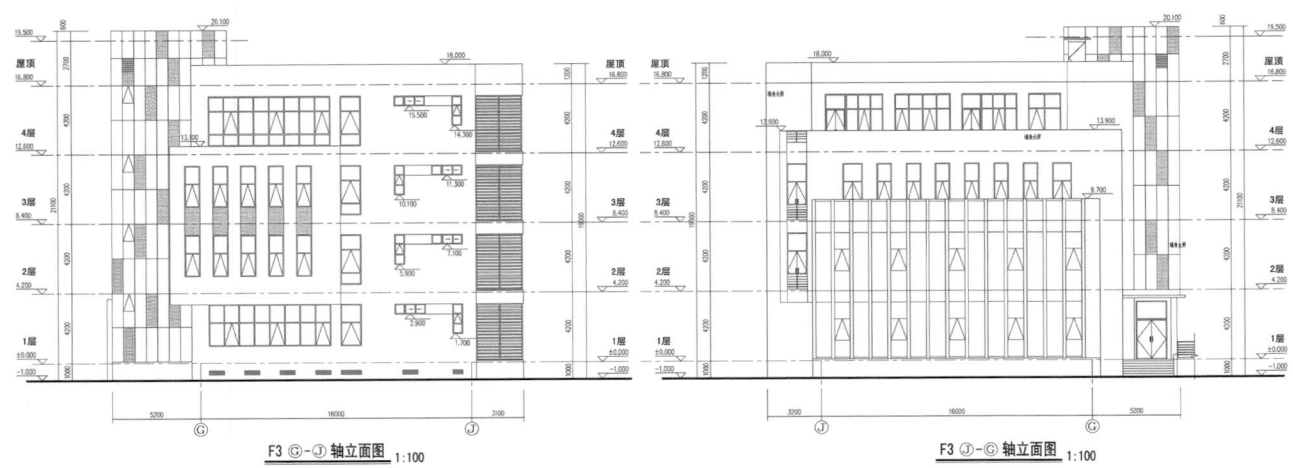

F3 ⒢-⒥ 轴立面图 1:100

F3 ⒥-⒢ 轴立面图 1:100

项目位于上海市规划的国际医学园区的核心区域，设计利用生命、健康、中西医学概念等丰富多元的文化形态，充分体现健康、舒适、生态、现代、快捷、高效、人性化的办公工作理念，将具有时代感的文化工作气氛作为整个园区的空间精神来加以强化。

The project is located at the core of Shanghai International Medical Park. The life, health, and science culture are the main subjects of the project.

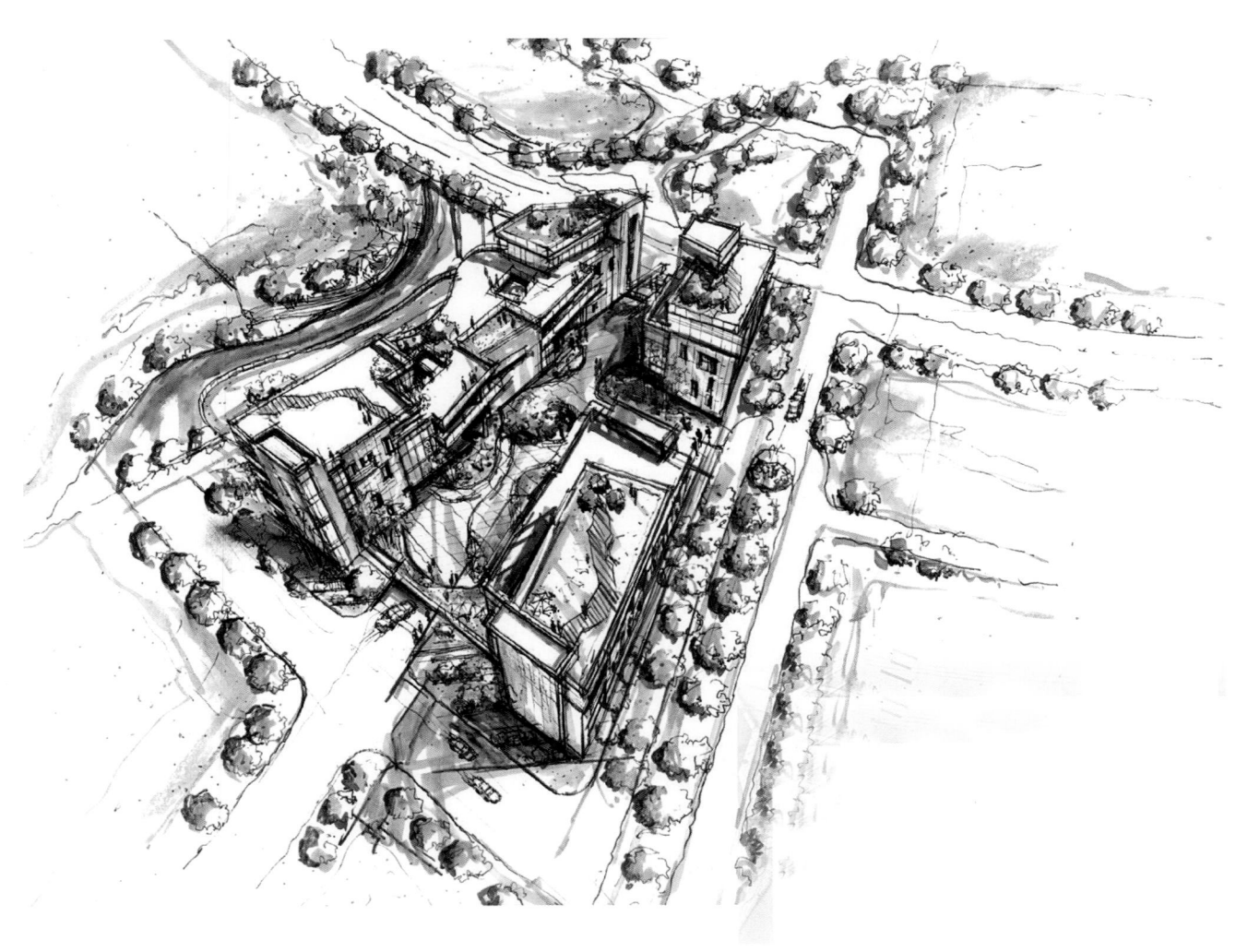

天津陈塘庄科技产业园改造
Tianjin Chentang Village Scientific Industry Park Reconstruction

项目名称：天津陈塘庄科技产业园改造
用地规模：10.33 万平方米
建筑面积：8.86 万平方米
容积率：0.86
项目地点：中国 / 天津
编制时间：2006 / 06

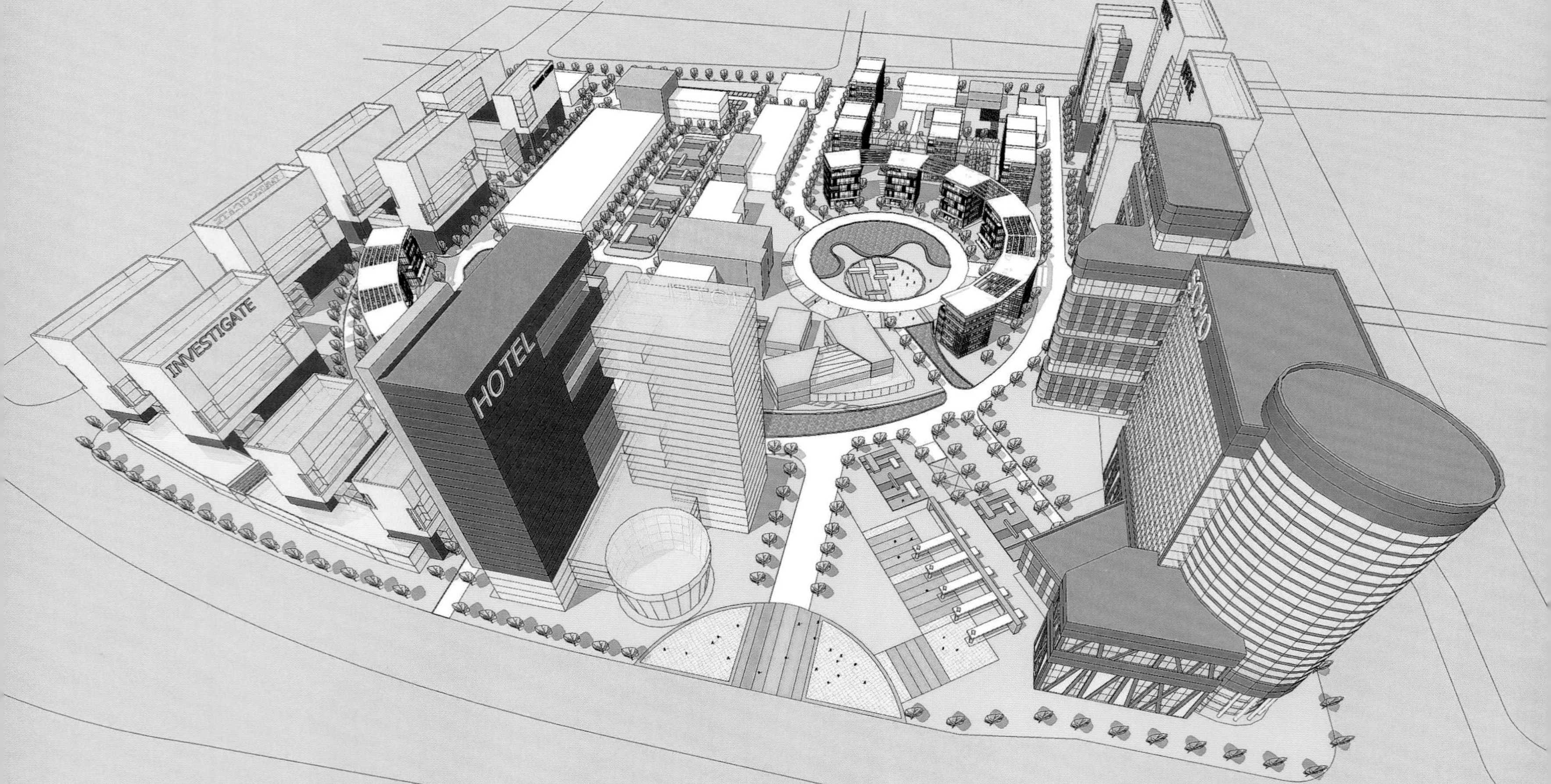

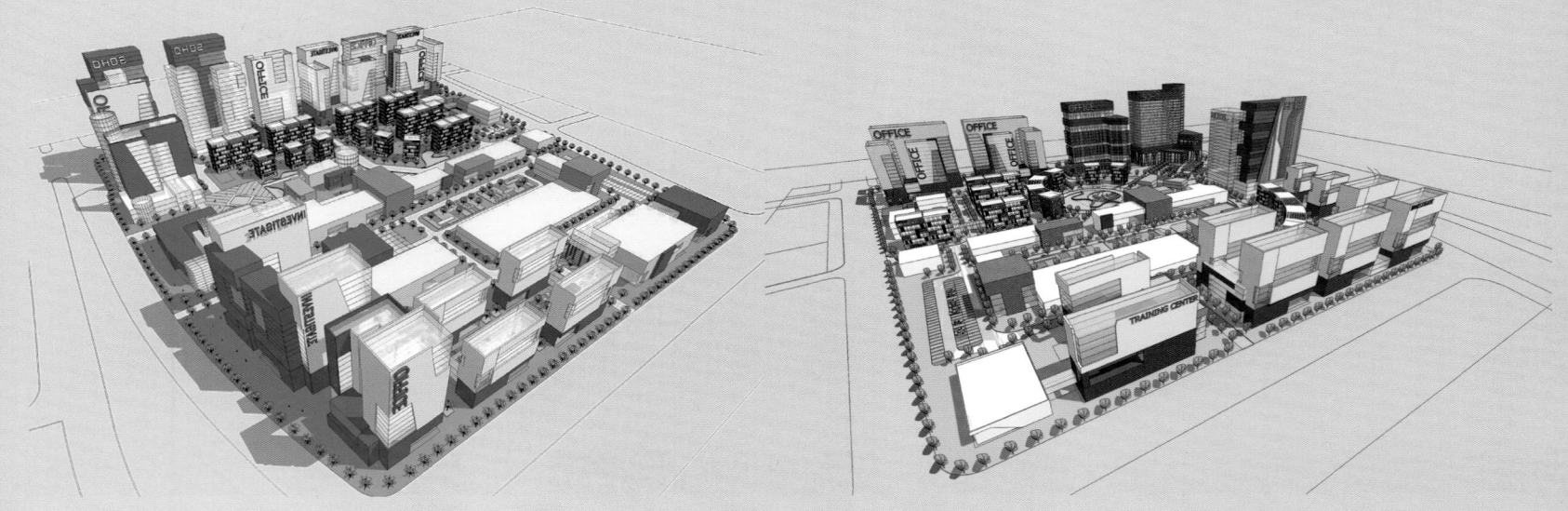

设计本着功能优先理念，创造出现代感极强的新社区，营造了良好的人文及生态环境，充分体现出人与建筑、科技、环境、文化的高度融合与互动。设计师引入先进的建筑理念和建筑形态，以后现代主义建筑为指导思想，充分展现了国际化高新科技园区的文化特点，给人以耳目一新的感觉和人文关怀。

"Functionality" is the principal concept. The innovated community provides good civilization atmosphere and nature environment which also emerges the integration and reaction of human, architecture, science and technology, environment and culture. The post-modernism is the guideline for this refreshing international high-tech park.

天津海泰产学研基地
Tianjin Haitai Research Base

项目名称：天津海泰产学研基地
用地规模：11.87 万平方米
建筑面积：17.8 万平方米
容积率：1.5
项目地点：中国 / 天津
编制时间：2007 / 06

建筑设计的风格必须与园区规划中所引导的形态一致，同时强调简洁、明快、极强的时代气息，建筑设计力求自身的完整和群落之间的整体性，建议以简洁的建筑形体张弛关系创造丰富流畅的外部空间，强调建筑节奏。组合内部空间巧妙多变。沿街建筑要体现街道界面的连续性，在高度、色彩、材质标识等方面取得统一协调，街面层次感强、富有变化。

建筑材料在统一的前提下多样化，立面设计和建筑色调以现代简洁为主，材质考虑玻璃幕墙、金属、石材等，并注意材料本身的软硬对比、虚实关系对比。

为避免本项目在开发建设中建筑布局无序、天际线单调的情况，在规划过程中建议采用架空，退台，高、中、低建筑形态结合等多种手法构筑高低层次错落、疏密有致的总部花园天际线，以形成移步换景的活动路线。

The building style must be consistent with the base planning form and at the same time, emphasize simple and straightforward feeling as well as the modern flavor and strive to ensure the own integrity and the coordination with the building group. The concise building shape is recommended to create a rich and smooth external space and to emphasize the building architectural rhythm. The internal space is variable; building along the street should reflect the continuity of the street interface, achieve coordinated effect of height, color, management and identification and create strong hierarchy and rich variation.

Under the premise of unity, diversified building materials are used. Simple elevation design and the building tone are adopted. Glass curtain wall, metal, stone and others are considered to be used, paying attention to the contrast of soft and hard material and the comparison between the reality and unreality.

In order to avoid the disorder of building layout and dull skyline of the project, it is recommended to use a variety of techniques to create the headquarter garden skyline with good hierarchy and well-conceived composition, such as elevating, set-back model, high, medium and low buildings and so on. Changeable scenery along with moving steps can be created hereby.

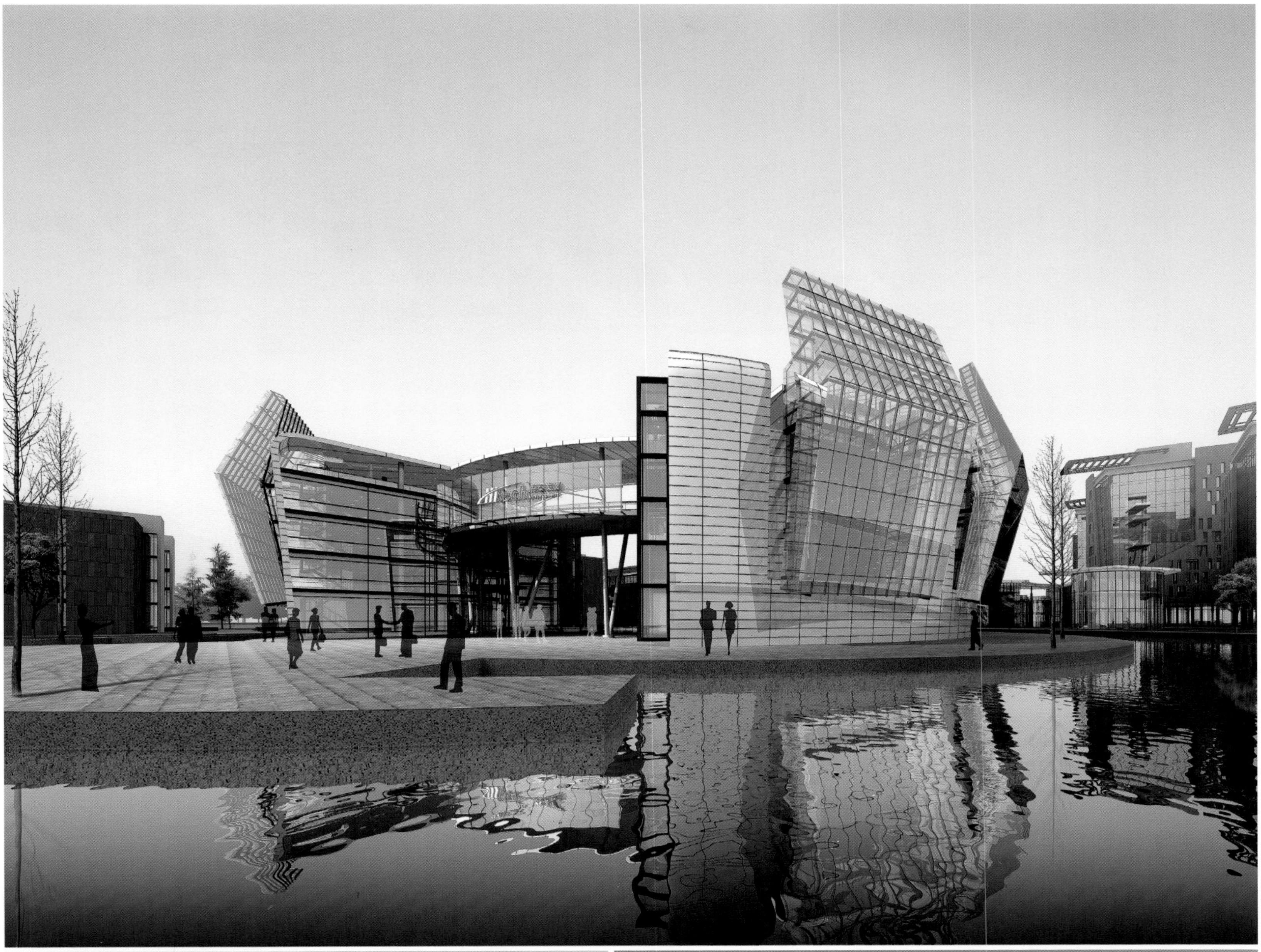

H AI TAI INDUSTRY INSTITUTE & RESEARCH CENTER
Flat perspective

H AI TAI INDUSTRY INSTITUTE & RESEARCH CENTER
Villa perspective

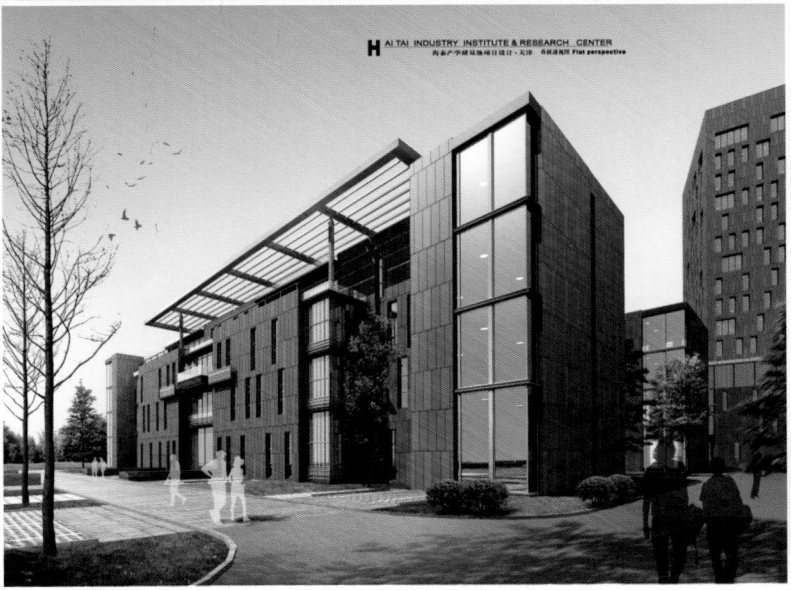

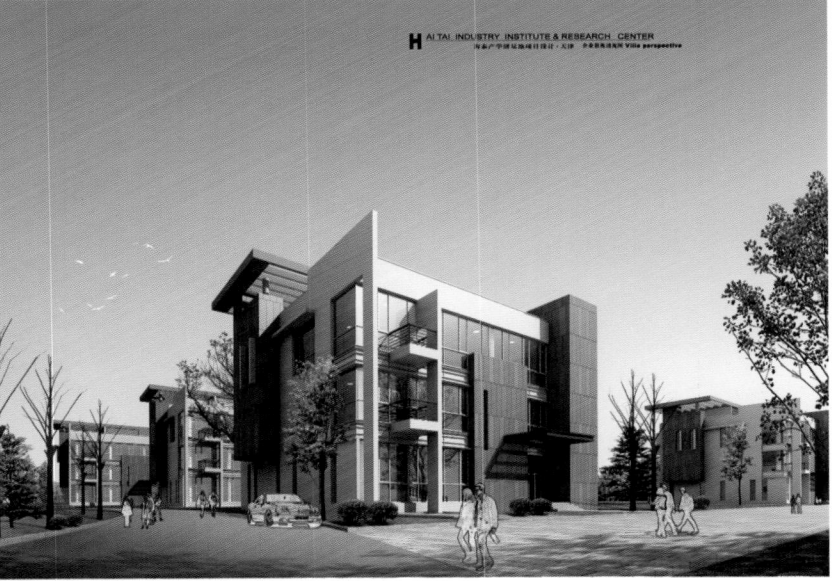

天津洪泽制造
Tianjin Hongze Manufacturing Project

项目名称：天津洪泽制造
用地规模：27.10 万平方米
建筑面积：69.12 万平方米
容积率：2.55
项目地点：中国 / 天津
编制时间：2007 / 03

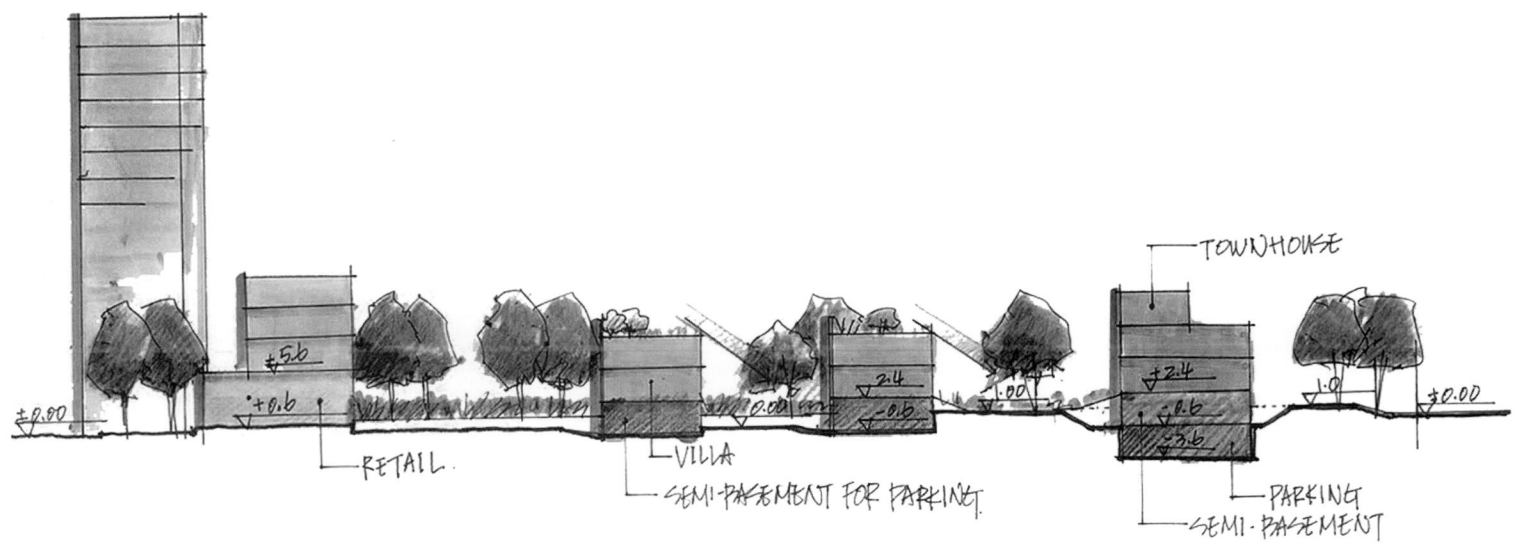

项目是位于天津市经济技术开发区内的高级住宅综合项目，是一个包含有高层、多层、联排、别墅等各种形式的居住、展示和演绎等公共建筑的组合式大型社区。设计师希望建立一个充满生机并且可以交流的场所，为社区的全体居住者提供一个全新的生活和社交的平台，以满足高层消费群体的各种需求。设计师通过运用对地形的处理和简洁的建筑形态与精致的建筑细部来体现该建筑的价值感和应对城市核心土地资源相对紧张所造成的空间压抑感。

The project is a senior residential complex project located in Tianjin Economic and Technological Development Area, which is a combined type of large-scale community including various types of high-rise, multilayer, terrace constructions and villas and other public buildings used for living and exhibitions. Designers expect to establish a place which is full of vitality and can be used as a site for public communication, and create a new life and social platform for all community residents to meet the diverse needs of high-level consumer groups. Designers apply the ingenious terrain, concise architectural form and exquisite architectural details to embody the value of the building and reduce the sense of space oppression caused by relatively insufficient urban central land resources.

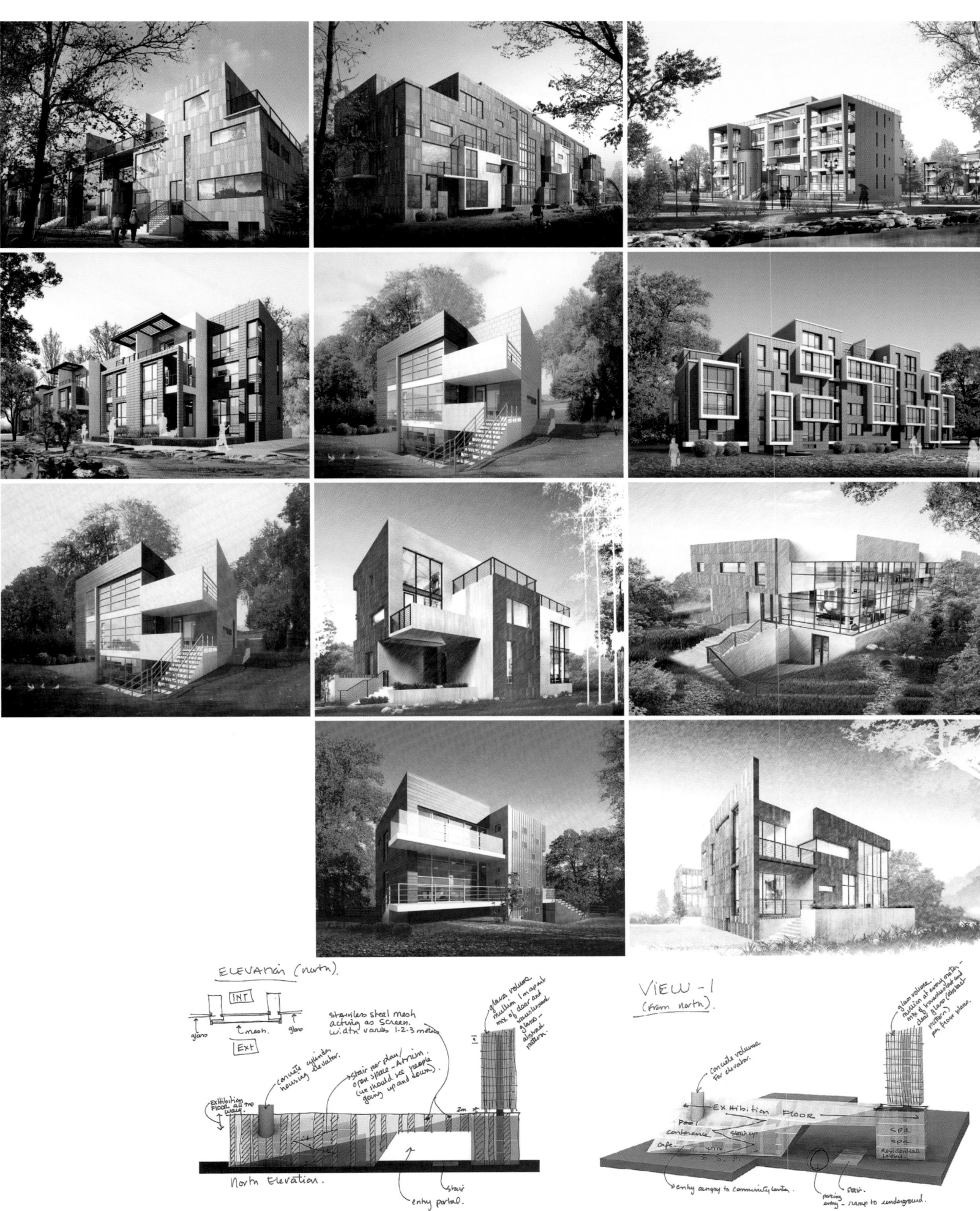

亚泰樱花苑
Yatai Sakura Garden

项目名称：亚泰樱花苑
用地规模：15.78 万平方米
建筑面积：28.41 万平方米
容积率：1.8
项目地点：中国 / 吉林 / 长春
编制时间：2005 / 07

整体建筑风格设计以体现简约德式建筑风情为主题，环境配置以"步移景异"的规划理念创造居住的诗意格局。运用现代化的材质和先进设施，传达传统的细节和环境要素，使人们感受家园的温馨，从而减轻城市建筑的陌生感。建筑立面造型通过建筑体块的穿插和色彩的大胆运用，体现其独特的区域风格特征和特色。

小区沿吉林大路的沿街商业区，结合小区人行主入口，形成商业内街的组织形式，汇集各具特色的代表性建筑，营造极富人情味的商业、休闲、社交、服务的公共空间。每栋房子都有自己的独特的"脸孔"，在时间和历史的长河中磨砺自己的性情，书写自己的故事，传递、融合甚至创造着人文的信息，不再是单调统一的仅供居住的"盒子"或者"机器"。传统的回归、文化的特征，是过度工业化之后人们必然的心理需求。

Simple German architecture style is adopted as the overall theme. Poetic pattern of the residence is created according to the planning idea of "scenery changing along with moving steps". Modern materials and advanced facilities are applied to convey the traditional details and environment requirements. It is expected that people living here can find the warmth of home and reduce strangeness to urban buildings. Block elements and bold colors are applied for facade shaping so as to display its unique style and characteristics.

Business area of the community along Jilin Road stretches to the main entrance to form a business street within the community. Typical buildings with different features here make the public space more humane. Business, leisure, social exchange and services functions are provided. Every building has its own "face" and builds its temper and composes its story along time. Buildings here are not just the "case" or "machine" for residence only; however, they shoulder the mission to deliver, integrate and even create humanistic message. Returning to tradition and cultural characteristics are the inevitable psychological needs of people after over-industrialization.

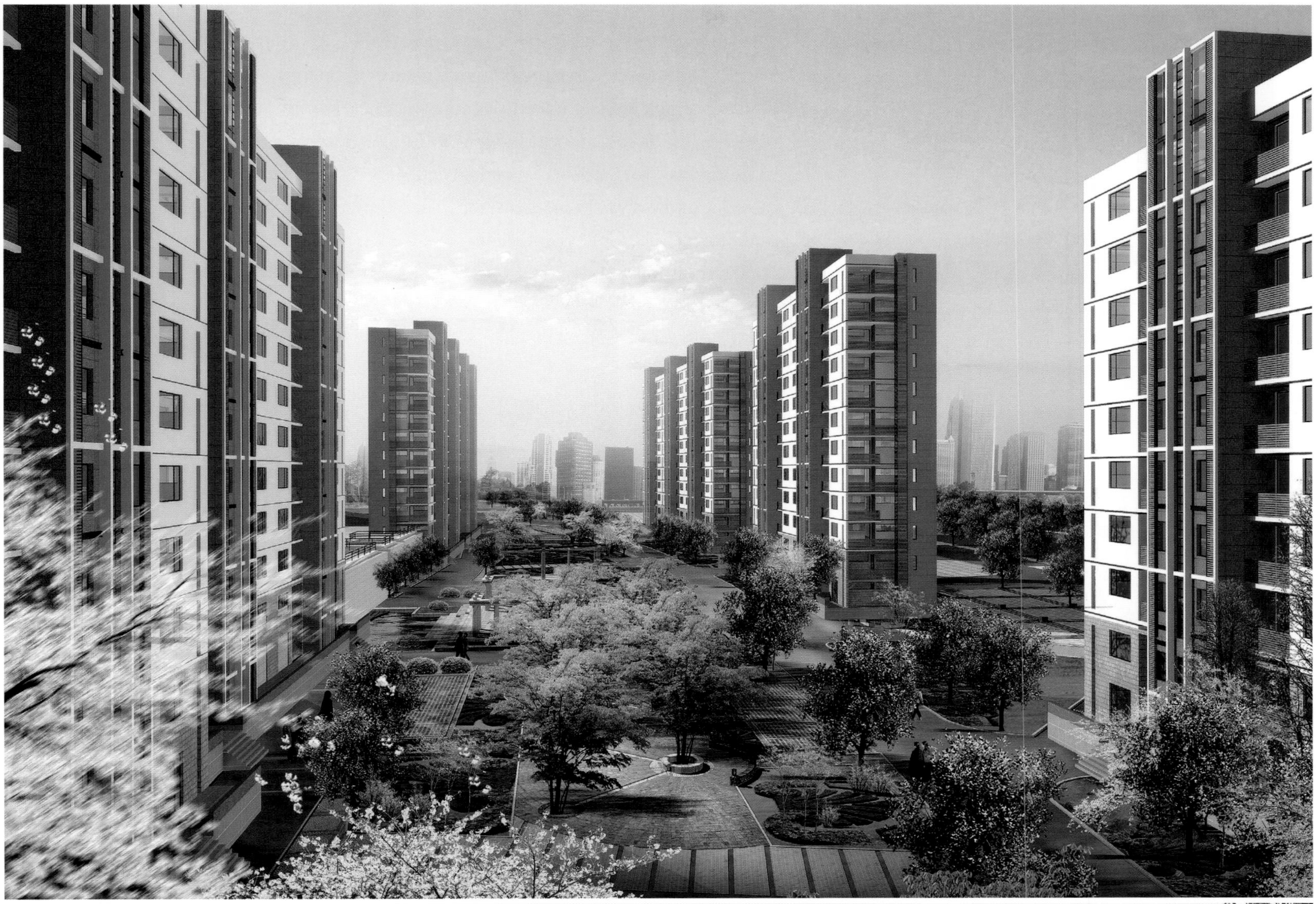

城市理想
Urban Ideals

项目名称：城市理想
用地规模：0.63 万平方米
建筑面积：8 万平方米
容积率：12
项目地点：中国 / 四川 / 成都
编制时间：2006 / 11

对于城市中心 CBD 区域的大型复合项目，设计师给予业主的是一个具有高度市场"弹性"的选择答案。而建筑造型与产品经济价值的关联性是本案设计中特别值得关注的"焦点"。作为商业和建筑塔楼的分隔和过渡，"空中会所"的概念感性地得到艺术化的放大。

Regarding the large-scale composite project in the city center CBD, the designer provides owners with an optional answer of high market "elasticity". Moreover, the relativity between the building model and the product's economic value is the "focus" especially worthy of attention in this design. As the division and transition of the commercial and building towers, the concept of the "aerial club" is artistically and sensibly magnified .

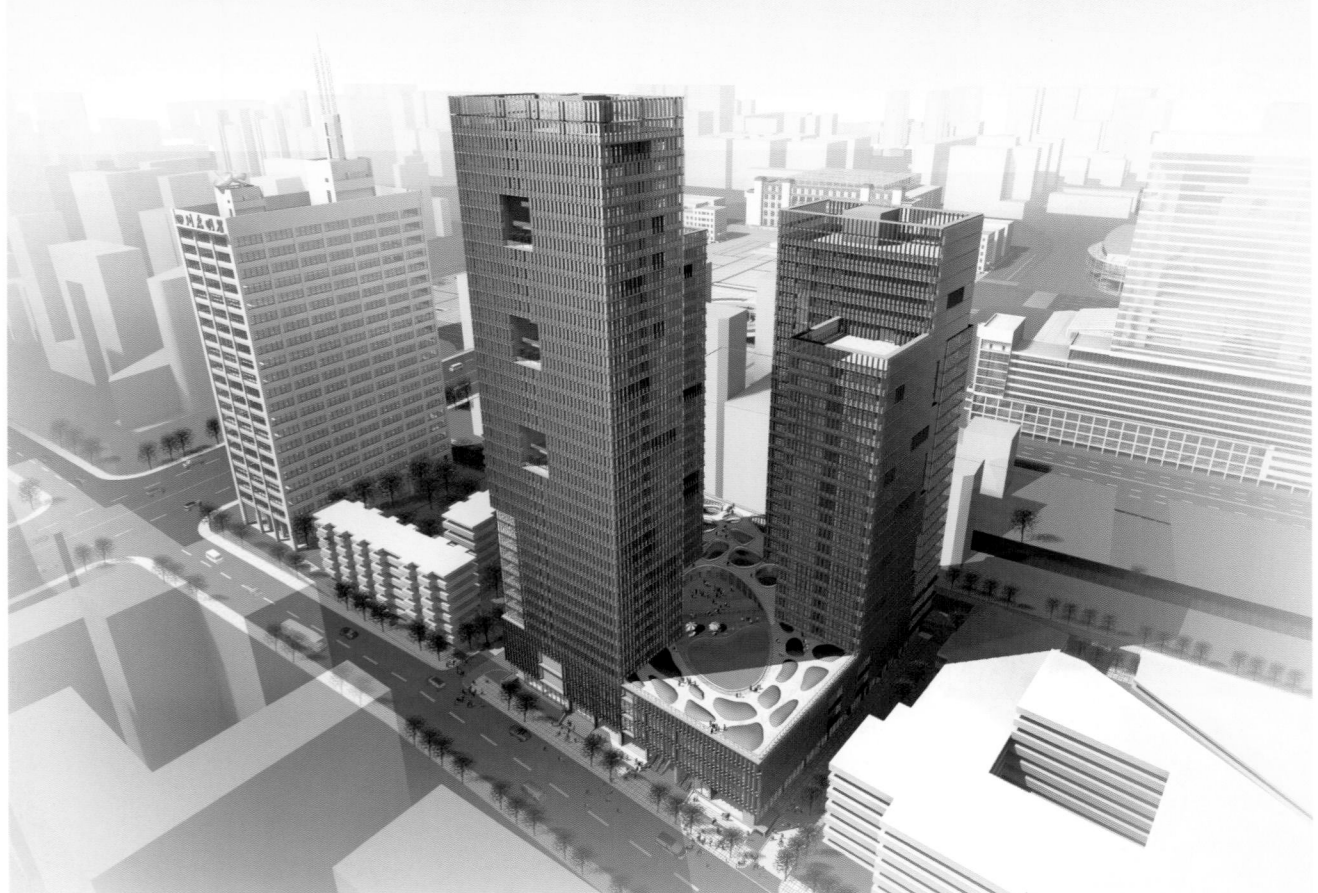

松江科技绿洲综合体项目
Shanghai Songjiang Technology Oasis Urban Complex Project

项目名称：上海松江科技绿洲城市综合体项目
用地规模：2.72 万平方米
建筑面积：8.17 万平方米
容积率：3.0
项目地点：中国／上海
编制时间：2010／07

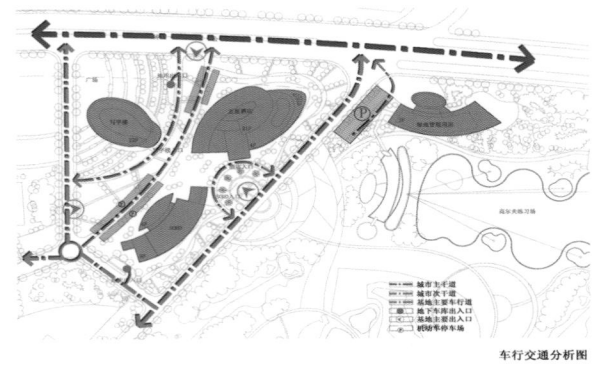

车行交通分析图

人行交通分析图

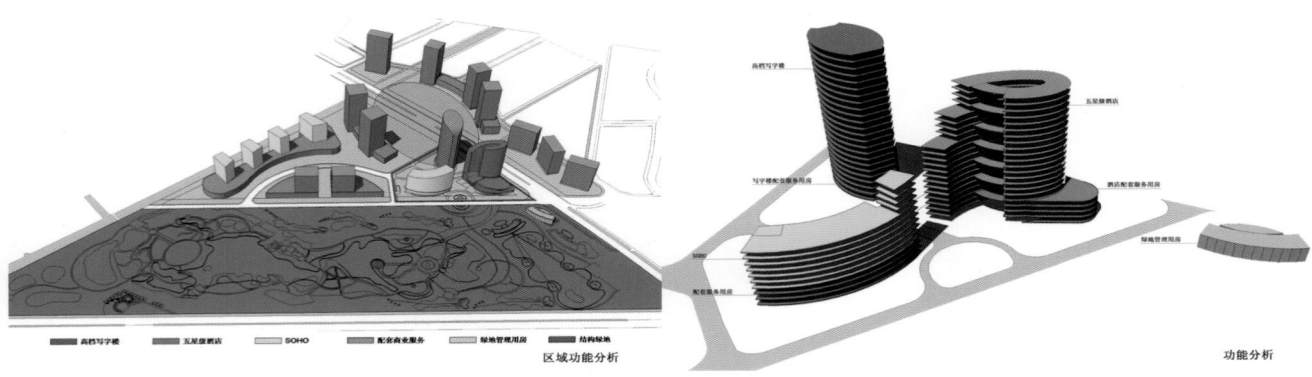

区域功能分析   功能分析

项目位于漕河泾开发区松江园区内，占地面积约 40 亩，综合容积率 3.0，建筑面积 8.3 万平方米，其中包含一座五星级酒店，两栋塔式办公楼和附属商业，另外还包括一个高尔夫球练习场和球会会所。

本项目的位置是由市区进入漕河泾松江园区的桥头堡，所以设计将如何营造区域良好的入口形象作为重点，注意保持建筑之间的统一和协调关系，同时尝试用两种不同形体与材料构成的组合，形成两个完全不同的流动的建筑形象，建筑强调了科技时代的流动、引入、发展的个性，同时更加突出门户建筑的个性。

The project is located in Songjiang Park of Caohejiang Developing Area, covering an area of 40 acres with the FAR of 3.0 and the building area of 83,000 square meters. The project includes a five-star hotel, two tower office buildings and their auxiliary commercial buildings as well as a golf club.

As it is the bridgehead of the way from the urban area to Caohejing Songjiang Park, the design focuses on creating the good regional reputation, maintaining the unity and coordination among buildings, And at the same time, attempts to create two completely different building images by two completely different shapes and different materials. It emphasizs the science and technology and the simple and stable personality.

天津滨海高新区科研孵化中心和综合服务中心
Tianjin Binhai High-Tech District Research Incubation Center and
Integrated Service Center

项目名称：天津滨海高新区科研孵化中心和综合服务中心
用地规模：3.67 万平方米
建筑面积：4.90 万平方米
容积率：1.33
项目地点：中国 / 天津
编制时间：2007 / 05

天津滨海高新区科研孵化中心和综合服务中心项目是天津
滨海高新区的又一高起点、高起步的重要项目，设计规划
全面地体现滨海新区科技创新领航区和高新技术研发转化
中心的特点，将建成全国一流的集科技研发和成果转化于
一体的应用基地，同时本项目也将建成标志性的绿色生态
建筑，多种绿色生态建筑理念在设计中得到运用，其中包
括能源的节约利用、生物技术、风能和太阳能的利用、零
碳排放等。

The project is the design for Tianjin Binhai High-Tech District
Research Incubation Center and Integrated Service Center in
another important high starting point project. High starting point
is required. The design should also reflect the feature of Binhai
High-Tech District as a technological innovation leading area and
a high-tech R&D and conversion center. According to the plan, a
national-class base for technology R&D and results conversion and
application will be built. At the same time, landmark green buildings
will also be built in the project.

项目名称：天津滨海高新区科研孵化中心和综合服务中心

SMITHGROUP cna

Architecture · Urban Planning · Landscape Design
Interiors Design · Art Consultant · Lighting Design
Since 1853
www.cna-group.com

Design Determines ...

设计决定……

04 商业娱乐
Commercial Entertainment

上海民生路码头改造项目
Shanghai Minsheng Dock Reconstruction

Architecture Design | 2010·05

上海民生路码头改造项目
Shanghai Mingshen Dock Reconstruction Plan

项目名称：上海民生路码头改造项目
用地规模：9.41 万平方米
建筑面积：13.98 万平方米
容积率：1.48
项目地点：中国 / 上海
编制时间：2010 / 05

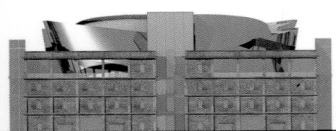   

民生码头，诞生于百年前，名曰"蓝烟囱码头"。蓝烟囱，源于十九世纪欧洲至中国第一批远洋船运航线之名。民生路码头，用地性质为商业办公娱乐综合体，规划要求为浦东及杨浦大桥地区有文化特色的滨江休闲目的地。本案作为一个承担起城市岸线坐标、后世博城市专场和城市商务平台三大功能的高起点项目，在其城市建筑规划的布局上，定义为"三区一带"的空间系列组织形态。

对比连接着过去和未来，创造出一个动态和谐奇妙的空间。我们保留老建筑原始的符号，如保留筒仓的符号使其内部作为酒店。我们找出所有有趣的细节，打造一个简洁高端的亲水码头，作为整个规划的宝石。旧的结构在城市广场得到重新利用。粮仓的桶是最重要的细节，把桶作为一种符号。老的临江仓库布局为工业尺度，不适合人的活动尺度，人会感到不舒服。我们的设计改变了这一切，创造出了亲人尺度的商业绿色广场空间。整个项目还有几个绿色屋顶，在屋顶上有咖啡馆和餐馆，绿色屋顶增强了整个项目的可持续性和吸引力，带给人们更加舒适高端的感受。

Minsheng Dock was built a hundred years ago and was known as "Blue Chimney Dock ". Blue Chimney is the name of the first ocean shipping lane from Europe to China in the 19th century. The project is a complex including commerce, office, entertainment and complex. It is required to plan this region to be a riverside leisure destination with cultural feature in Pudong and Yangpu Bridge area. This is a very important project regarded as the urban landmark, post-EXPO urban space and urban commercial platform. Its organization shape is determined in the urban construction planning as "three-district and one belt ".

We intend to connect the past and the future by comparison and to create a dynamic, harmonious and wonderful space. We reserve original symbols of the old building and reserve the silo and transform it to be a hotel. We find out all interesting details to create a simple and high-end dock. The old structure is reused in the urban square. Barrel of the granary is the most important detail. Trashes are removed and the barrel is remained as an important symbol. Warehouse near the river was originally an industrial place unsuitable for people's activity. People going here would feel uncomfortable. Our design changes this and creates an amiable green commercial square. Several green roofs are designed on which coffee shop and restaurants are arranged. The green roof enables the entire project to have the long-lasting attract and presents its more comfortable and advanced feelings.

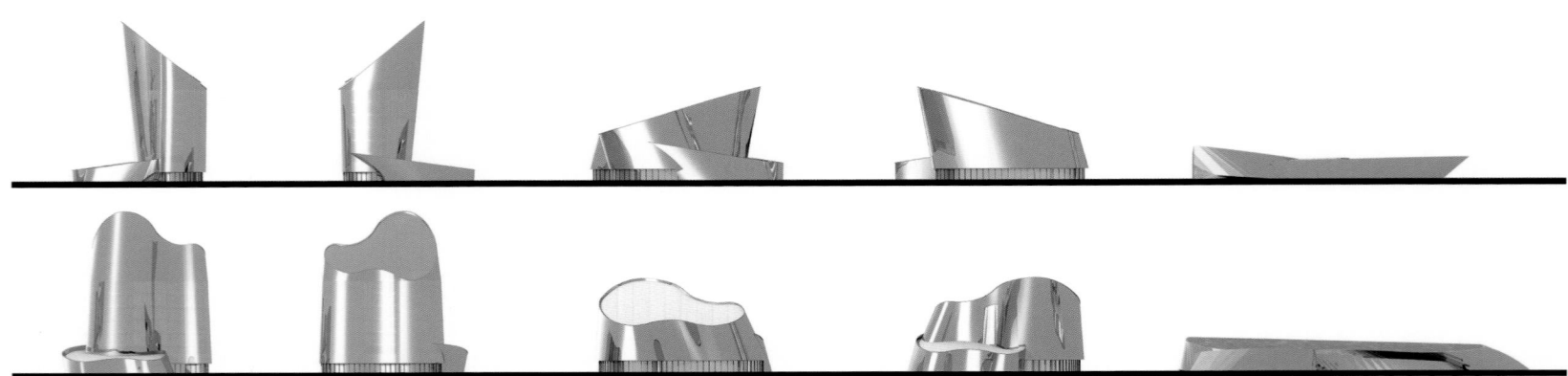

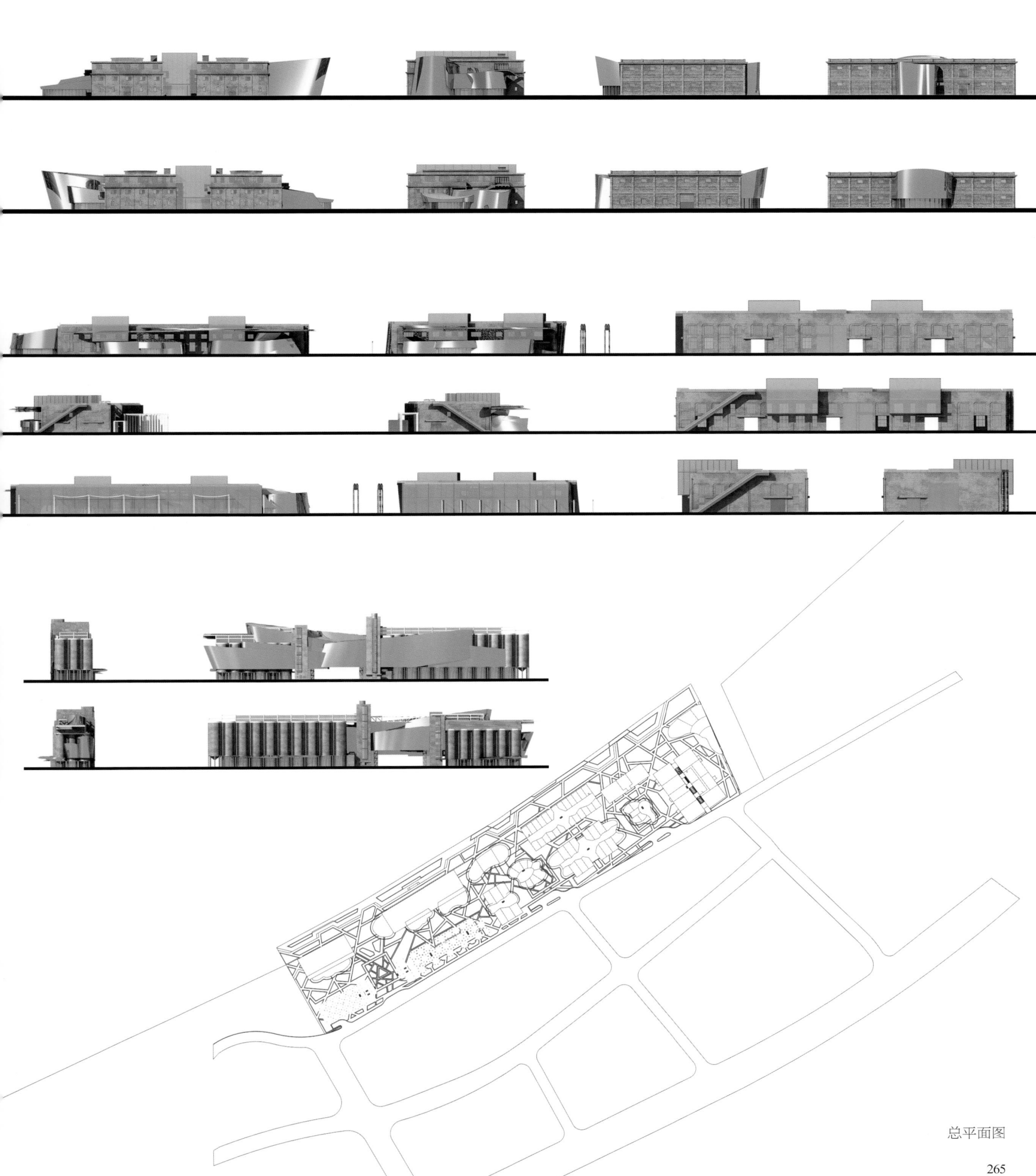

总平面图

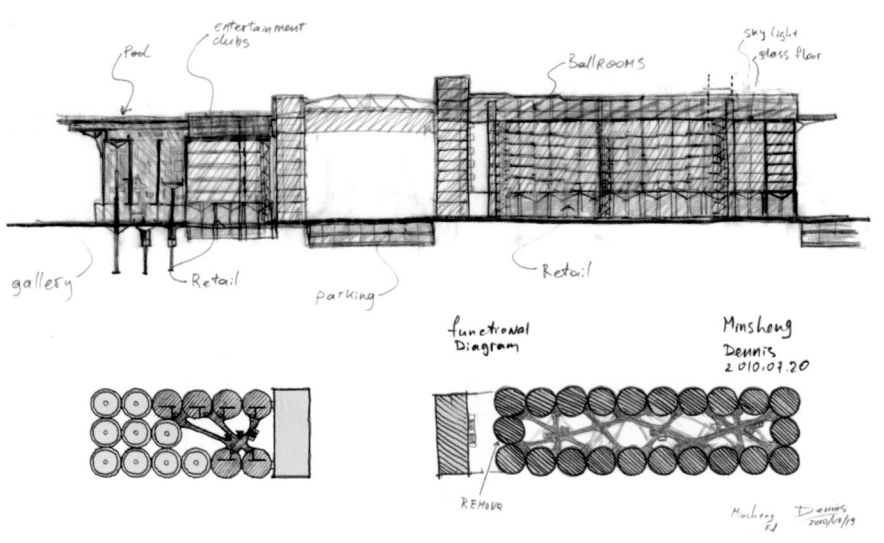

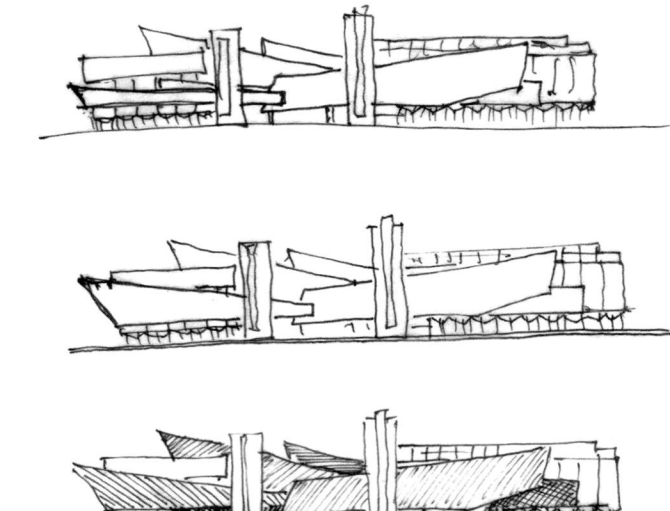

gallery   Retail   parking   Retail   pool   entertainment clubs   BALLROOMS   sky light glass floor

functional Diagram

Minsheng
Dennis
2010.07.20

REMOVE

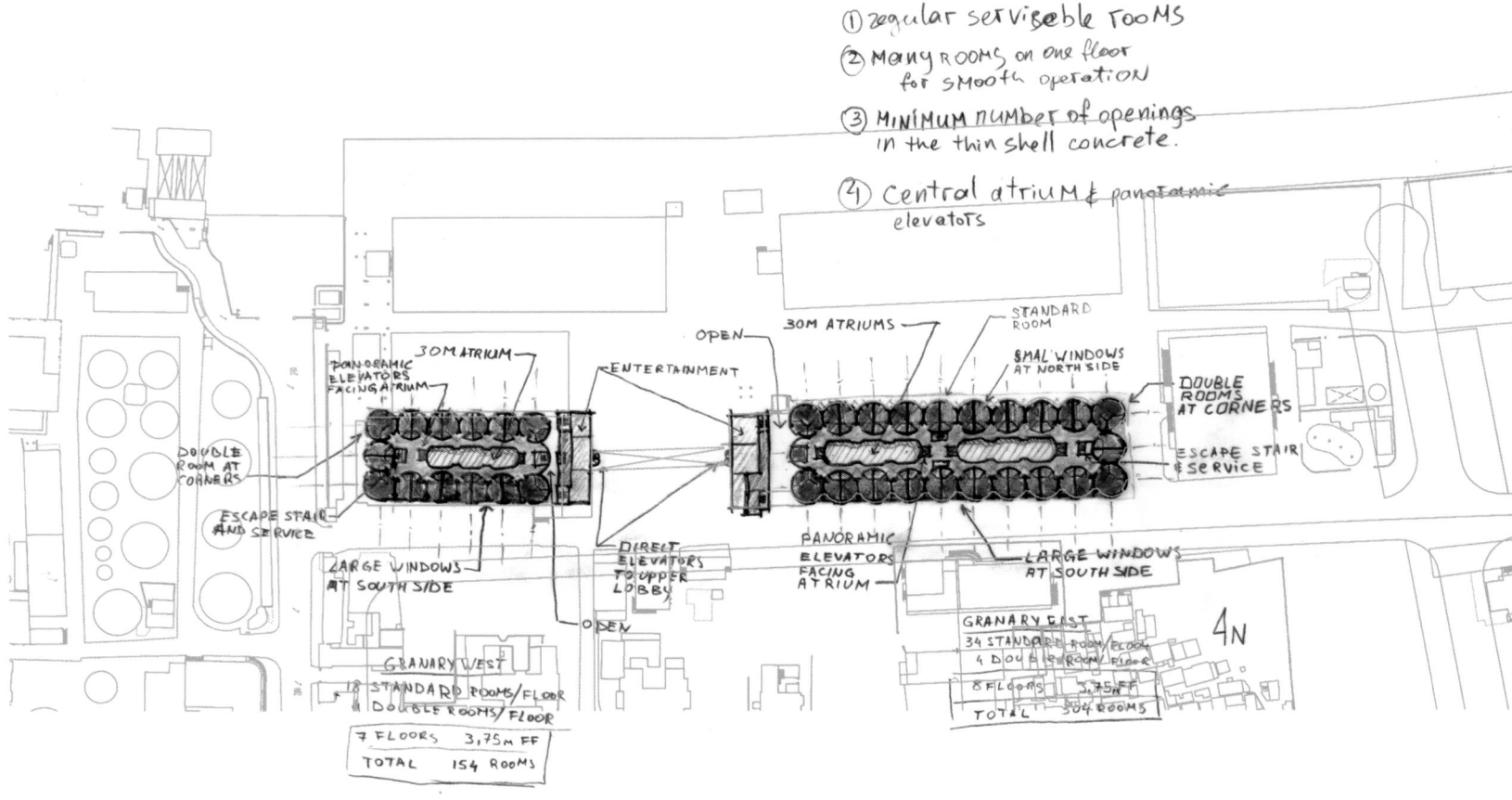

① regular serviceable rooms

② many rooms on one floor for smooth operation

③ minimum number of openings in the thin shell concrete.

④ central atrium & panoramic elevators

OPEN   30M ATRIUMS   STANDARD ROOM
PANORAMIC ELEVATORS FACING ATRIUM   30M ATRIUM   ENTERTAINMENT   SMALL WINDOWS AT NORTH SIDE   DOUBLE ROOMS AT CORNERS
DOUBLE ROOM AT CORNERS   ESCAPE STAIR & SERVICE
ESCAPE STAIR AND SERVICE   DIRECT ELEVATORS TO UPPER LOBBY   PANORAMIC ELEVATORS FACING ATRIUM   LARGE WINDOWS AT SOUTH SIDE
LARGE WINDOWS AT SOUTH SIDE   OPEN   GRANARY EAST
GRANARY WEST   34 STANDARD ROOM/FLOOR
8 STANDARD ROOMS/FLOOR   4 DOUBLE ROOM/FLOOR
2 DOUBLE ROOMS/FLOOR   8 FLOORS   3.75M FF
7 FLOORS 3.75M FF   TOTAL 304 ROOMS
TOTAL   154 ROOMS   4N

FF 3.75M   TOTAL 458 ROOMS
FF 5.0M   TOTAL 338 ROOMS

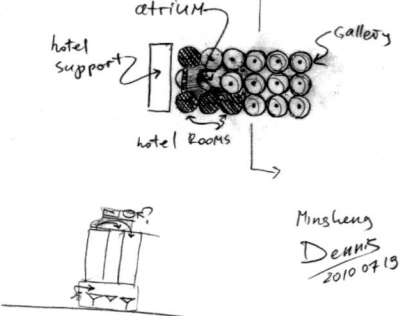

hotel atrium
hotel support   Gallery
hotel Rooms

Minsheng
Dennis
2010 07 19

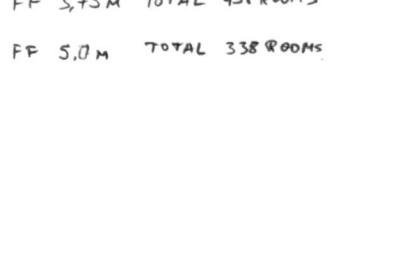

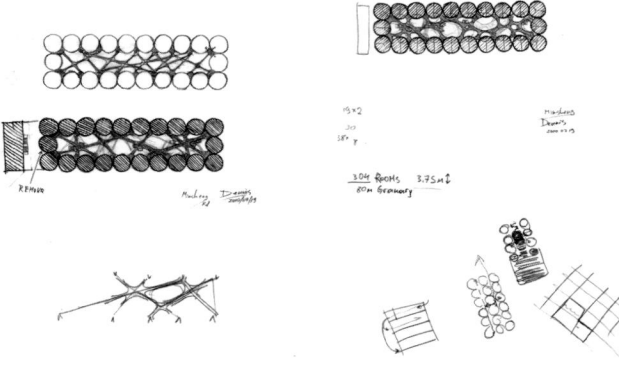

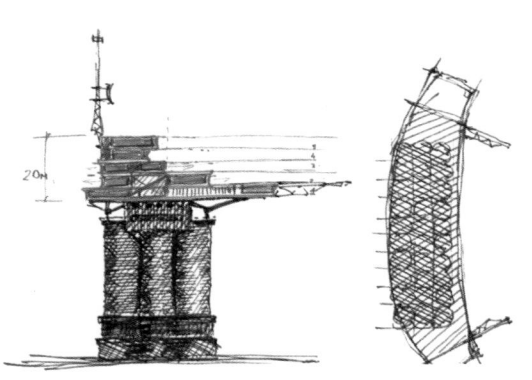

20m

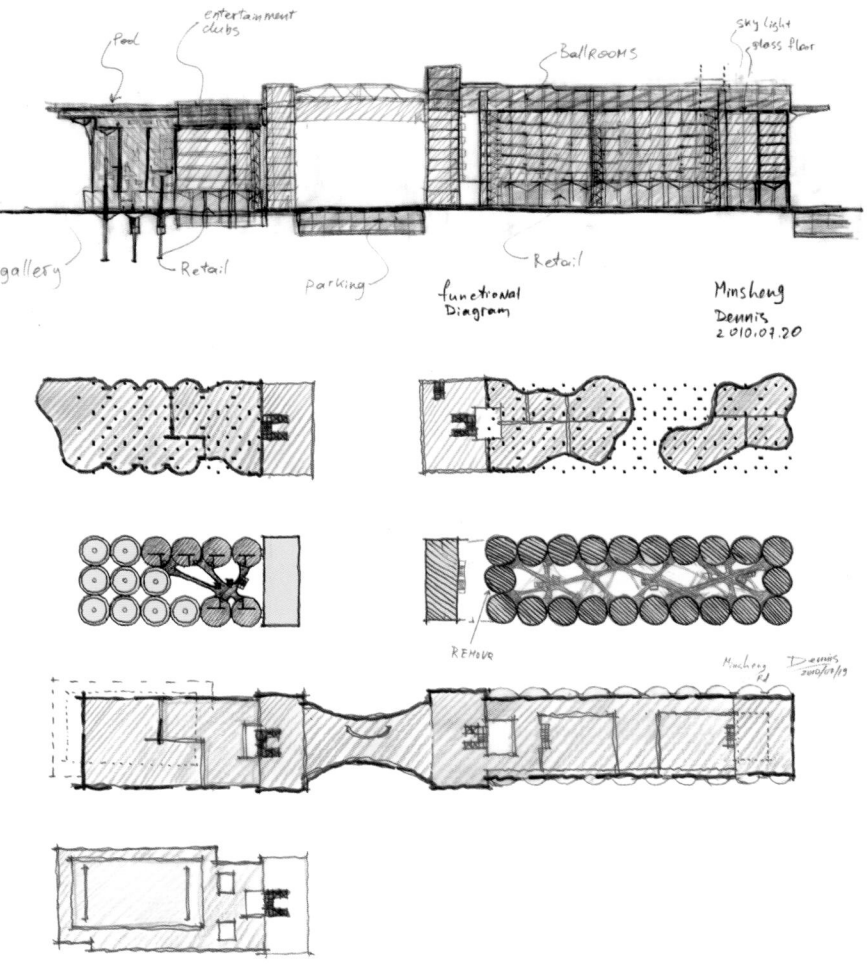

Pool
entertainment clubs
BALLROOMS
Sky light glass floor

gallery
Retail
parking
Retail

functional Diagram

Mmsheng
Dennis
2010.07.20

REMOVE

Mmsheng Dennis
2010/07/19

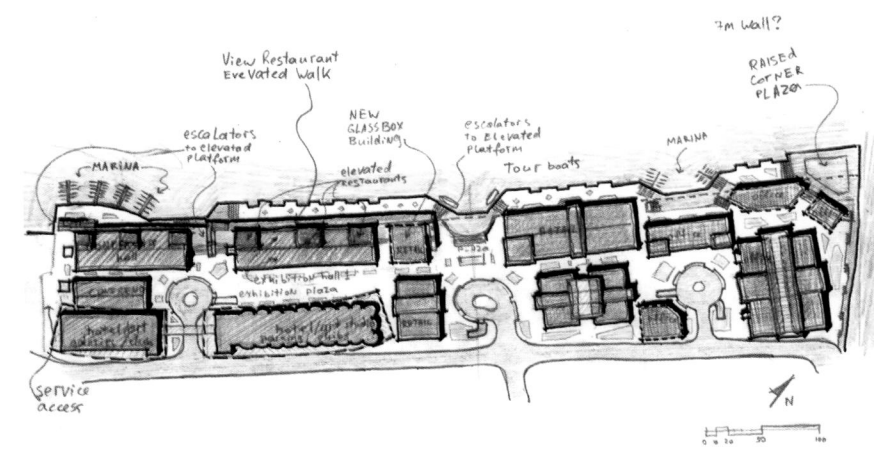

7m wall?

View Restaurant
Elevated Walk

escalators to elevated platform
NEW GLASS BOX Building
elevated restaurants
escalators to Elevated Platform
Tour boats
MARINA
RAISED CORNER PLAZA

MARINA

exhibition plaza

service access

N
0 10 30 50 100

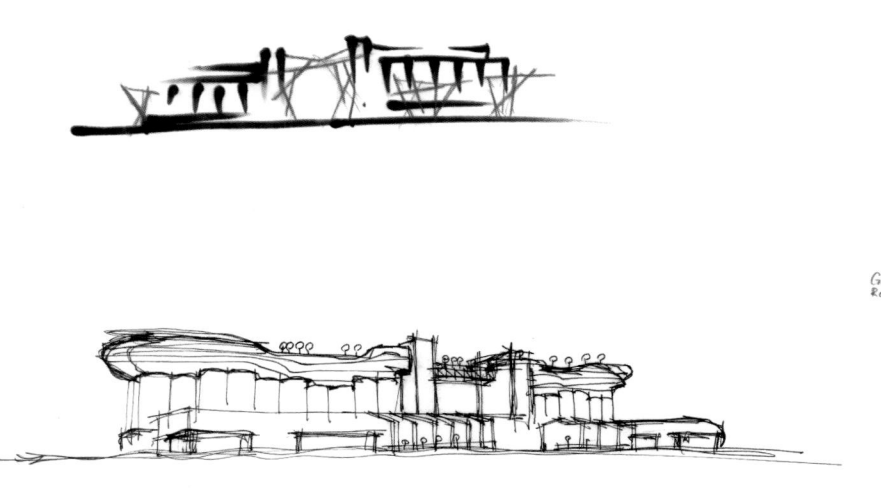

Landscape Diagram

Floating Pier

<-> Sky walk
elevated platform over the flood wall

Green Roof

waterfront walk (wood imitation)
Green Roof

Green Roof

Green Roof

large number of trees providing park atmosphere

welcome plaza

EAST Exhibition Plaza

variety of stone walkways & tree planters

(NOTE: Site → 3D rendering,
Buildings → white)

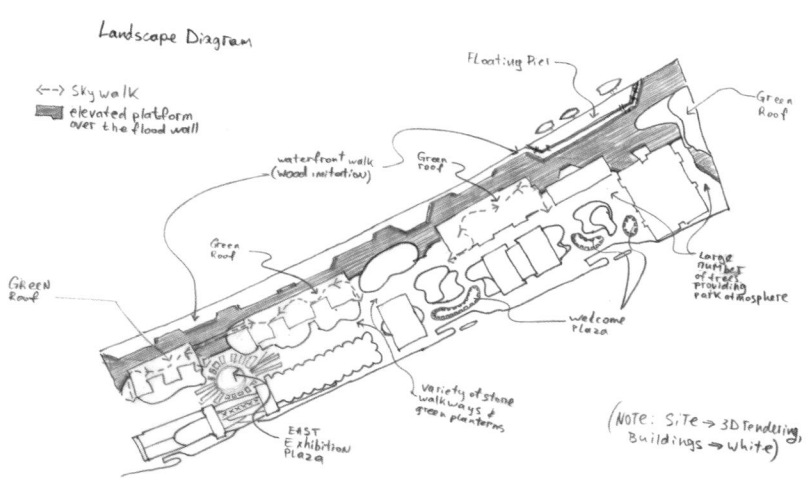

Architecture · Urban Planning · Landscape Design

上海东方渔人码头项目
Shanghai Oriental Fisherman's Wharf Plan

Architecture Design | 2009·04

上海东方渔人码头
Shanghai Oriental Fisherman's Wharf Plan

项目名称：上海东方渔人码头
建筑面积：4 279.43 平方米
项目地点：中国 / 上海
编制时间：2009 / 04

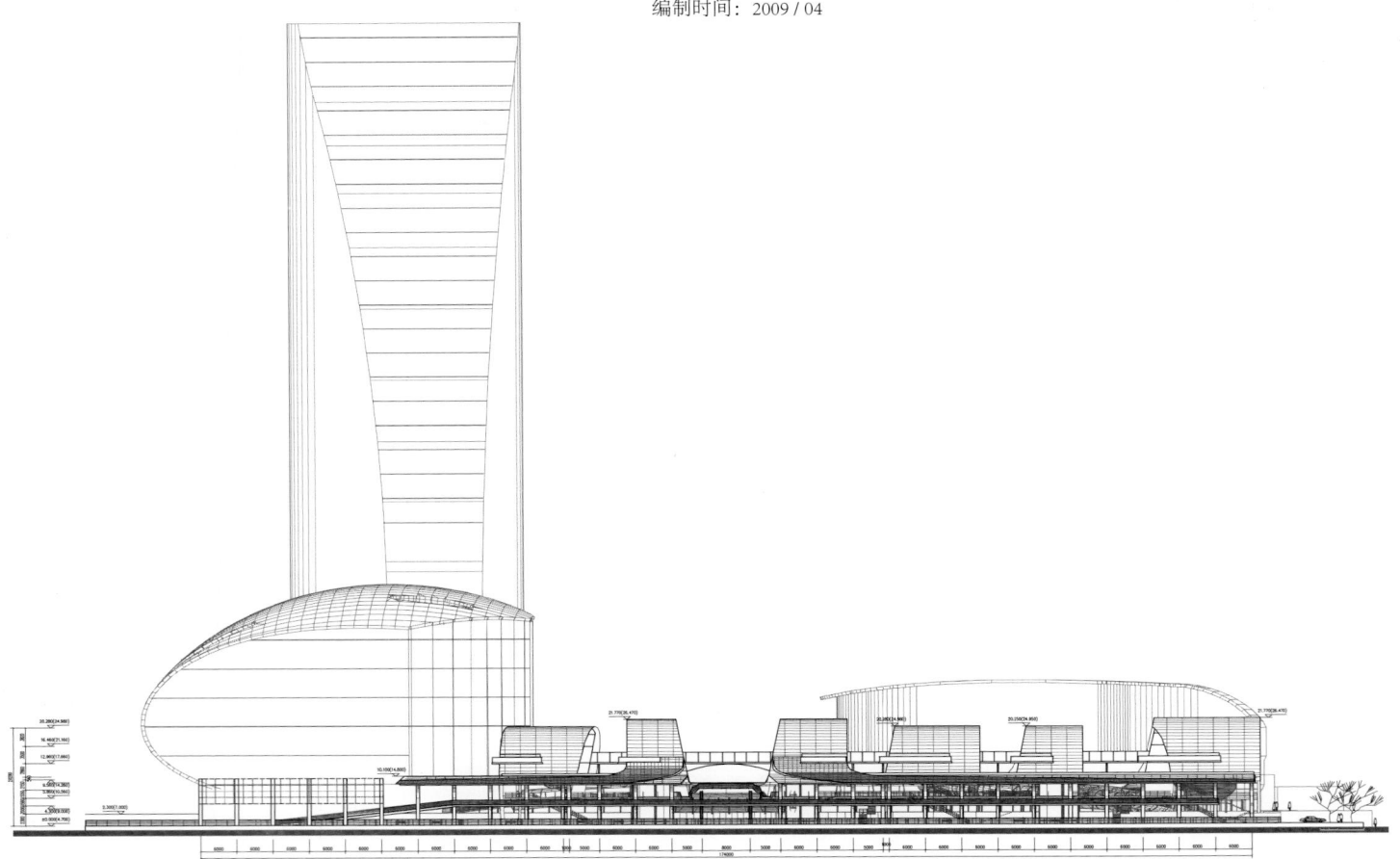

1. 方案用一条全天候步行风雨走廊将一期地块内整组建筑，以及地铁出口无缝连接。在交通上保证了相互之间的可达性和方便性，完善了一期步行系统的整体性。另外，此走廊在行进路线上的竖向高度经过精心设计，使行者在其间可以从不同视角观看景色。

2. 将建筑的屋面与连廊的屋面连为一个整体，连接部位作为入口的雨篷，充分考虑了整个体系的整体性和系统性，更创造出自由多变的灰空间。

3. 将原来单一的长条形体量打散，将其不等分为六段，使得江景可以通过其缝隙渗透到一期地块内；同时利用打开面，增加了观景的界面。

4. 打开的方式充分考虑了西南方向陆家嘴的优势江景朝向因素，使得建筑内更多的空间可以看到此方向。

5. 重视建筑与景观的融合。利用取消第五立面的手法，将屋顶面与建筑的南北立面融合，仿佛建筑为自然地理变迁的产物，使它做到视觉意义上的消隐存在。

6. 建筑物造型上的柔和曲线隐喻波浪，正立面上的横向长廊又仿佛是水中升起的龙门，与地块内的"卧鱼、立鱼及珍珠"相映成趣。同时又确立了 该建筑在整个建筑组群中的从属地位，丰富和完善了整个渔人码头一期建筑群体的整体意象。

1. An all-weather pedestrian corridor is designed to connect all buildings in the Phase I plot with the metro exit in a seamless way so as to ensure the accessibility and convenient transport and improve the walking system in Phase I. In addition, heights along the corridor have been carefully designed so that walkers can enjoy different sceneries from different points of view.

2. Connect the roof of the building with the roof of the corridor to be a whole. The connection place is designed to be the entrance canopy. The design ensures the integral and systematic feeling and creates a free and variable gray space.

3. Divide the original strip space into six sections to introduce the river landscape into the phase I plot through gaps among these sections; in the meanwhile, increase viewing space through the open face.

4. Open a space to introduce the river sight of Lujiazui in the southwest and ensure that more space in the building can have a view of this direction.

5. Emphasize the integrating of buildings and the landscape. By means of canceling the 5th facade, integrate the roof with the south and north facades of the building. It seems that the building is the product of the natural geography change to be a blank in the visual sense.

6. The soft curve of the building is just like the wave and the horizontal long corridor on the front facade is like the door rising from the sea. These elements just cope with "fish and pearl" in the plot. At the same time, determine subordinate position of the building among the building group, enriching and improving the overall image of phase I of the entire wharf.

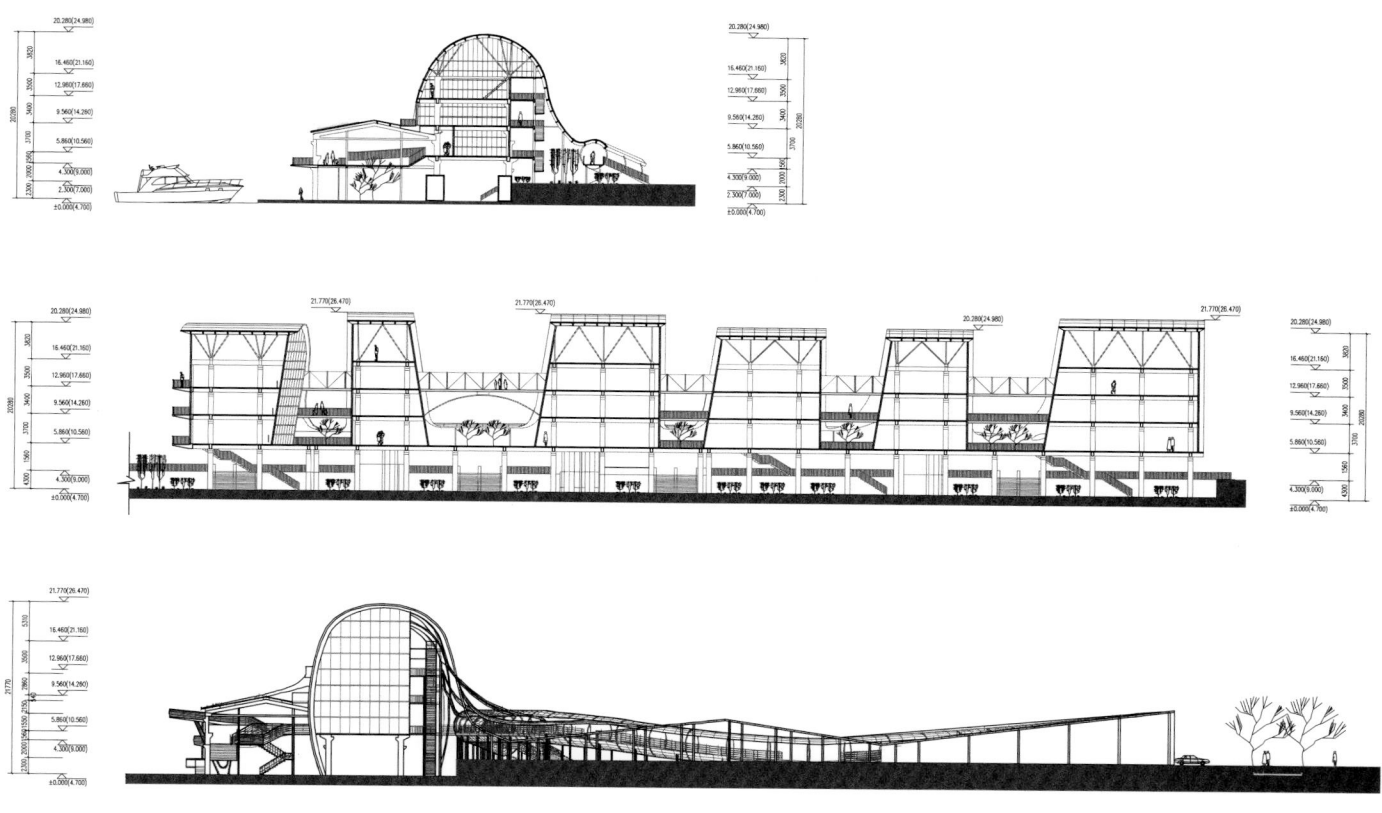

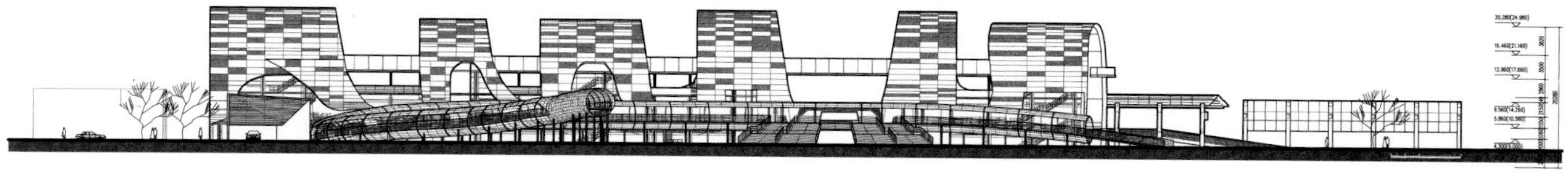

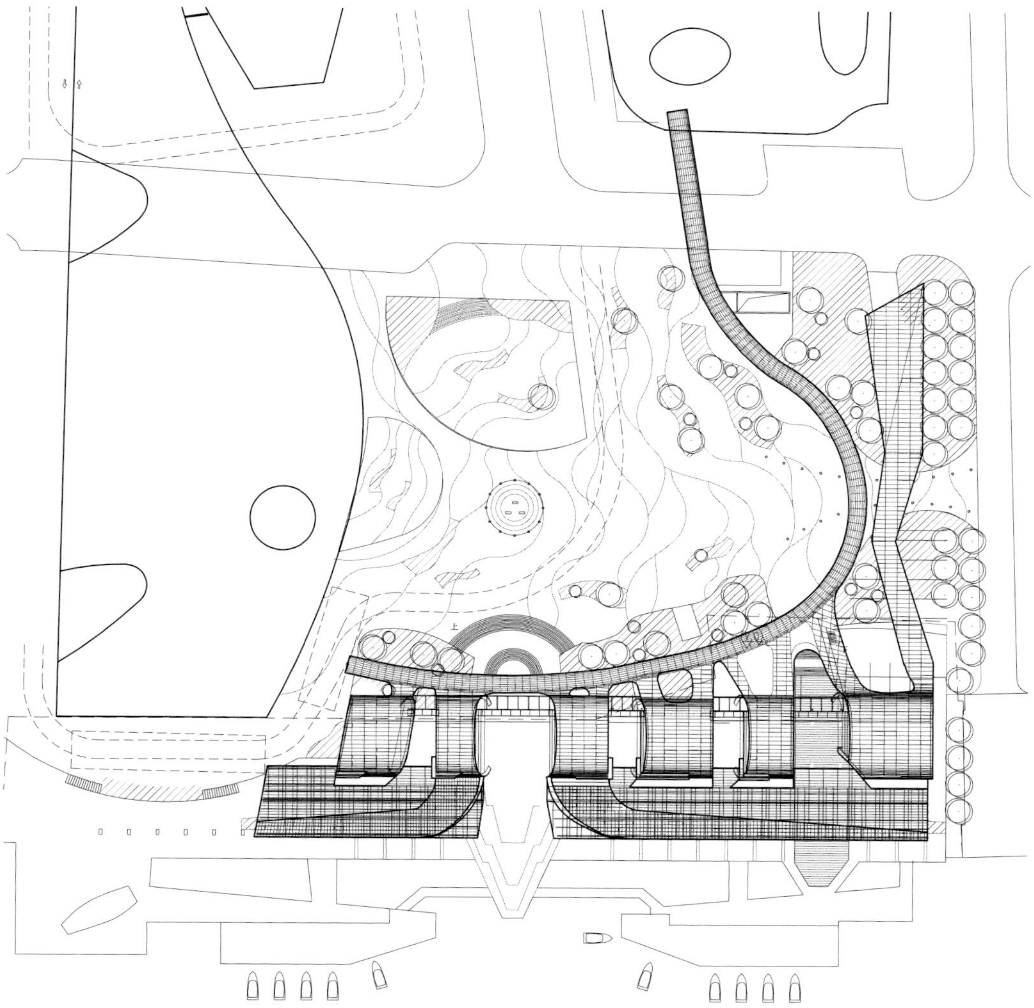

总平面图

Terminal 1A

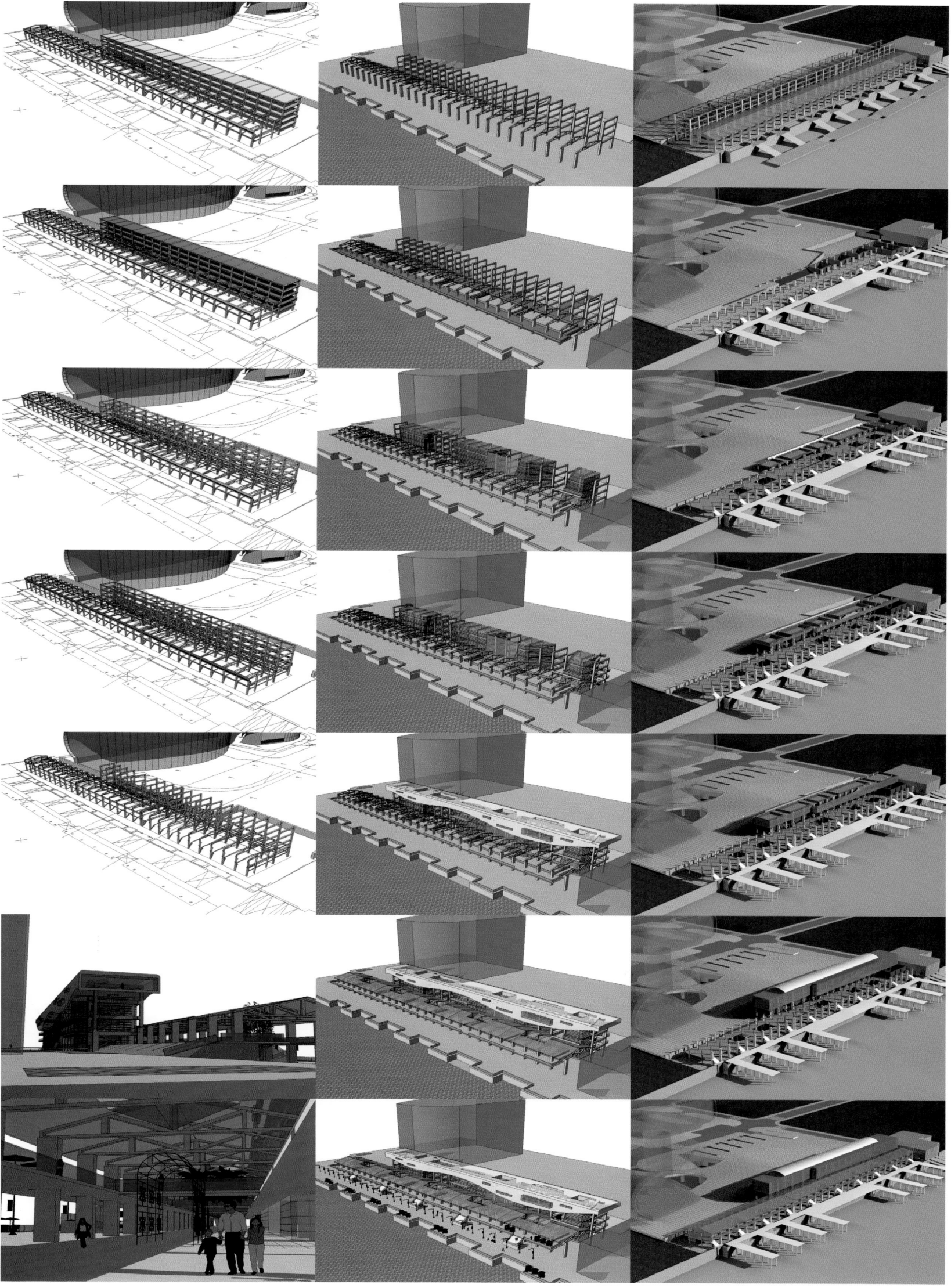

277

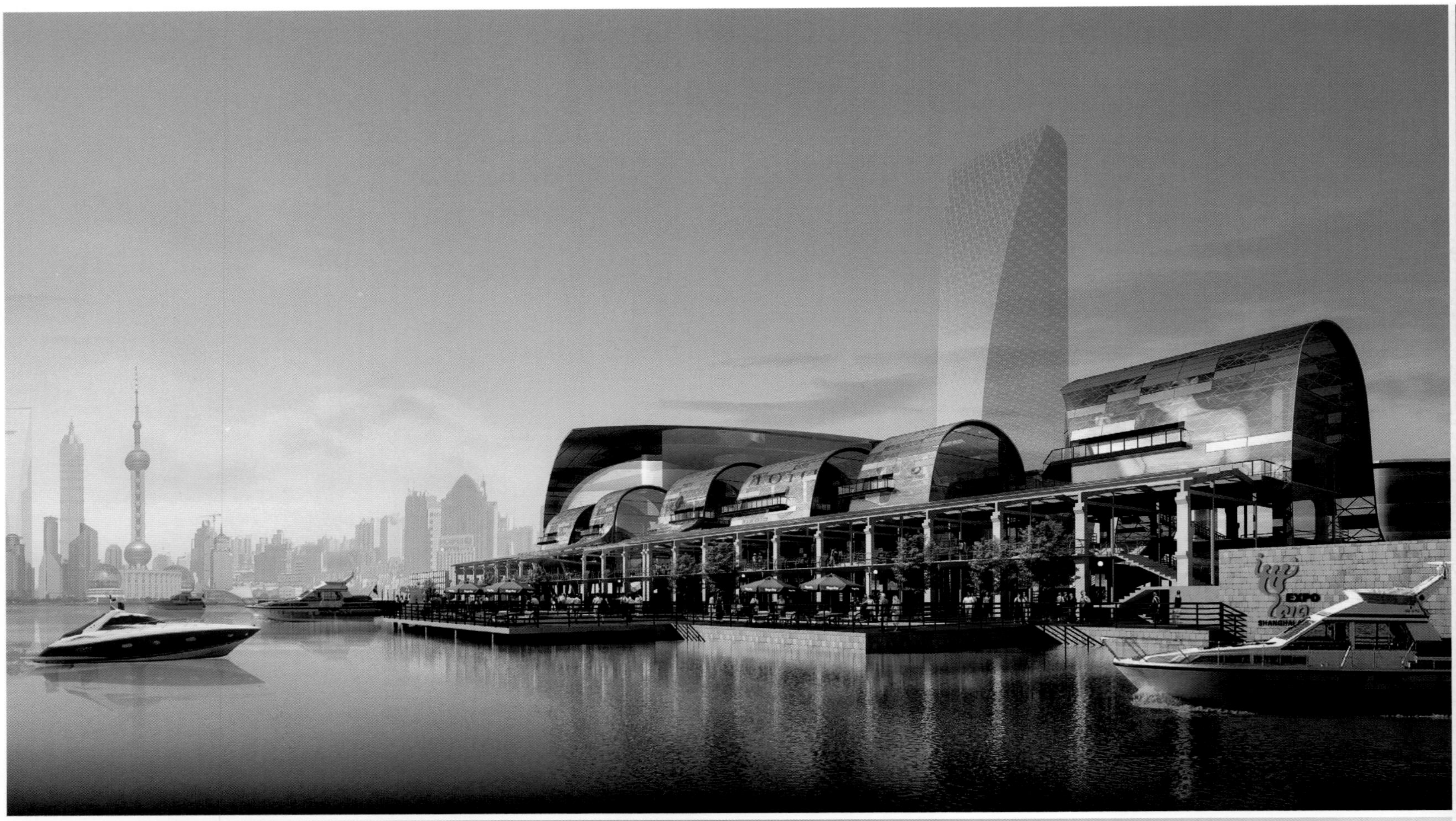
EXPO

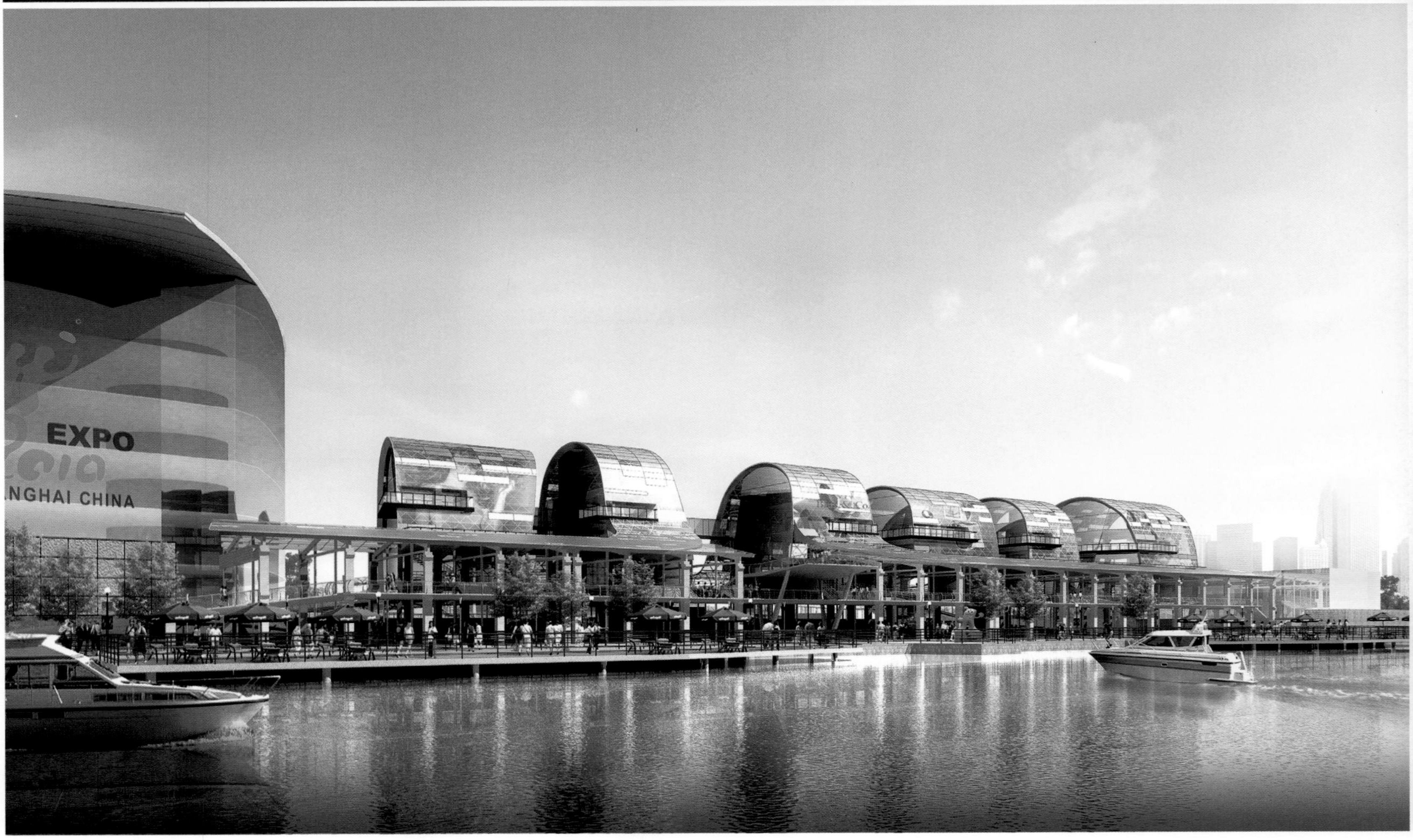
EXPO
NGHAI CHINA

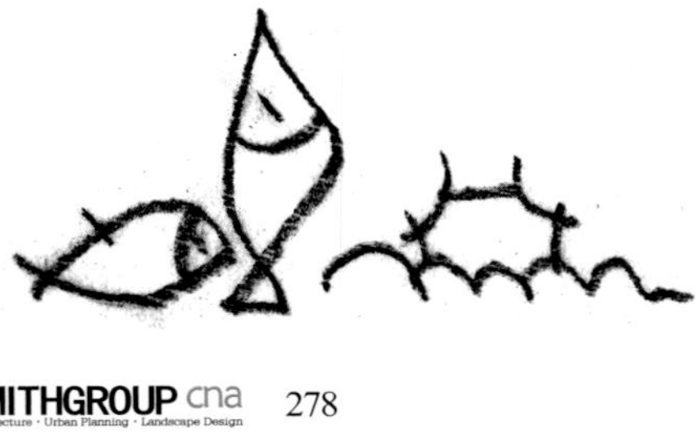

成都极地海洋世界
Chengdu Polar Sea World

项目名称：成都极地海洋世界
用地规模：133.33 万平方米
建筑面积：100.00 万平方米
容积率：0.75
项目地点：中国 / 四川 / 成都
编制时间：2006 / 06

成都极地海洋世界项目位于华阳镇天府大道南端，占地 600 余亩，总投资数十亿元。位于成都市南延线，是目前中国西部最大的展示极地动物、娱乐、风情的生态旅游、商业集群和生活大社区。

Chengdu Polar Sea World Project is located in the south end of Tianfu Avenue, Huayang Town. Covering an area of more than 600 mu, and a total investment of several billion. Designing an effective combination of street and other entertainment, residence regions, it is the biggest eco-tourism business cluster and living community in western China with the function of displaying polar animals, marine entertainment and tropical coquette.

上海金山嘴美食大世界
Shanghai Jinshanzui Food Plaza

项目名称：上海金山嘴美食大世界
用地规模：3.47 万平方米
建筑面积：5.02 万平方米
容积率：1.4
项目地点：中国 / 上海 / 金山
编制时间：2008 / 08

地块本身与金山区发展有着密切的联系，并且作为一个重要的区域融入到金山区的城市建设之中，这些直接影响到地块内部的规划布局和周边建设的呼应，并促使我们在规划中充分考虑金山滨海地区的远期发展规划并与之建立充分的对话关系，将美食广场、商业配套和娱乐业消费按照消费模式和心理模式去组合，使其成为周边人群可以停留一整天的场所。

The site has a very close relationship with development of Jinshan District, which pushes us to consider the future development of the whole district in the planning and establish an ineraction with it. The restaurant hall and other retail supports are set up according to the consumer model and psychology reason. These could be the commercial destination of the future habitants around the site.

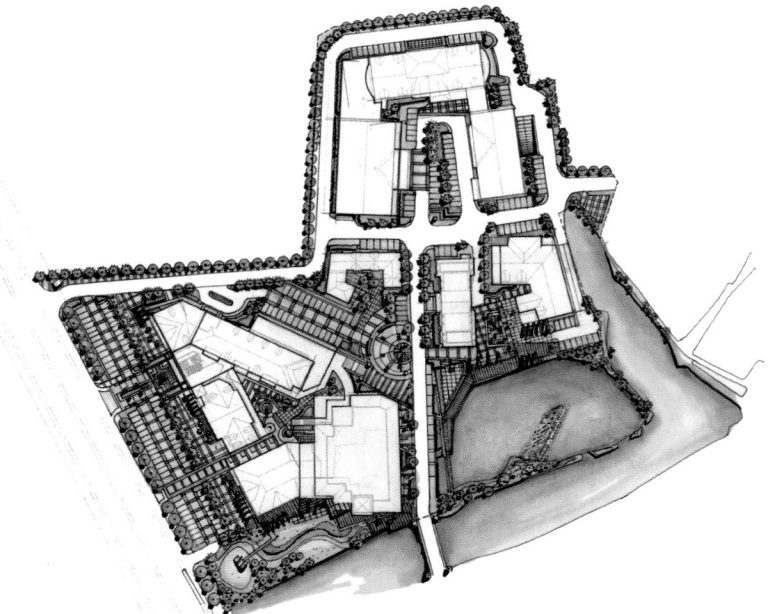

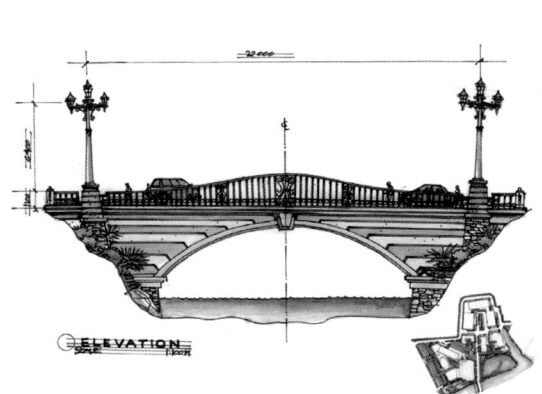

大理古城休闲商业改造
Dali Ancient City Leisure Business Transformation

项目名称：大理古城休闲商业改造
用地规模：1.5 万平方米
建筑面积：1.28 万平方米
容积率：0.85
项目地点：中国 / 云南 / 大理
编制时间：2006 / 06

"中国街"位于人民路（博爱路至复兴路段）的两侧，规划面积大约为 1.5 万平方米。根据大理古城的旅游特色和对该地块独到的功能定位，将"中国街"整个区域分为三大功能区块：市井商贾、文化休闲、展览会务。

市井商贾：商贾交易是古城生活的一个重要组成部分。将沿街四合院的倒座开敞出来，设计成通透的商铺，店面的门面可自由组合和变化。穿过沿街的铺位，游人可方便地进入四合院中，客商双方在四合院中交易，避免了古城旅游开发中商业气氛过重而破坏古城整体风貌的问题。同时方便了特色商业制作和传统手工艺表演活动的开展。

文化休闲：针对旅游者设计的四合院建筑，在保留传统四合院基本形制的前提下，开创了更广阔的活动空间和交流空间。可以将传统的四合院建筑的一面开敞，作为小型休憩广场的入口；或将几个四合院之间沟通起来，组合嵌套的四合院的形制，灵活组合，充分满足了旅游者娱乐、观演、休闲、居住等复合功能需求，最大限度地创造体验异地特色风情的场所。

展览会务：古城整体风貌与合院式建筑布局，是大理古城旅游的特色与亮点。为充分体现古城旅游的特色，营建几处档次、规格和环境都相对高档的四合院落组群，供展示和会议使用。局部采用玻璃中庭手法联系建筑空间，提高合院建筑的利用率，不仅有利于提升古城合院建筑的建设水平和档次，还能促进古城旅游形象的树立，扩大知名度和影响力。

"Chinese Street" is located on both sides of Renmin Road (from Boai Road to Fuxing Road) with the planning area of about 1.5 hectares. According to the tourism features of the ancient Dali City and the special functional positioning of this plot, the "Chinese Street" is divided into three functional blocks: merchants' marketplace; culture and leisure; exhibition services.

Merchants' marketplace: merchant trading is an important part of the life of ancient city. Open the front line of courtyards along the street to be shops. The store envelopes can be freely combined and changed. Visitors can easily enter the courtyard through these stores. Trading in the courtyard can avoid the excessive business atmosphere created in the ancient city tourism development process which will damage the overall style here. At the same time, this design can facilitate the special commercial fabrication and traditional arts and crafts performances.

Culture and leisure: while retaining the basic shape of traditional courtyard, courtyard buildings specially designed for tourists provide broader space for activities and communication. Open one side of traditional courtyard to be the entrance of small recreational square; for another choice, connect several courtyards together to be flexibly combined building which can fully meet visitors' needs of entertainment, performance watching, recreation, residence and others, thus creating an ideal place for experiencing special characteristics of the local place.

Exhibition services: the overall ancient city appearance and the courtyard building layout are features and highlights of the travel in Dali. To fully reflect the tourism characteristics of the ancient city, a number of high-end courtyard groups with large size and favorable environment will be built to provide exhibition and conference services. Glass atrium will be partly used to connect the building space and to increase the building use ratio. This will not only help enhance the courtyard construction level and grade, but also promote the city's tourism image and the expansion of reputation and influence.

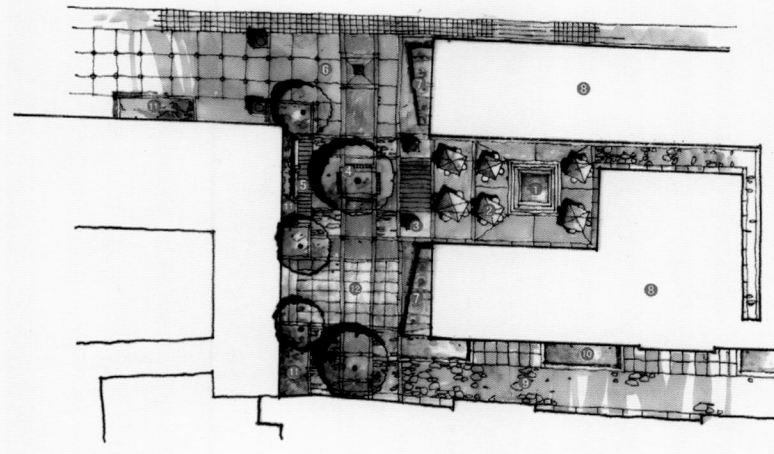

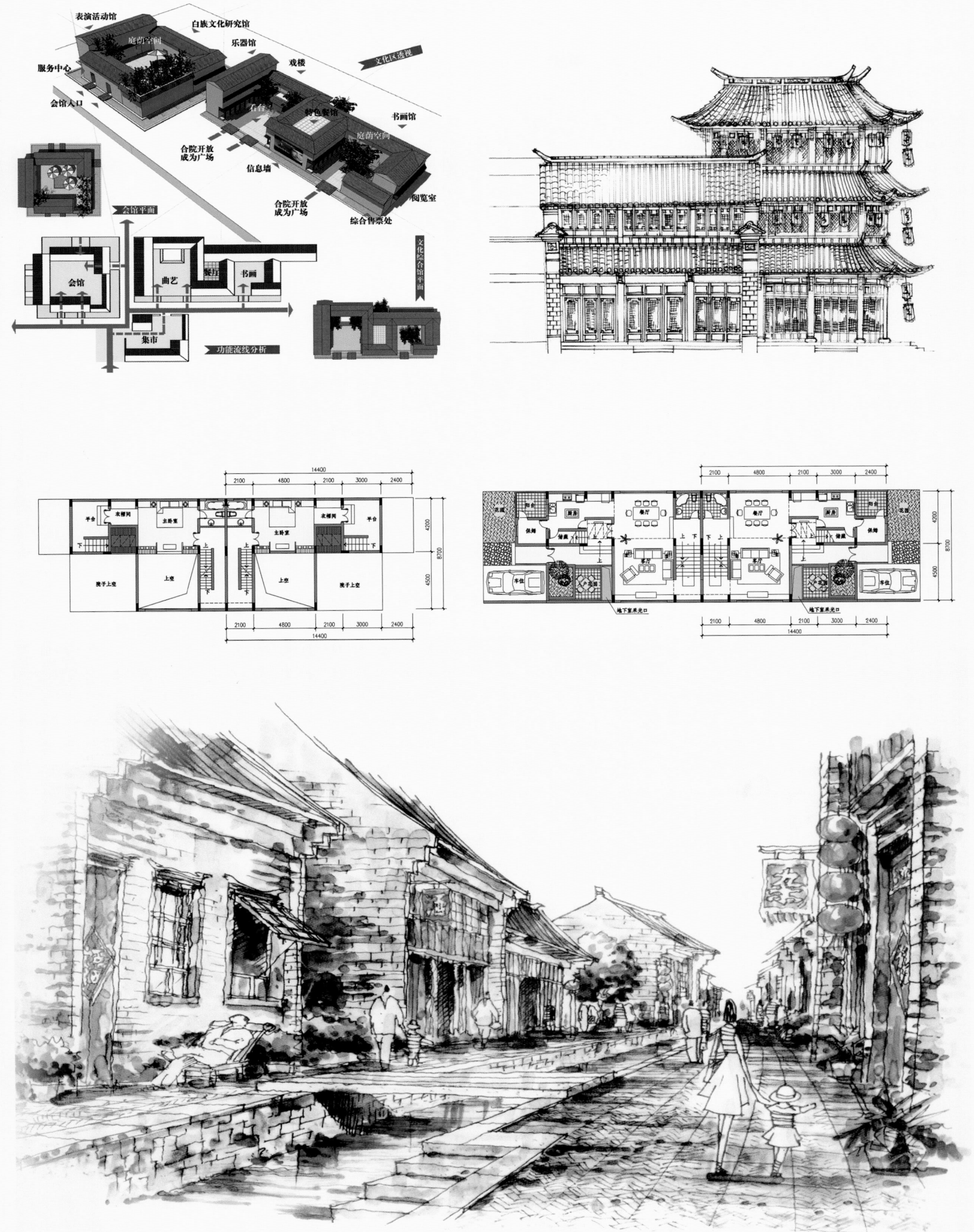

表演活动馆　白族文化研究馆
庭荫空间　　乐器馆
服务中心　　　　戏楼　　　　　　文化区透视
会馆入口

会馆平面

舞台　特色餐馆　　书画馆
合院开放
成为广场　　庭荫空间
信息墙　　　　阅览室
合院开放
成为广场
综合售票处

文化综合馆平面

会馆　曲艺　庭院　书画
集市
功能流线分析

14400
2100　4800　2100　3000　2400
平台　衣帽间　主卧室　　　主卧室　衣帽间　平台
下　　　　　　上　　下
院子上空　上空　　上空　院子上空
2100　4800　2100　3000　2400
14400
4200　8700　4500

2100　4800　2100　3000　2400
阳台　　　厨房　　　餐厅　　　　　　餐厅　厨房
保姆　　储藏　　　　　　　　　储藏　　　　保姆
上下　上　　　　　上下　上
客厅　　　　　　客厅
车位　花园　　　　　　花园　车位
地下室采光口　　　　　　　　地下室采光口
2100　4800　2100　3000　2400
14400
4200　8700　4500

成都壹购潮流广场
Chengdu E-go Trend Plaza

项目名称：成都壹购潮流广场
用地规模：0.45 万平方米
建筑面积：2 万平方米
容积率：4.4
项目地点：中国 / 四川 / 成都
编制时间：2007 / 07

本项目是成都市区年轻时尚的消费人群考虑的最好去处，设计师采用了极简约而又有视觉冲击力的里面表现，在红色染色玻璃上凸显出铝镁合金饰面的项目名字体，简洁大方而又有雕塑感的整体建筑体量质感强烈，使本项目明显有别于现有的周边建筑，为城市增添了活跃而又快乐的新元素。

The project is the best destination for young consumers in Chengdu. Designers adopts the simplest and visually powerful elevation. The metal projecting title protruding on the red colorful glass makes the whole building become a solid sculpture, and the distinction adds more active and happy elements for the city.

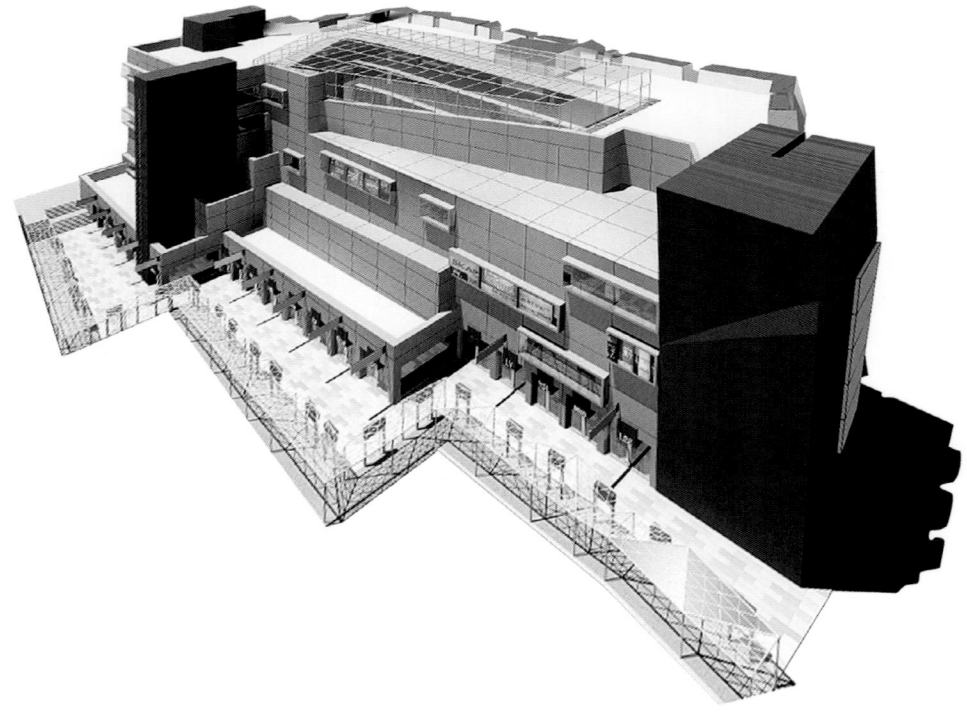

上海金山卫零路商业街改造项目
Shanghai Jinshan Weiling Road Commercial Street Reconstruction Plan

项目名称：上海金山卫零路商业改造项目
用地规模：3.6 万平方米
建筑面积：9 万平方米
容积率：2.5
项目地点：中国 / 上海 / 金山
编制时间：2006 / 05

这是对 70 年代老城区的一次实验性改造，将原有的居住、商业零售店面以及一些过时的文化设施进行重新的商业策划定位，并对街道进行立面、材料和铺装景观的调整，新建的主体建筑在设计中追求最大的商业沿街面，提升单位商业面积的使用率，增加商业空间的趣味性。利用空间步行系统和地下商业系统建设立体化步道系统，改善各层可达性，同时利用功能的混合提升建筑的商业活力，并将这种动力向外发散到整个卫零路老商业街区上，最终成为一个区域的商业引擎。

This is an experimental transformation for the 1970's old city, to relocate the business plan for the existing residential, commercial retail stores and some outdated cultural facilities, and to adjust the elevation, materials and landscape of the street pavement. The new main building is designed to ensure the largest commercial area along the street, enhancing the commercial space use ratio and increasing the interesting of commercial space. It adopts space pedestrian system and undergrounds business system to build three-dimensional walking system, improving the accessibility of all layers and at the same time, it enhances the commercial vitality of the building by function mixing. Spread the vitality to the entire business street on Weiling Road to finally form a regional commercial engine.

天津海洋世界
Tianjin Sea World

项目名称：天津海洋世界
用地规模：16.00 万平方米
建筑面积：63.96 万平方米
容积率：3.94
项目地点：中国 / 天津
编制时间：2006 / 06

在欢乐的主题公园的积极渗透下，居住为时尚、简约、充满现代美感的"都市休闲居所"，运用现代浪漫主义的设计手法，着重追求建筑构造本质的美感，融入现代技术美与人情味，使居住者的情感在经历充分的激励与想象后回归宁静与自然。

Influenced by the happy Marine Theme Park and featured business, the design idea is to create a fashionable and concise "urban leisure residence" full of modern beauty. We apply the design techniques of modern romanticism, focus on the natural building beauty and introduce modern technology and human touch to ensure that residents here can return to peace and nature after experiencing emotional inspiration and imagination.

武汉极地海洋世界
Wuhan Polar Sea World

项目名称：武汉极地海洋世界
用地规模：69.82 万平方米
建筑面积：59.32 万平方米
容积率：0.85
项目地点：中国 / 湖北 / 武汉
编制时间：2006 / 06

通过水公园、古镇、英伦风情商业建筑群，全面归纳和整合汉口特有的商业文化、古镇文化、两江文化，并使之与海昌特有的国际水公园全面接轨，将项目建设成为武汉文化新式娱乐海洋公园的综合体验区，成为武汉文化向往者的必游之处。

Comprehensively induce and integrate special Hankou business culture, ancient town culture and the culture of the two rivers by the water park, the old town and the classical British style commercial buildings. Integrate with the owner's special international water park form and build the project to be an entertainment sea park and comprehensive experience zone with Wuhan cultural form so as to make it a place which must be visited by those in favor of Wuhan culture.

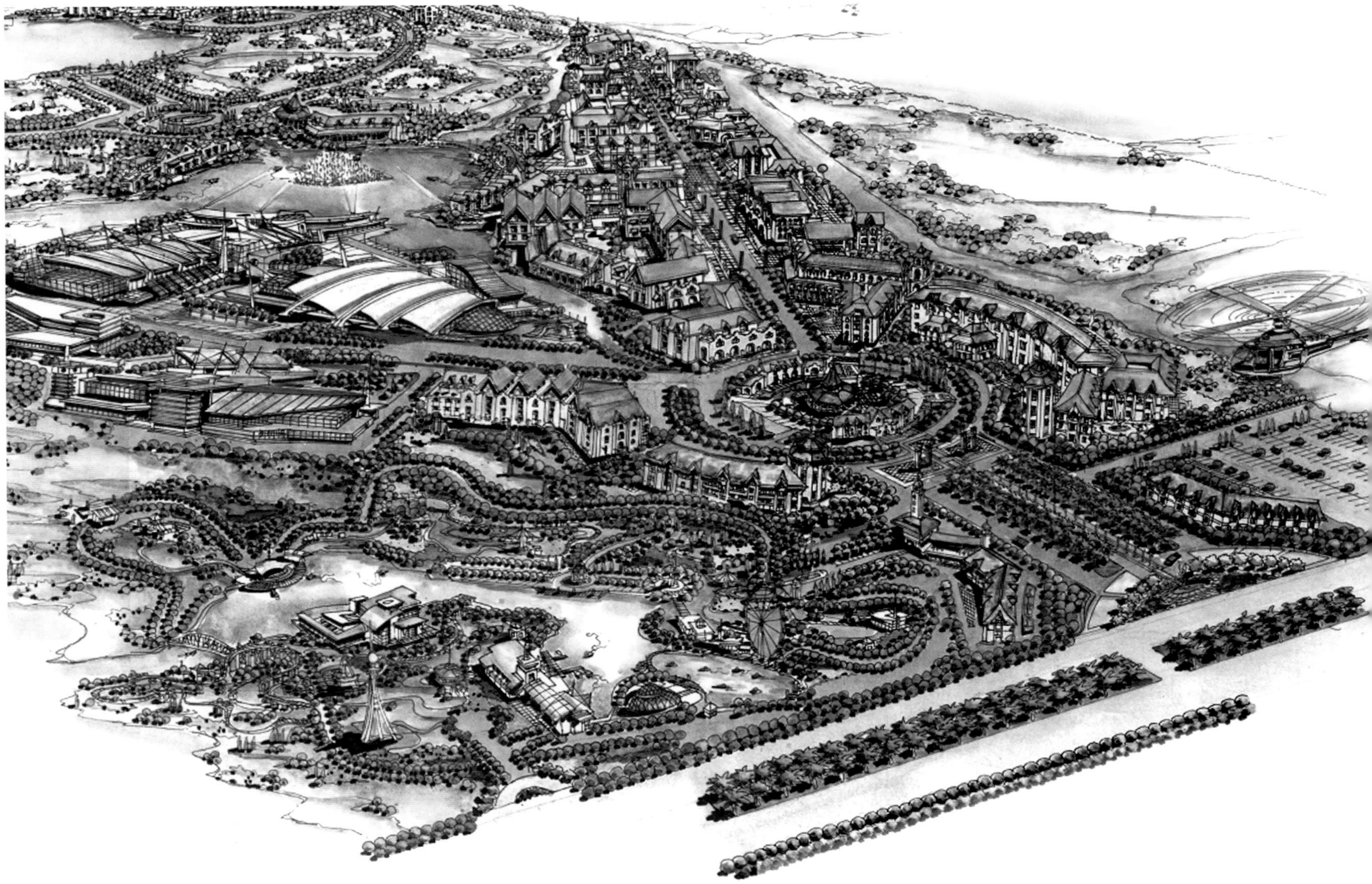

沪宁城际无锡新区站腾房区改造
Shanghai Nanjing Railway Wuxi New District Station Tengfang Area Reconstruction

项目名称：沪宁城际无锡新区站腾房区改造
用地规模：9.47 万平方米
建筑面积：9.32 万平方米
容积率：0.98
项目地点：中国 / 江苏 / 无锡
编制时间：2008 / 05

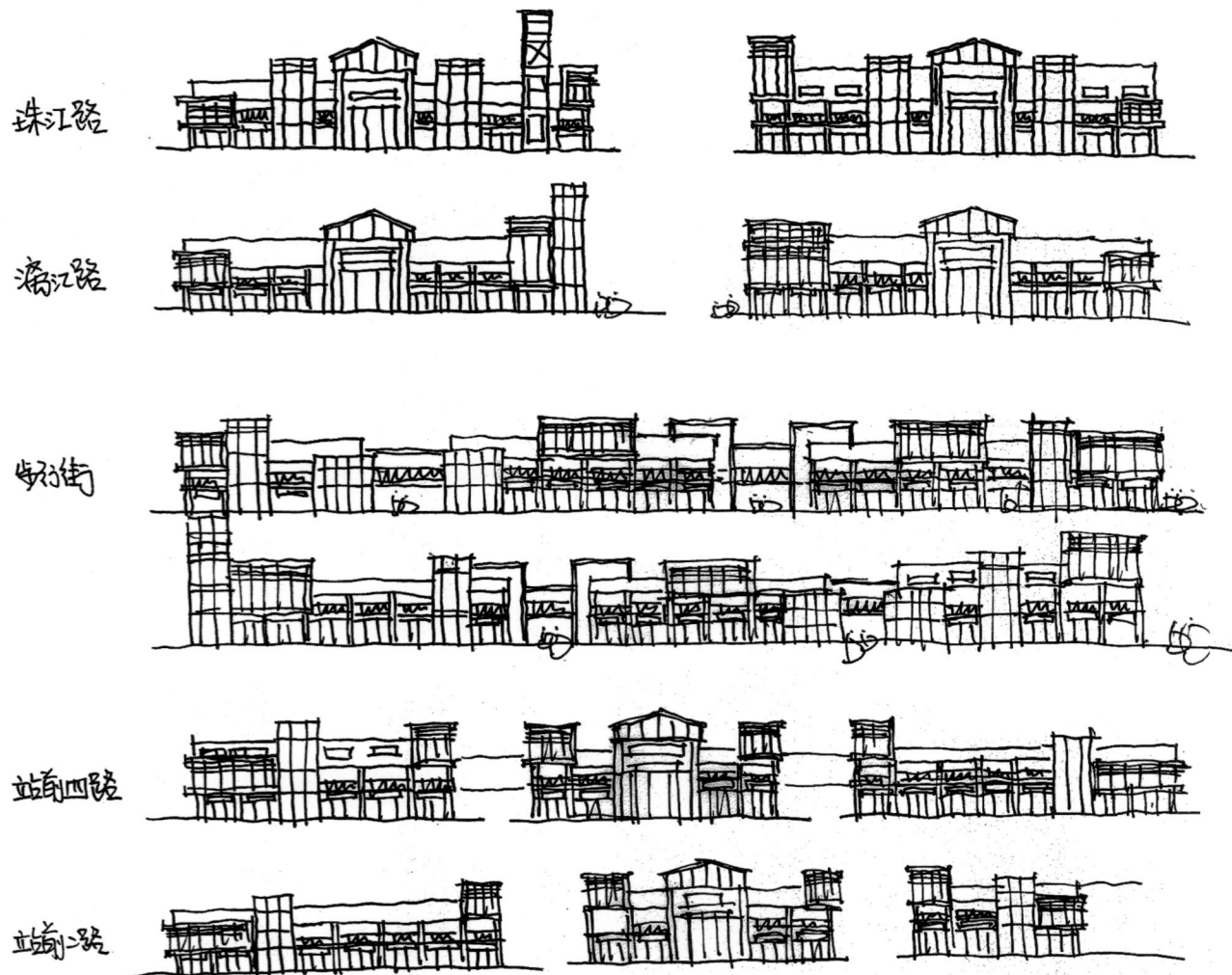

珠江路

滨江路

纯功街

站前四路

站前二路

沪宁城际铁路无锡新区站站前广场的大开发是当地政府希望对周边区域进行一次彻底改造和规划升级的工程，依规划的需要，设计计划拆除一期腾房区内临珠江路的部分厂房，并沿珠江路北侧改建为临时商业区，在前期改造阶段，先利用"轴线步行街"所处位置兼作机动车道。

其创作理念是展现中国最美的城市表情——打造 WALK CITY 体验型商业街区，并围绕"生活、色彩"来寻找规律和探索可行性的改造手段。

The square in front of Wuxi New District Station of Shanghai-Nanjing Intercity Railway is under great development. Local government hopes to transform and upgrade the surrounding region. According to the design, some plants of phase I near Zhujiang Road in Tengfang District can be dismantled and rebuilt to be temporary commercial street along the north side of Zhujiang Road. In the pre-transformation stage, the "axis pedestrian street " will also be used as traffic lane temporarily.

The creation idea is to show the most beautiful city expression in China-to create WALK CITY experiential business district and analyze and explore feasible means of transformation around "life and color ".

西宁火车站站前商务区
Xining Railway Front-Station Business District

项目名称：西宁火车站站前商务区
用地规模：7.30 万平方米
建筑面积：7.53 万平方米
容积率：1.03
项目地点：中国 / 青海 / 西宁
编制时间：2010 / 05

1. 以水为媒——汲取河湟文化之底蕴

建筑单体平面以水滴状造型自由灵活地沿北侧景观湖布置，平面肌理生动活泼，立面造型轻盈剔透，极具建筑张力，与湖岸景观形成较好的视觉冲击，提升了站前广场的商业形象，具有时代气息，同时在商业中还注重了下沉式广场及商业水景的处理，形成丰富的商业空间。

2. 取绸为意——隐喻丝绸之路的内涵

在项目总体的内部道路、花径等设计上，以曲线路径为设计元素，隐喻丝绸的柔美，取丝绸之路的意象，同时在主要景观节点及小品等处也围绕这一主题，这样的设计不仅与建筑的造型相得益彰，而且与建筑的串联也十分舒展，表现出项目所处特定地域的内涵与文化。

3. 以山为幕——倚仗祁连山脉之雄伟

主体建筑根据商业业态的形式，建筑高度为三层，体量轻巧，建筑外立面材质以玻璃幕墙及轻钢结构为主，部分立面饰以当地盛产的砂岩及花岗岩材质，注重当地建筑元素及本土化建筑材质的运用，同时形成现代时尚的商业风格，为北侧极具视觉冲击力的祁连山脉形成互为依托、互为映衬的效果，展示了西宁独特的山水格局，具有高原城市的独有风光。

1. Water as the Media-Learn from Hehuang Cultural Heritage

The individual buildings in drop shape are arranged freely and flexibly along the landscape lake in the north. The graphic texture is lively and the elevation shape is light and clear with great tensile force. Great visual impact can be formed together with the lake shore landscape. The commercial image of the front-station square is enhanced to be with flavor of times. Sinking-mode square and commercial waterscapes are also designed to form rich commercial space.

2. Silk-the Connotation of the Silk Road

Internal roads and flowery paths are designed in curve style, reflecting the beauty of silk. Main landscape places and other details are also designed with this theme. Such design can not only bring out the best in each other with the building shape but also form conformable connection of buildings, reflecting the local connotation and culture.

3. Mountain as the Curtain-rely on the Majestic Qilian Mountain

The main building is for commercial purpose. The 3-storey light building uses glass curtain wall and light steel structure for the envelope construction. Sandstone and granite are also used for some parts of the facade. We focus on the local architectural elements and the use of local building materials and at the same time, make efforts to form a modern and fashionable business style. Qilian Mountain in the north with great visual impact can better set off the project. It is designed to display the unique landscape pattern of Xining with unique scenery of the plateau city.

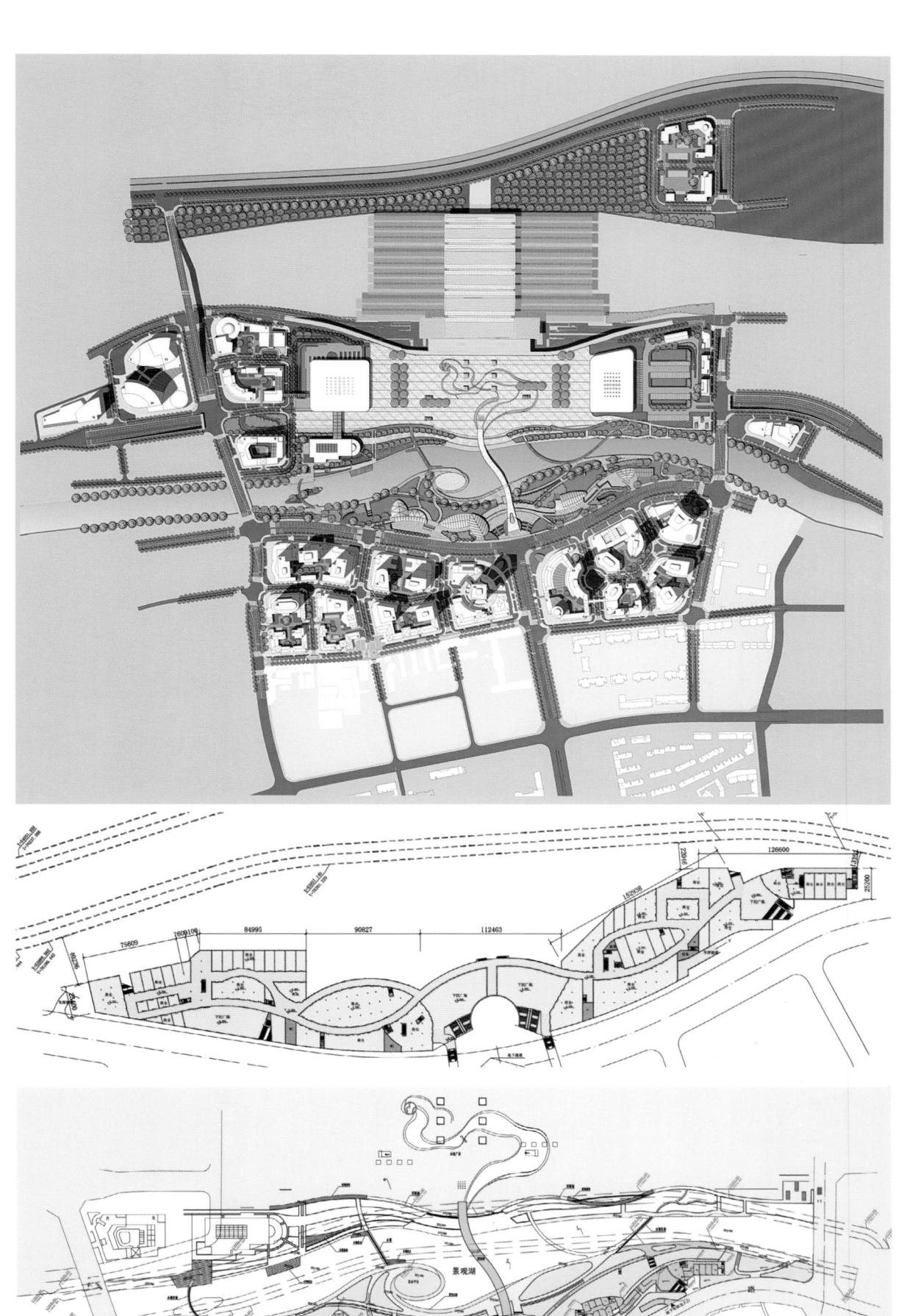

扬州京杭之心商业水街
Yangzhou Beijing Hangzhou Centre Commercial Water Street Project

项目名称：扬州京杭之心商业水街
建筑面积：1.70 万平方米
用地规模：1.39 万平方米
容积率：1.22
项目地点：中国 / 江苏 / 扬州
编制时间：2008 / 09

不是再造一个传统的江南水乡，也不是传统江南水乡的复原。

传统意义的江南水乡，水街正在逐步消失……

仅仅站在怀旧的角度上进行水乡的复原与复制，是一种完全不负责的态度！

这种城市形态的变更，更大尺度上是一种城市机体的新陈代谢。

汲取西方城市的布局，加强了其交通的可达性、独立性，使我们的水街别具一格，更具有丰富的娱乐性和休闲浪漫氛围。结合扬州当地民居特色、引用格网结构和城市肌理构成我们的建筑排布。

西方规划思想的引入，中国文人水墨的再现……

水乡古韵肌理，时尚元素变异……

It is neither to create anther nor to restore and reproduce a traditional water town in the south of the Yangtze River.

The traditional water town in the south of the Yangtze River is gradually disappearing at Water Street.

It is a completely irresponsible attitude to restore and reproduce a water town!

The change of the city form can be regarded as more of the metabolism of the city.

We learn from the layout of western cities to enhance the accessible and independent transport so as to make our Water Street unique with rich entertainment and romantic atmosphere. Combining local characteristics of Yangzhou, grid structure and urban fabric are introduced for the building arrangement.

Introduce western planning thought and reproduce China ink culture.

Old rhyme and texture of the water town; transformed fashion elements…

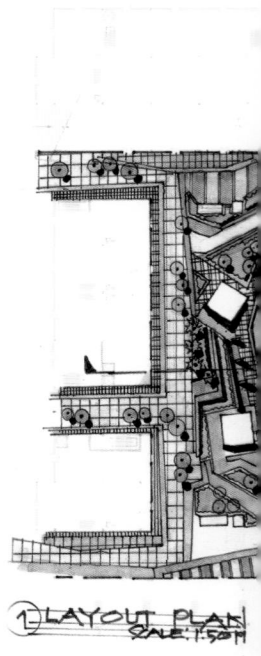

LAYOUT PLAN
SCALE 1:50 H

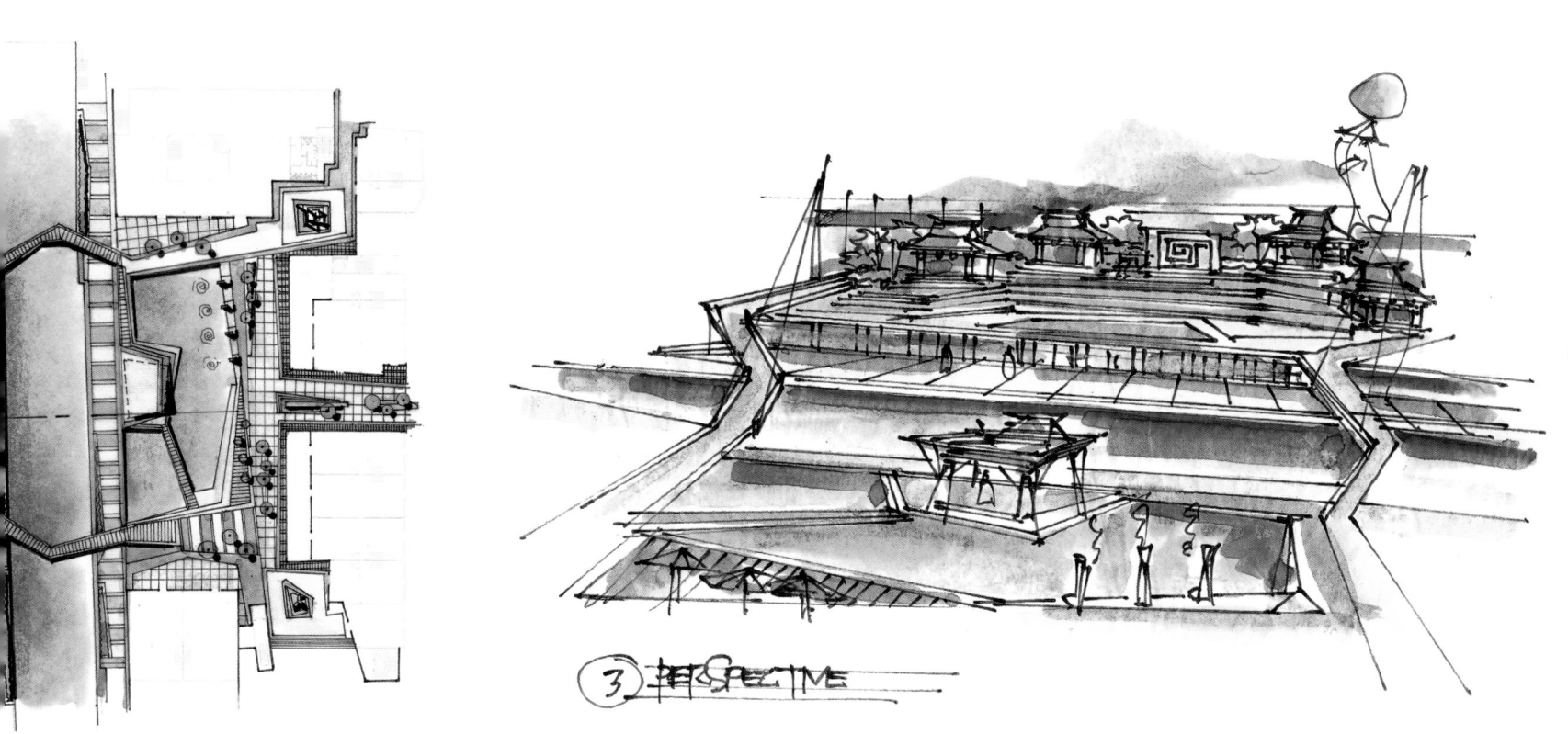

在这里我们需要重视新中式商业街的单元体量和在形成隔河相望的格局后人流的便利性问题。
细微的景观和小街区组团将商业的自然生长性体现出来。

③ PERSPECTIVE

银川绿地物流中心
Yinchuan Greenland Logistics Centre

项目名称：银川绿地物流中心
用地规模：13.73 万平方米
建筑面积：12.76 万平方米
容积率：0.93
项目地点：中国 / 宁夏 / 银川
编制时间：2008 / 05

项目位于银川市兴庆区科技园，设计师以强调功能的多样性为原则，将不同业态混合设置，帮助提升城市特色和活力。同时强调人的多样性和休闲居住人流的多样性，强调互动以在空间上加强不同主题的社会交流与联系。设计中部分承袭了当地的伊斯兰风格，采用提炼过的装饰符号，融合现代设计手法，构建具有当地特色的商业景观。

The project is located in Yinchuan Science and Technology Park Xingqing District City. Designers emphasizes the diversity of functions, integrates plenty of functions and modern large-scale integrated logistic park's construction to improve the activity of city. At the same time the design emphasizes the diversity of people and the interaction on space to reinforce the connection between different parts of society. The Islamic elements are well involved in the design so that the building is full of local environment features.

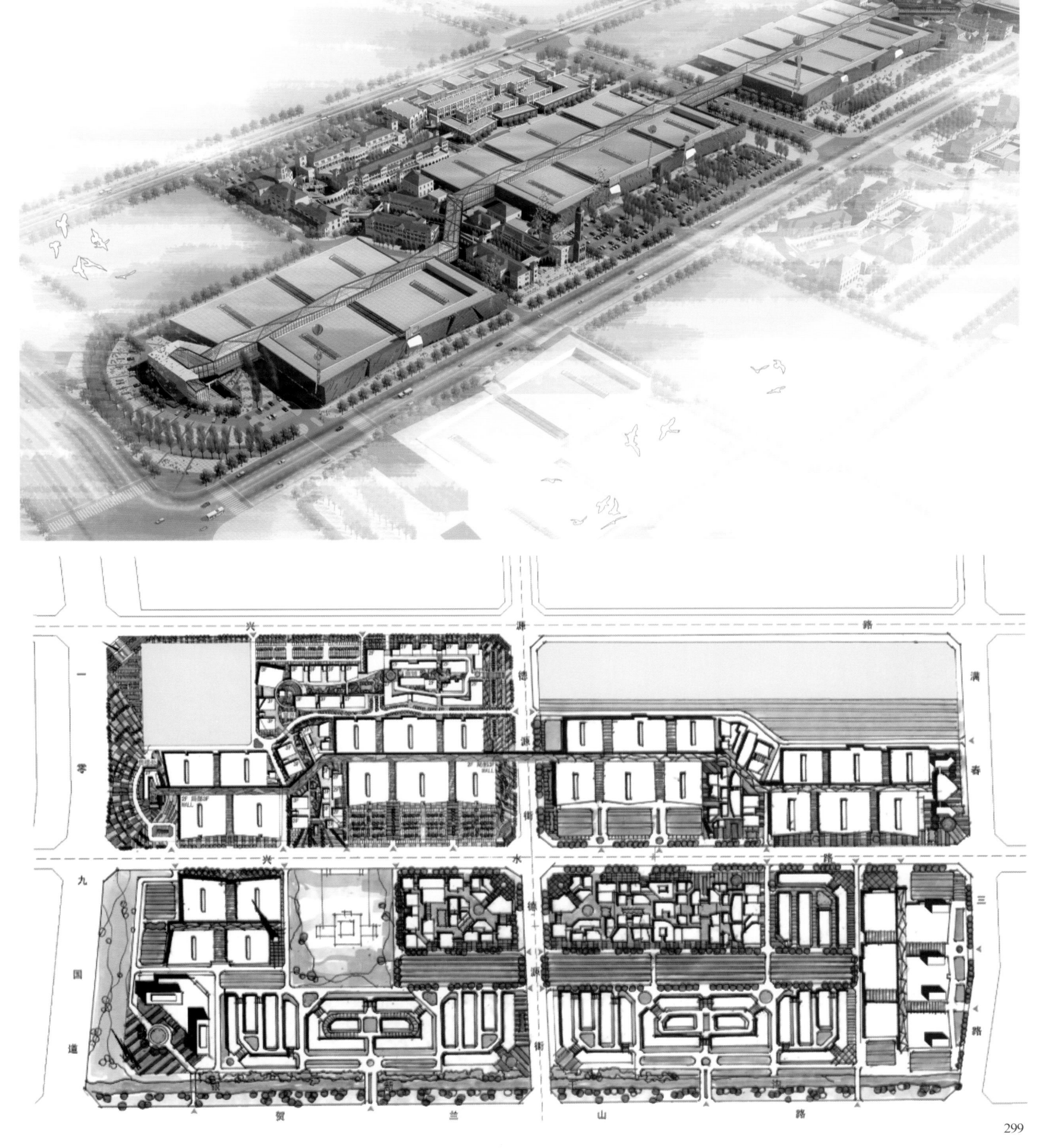

一个超百万平米的综合性商业中心已经远远超出单纯商业的范畴，
成为集合性的综合体，它包括对人在商业活动中各种行为的支持，
个人消费、批发、休闲、娱乐、旅游等等，
对业态安排和交通问题的综合考虑是其中重要的环节。

重庆加勒比海洋世界
Chongqing Caribbean Sea World

项目名称：重庆加勒比海洋世界
用地规模：13.25 万平方米
建筑面积：6.89 万平方米
容积率：0.52
项目地点：中国 / 重庆
编制时间：2006 / 05

作为重庆的一个高质量的大型综合体验区，将居住、文化、娱乐、休闲等功能与自然景观相结合，除使用功能之外，赋予园区更多的趣味性、知识性和独特的空间遐想，通过多样的形式及生动的动感空间来反映中国、重庆及南山的未来。

As a high-quality comprehensive experience community, the project integrates residence, culture, entertainment and other leisure functions. Additionally, the community reflects the local culture through many active methods.

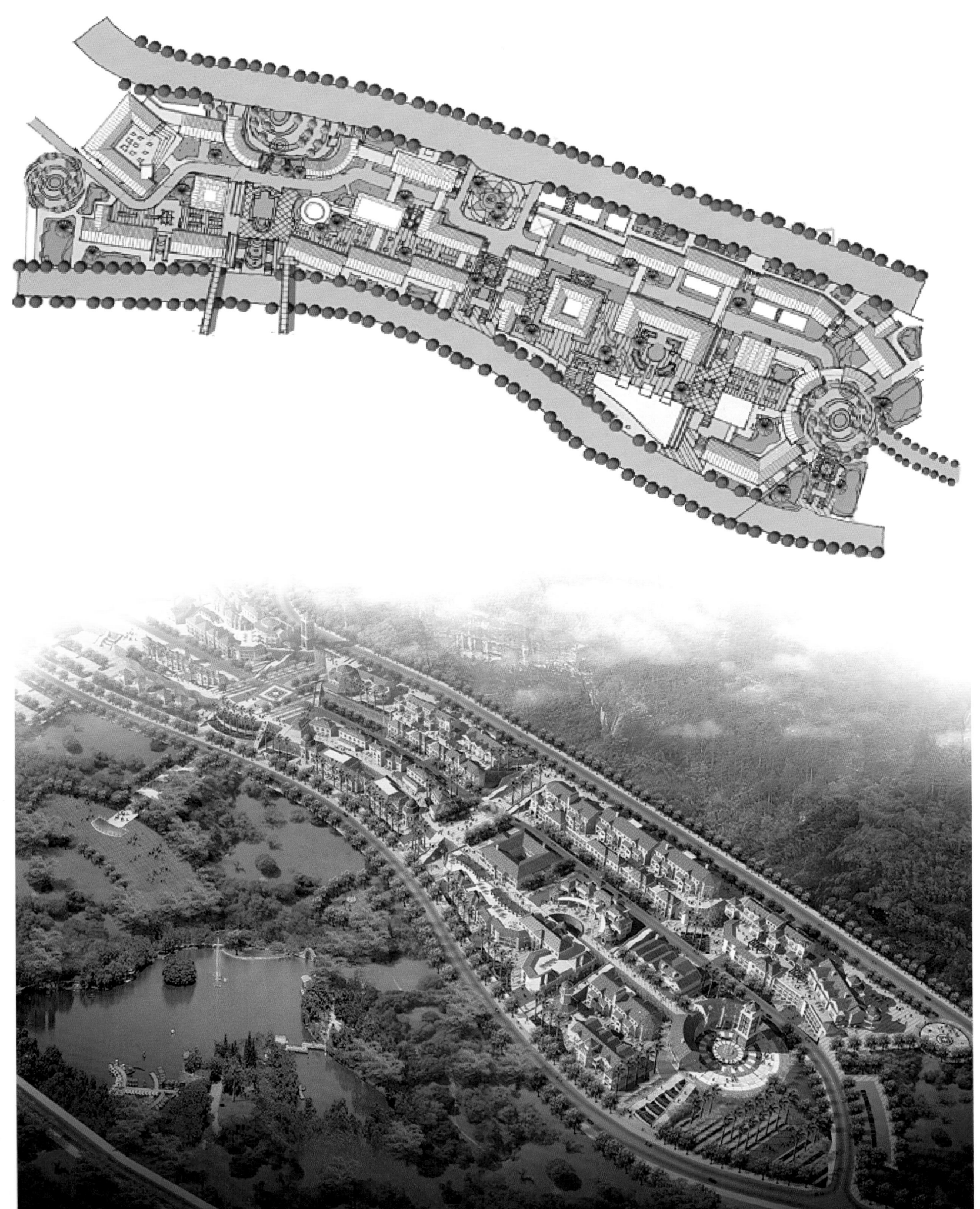

重庆金佛山天星小镇概念规划及建筑方案设计
Chongqing Jinfo Mountain Tianxing Small Town Conceptual Planning and Building Plan Design

项目名称：重庆金佛山天星小镇概念规划及建筑方案设计
用地规模：200 万平方米
建筑面积：10 万平方米（一期）
容积率：小于 1.0
项目地点：中国／重庆
编制时间：2009／01

金佛山位于重庆市南川区境内，具有明显的喀斯特地貌特征，融山、水、石、林、泉、洞为一体，集雄、奇、幽、险、秀于一身，是极具开发价值的自然风景区。随着我国"体验式"旅游的快速发展，金佛山将成为国内知名的都市近郊型山地休闲度假旅游胜地。

天星小镇紧邻金佛山西大门，属于景区的商业配套项目，它占地约 1.8 平方公里，建筑面积约七十万平方米，共分为三大部分：第一，天星商业街；第二，五星级酒店及高端度假别墅区；第三，山地生态主题度假区。建成后天星小镇将成为集"享受自然、休闲度假、文化风情"于一体的、具有健康、养生、休闲和文化品牌效应的——原乡生活小镇。

"巴渝文化"是长江上游最富有鲜明个性的民族文化之一，它起源于大山大川之间，大自然的熏陶和蜀国的历史沿革共同造就出了一种顽强、坚韧、率真的地域性文化。天星小镇的建筑设计体现了巴渝"乡土聚落"所特有的"多样统一、别开生面、诙谐生动、形态飘逸"的建筑特色。借鉴巴渝古镇传统商业空间形态，设置了"街、巷、楼、场"等公共活动场所；同时，为了适应现代社会度假生活的需要，在传统古镇的线性空间形态上又加入了"院、径、廊、台"等多种休闲空间模式。

天星小镇的文化内涵是古典的、它的空间导向是舒缓休闲的、是让人恍惚时光倒流的……

Located in Nanchuan District of Chongqing, JinfoMountain enjoys obvious features unique to Karst landform that integrates mountains, waters, stones, forests, springs and caves with their own magnificence, uniqueness, boldness and beauty. It is a valuable natural scenery spot. With the development of "Experience Tourism " of China, JinfoMountain will soon become a domestically famous tourism spot that is not far away from urban areas.

Tianxing Small Town adjoins to the west of JinfoMountain, serving as a commercial supporting project of the scenic spot. Covering an area of about 1.8 sqkm and occupying a construction area of about 700,000 sqm, it is divided into three parts. First,Tianxing Commercial Street; second, Five-star Hotel and high-end holiday villa zone; third, mountainous ecological entertainment zone. After completion, the Tianxing Small Town, which enjoys such functions as healthcare, entertainment and culture, will become a livable town with original features that integrates beautiful natural conditions, entertainment facilities and unique cultural environment.

"Ba-yu Culture" is one of the most unique national cultures on the upper reaches of Yangtze River. It originates from mountains and rivers and features stubbornness, toughness and frankness under the influence of natural conditions and history of Shu kingdom(211-263 AD). The construction design of the Small Tianxing Town manifests a diversified, innovative, vivid and smooth style unique to rural tribe. By taking advantage of its traditional commercial space design, streets, alleys, mansions and squares and other similar public spaces are set. At the same time, yards, tunnels, corridors and platforms and the like are added to the linear space unique to traditional town.

The culture of Tianxing Town is classical; the spatial orientation of it is smooth and comfortable and reminiscent.

上海国际汽车博览公园
Shanghai International Automobile Fair Park

项目名称：上海国际汽车博览公园
用地规模：78.05 万平方米
建筑面积：151.45 万平方米
容积率：1.95
项目地点：中国 / 上海
编制时间：2003 / 06

上海国际汽车博览公园是为国际著名汽车娱乐投资商发展汽车娱乐项目而提供的舞台；是上海国际汽车城和安亭新镇未来的城市中心绿岛；是人工和自然交错的充满情趣的休闲所在；是尽最大可能对当地生态环境和自然资源的还原；是地区性防洪蓄洪的重要设施。

建设上海国际汽车博览公园对改善安亭新镇的城市环境，提升居民的生活质量，促进土地开发和升值都具有十分重要的意义。

公园的设计将把最具吸引力的汽车文化的主题展示蕴藏于自然生态公园的绿色休闲之中。将技术、艺术与科普、娱乐相结合，最大限度地满足人们对过往自然痕迹的留恋和对新技术的向往，通过一系列经过精心安排的观赏性、参与性活动全面了解汽车文化、汽车结构以及各种驾驶知识。不仅如此，公园的设计师和建造者在满足公园整体规划的同时，预留最具灵活性的发展用地用于汽车博览会展中心的整体规划和建设，将充满情趣和现代气息的建筑和丰富的植物生态组织相结合而创造一个活力四射、充满情趣的集娱乐、休闲、科普为一体的专业化主题综合公园。这个公园的建成将进一步优化上海国际汽车城的城市空间结构，促进和完善上海国际汽车城的功能结构，创造良好的综合投资环境，有效地改善当地的生态环境。

Shanghai International Automobile Fair Park is invested by the government of Shanghai Jiading District Shanghai International Automobile City and Anting United Development company. It is located in the core area of Shanghai International Automobile city and is the future green Island of Shanghai International Automobile city and Anting new town, It interlaces artificial and natural leisure and it is the maximum extent to restore local ecological environment and natural resources. It is also a critical facility of regional flood control.

Park design combines the most compelling topics-car culture with the green natural ecological park. The combination of science and technology, art, entertainment, best satisfies people's memory of the past and yearning of new technology. Through a series of carefully arranged ornamental and participatory activities, people will be able to have a comprehensive knowledge of car culture, car structures, as well as driving skills. At the same time, to meet overall planning of the park, designers and builders have reserved the most flexible development space for overall planning and construction of Automobile Exhibition Centre, combining interesting and modern architecture with rich ecology organizations, creating a vibrant and interesting comprehensive park with entertainment, leisure, science. The park is built to further optimize urban spatial structure of Shanghai International Automobile city, promoting the function and structure of Shanghai International Automobile city, creating a good investment environment and effectively improving the local environment.

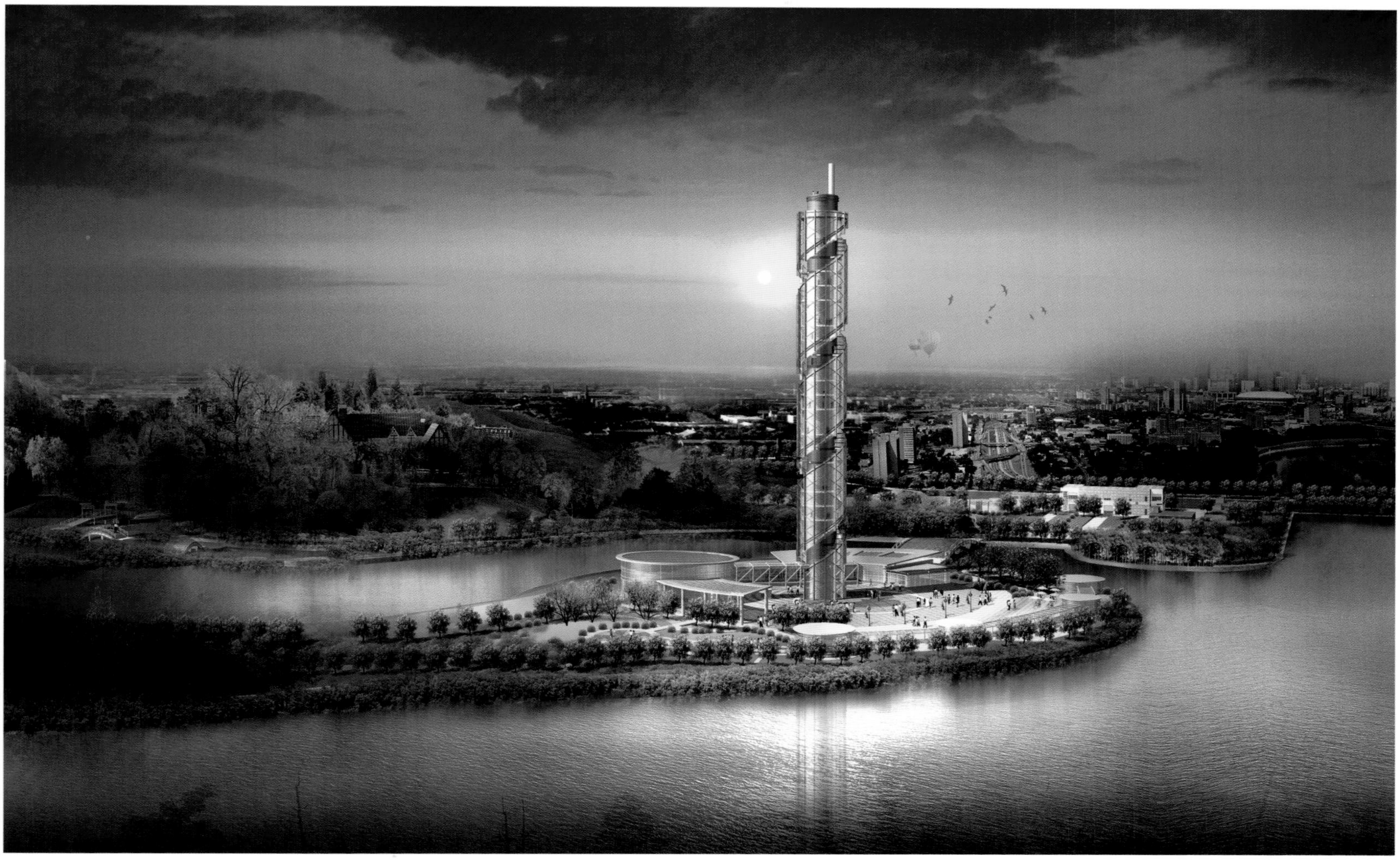

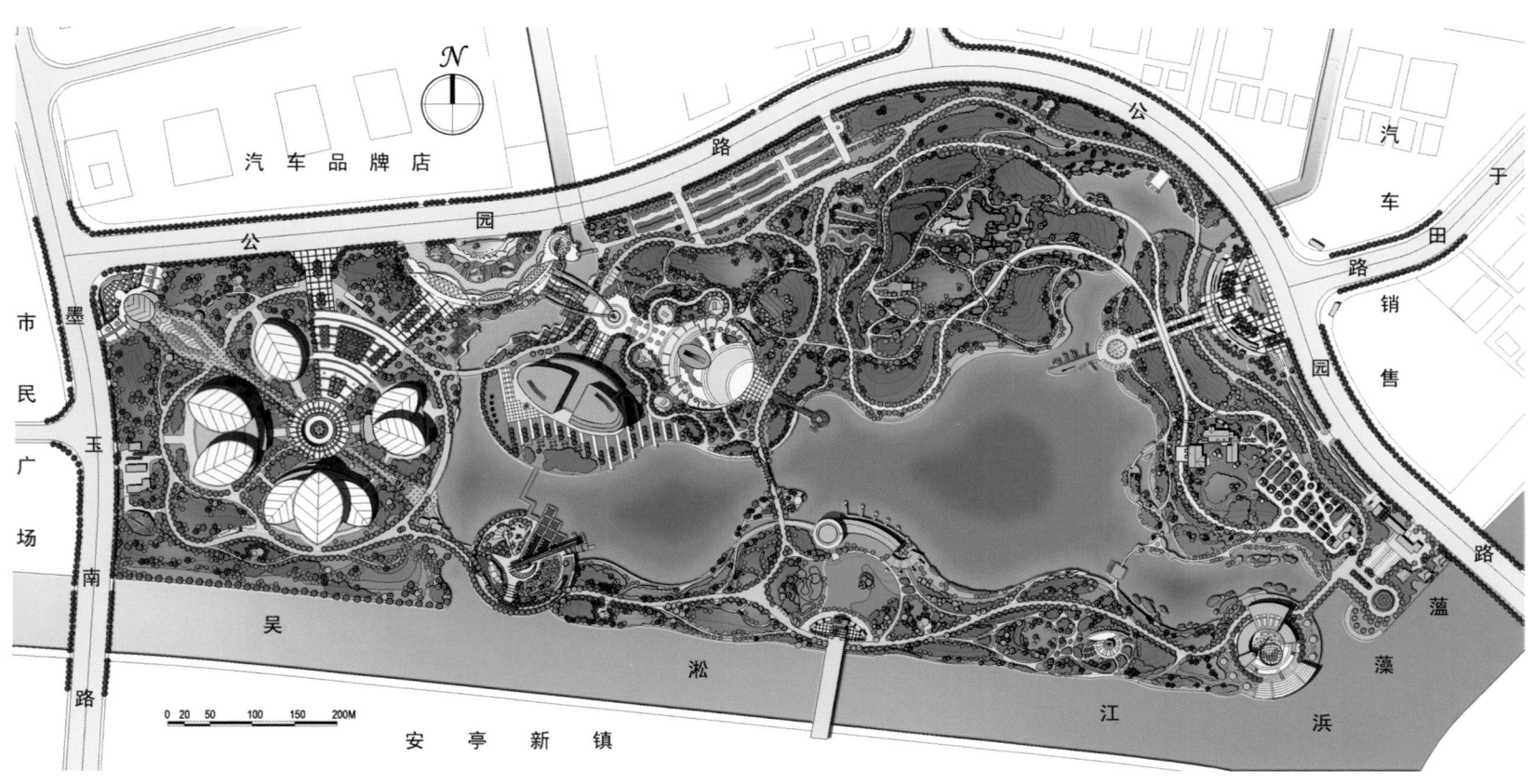

N
汽车品牌店
公园路
市民广场
墨玉南路
吴淞江
安亭新镇
0 20 50 100 150 200M
公路
汽车销售
于田路
园路
蕴藻浜

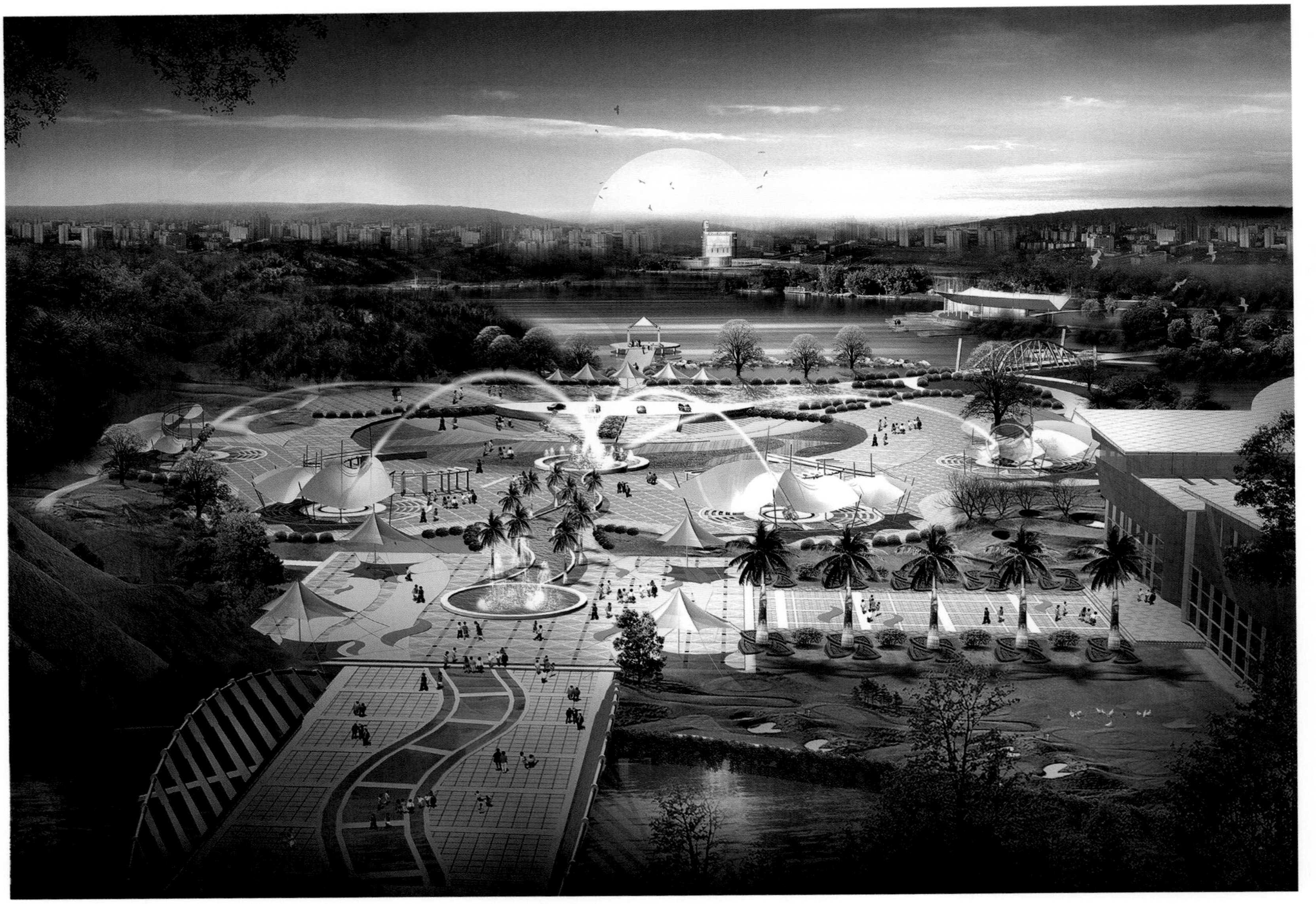

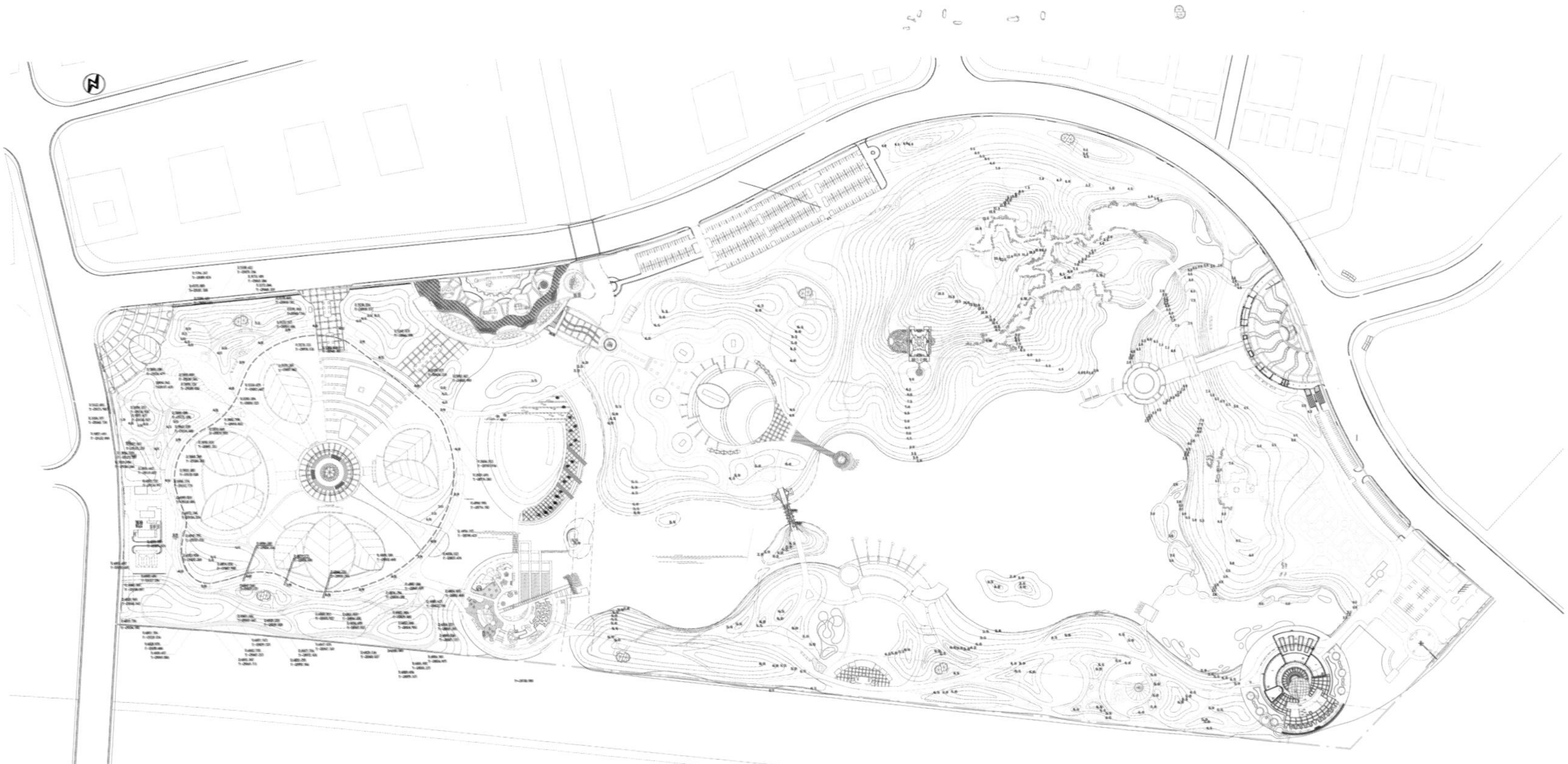

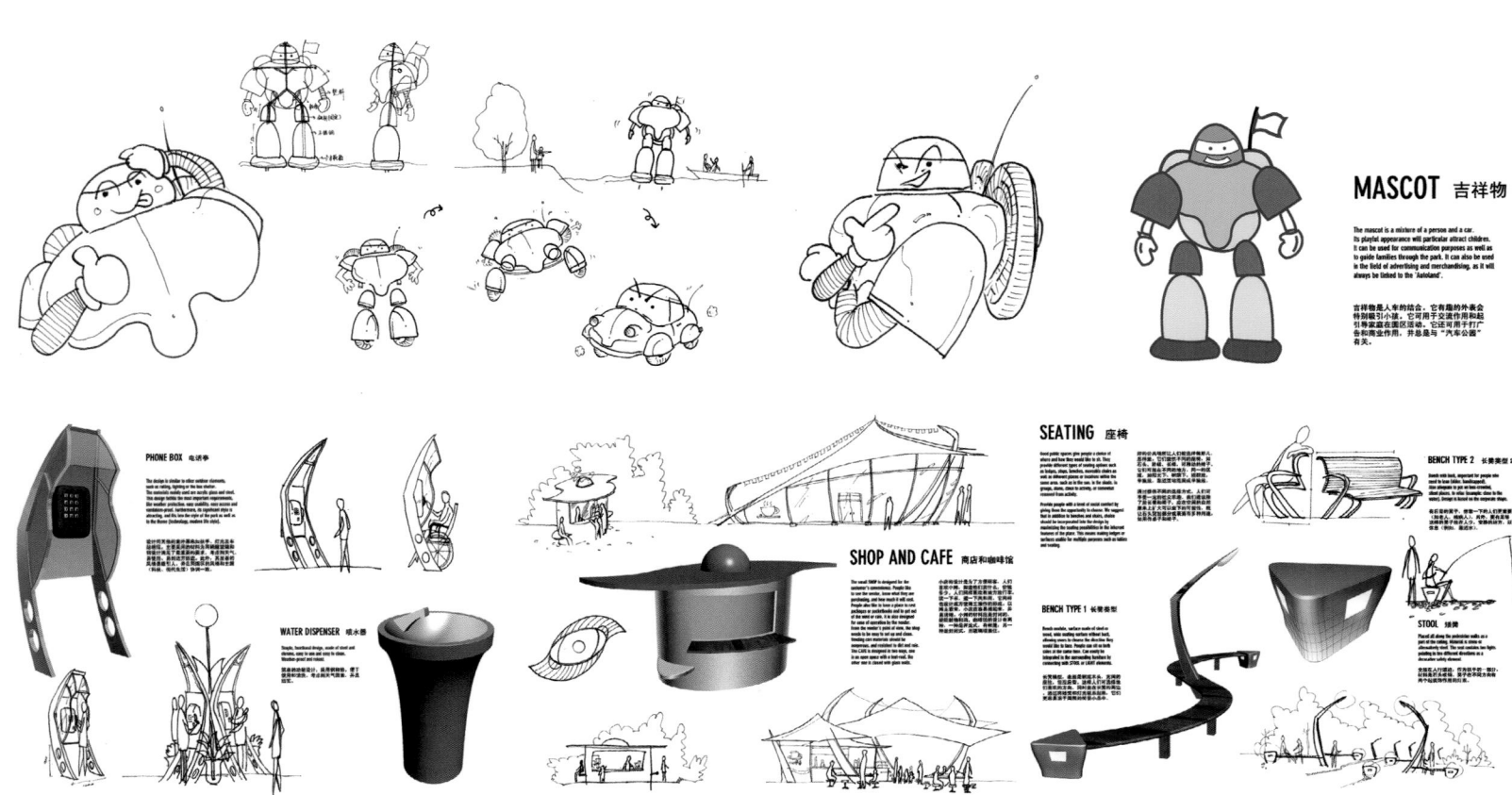

MASCOT 吉祥物

The mascot is a mixture of a person and a car. Its playful appearance will particular attract children. It can be used for communication purposes as well as to guide families through the park. It can also be used in the field of advertising and merchandising, as it will always be linked to the 'Autoland'.

吉祥物是人车的结合。它有趣的外表会特别吸引小孩。它可用于交流作用和起引导家庭游园区活动。它还可用于打广告和商业作用，并总是与"汽车公园"有关。

PHONE BOX 电话亭

The design is similar to other outdoor elements such as railing, lighting or the bus shelter. The materials mainly used are acrylic glass and steel, this design follow the most important requirement, the weather protection, easy usability, easy access and vandal proof, the phone box is attractive and characteristic.

WATER DISPENSER 饮水器

Simple, functional design, made of steel and stainless, easy to use and easy to clean. Weather proof and robust.

SHOP AND CAFE 商店和咖啡馆

The small SHOP is designed for the customer's convenience. People like to get the service, sauna and the relaxing and watching, and how much it will ask. People also like to look more at the same time they'll fetch the cinema, restaurant from the water's point of view, the shop needs to be easy to set up and clean. Seating are materials should be waterproof, and resistant to the weather, and is an open space with a hard wall, the other wall is closed with glass walls.

SEATING 座椅

bench public spaces give people a choice of where and how they would like to sit. They provide different types of seating options such as lumps, steps, benches, moveable chairs as well as informal pieces or furniture within the open area, such as in the sea, in the steps, groups, stairs, close to activity, or connected to another from activity.

Provide people with a trend of social comfort by giving them the opportunity to choose. We suggest that in addition to benches and chairs, choice should be incorporated into the design by maximizing the seating possibilities in the relevant features of the place. This means making ledges or surfaces usable for multiple purposes such as surface and resting.

BENCH TYPE 2 长凳类型 2

bench with back, exposed for people who need to take it easy, back supports, the designer to put on bent wooden seat places, in story, benches close to the water, under trees, in the restaurant, ...etc, Design is based on the corporate idea.

BENCH TYPE 1 长凳类型 1

Bench wooden, surface made of steel or wood, with lasting surface without back, allowing users to choose the direction they want to sit in, or use less, People can sit or both side at the same time, Can easily be integrated in the surrounding furniture in continuous parts of the pond, or child areas.

STOOL 凳

Placed all along the pedestrian walks as a part of the railing, intended to store a alternative seat. The seat contains low lights pointing in two different directions as a decorative safety element.

SMITHGROUP cna

Architecture · Urban Planning · Landscape Design
Interiors Design · Art Consultant · Lighting Design
Since 1853
www.cna-group.com

Design Determines . . .

设计决定······

05 商务办公
Business Office

上海绿地嘉创国际商务广场
Shanghai Greenland Jiachuang International Business Plaza

Architecture Design | 2007·06

上海绿地嘉创国际商务广场
Shanghai Greenland Jiachuang International Business Plaza

项目名称：上海绿地嘉创国际商务广场
用地规模：1.37 万平方米
建筑面积：5.48 万平方米
容积率：4.0
项目地点：中国 / 上海
编制时间：2007 / 06

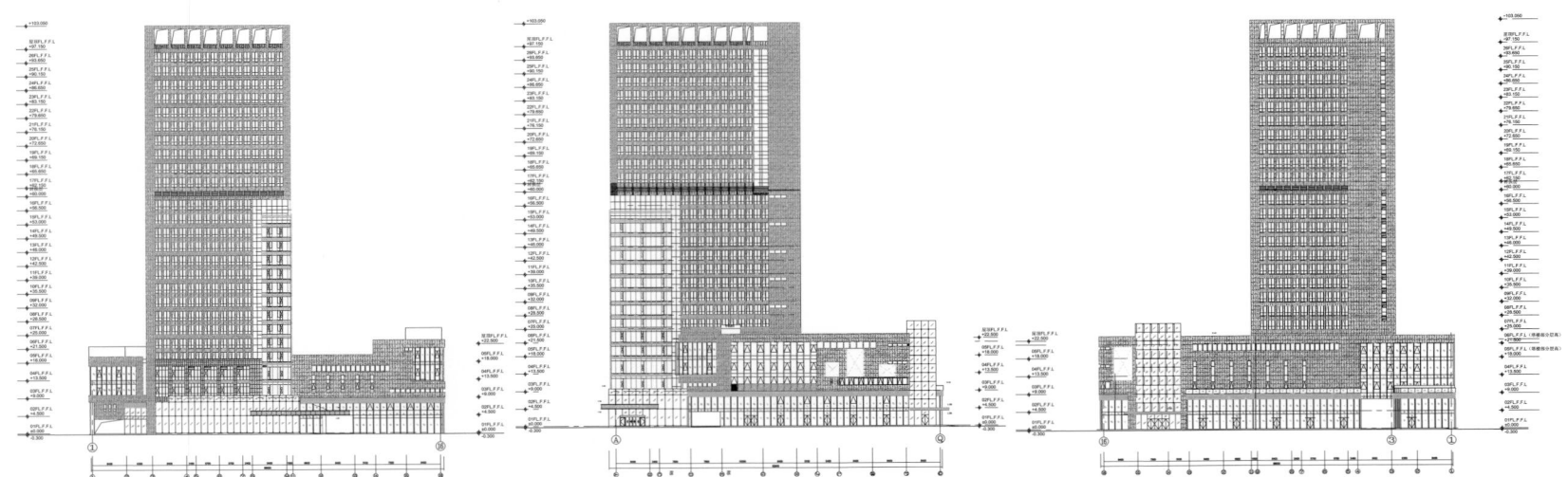

项目地块地处嘉定沪宜公路及叶城公路的交叉口。场地地形平坦，附近地块为多层住宅以及高层酒店。

塔楼成为地段内的标志性建筑群体。塔楼与商业辅楼在体量与建筑形式上共同起到了围合的作用。"L"形商业辅楼串联南面入口广场及西北面车行广场，在建筑内形成了清晰的流线，增加了商业面及商业价值。

1. 空间：裙房商业部分通过商业内街和放大的室外公共广场营造出舒适而丰富的空间变化。塔楼的竖向线条强化了主楼的标志性形象，无论在其基地中间还是地块外都从各个角度呈现出不同的变化。

2. 立面：裙房部分采用玻璃幕墙与石材挂板和铝板等高档材质，结合立面广告，营造一种现代、休闲、热闹的商业气氛。酒店主楼以干挂石材、铝板为主要材质，结合竖向条窗，强调竖向线条，突出主塔楼的标志性，整体效果优雅高贵，体现高档商务区的形象。

The project is located at the intersection of Shanghai-Yixing Road and Yecheng Road. The site is flat with multi-storey residential buildings and high-rise hotels around.

The tower building is the landmark of this region. The tower building and commercial building are designed in enclosed form. The "L-shape"commercial building connects the entrance square in the south and the vehicle flow square in the northwest. Clear flow lines are created inside the building to enhance business area and value.

The annex commercial buildings are designed to be comfortable with rich spacial change by the indoor atrium and the enlarged outdoor public plaza. The tower's vertical lines strengthen the iconic image of the main building. Different views can be obtained from different points no matter inside or outside it.

The annex buildings use glass curtain wall, stone, aluminum plate and other premium materials which create the modern, casual and prosperous business atmosphere together with facade advertisements. Stone and aluminum plates are main materials of the main buildings of the hotel. The vertical bar window emphasizes the vertical lines, thus highlighting the landmark of the main tower building. The overall effect is delivered to be grace and elegance, reflecting the image of the upscale business district.

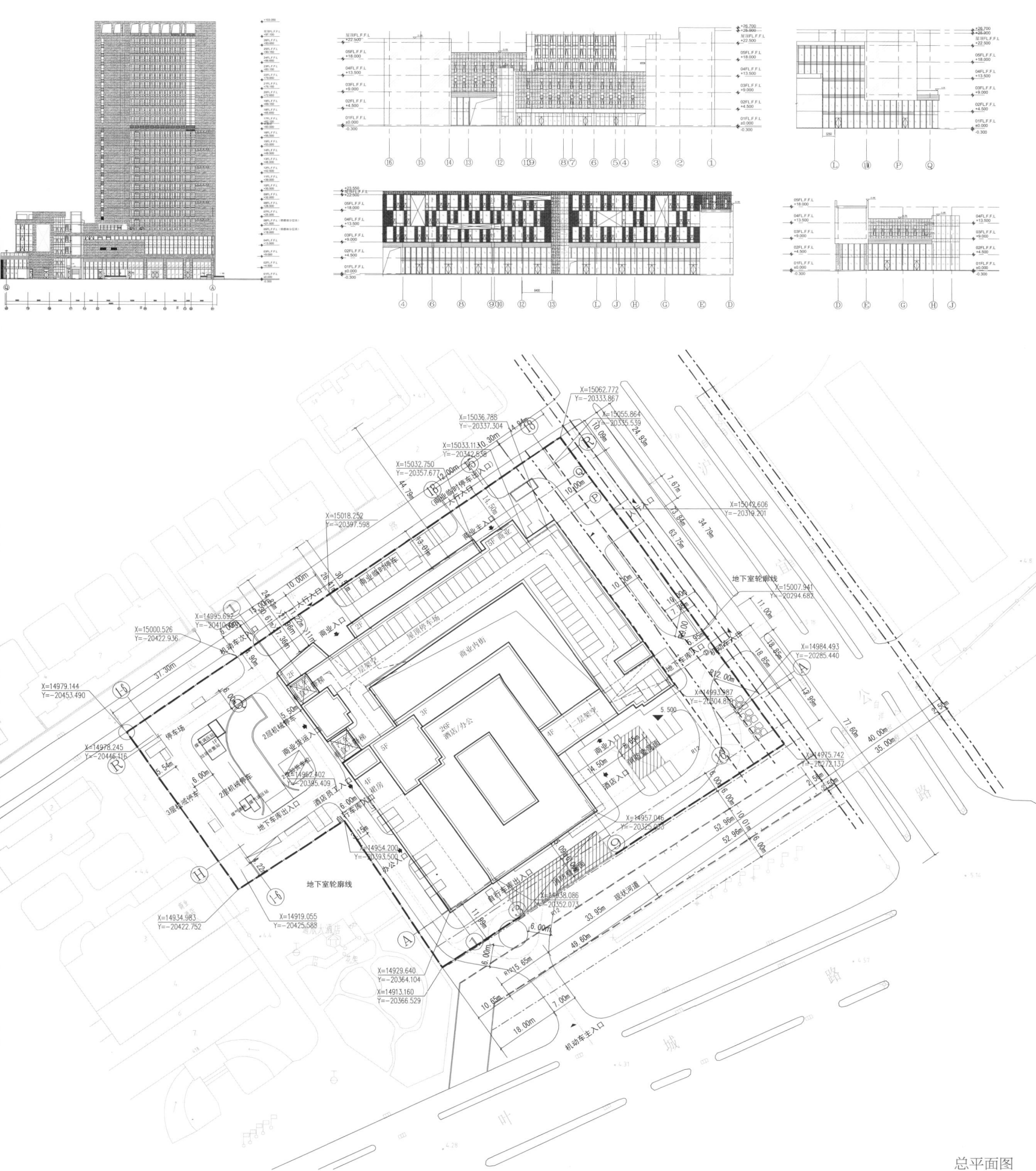

总平面图

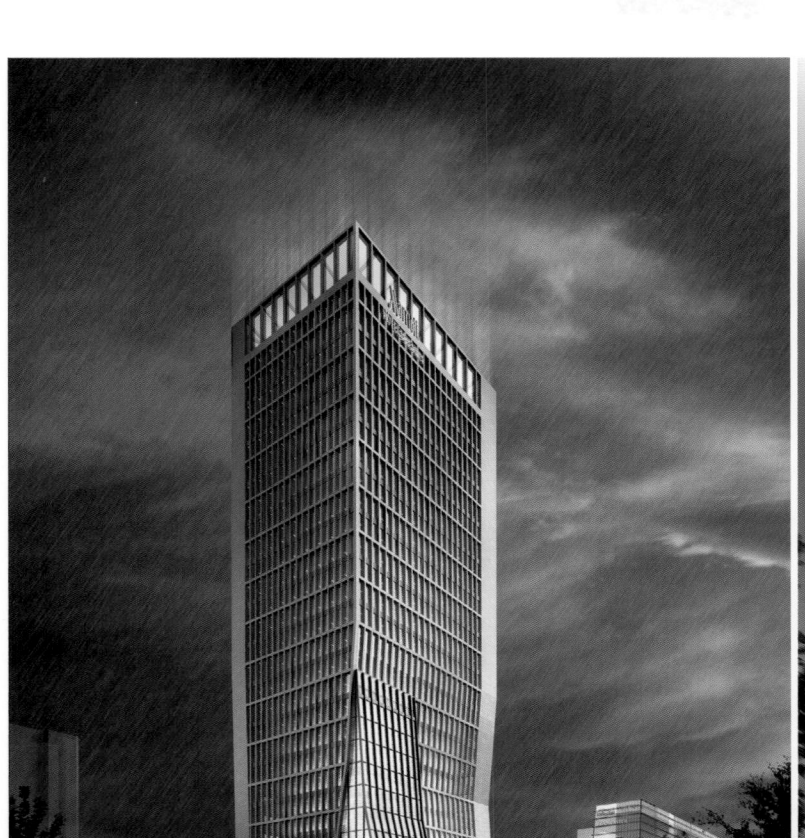

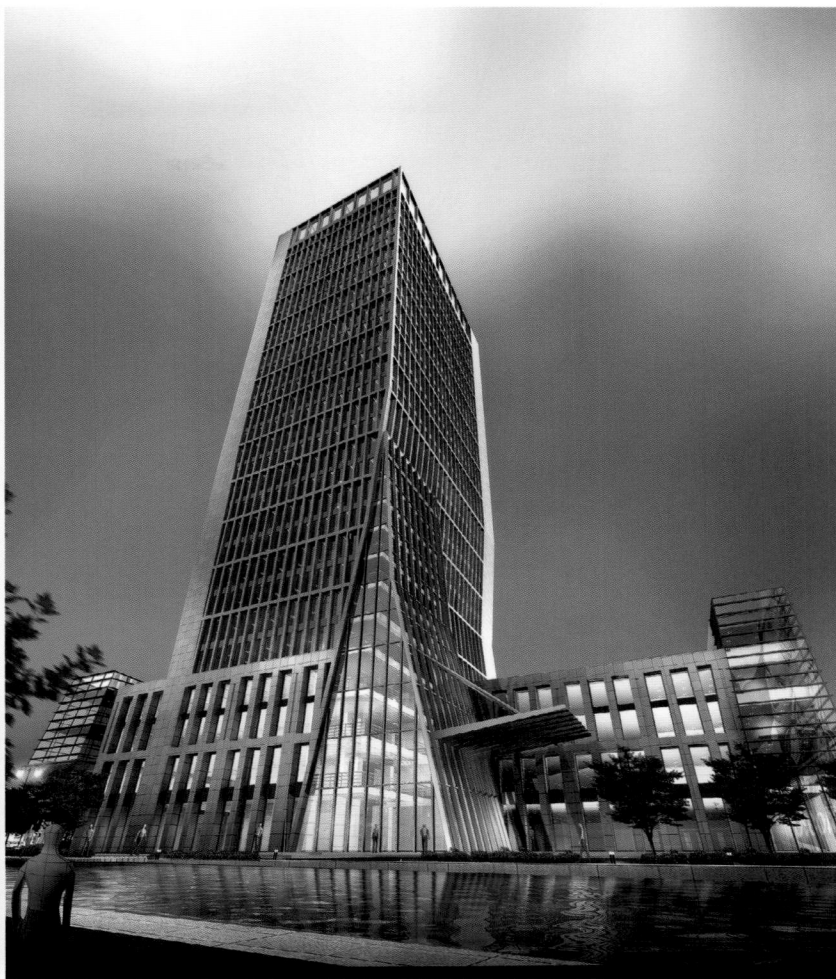

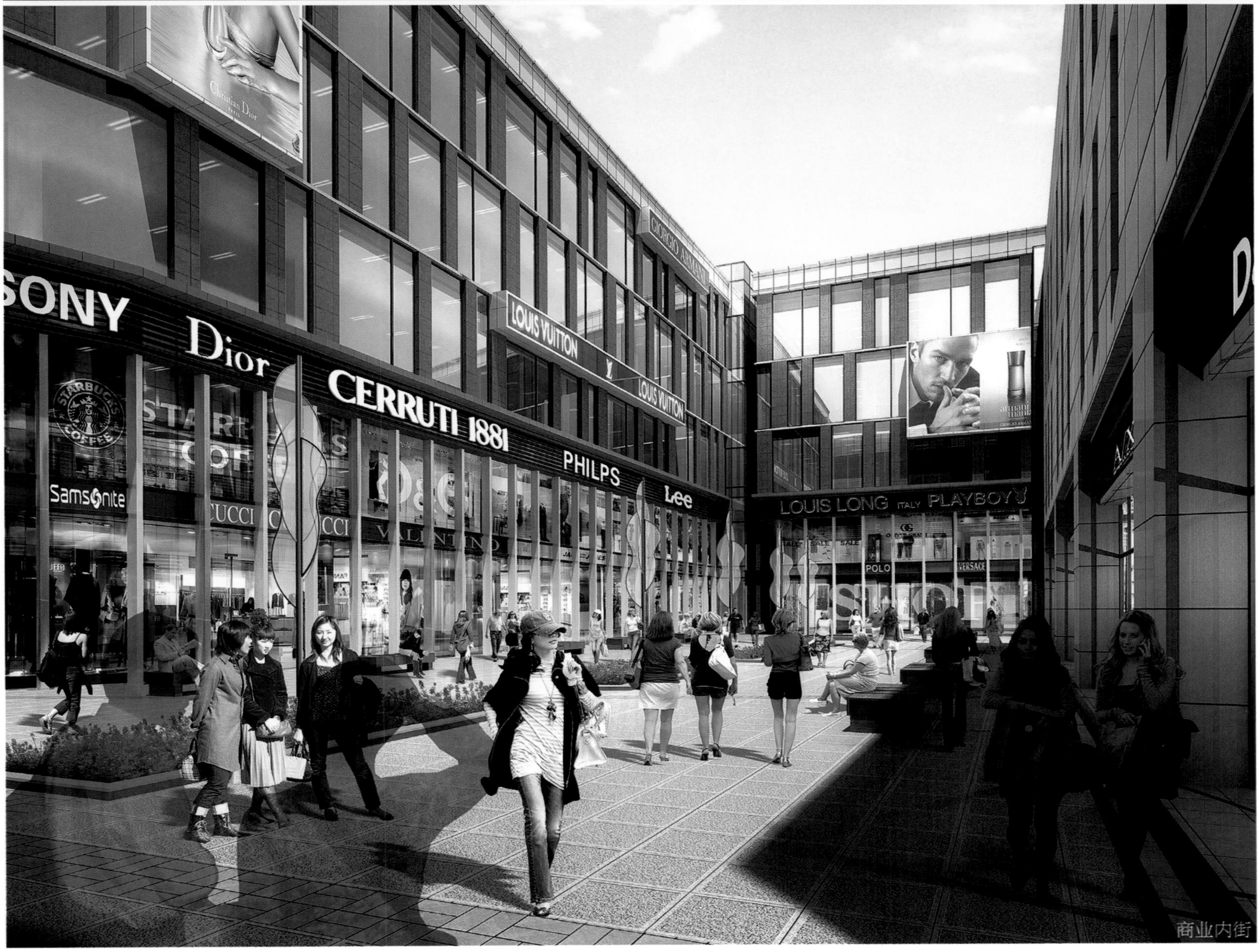
SONY
Dior
CERRUTI 1881
PHILPS
Lee
LOUIS VUITTON
LOUIS VUITTON
GIORGIO ARMANI
Samsonite
LOUIS LONG ITALY PLAYBOY
POLO
VERSACE

商业内街

中国农业银行重庆分行
Agricultural Bank of China Chongqing Branch

项目名称：中国农业银行重庆分行
用地规模：1.14 万平方米
建筑面积：4.95 万平方米
容积率：4.36
项目地点：中国 / 重庆
编制时间：2009 / 07

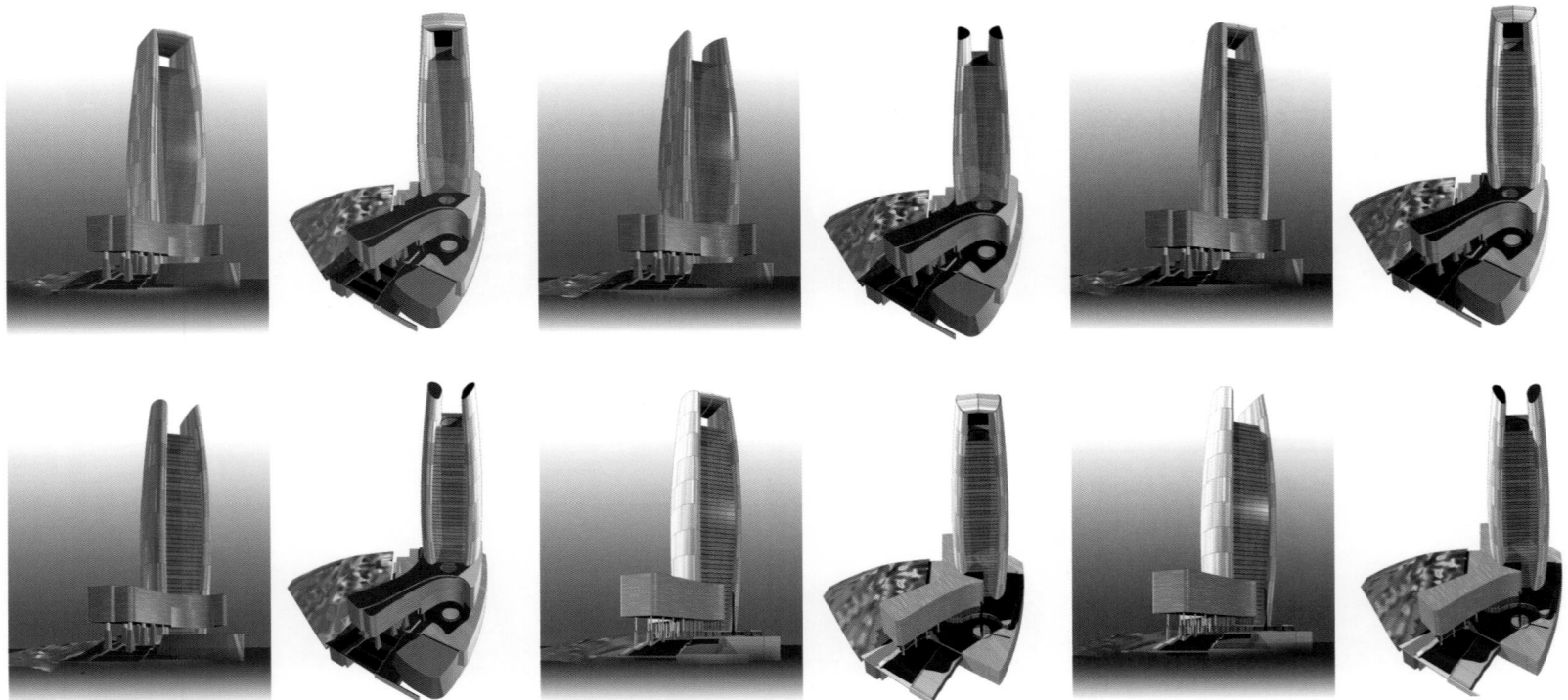

本项目基地位于江北区 CBD 的核心门户位置，地理位置极其优越。我们的设计理念是建造蕴含现代银行建筑理念与重庆文化特色的功能载体，使之成为 CBD 的一个新的带动点和标志性建筑，为城市建设增加新的亮点，并带动周边经济的发展。

设计结合对城市地理、城市规划、建造场地地形等的综合分析，从视线分析、采光通风、基地的整体感、功能分区与规划以及出入口等各方面，对建筑的整体布局和主要朝向进行深入的研究。

在全球金融系统处于动荡的恶劣环境时，银行尤其需要创造稳固而强大的形象。银行建筑必须传递出这样的信息：简洁而透明，安全而稳定，自信而负责任，可靠而持久；同时，海纳百川，平易近人，而又令人充满希望。

The project is located at a core portal place of Jiangbei CBD and enjoys extremely advantageous location. We design to create a functional carrier with both modern bank architectural idea and Chongqing cultural features and make it a new driving point and landmark in Jiangbei CBD. It is expected to add a new urban highlight and promote the surrounding economic development.

Through comprehensive analysis of urban geography, planning, construction site terrain and others, we make a careful study on the building's overall layout and major direction in combination of the line of vision, lighting, ventilation, overall sense of the base, function zoning and planning, as well as entrances.

While the global financial system is in a harsh state, the bank should especially create a solid and strong image. Bank building must deliver the following messages: simple and transparent, safe and stable, confident and responsible, reliable and durable; at the same time, comprehensive, amiable and presenting people with hopes.

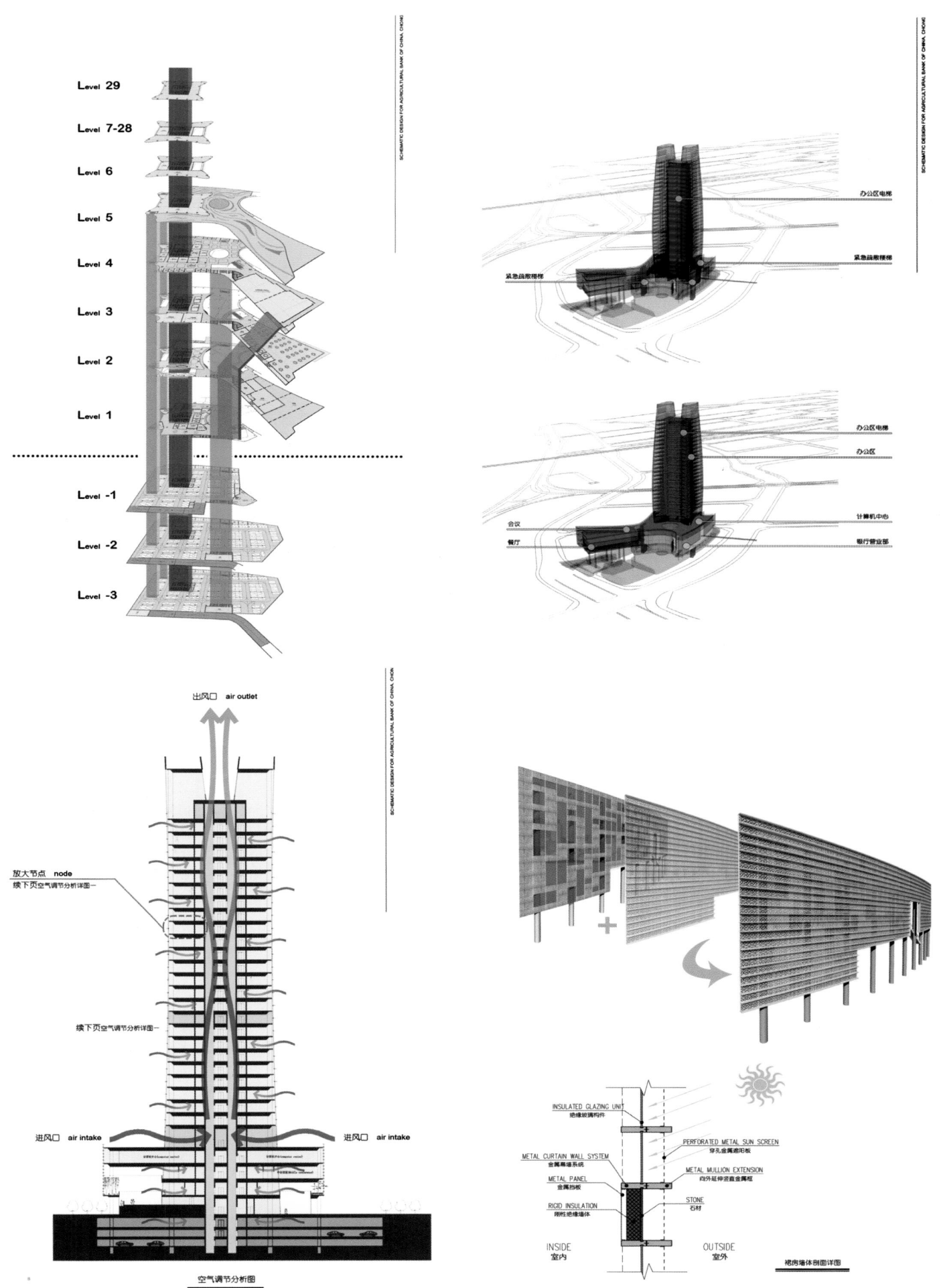

Level 29

Level 7-28

Level 6

Level 5

Level 4

Level 3

Level 2

Level 1

Level -1

Level -2

Level -3

办公区电梯

紧急疏散楼梯

紧急疏散楼梯

办公区电梯

办公区

会议

计算机中心

餐厅

银行营业部

出风口 air outlet

放大节点 node
续下页空气调节分析详图一

续下页空气调节分析详图一

进风口 air intake

进风口 air intake

空气调节分析图

INSULATED GLAZING UNIT
绝缘玻璃构件

PERFORATED METAL SUN SCREEN
穿孔金属遮阳板

METAL CURTAIN WALL SYSTEM
金属幕墙系统

METAL MULLION EXTENSION
向外延伸密直金属框

METAL PANEL
金属挡板

STONE
石材

RIGID INSULATION
刚性绝缘墙体

INSIDE
室内

OUTSIDE
室外

裙房墙体剖面详图

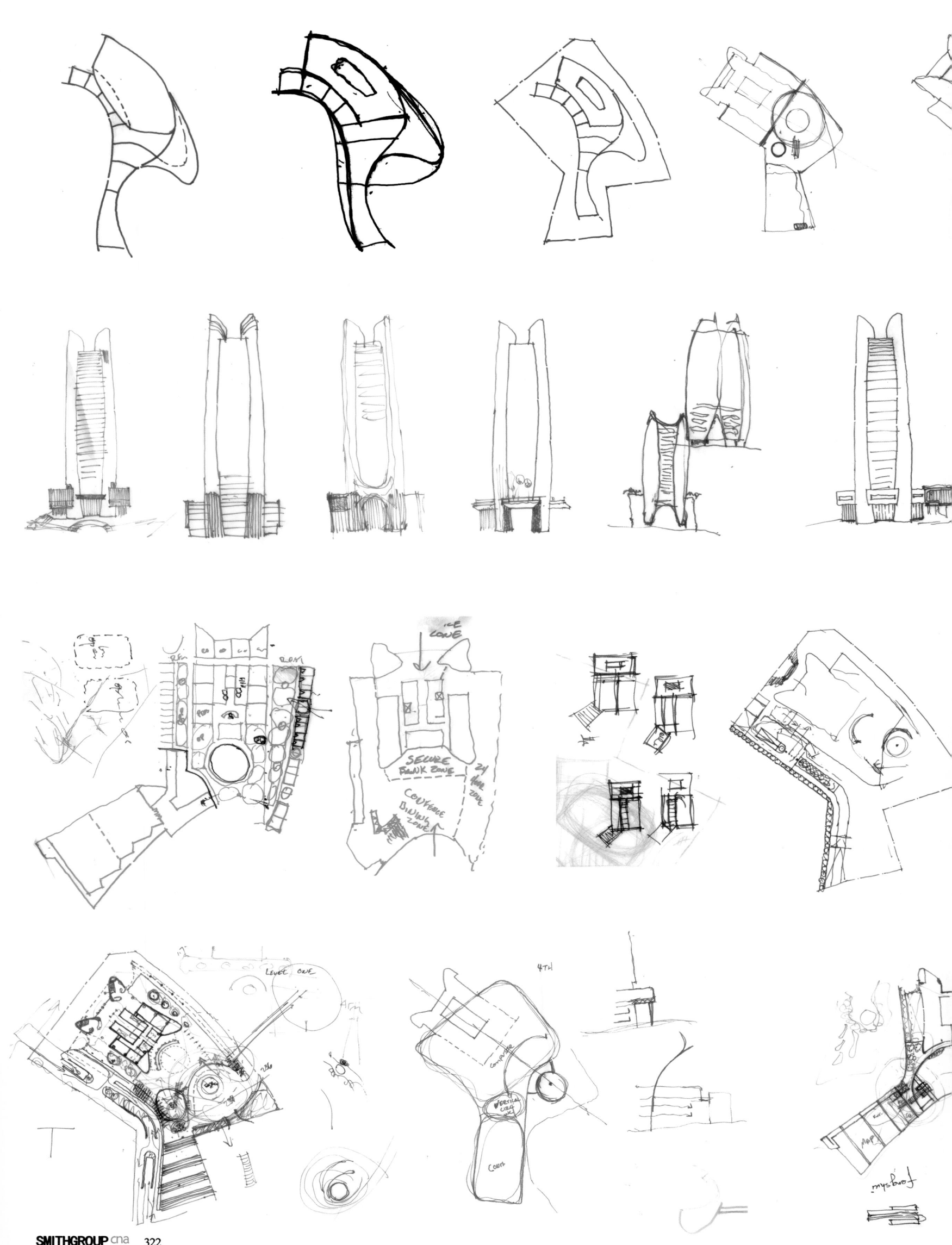

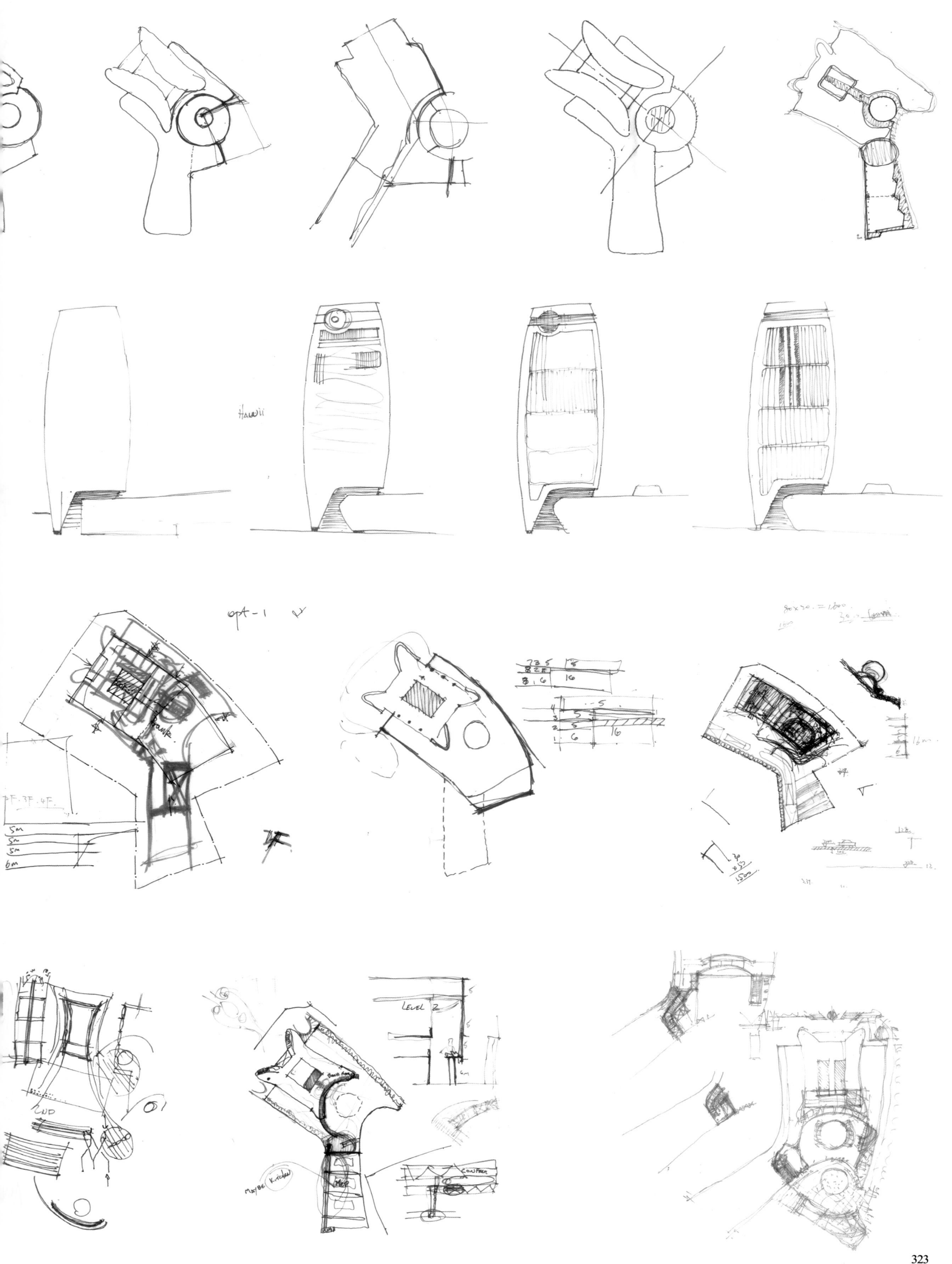

Hawaii

opT-1

LEVEL 2

323

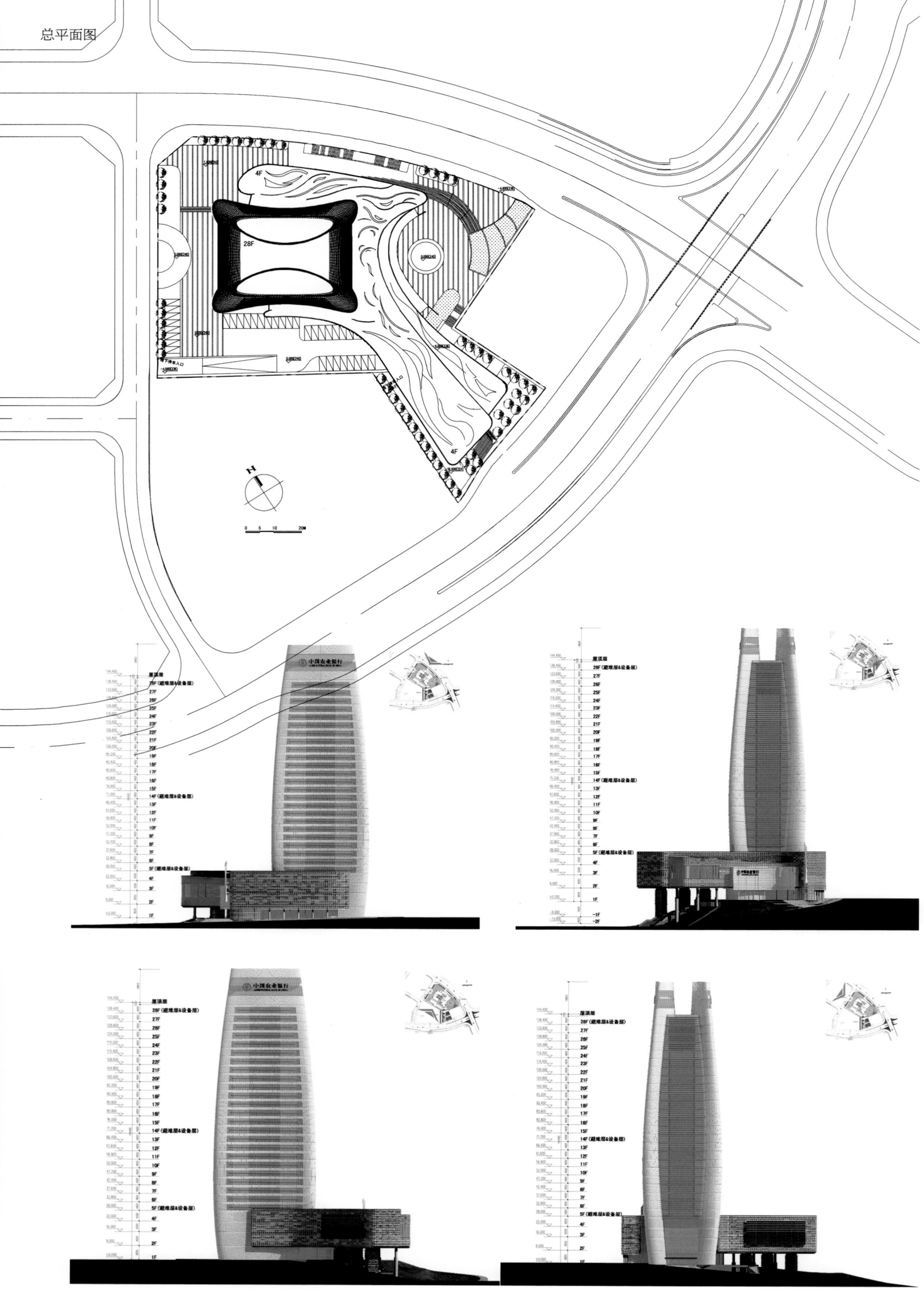

屋顶层
18F（避难层&设备层）
5F（避难层&设备层）
桂花街

屋顶层
18F（避难层&设备层）
5F（避难层&设备层）
江北城大街南路

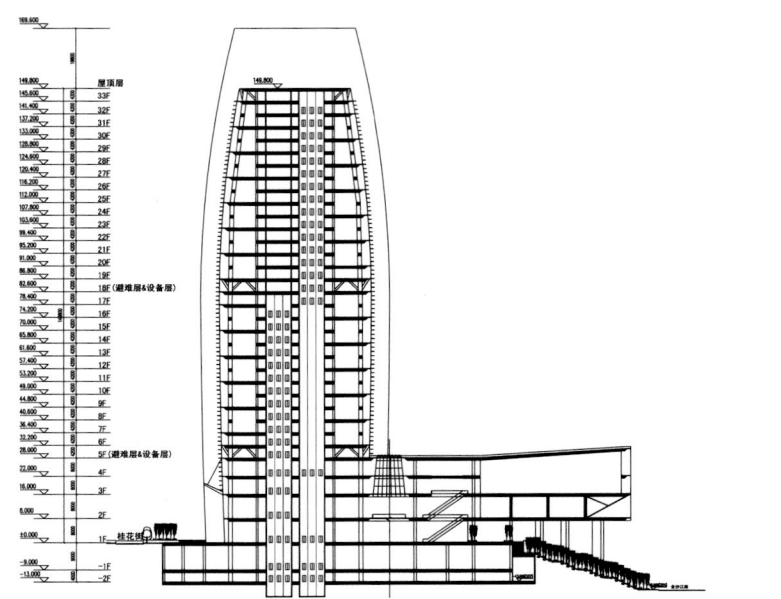

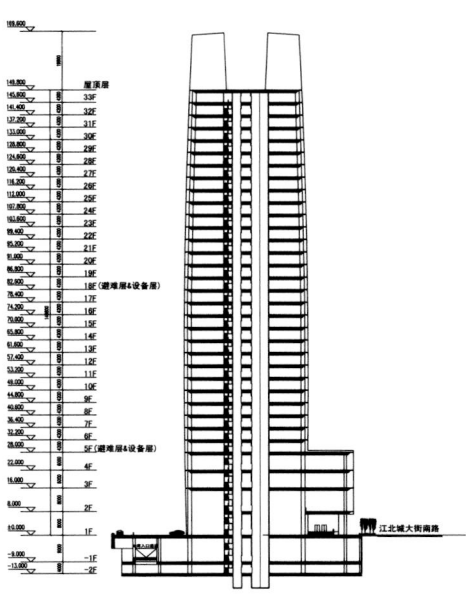

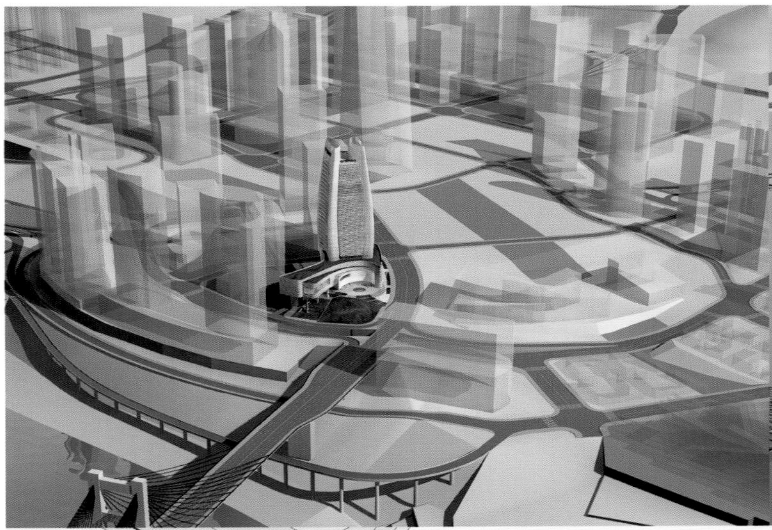

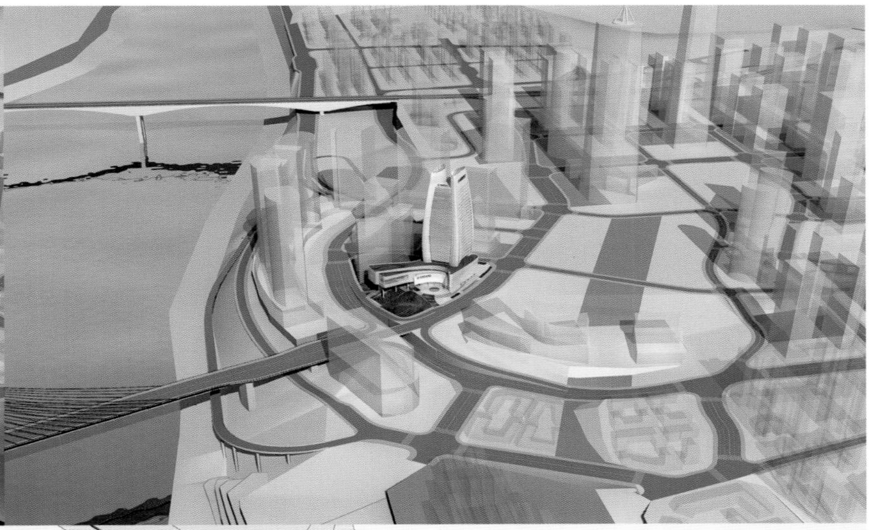

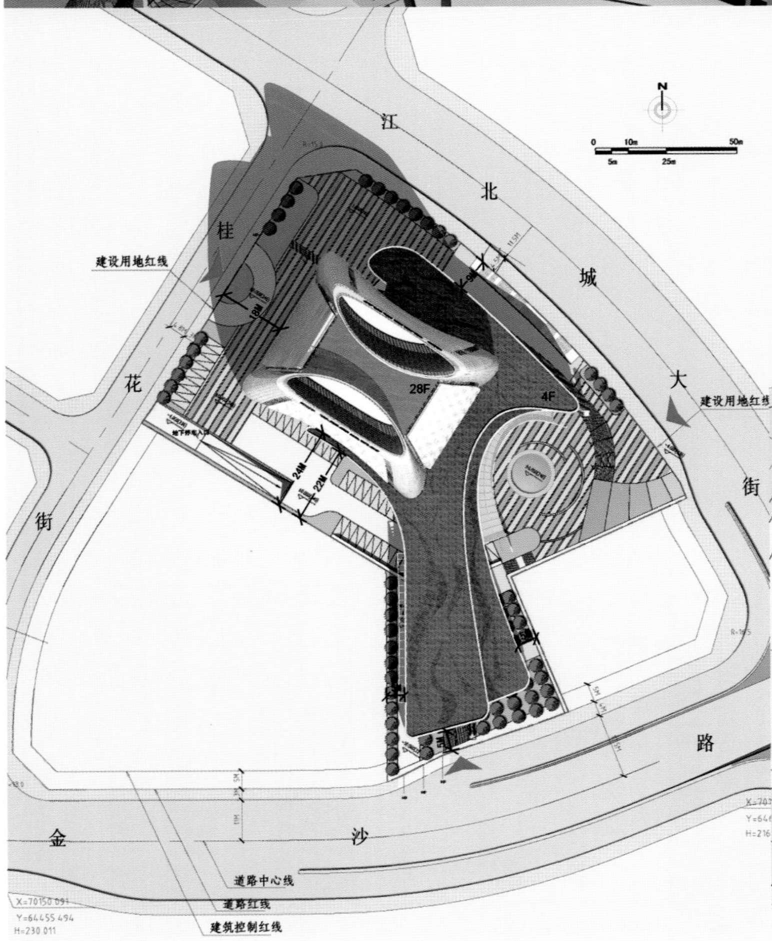

江
北
城
大
街
桂
花
街
金
沙
路
建设用地红线
28F
4F
道路中心线
道路红线
建筑控制红线

江
北
城
大
街
南
路
桂
花
街
金
沙
路
建设用地红线
道路中心线
道路红线
建筑控制红线

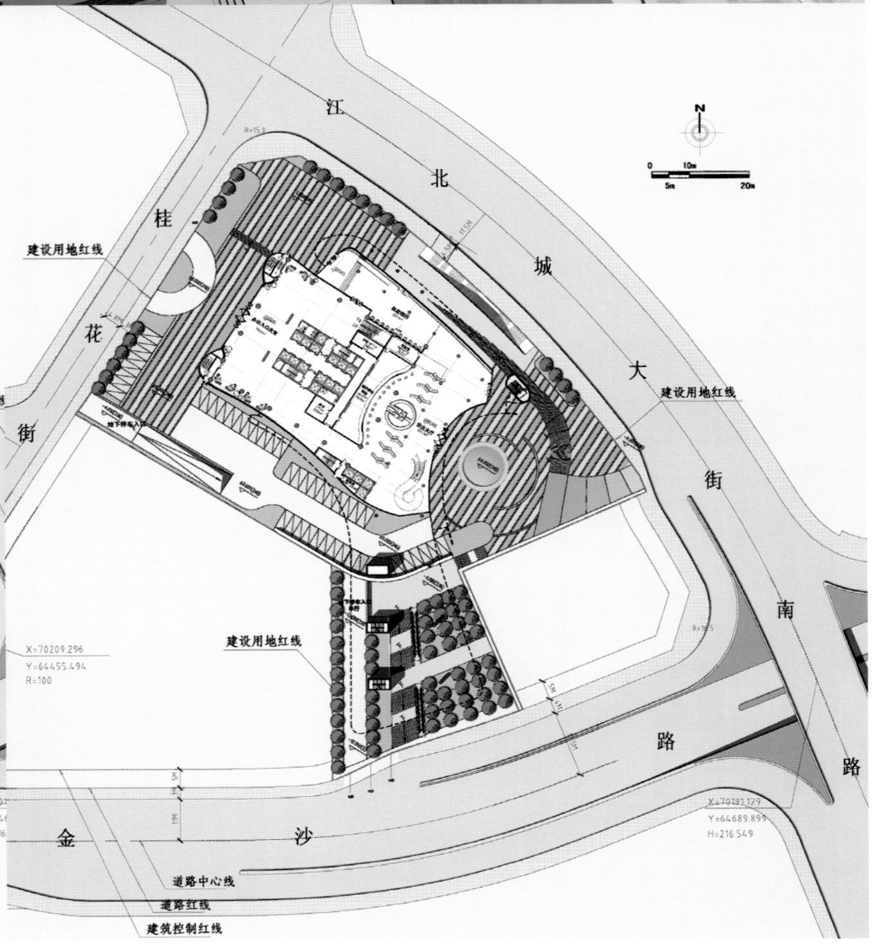

南京江宁天元路新城商业项目
Nanjing Jiangning Tianyuan Road New City Business Project

项目名称：南京江宁天元路新城商业项目
用地规模：1.20 万平方米
建筑面积：3.82 万平方米
容积率：3.2
项目地点：中国 / 江苏 / 南京
编制时间：2009 / 09

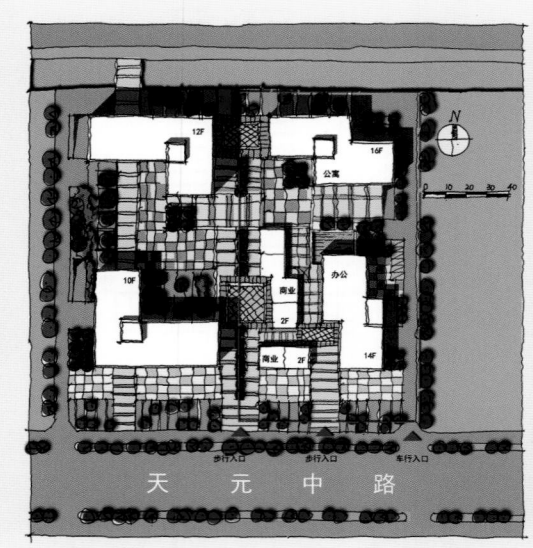
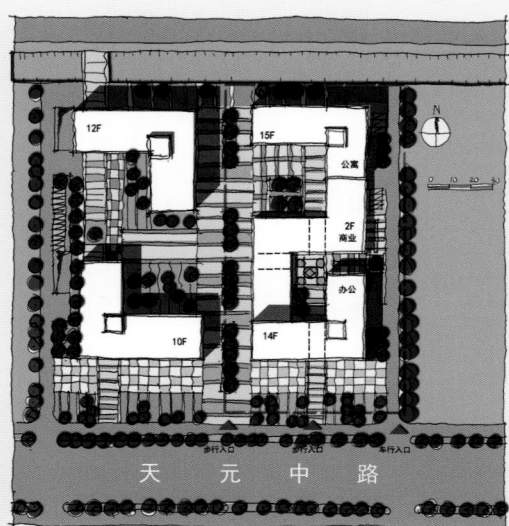
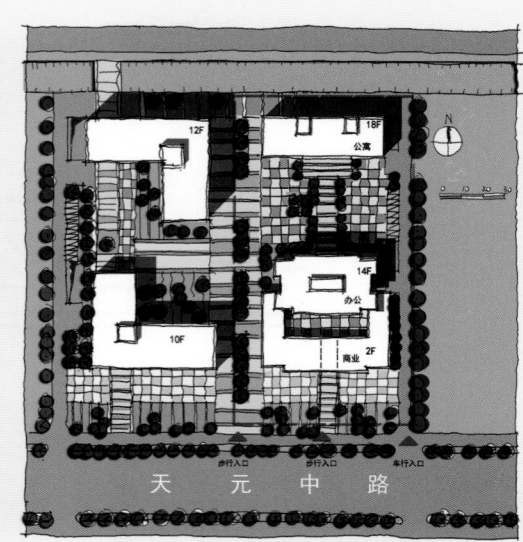

项目处于南京江宁开发区的天元中路，整个项目开发为复合业态的酒店式公寓、办公、商业等几部分，设计既需要满足基地内部大量办公和居住人群的消费需求，同时又要形成具有吸引力的区域形象，发挥区域主导作用，引导周边产业共同发展。

The project is located in Nanjing Jiangning Development Zone on Middle Tianyuan Road. The entire project is positioned as a complex format, including hotel-style apartments, office and commercial parts. The design can not only meet the consumer demands of a large office and residential population inside the base, but also can form attractive regional image so that the region can play a leading role and guide the common development of peripheral industries.

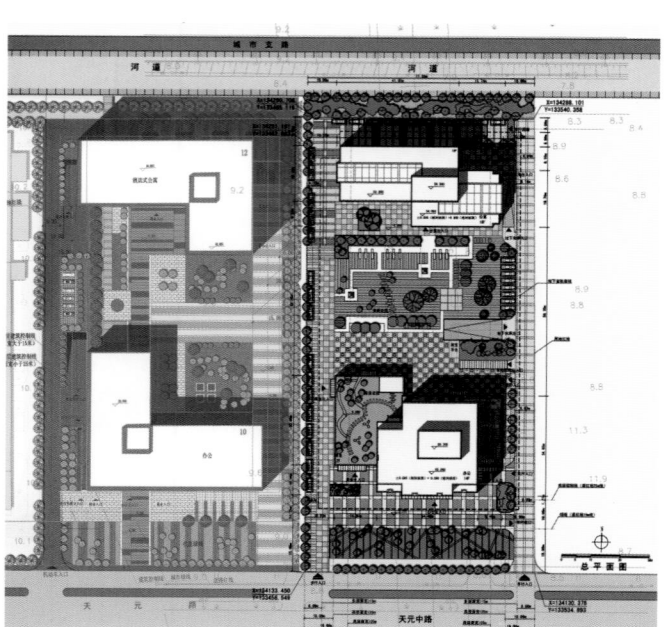

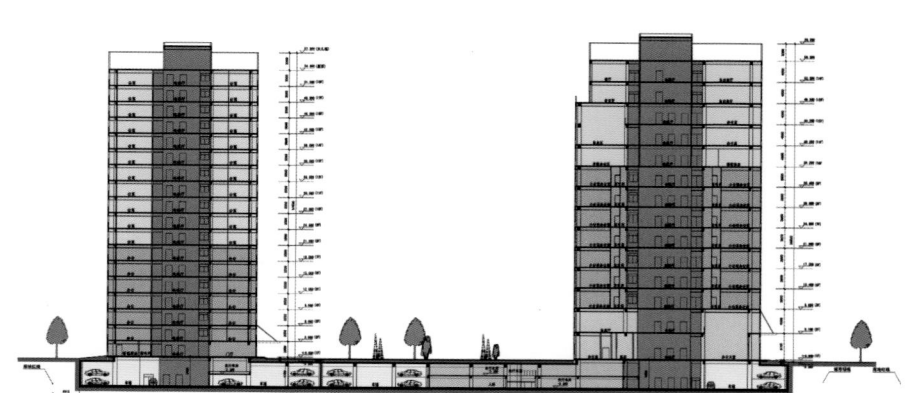

南通益兴集团大厦
Nantong Yixing Group Building

项目名称：南通益兴集团大厦
用地规模：1.01 万平方米
建筑面积：4.75 万平方米
容积率：3.6
项目地点：中国 / 江苏 / 南通
编制时间：2010 / 04

项目地处江苏南通市开发区能达商务区，位于规划经九路、纬十路道路交叉口的东北处，用地西侧为绿地公园，东侧为市政广场。建筑以自然、清新的现代风格为主，体现现代性、前瞻性；设计需要考虑作为一幢多功能建筑内部的功能使用合理，各部分交通流线清晰，以及建筑造型与城市环境形成的对话等等。

The project is located in Nengda Business Zone in Nantong Development Zone, Nantong, Jiangsu. It is in the northeast of the intersection of Jingjiu Road and Weishi Road with a green land park in the west and the municipal plaza in the east. The building is in natural and fresh modern style, modern and forward-looking; rational building functions and clear flows are ensured to achieve harmony between the building shape and the urban environment.

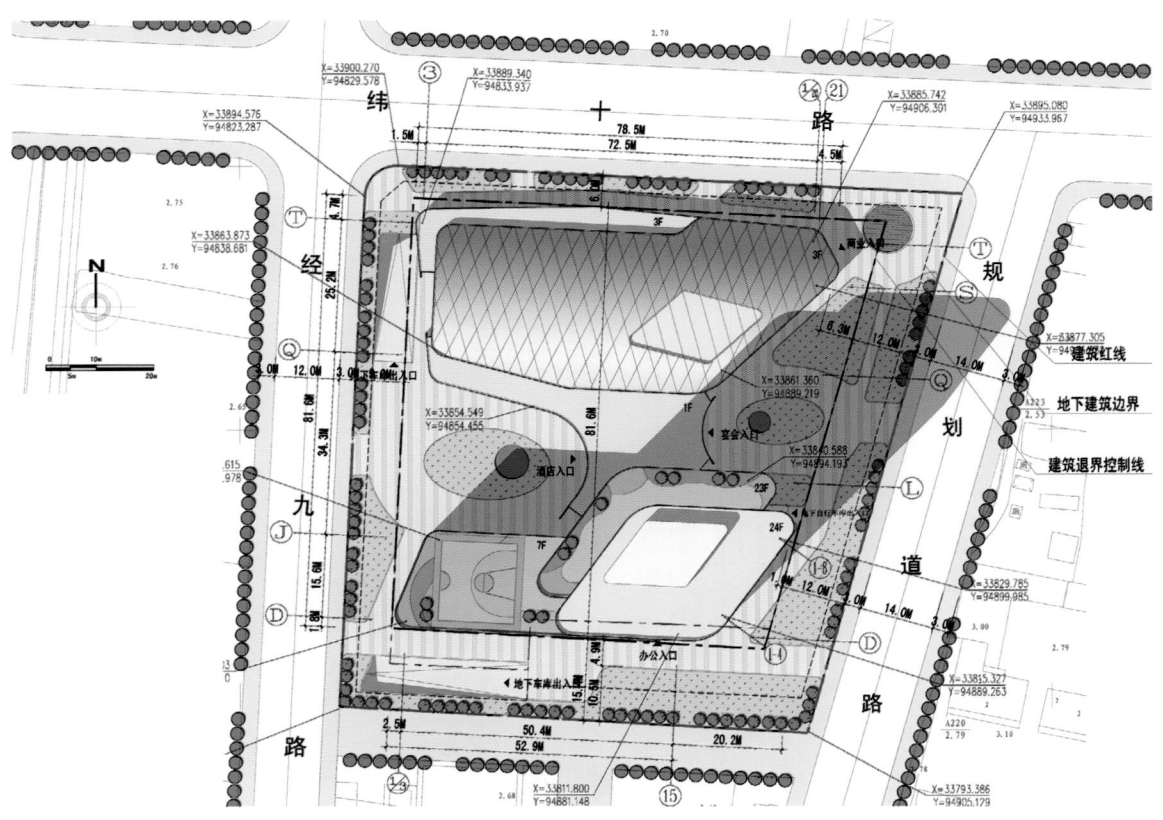
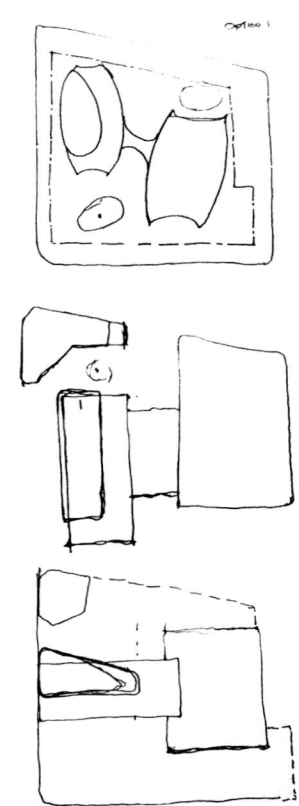

山西煤炭大厦
Shanxi Coal Building

项目名称: 山西煤炭大厦
用地规模: 6.24 万平方米
建筑面积: 12.42 万平方米
容积率: 1.99
项目地点: 中国 / 山西 / 太原
编制时间: 2007 / 05

项目位于太原市国家高科技产业区的火炬创业园主轴线的东部端点，是集会展、办公、酒店、研发、交易等多功能为一体的综合性大厦。项目由两个塔楼及一个裙房三个部分组成，通过对酒店、办公、交易大厅及其他配套功能整合轴线控制，达到主塔楼环抱裙房并对整个开发区主轴线进行端点控制的效果。

The project is at the eastern endpoint of the main axis of Torch Pioneer Park of Taiyuan National High-Tech Industrial Park. It is an integrated building integrating convention, office, hotel, research and development, trade and other functions in one. The project consists of two tower buildings and an annex. Control the axis of the hotel, office and trade hall as well as other supporting functions to achieve the effect that the main tower building encircles the annex and has an end-point control over the main axis of the entire development zone.

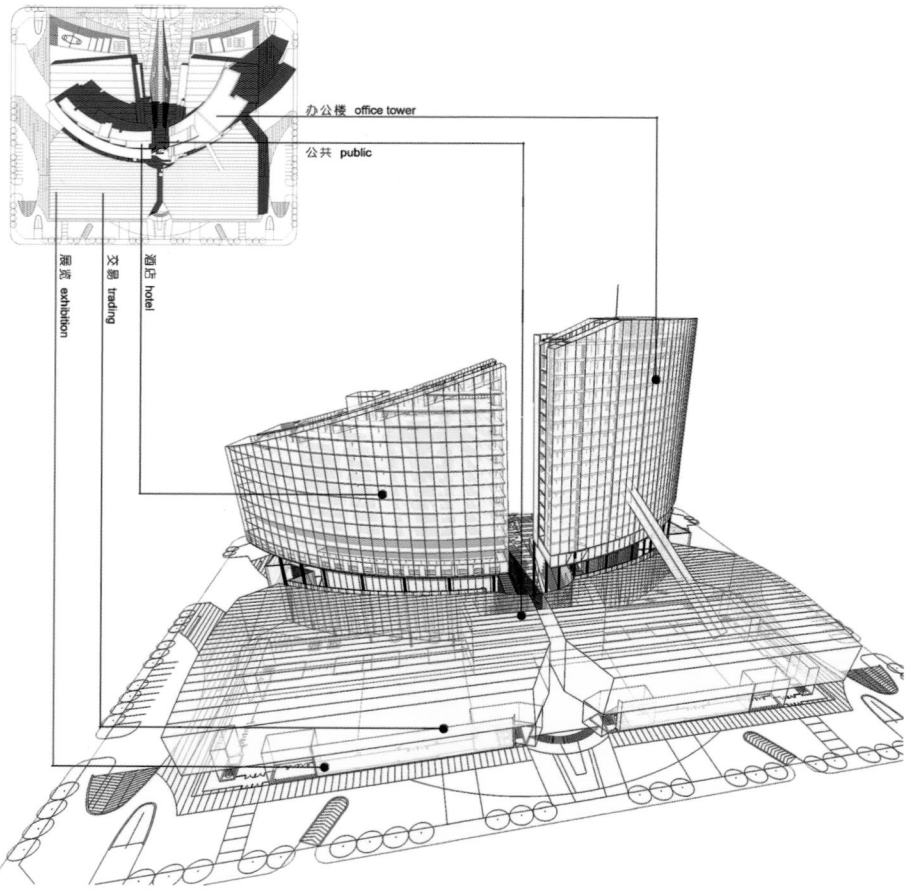

上海五角场 316 地块
Shanghai Wujiaochang 316 Plot

项目名称：上海五角场 316 地块
用地规模：1.26 万平方米
建筑面积：3.23 万平方米
容积率：2.55
项目地点：中国 / 上海
编制时间：2008 / 06

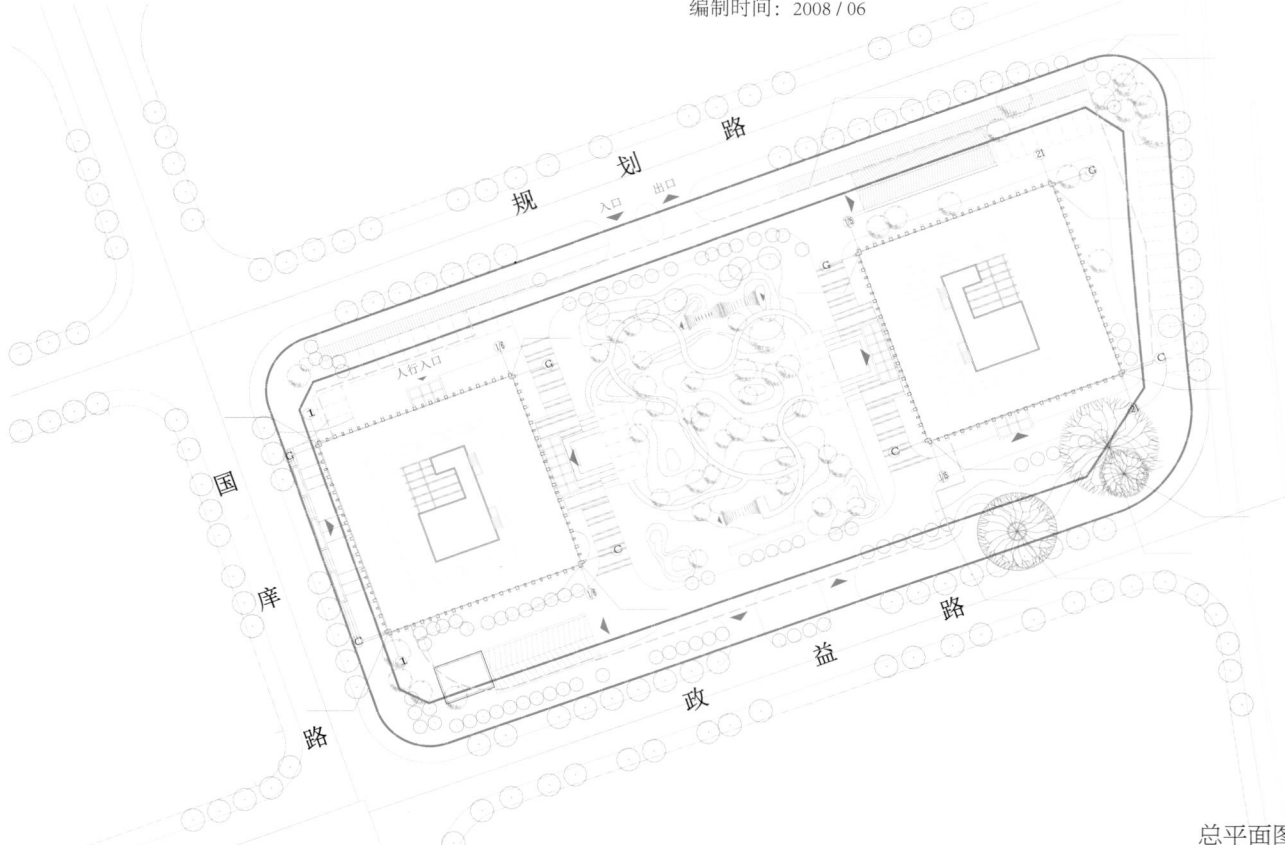

总平面图

本项目位于上海市五角场城市副中心的核心商业区，设计以简洁的手法处理建筑立面，但同时注重建筑的细节需求，用纯净的四边形体块和凹凸、排比构成的富有韵律的肌理效果，来创造优雅大方的艺术格调，构成完整高效的现代办公建筑群。

The project is located at the core business district of the sub-CBD in Wujiaochang, Shanghai. To cope with the dynamic architectural environment around Wujiaochang, simple facades are designed. Besides, the design also pays particular attention to details so as to create elegant artistic style by the clean quadrilateral body and the rhythmic texture effect. Thus, complete and efficient modern office buildings are created.

深圳移动生产调度中心大厦
Shenzhen Mobile Production Scheduling Center Tower

项目名称：深圳移动生产调度中心大厦
用地规模：0.56 万平方米
建筑面积：7.95 万平方米
容积率：14.20
项目地点：中国 / 广东 / 深圳
编制时间：2009 / 08

本项目位于深圳市中心中心广场东片区、深南大道与金田路交汇处东北角，用地东、北至市政规划路，西至金田路，南至深南大道，占地面积约 5 630.72 平方米。

设计定位

建造蕴含中国移动和深圳城市精神的功能载体，使之成为中心区新的标志性建筑。

设计策略

结合对城市地理、城市规划、建造场地地形等的综合分析，从规划关联、基地的整体感、功能分区、视线分析、以及出入口等各方面，对建筑的整体布局和主要方位进行深入的研究。

设计理念

百年基业，始于正气。无论是深圳鹏城这座伟大的城市还是中国移动这一企业巨擘，都应彰显出卓越、而非奇异的精神气质，大气方可经典恒久，潮流一时钟将湮没于时空。大厦的形象应与城市环境充分呼应，着重凸现企业文化。建筑由地上和地下二部分构成，其中地上建筑包括一座 36 层的塔楼及 7 层裙房；地下建筑为 5 层，主要为机动车停车库（部分兼人防工程）和设备用房。

The project is located in the eastern area of the Central Square in Shenzhen, at northeastern corner of the junction of Shennan Road and Jintian Road, reaching municipal planning roads on the north and the east, Jintian Road on the west and Shennan Road on the south. It covers an area of 5,630.72m².

Orientation

A carrier of comprehensive function combining the spirit of China Mobile and Shenzhen, becoming a new landmark in the central area.

Strategy

Combining all-round analysis of urban geography and planning with topography of construction site, in-depth investigation of buildings' integral layout, main position shall be commenced on the basis of relevance of planning, integrity, functional zoning, vision analysis and entrance and exit.

Concept

Hundred years of inheritance has begun with uprightness. Both Pengcheng and China Mobile reveal spirit of excellence, which makes them great and eternal. The building is the symbol of enterprise culture, which is in great harmony with the city. It is composed of two parts: one above the ground and the other underground, the former including a 36-storey main tower and a 7-storey podium.The 5-storey underground building is mainly used for vehicle parking garage (partly civil air defense works) and equipment room.

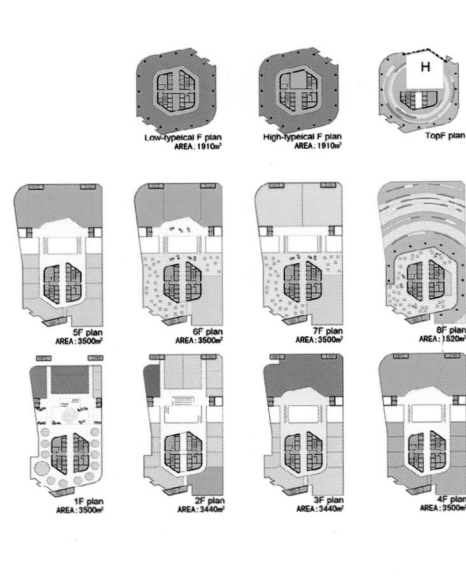

太原海鑫集团总部大厦
Taiyuan Haixin Group Headquarters Building

项目名称： 太原海鑫集团总部大厦
用地规模： 6.06 万平方米
建筑面积： 21.21 万平方米
容积率： 3.5
项目地点： 中国 / 山西 / 太原
编制时间： 2007 / 09

海鑫集团总部大厦项目位于太原市国家高新科技产业区的火炬创业园内，设计师用"一扇通往未来的门"的意念将办公，培训中心，展示中心，研发中心和五星级酒店等多功能集合为一体，筑成一栋崭新的综合性智能化大厦。

Haixin Group Headquarters Building is located at the Taiyuan High-tech Industry Park. The concept is a half-open door to the future. The building is a complex intelligent tower with various functions such as administration, training, exhibition, research and five-star hotel.

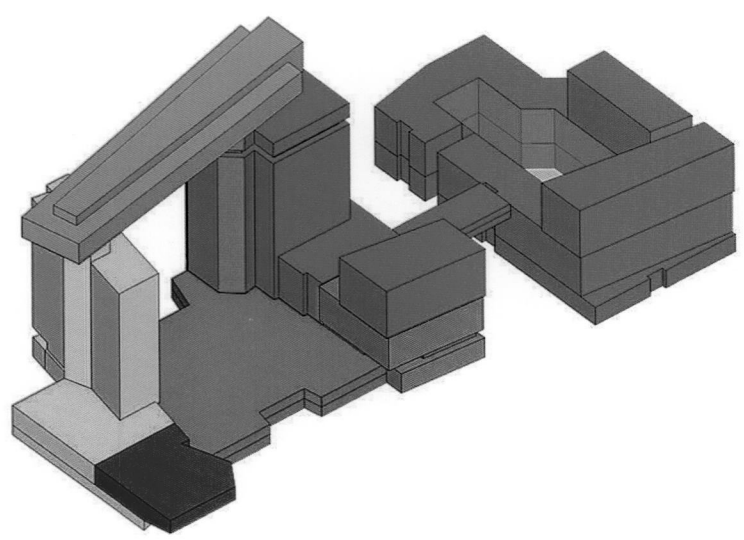

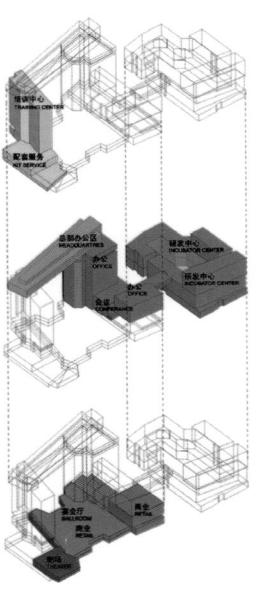

04
DESIGN EXPRESSION
PRADA
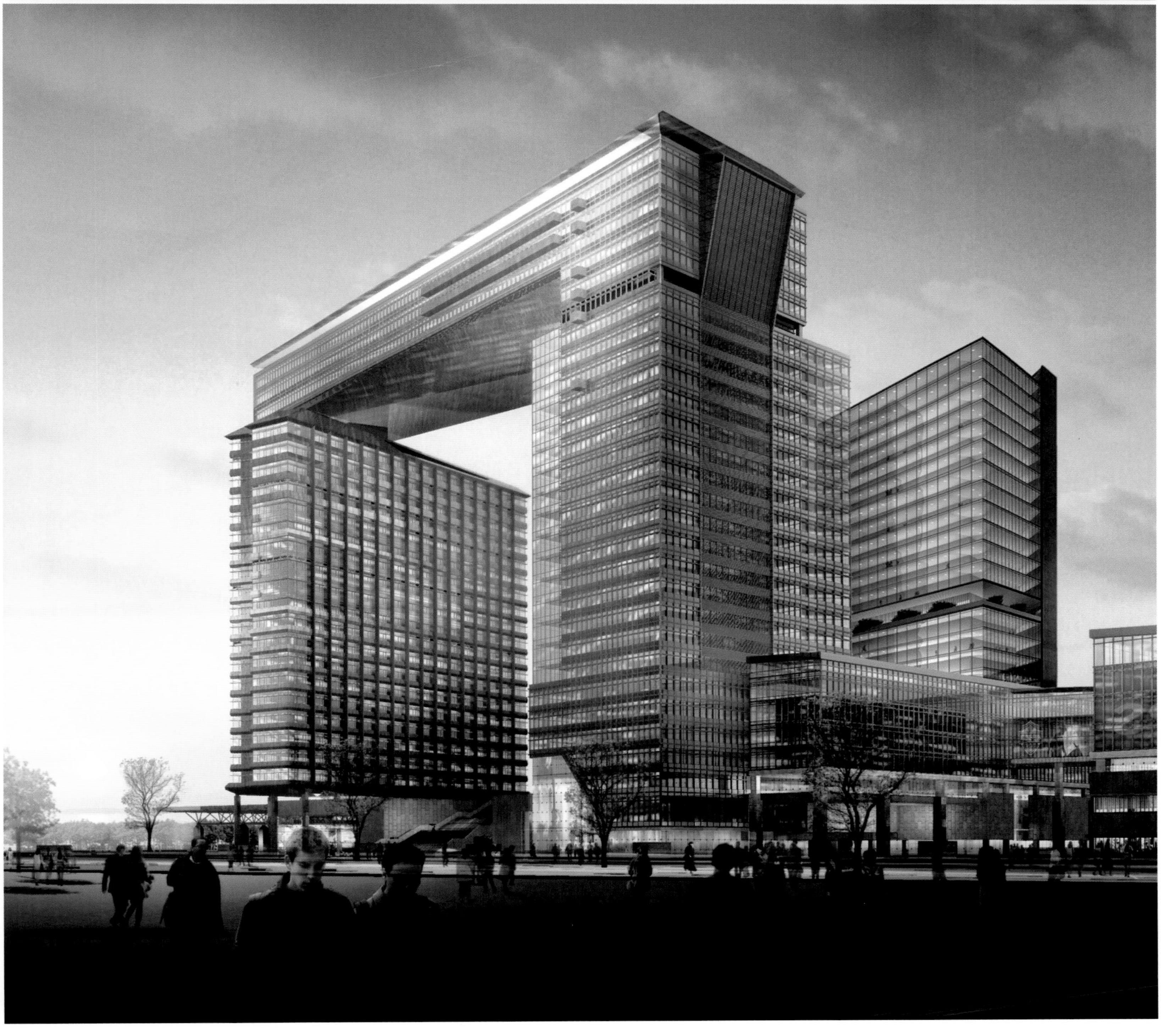

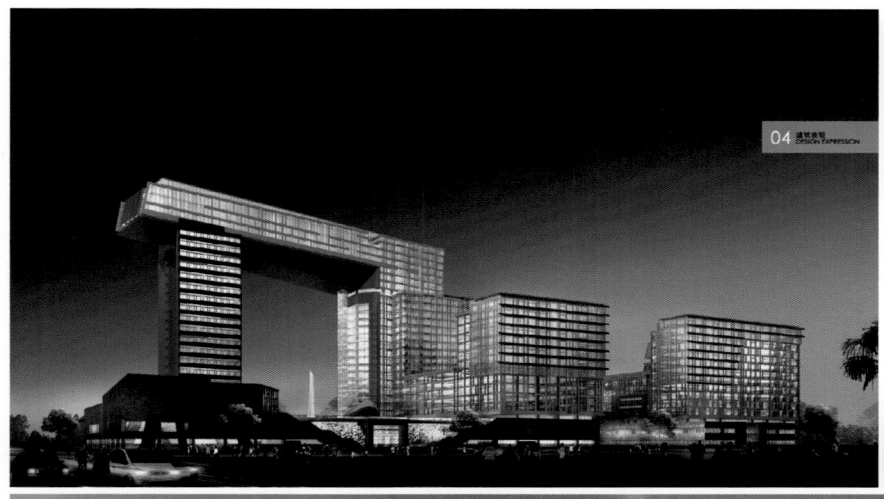

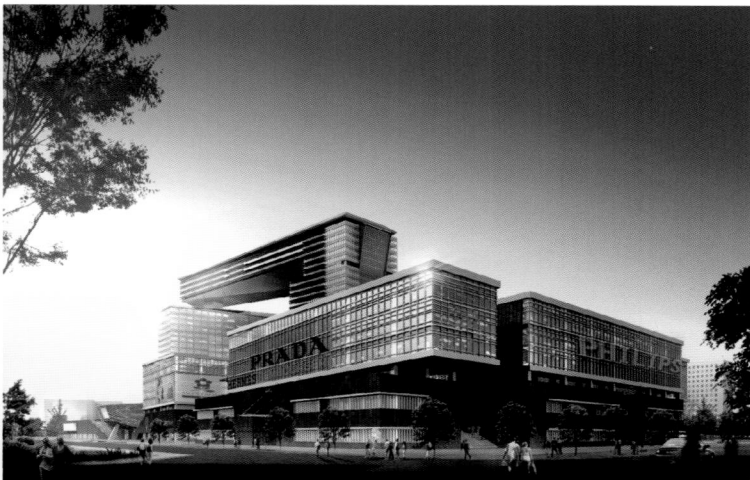

天津温州大厦
Tianjin Wenzhou Building

项目名称：天津温州大厦
用地规模：1.81 万平方米
建筑面积：17.17 万平方米
容积率：9.49
项目地点：中国 / 天津
编制时间：2006 / 05

设计哲学

设计哲学 1——多功能复合的充满活力的新城市中心区

仅仅是孤立的商贸、办公、酒店……，还是真正的市民中心？如果仅仅是孤立的功能的相加，其结果将是城市的失败，最终也将是经济的失败。如何对不同功能做出空间上的安排并超越任务书，做出新的设想，激发出强大的活力，使其成为天津独一无二的场所，是我们考虑的关键点之一。

设计哲学 2——将传统城市商务中心人性化空间处理方式引入现代中心区的规划已逐渐成为当代城市中心区建设的趋势。遵循可持续发展的原则，逐步实现人文环境与自然环境的协调。在塑造城市景观空间的过程中，不但充分强调空间景观的视觉感受，而且注重空间景观的人性化和可用性，使得新中心区的空间景观设计成为视觉优先、市民易享、人性化和感知性完美结合的空间设计典范。

设计哲学 3——与周边地块的充分融合

理解整个规划项目与周边地区发展的脉络关系，是使该项目的规划发展与整个地区的总体建设相协调的关键。地块本身与中心区的发展有密切的联系，并且作为一个重要的区域融入新的城市建设之中。这些周边条件直接影响到地块内部的规划布局和与周边建设的呼应，并促使我们在规划中充分考虑规划与地区性影响相融合，使之与未来周边的发展项目建立充分的对话关系。

Design Philosophy

1. Design Philosophy 1-energetic multi-functional complex new city center

The mere stacking of functions will be inevitably a failure of the city and ultimately will be an economic failure. We focus on the spatial arrangements for different functions and stimulate a strong vitality by new ideas so as to create a unique place in Tianjin.

2. Design Philosophy 2-people-oriented space landscape

It has become a trend to introduce traditional urban business center people-oriented space design into the planning of modern center. We should follow the principles of sustainable development and gradually realize the coordination between human environment and natural environment. For shaping urban landscape space, it is necessary not only to fully emphasize the visual feeling of the spatial landscape, but also to attach importance to humane and availability so as to design the spatial landscape in the new center to be visual-prior and public accessible space design model with the perfect combination of humane and perception.

3. Design Philosophy 3- full integration with the surrounding land
Understanding the relationship between the entire planning project and the surrounding area is the key to ensure the coordination between the project planning and the overall development of the region. This plot is in close relationship with the development of the center and has been integrated into the urban construction as an important region. The surrounding conditions will directly affect the internal planning of the plot and will promote us to fully consider the integration between the regional influence and the planning. Our purpose is to achieve harmonious relationship between the project and the future development projects in surrounding areas.

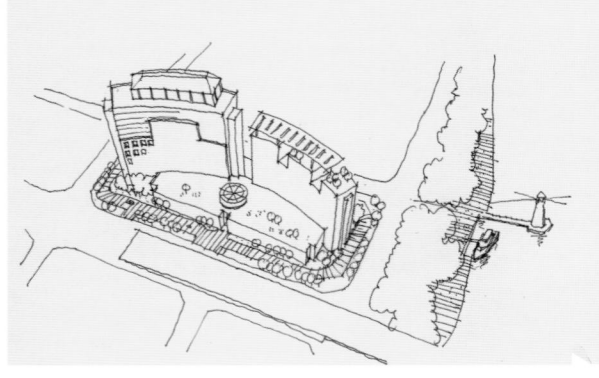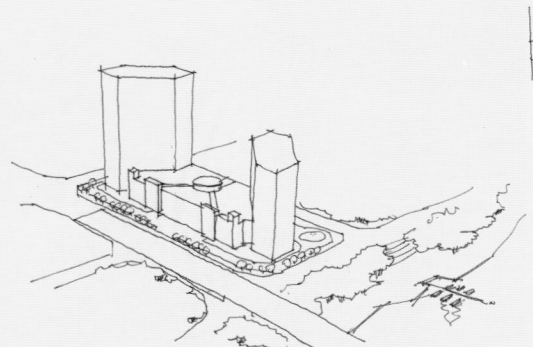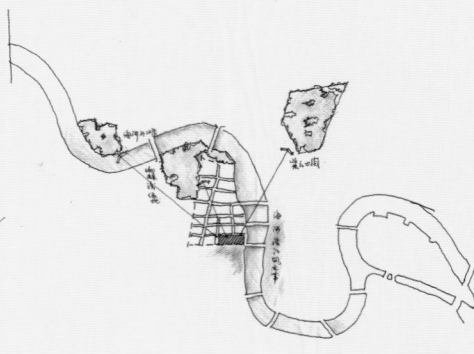

传媒大厦
The Media Mansion

项目名称: 传媒大厦
用地规模: 2.3 万平方米
建筑面积: 9.3 万平方米
容积率: 4.0
项目地点: 中国 / 四川 / 成都
编制时间: 2009 / 09

"L"形的体量与已建建筑在场地中围合形成了开阔的中庭景观,在内部景观上我们设计了小型广场,以及一个张拉膜景观小品,配合低矮的灌木及四季花卉等植物,将塔楼的集合造型充分反映在绿化景观上。在红星路与布后街的转角处,建筑的退让充分形成开放的城市节点,并设计了能反映建筑意向的喷水池及绿化带,营造了建筑沿街面的气氛及气势。

The L-shaped volume and the constructed buildings close in to form an open atrium landscape on the ground. In the internal landscape we design a small-scale square, and a tensioned membrane landscape sketch, in addition to the low shrubs and the plants of flowers of the four seasons, so that the aggregative model of the tower building is fully reflected on the green landscape. In the corner of Hongxing Road and Buhou Street, the recession of building fully forms an open urban node, and we design the fountain and greening that can reflect the building intention, to create the building's atmosphere and imposing posture along the streets.

金沙万瑞中心
Jinsha Winera Plaza

项目名称：金沙万瑞中心
用地规模：1.6 万平方米
建筑面积：14.7 万平方米
容积率：9.2
项目地点：中国 / 四川 / 成都
编制时间：2007 / 11

沿用地周边内置三栋造型相似、简洁大方的独立塔楼，整洁的竖向构成、干净的外立面以及整齐的城市轮廓线共同构成富有视觉冲击力的城市空间，使建筑物的体量和特征得到强烈而深刻的展示与表达。

The three concise and tasteful free-standing tower buildings of similar model are inwards laid alongside the usable land. The neatly vertical composition, the clean exterior facade and the orderly city outline together form an urban space rich in visual impact, making the volume and feature of the buildings strongly and profoundly exhibited.

SMITHGROUP cna

Architecture · Urban Planning · Landscape Design
Interiors Design · Art Consultant · Lighting Design

Since 1853

www.cna-group.com

Design Determines . . .

设计决定······

06 酒店建筑
Hotels and Resorts

哈尔滨华鸿洲际皇冠假日酒店综合体
Haerbin Huahong Hotel Comprehensive

Architecture Design | 2010·07

哈尔滨华鸿洲际皇冠假日酒店综合体
Harbin Huahong Hotel Comprehensive

项目名称：哈尔滨华鸿洲际皇冠假日酒店综合体
用地规模：3.95 万平方米
建筑面积：31.59 万平方米
容积率：7.99
项目地点：中国 / 黑龙江 / 哈尔滨
编制时间：2010 / 07

本项目位于哈尔滨市南岗区，东临城市主干道南直路，南邻闽江路，西北侧紧邻金色莱茵一、二期住宅，场地周边均为新建及在建居住社区，北部相隔一个街区为国际会展中心。因此区域条件非常优越，丰富的居住人群为本项目的商业部分提供了充足的客源支持。

五星级酒店分为两部分：下部为五星级酒店，上部为高端公寓式酒店。通过高端服务和齐全的配套设施，为游客和市民提供了一个趣味盎然、舒适便捷的临时居所，同时在功能上也与商业、娱乐设施以及住宅相辅相成。

设计理念

哈尔滨，一座美丽的冰城，阐述着一个个化冰雪为神奇的故事，它用晶莹剔透的冰块展现给人们熠熠生辉、明亮耀眼的冰雕杰作，同时也更多地赋予了这座城市激情、感悟与活力。

建筑是凝固的音乐，更是建筑师创造力与想象力的载体，本项目借鉴并采用冰雕工艺手法，强调建筑体量关系，突出形体基本特征，力求建筑造型轮廓鲜明，强调建筑细部，使建筑具有近观远瞩的绝佳效果。

Surrounded by newly-built and under-construction communities, the project is located in Nangang Area in Harbin, reaching the main city road Nanzhi Road on the east, Minjiang Road on the south, Phase I and II residential buildings of Golden Rhine on the north west and the international exhibition center one block away on the north. Therefore, the large community-living population is the target client resource of the project.

The five-star hotel is composed of two parts: the five-star hotel on the bottom and high-end apartment hotel on the top. Having functions like commercial and entertainment facilities and residence, its high-end service and associated facilities provide business tourists and citizens with temporary dwelling of interest and comfort.

Design Concept

Harbin, a beautiful ice city, has been telling stories of miracles. Not only does it bring people marvelous ice sculptures, but it also demonstrates passion, inspiration and vitality.

Architecture is the solidification of music and the carrier of architect creativity and imagination. Adopting ice sculpture technique, the project highlights architectural volume and basic shape features in order to make sharp outlines and outstanding visual effects.

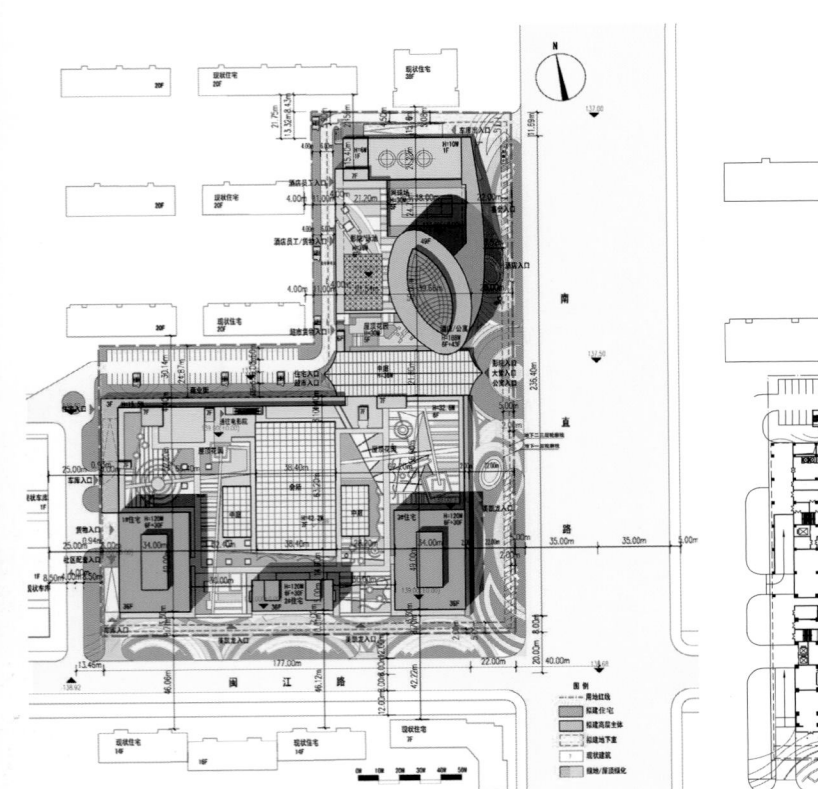

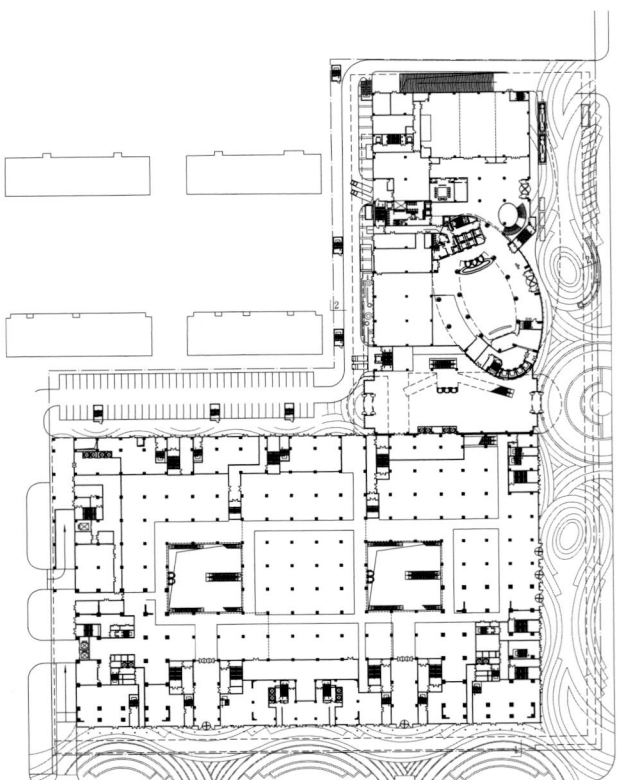

Golden Rhine
Macalline

Golden Rhine
Macalline

酒店客房
公寓式酒店
酒店配套功能
中央大堂
住宅
红星美凯龙商业

37400
6400
电梯厅
中庭上空
服务间
117,500
121,000

南
直
路
闽
江
路

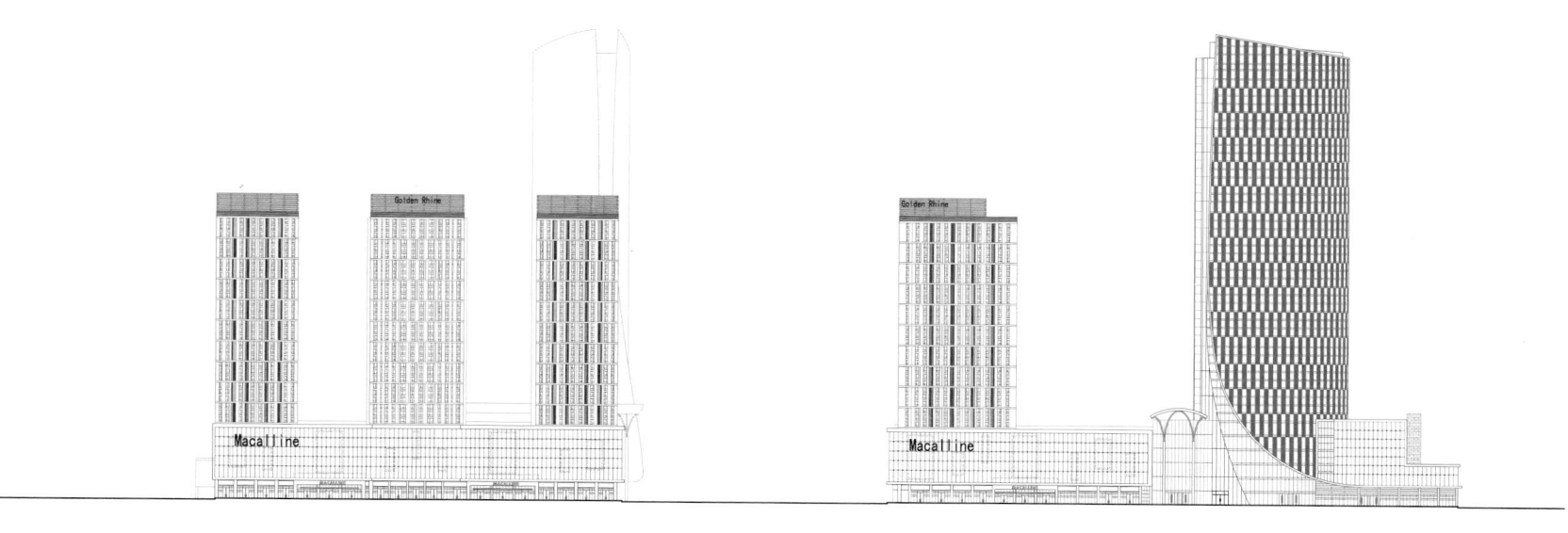

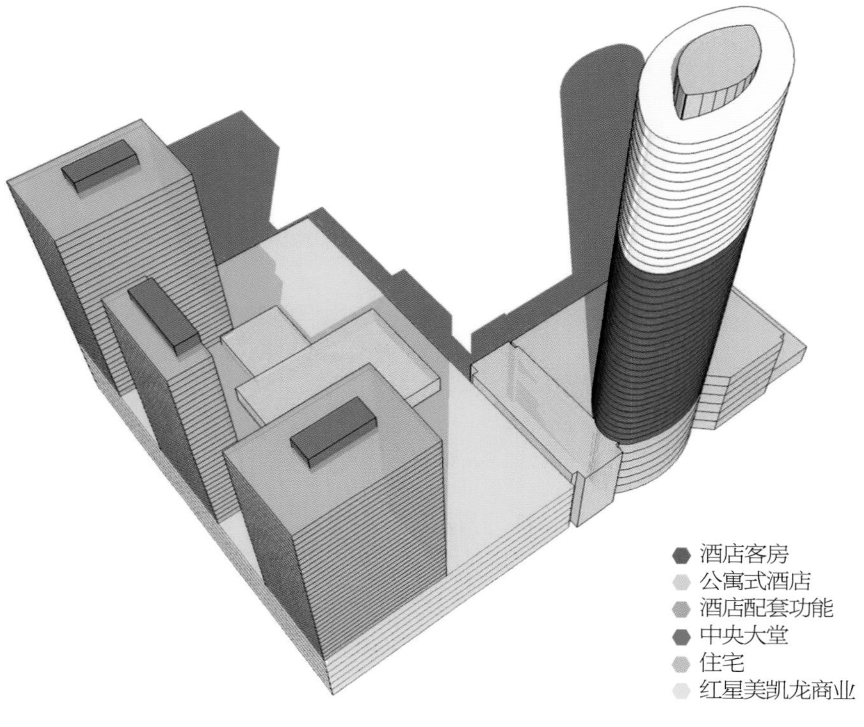

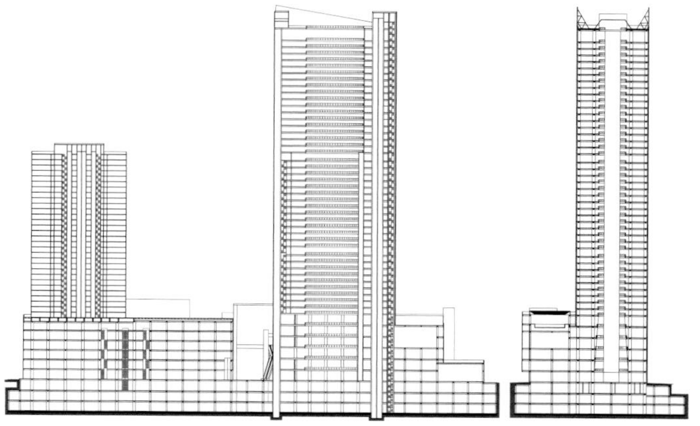

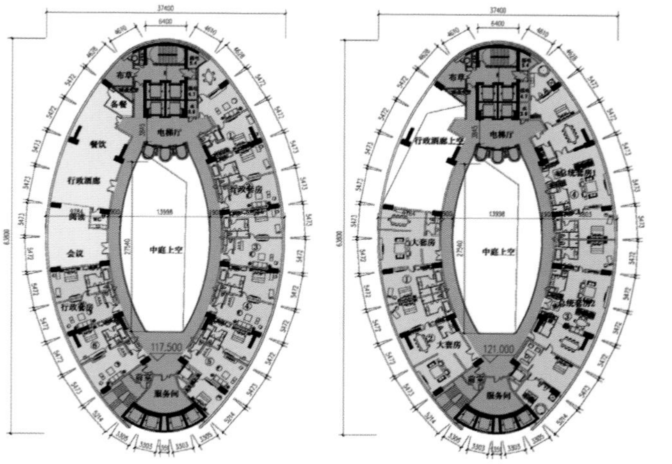

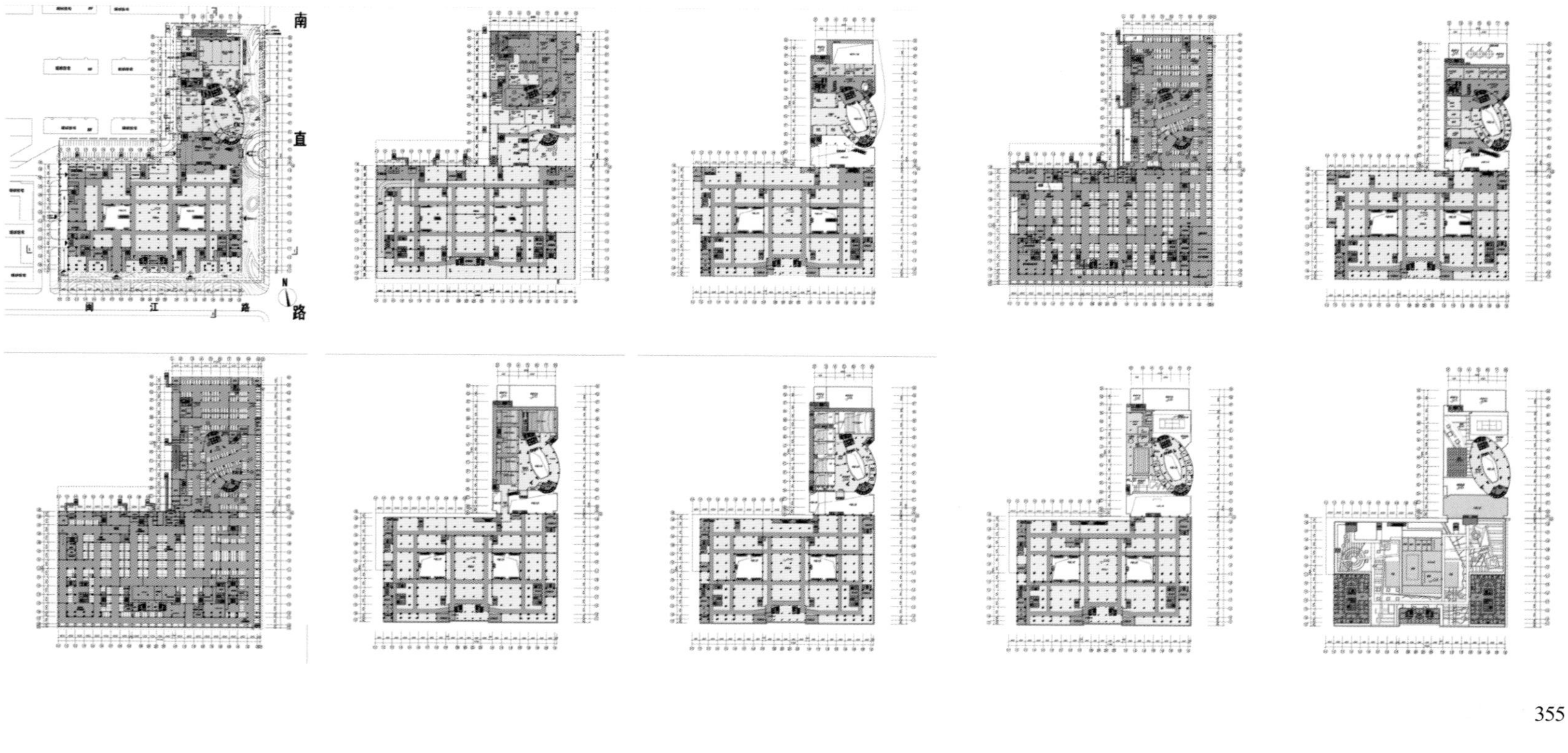

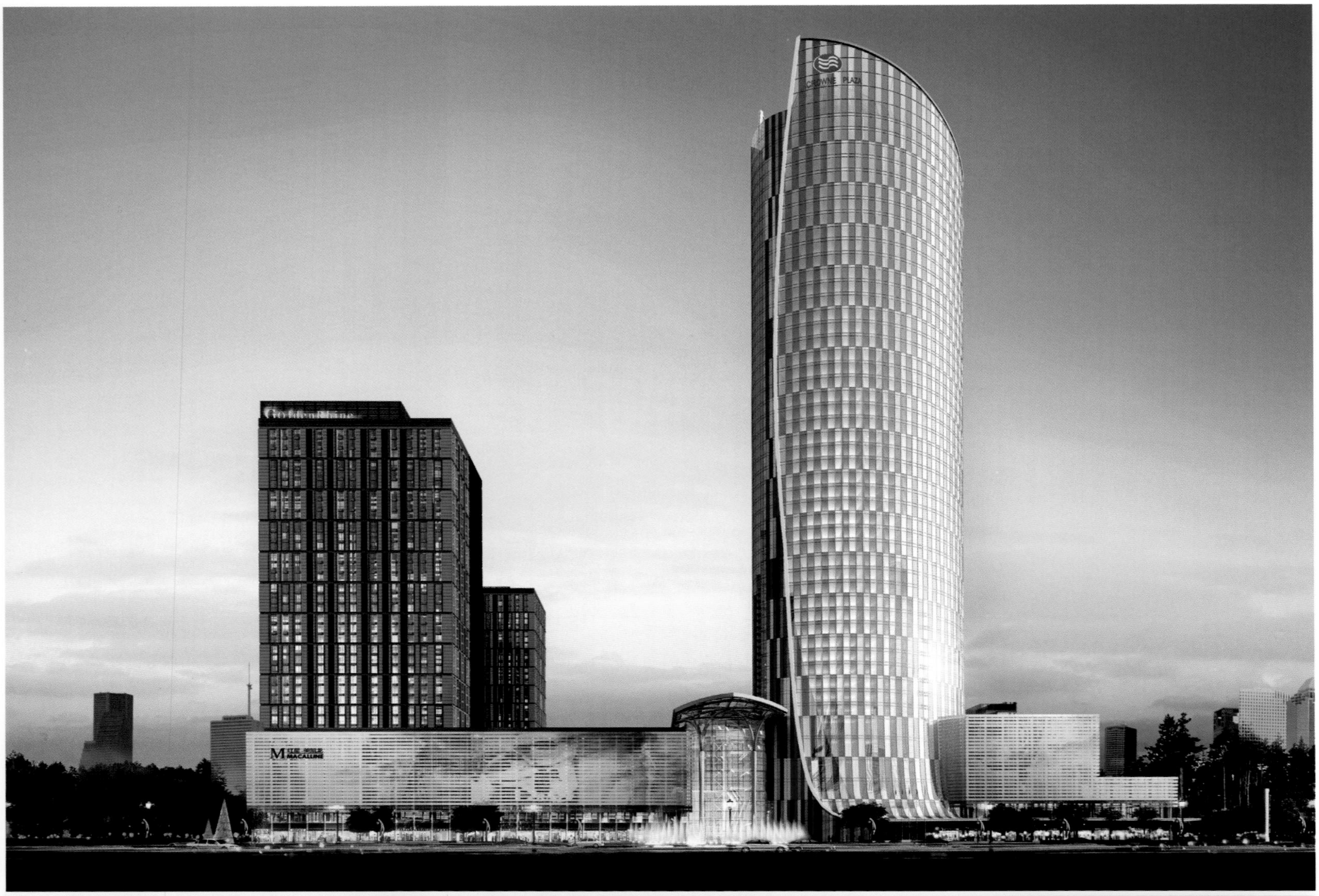
CROWNE PLAZA
MACALLINE

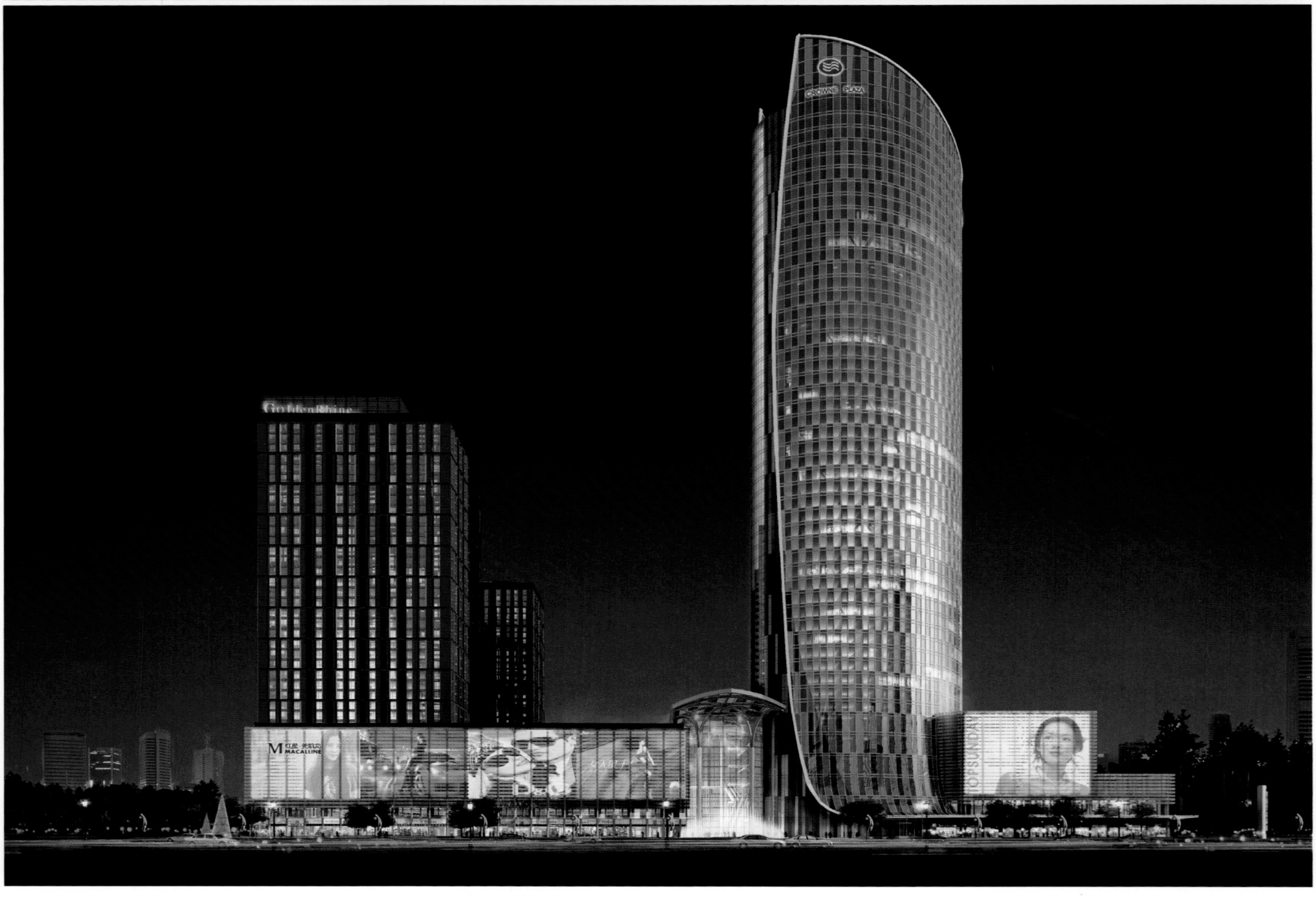
CROWNE PLAZA
MACALLINE

359

北京北湖九号
Beijing Beihu No.9

项目名称：北京北湖九号
用地规模：8.97 万平方米
建筑面积：2.16 万平方米
容积率：0.29
项目地点：中国 / 北京
编制时间：2009 / 09

知行合院——一个关于文化复兴的房子

1. 李约瑟难题与隔代传承

历史在明代转了一个弯。为何如此辉煌的明代却没有继续生成近现代的高势能文明？

这个疑问被人称为"李约瑟难题"。这个难题被无数人解过，有许多不同的答案。在这里我不是要做理论上的讨论，而是用一个由房子产生的案例做一个当代版的解释。我希望用隔代传承的方法，将集中国文明之大成的明代文明的势能直接转接到现代，从而隔开近代两三百年那些曲折的故事达到新的文化复兴。

2. 格物穷理与知行合一

中国的历代志士一直都在求"道"。明代的思想家王阳明用的求"道"或"理"的方法就是不断去"格"物然后穷其理。而今天我们要解决当代版"李约瑟难题"的方法就是去"格"或者"研究"院子来求道。那又怎么样去"格"呢？王阳明用的方法是"知行合一"，最后发现"理在心中"，有点太形而上了，解决不了具体的问题。

后来在明代有一个名叫徐阶的内阁大学士提出并实行了一个新的方法，叫"知行合利"。这是一个伟大的说法。若干年以后才有了西方的《国富论》，建立了资本主义的思想基础。《国富论》指出：每个人都有追求利益私有化的天性，而这种天性的发挥就带来了国富民强。在这里，我们有点明白历史为什么会在明代转那个弯了。简言之就是：我们传承的是"知行合一"而非"知行合利"。在今天，我们要做的是"知行合院"。

3. 三个合一

我们要做的是三个"合一"，即"天人合一""工作和生活合一""健康和娱乐合一"。通过这三个"合一"最后达到"知行合院"。

Courtyard House of Knowledge and Action-a House about Cultural Renaissance

1. Needham Puzzle and Atavistic Inheritance

Chinese history made a great turning in the Ming Dynasty. How didn't such a prosperous dynasty end with a modern high-potential civilization?

The question is the so-called "Needham Puzzle". The question has been solved in thousands of ways with diversified answers. Instead of theoretical discussion, we use a house as a case to give it a contemporary interpretation.

As a modern solution, we wish to adopt the method of atavistic inheritance to transfer the Ming Dynasty's civilization directly to the modern world in order to reach a new cultural renaissance without those winding stories in the recent two or three hundred years.

2. Exploring Thoroughly Nature of Things and the Unity of Knowledge and Action

Chinese patriots have been pursuing "Tao" all the time. Wang Yangming, an ideologist in the Ming Dynasty, studied and explored the nature of things ceaselessly for "Tao" or "truth". Nowadays, we ought to study courtyards for truth to find out the solution to the modern "Needham Puzzle". The problem is how Wang's way was "the unity of knowledge and action" and finally led to "the truth in one's mind", which is too superorganic to solve problems.

However, Xu Jie, a school scholar in the Ming Dynasty, raised and performed a new method called "knowledge and action for profits", which is a great theory. Many years later, there came The Wealth of the Nations, establishing the theoretical foundation of capitalism. It points out that everyone has the nature of pursuing privatized profits, which creates wealthier nations and stronger people. Thus, we learn why the change happened in the Ming Dynasty: we have inherited "the unity of knowledge and action" instead of "knowledge and action for profits". What we are going to build is "a courtyard house of knowledge and action".

3. Three Unities

The three unities which-are "the unity of the nature and people", "the unity of work and life" and "the unity of health and entertainment" are to finally reach a courtyard house of knowledge and action.

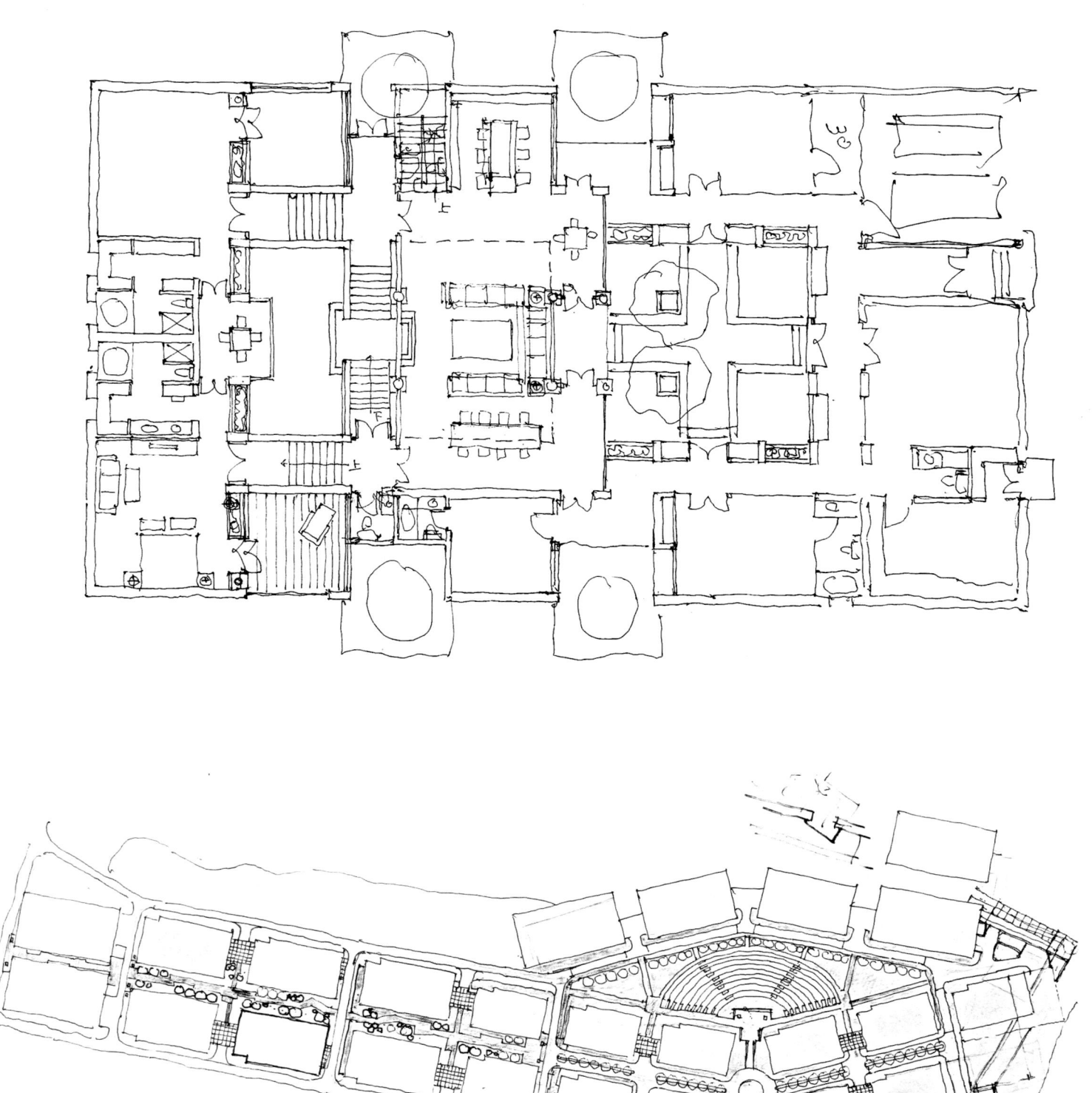

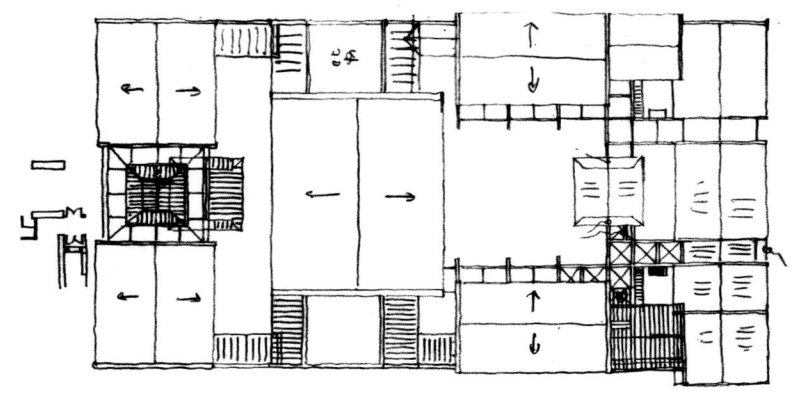
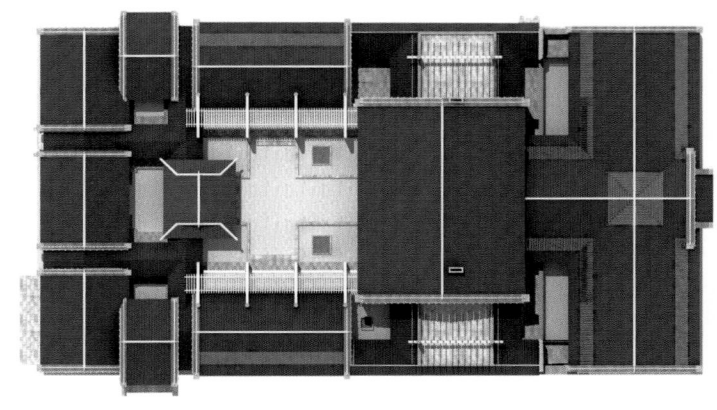

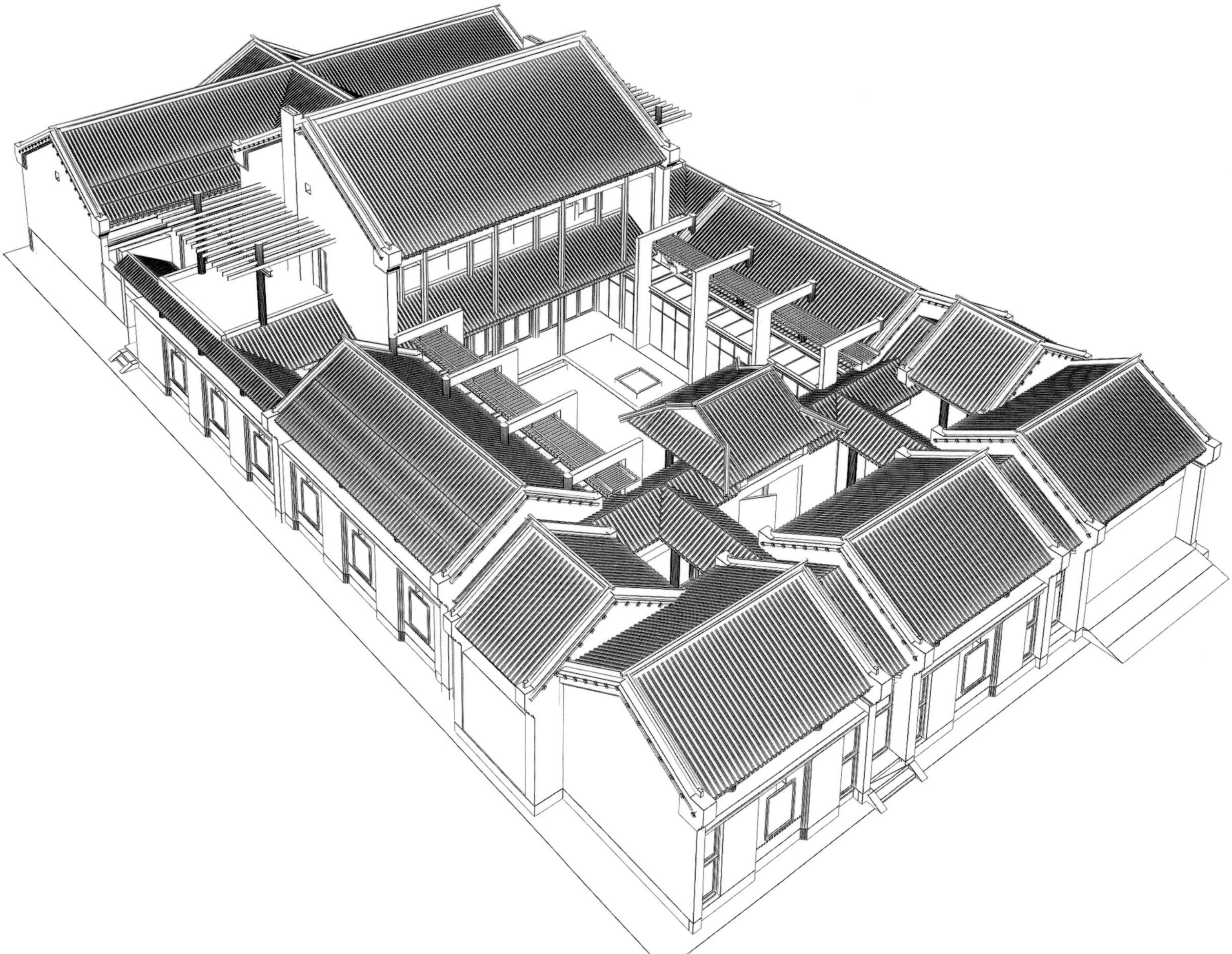

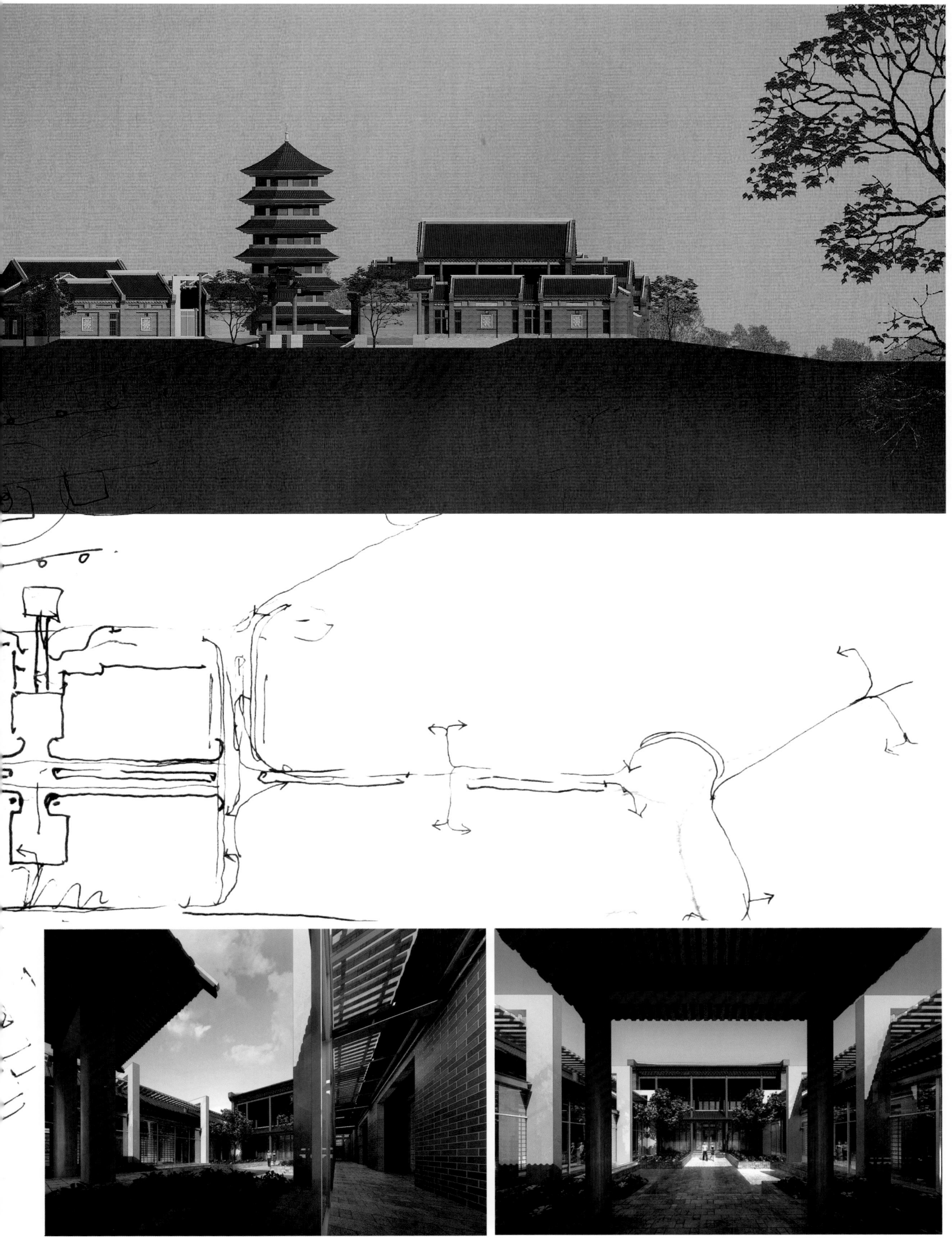

溧阳南山竹海温泉度假区
Liyang Nanshan Bamboo Sea Hot Spring Resort

项目名称：溧阳南山竹海温泉度假区
用地规模：11.73 万平方米
建筑面积：4.14 万平方米
容积率：0.35
项目地点：中国 / 江苏 / 溧阳
编制时间：2007 / 05

该项目紧邻 4A 级景区——江苏省天目湖南山竹海景区，依托周边良好的自然山水景色形成与自然共生的自然原生态系统，通过合理的规划，建立长久的生态保护与重构机制，创造山水体验景观的休闲度假生活，营造置身山水怡然自得的情景体验，实现山居生活的休闲享受。

The project is close to Tianmu Lake Nanshan Bamboo Forest, a AAAA scenic spot in Jiangsu Province. Depending on surrounding natural mountains, waters and ecosystem, it aims at establishing a lasting ecological protection and reconstruction system and creating a relaxing life style in the nature through reasonable planning.

总平面图

庐山天恒温泉度假酒店
Lushan Tianheng Hot Spring Resort Hotel

项目名称：庐山天恒温泉度假酒店
用地规模：12.53 万平方米
建筑面积：10.37 万平方米
容积率：0.75
项目地点：中国 / 江西 / 九江
编制时间：2007 / 09

设计导入生态的构筑手法，充分利用环境提供的通风和采光完成建筑的基本功能设置，整体规划通过引入不同性格特征的空间元素，结合功能的变化，演绎富于变化的视觉感受，与庐山山峰景观的轴线对应关系将新的酒店主体建筑塑造成山脚下的标志。

The design took advantage of the natural resources to provide ventilation and light function. The planning involved different spatial elements and manipulated a changeful visual effect so as to turn the main hotel building into a landmark at the foot of mountain.

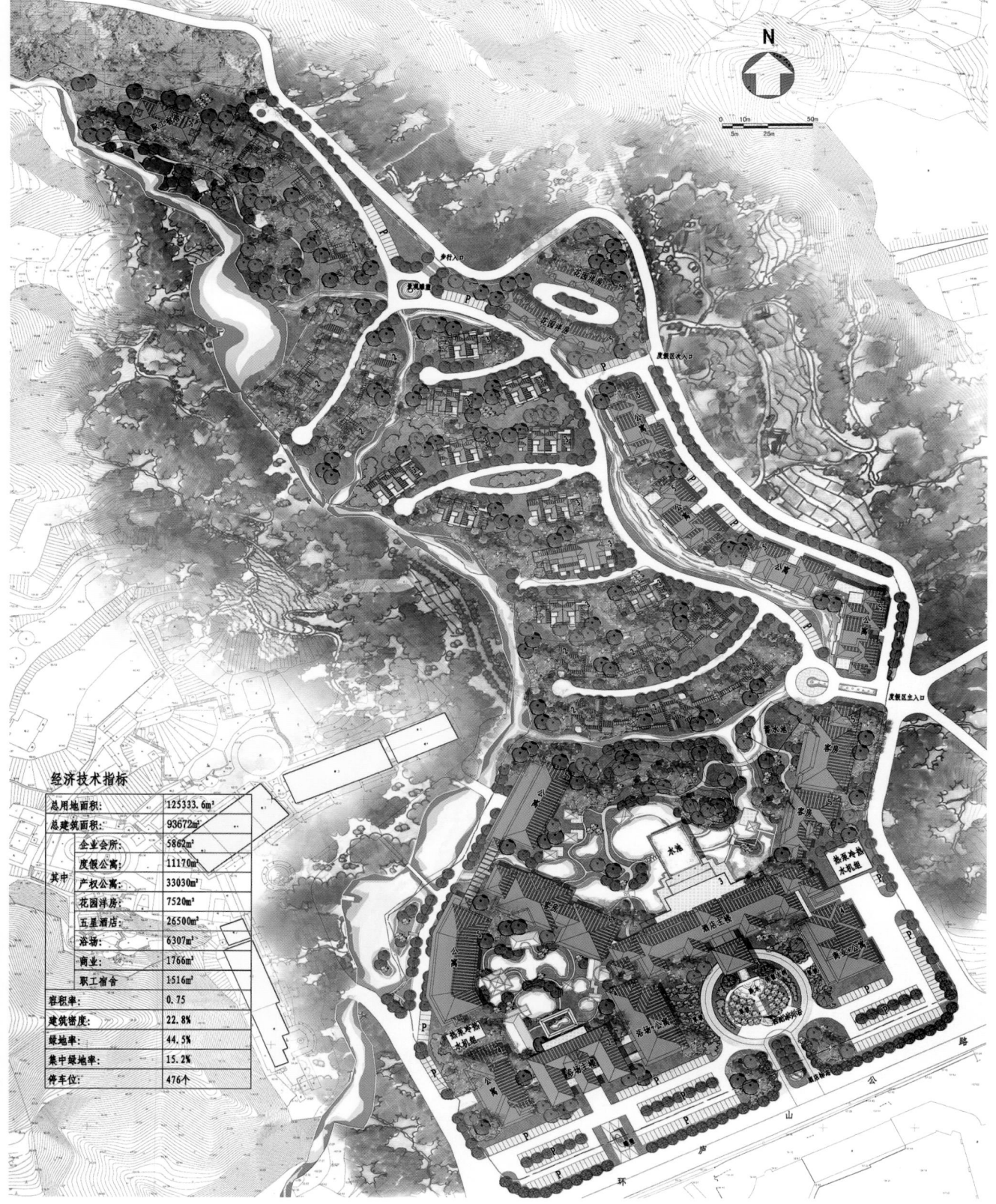

经济技术指标

总用地面积：	125333.6m²
总建筑面积：	93672m²
企业会所：	5862m²
度假公寓：	11170m²
产权公寓：	33030m²
花园洋房：	7520m²
五星酒店：	26500m²
浴场：	6307m²
商业：	1766m²
职工宿舍：	1516m²
容积率：	0.75
建筑密度：	22.8%
绿地率：	44.5%
集中绿地率：	15.2%
停车位：	476个

庐邑温泉度假区
LuYi Hot Spring Resort

项目名称：庐邑温泉度假区
用地规模：18.52 万平方米
建筑面积：7.93 万平方米
容积率：0.43
项目地点：中国 / 江西 / 九江
编制时间：2006 / 09

珍视清泉苑地块内的生态湿地与水畔生态绿带等资源，将自然环境与人工环境有机协调，创造生态化、自然化的社区，实现回归自然的追求，并使更多的建筑最大限度地融合到最好的环境中。

Preservating of natural green resources and integrating the natural environment and artificial environment, the pursuit of sustainable surrounding is the essential principle of the design. All the efforts allow the building to be involved into the surroundings maximally.

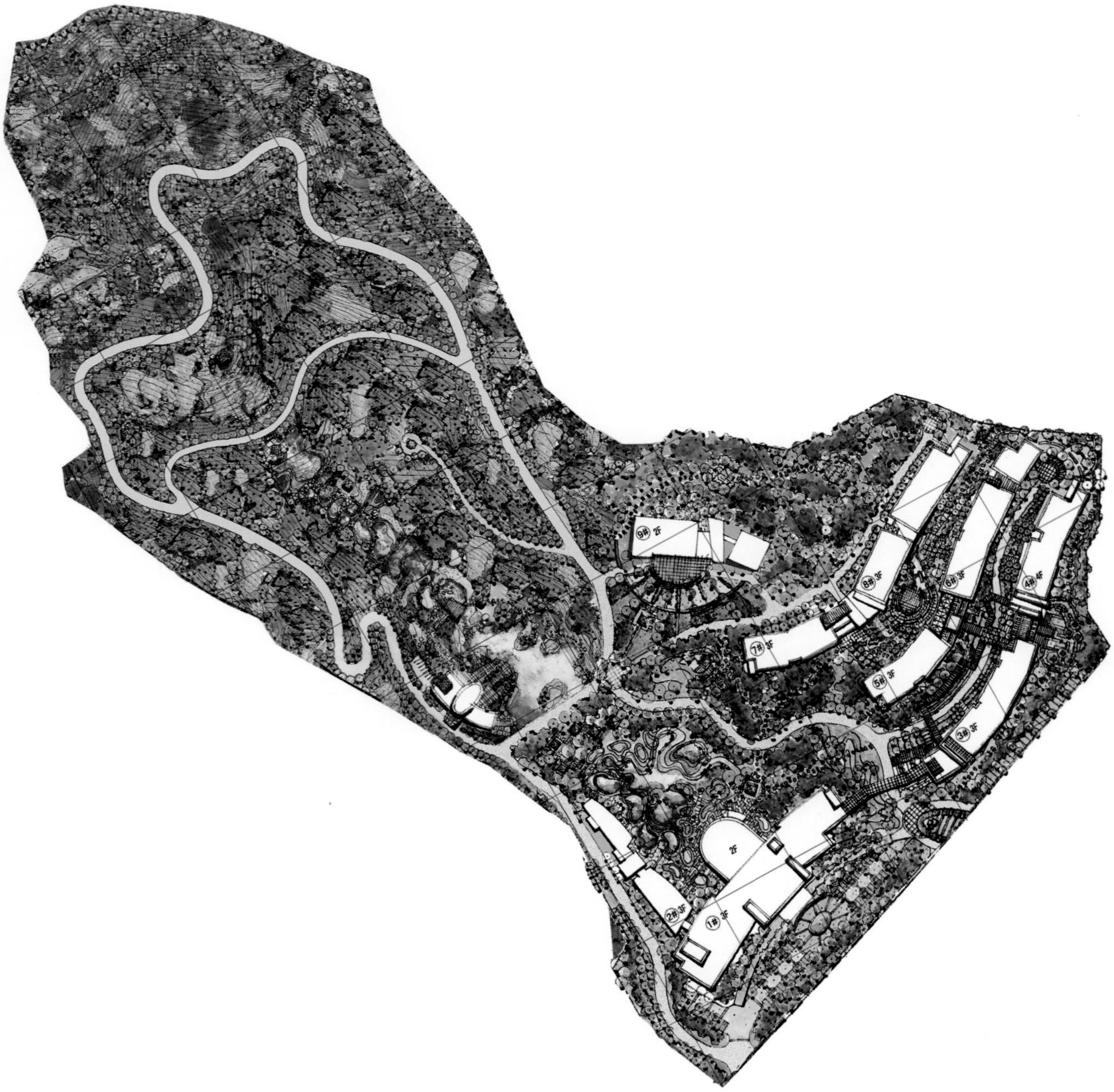

上海朱家角度假酒店
Shanghai Zhujiajiao Resort Hotel Project

项目名称：上海朱家角度假酒店
用地规模：2.66 万平方米
建筑面积：1.19 万平方米
容积率：0.45
项目地点：中国 / 上海
编制时间：2008 / 03

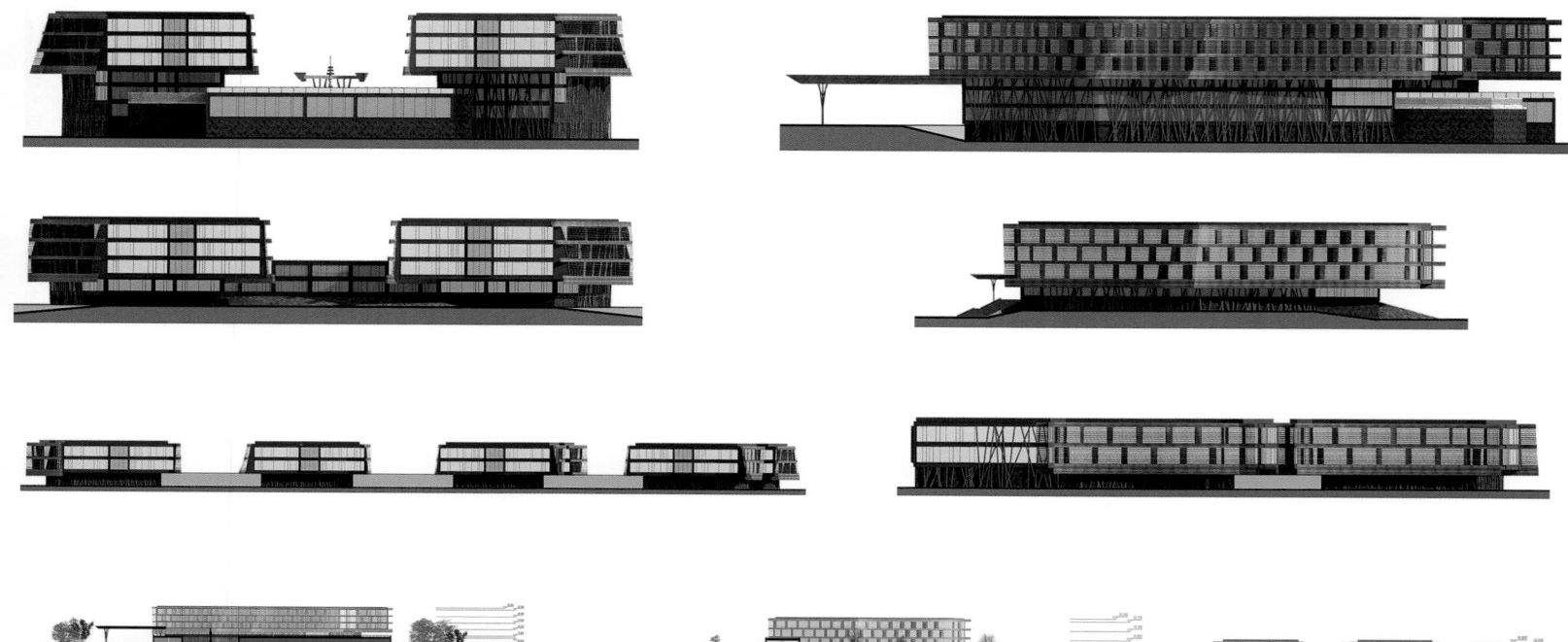

本项目位于上海历史文化名镇朱家角南镇区的南端，用地呈三角形，三面环水，规划用地面积约 2.7 万平方米。我们的设计理念吸纳江南水乡的灵气、大自然的精髓、古镇的人文情怀；重点展现设计师对大自然的感悟、对江南水乡文化的解读、对现代乡土建筑的构筑以及项目自身位于古镇旅游重要交通节点上的视觉感染力。

The project is located at the southern end of Nanzhen Area in Zhujiajiao, a historic town in Shanghai. It is triangular in shape and surrounded by water on three sides, covering a planning area of about 27,000 m² Having absorbed inspiration of the Yangtze River, natural essence and humanities in ancient towns, our design highlights designers' passion about the nature, their own understanding of the unique culture and visual appeal for modern vernacular architecture and the project's location in the tourist town.

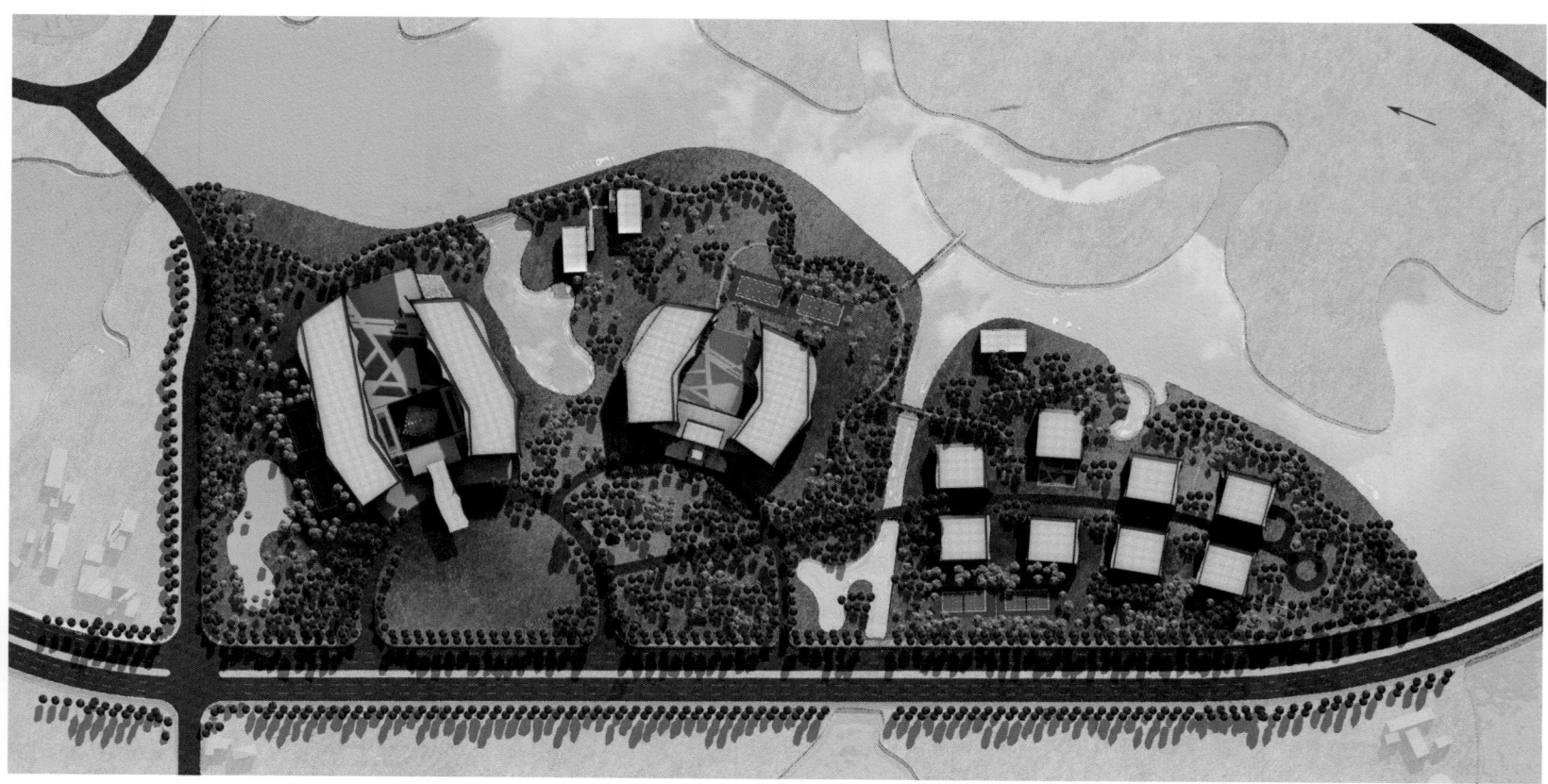

扬州京杭之心启动区酒店
Yangzhou Beijing-Hangzhou Centre Area Hotel

项目名称：扬州京杭之心启动区酒店
用地规模：2.39 万平方米
建筑面积：5.62 万平方米
容积率：2.35
项目地点：中国 / 江苏 / 扬州
编制时间：2009 / 02

古城是扬州凝固的历史， 运河是扬州流动的文化。作为中国大运河申遗的牵头城市，扬州着手建造一个"物化"的永久性会址，这里将成为与世界运河文明对话的永久性平台。唐风辉映大运河，整个建筑群落将成为"运河名城、精致扬州"的新地标。

扬州京杭之心酒店项目位于扬州 CBD 广陵新城一期启动区，西临京杭大运河，东临运河风情带。其中酒店由国际著名的洲际 IHG 集团管理。

Ancient city is the solidified history of Yangzhou; canal is the flowing culture of Yangzhou. As the leading application city of Grand Canal World Heritage, Yangzhou is going to construct a "materialized" permanent venue. Here is a permanent platform for dialogues with world canal civilizations. Tang style influences Grand Canal, and the main target of building community will be "famous canal city, refined Yangzhou ".

YangZhou BeiJing-Hangzhou Centre Hotel is located in the initiating zone of Yangzhou CBD Guangling New City, Phase I, adjacent to Beijing-Hangzhou Grand Canal in the west and canal landscape belt in the east. The hotel is under the management of the internationally-renowned IHG Group.

大邑凤凰岛国际旅游度假区
Dayi Phoenix Island International Tourism Holiday Resort

项目名称：大邑凤凰岛国际旅游度假区
用地规模：1 333.3 万平方米
建筑面积：9.6 万平方米
项目地点：中国 / 四川 / 成都
编制时间：2009 / 10

本项目为大邑县城北部森林溪谷里高端的旅游度假住宅社区，其中包括商业、商务、超五星级酒店、娱乐、度假、旅游、康体疗养以及高端居住，是一座环境优美的国际旅游度假区。

As a high-end tourism holiday residential community in the forest valley north of the county seat of Dayi County, it incorporates commerce, business, super 5-star hotel, entertainment, vacation, fitness recuperation and high-end inhabitation. It is an international tourism holiday zone with graceful surroundings.

丰德万瑞中心 & 奥克伍德酒店
Fengde Wanrui Center & Oakward Hotel

项目名称：丰德万瑞中心 & 奥克伍德酒店
用地规模：8.0 万平方米
建筑面积：3.04 万平方米
容积率：0.38
项目地点：中国 / 四川 / 成都
编制时间：2009 / 10

这是成都市场上第一个提出"单元式"写字楼的案例，超越了中心筒体结构和"回"字走廊的局限，更浅的建筑进深为办公空间带来充足的阳光和清新的空气。建筑公摊面积的压缩和使用的灵活性也为业主创造了可观的经济效益。

This is the first case which the "unit-typed office building" is presented in the Chengdu markets. It exceeds the limitation of the central tube structure and the "Hollow Square Shape" corridor. The shallower building depth allows adequate amounts of daylight and fresh air into office space. The compression and the use flexibility of the shared area also create considerable economic benefits for owners.

南京佛手湖万豪酒店
Nanjing Bergamot Lake Marriott Hotel

项目名称: 南京佛手湖万豪酒店
用地规模: 4.26 万平方米
建筑面积: 4.3 万平方米
容积率: 1.0
项目地点: 中国 / 江苏 / 南京
编制时间: 2011 / 04

佛手湖万豪酒店位于南京苏宁环球集团在该区域打造的一处高端别墅的中央，项目用地已经被建成的别墅挤压成一个狭小的三角地，但是投资人仍寄希望于通过高端酒店的建设提升整体地块的价值，在这个项目中我们的欧洲设计师对地形的综合利用和风格化形态的追求已经在方案中尽情表露，逐级而下的梯级花园以及梦幻般的酒店主体，每一个细节都希望体现出投资人对于价值的要求，我们认为这是一个成功的尝试，远远高于现在万豪给予的品牌定位。

Bergamot Lake Marriott Hotel is located in the center of high-end villa area built by Nanjing Suning Universal Group. The project site has been squeezed into a narrow triangle by the built houses, but investors still hope by the high-end hotel construction to enhance the overall value of land. In this project the comprehensive utilization of terrain and pursuit of style-oriented forms adopted by European designers have been fully presented. With step by step down the stairs of the hotel gardens and a fantastic body , every detail reflects investor's request for the value. We think this is a successful attempt, far higher than the brand positioning of Marriott Group.

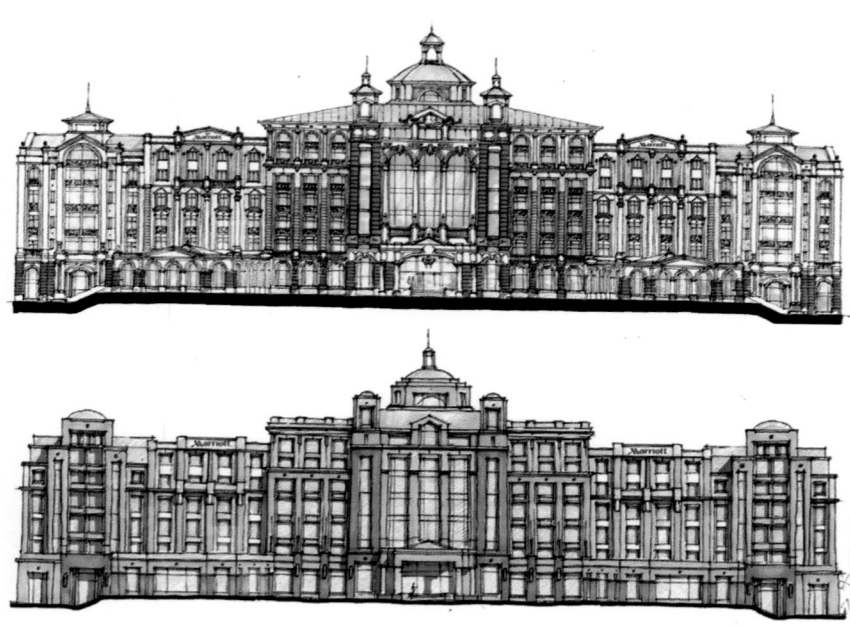

SMITHGROUP cna

Architecture · Urban Planning · Landscape Design
Interiors Design · Art Consultant · Lighting Design
Since 1853

www.cna-group.com

Design Determines . . .

设计决定……

07 居住建筑
Residential Building

南京汤泉东方温泉花园小镇
Nanjing Tangquan Oriental Garden Town

Architecture Design | 2010·04

南京汤泉东方温泉花园小镇
Nanjing Tangquan Oriental Garden Town

项目名称：南京汤泉东方温泉花园小镇
用地规模：61.2 万平方米
建筑面积：33.3 万平方米
容积率：0.54
项目地点：中国 / 江苏 / 南京
编制时间：2010 / 04

项目旨在以当地温泉和景观资源打造南唐风格的度假圣地和温泉养生天堂。在规划设计上充分结合原始自然坡地的地形地貌，依山就势、因地制宜，利用自然形成的老山山体，结合规划设计将项目打造成与山共栖、与水相融的温泉主题度假型别墅住宅社区。同时在户型设计上力求人性、舒适、生态、休闲、私密与安全，提供多样化生态人居选择。

延续南唐大气雄浑的规划格局，规划设计师法自然因地制宜，单体设计以双泉入户合院为特色度假产品。

The project aims at creating a holiday and health resort of the South Tang Dynasty style based on local hot spring and landscape resources. Its design is in full harmony with original land shapes, using natural mountains and waters to build a holiday villa dwelling community. The design of dwelling types pursues humanity, comfort, ecology, relaxation, privacy and safety with various choices.

Continuing marvelous planning style of the South Tang Dynasty, the planning design makes compliance with the nature and local features, and courtyards with two spring inflows are the featured vacation product.

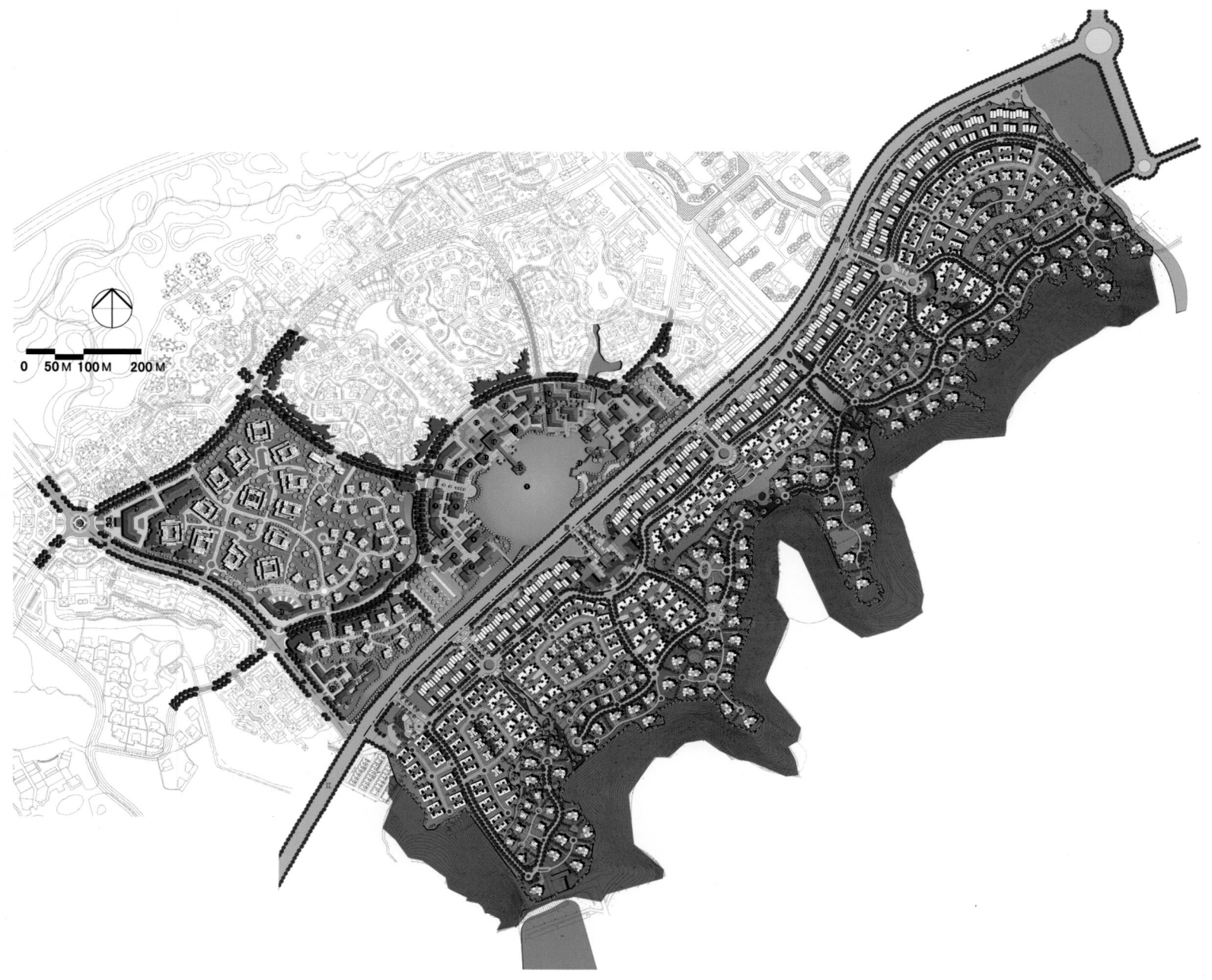

0 50M 100M 200M

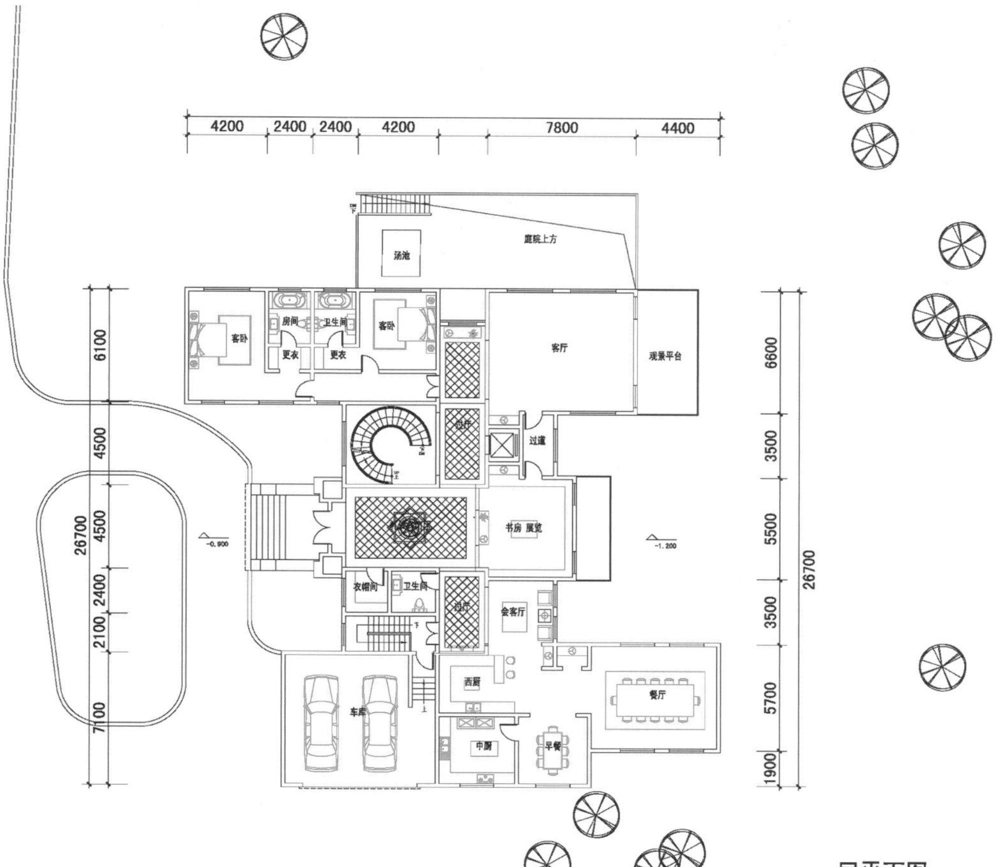

一层平面图 1:200
层建筑面积 500m²
地上总建筑面积 810m²

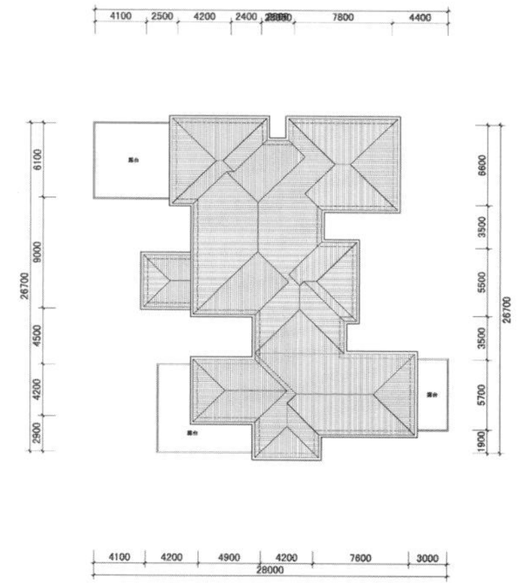

4200 2400 2400 4200   7800   4400

6100

4500

26700
4500

2100 2400

7100

6600

3500

5500

3500

5700

1900

26700

4500   8100   4200   3800   6800
27400

汤施
庭院上方

客卧   房间 卫生间 客卧
     更衣 更衣

客厅   观景平台

书房 展览

衣帽间 卫生间

会客厅

西厨

车库

中厨   平储   餐厅

-0.900
-1.200

4100 2500 4200 2400 2800 7800   4400
6100
9000

26700
4500
4500
7100
2500

4100 4200 4900   4200   7800   3000
28000

4100 2500 4200 2400 2800 7800   4400
6100
26700
4500
4500
7100
1900

6600
3500
3500
5700
26700

4100 4200 4900 4200 2000 5600 3000
28000

9.000

6.000

3.000

±0.000
-0.900
-1.500

正 立 面

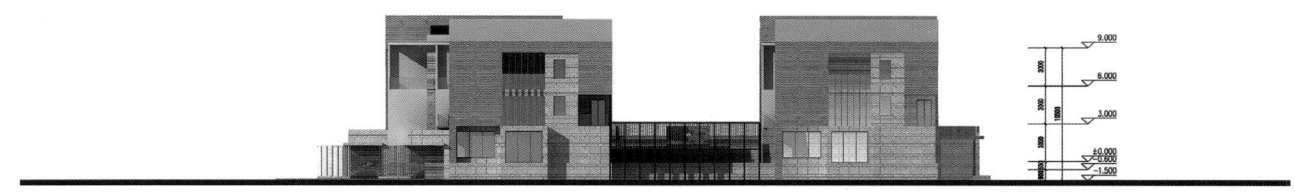

9.000

6.000

3.000

±0.000
-0.900
-1.500

侧 立 面

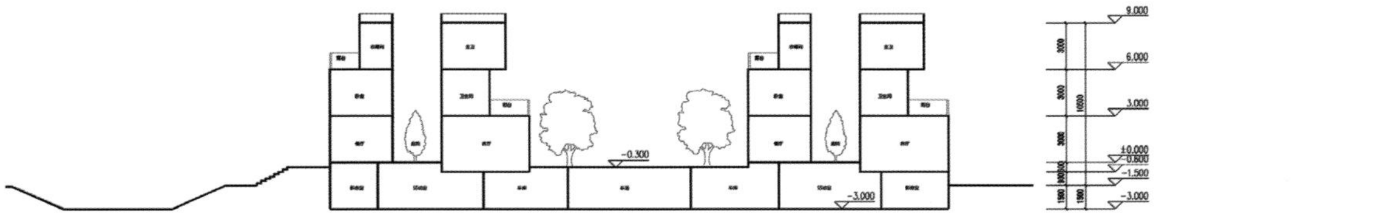

9.000

6.000

3.000

±0.000
-0.900
-1.500
-3.000

-0.300

-3.000

剖 面 图

403

双泉入户是每一个客户对这种类型的度假别墅的要求，
而我们考虑的问题除了住宅的娱乐功能之外，
还有安全性、舒适性等各种细节的要求。

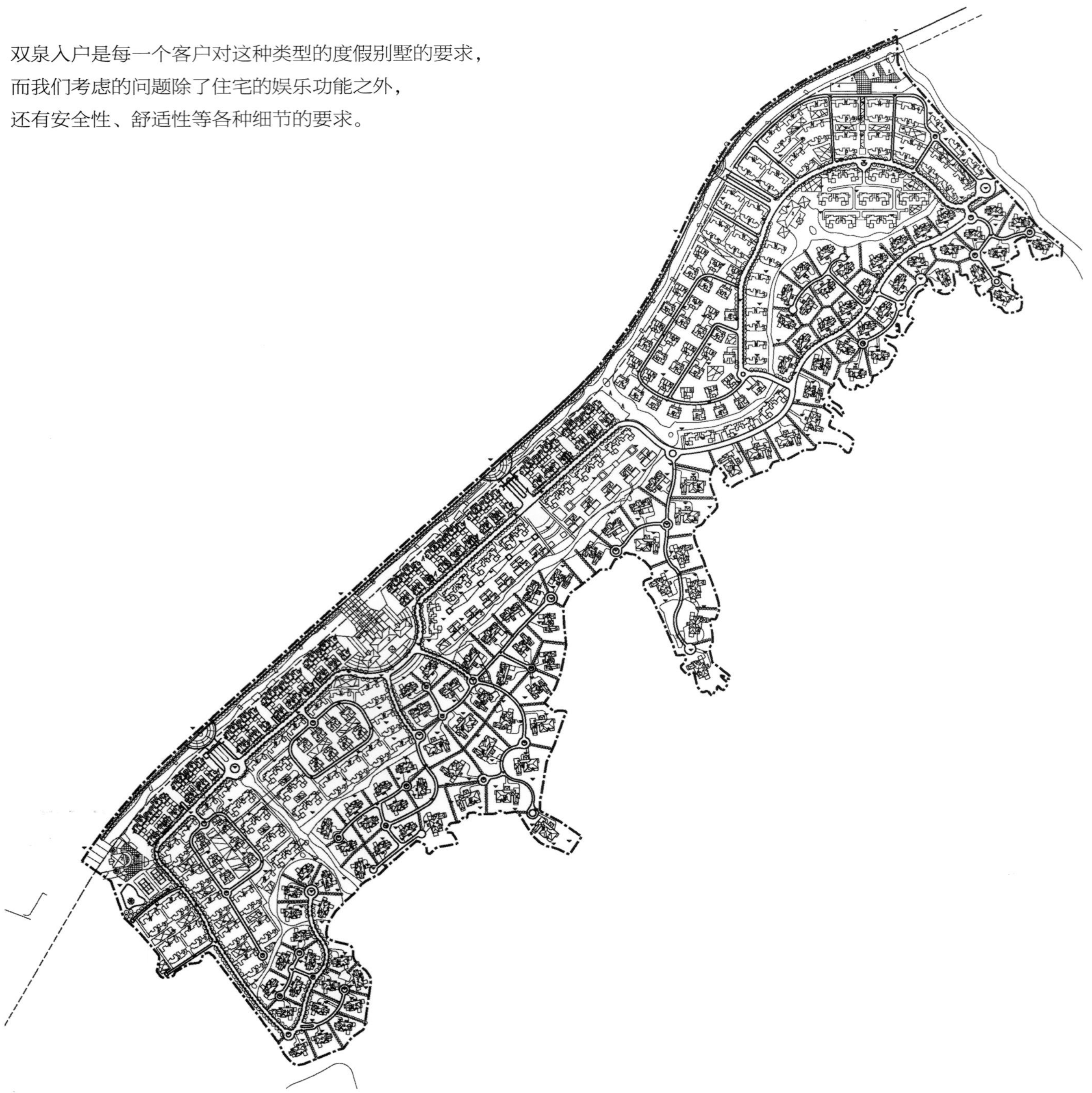

追寻该地区南唐文化和历史使之最终成为新建建筑语言的一部分。

上海大宁瑞仕花园
Shanghai Daning Ruishi Garden

Architecture Design | 2009·06

上海大宁瑞仕花园
Shanghai Daning Ruishi Garden

项目名称：上海大宁瑞仕花园
用地规模：5.13 万平方米
建筑面积：10.25 万平方米
容积率：2.0
项目地点：中国 / 上海
编制时间：2009 / 06

项目位于闸北大宁板块，紧邻大宁商业中心和大宁绿地。

设计追求把不动产项目本身作为一个社会、经济、人文问题来综合考虑，不追求单纯的设计概念或形式，我们谨慎地从细节出发，从技术、施工、销售、市场等多个角度入手，根据基地与市场的现状，采取最合理的手段，责无旁贷地配合发展商在本项目上获得最大的社会效益和经济效益。

如何发挥本地块的资源优势，打造最具市场销售竞争力的超值精品是设计首要考虑的核心议题，通过对市场及基地的分析评估，高附加值的产品，前瞻性的客群定位，差异化的竞争策略是设计需要考虑的重点。

因此，"建筑以景为先的质感生活，使人们对品质感、附加值的需求，差异化的满足达到极致！"是本项目设计的主要理念。

规划上引入"以景为先"的全新理念，依傍彭越浦的天然优势，将自然水景与中心绿地有机融合，打造人人均享自然生态景观的居住形态，无形间拉近居者与自然的距离，大大提升居住价值！

The project is located in Daning in Zhabei Area, close to Daning commercial center and green space .

Considering the real estate project itself as a social, economic and humanistic complex and having not only pursued pure design concepts or forms, we adopt the most reasonable means to help developers achieve maximum social and economic profits, from perspectives of technique, construction, sales and marketing.

The major issue of the design is how to take advantage of the land's resources to create the most competitive product. Key points are analysis and assessment of market and the site, high value-added products, prospective positioning of target customers and differentiated competition strategies.

Therefore, our design concept is "the landscape will satisfy people's demand for quality, added value and difference to the maximum".

A new concept of prioritize "the landscape" is brought into the planning, combining natural water landscape with central green space and creating a living style of enjoying natural ecological landscape equally.

重视空间的整合设计，景观资源不同的地方采用不同的户型，侧向有景的，采用横厅设计，北向有景的，采用北厅设计，南北向均有景的，采用南北通的格局。

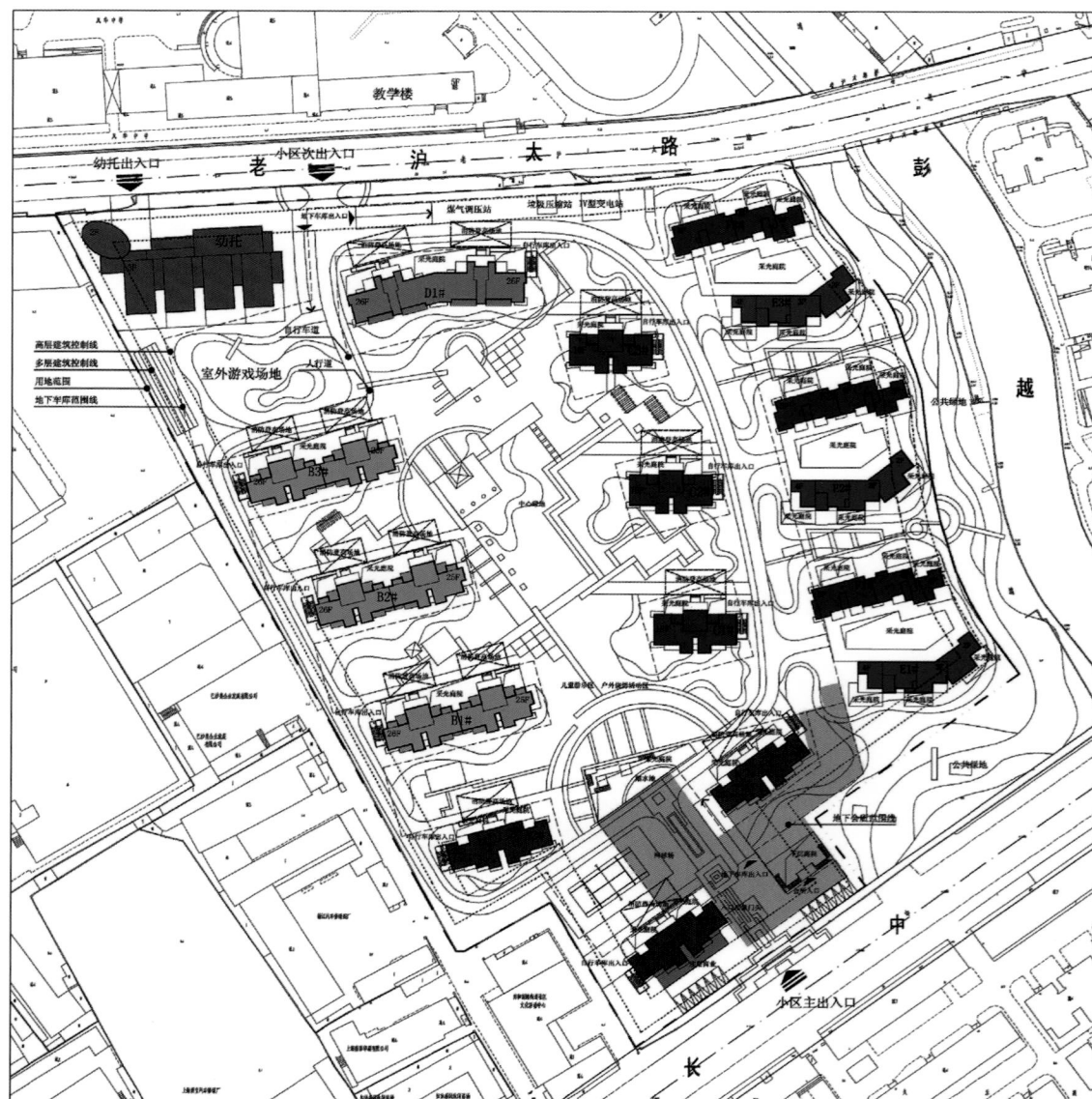

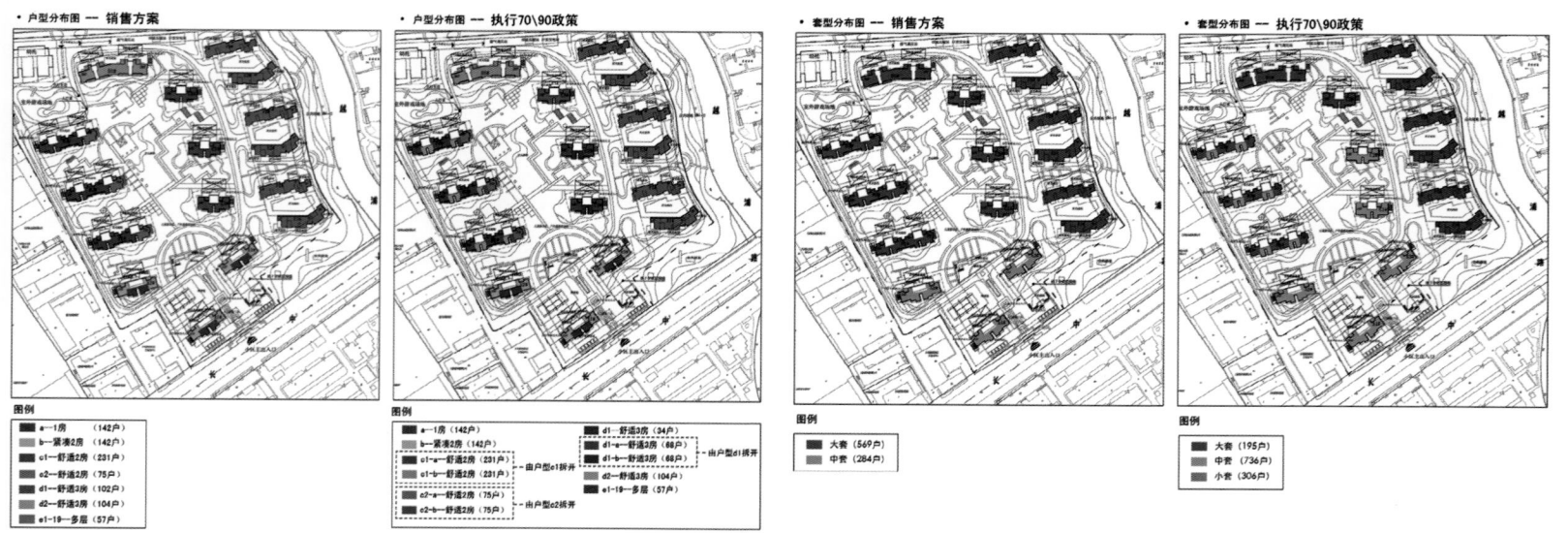

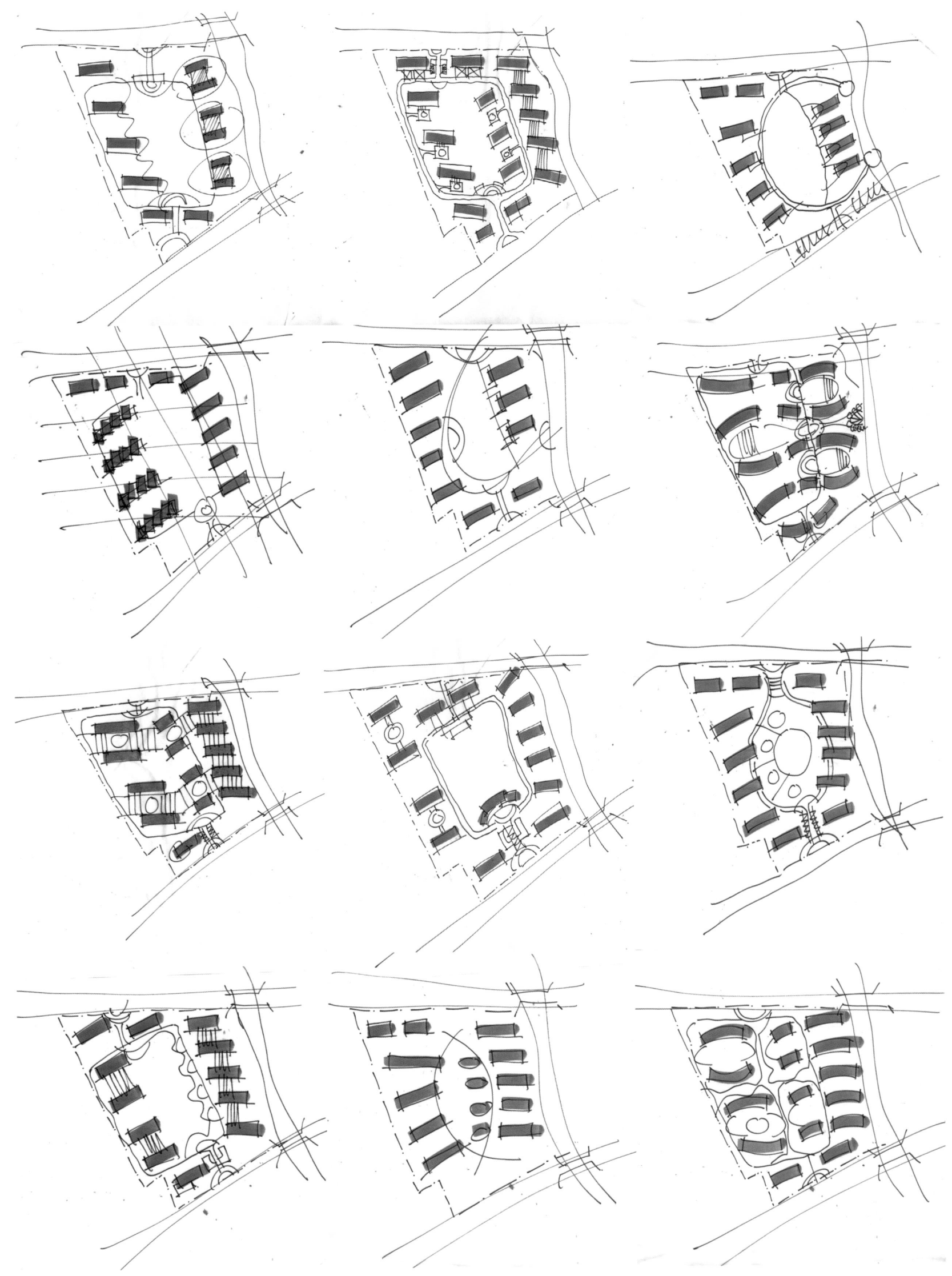

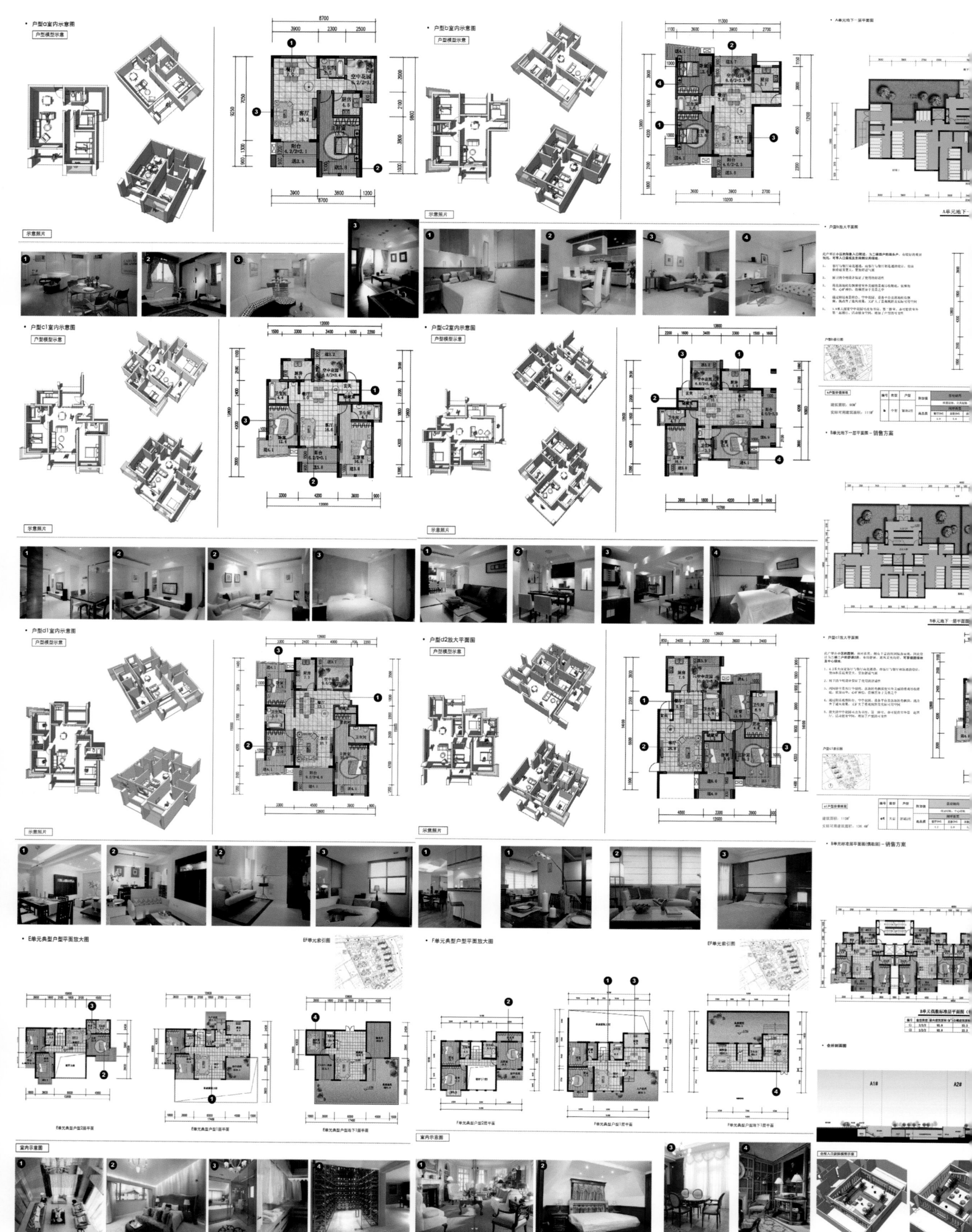

上海朱家角海源别墅项目
Shanghai Zhujiajiao Haiyuan Villa

Architecture Design | 2009·06

上海朱家角海源别墅项目
Shanghai Zhujiajiao Haiyuan Villa

项目名称：上海朱家角海源别墅项目
用地规模：19 万平方米
建筑面积：2.66 万平方米
容积率：0.14
项目地点：中国 / 上海 / 青浦
编制时间：2009 / 06

项目处于青浦朱家角行政区西北角，与"十五"期间市政府重点建设的朱家角新镇区一路之隔，相互依托，地理区位十分优越。它将成为长江三角洲地区最具有国际水准的高级社区，吸引国内外人士入住，是上海目前容积率最低、综合品质最高，完全按每户不同客户的要求定制的奢华级别墅。

设计原则

结合东方文化价值观与西方成熟的生活方式，形成新世纪上海顶级居住理念。追求个性、展现其文化价值，建设长江三角洲地区最具国际水准的高档小区，形成新时尚风格的别墅建筑。

建立一个平缓、统一的小区地形，从江南水系的自然形态出发，建立以平缓地形为主的基本地貌，调整河道走向，尽可能使大部分别墅临水，小区内基本保留了自然河道，并根据水资源调度规划和正常期河道的水流方向，将新挖河道与湖面相互串接，形成一个主次分明、流线通畅的动态水系，使小区内的排水更加顺畅自然。体现以水景为主题，大部分别墅临水的特色与品质。

This project is located in Zhujiajiao canton of Qingpu District, only a street away from the Zhujiajiao New Township which was under the government's key construction during "the 10th five-years". They support each other's development, which gives the project a good geological location. It aims at building a high-class community with the top international level in Yangtze River Delta region while attracting people at home and abroad to move in. It is a upscale villa with the lowest plot ratio and the highest quality in Shanghai. It is close to Fuxing Road to the east.

Principles of Design

Combining oriental cultural values and occidental mature life style, the project shall establish the top residential concept in the new century of Shanghai, to pursue individuality and show its cultural value. The project aims at building the top international level upscale residential area in Yangtze River Delta region and constructing villa buildings with new fashion.

Establish a flat and unified terrain in the residential area on the basis of natural features of water system ; establish the basic terrain mainly consisting of flat terrain; adjust the path of stream ways, so as to make most of the villas close to water. The project also connects the newly completed stream ways with lakes according to the planning and scheduling of water resources and the water direction at usual time, which is to form a dynamic unobstructed water system with a clear distinction between the major water and the subordinate one and make the disposal of water in the residential area smoother. The project shall represent the theme of waterscape as well as the feature and quality that most of the villas are near water.

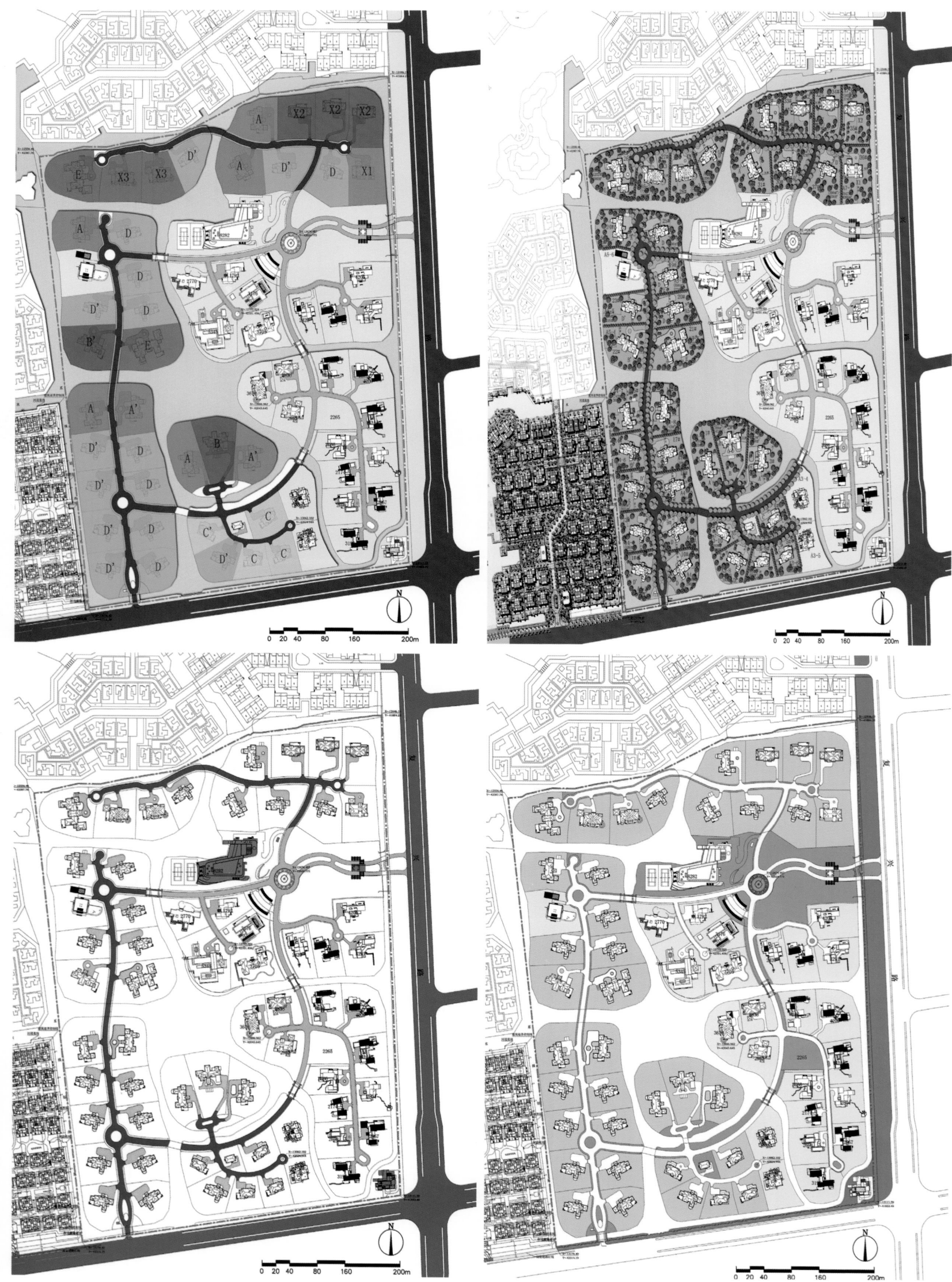

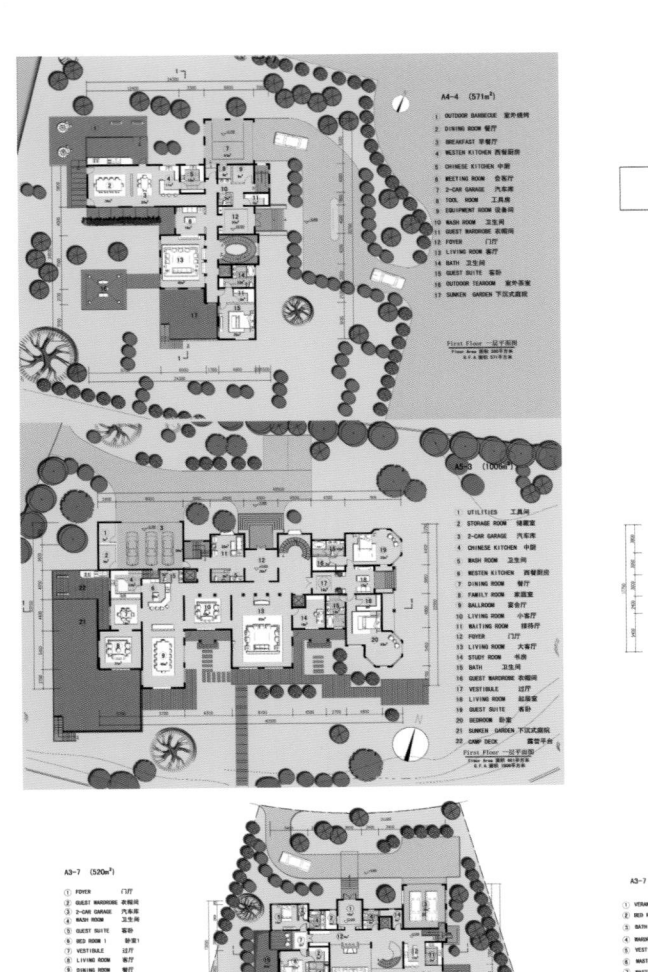

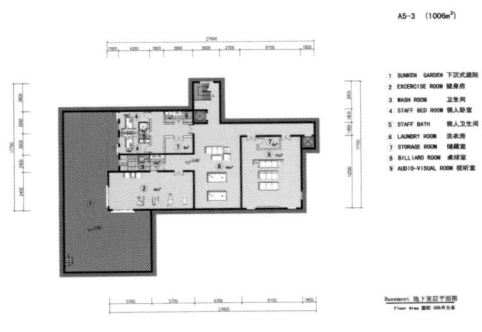

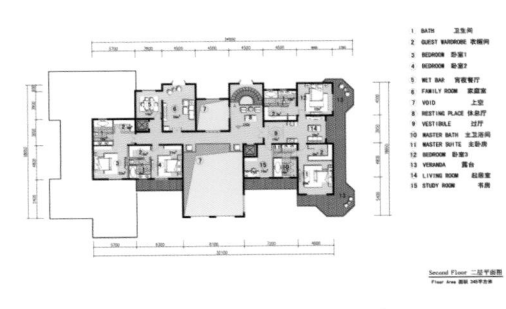

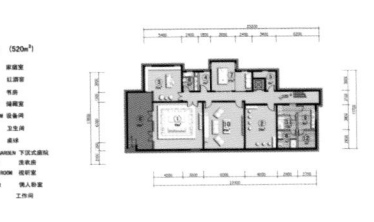

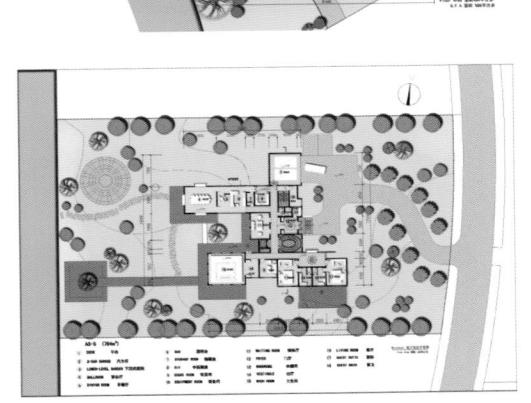

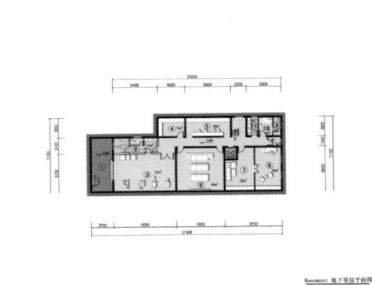

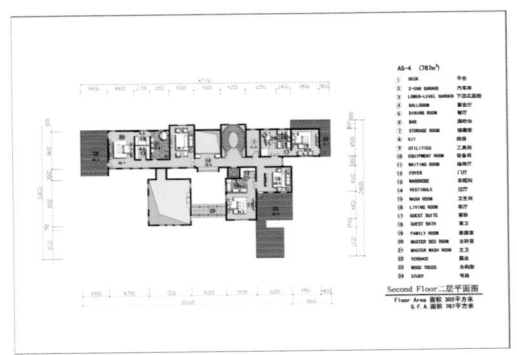

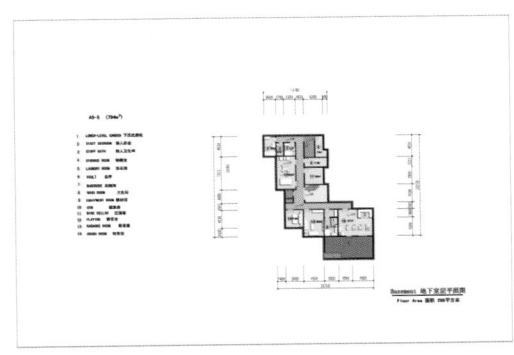

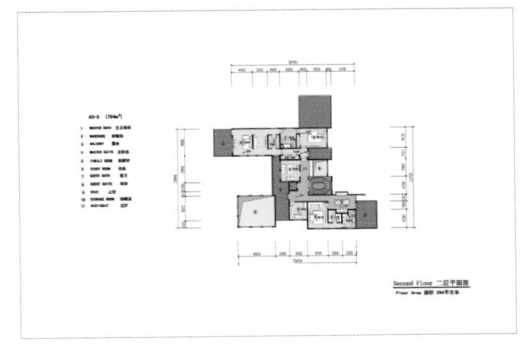

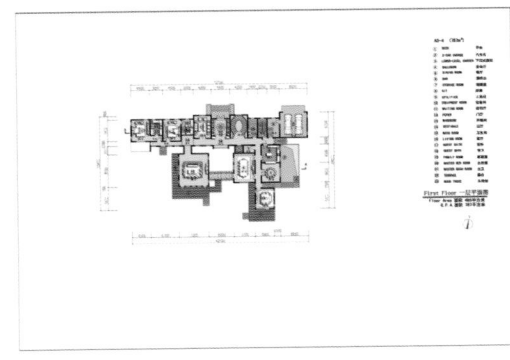

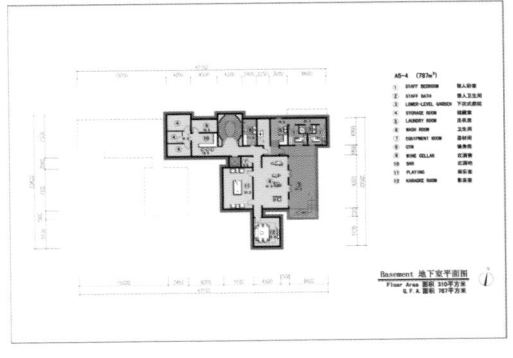

保利花园
Poly Garden

项目名称：保利花园
用地规模：23 万平方米
建筑面积：27 万平方米
容积率：1.2
项目地点：中国 / 四川 / 成都
编制时间：2009 / 10

位于成都的大型社区的成熟案例，新古典稍带现代风格的设计风格，严谨实用的总图规划，以及对建筑单体户型和整体均衡性的反复研究，将一个开发项目对市场品质的要求和利益追求平衡的表现出来，为中高档大型社区的开发模式提供了又一个坚实的证明。

The typical large-scale Community project in Chengdu, new-classical intergrated with modern style, the rigrous master plan and the repeated studies of the architectural layout and general balance ,represent the harmony between marketing quality and benifit pursuing. It also provide another development example for large-scale high class community .

成都鹏程金沙
Chengdu Pengcheng Jinsha

项目名称：成都鹏程金沙
用地规模：1.60 万平方米
建筑面积：9.75 万平方米
容积率：6.1
项目地点：中国 / 四川 / 成都
编制时间：2007 / 08

项目地处城西中房蜀都花园西侧，金沙遗址公园正西，是以居家为主的高品位住宅项目，现已成为城西标志性的居住亮点，在造型和外墙材料处理上，采用现代中式风格，以高档中空玻璃、面砖和花岗石为主，竖向垂直空调矩管及柱似线条使建筑显得俊秀挺拔，表现出时尚气派、端庄雅致的视觉效果。

The project is located just in front of the Jinsha Heritage Park. It's a high quality residential project and becomes a shinning landmark in the west of city. On the elevation and exterior materials, the designer adopts the modern Chinese style, mainly uses glass, brick and granite as materials. The vertical air condition pipe shapes straight building with elegant manner.

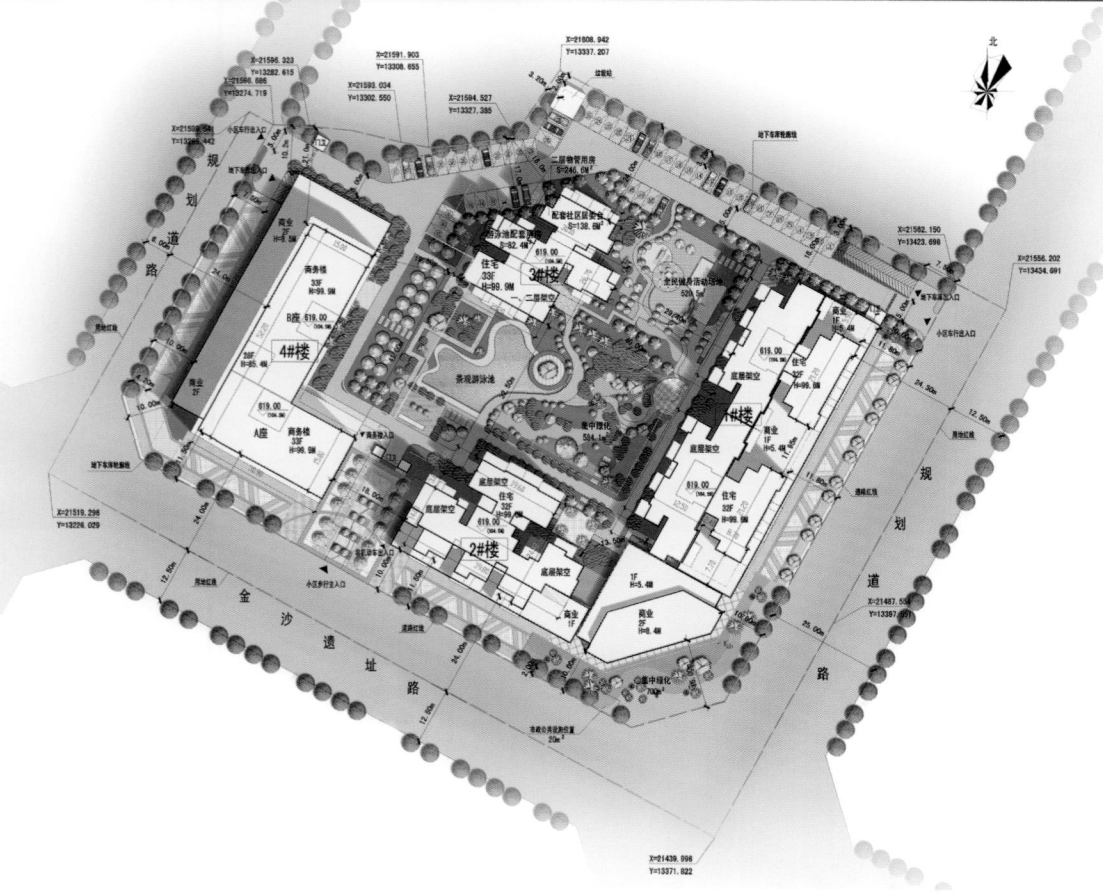

成都蜀郡住宅区
Chengdu Shujun Residential Area

项目名称：成都蜀郡住宅区
用地规模：12.63 万平方米
建筑面积：14.75 万平方米
容积率：1.17
项目地点：中国 / 四川 / 成都
编制时间：2006 / 09

设计完全承接了中国传统古镇老街的特色文化，如传统手工艺坊、茶馆、戏台等；运用大面积绿化分割了仿古建筑与道路，避免了仿古建筑与现代城市道路的格格不入，更达到了古镇老街的空前艺术效果。

The design carries on totally traditional old street features such as traditional handicraft, tea house, theatre stage etc; the large area green land divided the old style architecture and road, so as to avoid the large difference of archaistic building and modern urban road.

福建融侨水都五 / 六期方案设计
Fujian Rongqiao Water City Phase 5/6 Schematic Design

项目名称：福建融侨水都五 / 六期方案设计
用地规模：36.24 万平方米
建筑面积：74.51 万平方米
容积率：2.86
项目地点：中国 / 福建 / 福州
编制时间：2006 / 05

本项目特别强调进行居住建筑的空间实验，利用 15 度的概念，在平面上建立有规律的围合，在立面上形成律动的天际线；层层退台和多元的空间魔方既有规律地呼应菱形组团，又错开视野，强调了每户的均好性。

The project places special emphasis on space experiments in residential building. It strives to build regularly closed spaces on the master planning by using the concept of 15 degree. The project also adopts special method to design the fifth envelope of the building to expand the skyline of the high-rise residential building. The "step type" building and its multiple space cubes show the diamond group supported by the existing principles, stagger the residential horizon, and ensure equally good quality of each household.

哈尔滨绿地文景公馆
Harbin Greenland Wenjing Residence

项目名称：哈尔滨绿地文景公馆
用地规模：7.27 万平方米
建筑面积：35.62 万平方米
容积率：4.9
项目地点：中国 / 黑龙江 / 哈尔滨
编制时间：2010 / 05

由于规划用地及其周边环境的独特性，设计中既要对建筑进行统一合理的规划，充分考虑空间视线和住宅采光等需求，更要对这一区域进行整体的环境规划设计，以体现当地特有的地域特征及统一的城市环境。尊重历史，继承传统，延承城市文脉，重视人和生态环境的完美融合，将哈尔滨的城市自然环境特色和文化积淀有机结合，赋予文林社区住宅小区独特的文化内涵和人文关怀。

本社区采用经典高贵的新古典主义风格，并将其作为一种尊贵生活的代表符号，在融入了更多精神层面的东西之后，这些建筑也因此显示了极强的适应性和极好的融入性，在满足东方人的居住基础上，更产生了精神上的沟通和情感的共鸣，在传承中获得了新生的力量。新古典建筑散发的是具有厚度的形式美，摒弃繁复的艳媚装饰，以最经典的古典元素，最简约的表现手法，展现历史感和文化纵深感。另外，新古典主义建筑更加强调协调人与人之间，人与社会之间的关系和改善建筑的亲和性，把建筑从一个冷冰冰的物体，变成一种富有人情的空间：在建筑外立面色彩上揉和了明丽的建筑色彩，使用门窗、露台等经典流传的建筑表现元素，使得建筑外形洋溢着轻松、浪漫的生活气息，力图表现更加自然、更具时间感、更具人文精神的氛围，建筑与生活一同生长、变化，历久弥新。

Due to the uniqueness of the planned land use and its surroundings, the design should not only make the unified and rational planning for buildings taking demands of space, view and lighting into account, but also should make the comprehensive planning and design of environment for this area in order to represent the unique regional feature and unified urban environment. The design shall respect history, inherit traditions, continue the cultural context of this city, and value the perfect fusion of human and nature, to combine the natural environment of Harbin with its cultural accumulation and give the unique cultural connotation and humanistic care to Wenlin Community Residential Area.

This community adopts the new-classical style. As a symbol of honorable life, the style not only represents a form of architecture, but also contains more spiritual content,with which, those buildings have been given a great flexibility and harmony. Besides satisfying oriental residential demand, this community also establishes the spiritual communication and emotional resonance; it has gained a new life in the heritage. The new-classical architecture represents the rich beauty in form while abandoning those excessive decorations, to express the historical and cultural feeling through the most classical elements with the most contracted techniques. In addition, the new-classical architecture style emphasizes more harmony of between people and between human and the society as well as the improvement of the compatibility of architecture. Such style wants to turn architecture from a non-life object to a space full of spirit: it uses bright and beautiful architectural colors in elevation color usage; and it also uses other classical architectural expression elements such as doors, windows, and terraces which can give the architectural appearance the vitality of relaxation and romance. It tries hard to make its expression more natural, more historical and have stronger atmosphere of humanistic spirit. The architecture here shall grow, change and have its lasting charm along the life.

杭州和家园项目 A 组团
Hangzhou Harmonious Homes Project Section A

项目名称： 杭州和家园项目 A 组团
用地规模： 6.22 万平方米
建筑面积： 10.1 万平方米
容积率： 1.62
项目地点： 中国 / 浙江 / 杭州
编制时间： 2008 / 09

东方剑桥——融合了剑桥学院和东方园林精髓，具有纯正的英伦风情的住宅高档社区。

和家园 A 组团拥有杭州得天独厚的位置，以及原生态景观优势，这样一个幽静环境将酝酿出具有纯正的英伦风情与东方庭院意境相融合的围合院落空间。

住宅建筑主要以院落围合布置，使住宅空间更加有趣味性和生动性以及围合感，住宅之间形成相对集中的绿地及精致的英伦景观小品，极大地提升了小区的生活品质。建筑设计与园林景观设计统一考虑、有机结合，在承袭纯粹英伦的贵族和人文气质的同时，融入东方庭院的意境和精神真谛，使这个区域的建筑成为全区的视线焦点，当置身其中的时候，可以感受到英伦的幽雅与风情。

和家园 A 组团拥有绝版稀缺的自然与人文条件：生态环境优越，有挂牌山系环绕、天然水系、树木葱茏的生态大盘，这些唯和家园独有，因此它具有稀缺而不可取代的地段。

和家园是人文生态与自然生态并存的大型社区，通过合理的规划，使每户的窗口及露台都看得到山，"户户见山景"。凭借卓越的自然景观条件，其园区景观设计浑然天成，外围山景蔓延到小区内部形成拥有丘陵地形的疏林草地景观，并在周边山系设计登山游步道，与西湖风景名胜区直接相连。

凭借自然的恩赐，和家园的建筑就像是长在山上的房子。充分利用周边山系及坡地，将山景景观资源最大限度引入户内，无论 90 平方米的公寓还是大户性豪华住宅，户户可享山景，在自然中体味居住本色；依靠自然的坡地，结合英伦风情式的经典建筑，融入东方庭院的意境，充分满足业主对居住品质的追求，也保证了山景健康的生活品位。

The oriental Cambridge-An upscale residential community with pure British style blending the essence of Cambridge College and oriental courtyard.

The Group A of Harmonious Homes Project occupies an advantageous location in Hangzhou and possesses a great advantage of original natural landscape. Such a peaceful scene shall make this land an enclosing courtyard space blending both pure British charm and the artistic conception of oriental courtyard.

The residential buildings are mainly distributed according to enclosing courtyard, which makes the residence space more interesting and lively, forming a relatively concentrated green area and delicate British scenery and greatly upgrading the quality of life in the residential area. The architectural design and landscaping design are considered as a whole and combined with each other, blending the artistic conception and true essence of oriental courtyard while adopting the pure British noble and cultural elements. This shall make architectures in this area the visual focus within the whole region. Being situated in such a scene, you can feel the grace and charm of Great Britain.

Group A of Harmonious Homes Project has the scarce natural and cultural advantages: it is a residential area with excellent surroundings of listed hill system, natural water system and lush forest. Such an irreplaceable plot can only be found in Harmonious Homes.

The Harmonious Homes is a large community with both human ecology and natural ecology. Through the rational planning, the hills can be seen from every house, achieving the goal of "Every house shares the hill scene". Depending on the excellent natural landscape advantage, the scene design inside is like nature itself; the surrounding hill scenes extend into the residential area which forms a scene of open forest and grassland with undulating topography. The design also arranges hill-climbing travel trails on the surrounding hill chain, which connect with scenic spots and historic sites of Xihu directly.

By the gifts of Nature, buildings of Harmonious Homes are just built in the hill. These buildings take full advantage of the surrounding hill chain and hillside field, giving every house the hill scene as much as possible. No matter the 90 m² apartments or the large dwelling-size luxurious residences, every house can share the hill scene and enjoy the natural residence. With the natural hillside fields, it combines the classical architectural style with British cultures and customs and blends the artistic conception of oriental courtyard, which shall fully satisfy the owners' high-quality expectations and also ensure the healthy mountain life taste.

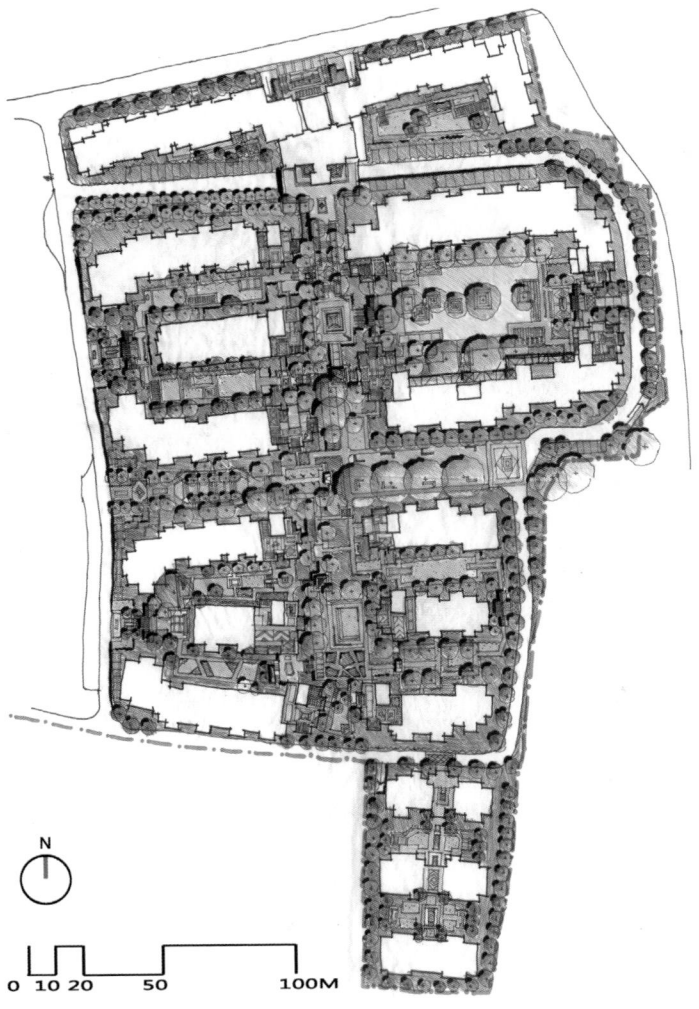

红蜻蜓机场镇 C 地块别墅
Red Dragonfly Airport Town Villa in Block C

项目名称: 红蜻蜓机场镇 C 地块别墅
用地规模: 14.85 万平方米
建筑面积: 4.78 万平方米
容积率: 0.32
项目地点: 中国 / 上海
编制时间: 2004 / 03

针对红蜻蜓项目总体 854 亩地块的特征及特点，我们将其定位为"乡村里的休闲都市"。将乡村的休闲生活和小城市的便捷生活有机地结合起来，以丰富而具层次感的空间舒展的田园生活，以规范而具情趣感的构图创造方便的都市居住环境，并以此作为我们的设计主旨，将这个主导思想贯穿于我们的设计中，形成本地块的特点；使景观和商业以本项目为中心贯联起来，对周边地区形成影响，乃至改变周边地区现有的氛围。大片的景观布置在组团空间的主轴上，小城周边设计环区景观带，将整个小城包围在绿色之中，同时也改善了交通繁忙的省道和基地周边农村居点对基地的不良影响。景观设计最大限度地利用原有资源，保留原有地貌，一切都以自然为主题，精致而毫不张扬。

Depending on features of 854-acre block, we position it as "a city of leisure in rural area". Combining rural life with convenient life in the city, the concept of rural community is thus formed on the basis of space of depth to create a city living environment of convenience with drawings of standards and interest, which is also our design concept that connects landscape with commerce. The integral project is surrounded by rings of green landscape, reducing adverse influence by surrounding traffic and village communities. Its landscape design makes utmost use of natural resources and preserves original land shapes. Slopes, springs, forest and path seem so natural that people walking among them can feel so close to Nature.

淮安富仕达国际花园
Huai'an Fushida International Garden

项目名称：淮安富仕达国际花园
基地面积：6.53 万平方米
建筑面积：18.92 万平方米
容积率：2.9
项目地点：中国 / 江苏 / 淮安
项目制作时间：2009 / 10

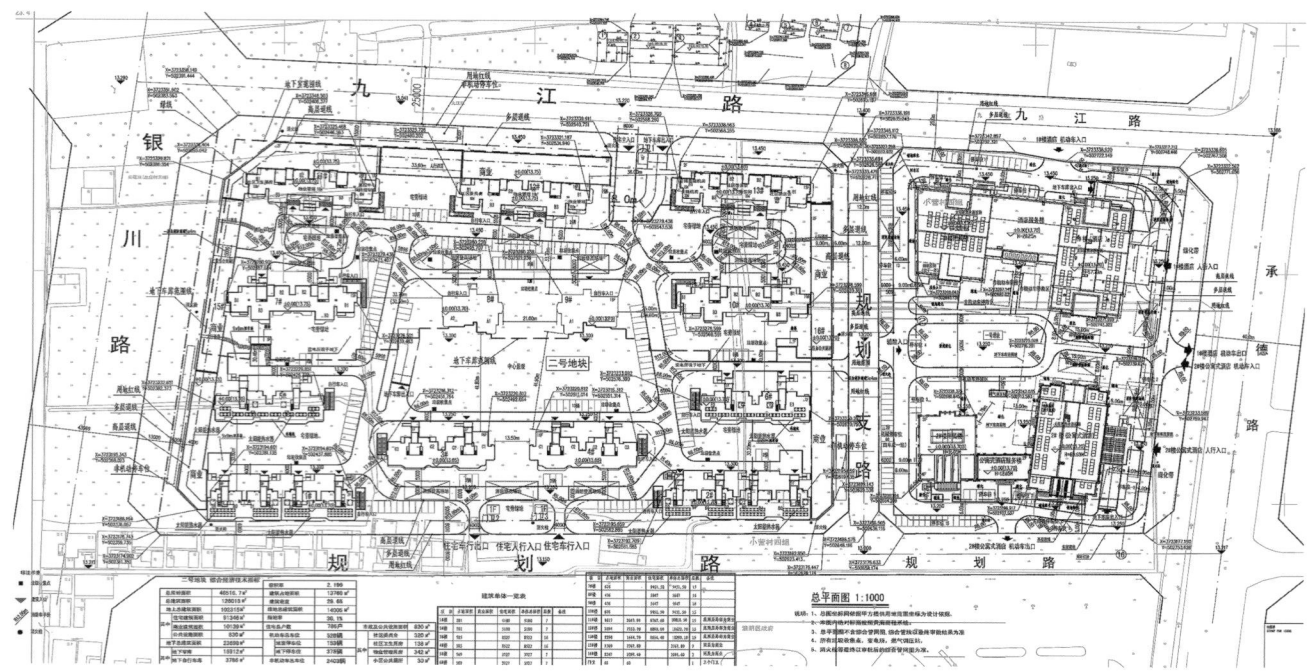

商业酒店地块作为淮安的"城市名片"及淮安北部新城北大门地标，设计充分考虑人性化的空间景观和满足管理公司要求的内部使用空间，及与周边地块的充分融合。

住宅社区地块与淮阴区政府大楼隔楼相望，为项目提供开放的公共环境景观，同时为淮阴新区建设高品质住宅小区提供优越条件。

住宅规划以短板和点式相结合的平行四排布局，高度、疏密有致，各组团公共绿地相互渗透，形成"一中心两轴四组团"的模式，"两轴"既南北向贯穿小区，联系南面区政府及背面火车新站的景观主轴，又在东西方向联系二号地块的园区景观渗透轴线。

The commercial hotel plot and Huai'an North New City fully integrates with humanistic landscaping and fulfills the internal management requirement, at the same time is integrated with the surrounding plot.

The residential plot is opposite to HuaiYin District government building, providing open public view, at the same time providing good condition for high-quality residential community.

The plan adopts the parallel 4-row layout combining short slab and point mode with the orderly height and concentration. The public green areas of various groups blend with each other, forming the mode of "one center, two axis and four groups". One of the "two axis" represents the north-south landscape major axis which passes through the residential area and connects the district government in the south as well as the new railway station in the north; and the other of the "two axis" represents the east-west park landscape blending axis which connects the plot No.2.

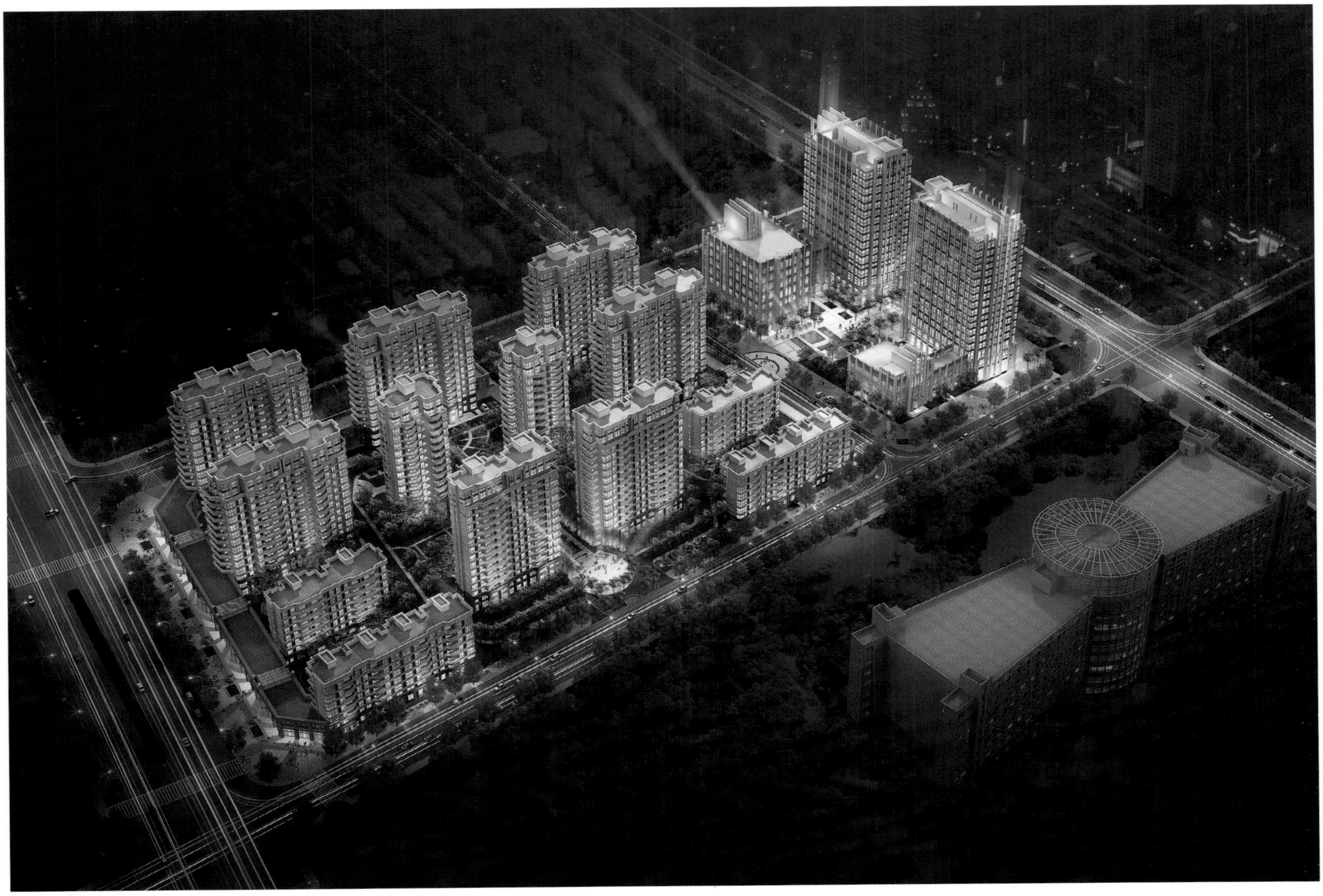

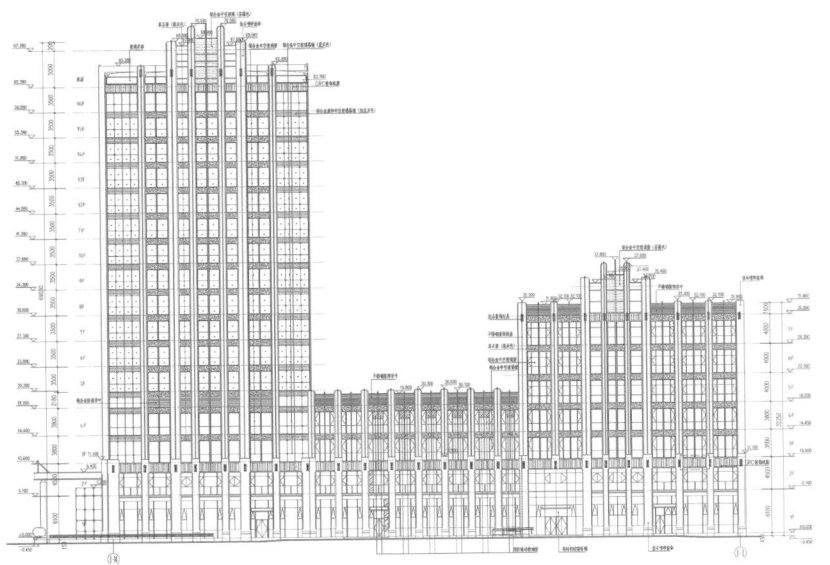

昆山东森花园
Kunshan Dongsen Garden

项目名称：昆山东森花园
用地规模：12.11 万平方米
建筑面积：14.21 万平方米
容积率： 1.17
项目地点：中国 / 江苏 / 昆山
编制时间：2007 / 12

昆山东森花园位于昆山市西部居住生态区，其设计的精髓在于营造一种充满英伦风情的小城居住模型，让人们在忙碌的生活中寻找到轻松的居住感受。通过塑造典雅庄重的英伦小镇的建筑风貌和交通流线，高绿化低密度的田园风情，成为体现 21 世纪新城市生活的特色生态居住示范区。

Located in the ecological residential zone of west Kunshan, Kunshan Dongsen Garden stretches to Hongqi Rd in the east and Huayuan Rd in the south. It occupies a cubic area of 121,077 square meters and enjoys a flat terrain. The essence of the project lies in the creation of a small town full of British charm so as to provide busy people with ideal residential joy. It presents a unique ecological residential model zone of city life in the 21st century by adopting an elegant and solemn architectural style with British charm and building a pastoral amour with high greening and low population density.

昆山龙隐水庄
Kunshan Longyin Water Village

项目名称：昆山龙隐水庄
基地面积：6.39 万平方米
建筑面积：8.27 万平方米
容积率：1.3
项目地点：中国 / 江苏 / 昆山
编制时间：2007 / 06

整个小区包括住宅、商业和酒店式公寓，设计考虑住宅、空间、环境三位一体，商业以及酒店式公寓立面沿大学路商业立面，完全结合了住宅立面新中式主义风格，在街角以及住宅入口处设置塔楼以增加项目的可识别性和标志性。酒店式公寓立面利用开敞式柱廊以及底层增添的细部小拱圈丰富了与内院空间的交流。

The whole residential area includes residences, stores and apartment hotels. Its design takes residence, space and environment together into account and makes these three complement each other. According to the elevation design features of commercial and hotel apartments, the planning of commercial elevation along Daxue Road completely combines with the new Chinese residential style. Towers are established in street corners and entrances of residences in order to promote the identifiability and significance of the project. The elevation of hotel apartments uses open type of colonnades and adds arch rings to the bottom which enrich the communication with inner courts.

朗基望今缘
LangJi WangJinYuan

项目名称：朗基望今缘
用地规模：1.46 万平方米
建筑面积：5.98 万平方米
容积率：4.0
项目地点：中国 / 四川 / 成都
编制时间：2003 / 05

上海奥林匹克花园三期
Phase III of Shanghai Olympic Garden

项目名称：上海奥林匹克花园三期
用地规模：14.6 万平方米
建筑面积：21.9 万平方米
容积率：1.5
项目地点：中国 / 上海
编制时间：2005 / 01

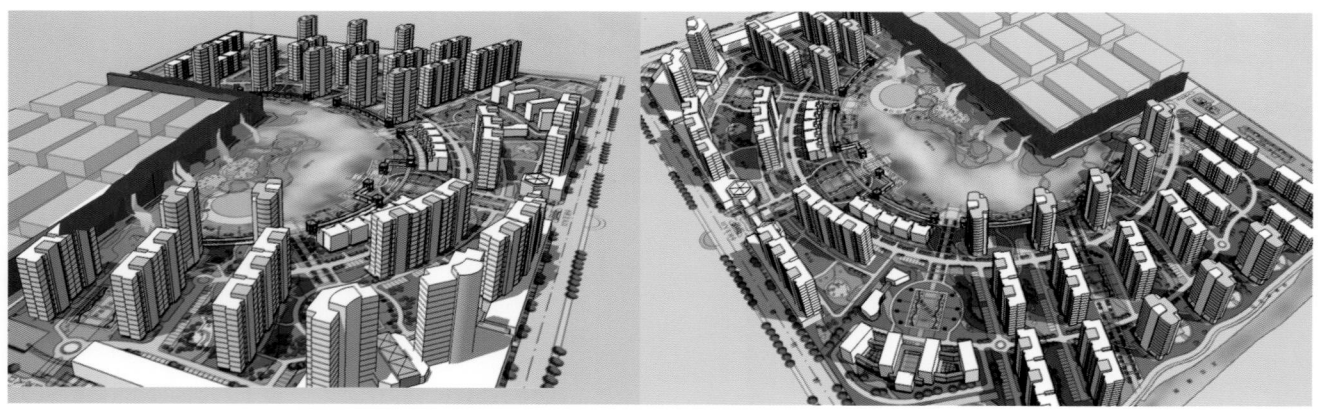

上海奥林匹克花园三期住宅位于上海市松江新区，美丽的自然河道，清新的田野空气和一二期已经提供的丰富的体育休闲娱乐生活设施将为三期的人们实现生活理想提供无与伦比的机会。

方案将体现新时代生活的特点，同时延续奥林匹克的文化底蕴与体育健康精神。整个社区将具有卓然不同的社区认同感，它安享宁静，呵护家庭的成长。以现代的建筑风格，轻松简洁的建筑形态和色彩，灵活的布局和松弛的结构（大绿地、多高层），集中展现活力、激情、健康生活的本质。

Phase III of Shanghai Olympic Garden is located in Songjiang New District of Shanghai and enjoys beautiful natural watercourses, pure and fresh field air and colorful sports and leisure entertainment facilities provided by Phase I and II projects, which offer people unparalleled opportunities to pursue ideal lives.

The design concept is supposed to project the characteristics of the new age life and meanwhile continue to cherish the cultural deposits and sports and health spirit of the Olympics. The whole community boasts a quite brilliant community identity and cherishes quietness and family growth. By adopting a modern architectural style, it well presents the essence of vigorous, energetic and healthy life with easy and simple architectural patterns and colors, flexible layout and loose structure (large areas of greenbelt and many high-rises).

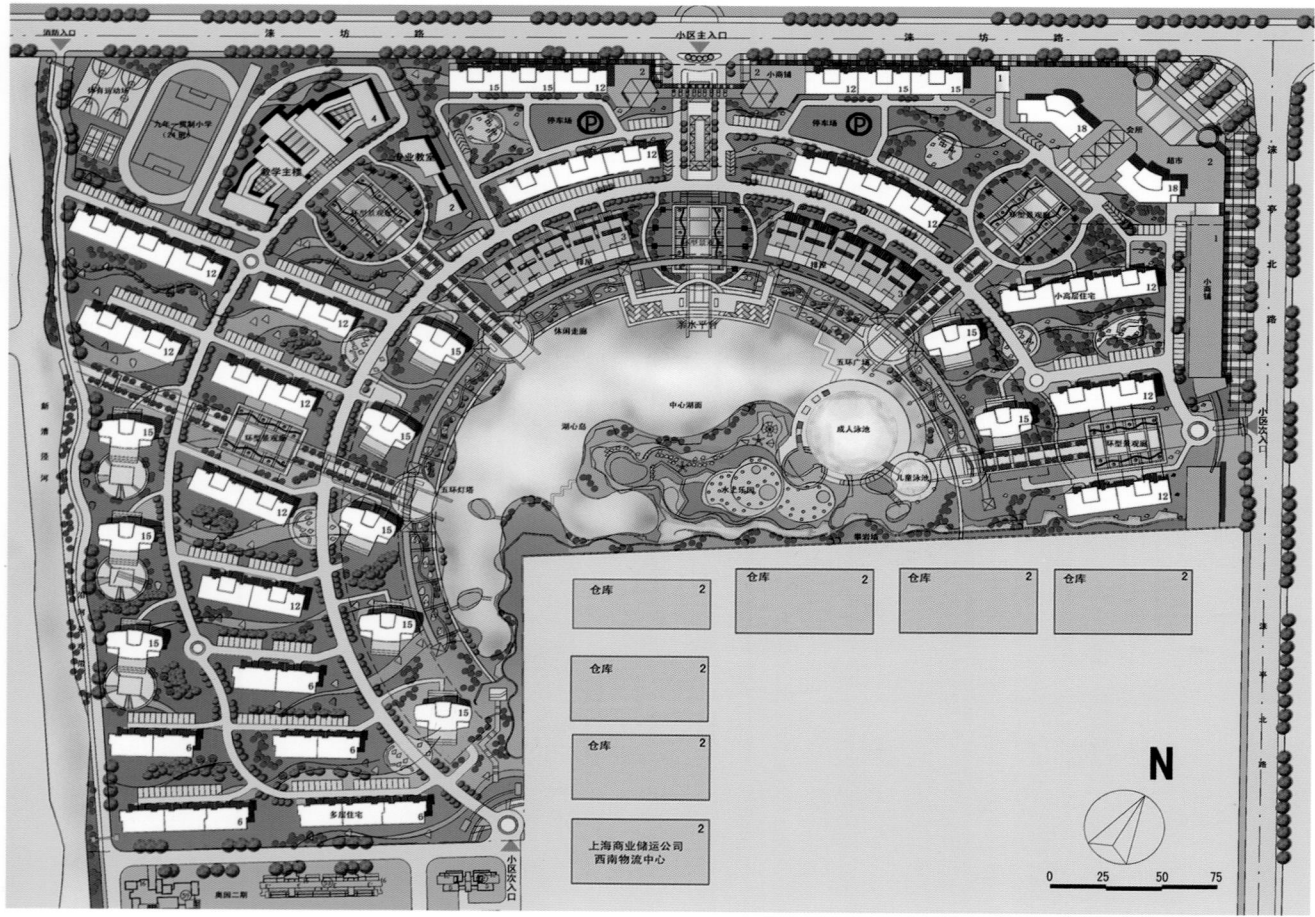

乐清上海花园
Yueqing Shanghai Garden

项目名称：乐清上海花园
用地规模：23.17 万平方米
建筑面积：11.66 万平方米
容积率：0.5
项目地点：中国 / 浙江 / 乐清
编制时间：2004/ 05

针对上海花园地块的特征及特点，我们将其定位为"都市中心的休闲社区"。将乡村的休闲生活和小城市的便捷生活有机地结合起来，以丰富而具层次感的空间形成田园社区理念，以规范而具有情趣的构图创造方便的都市居住环境，并以此作为我们的设计主旨，将这个主导思想贯穿于我们的设计中，形成本地块的特点：使景观和商业以本项目为中心串联起来，对周边地区形成影响，乃至改变周边地区现有氛围。大片的景观布置在组团空间的主轴上，用地周边设计环区景观带，将整个小城包围在绿色中，强调"以人为本"，加强人性感受。

In accordance with the features of Shanghai garden, we position it as "a community of leisure in central city". Combining rural life with convenient life in the city, the concept of rural community is thus formed on the basis of space of depth to create a city living environment of convenience with drawings of standards and interest, which is also our design concept that connects landscape with commerce. The integral garden is surrounded by rings of green landscape, highlighting "people orientation".

南立面图

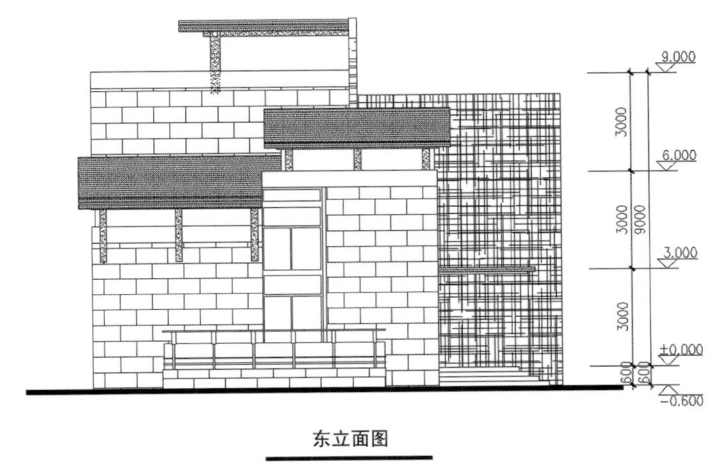

东立面图

成都礼顿山
Chengdu Leighton Hill

项目名称：成都礼顿山
用地规模：3.2 万平方米
建筑面积：12 万平方米
容积率：3.75
项目地点：中国 / 四川 / 成都
编制时间：2010 / 10

基地的利用实现最大化，围而不和的庭院带来的是灵动而富于变化的空间，丰富多彩的立面效果诠释着建筑的价值。

The base is utilized maximally. What the surrounding but unenclosed courtyard brings about is an animated and changeful space, and the colorful facade effects expound the values of the buildings.

洛阳顺驰城
Luoyang Shunchi Center

项目名称：洛阳顺驰城
用地规模：11.2 万平方米
建筑面积：25.44 万平方米
容积率：2.27
项目地点：中国 / 河南 / 洛阳
编制时间：2005 / 12

作为洛阳市大型高档住宅区将运动的深层次内涵，即"乐观、健康、公平、和谐"作为主线，同时以"健康住宅"的概念及标准来规范设计，强调健康与和谐，将运动文化与社区文化、居住文化有机地互相融合。坚持"以人为本"，巧妙地在规划设计中诠释运动的精神，并进一步地延伸健康住宅的深刻含义，使住宅、商业、体育设施、生活配套以及休闲景观之间相互依存且动静分明，共同构筑出一个现代化的运动社区。

健康住宅在中国是一个全新的居住概念，有别于绿色生态住宅及可持续发展住宅的概念，不仅仅重视环境，重视山、水、土、石、绿地、阳光、空气等组成的要素，同时也重视居住者生理及心理的健康，一切从居住者出发，满足居住者生理和心理健康的需求，使之生存在健康、安全、舒适和环保的室内和室外的居住环境中。好的住宅不仅提供人们基本的生活居住条件，而且对人们修身养性、培养情操、提高卫生文明有着重要的作用。

Having held the in-depth connotations, i.e. "optimism, health, justice and harmony" as its main purpose, the buildings' design complies with health and harmony and combines sport culture, community culture and living culture together. The design describes sport spirit subtly with people-oriented concept and further extends the profound implications of healthy residence, making residence, commerce, sport facilities, living accessories and leisurely landscape exist in great harmony.

Residence of health is a new concept in China, which differs from green eco-residence and sustainable residence. Not only does it pay much attention to the environmental factors such as mountain, water, soil, stone, green land, sunshine and air, but it also cares about people's physical and psychological health. The concept is of people-orientation, satisfying their physical and psychological health demands and providing them with both internal and external living environment of health, safety, comfort and environmental protection. It is convinced that a quality residence can do good to people's health, mind and civilization in addition to basic living conditions.

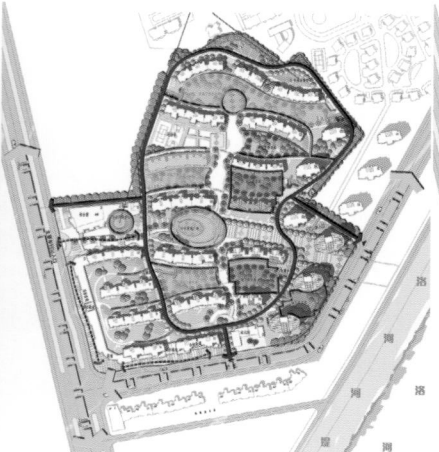
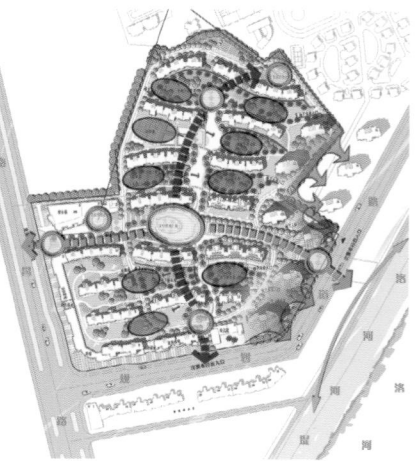

牧山书院
Mushan Academy

项目名称：牧山书院
用地规模：1.03 万平方米
建筑面积：1 674.3 平方米
容积率：0.162
项目地点：中国 / 四川 / 成都
编制时间：2009 / 03

我们的设计理念来自 "牧牛图"， "明心见性牧牛处，开坛论道山郎庐"。唐代的大禅师们——诸如马祖道一、百丈怀海——很喜欢用 "牧牛" 比喻 "治心"，即将牧童比作 "人"，将 "牛" 比作 "心"，或是将牧童比作 "心"，而将牛比作 "性"，以十牛比作修心的十个阶段。普明的《牧牛图颂》所绘之牛由黑变白，分成未牧、初调、受制、回首、驯伏、无碍、任运、相忘、独照、双泯十个阶段，其着眼点在于调心证道，以人牛不见、心法双亡为最高境界。

Our design conception is derived from Chinese classical painting "Ox-herding Pictures". The real heart and essential nature are discovered and realized where the ox is herded, a forum is opened and the truth discussed in a mountaineer's cottage". Great Zen masters of the Tang Dynasty, such as Mazu Daoyi, Baizhang Huaihai, liked very much to compare "herding the ox" to "governing the heart", which is to liken the cowboy to the man, and the "ox" to the "heart", or to liken the cowboy to the "heart", and the ox to the "character", further to liken the ten oxen to the ten phases of cultivating the heart. The oxen described in Puming's Ox-herding Illustrations and Verses, becoming white from black, are divided into the ten phases of having not been herded yet, being preliminarily regulated, being controlled, turning the head, taming, being not hindered, moving at will, forgetting each other, seeing the true nature with a clear mind, and achieving Nirvana, the focus thereof is on adjusting the heart and demonstrating the truth, and the supreme realm is that neither the man nor the ox is seen, and both the heart and methodology are extinct.

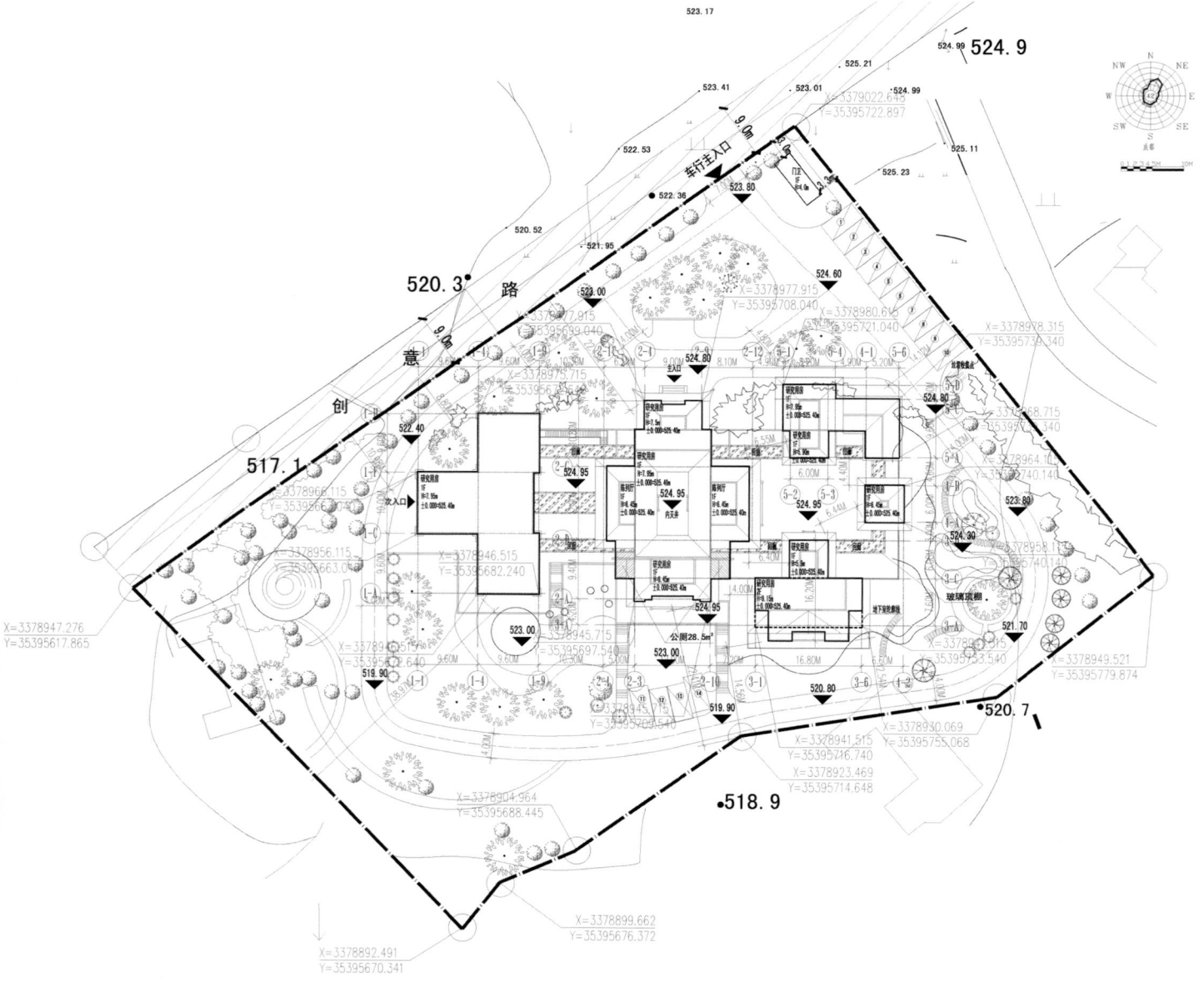

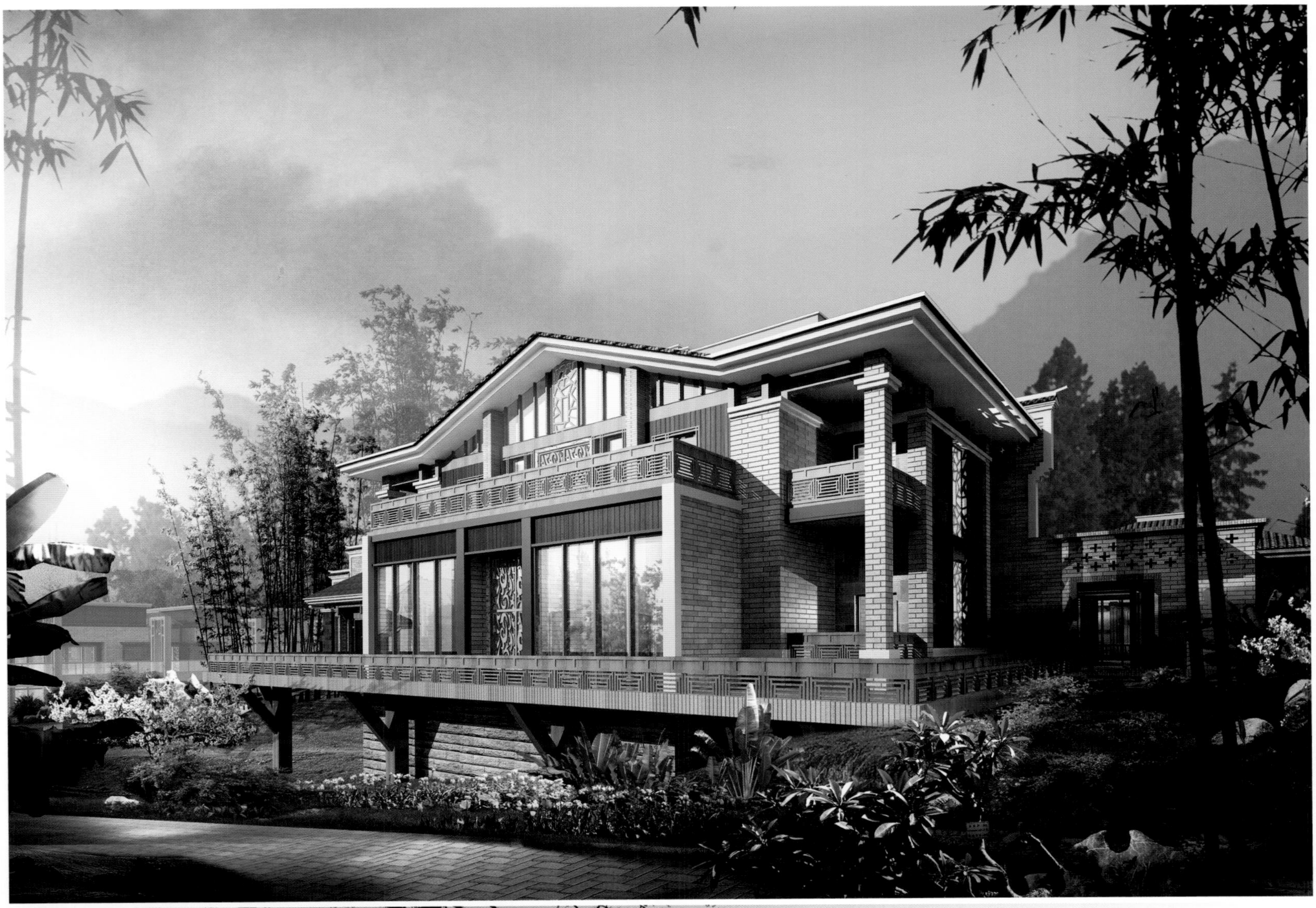

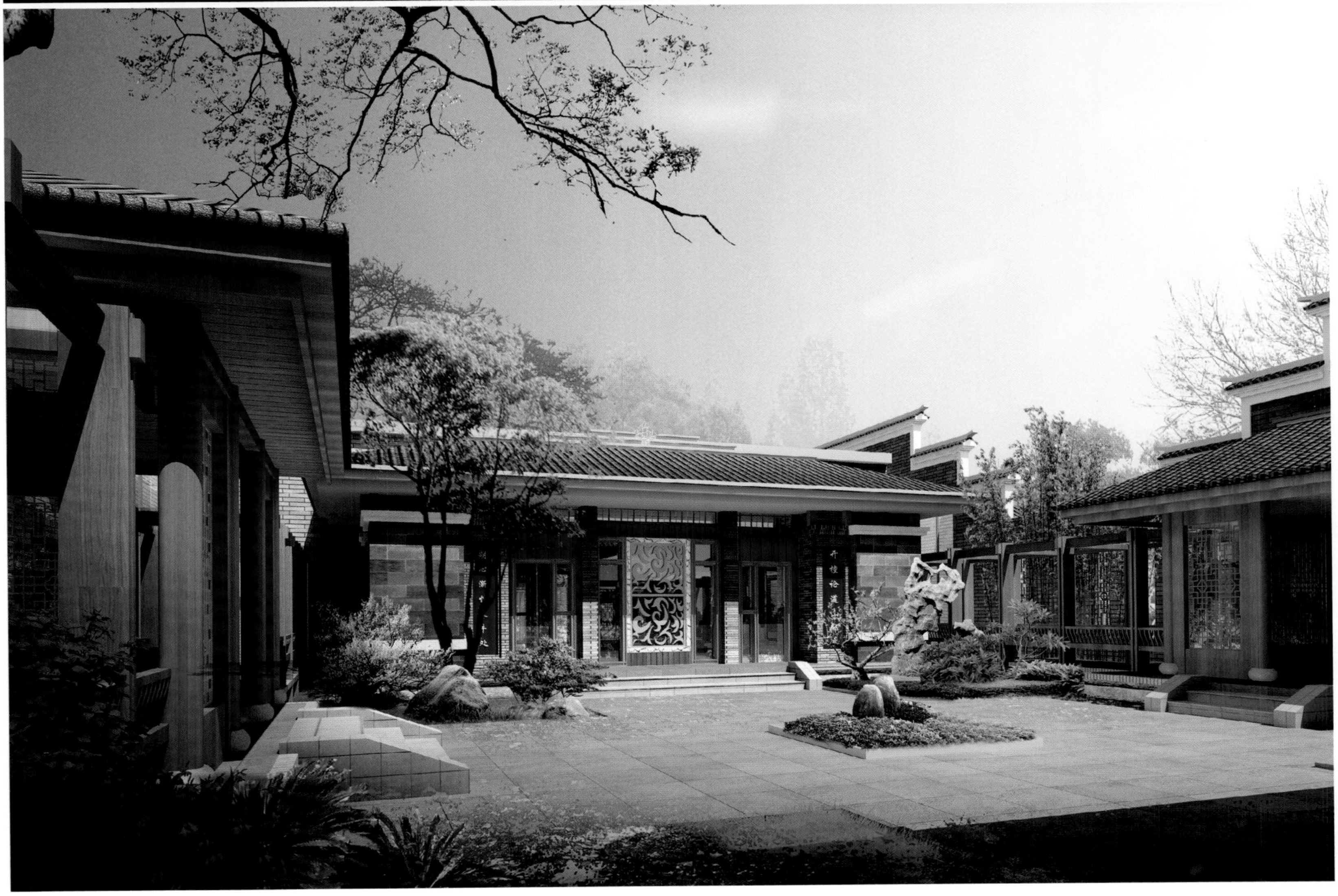

靈犀一點通古道自新星
猶勞徙筆暖毫襌意蜚糊
天青驟巨峰地镜化奇峰
細寫诗出喜半扣最勤慎

南京苏宁环球朝阳山
Nanjing Suning Global Zhaoyang Mountain Project

项目名称：南京苏宁环球朝阳山
用地规模：6.98 万平方米
建筑面积：5.57 万平方米
容积率：0.8
项目地点：中国 / 江苏 / 南京
编制时间：2009 / 10

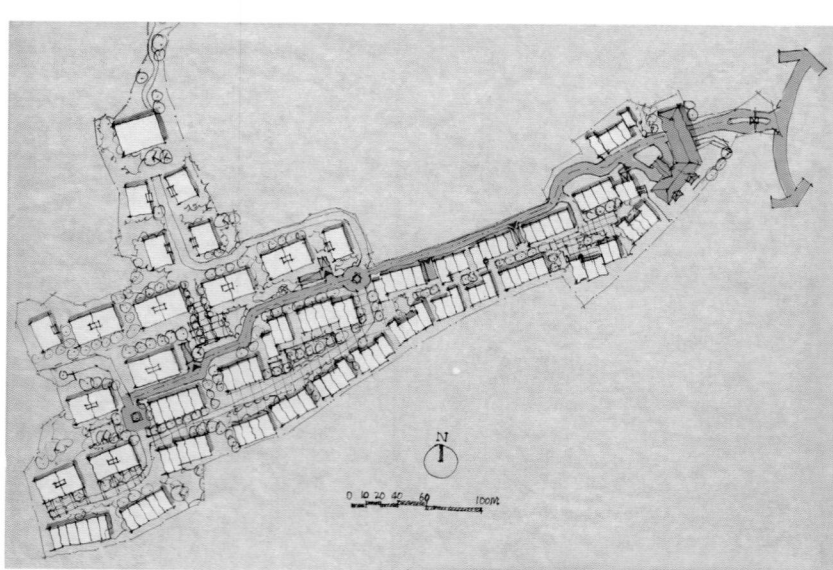

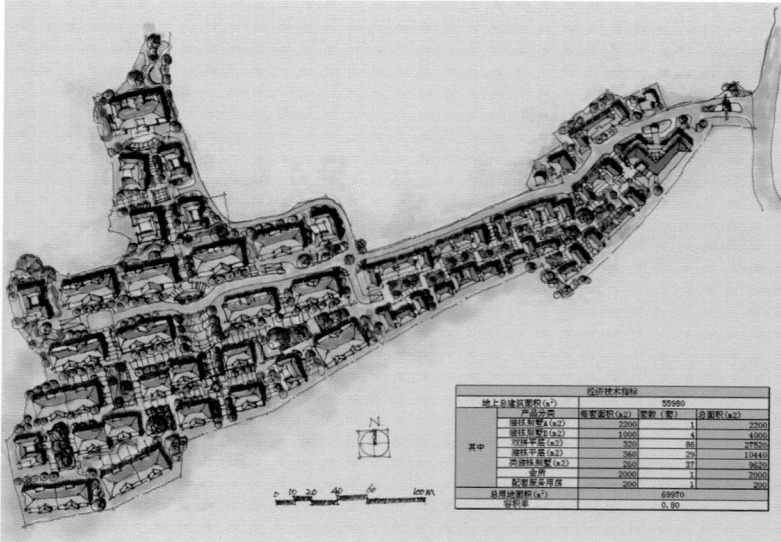

本项目基地内最大高差约 20 米，整体上南低北高，东低西高，并且微地形复杂，存在大量的陡坎和护坡。

产品开发类型以类独栋别墅和大平层豪宅为主，类独栋别墅近似于传统的联排别墅产品，但在私密性和独立性上更优于联排产品，同时通过邻里单元围合创造出内向性坊间合院空间。

大平层豪宅每户独占一层，270 度以上景观视野，同时每户均有独立的入户大堂和私家电梯。

整体布局上与地形地势紧密结合，形成层叠错落的整体效果，并且实现人车分离的交通系统，车行道路低于建筑首层地面，最大可能减少干扰。

There is obvious altitude difference in the project, about 20m to the maximum. Generally, it is low in the south and the east and high in the north and the west with complicated micro topography and a plenty of steep slopes and protection slopes.

Its products are mainly detached villa and large flat mansion. The former is similar to row villa other than with more privacy and independence, which creates internal courtyard space through neighbor enclosure. Its architectural style is new classics of new features with full stone materials on the elevation and high-efficiency and energy-saving intelligence system as highlights to improve integral quality.

Large flat mansion indicates one storey for each family with above 270° landscape view and there are independent lobbies and private escalators for each family.

Its integral layout is closely combined with land shape, forming cascading expression and realizing a communication system separating people from vehicles. Driveways are lower than the ground floor, which can reduce interruption as much as possible.

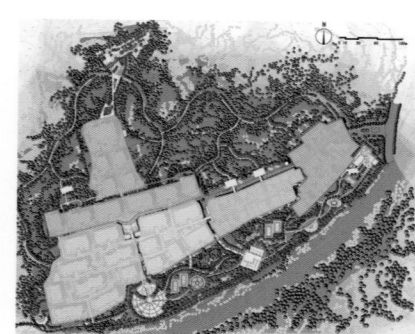

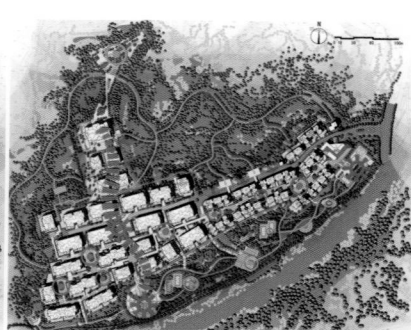

内蒙古呼和浩特范家营项目
Inner Mongolia Hohhot Fanjiaying Project

项目名称：内蒙古呼和浩特范家营项目
用地规模：30.38 万平方米
建筑面积：51.86 万平方米
容积率：1.7
项目地点：中国 / 内蒙古 / 呼和浩特
编制时间：2006 / 04

整个项目区内设有动迁安置房、公寓、城市联排、情景洋房、双拼别墅、社区大卖场、小学幼儿园、商业街、独享私密会所、星级酒店及一块开放式集中绿地。规划设计通过因地制宜合理配置将该地块打造成未来呼和浩特市的高尚居住中心的核心。

This project is located in the central area of the city in Hohhot. There are many kinds of facilities in the community such as allocation apartment, normal apartment, townhouse, supermarket, primary school, commercial street , private club and high-class hotel. The project will become the residential center of Hohhot.

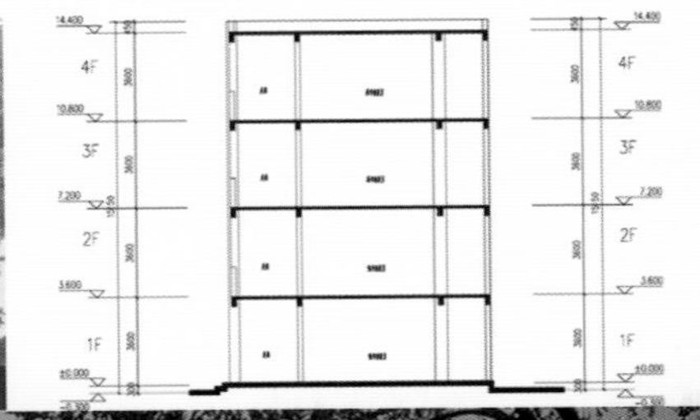
14.400
4F
3600
10.800
3F
3600
7.200
2F
3600
3.600
1F
3600
±0.000
-0.300

上海绿地花桥青年城以东住区
Shanghai Greenland Huaqiao Youth's City and Residential Area

项目名称：上海绿地花桥青年城以东住区
用地规模：36.10 万平方米
建筑面积：101.07 万平方米
容积率：2.8
项目地点：中国 / 江苏 / 昆山
编制时间：2007 / 08

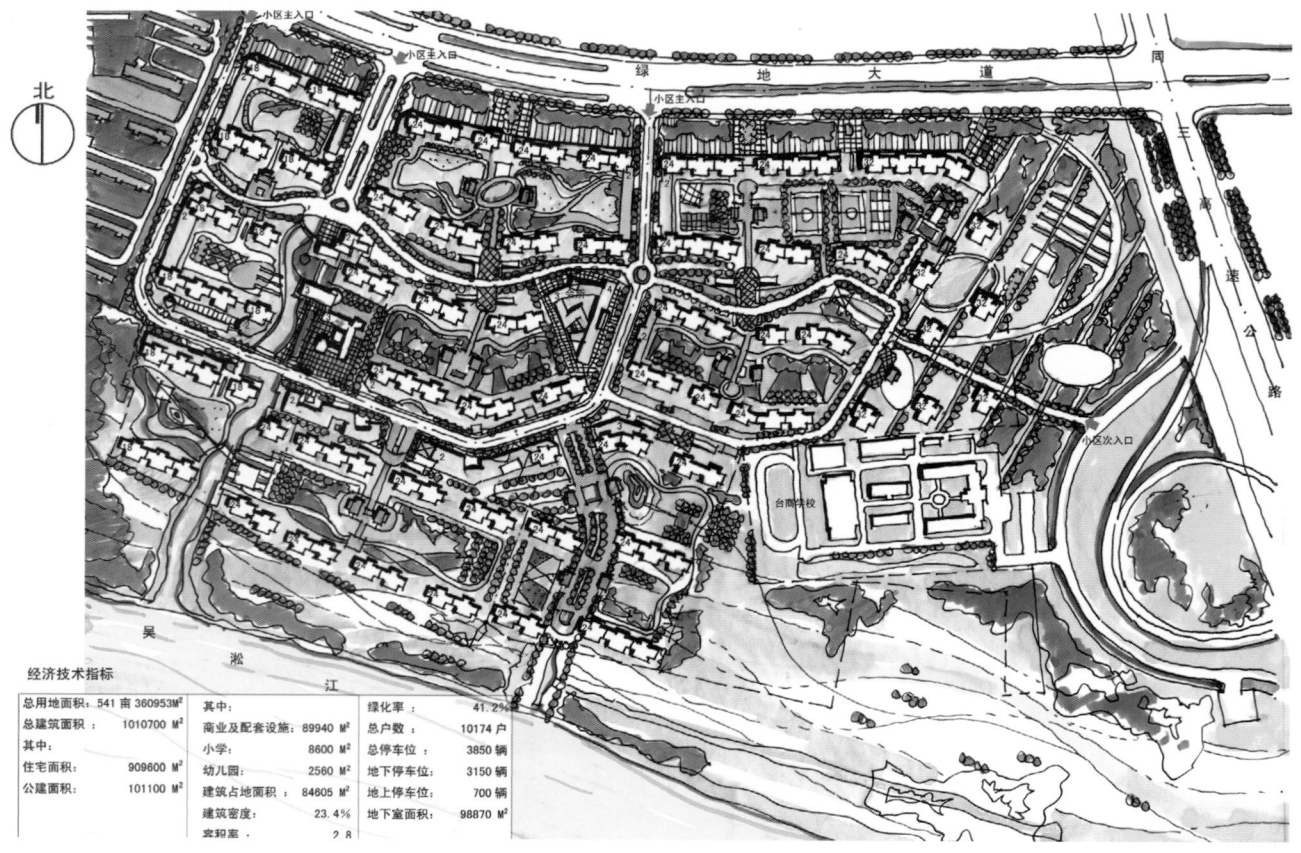

经济技术指标

总用地面积：541 南 360953 M²	其中：	绿化率：41.2%
总建筑面积：1010700 M²	商业及配套设施：89940 M²	总户数：10174 户
其中：	小学：8600 M²	总停车位：3850 辆
住宅面积：909600 M²	幼儿园：2560 M²	地下停车位：3150 辆
公建面积：101100 M²	建筑占地面积：84605 M²	地上停车位：700 辆
	建筑密度：23.4%	地下室面积：98870 M²
	容积率：2.8	

设计之初，我们将自己定位成一个普通的居住者，设身处地地去思考我们需要的空间，并引导我们设计的方向。通过创造性的设计和丰富的生活细节来提高居住的情趣。

At the beginning of our design, we imaged that we were the habitants, and tried to think about the real lives we wanted. Following this principle, the creation is improved with the rich living details and we are successful in forming the final products we wanted.

中铁置业·世纪山水
China Railway Real Estate • Century Landscape

项目名称：中铁置业·世纪山水
用地规模：29 万平方米
建筑面积：80 万平方米
容积率：2.7
项目地点：中国 / 湖南 / 吉首
编制时间：2008 / 09

大自然的鬼斧神工，造就了湘西这块神奇的土地，山水奇异，人文独特。湘西的美，在于山、在于水、在于人。对项目原始地貌的充分尊重形成了依山傍水、得天独厚的平面格局，而"三山两湖一环"的总体规划构架，则在将"湘西地王"打造成兼具地域性和独创性山水大盘的同时，让建筑与环境的包容展示着人文的延续与变迁，在空间和细节的刻画中传承着湘西独特的人文魅力。

The magic craft of Nature has created this miraculous land of western Hunan. There are fantastic landscapes of mountains and rivers, as well as unique human cultures. The beauty of western Hunan is in its mountains, rivers and human cultures. The full respect for the original landforms of the project gives shape to the mountain-front and riverfront plane pattern rich in gifts of Nature, while the "three mountains-two lakes-one ring" general planning frame, at the time of making the "Western Hunan Land King" into a landscape project with both regionality and originality, lets the building and environment harmoniously exhibit the continuation and vicissitude of human cultures, to inherit and pass on the peculiar human cultural charms of western Hunan in the depiction of space and details.

上海紫园 8 号
Shanghai No.8 Purple Apartment Project

项目名称：上海紫园 8 号
用地规模：0.93 万平方米
建筑面积：0.3 万平方米
容积率：0.14
项目地点：中国 / 上海
编制时间：2002 / 06

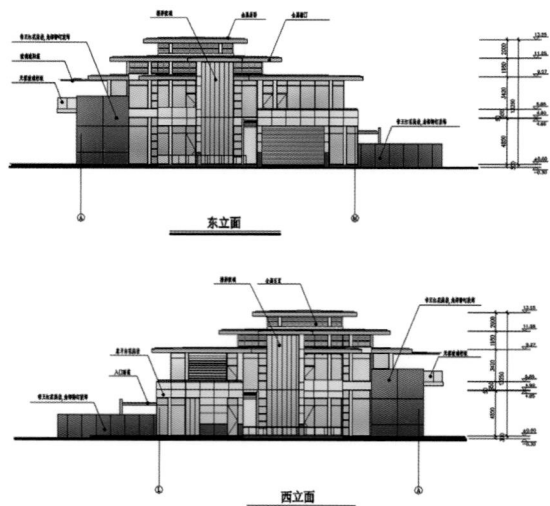

东立面

西立面

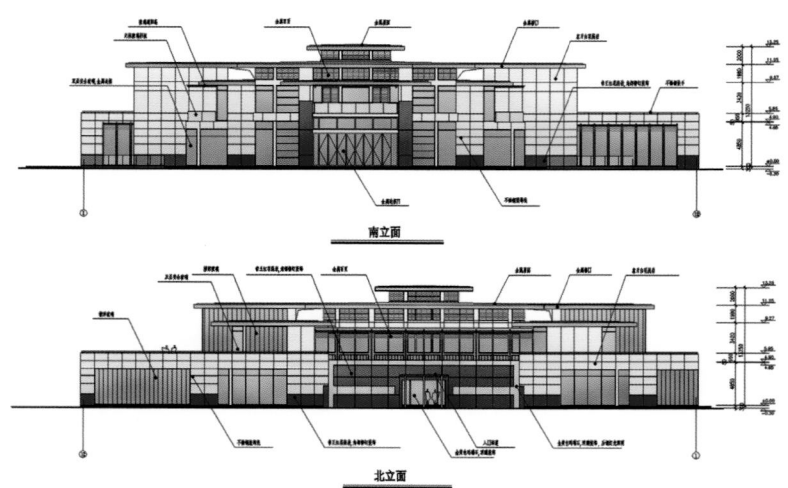

南立面

北立面

建筑总体呈现厚重、沉稳、现代、价值、私密、个性的居住体验和视觉体验，基地依山傍水，北面为青翠秀丽的余山，四周碧水环绕，设计希望将别墅融于青山绿水的景色之中，在基地中留出大片绿地，使别墅非常协调地掩映于苍翠欲滴的绿树丛中。

设计物超所值地体现"现代与尊贵"和没有脂粉气的奢华，以中国传统造园手法为基础的曲径造就深宅的诡秘，厚重的花岗岩塑就的粉墙在浓荫中时隐时现，大型的钢构件和室内玻璃泳池无穷变化的水景与光影产生的效果变无装饰为有装饰，处处体现一种现代空间的简单和奢华之间的关系。

The overall design presents a residential and view experience which is dignified, profound, modern, valuable, private and unique. The plot is surrounded by the mountains and rivers, with verdant and beautiful Sheshan at its north and waters around. The design aims at integrating the villas into the scenery of mountains and waters and leaving a large land area for greenbelt where grasses and plants can be planted layer after layer so that the villas can be shaded by fresh green trees harmoniously.

The design projects a kind of "modern nobility" with a well worth price and successfully avoids extravagance. It brings up the mystery of grand villa by building a labyrinth with the Chinese traditional gardening architecture as its basis. The whitewashed walls built by the stately granite flicker among the shades and the grand steel frames and the changing waterscape inside and outside of the room serve as a kind of decoration themselves, which presents a modern spatial luxury.

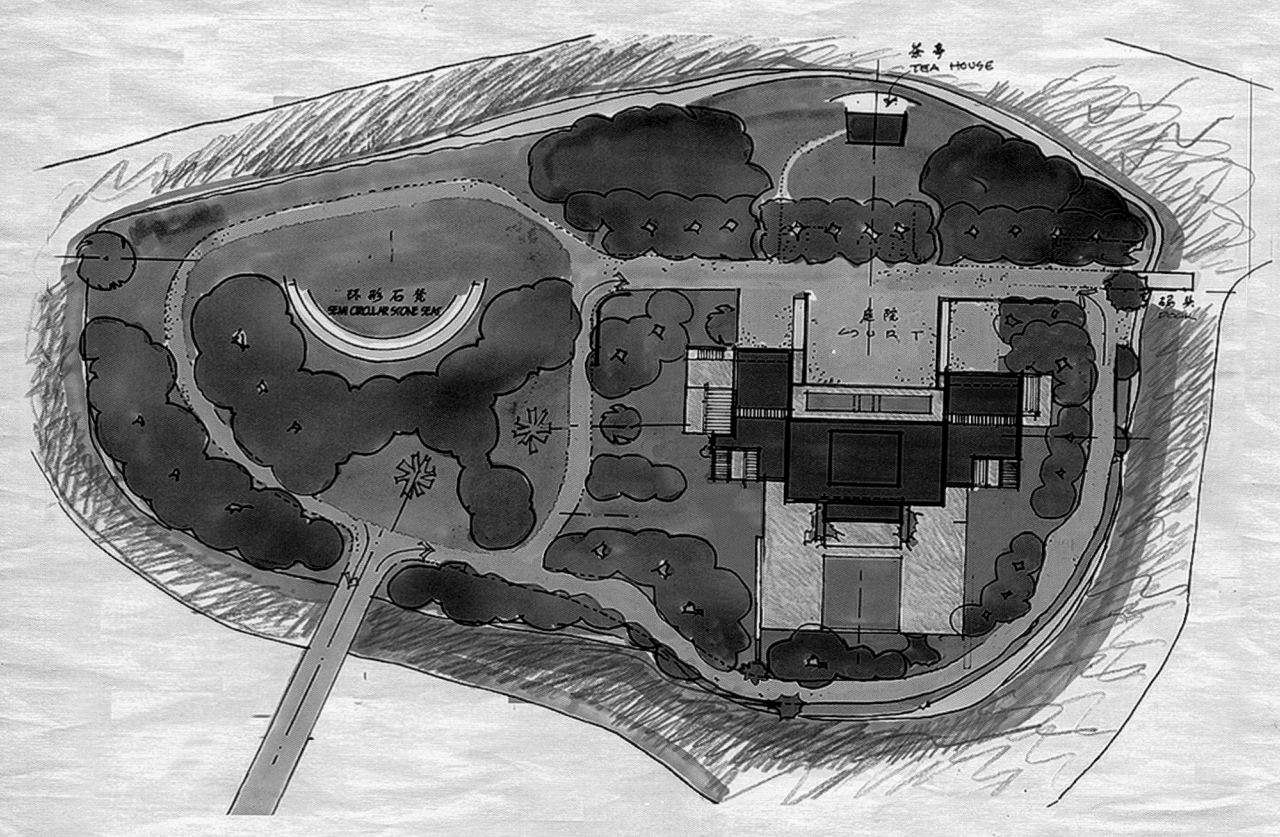

茶亭
TEA HOUSE

环形石凳
SEMI CIRCULAR STONE SEAT

庭院
COURT

SITE PLAN
总平面

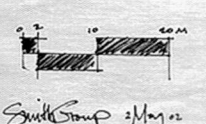

0 5 10 20M

SmithGroup 2 May 02

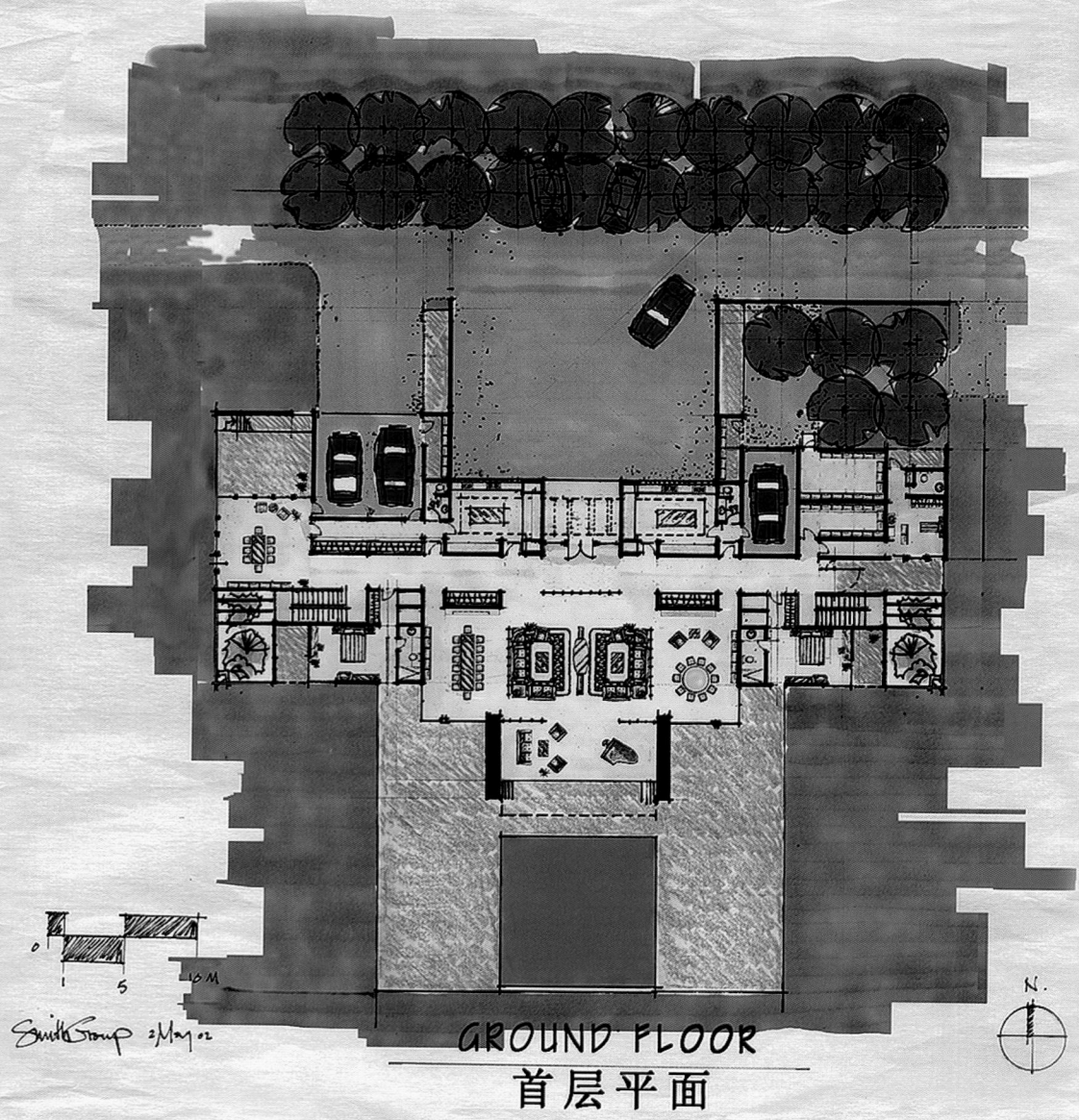

0 1 5 10M

SmithGroup 2 May 02

N.

GROUND FLOOR
首层平面

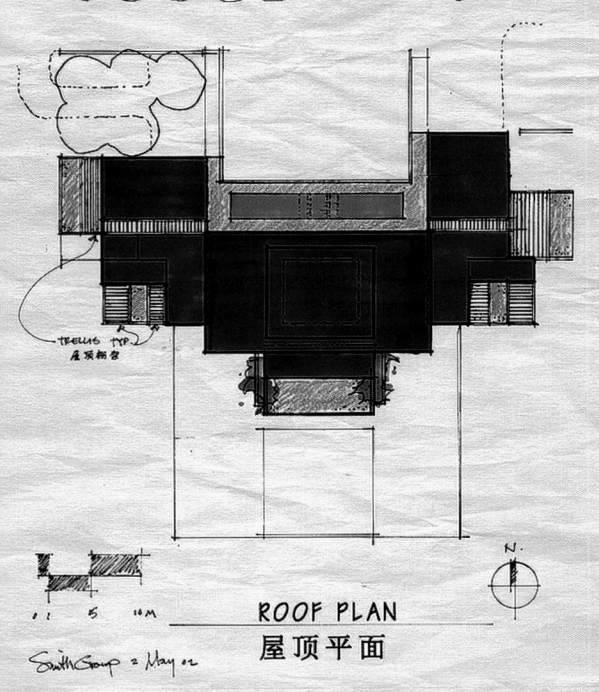

ROOF PLAN
屋顶平面

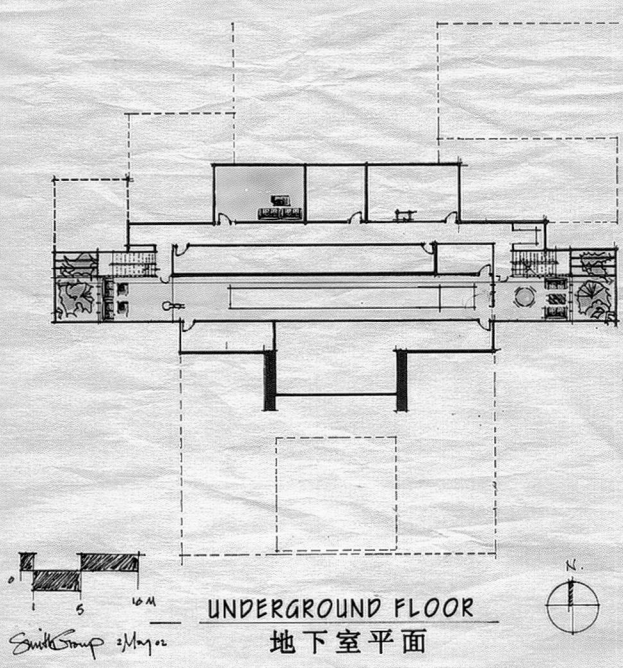

UNDERGROUND FLOOR
地下室平面

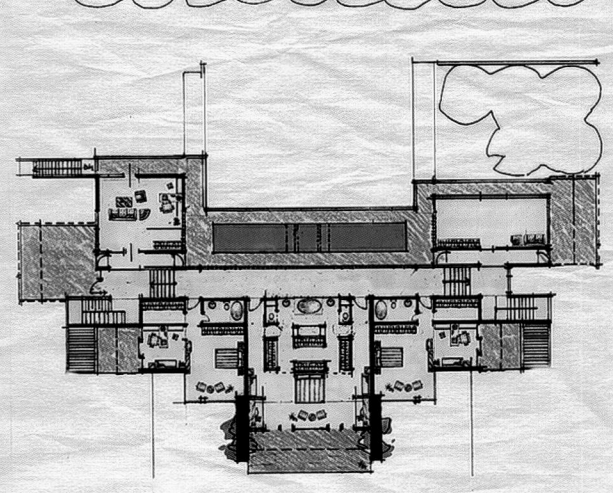

SECOND FLOOR
二层平面

NORTH ELEVATION
北立面

SOUTH ELEVATION
南立面

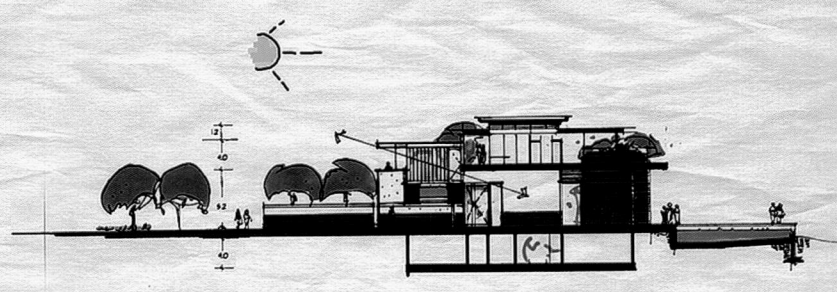

NORTH / SOUTH SECTION
剖面

南京苏宁环球和燕路项目
Nanjing Suning Global Heyan Road Project

项目名称：南京苏宁环球和燕路项目
用地规模：3.78 万平方米
建筑面积：6.15 万平方米
容积率：1.63
项目地点：中国 / 江苏 / 南京
编制时间：2009 / 02

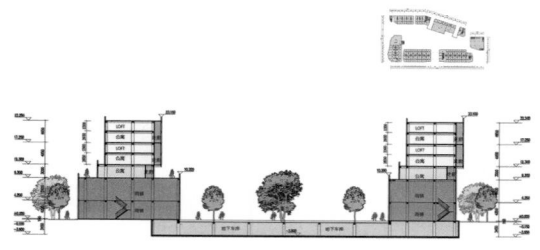
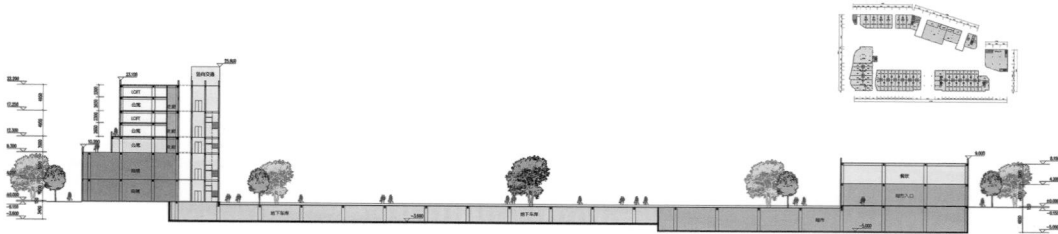

地块距地铁迈皋桥站一公里。西临南北向规划道路，东接高力国际家具港，北面为汽车 4S 店，南面为住宅区和化工厂。一条 35 米宽东西向市政规划道路横穿地块，将整幅地块分为两部分。其中北地块为底层商业加公寓综合体。南地块为高层住宅建筑。整个建筑以现代欧式风格为主，彰显高端商业与住宅的品质与厚度。

北地块商业部分立面设计采用稳妥的现代经典式手法，在简洁的体量关系中体现出细节的韵味，局部建筑细部尺度宜人，充分展示对使用者的关怀，建筑整体形象优雅大方，建筑语言丰富而不繁琐。南地块住宅建筑风格与北地块相互呼应。在建筑材料选用上以暖色调为主，经济实用的同时提升建筑外观品质。

The building land of this project is 1 kilometer away from subway. Maigao Bridge Station. A 35-meters-wide municipal planning road goes across the building land, and divides the area into two irregular blocks. The northern part is the comprehensive building for underground business district and apartments. The entire southern part is used for high-rise residential buildings.

The northern commercial part adopts modern classical style to design its envelope, which reflects the lasting appeal of particulars in the succinct dimension relationship. The design emphasizes the detail scale that fully demonstrates the caring and concerning about the users. The building boasts of graceful and grand overall image. The architectural language is rich but not complicated. The architectural style of the residential building in the south is in line with the northern part in a harmonious manner.

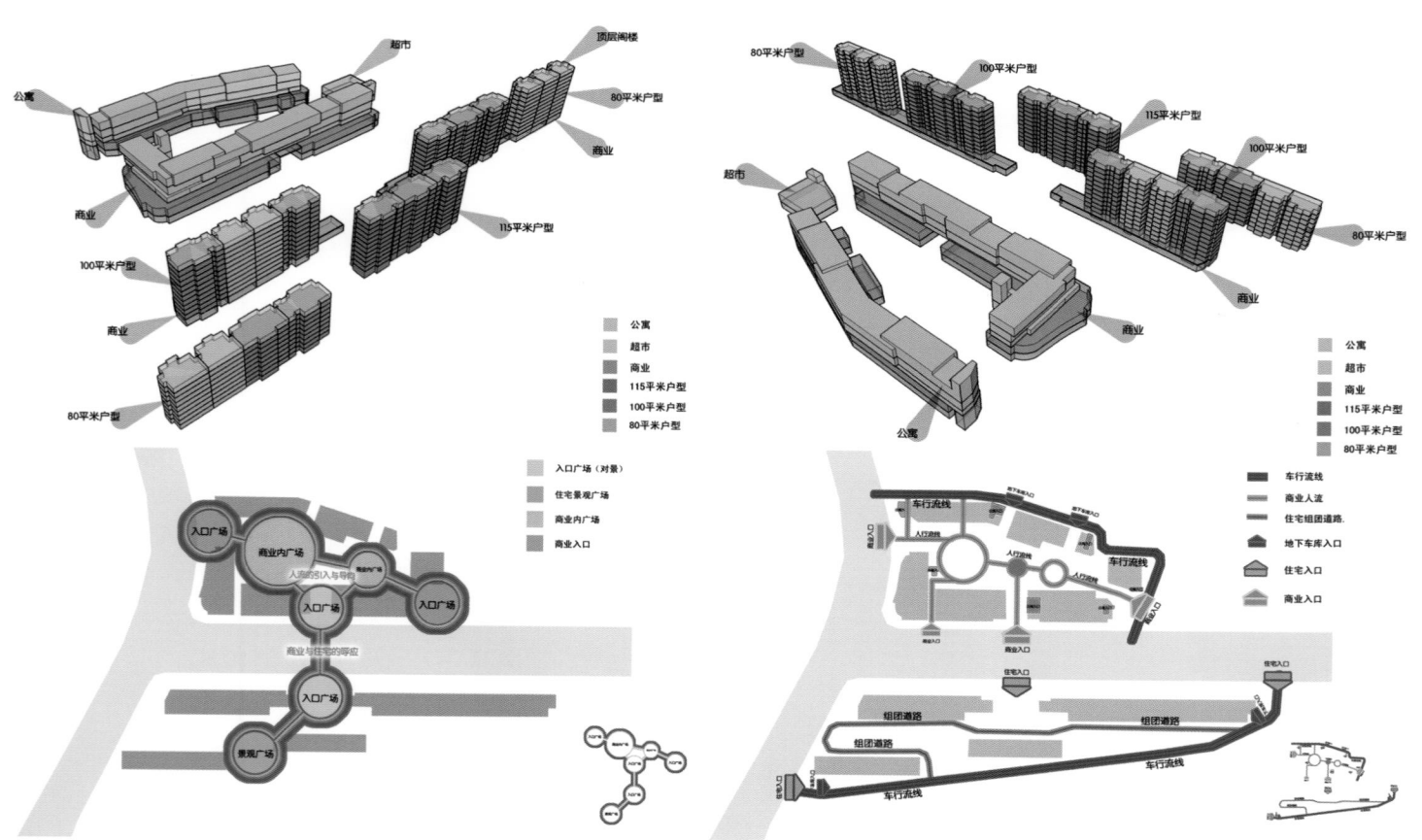

地下室范围线
人行通道
地库出入口
停车
机动车出入口
建筑红线
地库出入口
5F
5F
5F
2F
2F
人行出入口
5F
2F
用地红线
人行通道
人行出入口
机动车出入口

人行出入口
机动车出入口
地下室范围线
建筑红线
用地红线
1F
1F
11+1F
11+1F
11+1F
11+1F
11+1F
11+1F
11+1F
11+1F
11+1F
11+1F
11+1F
11+1F
8+1F
8+1F
8+1F
8+1F
地库出入口
停车
机动车出入口
地库出入口
停车

N

0 10 30 50m

住宅沿街透视图

天津檀府
Tianjin Tanfu

项目名称：天津檀府
用地规模：6.4 万平方米
建筑面积：6.5 万平方米
容积率：1.01
项目地点：中国 / 天津
编制时间：2006 / 09

本项目位于天津市区老城厢的中心地带，是高密度的城市联排和独栋建筑，在构建城市肌理的理念指导下，通过不同方向的建筑组合，形成丰富的街区与胡同的灵活布局，是居住、休闲和商业共同组成的充满回忆的社交建筑空间。

设计要点

院子建筑的理念是经过提炼的中式宅院建筑的院落精神，将院子所能形成的空间围合感和现代建筑的空间变换原理结合起来，可以形成现代的中式空间建筑。

平面设计中的加减法是 N+1 状态建筑和 N-1 状态建筑，N 就是建筑空间的正常状态，N+1 就是在提供人们正常的居住功能以后，另外增加阳光、空气和花园，让人们感到物业的超值。同样的 N-1 就是去掉现在习惯性多余空间，包括那些习以为常的多余的功能和多余的面积。

研究空间设计与气候的关系，设计可以内透的空间，把应作为住宅的外院尽量地开敞。同时在墙体和窗户的设计上引进西方特别是北欧的先进技术，用中国的方式建造符合老城建筑的节能工艺。

整体建筑风格设计以现代中式风格和老城厢的韵味相结合形成"新建筑群体"，环境配置以"步移景异"的规划理念，创造居住的诗意格局，在集体记忆与现实生活的交织中得到更高层次的心理享受和精神依托。

This project is located in the heartland of Tianjin old downtown, which contains highly concentrated urban flat buildings and stand-alone buildings. Under the idea of building skin texture of the city, the project forms the flexible layout of streets and alleys through the combination of buildings with different directions. It is a social architectural space full of memories, performing residential function, leisure function, and commercial function.

Essentials of the Design

The idea for courtyard architecture can be expressed as a refined Chinese courtyard spirit, which is to combine the feeling of enclosure created by the yard with the spatial alternation principles of modern architectures, to form a kind of modern Chinese space architecture.

The addition and subtraction in plane design mean the architecture of N+1 and the architecture of N-1, in which the N represents the normal state of architectural space; the N+1 represents the sunshine, fresh air and garden added to the normal residential functions, which will make people feel the value of their real estate; and the N-1 represents cutting off the currently customary redundant space, including familiar redundant functions and unnecessary area.

The design studies the relationship between space design and climate to design the space of inner-open. It makes the outer court which is supposed to be residence as open as possible. It also uses advanced technologies from the West (especially Northern Europe) into the design of wall bodies and windows, and adopts Chinese methods to meet the architectural energy-saving process in old town.

The general architectural style is the "new building group" combined with modern Chinese style and the charm of old downtown. Its environment configuration uses the planning idea of "different views in different positions" to create the poetic layout for residence. People shall enjoy higher level of mental experience and inner sustenance in the intertexture of collective memories and the reality.

外高桥新市镇配套商品房
Waigaoqiao Area Xinshi Town Supporting Commercial Housing

项目名称：外高桥新市镇配套商品房
用地规模：6.68 万平方米
建筑面积：12.02 万平方米
容积率：1.8
项目地点：中国 / 上海
编制时间：2006 / 04

本项目属于上海市配套商品房项目，但并不意味着设计起点和标准的降低，不仅要按照"四高"小区的要求规划建设，同时要在有限的条件下创造出富有特色的居住环境。

设计以建设生态型居住环境为规划目标，创造一个布局合理、功能齐备、绿意盎然、生活方便，具有文化内涵的住区。注重居住地的生态环境和居住的生活质量，合理分配和使用各项资源，全面体现可持续发展思想，把提高人居环境质量作为规划设计、建筑设计的基本出发点和最终目的。而绿化系统的规划及绿化覆盖率的提高，是现阶段改善居住地生态环境的有效手段。完善的配套设施，方便的交通系统，宜人的空间设计以及休闲、娱乐场所的设置，又将有助于居民生活质量的提高。

This project is part of the Shanghai supporting commercial housing project. But this does not mean the lowering of starting point and standards for design. Not only must the planning and construction accord with "four-high" housing requirements, but it must also create a dwelling environment with distinctive characteristics under the limited conditions.

It takes the construction of an eco-living environment as the planning objective, creates a residential area with rational layout and complete functions and smiling green and convenient livelihood and profound cultural connotation. We focus on the ecological environment of the residential area and the living quality, rationally allocate and utilize various resources, comprehensively reflect the thinking of sustainable development, and take the improving of inhabitation environment as the basic starting point and ultimate aim of planning design and building design. To improve the ecological environment of residential area, the effective measures are to map out the greening systems and to raise the percentage of greenery coverage, while complete supporting facilities and convenient traffic systems and delightful space design and the setting-up of leisure recreation spots will help to improve residents' living quality.

项目名称：外高桥新市镇配套商品房
Waigaoqiao Area Xinshi Town Supporting Commercial Housing

温莎花园
Wensha Garden

项目名称：温莎花园
用地规模：11.9 万平方米
建筑面积：7.79 万平方米
容积率：0.65
项目地点：中国 / 上海
编制时间：2003 / 05

项目名称：温莎花园
用地规模：11.9 万平方米
建筑面积：7.79 万平方米
容积率：0.65
项目地点：中国 / 上海
编制时间：2003 / 05

温州 A19 地块
Wenzhou Plot A19

项目名称：温州 A19 地块
用地规模：0.3 万平方米
建筑面积：1.0 万平方米
容积率：3.3
项目地点：中国 / 浙江 / 温州
编制时间：2004 / 08

沿街商铺最大限度地把"黄金立面"向主要干道开敞，以获得最大的经济效益。住宅面向已规划的住宅小区，既营造大自然休闲的生活氛围，也形成了景观的资源共享。在住宅的布置上除了三个单元的板式小高层的布置方式外，本设计挖掘了另一种可能的布置方式：将通常情况下连成整体的板式小高层住宅体量打散为两部分，一部分成为层数较低的独立点式住宅，另一部分则是两个单元的板式小高层住宅。

Roadside shops open its "golden facade" towards the main road to the greatest extent to gain the maximal economic benefit. The dwelling houses face a planned residential community, creating a living atmosphere of nature and leisure, as well as facilitating the resource sharing of the landscape. As to the layout of the houses, besides the 3-unit layout of the medium-height flats, we dug another mode of layout in our design: dividing the volume of the medium-height flats which are usually connected as a whole into two parts; turning the lower floors into independent point blocks, and the higher storeys into 2-unit slab blocks.

紫阳美地山庄
Ziyang Meidi Village

项目名称：紫阳美地山庄
用地规模：10.85 万平方米
建筑面积：13.30 万平方米
容积率：1.2
项目地点：中国 / 江苏 / 江阴
编制时间：2007 / 04

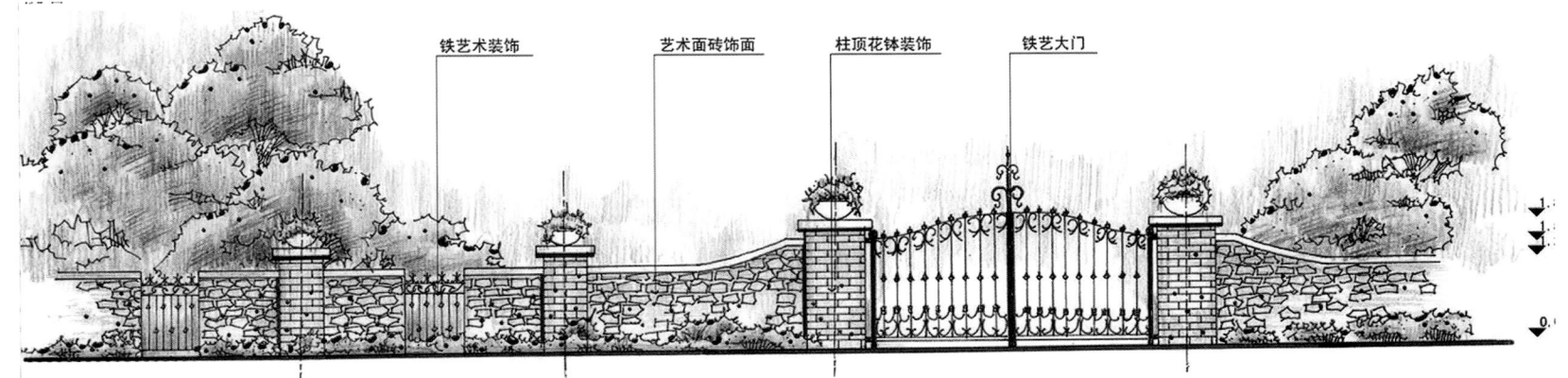

铁艺术装饰　　　艺术面砖饰面　　　柱顶花钵装饰　　　铁艺大门

该项目为联排别墅、六层叠加别墅、小高层、会所的常规组合住宅区，小区内建筑布置多采用四合院的围合空间形式，在常规的建筑规划项目中创造多变活跃的空间效果，融合建筑材料与色彩，共同强化居住环境与组团空间的识别性。

Residences in the community are mainly row houses, club house or middle high-rise apartments. The various cluster, material and color enhance the identification of the living environment and group space.

郑州五胡　城中村改造
Zhengzhou Wuhuzi Chengzhong Village Reconstruction

项目名称：郑州五胡旮城中村改造
用地规模：128.36 万平方米
建筑面积：398.00 万平方米
容积率：3.1
项目地点：中国 / 河南 / 郑州
编制时间：2007 / 09

设计通过对位于片区总体功能的分析，刻画出数个主题的分区，并在每一个分区进行单独的功能定位。从土地使用、生活风情、建筑特色和景观概念等方面赋予其鲜明的个性。大量并大胆地运用不同风格与尺度的开放空间，以鼓励公共集聚、文化活动、体育运动以及个体的交流，还原丰富的社会生活形态，拒绝单一地对大型区块进行简单的描绘。

Through the analysis of the whole district, each function has been basically distributed. The land use, living style, architectural features and landscape concept of each district are unique. Lots of activity spaces revert the colorful living scenery and promote the communication of each other.

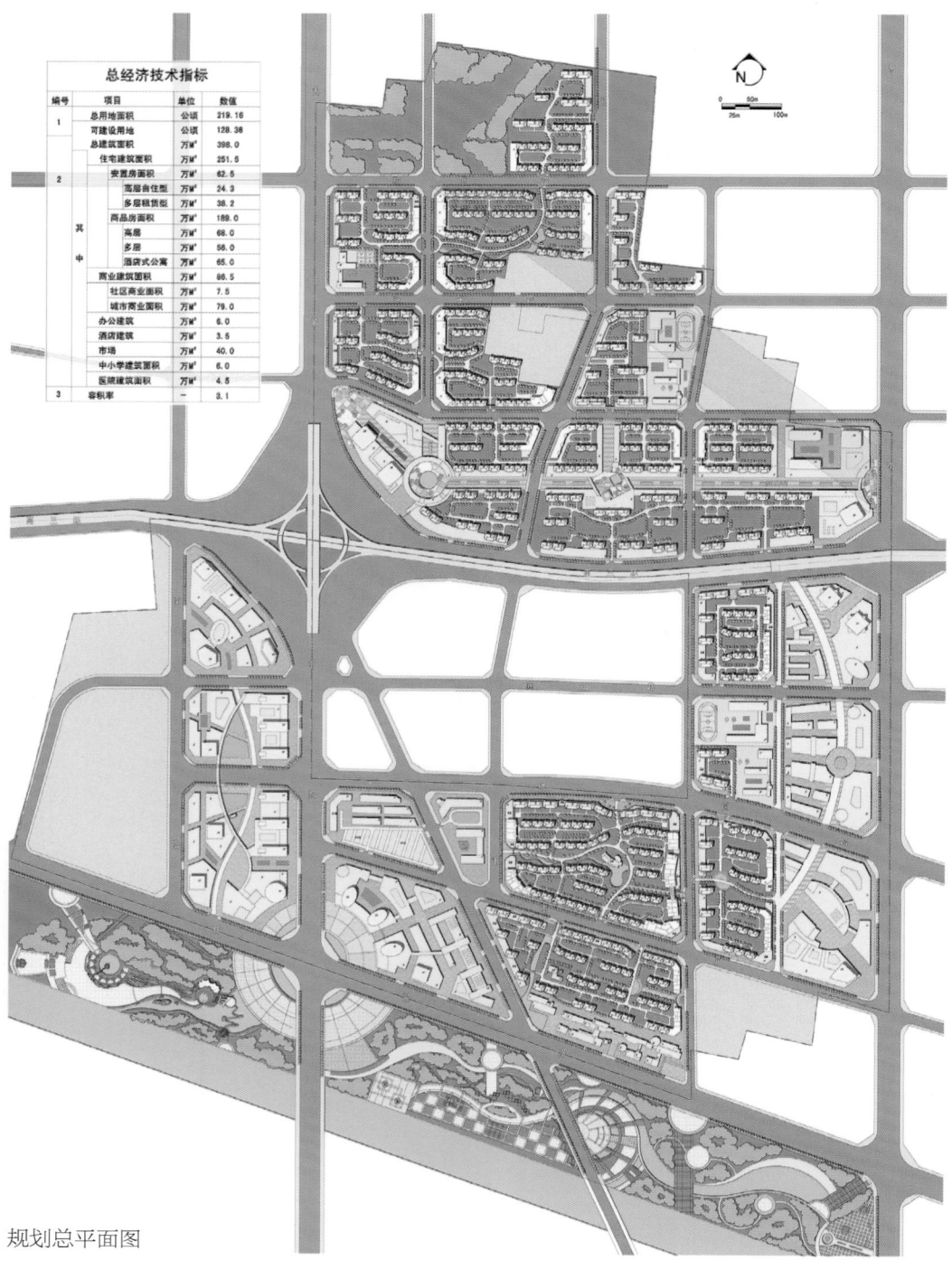

总经济技术指标			
编号	项目	单位	数值
1	总用地面积	公顷	219.16
	可建设用地	公顷	128.36
	总建筑面积	万m²	398.0
2	住宅建筑面积	万m²	251.5
其中	安置房面积	万m²	62.5
	高层安置型	万m²	24.3
	多层精装型	万m²	38.2
	商品房面积	万m²	189.0
	高层	万m²	68.0
	多层	万m²	56.0
	酒店式公寓	万m²	65.0
	商业建筑面积	万m²	86.5
	社区商业面积	万m²	7.5
	城市商业面积	万m²	79.0
	办公建筑	万m²	6.0
	酒店建筑	万m²	3.5
	市场	万m²	40.0
	中小学建筑面积	万m²	6.0
	医建建筑面积	万m²	4.5
3	容积率	-	3.1

规划总平面图

滨水景观商业街景

中科大学村五期项目
Zhongke University Phase 5 Project

项目名称：中科大学村五期项目
用地规模：4.6 万平方米
建筑面积：1.99 万平方米
容积率：0.43
项目地点：中国 / 上海
编制时间：2007 / 09

本项目地块位于上海市浦东康桥地区，以田园风格倡导"回归自然"。设想在当今高科技快节奏的社会生活中获取生理和心理的平衡。

住宅设计部分在认真研究市场需求的同时力求做到不同规格的房型多样化，以满足不同人士的需求。设计中功能分区明确，空间组织合理，做到动静分区、洁污分区、干湿分区。使起居厅、餐厅、厨房相对集中，卧室、卫生间相对集中，居住层高 3~3.5 米，充分满足人们家居生活的多方面要求。

平面设计中做到明厅、明厨、明卫、朝向、采光、通风良好。每栋均做到至少两到三间居住空间朝南，做到两间居住空间朝向主要景观。起居厅至露台为落地玻璃门，保证了良好的日照采光及优美的景观视野。每户进户处都有玄关作为过渡空间，避免开门见厅，营造了良好的室内空间层次效果。

Located in Kangqiao Area of Pudong New Area, Shanghai, this project plot boasts an integral pastoral style of architectural design. It advocates "returning to nature", highly praises "natural beauty" aesthetically and holds the view that the physical and psychological balance can only be achieved by upholding and integrating nature in today's fast-paced social life with highly developed science and technology.

To carefully study the market demand, develop buildings elaborately, strengthen the consciousness for the best and strive for various specifications of houses so as to satisfy different needs, the houses should be designed with clear function divisions, rational spatial organization, dynamic-static division, clean-dirty division and wet-dry division. The living room, dining room and kitchen should be relatively concentrated, so are the bedroom and the washing room. The height of each residential floor is between 3-3.5 meters so as to fully satisfy the various life needs of the residents.

The following aspects shall be guaranteed during the graphic design: bright hall, kitchen and washing room, good orientation, day lighting and ventilation. At least two to three sitting rooms of each building should face the south so that two sitting rooms can be welcomed by the major landscape. The space between the sitting room and the terrace is designed as landed glass doors which guarantee good sunshine, daily lighting and a beautiful view. A hallway at the entrance of each house serves as an in-between so as to avoid "taking in everything in a glance while opening the door", which helps create a good space dimension.

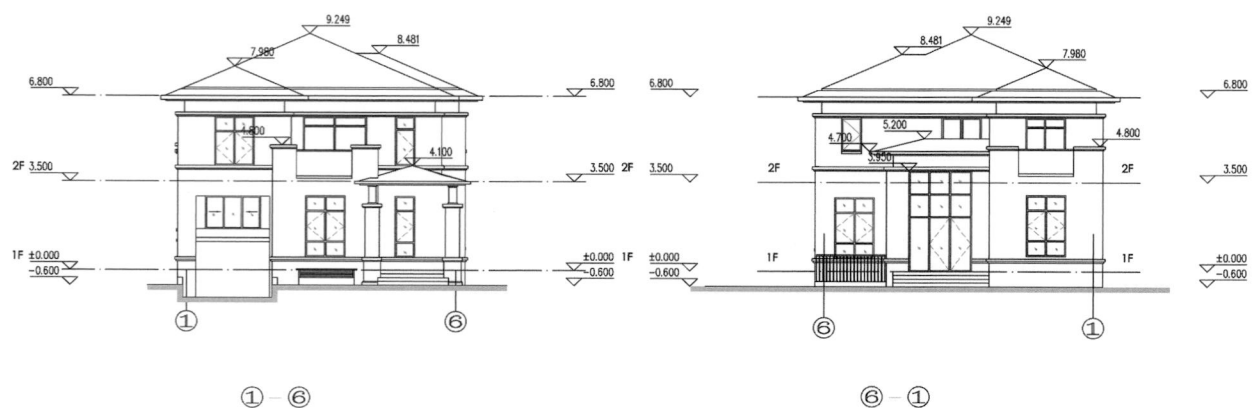

① - ⑥   ⑥ - ①

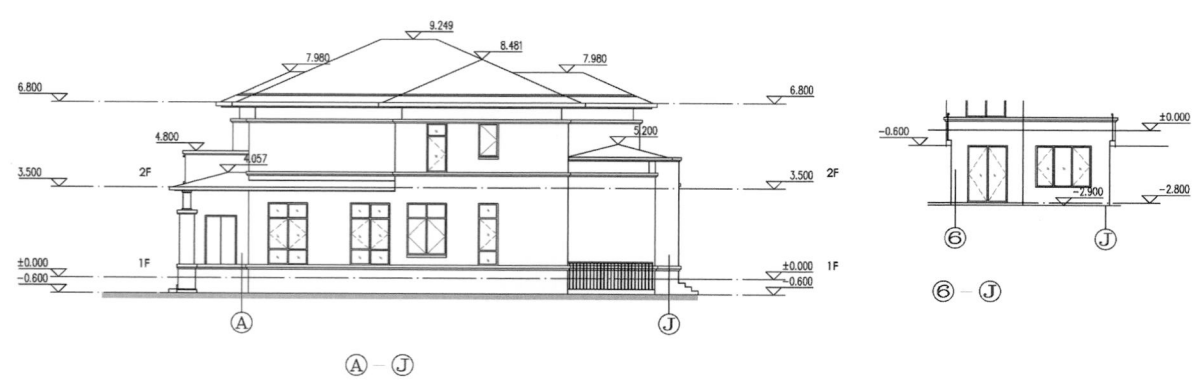

Ⓐ - Ⓙ   ⑥ - Ⓙ

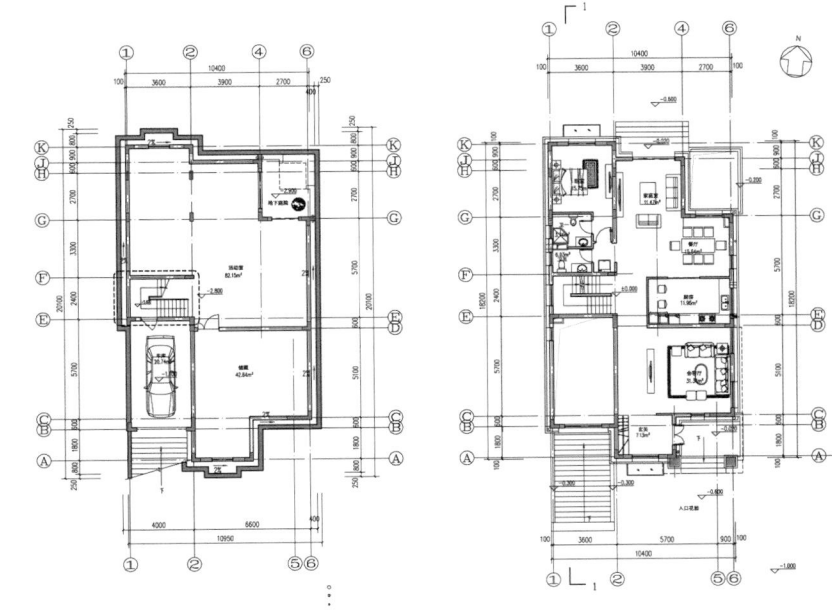

诸暨陶朱山壹号
Zhuji Taozhu Mountain No.1 Project

项目名称：诸暨陶朱山壹号
用地规模：23.24 万平方米
建筑面积：27.89 万平方米
容积率：1.2
项目地点：中国 / 浙江 / 诸暨
编制时间：2010 / 05

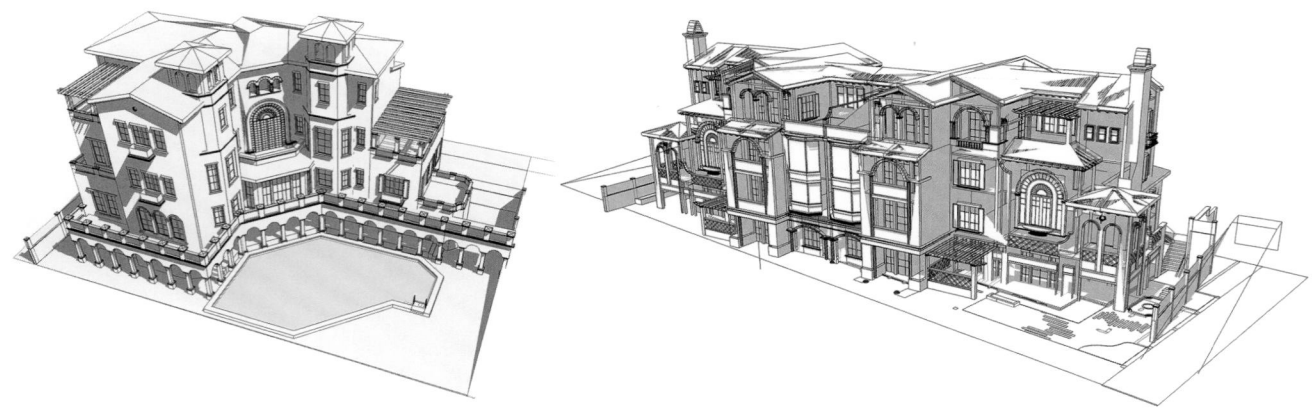

陶朱山壹号项目位于浙江省诸暨市南郊，定位于高档低密度住宅区，本项目将致力于打造诸暨顶级高端品牌住宅。本案运用南向阳坡的绝佳地理优势，利用自然的山与水，创造一个宜居的高档社区，迅速提升地块的最大价值。

本项目主体建筑分为低层住宅、联排住宅、高层住宅、中央会所、配套服务用房等，在整体布局中，充分结合地块地形特征，营造出各个相对私密的组团，以水系将低层住宅与联排住宅划分开，并以水系为载体，贯穿于整个小区中，分离成相对独立的生活区。

本项目以意式乡村托斯卡纳风格为主题，在总体的规划布局以及建筑物的细部设计上均尊重原汁原味的意大利风格主题，立面设计为红色陶土屋瓦，舒展而自然；墙面多为石灰泥涂层，以颜色自然质感的黄色或桔红的砖石为主，铺装材料多切割成规则的形状，适当使用彩色瓷砖；墙面用灰泥或者涂刷鲜艳的颜色，涂料均选用户外涂料，保证色彩的耐久性；彩色的木质结构的廊架等构筑物，与周边建筑和整体庭院风格相得益彰。

The Zhuji Taozhu Mountain No.1 Project is located in the southern suburb of Zhuji (City), Zhejiang Province. This project is positioned to be an upscale multi-storey residential area with low concentration. It is committed to build a residence of top high-end brand in Zhuji.

Taking the superexcellent geographic advantage of the southern sunny slope, the project makes its layout based on natural hills and waters, to create an upscale livable community environment and space form, and promote the value of this land quickly.

The main buildings of this project include low-rise dwellings, row houses, central chamber, buildings for adequate and systematic service and so on. Combining the terrain features of this land, the layout manages to build the project into several relatively private groups; the layout adopts water system to seperate the low-rise dwellings from row houses; and also uses the water system which passes through the whole residential area to divide the area into different residential areas which can be developed independently.

This project adopts Italian Tuscan rustic style as its theme. It totally adopts the original Italian style in both general layout and the detailing of buildings; its elevation design mainly uses the effuse and natural red clay tile as major symbol; its layout adopts lime slurry as the coating for wall surface which is painted with yellow or nacarat and added with natural bricks; the laying materials adopted include certain color tiles to ensure the durability of color; the color wood-frame corridors and other structures are harmonious with surrounding buildings and the whole courtyard style.

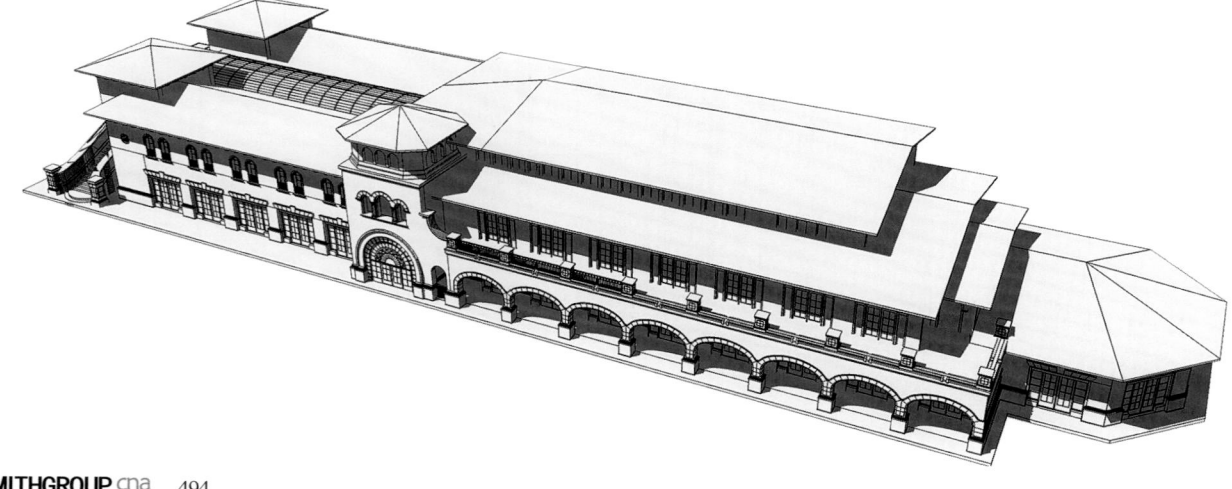

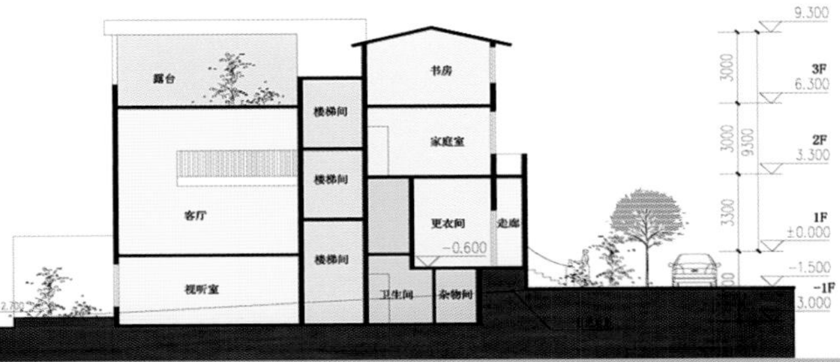
露台
书房
楼梯间
家庭室
客厅
更衣间
走廊
视听室
卫生间
杂物间
-0.600
9.300
3F 6.300
2F 3.300
1F ±0.000
-1.500
-1F -3.000
3000
3000
3300
9300

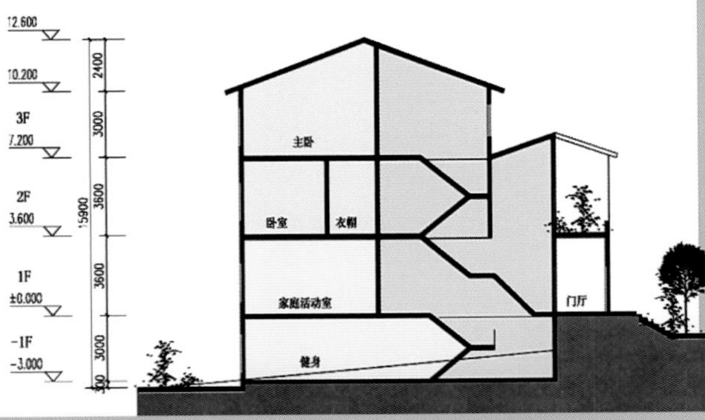
12.600
10.200
3F 7.200
2F 3.600
1F ±0.000
-1F -3.000
2400
3000
3600
3600
3000
15900
主卧
卧室
衣帽
家庭活动室
门厅
健身

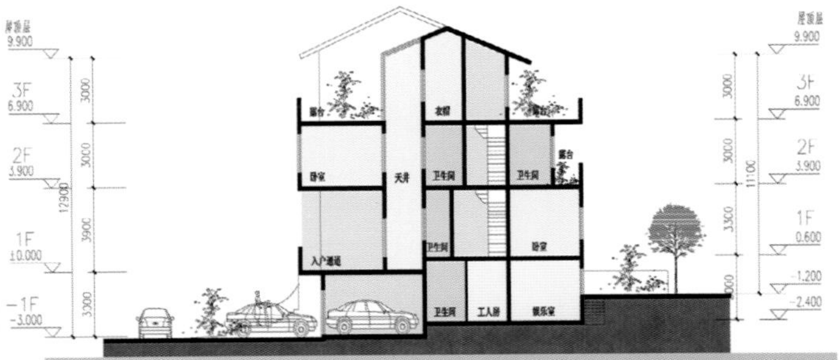
屋顶层 9.900
3F 6.900
2F 3.900
1F ±0.000
-1F -3.000
3000
3000
3900
3300
12900
露台
衣帽
卧室
天井
卫生间
入户通道
工人房
娱乐室
屋顶层 9.900
3F 6.900
2F 3.900
1F 0.600
-1.200
-2.400

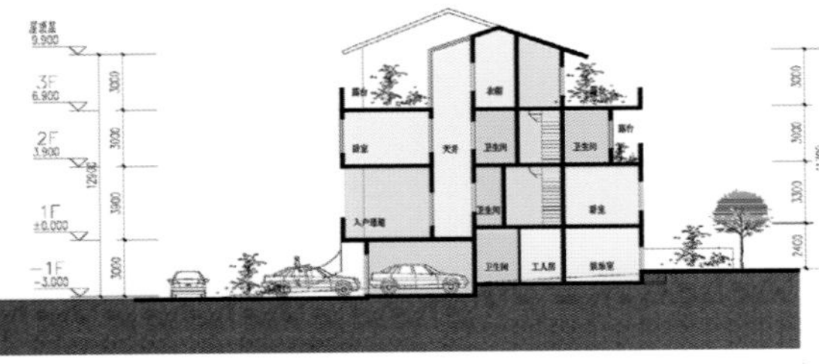

CRA®

规划 、景观设计类
Urban Planning & Landscape Design

∙∙

TenYears
SmithGroup CNA
Design Determines …

TenYears
SmithGroup CNA
Design Determines …

思纳熙三景观规划设计公司是思纳史密斯公司和加拿大熙三规划事务所优化组合的成果。思纳史密斯在 2000 年进入中国后，取得了长足发展，由一家专业的建筑设计咨询公司发展为含建筑设计、景观规划、室内设计的综合性公司。加拿大熙三规划事务所于 2000 年进驻中国上海，在城市街区设计、街道设计和城市更新设计方面有着独到的见解。2007 年思纳史密斯和熙三事务所合并，成立了思纳熙三景观规划设计公司，实现了优势互补、强强合作。

3 年多来，思纳熙三设计公司完成了公司架构、人员和资源的整合，形成了一支技术特色鲜明、高效快速、综合多样的景观规划团队，并在国内市场设计环境日益不景气的状况下，完成了一系列重要的规划景观项目。

思纳熙三景观规划公司分为规划和景观两个团队，规划团队以原熙三事务所的技术总监，资深规划师曹洁先生担当规划设计总监，带领规划小组，专业从事街区城市设计、核心地区详细规划、特殊地区专业规划等综合性强、条件复杂的规划项目。

回想思纳熙三景观规划公司的发展，10 年来虽然经历风雨，但总体还是顺利平稳的。这和我们的发展策略是一致的，即稳固中求变化，一年一个脚印。2000 年我们从北美进入中国，以上海为基地。以 5、6 个人的小型规划工作室起步，逐步扩大到以景观、规划为核心的综合事务所，人员也增长到 30 多人，设计和服务领域不断扩大，产值年年提升。

我们取得的一些成绩，从整个中国设计市场来看，是微不足道的。但作为其中的一分子，我们又是幸运的，因为我们参与了人类历史上规模最大的城市、地区和国家的急剧发展过程。这短短的 20 年，彻底改变了这个具有 5 000 年历史的、世界上人口最多的国家的面貌，我们见证了这一宏大的过程，并贡献了自己的力量。

伴随着思纳熙三的成长，我们的设计团队也经历了一个由稚嫩到成熟的过程，每个成员，不管是现在在思纳熙三的或曾经加入过的，都是值得赞扬和感谢的。因为没有你们的努力和奉献，就没有今天的思纳熙三。同样，希望每个思纳熙三人继续伴随公司一起走上新的台阶，当然这需要无比的勇气和无限的热情，因为我们选择的设计这个职业，是一片充满艰辛和挑战的天地，远没有当初在学校所想象的那样浪漫和轻松。中国的设计师是值得尊敬的，中国的设计公司也是值得称道的！

CNA + C3 Landscape Planning Design Company is the product of a merger between CNA Smithgroup and Canada C3 Urban Planning Company. CNA Smithgroup entered in China in the year 2000 and has since made considerable progress from a professional architectural design consulting firm to a multi-disciplinary company integrating architecture, landscape planning and interior design. Canada C3 entered Shanghai, China in 2,000, bringing a unique approach to urban planning, streetscape and urban renewal design. In 2007, CNA Smithgroup and Canada C3 formed a partnership to become CNA + C3 Landscape Planning Design Company, consolidating the capacities of both firms.

For the past three years, CNA + C3 has succeeded in strengthening its corporate structure, personnel and resources ,thus forming an experienced, distinctive, efficient, comprehensive and diverse team. The team has since then achieved completion of a series of important urban planning and landscape projects in spite of the unfavorable situation of the domestic market.

CNA + C3 is composed of the Urban Planning and Landscape Planning teams. The Urban Planning team is headed by the former technical director of Canada C3 and senior planner Mr. Cao Jie as director. He is responsible for leading the team in city planning, detailed core design, specialized urban planning and other comprehensive and complex planning projects.

Looking back on the development of CNA Xisan Landscape Planning Company, despite ten years experience of various difficulties, overall it is all right and smooth. This has been consistent with our development strategy-pursuing changes in steady going, and one footprint one year. We entered China from North America in 2000, based in Shanghai. A small-scale planning studio of 5 or 6 staff as the starting point, now has gradually grown into a comprehensive office focusing on landscaping and planning as the core business, with its personnel increasing to more than 30. The scope of design and service has been continuously expanding, and the production value is rising year by year.

We have made some achievements, which are minute and inconsiderable in terms of the entire Chinese design market. And yet, as one member in it, we are lucky, because we have participated in the largest-scale urban, regional and national rapid development process in human history. This short period of 10-20 years has thoroughly changed the appearance of this most populous country with a 5,000-year history, and we have witnessed this grand course, also contributed our efforts.

Along with CNA + C3 growth, our design team has also experienced a process of maturing, and every member, either now at or having once joined CNA + C3, is worthy to be praised and appreciated. It would be no today's CNA + C3 without your efforts and contributions. So, we hope every CNA + C3 member continue to accompany the company together, to walk onto a new step, and of course this requires unmatched courage and boundless enthusiasm, because this profession of design we have chosen is a field full of hardship and challenge, far from the romance and easiness we originally imagined at school. Designers in China are worthy of respect, and design companies in China are also worthy of commendation!

李超 ~ 思纳史密斯（集团）中国 (SmithGroup CNA) 上海区规划公司总经理，景观总监现代城市规划专业从建筑学中分离出来，成为单独的一门学科，尽管其时间不长，道路曲折，功过难断，但其发展始终伴随着两条主线。第一条主线就是大家有目共睹的，即人类的工业文明的进步和科技的迅猛发展，迅速地改变了人类在地球上的生存方式。回顾人类的文明史，我们可以发现，相比过去缓慢的进化，现在的发展可以说是在转瞬之间。我们突然之间面临着自己所制造的种种问题：人口的急剧膨胀、自然资源日渐枯竭、物种灭绝、环境污染、贫富加剧、社会动荡等等，比之以往的战争与饥饿，其复杂程度有过之而无不及。第二条主线常常被大家忽视，那就是人类的生存态度的转变，即人与人之间关系和人与大自然之间关系的思考和再认识。特别是发达国家从工业化社会转变为后工业化社会，或所谓传统产业向知识经济、信息产业转变之际，总结过去的发展经验，有许多教训值得吸取。这是区别于第一条物质主线层面和第二条精神主线的发展进步。

回到城市规划这门学科中来。作为一名城市规划工作者必须对这项专业有一个较为全面而又深刻的了解，这样才能把一些所谓先进理念落实到实践中去。城市规划能解决上述两条主线上的一系列问题吗？比如说人们通常很容易将交通堵塞归结为道路交通没有规划好；将房价上涨归结为城市规模不够、供地不足；将人口规模超标归结为城市规划预测不准、控制不力；将贫富相隔归结于城市规划上没有将富人区和穷人区规划在一起；从城市老龄化问题引申出规划养老设施不足等等。这些论调都直接或间接地反映了人们对城市规划专业及作用的误解。

城市规划是什么？简单地说，城市规划是以城市为研究对象，对城市空间未来发展作出的一系列安排。那么城市是什么？不管对城市的定义怎样千差万别，城市首先是一个巨大的容器或者说是物质载体，承载着人类有史以来辉煌的文明以及千千万万城市居民日日夜夜、平平凡凡的城市生活。以一个在时间、空间、内涵等诸多方面如此复杂而动荡不定的"容器"作为城市规划学科的研究对象，城市规划能否有效地把握城市发展的脉络和进程尚是一个问题，上述种种误解也就无可避免了。

对一个规划师而言，没有比认清行业的特质更加重要的事了。在这方面，我们应该始终把握上述城市规划学科发展的两条主线，缺一不可。要避免两种倾向，第一种所谓"虚无型"，即只重视战略层次面研究，忽视规划落位，脱离城市实践、纸上谈兵；第二种所谓"埋头拉车型"，即只知道排房子、做形态，对城市发展的大背景、大战略、大格局漠不关心。

从城市规划的具体研究对象来看，往往有一种否定物质规划的倾向，认为物质规划都是"墙头挂挂的东西"，说变就变。正如前面所述，由于城市的复杂多变，与时俱进十分重要。这要求城市规划专业者有能力在不同层面做不同的工作，制定不同的目标，编制不同的但却切合一定时间段需要的规划成果。综合而言，不论是宏观层面还是微观层面，对城市物质空间的研究应始终是城市规划的主要对象。

作为在一家综合甲级设计机构中的城市规划专业部门，相对而言，我们能够更加深切地体会到城市物质空间的质量对城市的发展起着多么大的作用。在此我们将规划成果编为城市空间、旅游休闲、城市街区、滨水地区四个主题，分别向大家展示我们一些阶段性的心得。这些成果分属于不同的地区，每个地区的自然条件、资源禀赋、人文特质、城市形态都不尽相同，但是对我们来讲，"重创意、必前瞻、求务实、求整合、求公义"之理念是一贯而坚定的。

Modern urban planning was separated from separates architecture and now is a separate discipline. Although it did not take a long time, its development road has been in amid twists and turns. It is hard to judge its merits and demerits; its development is always accompanied by two main lines. The first main line is obvious and is the rapid change of mode of existence brought by progress of human industrial civilization and the quick development of science and technology. Looking back on the history of human civilization, we can see that compared to the slow evolution of the past, the present development can be said in a few seconds. Suddenly we are faced with problems created by ourselves: the rapid expansion of population, impoverishment of natural resources, species extinction, pollution, widening income gaps, social unrest, etc. Compared with the previous war and hunger, the complexity is greater. The second main line often neglected is the changing attitude of human existence, namely, thinking and re-understanding of the relationship between people and the relationship between man and nature. Particularly the developed countries shift from the industrialized society to post-industrialized society, or so-called traditional industries to knowledge-based economy and information industry. To sum up, there are many lessons to be learned. This is progress and development different from the first main material level and the second spiritual main line.

Let us go back to the discipline of urban planning. As a city planner, we must have a comprehensive and profound understanding about this profession in order to implement the so-called advanced ideas into practice. Can urban planning solve a series of problems on the two main lines? For example, people always think that traffic jam is caused by poor traffic management; the rise of housing prices is due to insufficient city size and the land shortage; the excessive size of the population has been attributed to improper urban planning and control. The social problems that the rich and the poor are separated is due to non-separation of the rich and the poor districts. They extend the problem of aging to inadequate planning of retirement facilities and so on. These arguments directly or indirectly reflect people's misunderstanding about the role of urban planning.

What is urban planning? Briefly speaking, urban planning uses the city as the research object, and it makes a series of arrangements for the future development of urban space. So what is a city? Regardless of how diverse the definitions are diverse, a city is a huge container or a material carrier of the brilliant civilization in human history of city residents in thousands of days and nights. For such a complex container in the aspects of time, space and content, it is hard to tell the effective grasping of context and progress, so those misunderstandings can not be avoided.

For a planner, there is nothing more important than having a clear understanding of the nature of industry. In this regard, we should always take the two main lines of the development of the urban planning, either one of which is indispensable. To avoid two tendencies, the first one is so-called "null type", which is only the attention to the strategic level surface and ignores the plan, and is away from urban practices; the second one is so-called "buried pull model", of which the designers only know row of houses and do forms but they pay no attention to the background of urban development, major strategy and pattern.

From the point of view of specific urban planning study object, we often have a negative tendency of physical planning. Physical planning has been considered as changeable as "hanging things on the wall". As we mentioned earlier, changing with the times of the city is very important because of its complexity. This requires urban planners to have the ability to do different jobs at different levels, for different goals, and be prepared to meet the planning results of a certain period of time. Overall, the main target of urban physical space planning should always include either the macro level or micro level.

As a professional urban planning sector in an integrated Class A design firm, relatively speaking, we are able to realize the big role that quality of urban physical space plays during the development of the city in a deeper manner. Here we will show some stage experience to you compiled by the four themes of urban space-tourism and leisure, urban neighborhoods and waterfront planning. These results belong to different regions, each of which is different in natural conditions, resource endowments, cultural characteristics, urban form, but for us, the concept of "focus on creativity, forward-looking, seeking pragmatic, integration and justice " is consistent and firm.

曹洁～思纳史密斯（集团）中国 (SmithGroup CNA) 上海区规划总监

SMITHGROUP cna

Architecture · Urban Planning · Landscape Design
Interiors Design · Art Consultant · Lighting Design
Since 1853

www.cna-group.com

Design Determines ...

设计决定······

01 规划设计类项目
01 Urban Planning

滨水地区
Water Front

城市街区
City Block
城市空间
Urban Space
旅游休闲
Travel and Leisure

02 景观设计类项目
02 Landscape Design

保华常州玫瑰湖项目规划设计
PYI Changzhou Rose Lake Concept Planning

Urban Planning | 2010·10

保华常州玫瑰湖项目规划设计
PYI Changzhou Rose Lake Concept Planning

项目名称：保华常州玫瑰湖项目概念规划设计
项目地点：中国 / 江苏 / 常州
建筑面积：214.9 万平方米
编制时间：2010 / 10
业主：保华集团有限公司
规划类型：概念规划

常州地处江苏省南部，与上海、南京等距相望，是江苏金三角之一，拥有较强的经济实力和城市竞争力。戚墅堰区位于常州市东部，属常州承东启西的交通十字架区域，也是常州东部门户片区，区域发展潜力巨大。

本次规划项目位于戚墅堰区玫瑰湖畔地块，规划总面积约为 254.3 万平方米，其中玫瑰湖水体面积约 41.2 万平方米。总建筑面积约为 214.9 万平方米，容积率为 0.85。

项目以玫瑰湖生态环境恢复为核心，以打造以高尔夫景观为特色的生态型城市社区为目标，以恢复城市生态环境为基石，本着前瞻性和整体性原则，遵循国际化生态社区发展趋势，以高尔夫景观为特色，优化玫瑰湖自然生态环境，打造一个布局合理、功能复合、形态独特的城市社区，提升戚墅堰乃至常州区域服务等级和城市品牌感知力。

项目以"一轴多区，绿化渗透"为总结构，其中，一轴为沿大明路的开发拓展轴，是联系规划区内各个功能组团的空间组织主轴。多区为规划区内形态各异的功能组团，包括别墅组团、生态社区组团、休闲娱乐组团和特色商业街区等，并以高尔夫景观组团为内核，向外渗透，在空间上构筑生态、宜居的社区形象。

在未来的城市发展中，玫瑰湖地区将作为戚墅堰区未来发展新中心，其发展在戚墅堰起到旗舰作用，规划设计在打造戚墅堰区首个度假休闲社区以成为区域内高品质社区的典范的同时提出在玫瑰湖建设"生态环境型休闲度假社区"，做为戚墅堰城市发展战略的一部分。届时玫瑰湖将集居住、休娱、商务、酒店于一体，以度假、休闲为社区特色，引领戚墅堰区的生活潮流。

Changzhou is located in the southern part of Jiangsu Province, the same distance from Shanghai and Nanjing, one of the Jiangsu Golden Triangle in Jiangsu, with strong economic strength and city competitiveness. Qishuyan District is located in the east of Changzhou City, and is the eastern and western traffic gateway of Changzhou Area with regional development potentials.

The project is located in Rose Lake of Qishuyan District. The total planned area is about 2,543,000 square meters of which the body area of the rose lake is approximately 412,000 square meters. The total construction area is about 2,149,000 square meters with floor area ratio of 0.85.

Rose Lake project aims to restore the ecological environment as the core, build eco-city communities with golf landscape features as the goal, restore the ecological environment as the cornerstone, in the spirit of forward-looking and holistic principles, following the international trend of ecological communities, using golf landscape features to optimize Rose Lake natural environment, creating an urban community with rational layout, composite function, and unique to improve the service level and city brand perception of Qishuyan District and Changzhou City.

The project uses "an axis and multi-zone, green infiltration" as the total structure. One axis is the development and expansion axis along Daming Road, which is the main axis to contact various functional groups in the planning area. Multi-zone is a group of various functional spaces, including the villa group, ecological community groups, recreational groups and commercial district and other commercial blocks, using golf landscape as the core group, expanding to the outside, forming an ecological, livable community image.

In urban development towards the future, Rose Lake area will serve as the new future development center of Qishuyan District. Its development plays a leading role in the Qishuyan District. To create the first leisure and resort community within the region which is a high-quality model, as well as to build the Rose Lake as "resort-type environment community", it is part of Qishuyan city development strategies. The Rose Lake will integrate residential, sports and entertainment, business and hotel in one with features of vacation and leisure to lead the trend of the area, and enhance urban vitality of this area.

①②　①道路交通分析；
③④　②功能结构分析；
⑤⑥　③公共设施分析；
　　④水系分析；
　　⑤日照分析；
　　⑥景观分析

规划总平面图

511

扬州广陵新城"京杭之心"规划设计
Yangzhou Guangling New City "The Heart of Beijing-Hangzhou"

项目名称：扬州广陵新城"京杭之心"规划设计
项目地点：中国 / 江苏 / 扬州
建筑面积：47.80 万平方米
编制时间：2009 / 04
业主：泰达集团—南京新城发展股份有限公司
规划类型：概念规划

项目位于扬州广陵新城中央商务区。该项目的建成将打造广陵新城最具特色的运河主题景观，成为广陵新城的一颗璀璨明珠。

规划师将"京杭之心"作为扬州城市新的发展中心及扬州城市活力的又一激发点，将功能复合、集约的城市中心作为现代城市生活方式的反映，以增加城市中心运营的高附加值。

此设计以尊重和发扬扬州的历史文脉为模式，其运河文化、历史人文以及江南水乡的城市肌理在项目中充分得以体现。

我们在规划中以"活力、尺度、人性化"为理念，其中：

"活力"旨在融合 live、work、play 等复合功能；"尺度"重在打造精致扬州，宜人街坊；而"人性化"主要体现在构筑高品质的公共空间系统上。

在上述理念指导下，规划突出城市综合功能，将项目打造成一个集商贸金融、商务办公、生活居住、文化娱乐、旅游休憩等于一体的城市核心功能区，以引领扬州生活潮流，增强扬州城市活力。

项目总用地面积为 47.8 万平方米，包括一个约 10 万平方米的湖面。项目开发总建筑面积约 41 万平方米，平均容积率为 0.86。

The project is located in the central business district of Guangling New City, Yangzhou. It aims to create the most unique canal theme landscape of Guangling New City as a shining pearl in Guangling New City.

"Beijing-Hangzhou Heart" is a new development centre in Yangzhou city which is another excitating point in the city. Urban centers which are intensive and with complex function are reflection of modern urban life, to increase the added value from the operation of the city centre .

The development of "Beijing-Hangzhou Heart" is a project that should respect and carry forward the historical context of the development model. Yangzhou canal culture, history and culture should be reflected in the project. The urban fabric of south of Yangtze River should be able to be displayed.

Therefore, we propose the "vitality, scale, humanity" concept, in which:

"Vitality" aims to combine functions like live, work, play and others; "scale" focuses on creating exquisite Yangzhou and pleasant neighborhood; "humanity" is mainly reflected in building high-quality public space system.

Under the guidance of the above concept, the plan highlights the city's comprehensive planning capabilities to create an urban core functional area combining financial, business office, residence life, culture and entertainment, tourism, leisure, which leads the trend of Yangzhou life and enhances the urban vitality of Yangzhou.

The total land area of the plan is 478,000 square meters, including a lake of about 100,000 square meters. The total construction area of the project is about 410,000 square meters with an average floor area ratio of 0.86.

文昌大桥

文 昌 东 路

滨

京 杭 大 运 河

河

北

扬州运河大桥

运 河 东 路

路

规划总平面图

515

①② ①道路分析；
②交通分析；

→ 城市主干道
→ 城市次干道
→ 城市支路
→ 地块内道路
→ 滨河景观道
→ 坝顶防洪通道
→ 应急道路

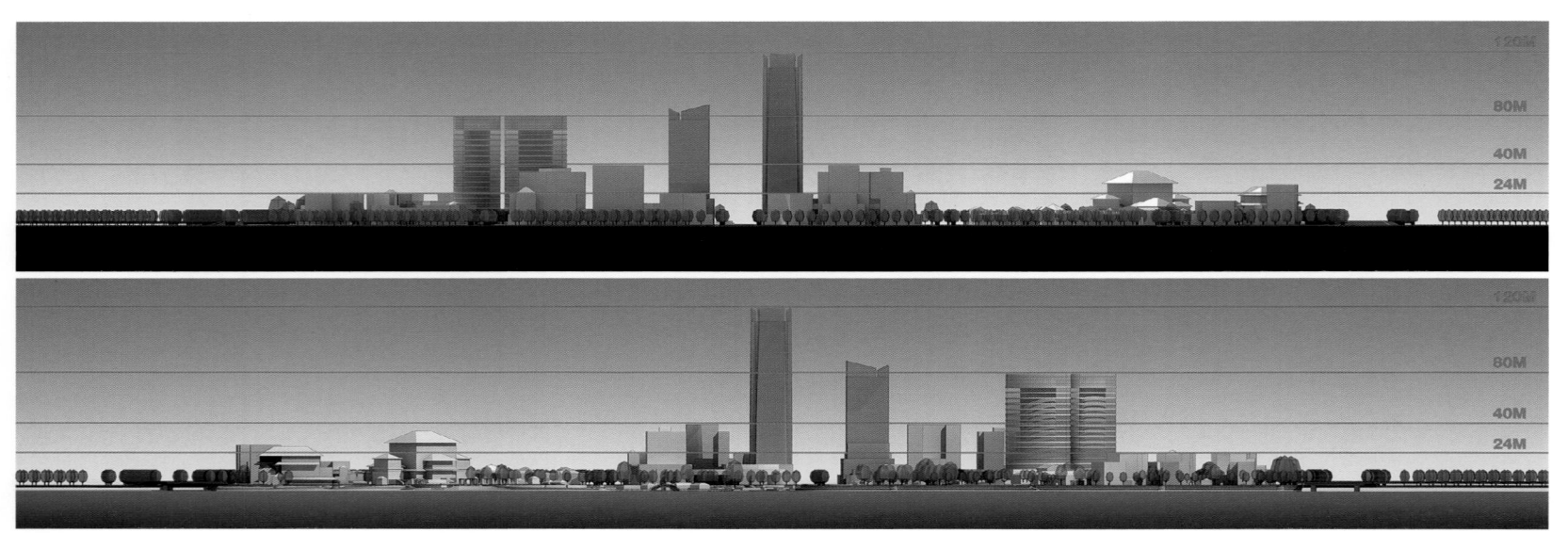

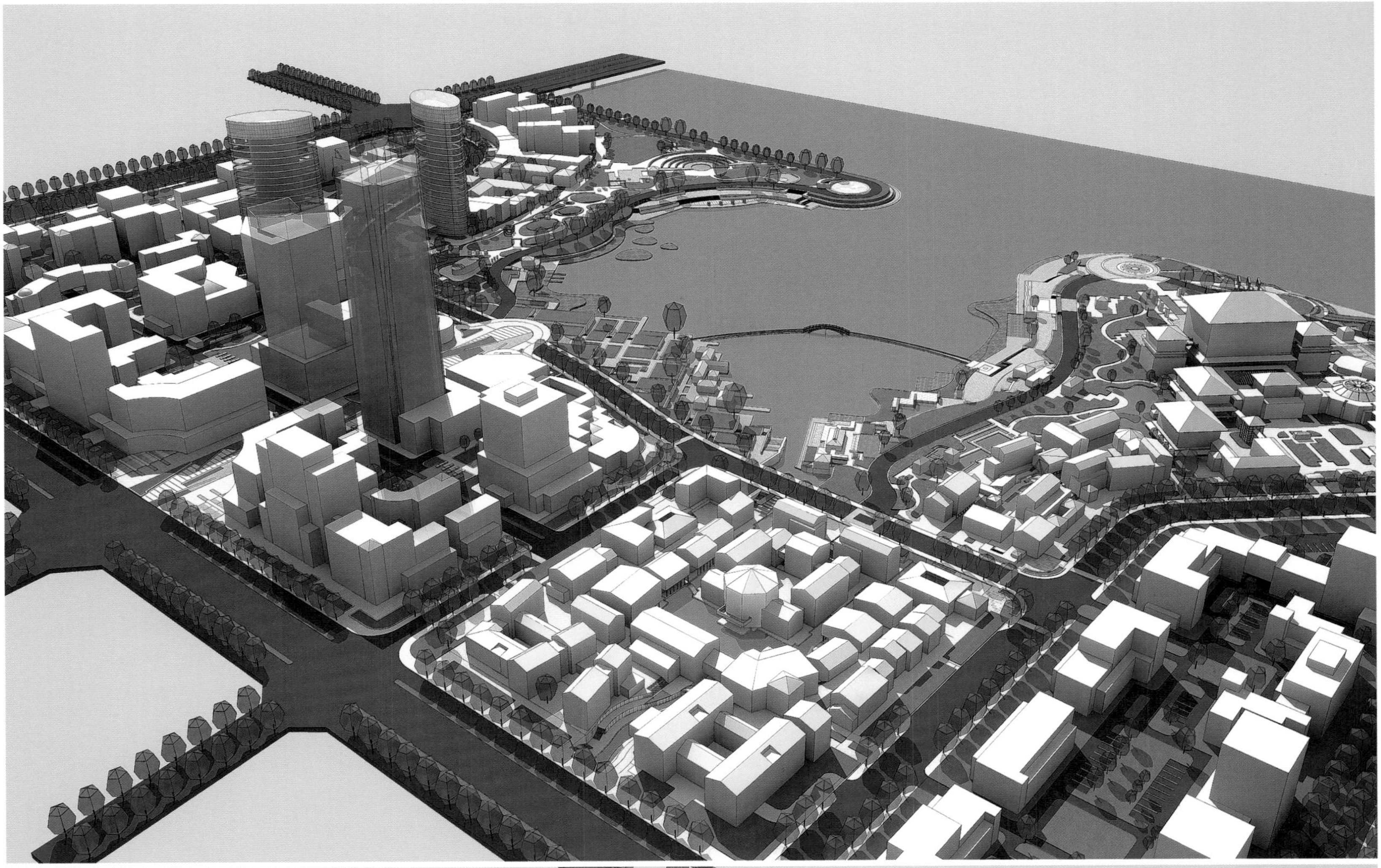

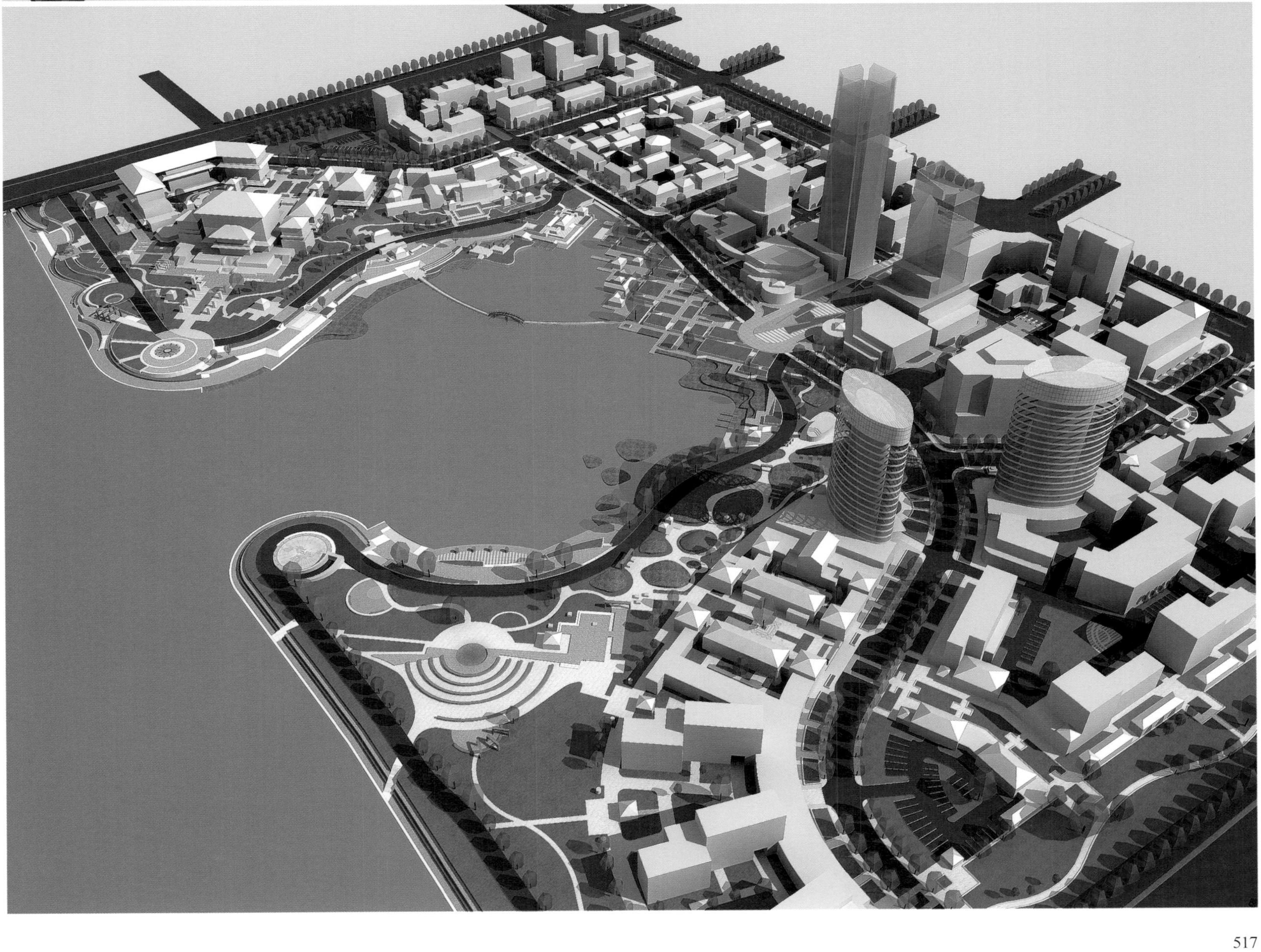

临港新城中心区城市设计
Lingang New City Center Urban Design

项目名称：临港新城中心区城市设计
项目地点：中国 / 浙江 / 临海
建筑面积：1160 万平方米
编制时间：2009 / 08
业主：临海市临港新城管委会
规划类型：城市设计

项目位于临海市滨海新城东部，南洋滩涂围垦区与北洋滩涂围垦区之间。区域内部地形地貌较为鲜明，除正对山脉及王峙山等之外多为滩涂围垦区。本次中心区规划范围约 1160 万平方米，启动区范围 340 万平方米。

规划对基地内建筑、道路、绿化、水体等基础条件进行了详尽的分析，最终确定了以拆、改、留、迁四种技术手段，对城市用地功能布局进行优化整合，力争将临港新城打造成富有魅力的城市新中心，营建亲切宜人的滨海地带，塑造充满活力的城市街道空间，形成生态为本的居住环境，使其成为一座拥有蓝天、碧海、青山、绿城的山水生态新城。

规划启动区可开发建设地块用地面积 136.5 万平方米，总建筑面积为 1160 万平方米。

The project is located in the eastern coastal Linhai city, between south and north beach reclamation areas. The region boasts a distinctive topography. In addition to currently facing the mountains, Wang Zhi Mountain and other areas, most of which are reclamation area. The central area of the plan is about 11,600,000 square meters with starting area of 3,400,000 square meters.

The plan conducts a detailed analysis of basic conditions like buildings, roads, forestation, water inside the base and finally determines to use four technical means of dismantling, changing, staying and moving to optimize the functional layout of urban land consolidation, striving to make Lingang New City a glamorous new city center, building a pleasant coastal zone, creating vibrant urban street space, forming eco-oriented living environment, making it a landscape ecological city with blue sky, clear sea, green hills and green city.

The exploitable development area is 1,365,000 square meters with a total building capacity of 11,600,000 square meters.

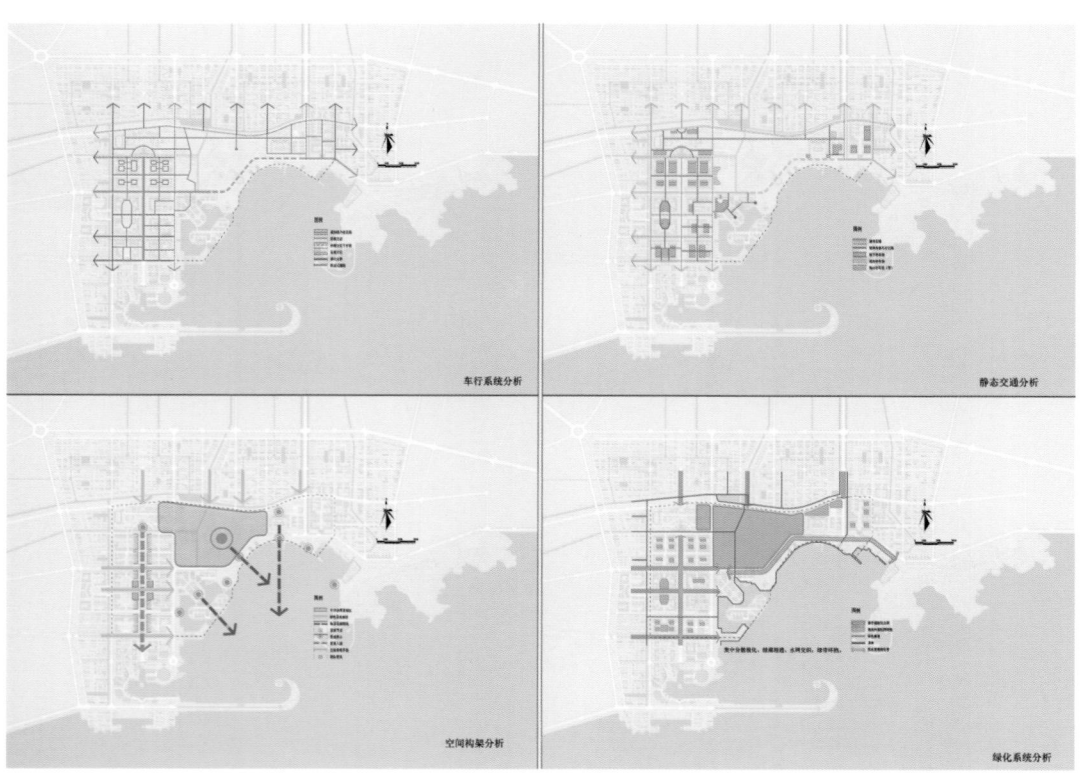

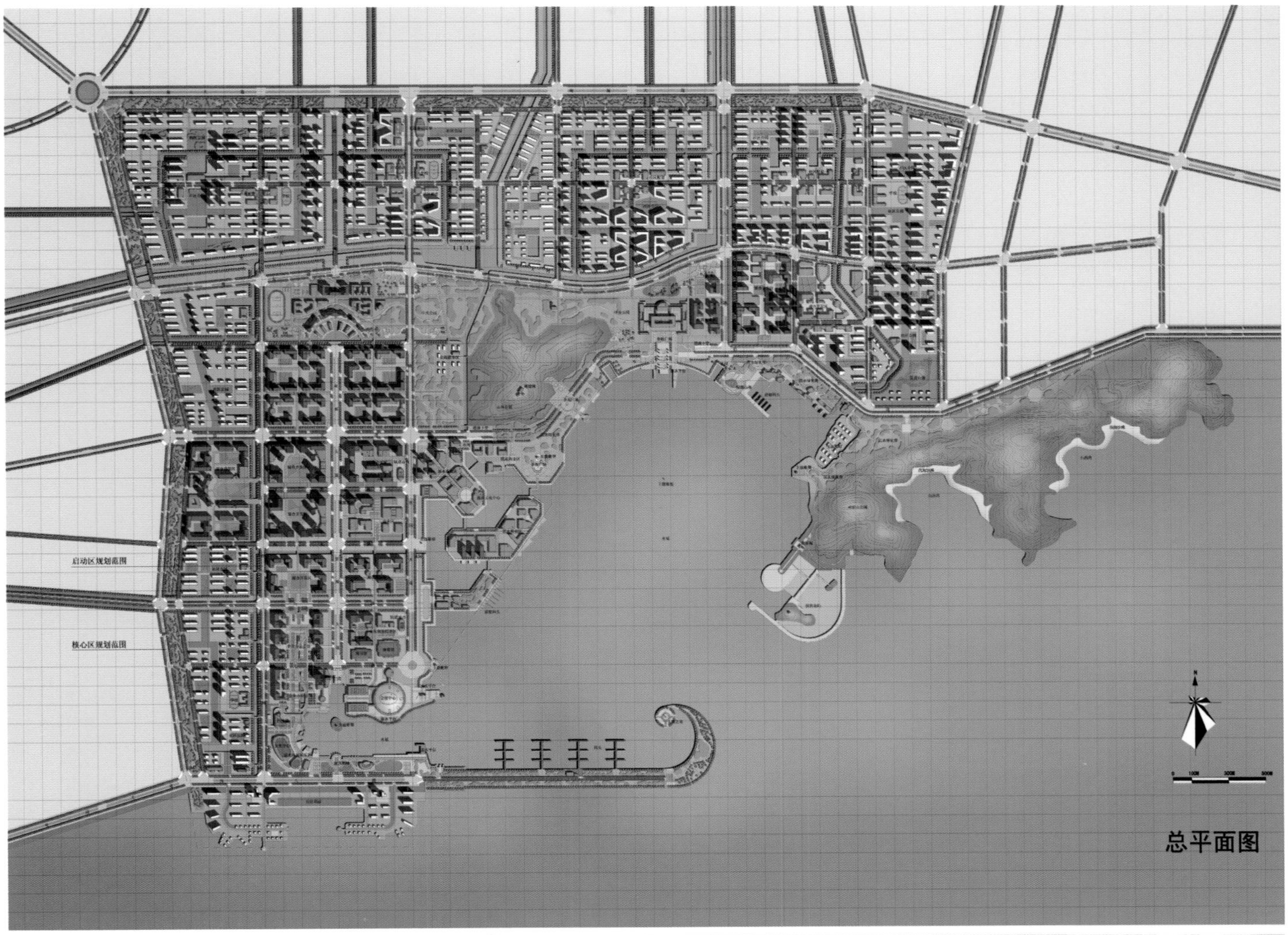

总平面图

核心区总平面图

上海十六铺地区城市设计
Shanghai Shiliupu Region Urban Design

项目名称：上海十六铺地区城市设计
项目地点：中国／上海
建筑面积：23 万平方米
编制时间：2001 / 09
业主：上海黄浦区城市规划管理局／上海港务局
规划类型：城市设计

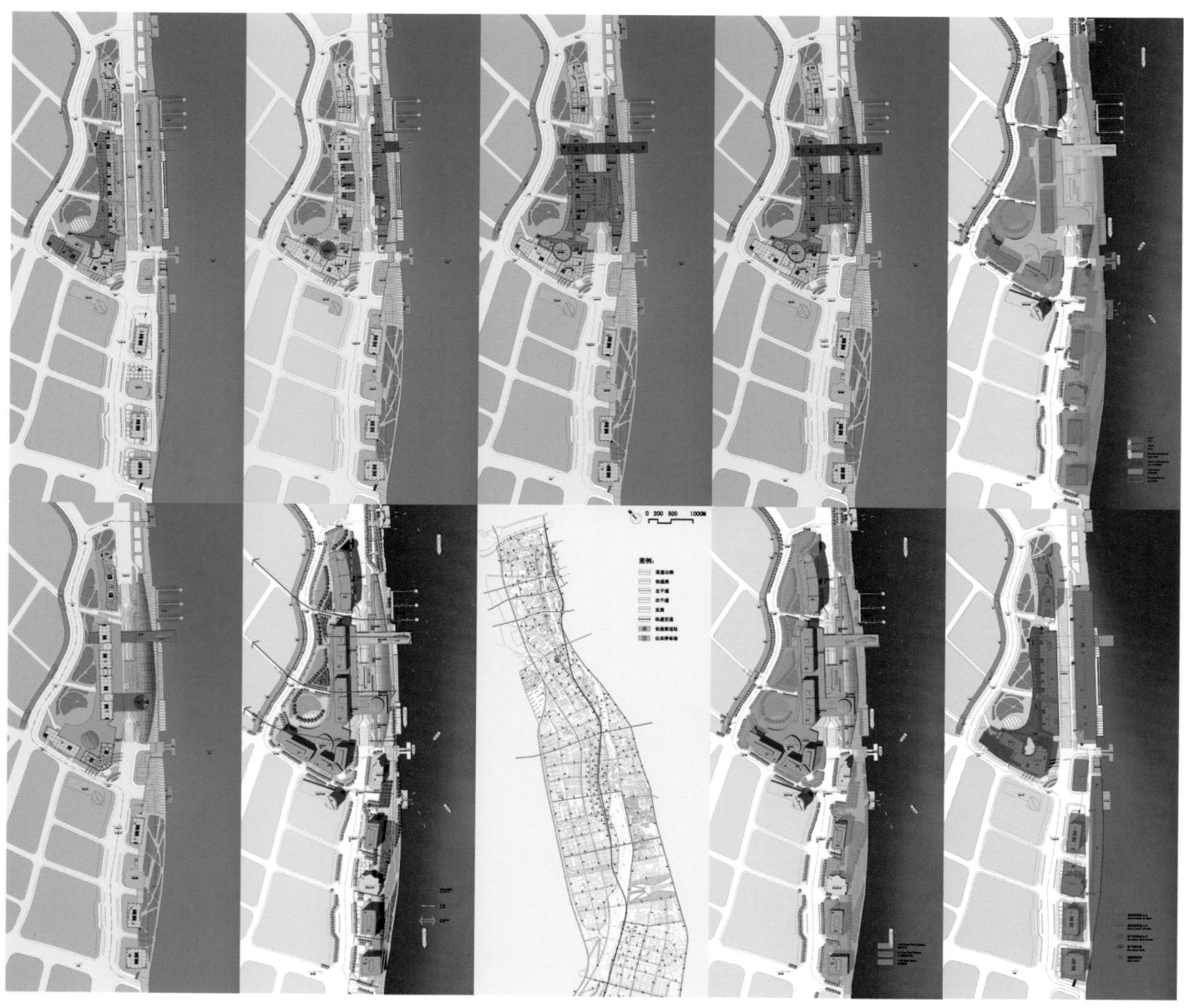

项目位于上海市中心黄浦江畔，北接外滩历史文化滨河带，南临南浦大桥，具有得天独厚的交通和景观资源。规划通过现状分析与未来预测，对十六铺地区的未来发展提出规划目标：1. 通过完善 CBD 的功能，延伸外滩步行带，使之成为旅游活动的中心；2. 建立浦东、豫园和黄浦江之间实体和视觉上的互相联系；3. 使该地区成为包括汽车交通（区域的与当地的）、货运、巴士、轮渡与水上巴士复杂联系的重要交通节点；4. 增强滨水区的可达性，提供市民活动空间，重建部分老城墙以保护老城肌理。

本次规划提倡多功能混合和高覆盖率的地块开发模式，达到以中高层建筑来满足较高容积率、绿地率要求，延续外滩城市风貌特色，并与对岸浦东新建筑的天际线形成对比，清晰地刻画出十六铺江岸的特点。

This project is located near the Huangpu River. It reaches Bund historical and cultural riverside in the north, Nanpu Bridge in the south. It boasts a unique transport and landscape resource. Through analysis and prediction, the plan puts forward future objectives on the development of the Shiliupu region : 1. improving the function of CBD, with an extension to the Bund which makes it the center of tourism activities; 2. establishing the relationship between Pudong, the Yu Garden and Huangpu River in physical and visual aspects; 3. making the region, including automobile traffic (regional and local), freight, bus, ferry and water taxi transport complex as an important node; 4. enhancing accessibility of the waterfront, providing public space for citizens, reconstructing part of the old city walls to protect old city texture.

The plan promotes land development mode of multi-function mixture and high-coverage land, meeting high floor area and green space ratio requirement of using high-rise buildings, sustaining the Bund's urban features. It makes a contrast with the other side of the Pudong skyline, clearly depicting the characteristics of Shiliupu riverbank.

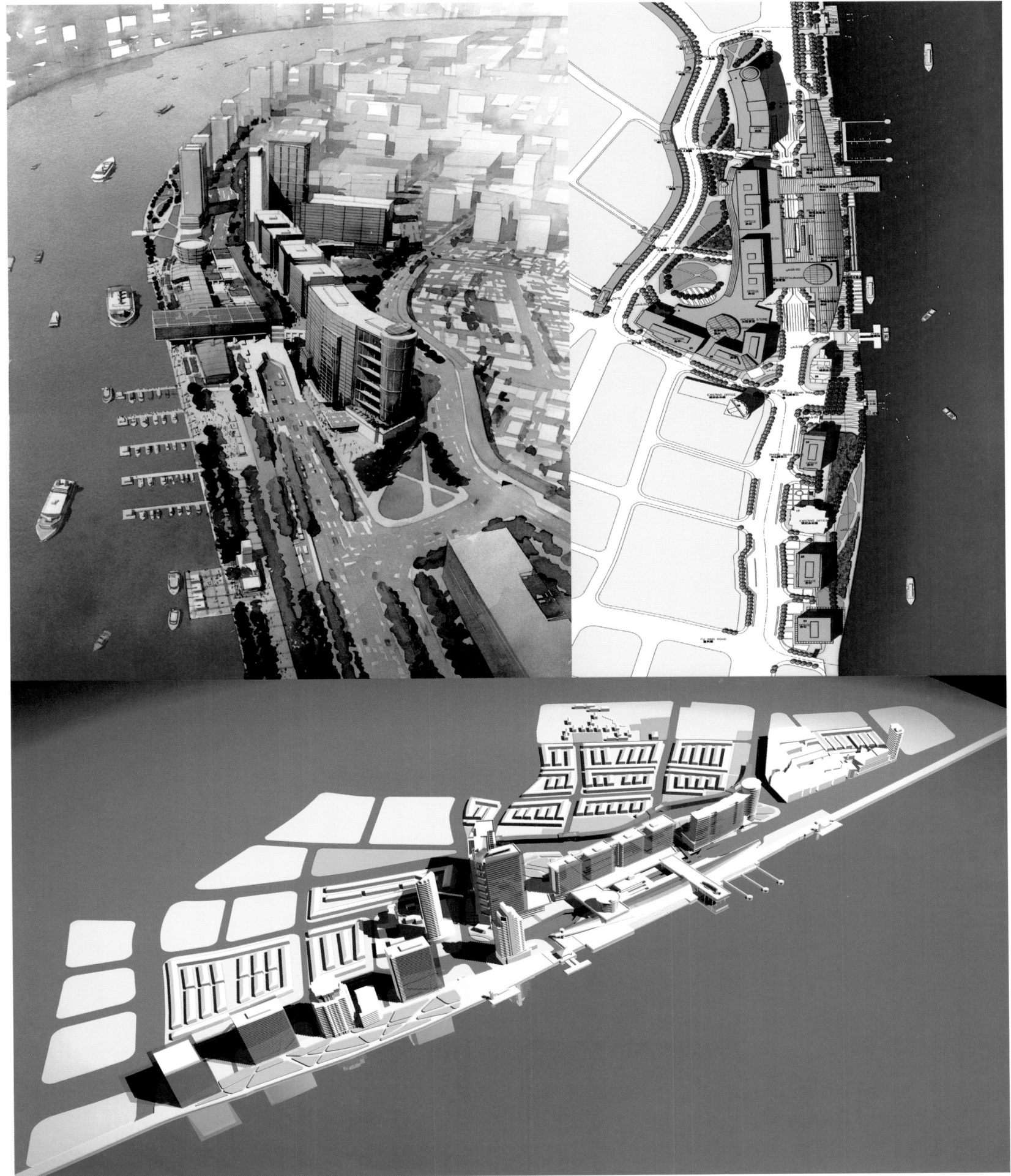

西宁市南川地区总体规划
Xining Nanchuan Regional Master Plan

项目名称：西宁市南川地区总体规划
项目地点：中国 / 青海 / 西宁
建筑面积：3 700 万平方米
编制时间：2010 / 07
业主：西宁市正华建设投资股份有限公司
规划类型：概念规划

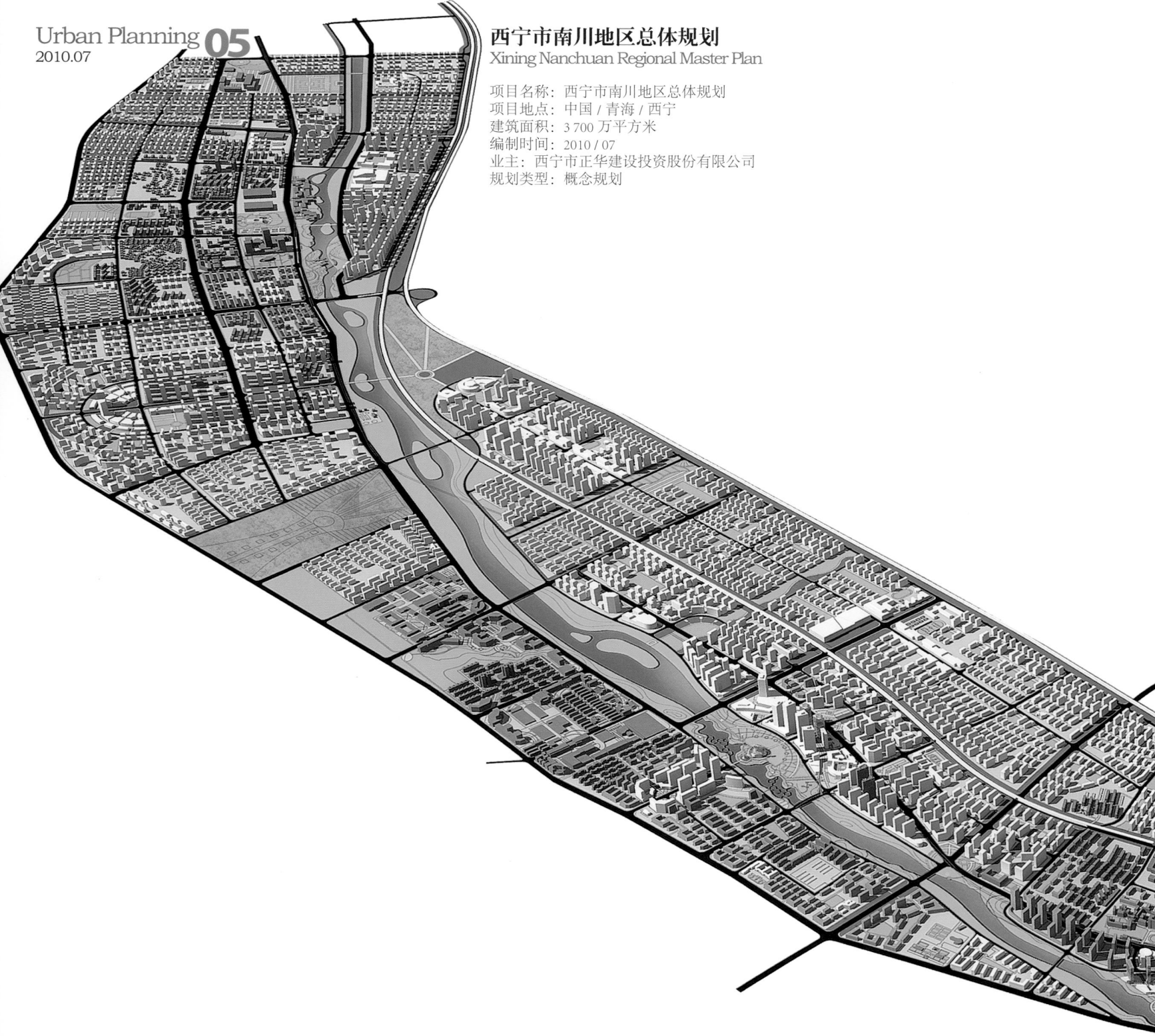

凭借西宁市城市发展契机，城南组团不断发展壮大，南川地区也将面临新的发展机遇。

规划南川地区北以湟水河及滨河路为界，南至总寨西路，西临规划的西塔高速，东为西久公路，用地较为狭长。

规划设计定位为夏都未来的黄金水岸，打造因河而荣的西宁城市南拓轴线，成为西宁第一条滨水公共活动走廊、第一片滨水特色生态居住地区以及第一个公共服务核心区，并将其整体形象定位为十里南川、都市金谷、生态绿带和魅力水廊。

规划区总建筑面积约 3 700 万平方米，其中居住建筑面积约 2 400 万平方米，可容纳的居住总人口为 68.4 万人。

With the development opportunity of Xining City, south group is flourishing. Nanchuan areas are facing new development opportunities.

Planning of Nanchuan has the long land area covering northern boundary to Huangshui river and Binghe Road, south to West Zongzhai Road, west to the planned West Tower highway, east to Xijiu highway.

The planning is positioned to be Xiadu future golden waterfront, building south wing extension axis famous for the Main River, making Xining the first public event waterfront corridor, the first piece of waterfront living areas with ecological characteristics and first public service core area and transferring its overall image an ecological green belt and a charming water gallery.

Planning has a total construction area of 37 million square meters, including residential construction area of about 24 million square meters for a total population of 684,000.

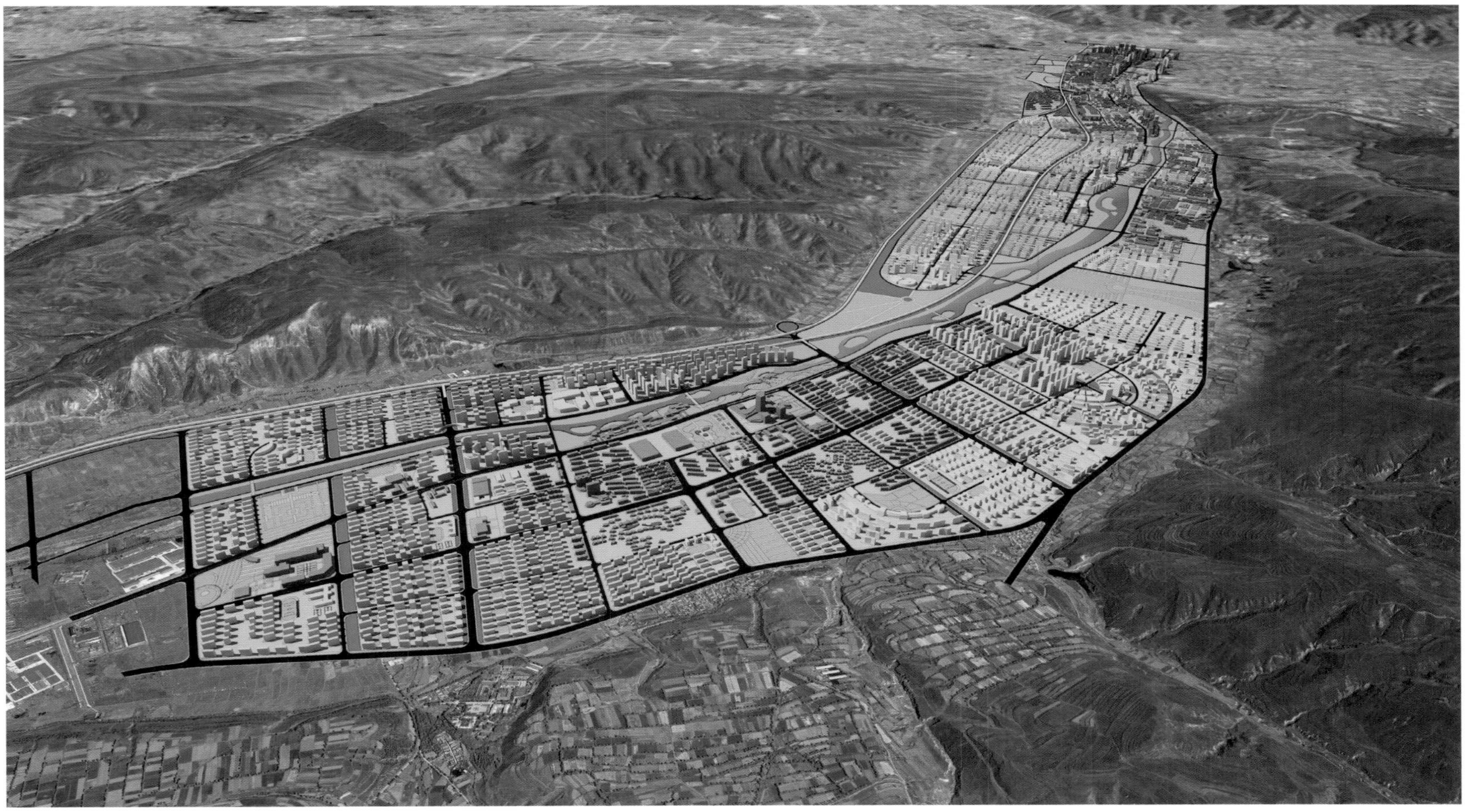

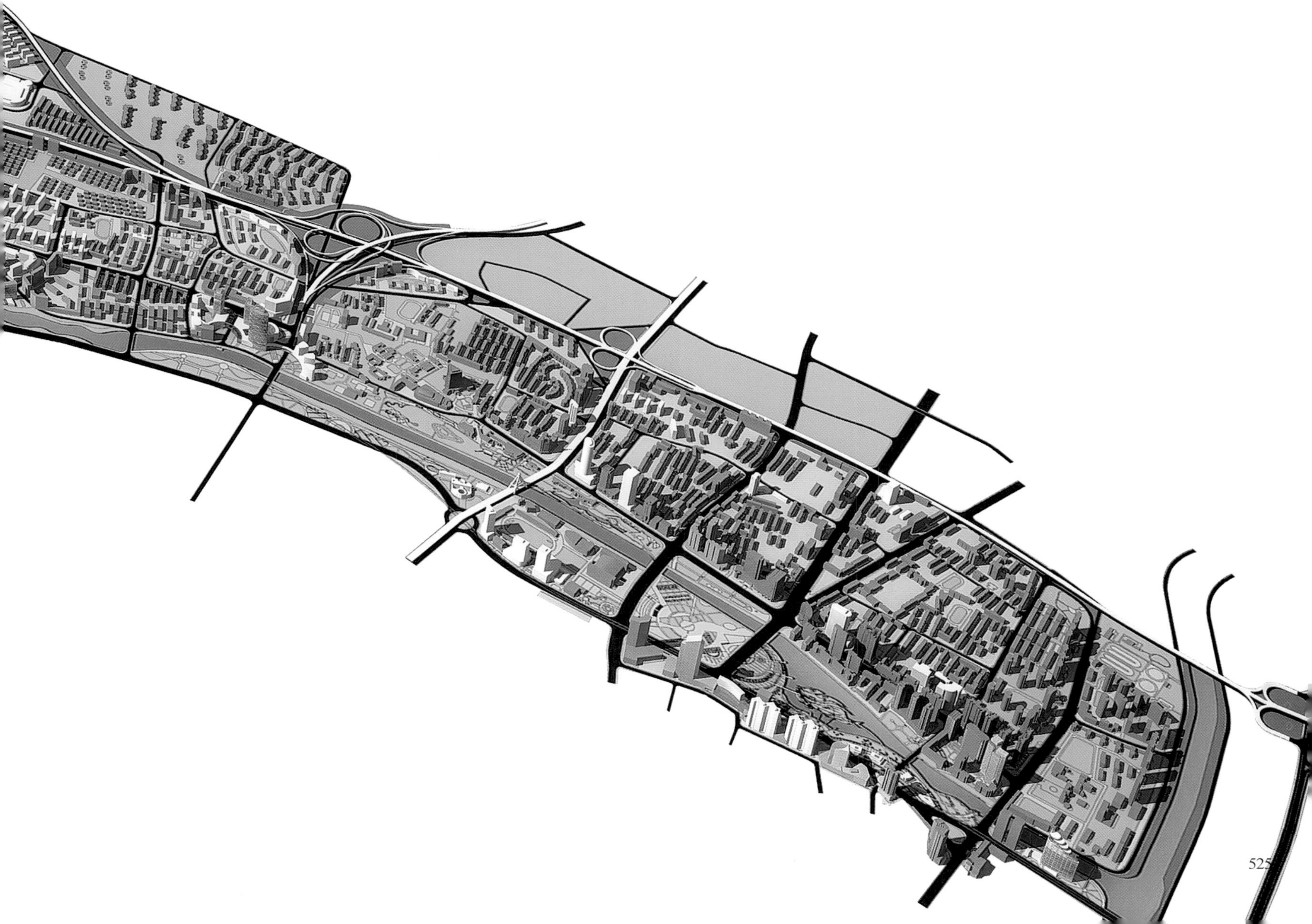

用地总平面图

SMITHGROUP cna
Architecture · Urban Planning · Landscape Design

图例：

一类居住用地（R1）
二类居住用地（R2）
中、小学用地（R22）
四类居住用地（R4）
行政办公用地（C1）
商业金融用地（C2）
文化娱乐用地（C3）
体育用地（C4）
医疗卫生用地（C5）
教育科研设计用地（C6）
商住混合用地（C/R）
工业用地（M）
仓库用地（W）
高速公路用地（T21）
长途客运站用地（T23）
道路用地（S1）
广场用地（S2）
社会停车场库用地（S3）
市政公用设施用地（U）
公共绿地（G1）
特殊用地（D）
水域（E1）
耕地（E2）
村镇建设用地（E6）

0 200 500 1000M

系列分析图：

图例：
建筑高度≤18m
建筑高度≤36m
建筑高度≤50m
建筑高度≤80m
建筑高度≤100m
建筑高度100m以上

城市中心区
（城西片区）

拓展延伸区域

公共

城南公共服务

图例：
南川河核心开放空间
滨水公共活动节点
步行景观桥
外围绿色空间渗透轴

0 200 500 1000M

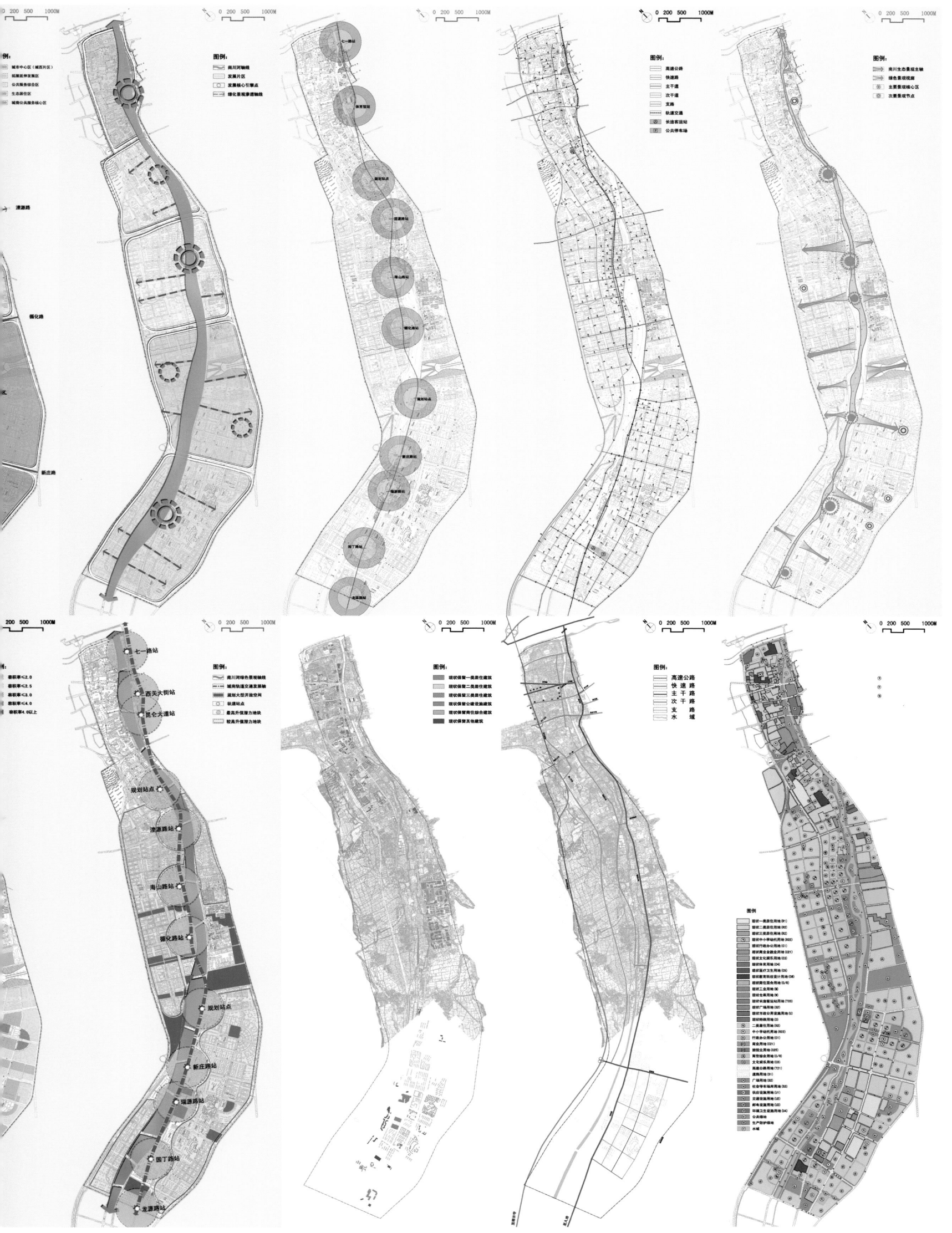

SMITHGROUP cna

Architecture · Urban Planning · Landscape Design
Interiors Design · Art Consultant · Lighting Design
Since 1853
www.cna-group.com
Design Determines . . .
设计决定……

01 规划设计类项目
01 **Urban Planning**

滨水地区
Water Front

城市街区
City Block

城市空间
Urban Space
旅游休闲
Travel and Leisure

02 景观设计类项目
02 **Landscape Design**

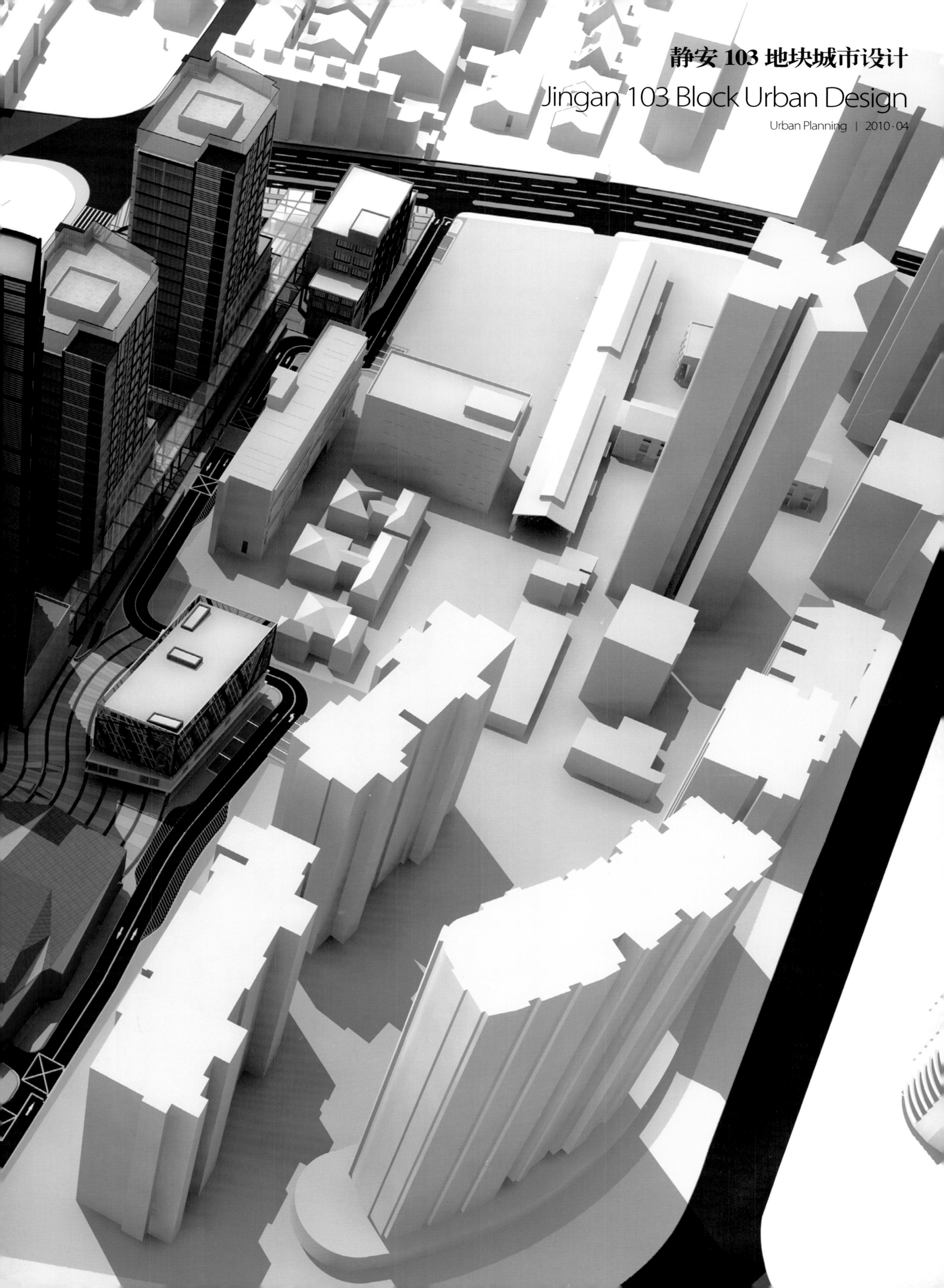

静安 103 地块城市设计
Jingan 103 Block Urban Design

项目名称：静安 103 地块城市设计
项目地点：中国 / 上海
建筑面积：14.33 万平方米
编制时间：2010 / 04
业主：上海静安城建配套发展公司
规划类型：城市设计

本次规划的 103 地块位于上海静安区北部，北近昌平路、南临康定路、西靠常德路、东临西康路。地块以里弄住宅为主，工厂分列北侧东西，零星有大宅。目前由于周边高楼拔地而起，103 地块日益成为由高层住宅形成的多岛多带之间的"谷地"，并且随着轨道交通 7 号线的开发及竣工，800 秀项目的竣工，103 地块本身的价值逐步显现，日益需要由一定深度和力度的开发来衬托这个地块。

基于这样的开发意愿和市场需求，且为了更好地控制和引导 103 地块的开发建设和整体环境品质，在上一轮概念方案研究的基础上，本次城市设计重点研究开发手段、业态配置、交通组织、整体空间形象和景观等关键问题，以提升 103 地块的环境品质、城市形象和经济价值。

该项目的总体定位为：以文化展示、商业商务、娱乐休闲为主导功能。

其开发思路抓两条主线：其一为考虑 103 地块的历史内涵，为了保留地块的历史回忆，重视历史建筑与地块文化特征，规划在开发思路上首先要强调"拆、改、留、建相结合"，而非推倒式的城市更新手段；其二为结合地铁 7 号线站点充分开发利用地下空间。

在空间塑造上，规划通过点、线、面相结合的设计手法将地块内不同功能区块和不同时期的建筑进行完美融合，创造出复合多元化的城市公共活动空间。

面：从功能上分为商业商务综合区和文化娱乐综合区，分居地块东西两侧，描绘出风格各异的空间景观。

线：多条景观通廊向周边地块辐射，使得人们能够方便快速地到达地块内部。

点：由多个集散广场构成的景观节点成为空间的点睛之笔。

本次规划地块总建筑面积为 14.33 万平方米，其中地上建筑面积为 10.42 万平方米；地下建筑面积为 3.91 平方米；平均容积率为 3.15。

103 plot is located in the northern part of Shanghai Jingan District, north to Changping Road, south of Kangding Road, west of Changde Road, east of Xikang Road. The lanes are the main part of the plot. The factory and mansion are in the north. At present, because of the grown up of the surrounding high-rise buildings, 103 plot has become a "valley" between the high-rise apartments and multi-island. With development and completion of metro line No.7 and 800 show project, the value of 103 plot has gradually emerged, increasingly needing to have a certain depth and strength to support the development of this land.

Based on the development will and market demand and in order to better control and guide the development and the overall environmental quality construction of 103 plot, on the basis of last round of the concept plan, the focus of urban design in this study is to develop tools, format configuration, traffic organization, the overall landscape image of space and other key issues to improve the quality of the environment, city image and economic value of 103 plot.

The overall positioning of the project is: functions like cultural show, commercial business, entertainment and leisure.

There are two main lines of development: one is to consider the historical connotations of 103 plot in order to preserve historical memory, focusing on historical buildings and cultural features of the plot. The plan first emphasizes" a combination of demolition, change, leaving, building ", as the development of ideas rather than urban renewal means of tearing down; the other is combined with Metro Line No.7, taking advantage of the underground space.

In the respect of space shaping, through combining design approach of the points, lines and planes, the plan blends buildings of different functions and different periods to create a complex variety of urban public space.

Side: from the function, it can be divided into commercial and business district and entertainment complex zone, separately located in eastern and western sides of the plot, depicting different styles of space landscape.

Line: a number of landscape corridors radiate to the surrounding plots so people can quickly and easily reach the land.

Point: the landscape node formed by multiple distributed square is a shining point.

The total construction area of the plan is 143,300 square meters of which the ground floor area is 104,200 square meters; underground floor area is 3.91 square meters; the average floor area ratio is 3.15.

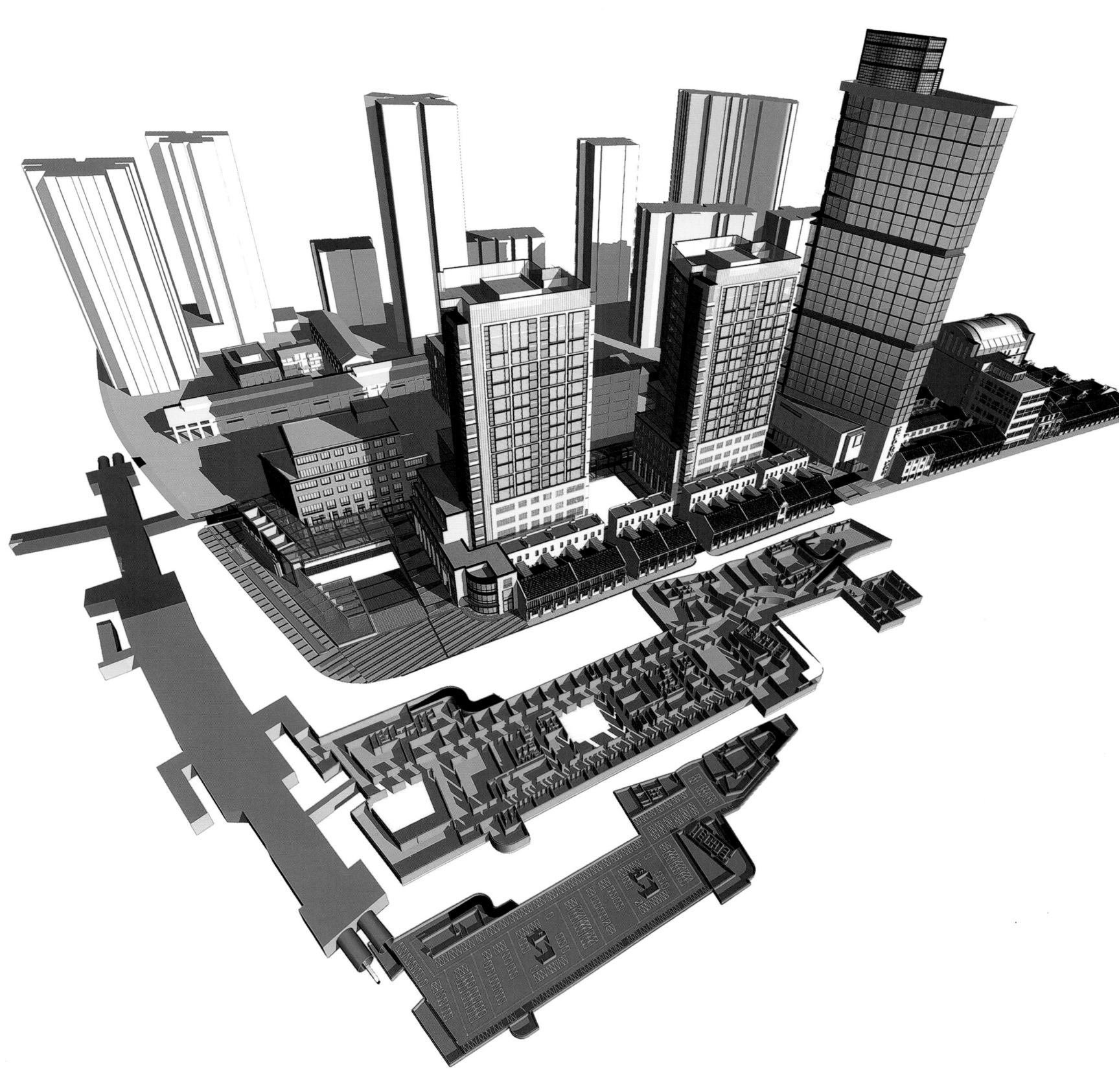

地下空间分析

方案一

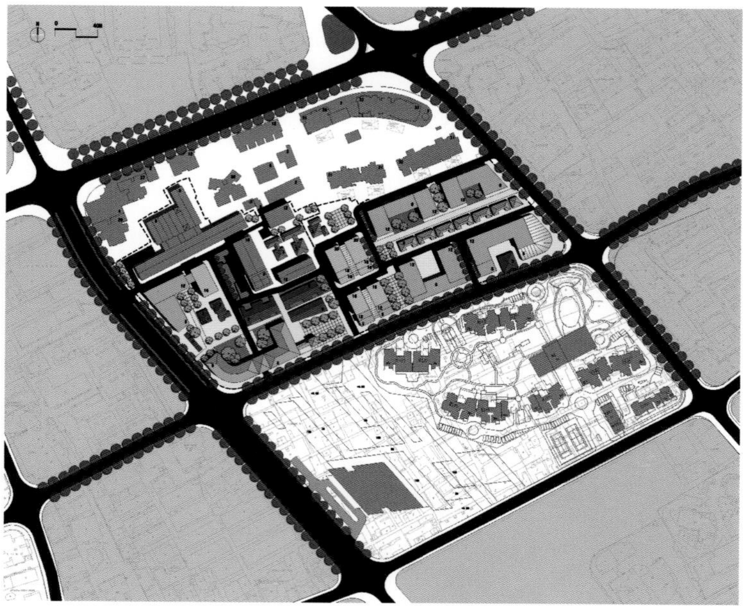

方案二

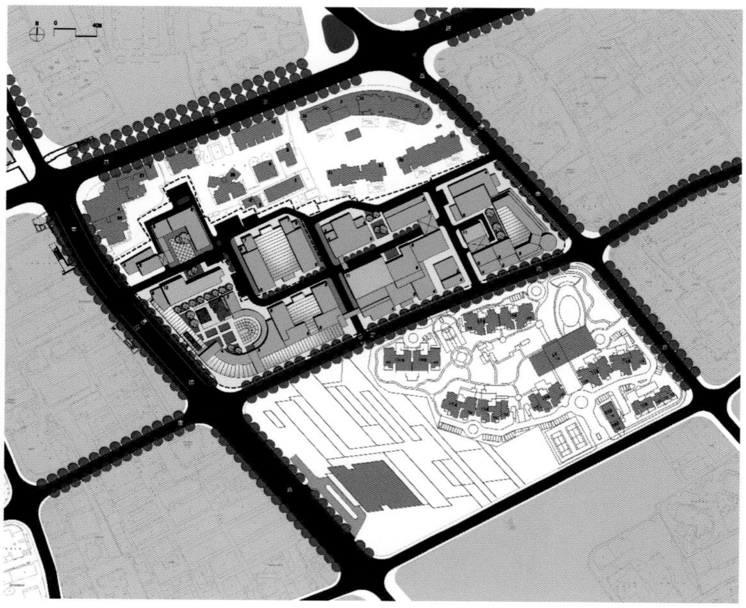

常德路立面

BLOCK 103

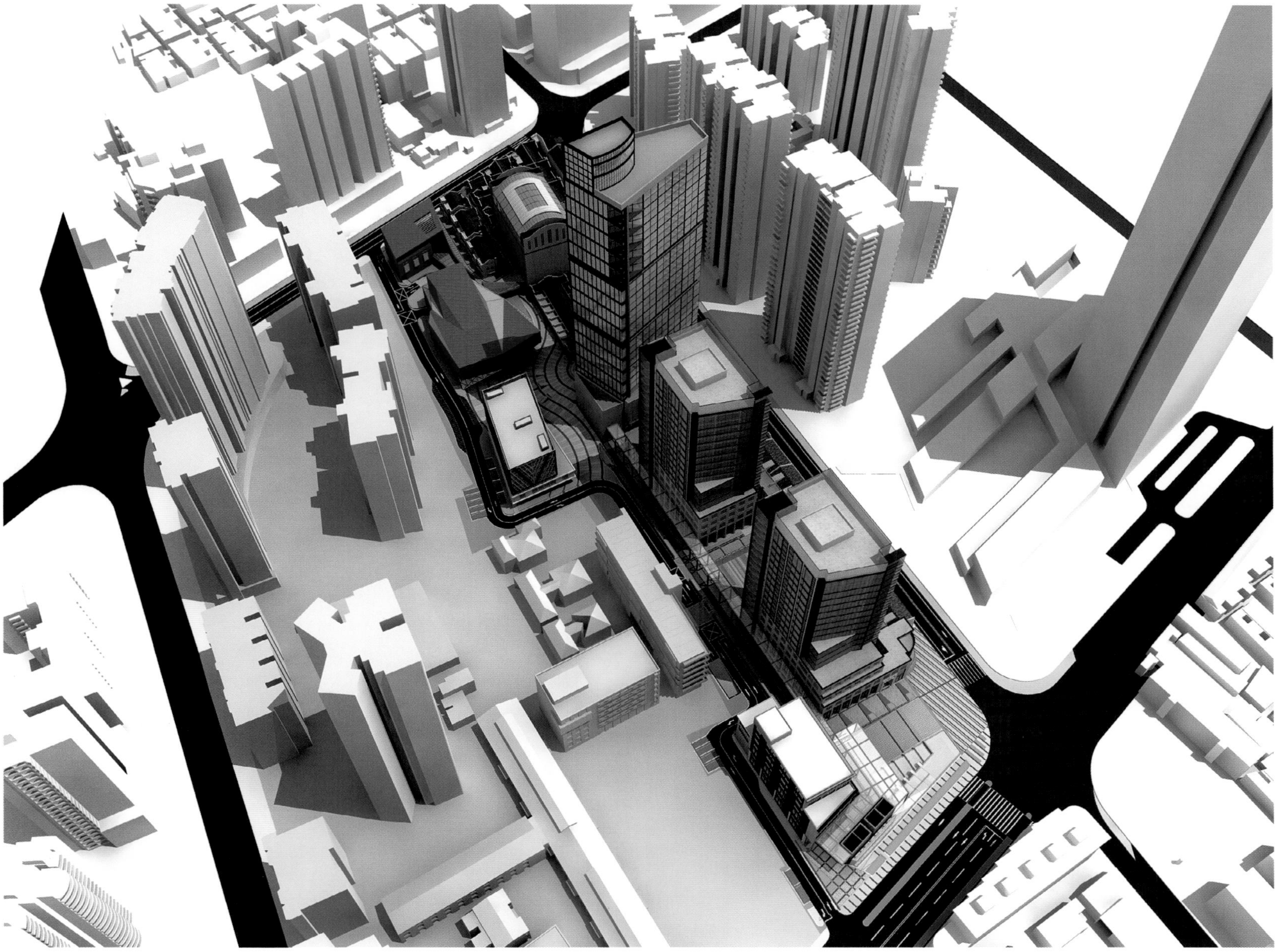

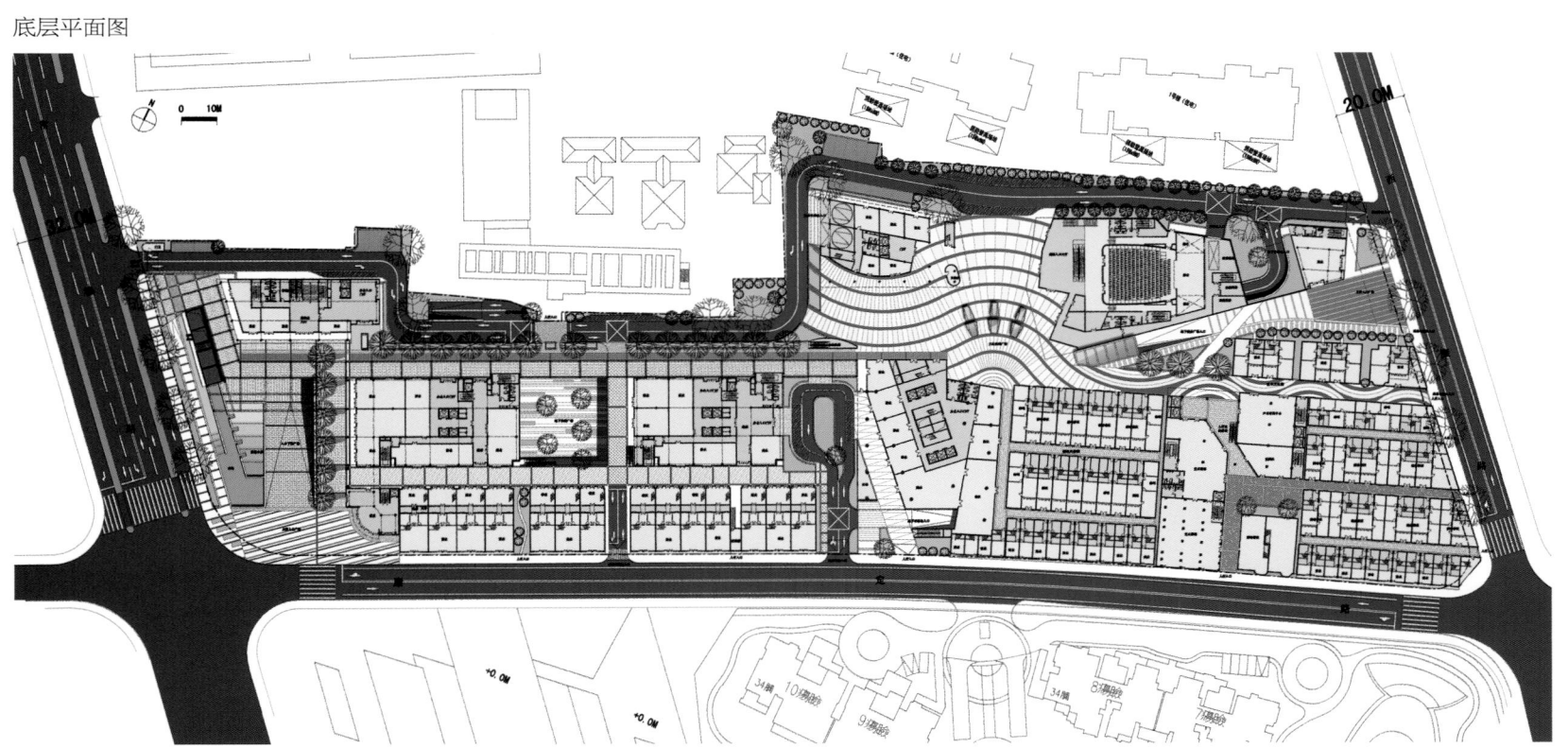

底层平面图

青海省西宁市解放路改造规划设计
Jiefang Road Reconstruction Planning, Xining City, Qinghai Province

项目名称：青海省西宁市解放路改造规划设计
项目地点：中国 / 青海 / 西宁
长度：465 米
编制时间：2009 / 06
业主：西宁正华建设投资控股有限公司
规划类型：详细设计

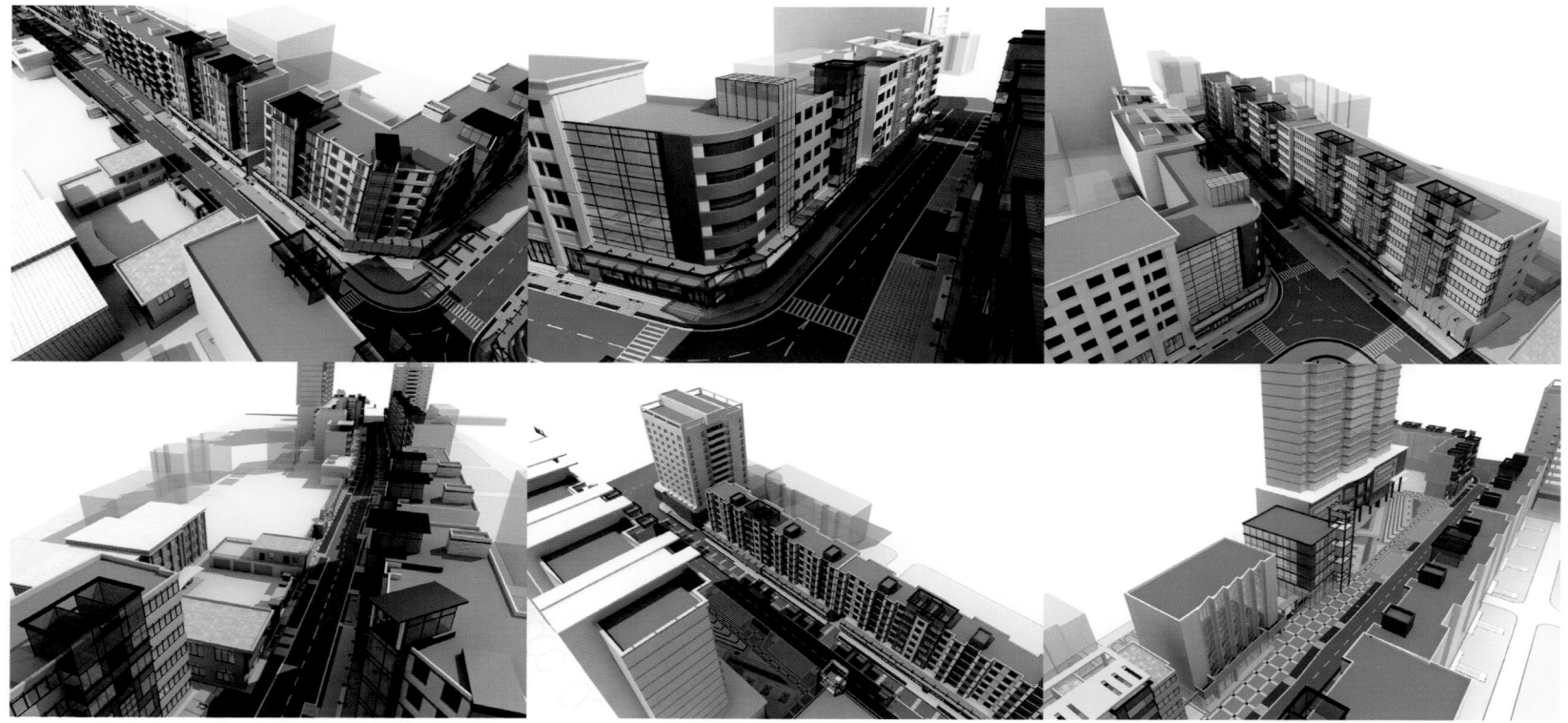

解放路位于西宁市中心区，东接北大街，西至礼让街及互助巷，南临省政府，北临城隍庙，为综合性的城市道路。

规划对其功能定位为集购物、休闲、时尚于一体的多功能街区，体现现代商业及生活气息。同时以步行休闲为主，是分时段限制穿越性机动车通行的综合性城市道路。力图将解放路打造成为：

城市之路——打造一条具有特色功能的混合型城市街区；

生活之路——形成一片生活氛围强烈、风光旖旎的生活社区；

时尚之路——创造一种时尚、高雅的生活氛围；

休闲之路——体现一种休闲、自在、具有活力的生活模式。

Jiefang Road is located in Xining city center, east of North Avenue, west to Lirang Street and Huzhu Lane, south of the provincial government, north of Chenghuang Temple and it is a comprehensive urban road.

According to its function, the plan's position is a multifunctional block integrated by shopping, leisure and fashion, reflecting the modern business and living atmosphere. The main style is casual; it is a comprehensive urban road that limits motor vehicles to pass through during some period. We are trying to build Jiefang Road as a: City road-to create an urban neighborhood with mixed features;

The road of life-to form of a community with strong atmosphere of life and beautiful scenery;

Fashionable road-to create a stylish and elegant life atmosphere;

Leisure road-to embody a casual, comfortable, vibrant lifestyle.

西宁市解放路透视（上图）

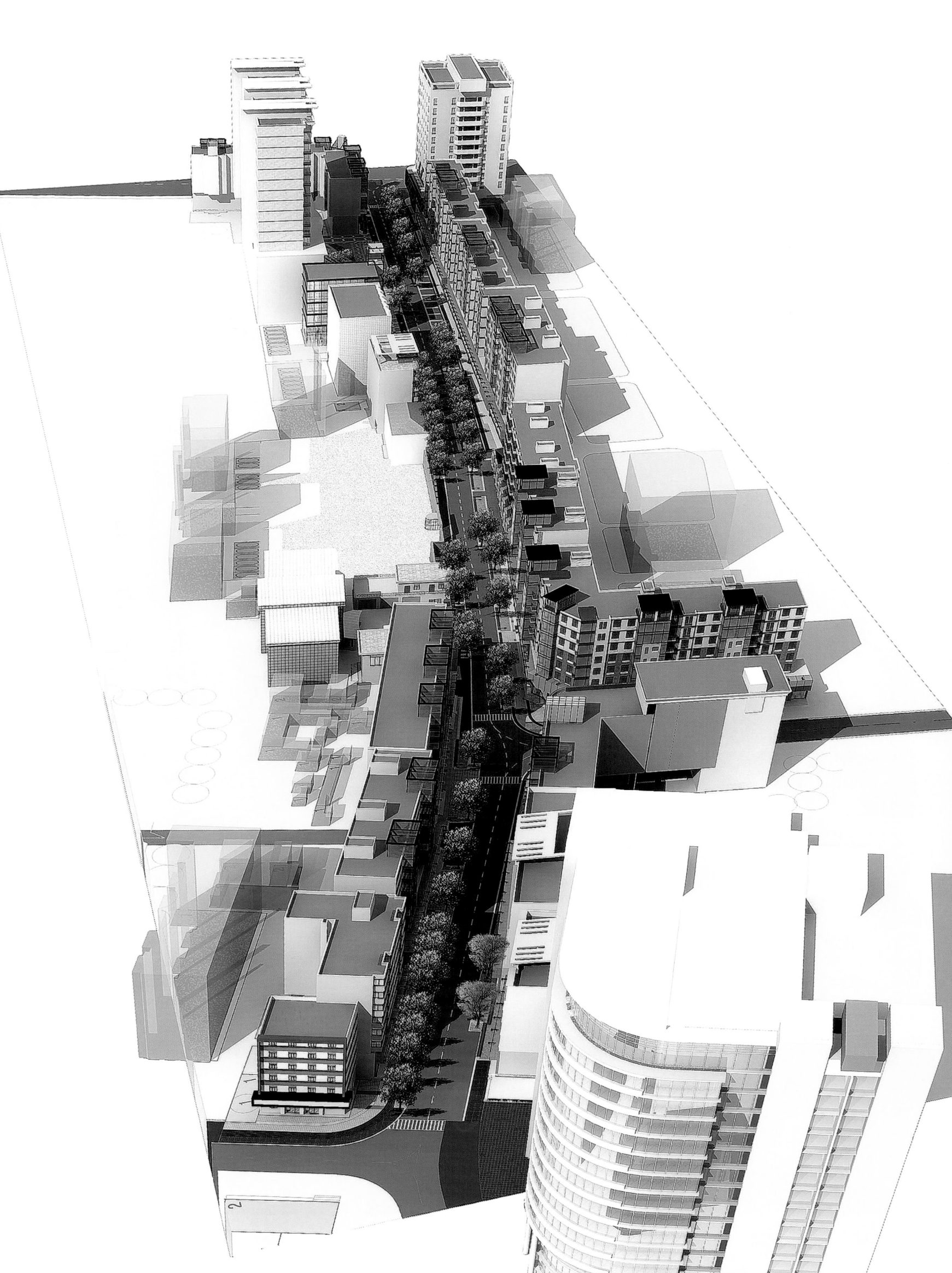

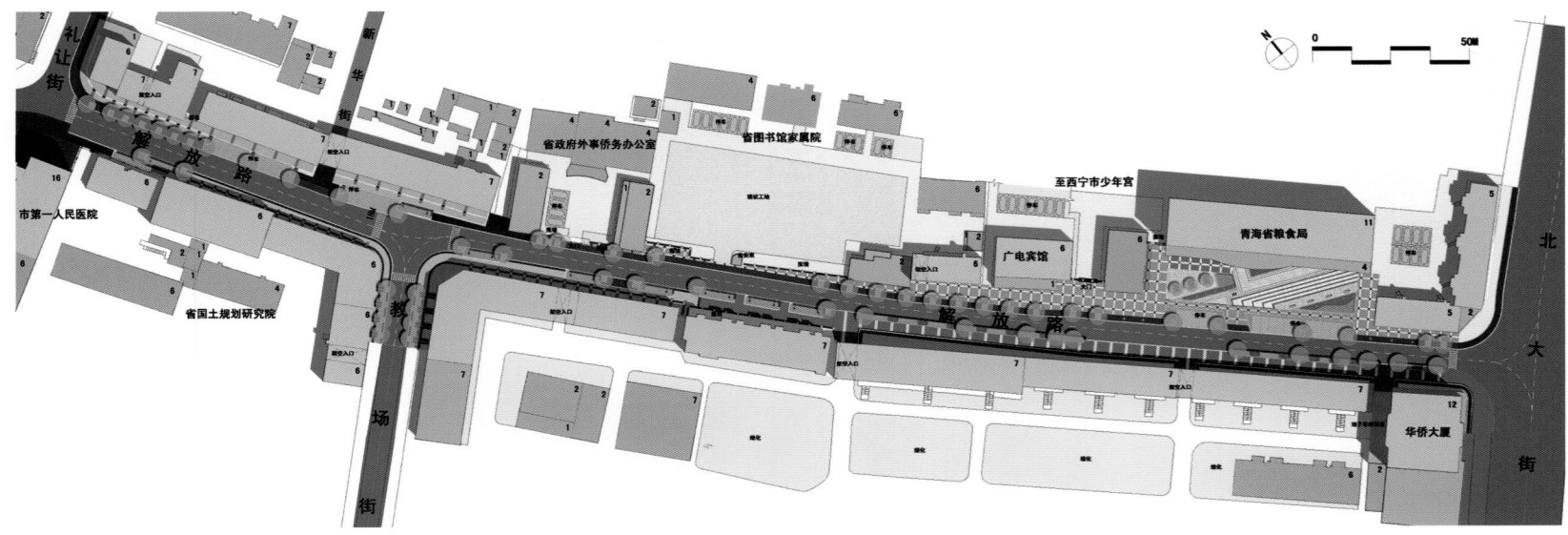

规划总平面

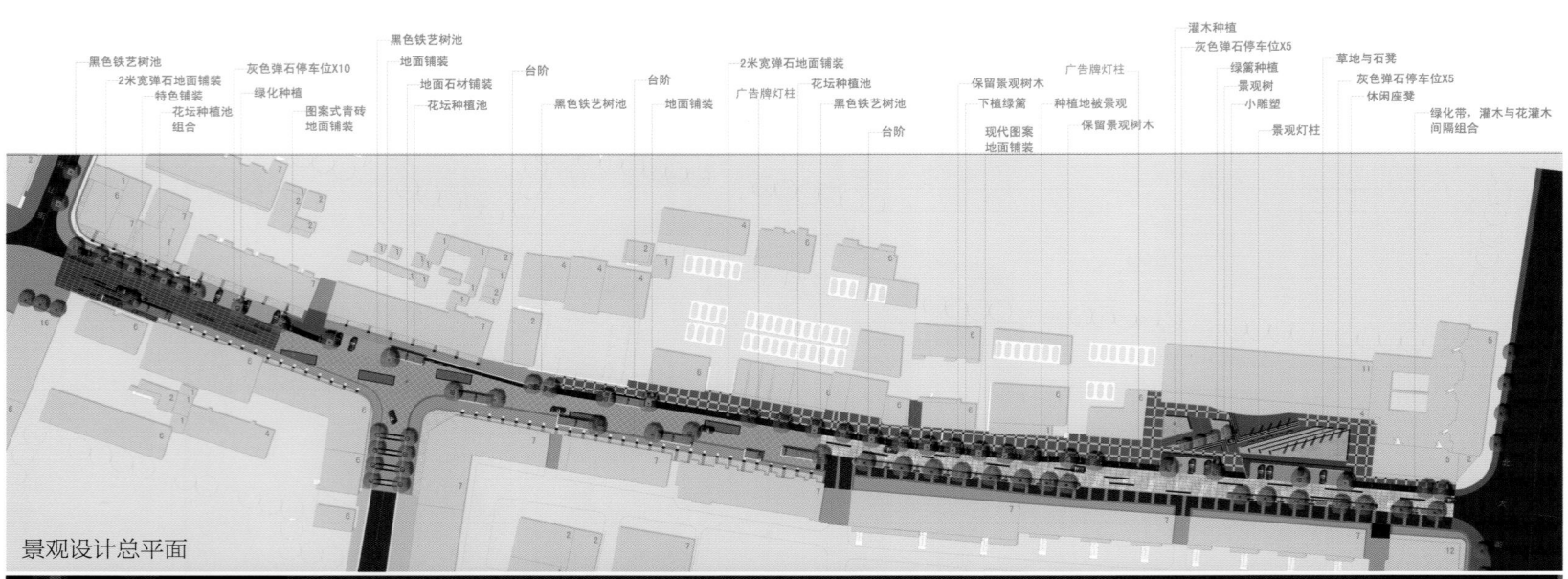

黑色铁艺树池
　　2米宽弹石地面铺装　　　　灰色弹石停车位X10
　　　　　　　　特色铺装　　　　　　　　绿化种植　　黑色铁艺树池　　　　　　　　　　　　　　　　　2米宽弹石地面铺装　　　　　　　　　　　　　　　　灌木种植
　　　　　　　花坛种植池　　　　　　　图案式青砖　　地面铺装　　台阶　　　　台阶　　　　　　　　　　　　　　　　灰色弹石停车位X5
　　　　　　　　　　组合　　　　　　　　地面铺装　　地面石材铺装　　　　　　　　　　　　　广告牌灯柱　　保留景观树木　　　　　草地与石凳
　　　　　　　　　　　　　　　　　　　　　　　　　　花坛种植池　　　　　黑色铁艺树池　　　　　　花坛种植池　　　　绿篱种植　　灰色弹石停车位X5
　　　　　　　　　　　　　　　　　　　　　　　　　　　　　　　　　　地面铺装　　　　　　黑色铁艺树池　　下植绿篱　　广告牌灯柱　　景观树　　　休闲座凳
　　　台阶　　现代图案　　种植地被景观　　小雕塑　　　绿化带，灌木与花灌木
　　地面铺装　　保留景观树木　景观灯柱　　　间隔组合

景观设计总平面

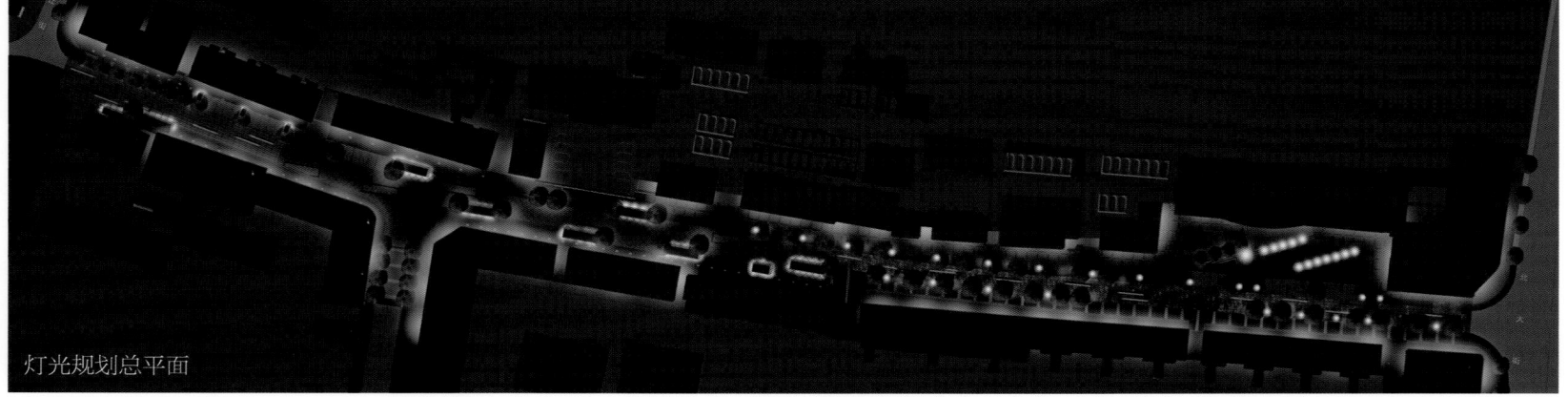

灯光规划总平面

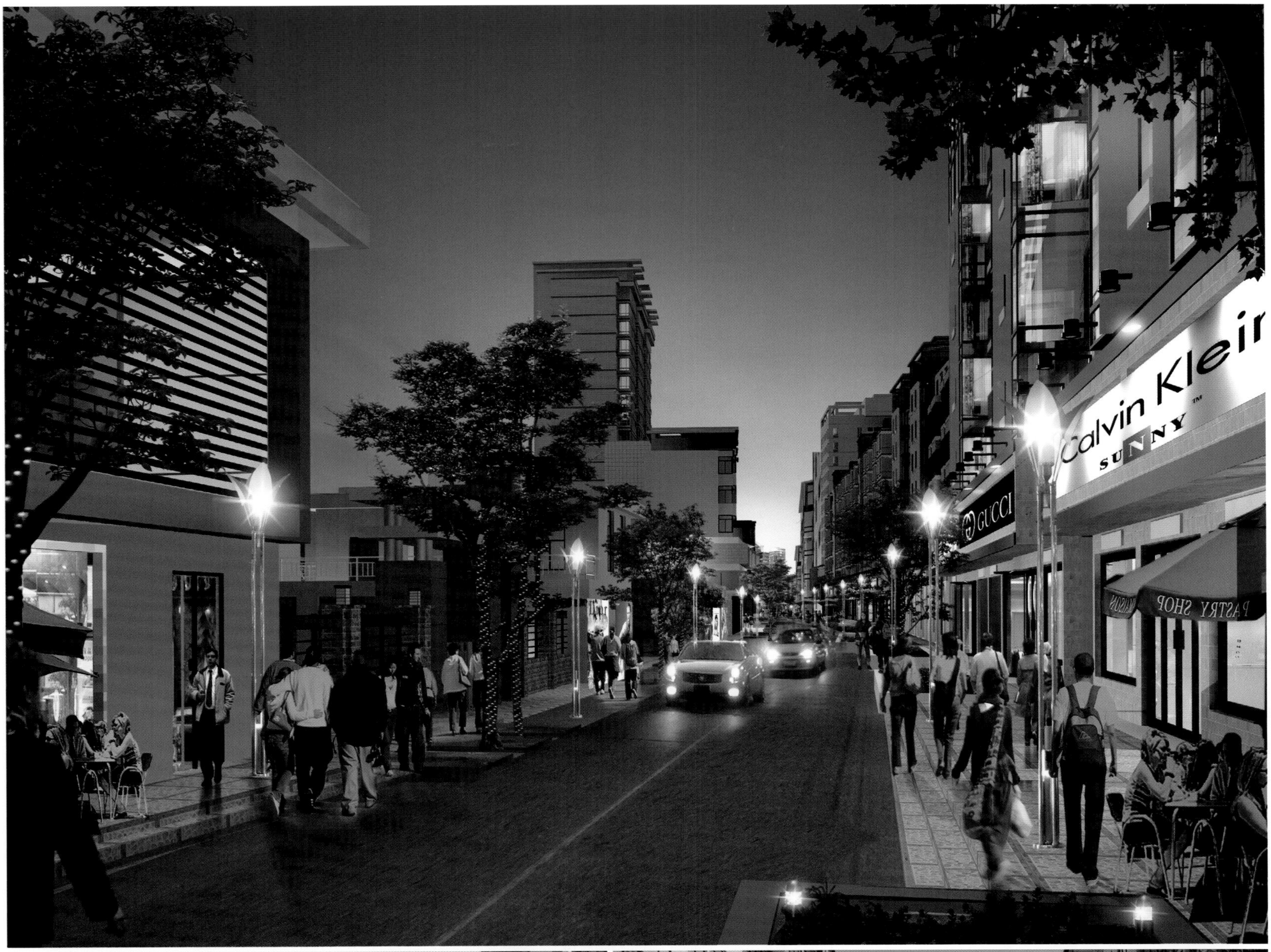
Calvin Klein
SUNNY
GUCCI
PASTRY SHOP

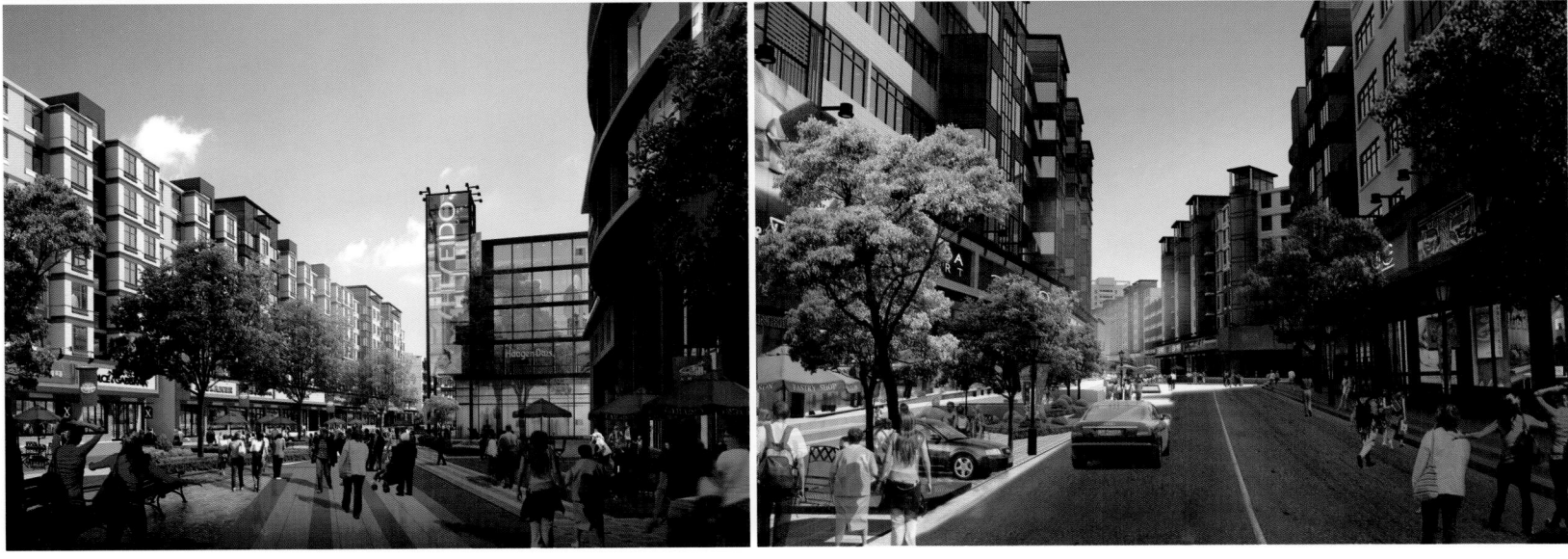

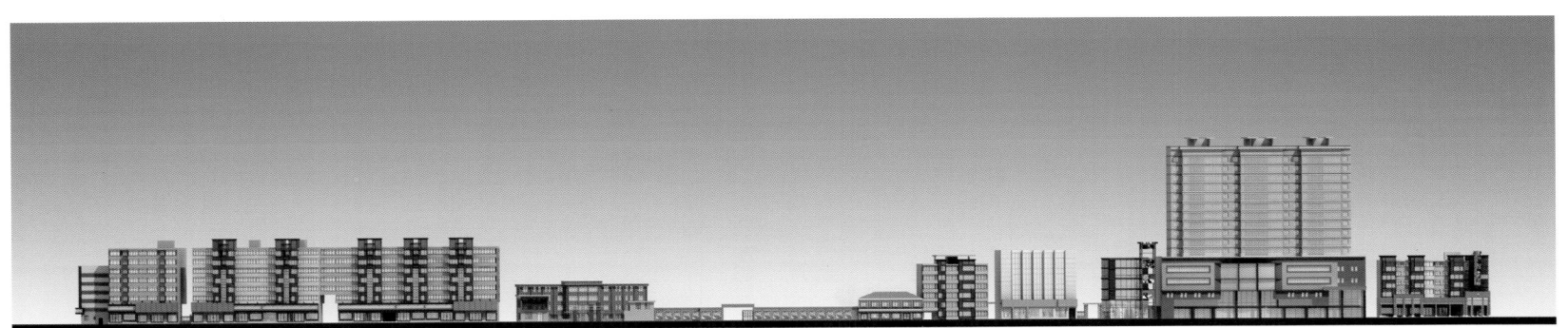

上海古北新区一、二区居住分区
Shanghai Gubei Phase1/2 Residence

项目名称：上海古北新区一、二区居住分区
项目地点：中国 / 上海
建筑面积：95 万平方米
编制时间：2000 / 12
业主：上海古北（集团）有限公司
规划类型：详细规划

规划基地范围北起红宝石路、南至古羊路、东起姚虹东路、西至古北路。规划以"创造国际都市一流的居住区，并成为未来社区模型"为目标，力求突出地域文化的特性、规划布局的整体性和延续性、开放空间的尺度感和丰富性、步行网络的安全感和便捷性、交通组织的次序感和导向性、建筑单体的超前感和科技性、公共服务设施的完善性和组织性。在规划建筑设计中，注重城市脉络和古北新区特有气质的延续，从而演绎都市高档住宅区的新理念。

规划总建筑面积 95 万平方米，居住人口 1.45 万人。

The planning project is in the north of Hongbaoshi Road, south of Guyang Road, east of Yaohong Road, west of Gubei Road. It aims to "create an international city-class residential area, and become future community model" , seeking to highlight the characteristics of local culture, integrity of planning the layout ,the continuity, scale sense of open space, the richness, security of pedestrian network, convenience, sense of traffic organization , orientation, order of sense, a sense of the monomeric building , the scientific and technological feature, the improvement and organization of public service facilities . In the planning of architectural design, we focus on specific urban context and continuation of Gubei New Area's quality so as to perform a new concept of urban high-class residence.

The total planned construction area is 950,000 square meters, and the population is 14,500.

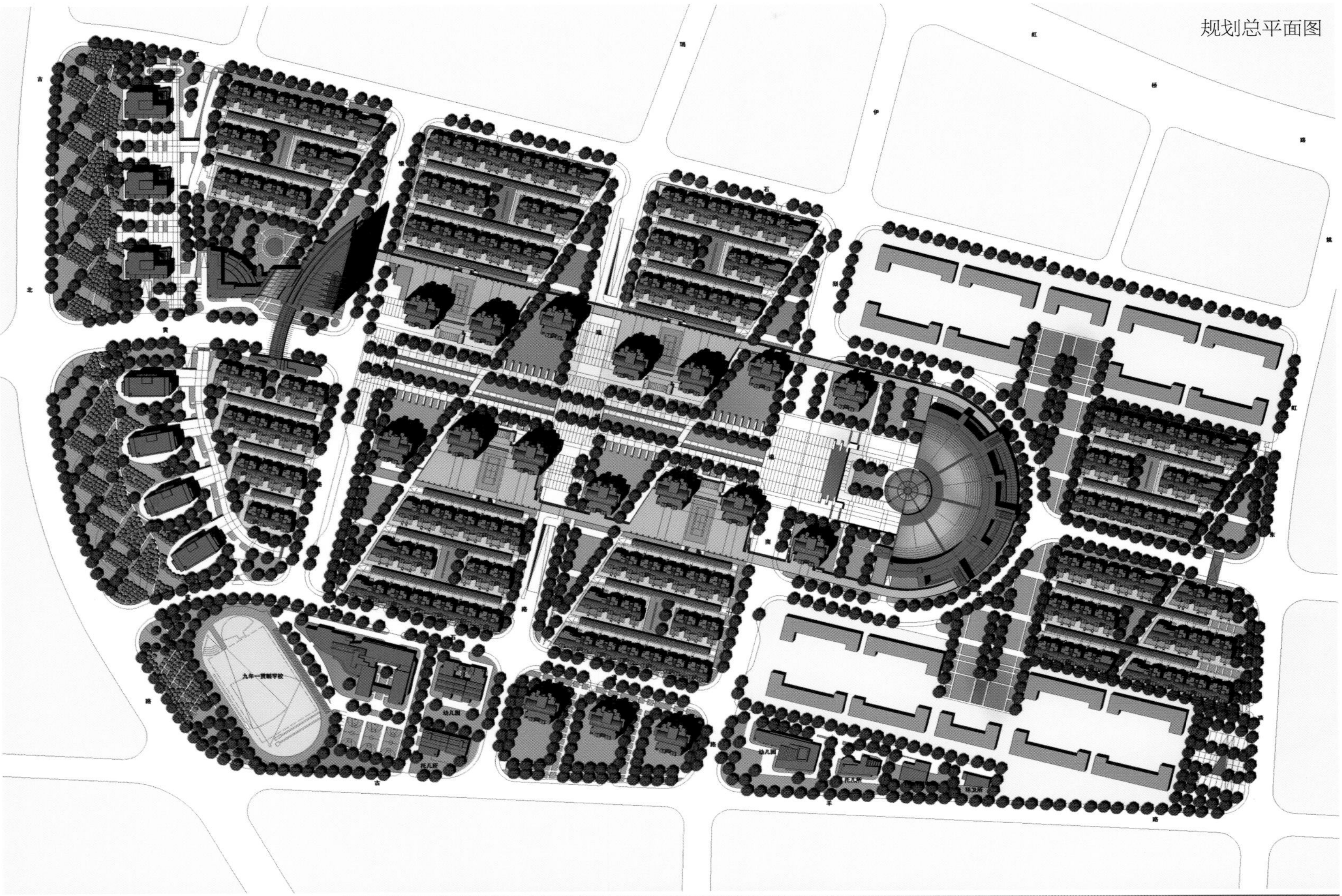

西宁市教场街改造规划
Xining Jiaochang Street Reconstruction Plan

项目名称：西宁市教场街改造规划
项目地点：中国 / 青海 / 西宁
长度：320 米
编制时间：2010 / 10
业主：西宁市正华建设投资控股有限公司
规划类型：详细设计

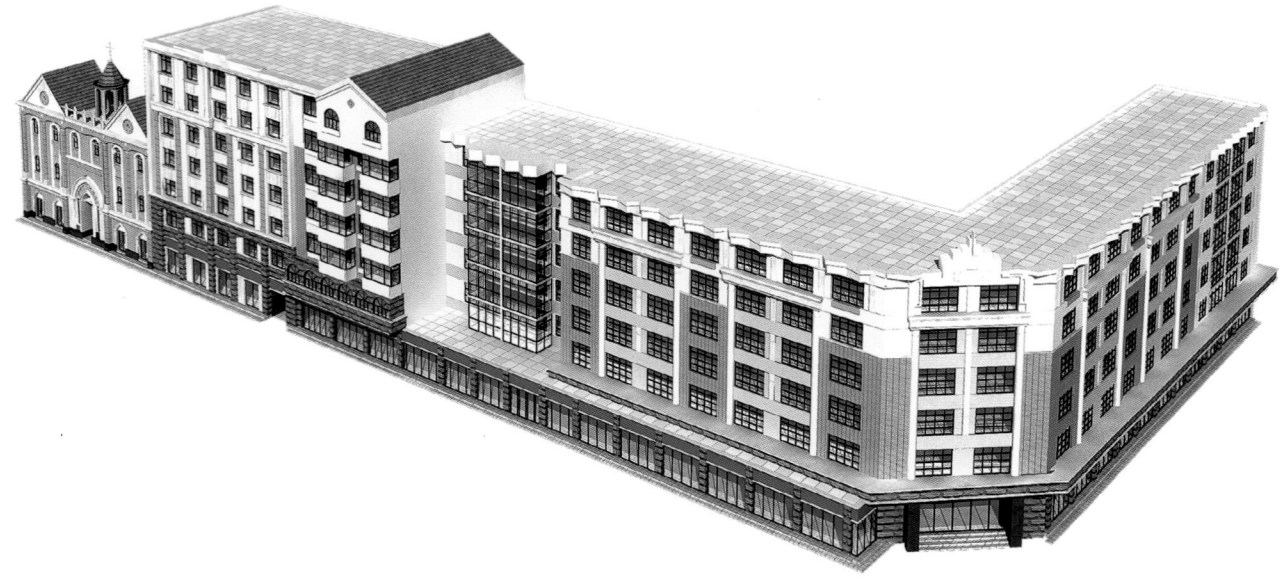

教场街位于西宁市城东区，南至西大街，北至解放街，为西宁市中心区内生活性城市支路。

本次规划改造将教场街整体定位为以现代简欧风格为主基调的生活性城市支路。

在功能上，规划保持现有生活性服务道路的特点，重点通过建筑立面和街道景观改造等手段，塑造宜人、舒适的生活性街道。

在形象上，规划采用现代简欧风格为基调对建筑立面和景观环境进行再塑造，形成风格鲜明而统一的街道空间环境。

Jiaochang Street starts from West Street to the south, Jiefang Street to the north and it is a life road of Xining city.

The planning aims to transform Jiaochang Street into a life road with modern European style as the main tone.

Functionally, the plan maintains the current characteristics of life services, and creates a pleasant and comfortable life street through the building facade, streetscape renovation and other means.

For the image, the plan uses modern simple European style for the tone to reconstruct the building facade and landscape thus forming a clear and consistent style of street space environment.

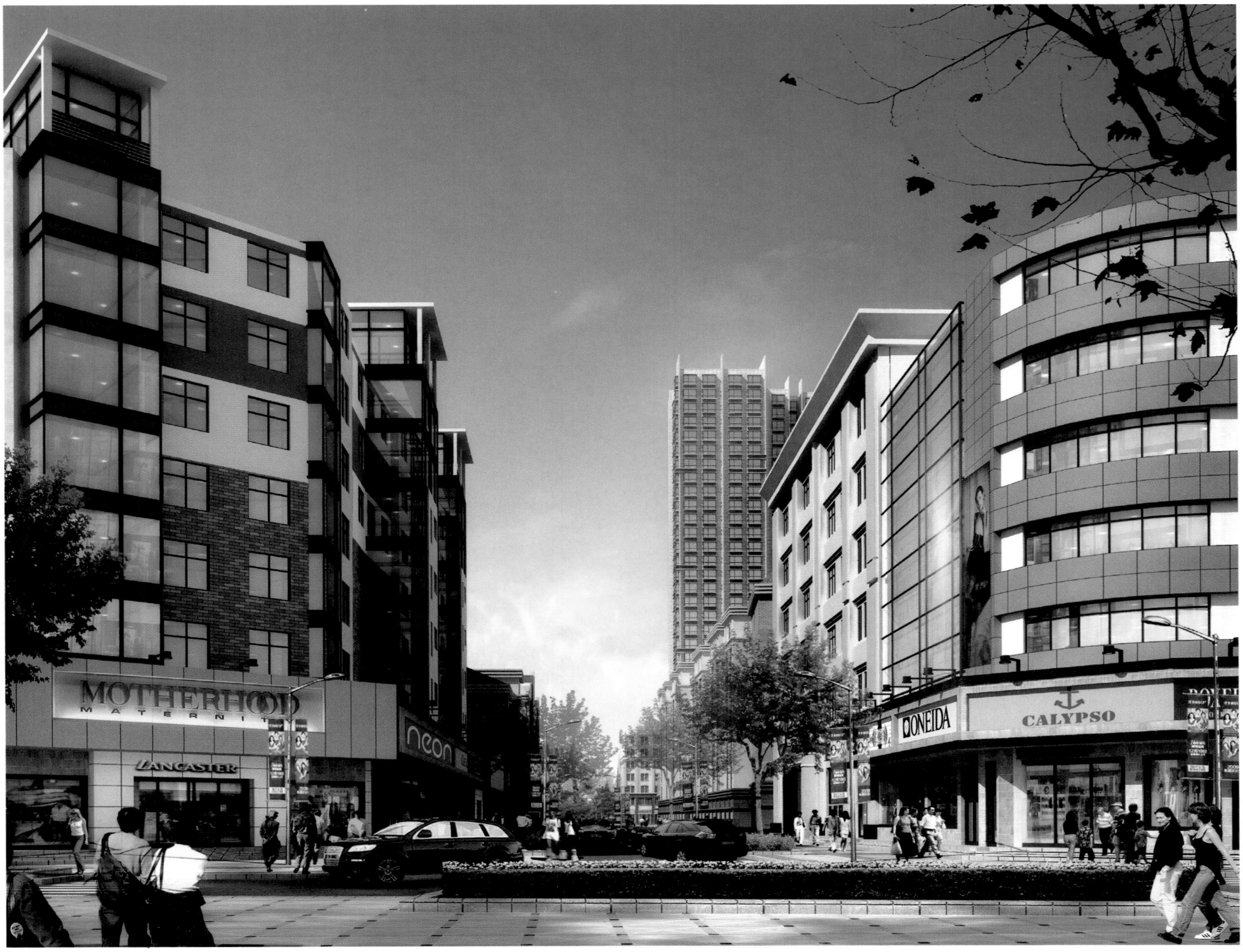
MOTHERHOOD
LANCASTER
neon
ONEIDA
CALYPSO

交通分析

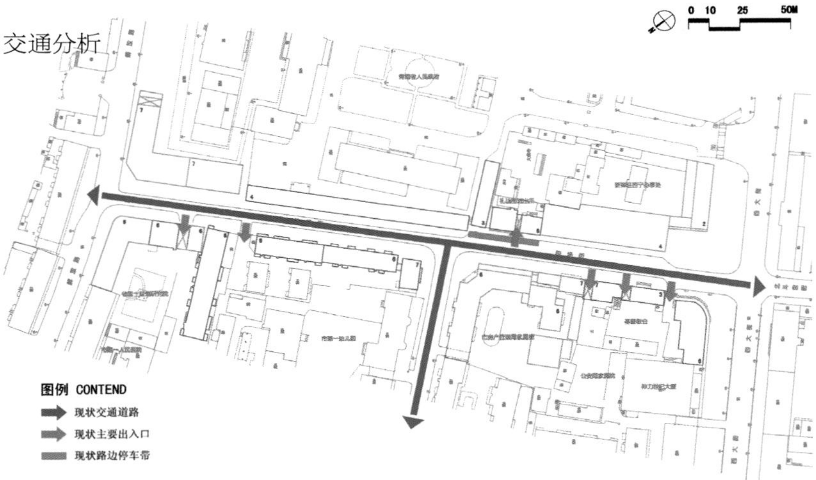
0 10 25 50M
图例 CONTEND
现状交通道路
现状主要出入口
现状路边停车带

建筑功能

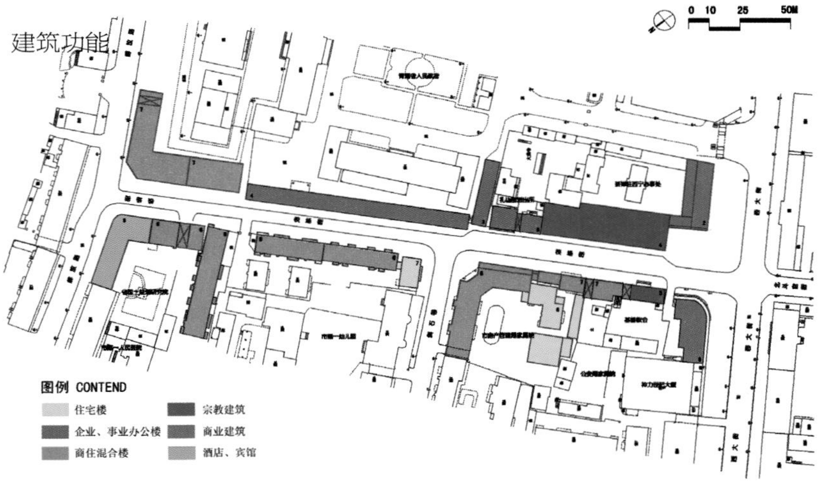
0 10 25 50M
图例 CONTEND
住宅楼
企业、事业办公楼
商住混合楼
宗教建筑
商业建筑
酒店、宾馆

建筑质量分析

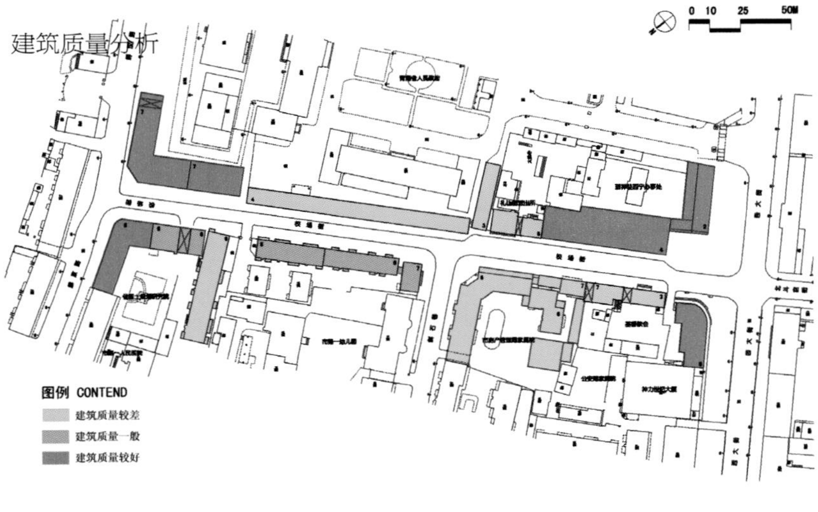
0 10 25 50M
图例 CONTEND
建筑质量较差
建筑质量一般
建筑质量较好

建筑改造分析

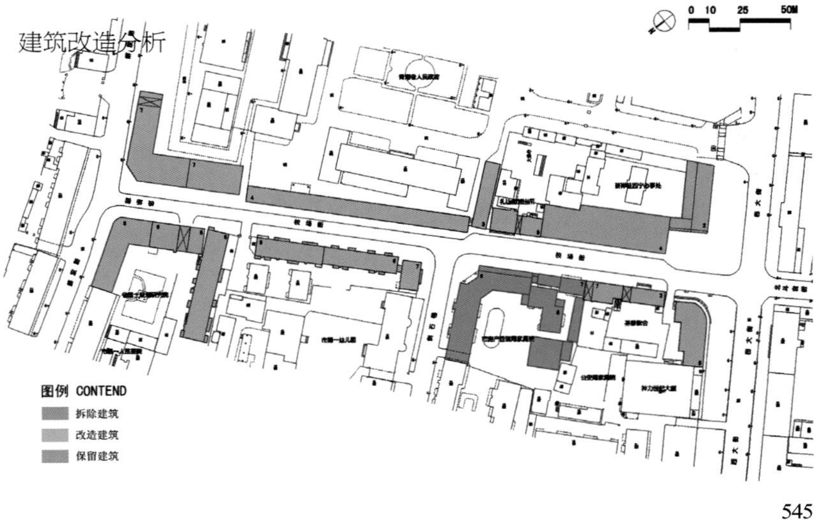
0 10 25 50M
图例 CONTEND
拆除建筑
改造建筑
保留建筑

西宁五四大街虎台地区控制性详细规划
Xining Wusi Avenue Hutai Area Controlled Planning

项目名称：西宁五四大街虎台地区控制性详细规划
项目地点：中国 / 青海 / 西宁
建筑面积：375.6 万平方米
编制时间：2010 / 07
业主：西宁市城乡规划建设局
规划类型：控制性详细规划

项目规划范围东起新宁路、西至海湖路、北到湟水河、南至昆仑路，是今后几年西宁政府着力开发建设的重点区域。基地现状城市空间饱和，环境压力大，用地混杂，高品质复合化公共空间场所较为匮乏。

规划定位：以五四大街为城市发展轴，形成以行政管理、教育科研和科技商务为主体功能的体现地方历史风貌特色的城市中心区。遵循资源节约型、环境友好型的发展理念，突出绿色、人文、环保、科技特色，体现行政管理、商务科技、教育科研和文化展示四大职能特点，规划重点将其塑造成省、市行政管理服务中心，市教育科研硅谷和地方历史风貌展示窗口。

在控制性详细规划基础上，规划进一步编制了城市设计导则，对每个地块的空间要素做出进一步的说明和指导，利于形成完整统一的城市风貌。

该项目总建筑面积约为 375.6 万平方米，可容纳 5.4 万人，地块平均容积率为 1.6。

The project starts from Xining Road to the east, Haihu Road to the west, north to the Huangshui River, south to Kunlun Road. It will be the key development and construction area of Xining Government in the next few years. Currently the base is facing environmental pressure, mixed land use, lacking of high-quality compound and public space.

Planning location: use Wusi Avenue as the urban development axis; form an urban center with main functions like administration, education, scientific research and technology business; embody local historical features and characteristics. Follow the resource-saving and environment-friendly development concept, highlighting the green, cultural, environmental, technological characteristics, reflecting four functions of administrative management, business technology, education and scientific research and cultural show to mold the area into a provincial and municipal administrative service center, the urban Silicon Valley of education and scientific research and local history display window.

On the basis of the controlled plan, the plan further develops the urban design guidelines, makes further instructions and guidance for the spatial elements of each plot which will help to form a complete and unified cityscape.

The total construction area is about 3,756,000 square meters which can accommodate 54,000 people. The average floor area ratio is 1.6.

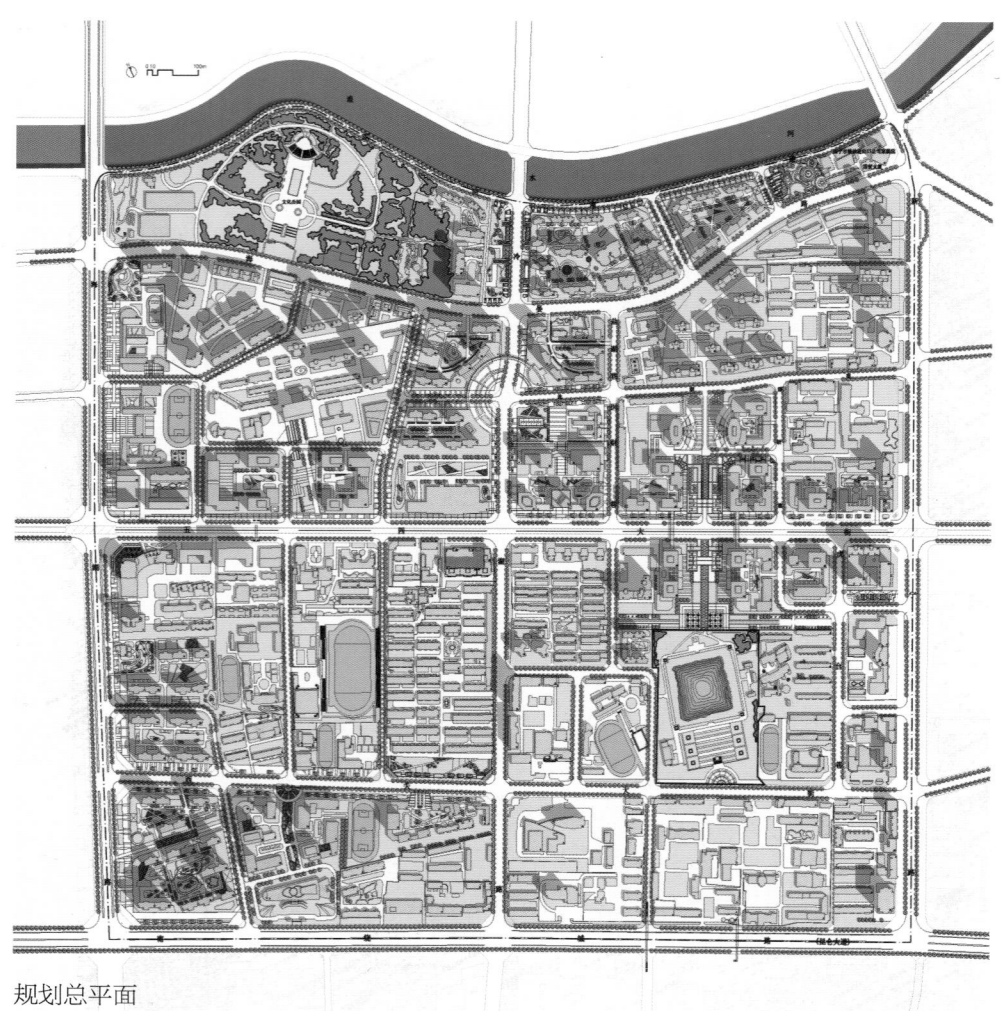

规划总平面

① ② ③　①功能结构分析；
④ ⑤ ⑥　②景观绿化分析；
　　　　③建筑高度分析；
　　　　④公共交通服务半径；
　　　　⑤容积率分配；
　　　　⑥规划用地分析

547

西宁中心广场北侧片区城市设计
Urban Design of Xining North Central Square

项目名称：西宁中心广场北侧片区城市设计
项目地点：中国 / 青海 / 西宁
建筑面积：29.5 万平方米
编制时间：2010 / 11
业主：西宁市南川综合开发项目管理委员会
规划类型：城市设计

规划基地位于西宁市城市核心区域，南侧为中心广场，西侧为湟水河之分流南川河，北与北山寺、南与南禅寺遥遥相望，是南川河沿线发展轴上的重要节点。

规划定位：打造成集城市公共文化空间、市民休闲活动、生态滨河景观和商业开发等于一体的西宁城市公共活动核心，并通过商务办公、星级酒店等功能植入和有机联动形成全新的城市金融商务区，使其成为人文与生态、历史与现代完美融合的富有生机和活力的城市综合区域，并努力打造成西宁城市标志性空间，缔造城市品牌！

在研究和讨论现状问题的基础上，规划提出五个核心对策，分别从城市功能、城市视线通廊、区域交通组织、历史遗址保护和标志性空间塑造等方面进行空间设计。

规划总建设面积约为 29.5 万平方米。

The planning base is located in the core area of Xining City, south to central square, the west side of the Nanchuan River, south of North Temple, north of Nanzen Temple which is the important node of development along Nanchuan River axis.

Planning Location: create a core public activities centre in Xining city with urban public cultural space, public recreation, ecological riparian landscape and commercial development and through implantation like the business office, hotel features, form new urban organic linkage the financial business district, making it dynamic and vigorous urban integrated region with humanity and ecology, history and the perfect fusion of modern city and strive to create space for the landmark in Xining city thus creating an urban brand!

Based on study and discussion of current problems, the plan addresses five core measures, respectively, from the urban functions, line of sight traverse the city, regional transportation organizations, historic sites landmark protection, and other aspects of space design.

The total planned construction area is approximately 295,000 square meters.

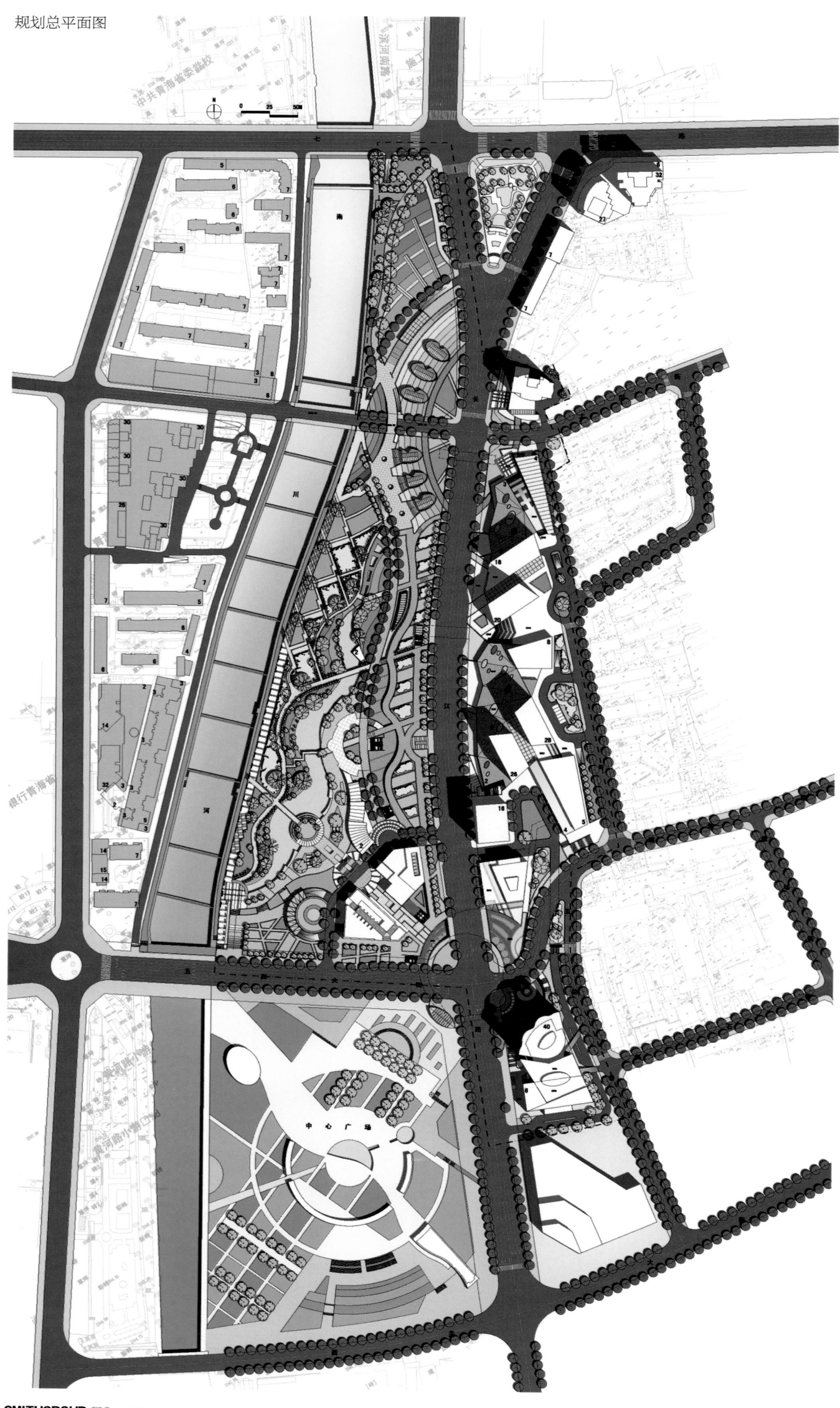

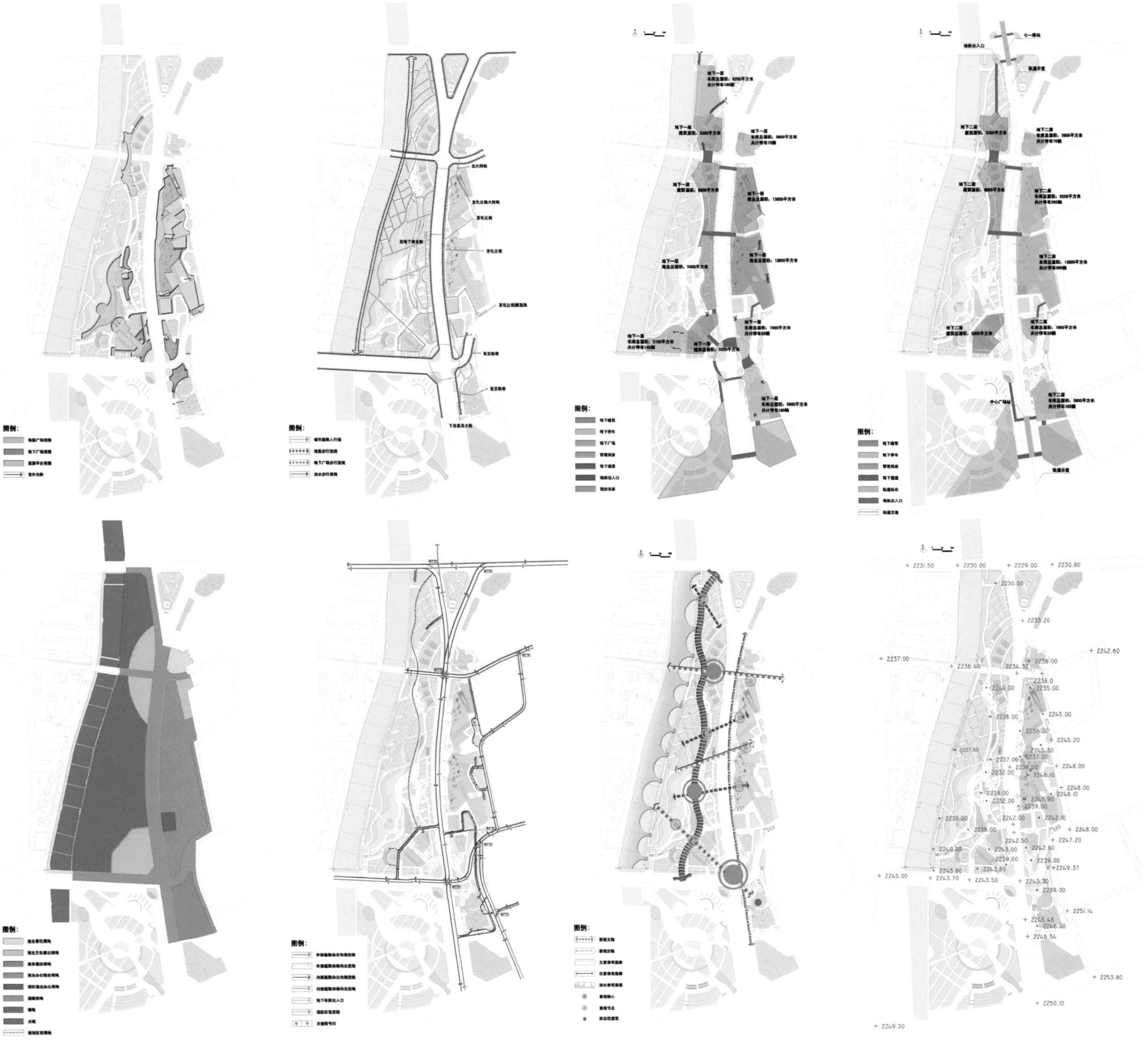

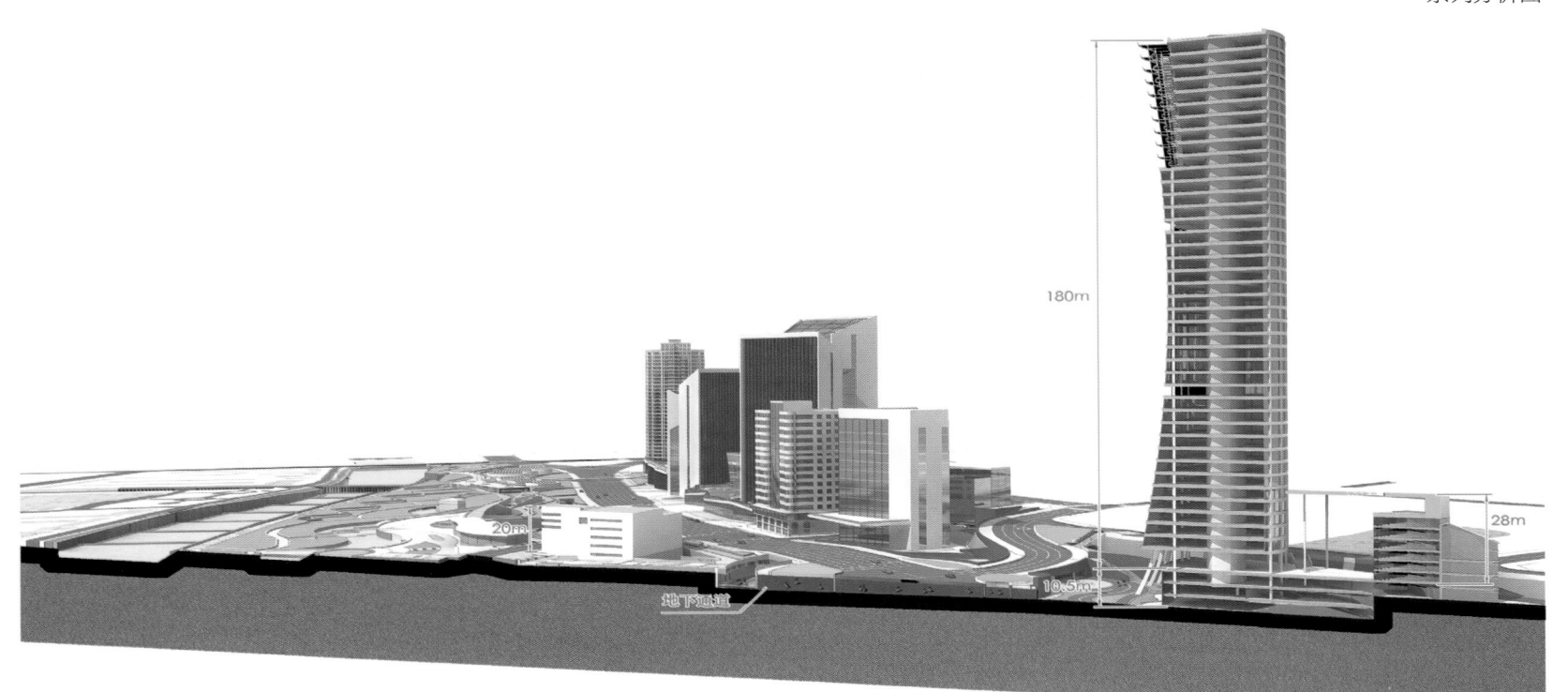

系列分析图

静安北部地区发展规划——滨江绿洲城市设计
Jingan Northern Development Plan-Riverside Oasis

项目名称：静安北部地区发展规划——滨江绿洲城市设计
项目地点：中国 / 上海
建筑面积：91 万平方米
编制时间：2003 / 05
业主：上海市静安区城市规划管理局
规划类型：城市设计

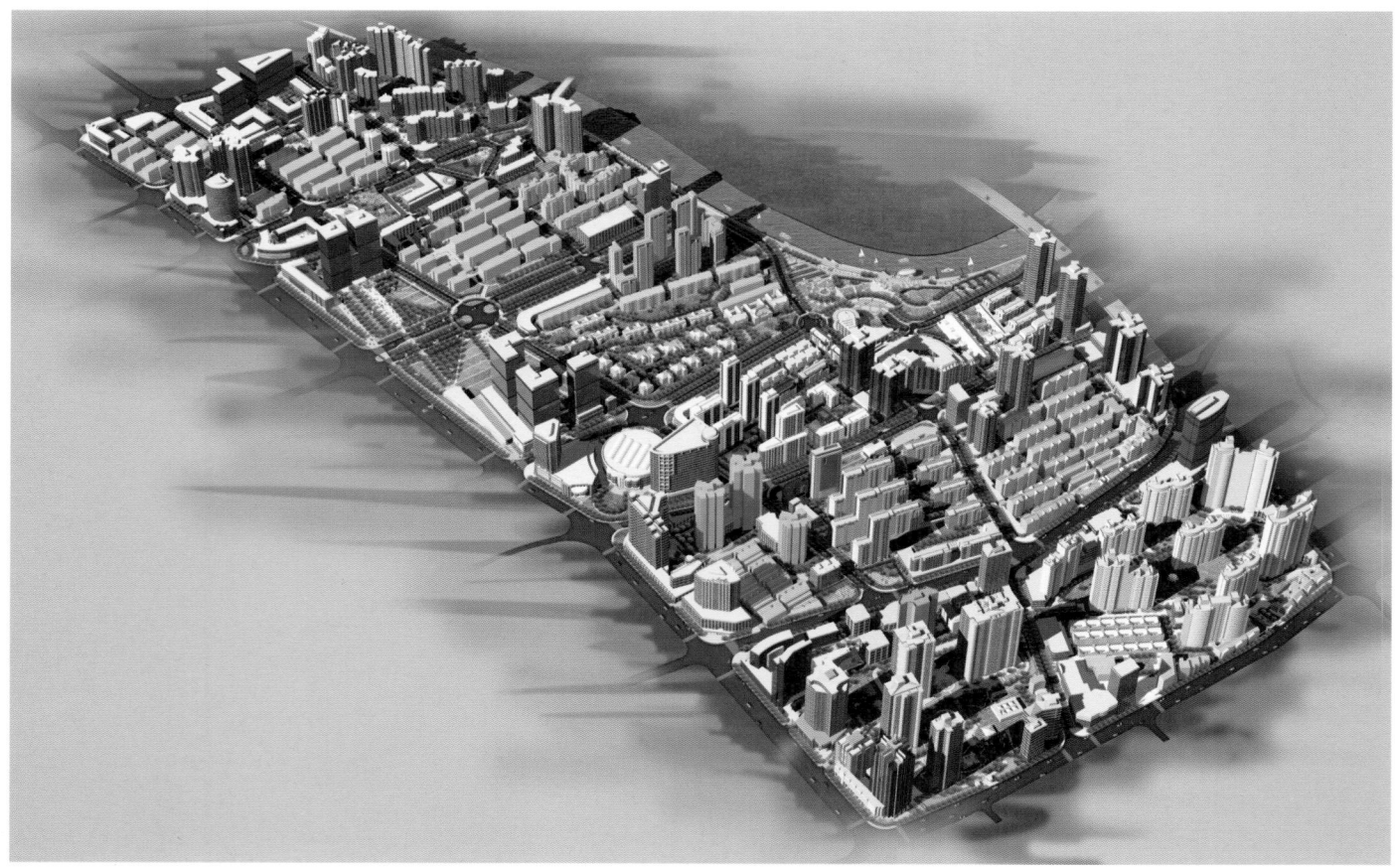

"滨江绿洲"位于静安区北部地区，东临苏州河及恒丰路，西临江宁路，北起安远路，南至北京西路。

本次规划总结前一段时期的开发经验教训，特点主要体现在：

1. 梳理、整理各个地块（包括现状、在建、有方案），重新建立地区空间发展次序；

2. 重点规划开放空间提炼外部环境体系；

3. 编制、控制、引导各地块的空间环境发展的图则。

"Riverside Oasis" is located in northern Jing'an District, which faces Suzhou River and Hengfeng Road in the east, Jiangning Road in the west, An Yuan Road in the north, West Beijing Road in the south.

The plan is made on the basis of concluding lessons learned from previous planning and developing experience. Its characteristics are mainly reflected in:

1. Sorting out each block (including current status, under construction, the one with programs), re-establishing the order of the space development;

2. Focusing on planing open space to refine the external environmental system;

3. Preparing, controlling and guiding the development of plans of spatial environment for each block.

0　　　　100M

静安常德路沿街街景设计
Street Design for Changde Road, Jingan District

项目名称：静安常德路沿街街景设计
项目地点：中国 / 上海
建筑面积：2 045 米长的两侧地区
编制时间：2004 / 10
业主：上海市静安区城市规划管理局
规划类型：城市设计

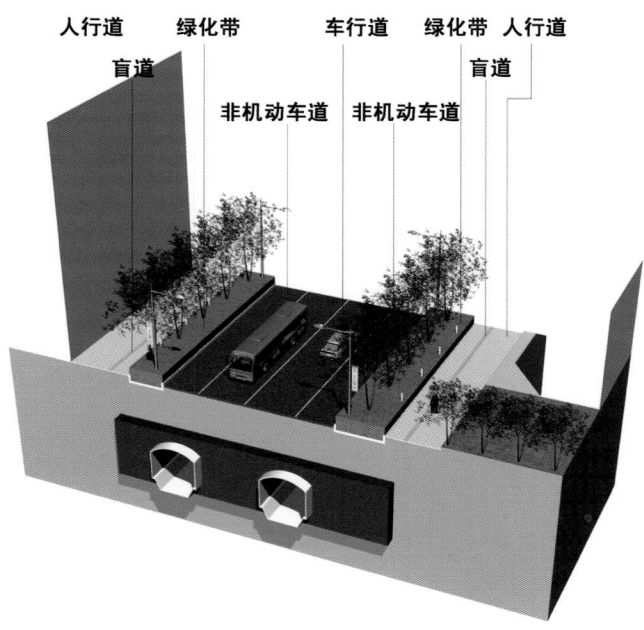

人行道　绿化带　车行道　绿化带　人行道
盲道　非机动车道　非机动车道　盲道

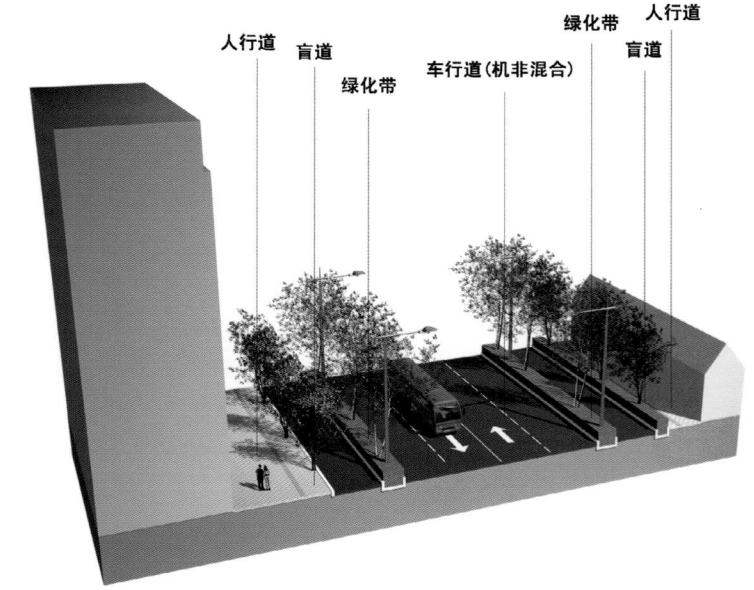

人行道　盲道　绿化带　车行道（机非混合）　绿化带　盲道　人行道

常德路是上海中心城区一条南北向城市道路。本次街景规划，通过对该道路两侧沿线现状道路交通、生活服务设施、绿地景观等系统的分析研究，对现有资源进行整合与重建，促进沿线城市景观及环境品质的提升，强化沿线道路交通功能的畅通，促进生活服务设施的完善，进一步增强静安北部现代都市住宅的活力，加强市级中心即南京西路的辐射效应。以街景塑造为核心将常德路沿线建设成为静安区中部一条兼具交通性功能的特色景观道路。

除了常规的广场绿地等景观设计、沿街建筑立面改造之外，规划还重点编制了沿街街景设计图则，既作为街景设计内容的一部分，又可独立成册，在指导具体街景建设中起重要作用。

Changde Road is a north-south road in Shanghai downtown. This streetscape plan, through analysis of status of the road transport along both sides, service facilities, green space and landscape system, integrates and reconstructs the existing resources, promoting the environment along the urban landscape and improving the quality and strengthening the smooth flow of the road traffic function, improving life services to further enhance the Jingan northern residential part's vitality, and strengthening the radiation effects of Nanjing West Road. Using the core of the streetscape construction, the design makes Changde Road a road featured with special parkway in Jingan District.

In addition to the conventional reconstruction of landscape design like square green space and other building facades along the street, the plan also focuses on the preparation of the street streetscape design plans. It can be either a part of streetscape design elements or a separate book, playing an important role in guiding the construction of the concrete street.

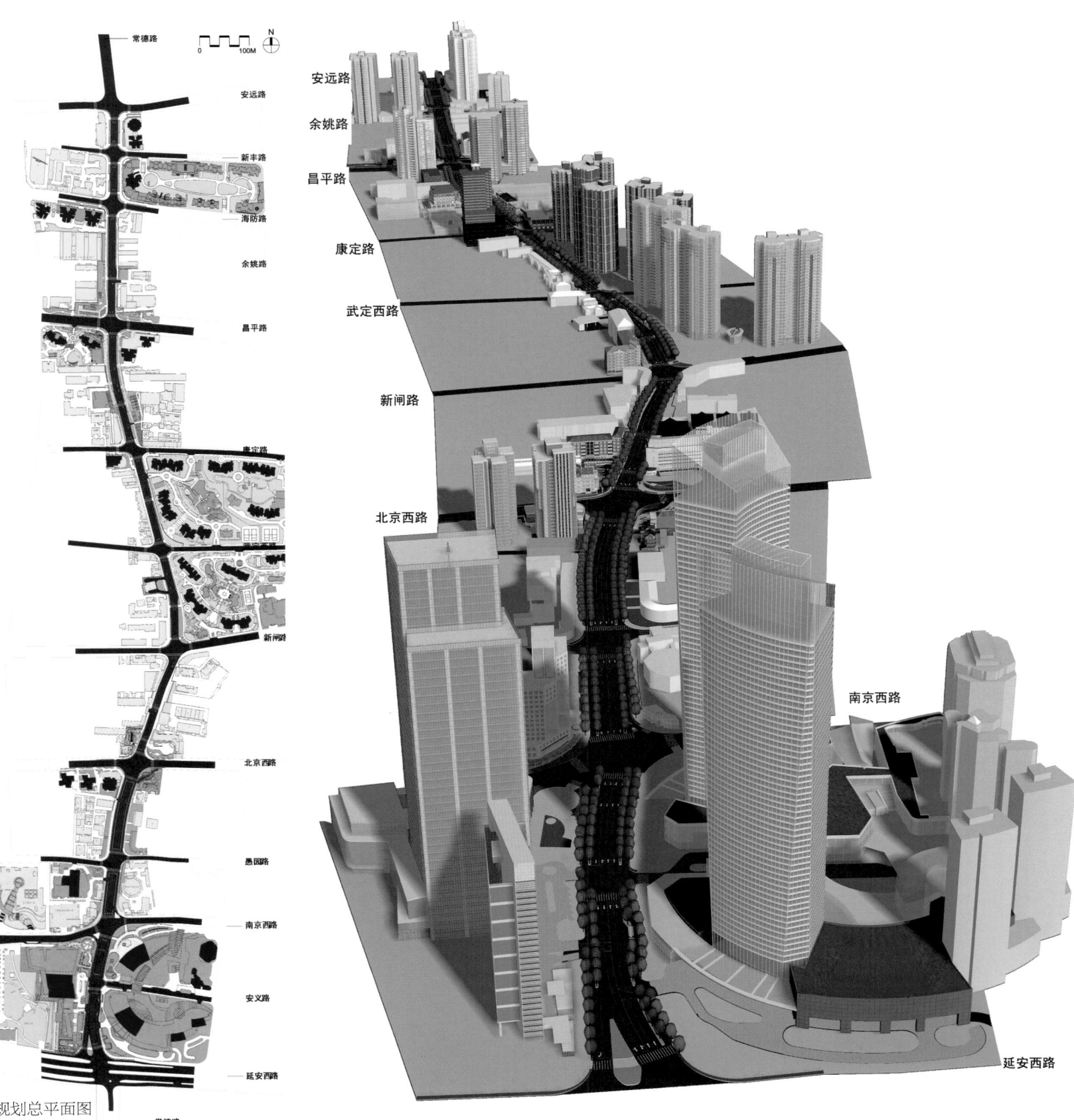

常德路

安远路

新丰路

海防路

余姚路

昌平路

康定路

新闸路

北京西路

愚园路

南京西路

安义路

延安西路

常德路

规划总平面图

0 100M

N

安远路

余姚路

昌平路

康定路

武定西路

新闸路

北京西路

南京西路

延安西路

上海市静安南京路城市设计
Nanjing Road Urban Design in Jingan District, Shanghai

项目名称：上海市静安南京路城市设计
项目地点：中国 / 上海
建筑面积：185 万平方米
编制时间：2002 / 06
业主：上海市静安区城市规划管理局
规划类型：城市设计

静安区南京西路东起成都北路，西至镇宁路，长约 3 260 米。本次城市设计内容包括现状分析、总体规划、交通分析、重要节点城市设计以及城市街道景观分析设计五大部分。

设计团队在规划设计的同时对现状用地、城市空间、街道个性、商业活动、地下空间、道路交通和历史文脉七个方面进行了广泛深入的调研和研究，全面地总结了这条上海最著名的街道的独特个性，并在此基础上确立了未来南京西路发展的五大板块，包括石门路文化传媒休闲区、"梅恒泰"商务办公购物区、会展区、静安寺购物酒店区以及协和世界主题商住办综合区。

West Nanjing Road in Jing'an district stretches from Chengdu Road to the east, west to Zhenning Road. It is about 3,260 meters. The urban design includes five parts: current situation analysis, overall planning, traffic analysis, urban design of important node, urban street landscape analysis and design.

The design team conducted a deep research and analysis on the status of land use, urban space, street character, commercial activities, underground space, road traffic and historical context while planning the design. It aims to establish five major sections of West Nanjing Road in the future, including Shimen Road cultural and media recreational area, "Mei Hengtai" shopping and business district, exhibition district, hotel shopping area in Jingan Temple and integrated commercial and residential areas of Concord World Office.

沿街建筑立面分析

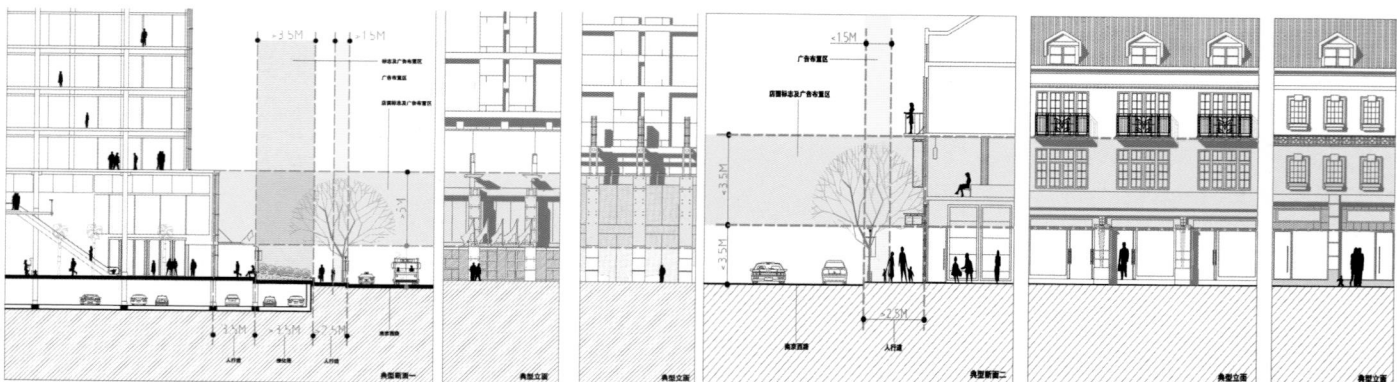

CIRCULATION DIAGRAM
GREEN DIAGRAM
DISTRICT DIAGRAM
HISTORIC PRESERVATION

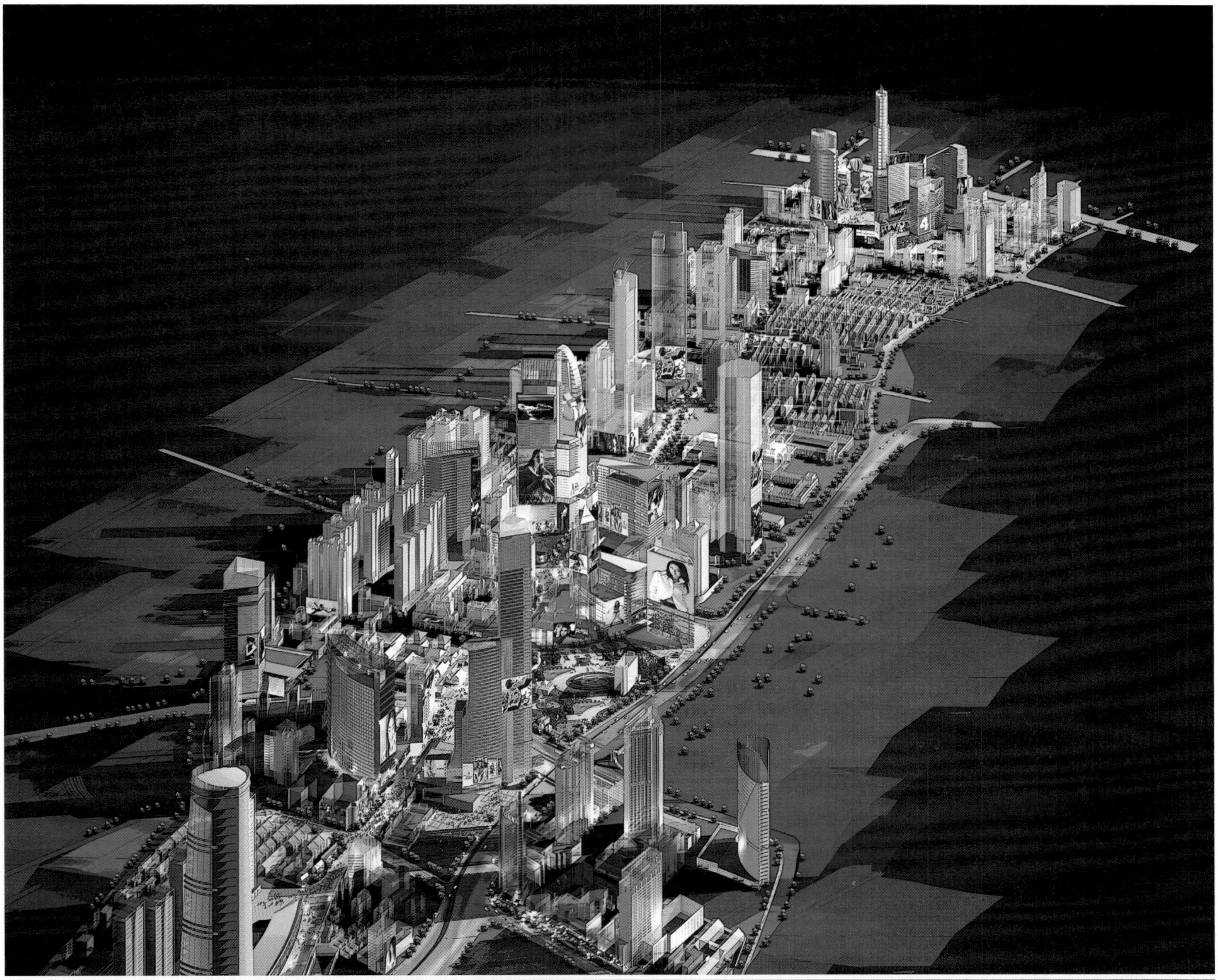

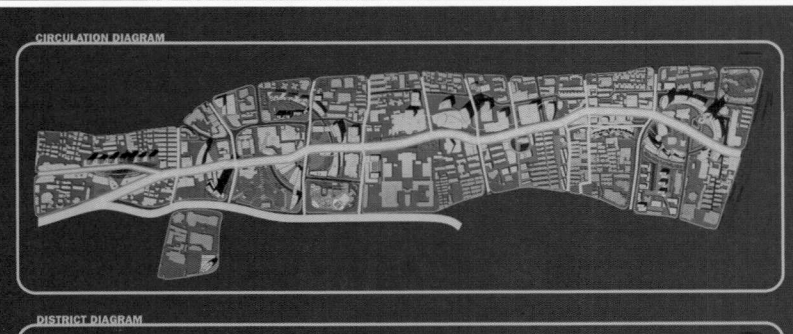

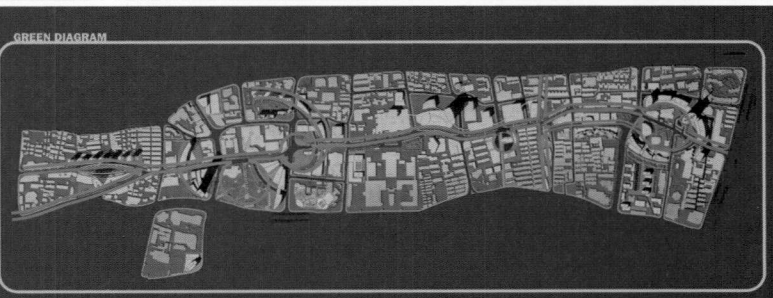

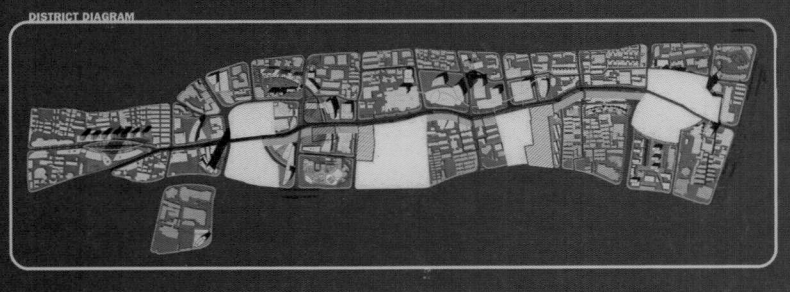

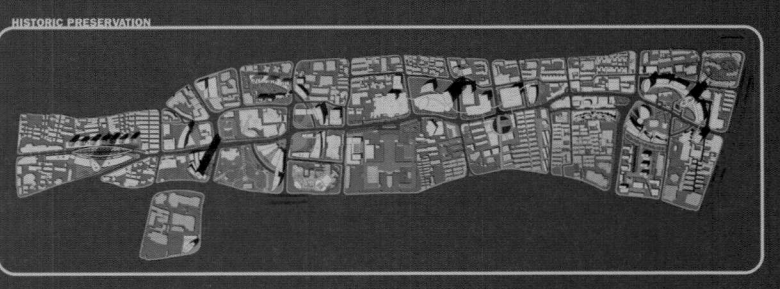

静安南京西路沿街改造
Jing'an West Nanjing Road Reform Planning

项目名称: 静安南京西路沿街改造
项目地点: 中国 / 上海
长度: 1 250 米
编制时间: 2004 / 02
业主: 上海市静安区商业委员会
规划类型: 详细设计

本次改造规划目的是通过发掘和梳理南京西路（南北高架道路至陕西北路段）沿街及周围街区资源，改造局部沿街立面，优化环境，提高南京西路整体形象和潜在商业价值，从而为新一轮南京西路商业改造和投资引入创造条件。

本次规划内容分三部分：商业发展总体布局；商业发展形象设计；商业环境设计。

The reform plan aims to identify and arrange resources along the street and the surrounding neighborhood of West Nanjing Road (north-south elevated road to north Shanxi Road section), through reformation of the local street facade, by optimizing the environment, to improve the overall image and potential commercial value of West Nanjing Road, thus creating conditions for investment introduction for the new round West Nanjing Road business reformation.

The content of the plan can be divided into three parts: the general layout of the commercial development; business development image design; business environmental design.

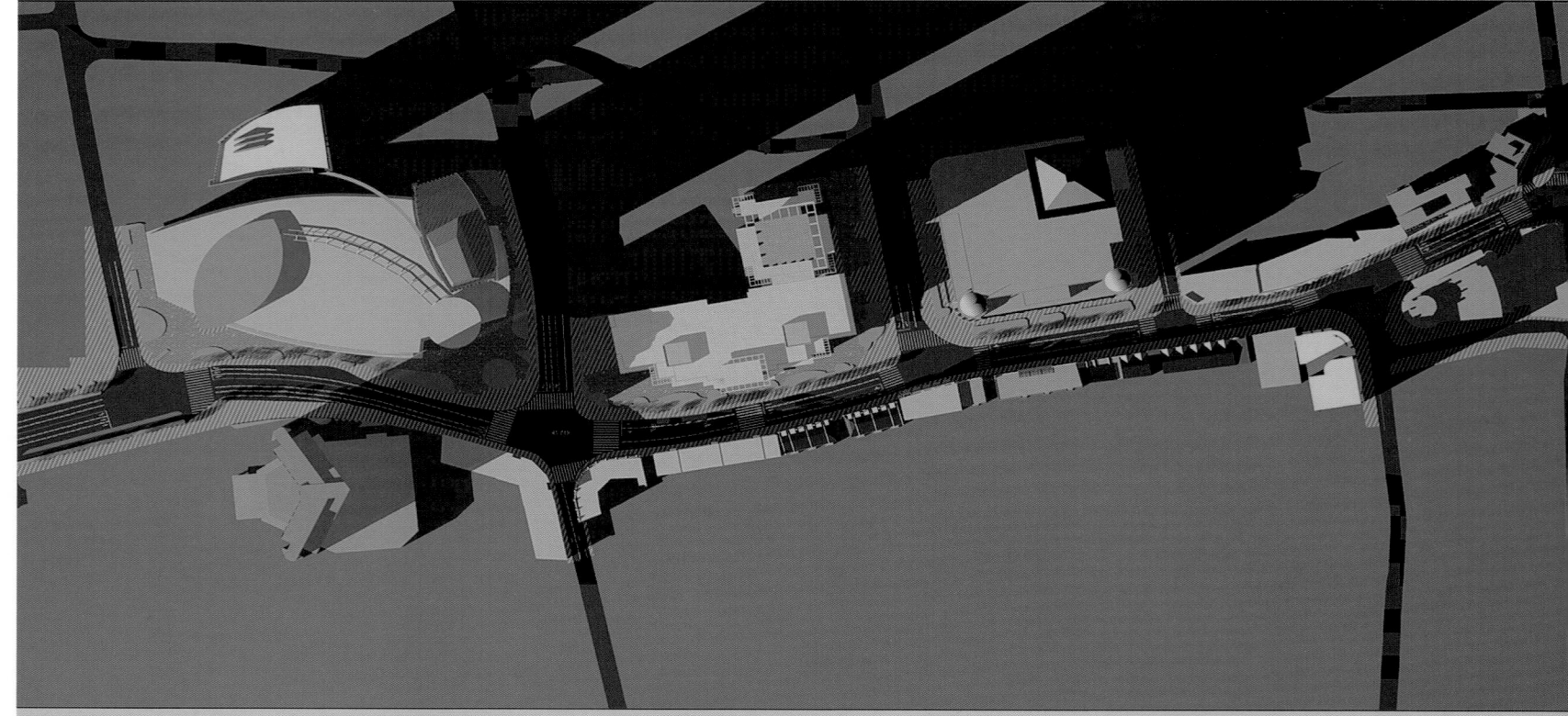

静安南京西路南北高架道路至陕西北路段沿街商业改造规划——北部立面

静安南京西路南北高架道路至陕西北路段沿街商业改造规划——南部立面

青城山研发社区规划设计
QingchengMountain Research & Development Community Planning

项目名称：青城山研发社区规划设计
项目地点：中国 / 四川 / 成都
建筑面积：67.5 万平方米
编制时间：2008 / 11
业主：成都泰达新城建设发展有限公司
规划类型：详细规划

项目地段为青城山镇发展区域，地处城市旅游最重要的"成都——青城山——都江堰干线"上，也是成都软件信息技术三大基地之一，南临软件园中央发展轴——东软大道，西靠106绕镇线，地理和交通条件十分优越。

规划总体定位：集研发、办公、酒店、商业、会议于一体，配以信息化的配套居住，形成生态型综合性研发社区。

规划核心内容：研发社区、动力地产、新型复合社区 5+2 生活模式。

规划特点：1. 工作、生活、娱乐综合一体化共享社区；

2. 度假休闲风格鲜明，处处体现人文精神。

The project is located inside Head Qingcheng Town development area, which lies on the most important tourist "Chengdu-Qingcheng Mountain-Du River-Weir Link" line and is also one of the three Chengdu information technology bases. It is located at the south of major Software Park axis- central way, west of the 106 line and has a favorable geographic and traffic condition.

The overall orientation of the plan: integrating research and development, office, hotel, business, meeting together with information supporting housing to form of a comprehensive research and development ecological community.

Core content of the plan: R & D community, dynamic real estate, New Compound 5 +2 lifestyle community .

Planning features: 1. integration community of work-life-entertainment;

2. The distinctive style of vacation and leisure, reflecting the humanistic spirit.

规划总平面图

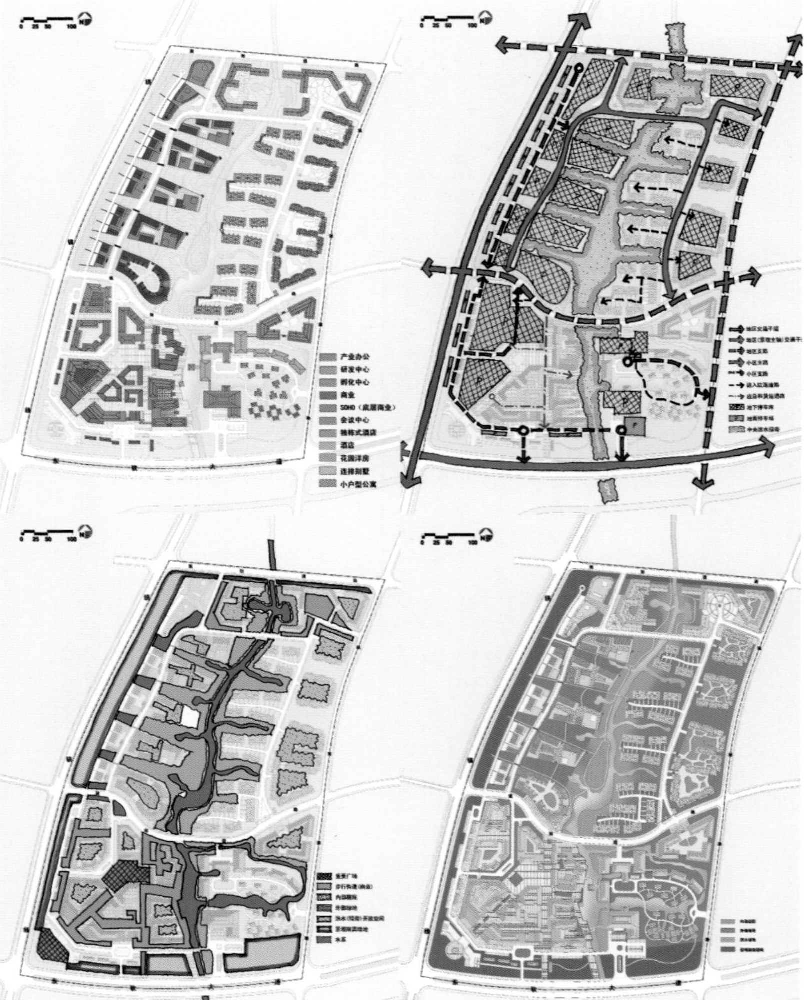

系列分析图

上海都市阳光花园规划与景观设计
Shanghai Sunshine Garden Urban Planning and Landscape Design

项目名称：上海都市阳光花园规划及景观设计
项目地点：中国 / 上海
建筑面积：13.3 万平方米
编制时间：2002 / 06
业主：上海颛桥房地产公司 / 上海颛盛房地产公司
规划类型：详细设计

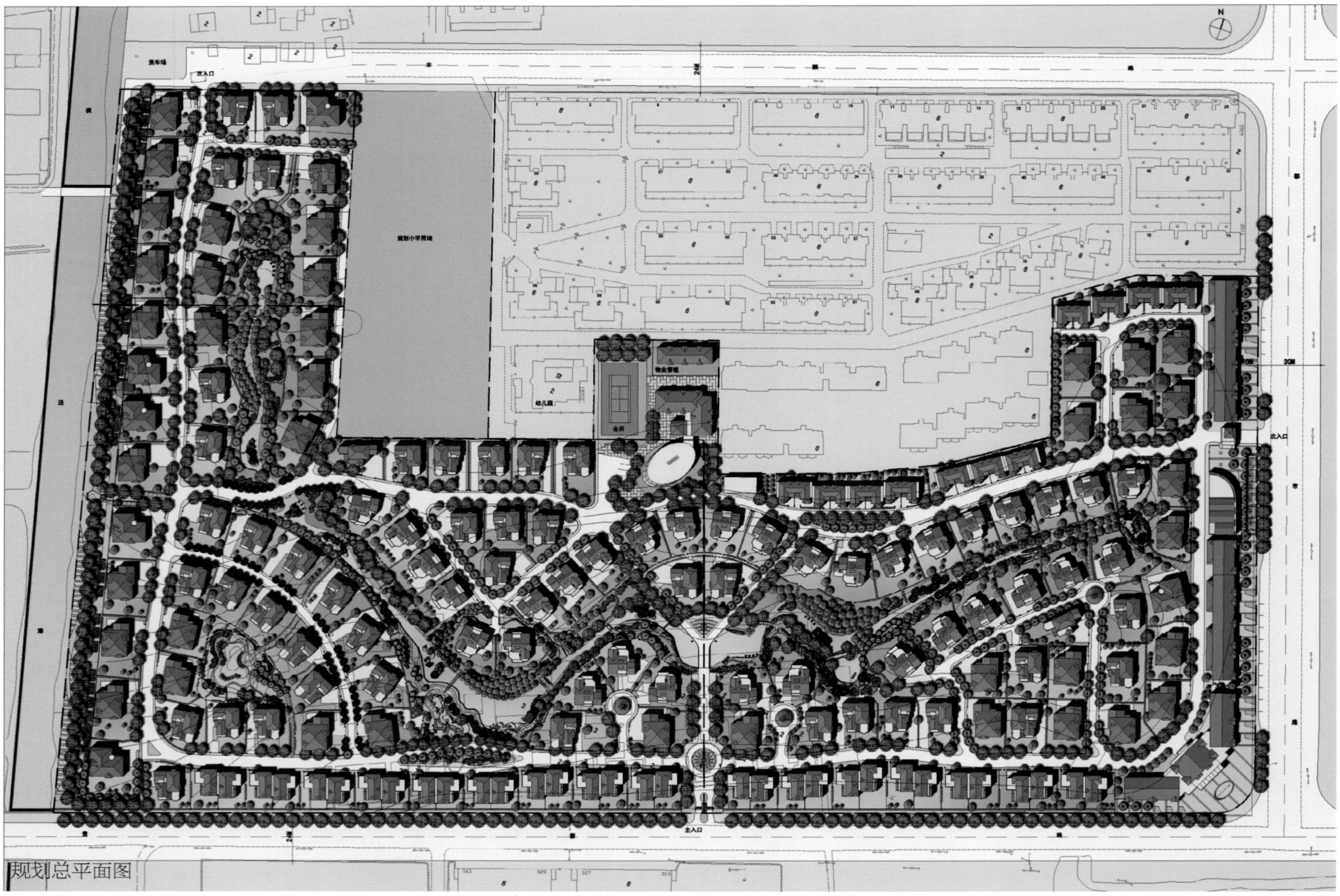

规划总平面图

上海都市阳光花园位于闵行区都市路、贵都路一侧。其规划和景观设计力求以欧美成熟的独立式别墅为特色，结合都市田园风情，以自然环境、人文环境为主线，通过总体布局、建筑设计和环境营造体现人与环境的和谐统一，切合现代人回归自然的居住需求。别墅区总体布局确立了"先造景后布置别墅"的规划思路，抓住中央绿带、滨河绿带和都市路景观带等重要景点深入研究，使都市阳光花园具备了完整而丰富的空间形象和细部特征。

Shanghai Sunshine Garden is located near Dushi Road and Guidu Road in Minhang District. Its planning and landscape design is featured by European and American mature style, combined with urban pastoral style, with the natural environment and human environment as the main line, through the overall layout, building design and environmental construction, reflecting the harmonious unification of man and environment, meeting needs of returning to nature. The overall layout uses planning ideas " first landscape, second layout of Villa ", conducting in-depth study of central green belts, green belts along the river and urban road landscape riparian zone and other important sights so that Sunshine Garden boasts a full and rich garden spatial image and detailed features.

中心水带断面 A

中心水带断面 B

中心水带断面 C

上海古北新区居住分区黄金城道城市设计
Shanghai Gubei New District Residential Area Huangjin City Road Urban Design

项目名称：上海古北新区居住分区黄金城道城市设计
项目地点：中国 / 上海
建筑面积：7.2 万平方米
编制时间：2001 / 11
业主：上海古北（集团）有限公司
规划类型：城市设计

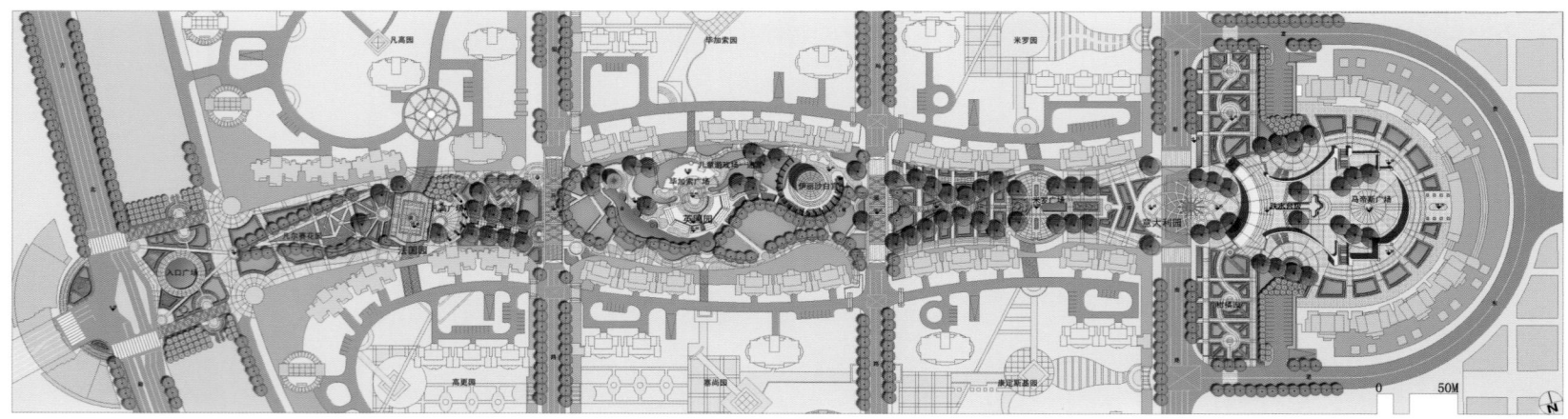

规划力图通过对古北新区城市空间主轴——黄金城道的分析形成城市设计导则，以引导整个地区的发展，有助于规划主管部门、土地管理者、地块开发商、设计师以及今后的使用者之间建立起一致、连续、协调的相互关系，从而为整个地区创造出充满活力的、有序的城市空间。其主要框架包括：城市中的街道、街道空间、街道生活、要素集合——城市设计方案四个部分。城市设计导则包括设计总则及细则，分别对黄金城道沿街裙楼和高层建筑的风格、材质、体量、用途，以及街道空间的性质、界面、地面铺砌、植物配置、广告、照明作了说明，以文字及图解为形式，对今后的设计、开发、管理加以控制和引导。

Through the analysis of Huangjin City Road which is the main axis of Gubei New District, the urban design guide book is formed for leading the whole area's development. Also, it helps Planning Office, Land Office, Developer, Designers, and Users to have a common, continuous and harmonious relationship among each other, in order to create an active and organized city space. The main frame includes street, street space, street life, elements combining. The Urban Design Guide Book with the general rules and details has the explanation on the podium building and high building's style, material, size, function, as well as the explanation on street space's character, interface, paving, greens, advertising, lighting, to control and guide future design, developing, and management.

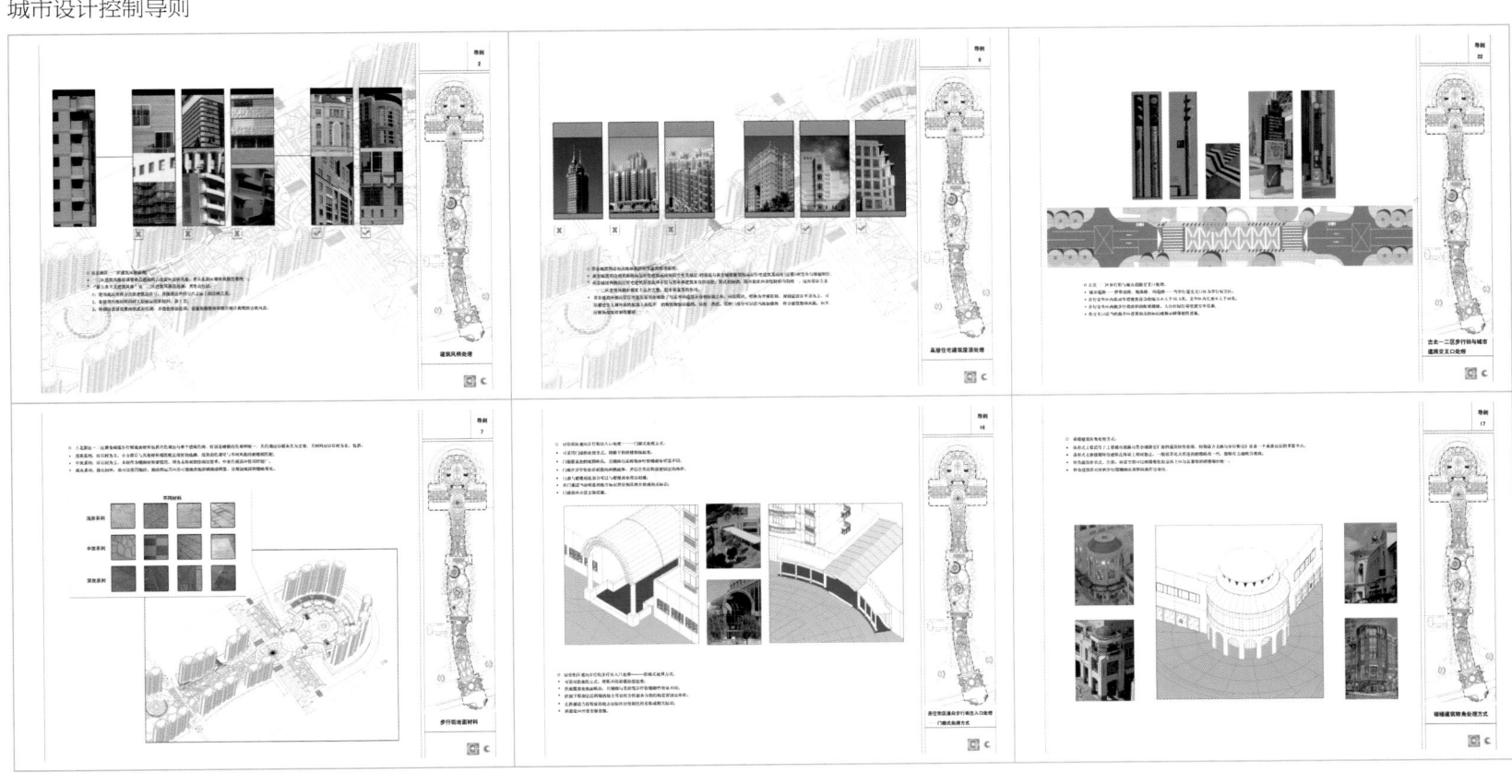

城市设计控制导则

青海省西宁市黄河路改造规划设计
Reconstruction Planning of Huanghe Road, Xining, Qinghai

项目名称：青海省西宁市黄河路改造规划设计
项目地点：中国 / 青海 / 西宁
长度：1 840 米
编制时间：2009 / 10
业主：西宁振华建设投资控股有限公司
规划类型：详细设计

图例
- 保留建筑（现状）
- 保留建筑（已批或已批在建建筑）
- 拆除建筑
- 整体改造
- 立面改造

黄河路位于西宁城西片区，南起青海宾馆，北至七一西路，是西宁市中心的南北向城市干道。

根据现状条件，规划将黄河路建设改造定位为：集公共服务与形象展示于一体的城市公共空间，现代、时尚、大气、生态、精致的西宁迎宾大道。

设计重点考虑道路改造实施过程中的可操作性，改造规划主要分近期和远期两个层面来进行。其中，近期重点改造黄河路北侧入口（与七一西路相交处）的街道景观形象；远期通过拆除、改造、保留相结合的手段对沿街建筑、街边绿化及街道设施等进行具体研究与设计，以塑造其作为迎宾大道的城市形象。

Huanghe Road is located in the west of Xining city, south to Qinghai hotel, north to West Qiyi Road, which is the north-south trunk road of Xining city.

According to the current conditions, the plan of Huanghe Road construction and renovation will be defined as an urban public space with public service and image display to create a modern, stylish, atmospheric, ecological, sophisticated Xining Welcome Boulevard.

Considering the feasibility during road rebuilding process, the plan will be divided into two main levels of short and long term. Among them, the short-term plan focuses on the transformation of street landscape image of Huanghe Road's northern entrance (the intersection with West Qiyi road). By means of demolition, renovation, retaining to conduct specific studies of a combination of street construction, street landscaping and street furniture,the design helps to shape an image of a Welcome Boulevard of the city.

① 规划总平面图
② 建筑改造类型分析

青海省西宁市七一路改造规划设计
Reconstruction Planning of Qiyi Road, Xining City, Qinghai Province

项目名称：青海省西宁市七一路改造规划设计
项目地点：中国 / 青海 / 西宁
长度：1 435 米
编制时间：2009 / 10
业主：西宁市城辉建设投资有限公司
规划类型：详细设计

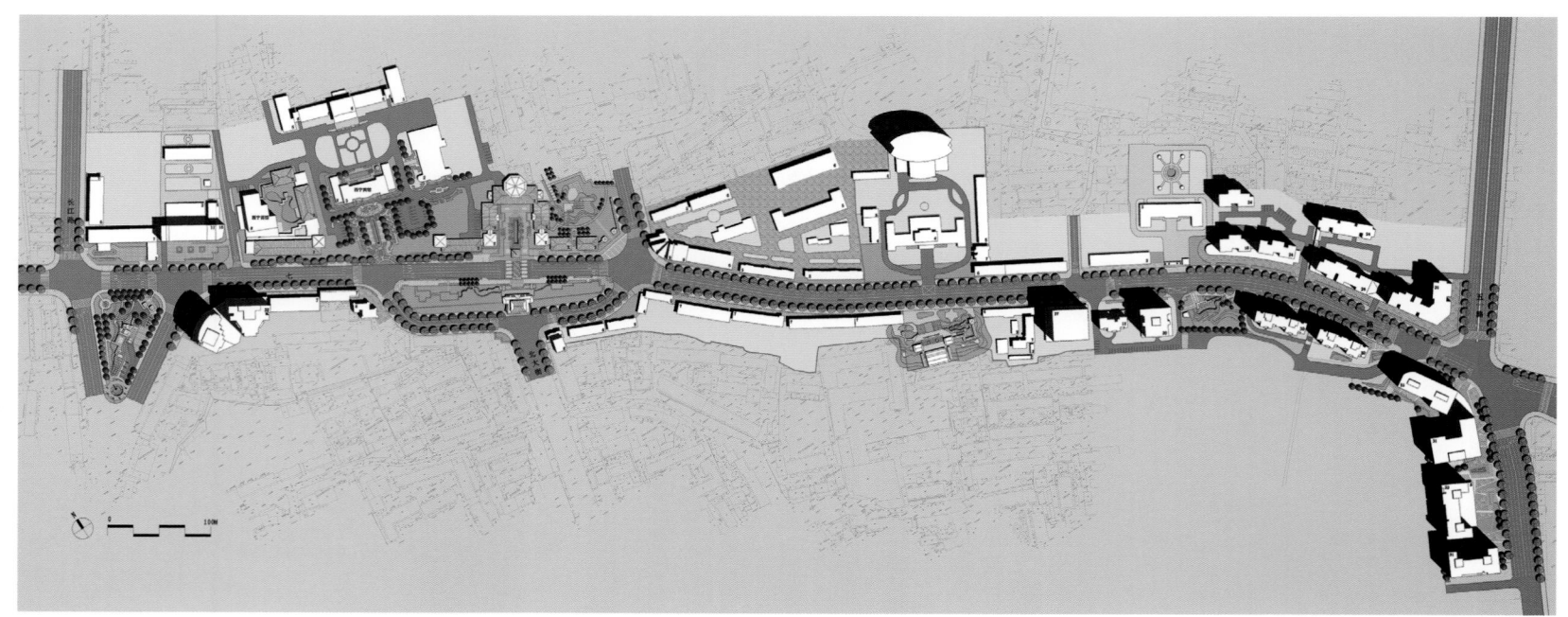

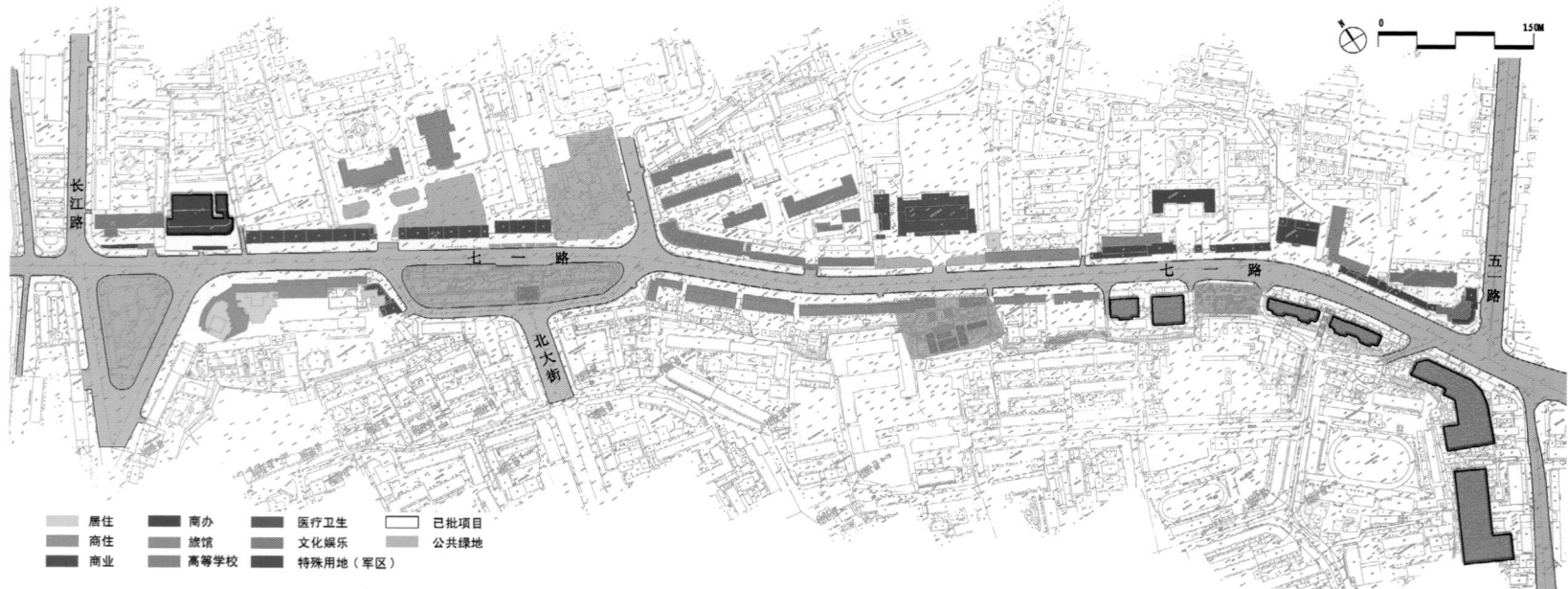

七一路位于西宁老城区，紧邻省委和省人大，西临南川河，南接十字商业中心，是西宁主城区一条东西向的城市干道。道路西起长江路，东至五一路，分别与长江路、北大街和五一路相交。

通过现状条件分析，规划将其建设改造定位为：融合了城市多元活动要素的功能复合型街区，将七一路打造成集城市历史文化与绿色生态于一体的城市景观大道。

Qiyi Road is located in Xining old city area, close to the provincial committee and provincial people's congress, west to Nanchuan River, connecting Cross Business Centre in the south, and is a main east-west road of Xining city. The road starts from Changjiang Road in the west, east to Wuyi Road, and intersects with Changjiang Road, North Avenue and Wuyi Road.

Through analysis of the current condition, the plan positions its construction and renovation as: integration of multiple city blocks with elements of city activities, making it a city landscape road with history, culture and green ecology.

① 规划总平面图
② 现状功能

青海省西宁市为民巷改造规划设计
Weimin Lane Reconstruction Planning Xining, Qinghai

项目名称：青海省西宁市为民巷改造规划设计
项目地点：中国 / 青海 / 西宁
长度：560 米
编制时间：2009 / 10
业主：西宁市城辉建设投资有限公司
规划类型：详细设计

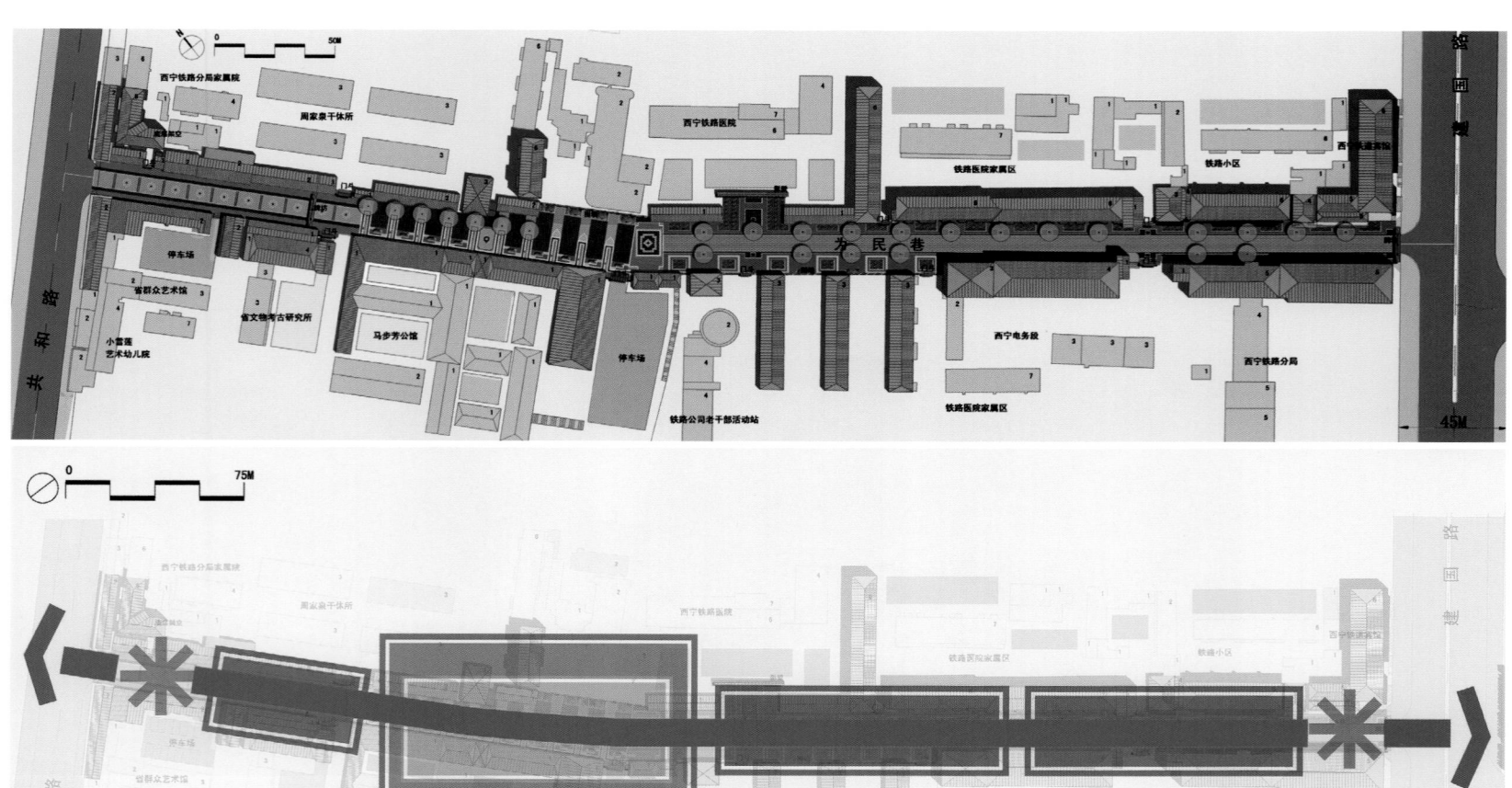

图例
核心景观区
核心路面改造区
延伸景观区
道路景观主轴
门户节点

为民巷位于西宁市中心区偏东地带，湟水以南，东接建国路，西至共和路，距离湟水约 400 米。道路现状为生活性城市道路，沿街以商业、餐饮、金融服务为主。街区内分布若干个单位，其中马步芳公馆（馨庐）是青海省级文物保护单位，建筑风格和院落形式均具有一定的民族特色和艺术水平。

鉴于现状街道两侧建筑均带有西宁民居的建筑元素，规划立足于弘扬中国传统文化，以促进传统旅游资源开发为落脚点，建议对街道两侧建筑进行整体性风貌改造，拟建成一条容特色餐饮、传统文化、旅游休闲等元素为一体的风情仿古街。

Weimin Lane is located in the east of Xining downtown area, south to Huangshui, connecting Jianguo Road in the east, west to Gonghe Road, with a distance of about 400 meters away from Huangshui. Currently it is a road with commercial, catering, financial services. A number of regional distribution units are located in the street including Ma Bufang Residence (Xin Lu), the cultural relic protection units of Qinghai, of which architectural style and form of the compound show national characteristics and artistic level.

According to the status that both sides of the street are building houses with Xining architectural elements, based on the promotion of Chinese traditional culture and traditional tourism resources, the proposal aimed to reform buildings on both sides of the street to build an antique street characterized by food, traditional culture, tourism and leisure.

① 规划总平面图
② 景观分析

青海省西宁市兴胜巷改造规划设计
Xingsheng Lane Reconstruction Planning, Xining, Qinghai

项目名称：青海省西宁市兴胜巷改造规划设计
项目地点：中国 / 青海 / 西宁
长度：330 米
编制时间：2009 / 10
业主：西宁振华建设投资控股有限公司
规划类型：详细设计

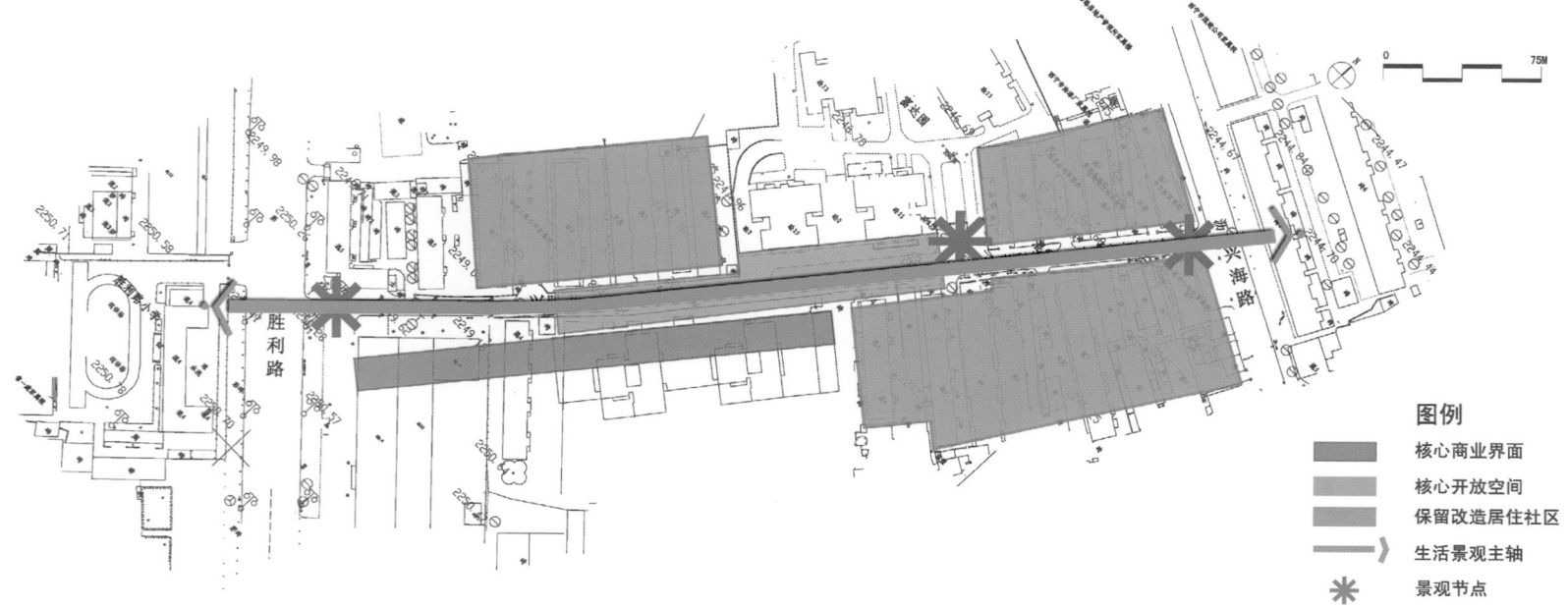

图例

核心商业界面
核心开放空间
保留改造居住社区
生活景观主轴
景观节点

兴胜巷位于西宁市中心区，人民公园东南，北接兴海路，南至胜利路。现状道路红线总宽度约为 12 米，为生活性的城市道路。

本次规划目的在于：1. 借助文化理念，塑造街区特色；2. 发挥城市活力，展现街区魅力；3. 利用现有配套，完善街区服务。以规划目的为指导，规划将兴胜巷定位为：集休闲、生活、社区服务于一体的综合型生活服务道路。在此基础上，规划确定了相应的改造策略：通过现状分析，分别对沿街建筑或构筑物进行拆除、改造或保留，并从整体空间形象出发对街道景观进行统一设计。

Xingsheng Lane is located in the downtown area of Xining, southeast to People's Park, connecting Xinghai Road to the north, south to Shengli Road. Current width of the road is about 12 meters. It is a city road.

The plan aims to: 1. use cultural ideas to create neighborhood characters; 2. play urban vitality to show neighborhood charm; 3. use existing infrastructure to improve neighborhood services. According to the purpose, the plan defines Xingsheng Lane as: road with integrated life services like leisure, life, community service. On this basis, the plan determines the corresponding transformation strategy: through current analysis, to remove, modify or retain the street building or structure, and conduct unified design for the streets.

① 规划总平面图
② 规划结构分析

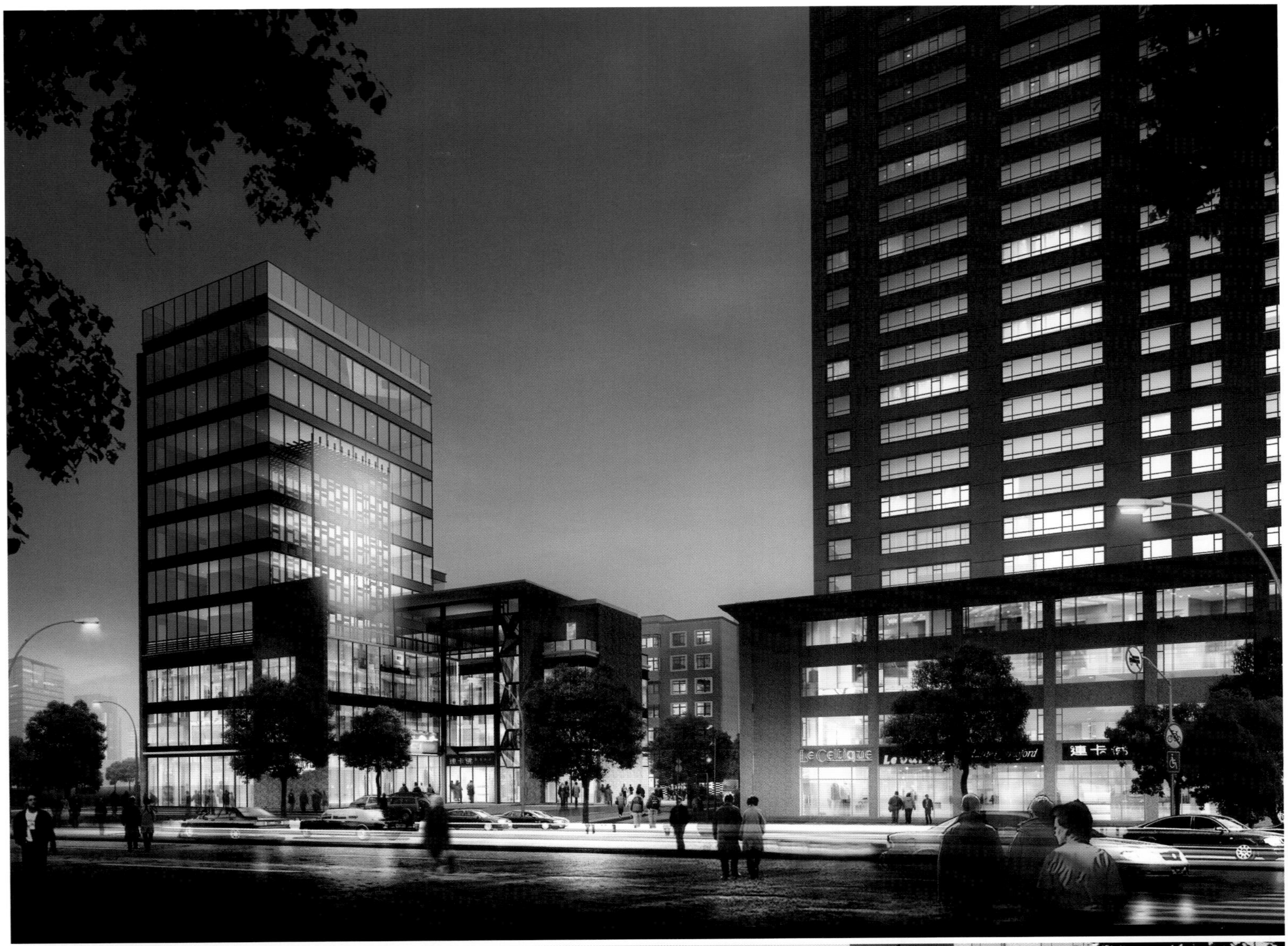
Le Celtique

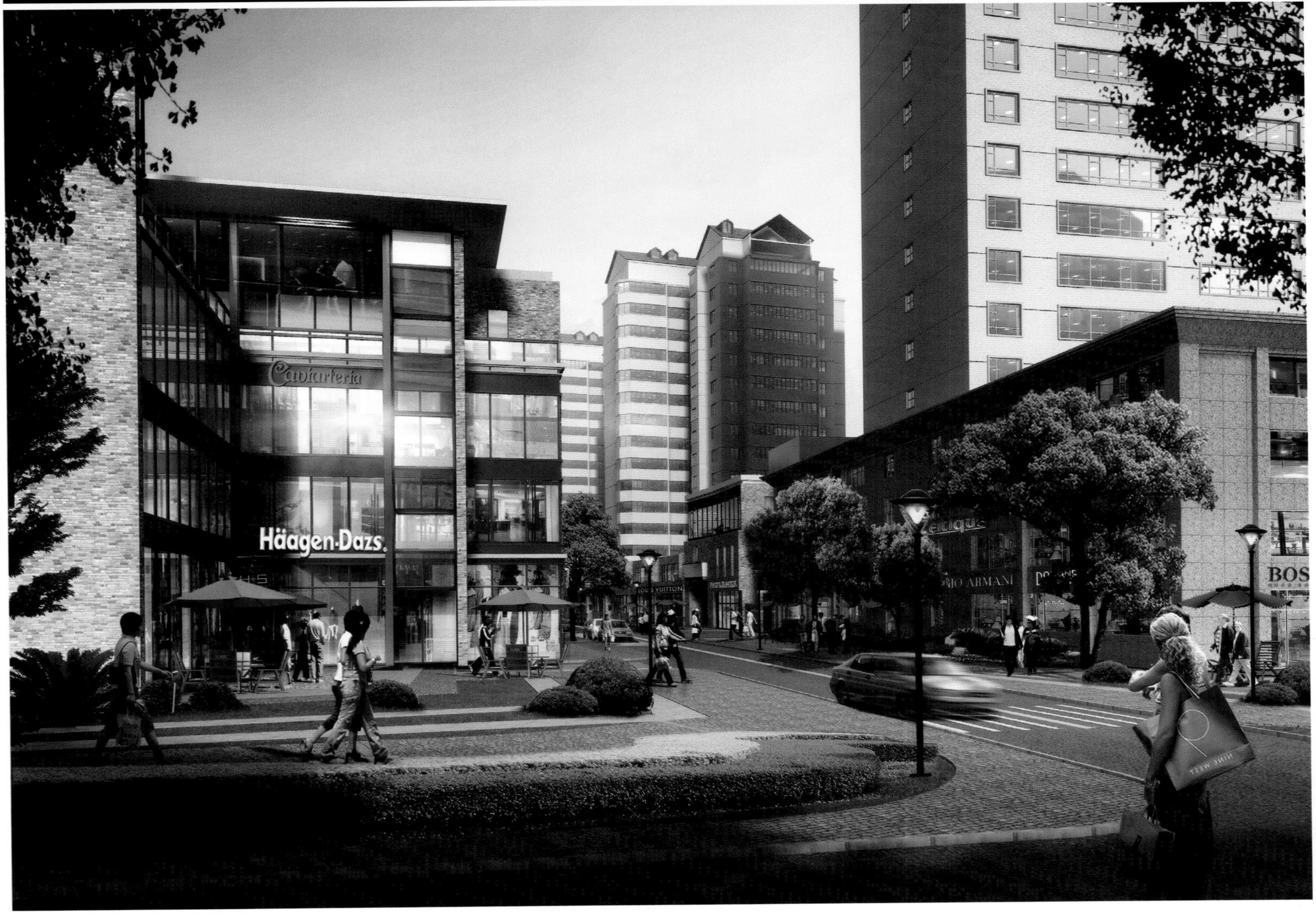
Caviarteria
Häagen-Dazs

西宁市中下南关街改造规划设计
Xining Zhongxia Nanguan Street Reconstruction Planning

项目名称：西宁市中下南关街改造规划设计
项目地点：中国 / 青海 / 西宁
长度：987 米
编制时间：2009 / 10
业主：西宁市城辉建设投资有限公司
规划类型：详细设计

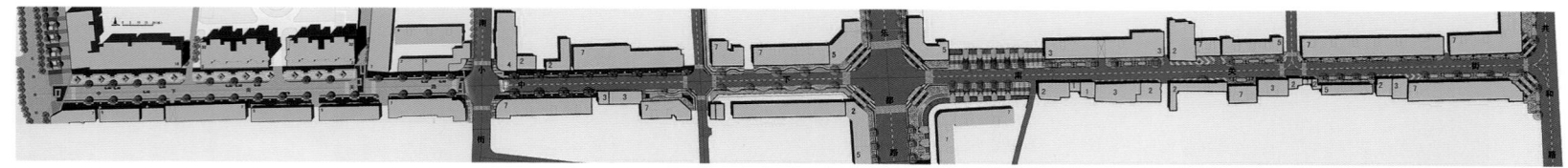

中下南关街位于西宁市城东区，东至共和路省儿童医院，西至花园街华联超市，为西宁市少数民族聚集区。现状道路规划红线宽度为 18 米，为生活性城市道路。

在功能上，规划基本保持现状生活性服务道路的特点，力求通过拆、改、留、建等手段改变现状较差的空间环境，塑造宜人、舒适、具有一定民俗特点的街道空间环境。在形象上，规划萃取传统民族建筑语言符号为母体，形成能适度体现民族建筑风格的特色民俗风情街。

为形成"民俗风情街"的总体规划定位，同时又考虑到规划实施的诸多限制条件，规划确定一、二、远期实施计划，并分别给予明确的规划目标。规划确定一期改造花园南街至南小街路段，形成民俗风情街的样板路段。二期改造南小街至共和路路段，完成民俗风情街的整体环境整治。此外，规划建议在中期实施完成的基础上，整体改造清真寺南侧地块，完善街道功能配套，提升环境品质。

Zhongxia Nanguan Street is located in the Chengdong District, Xining City, east to the Provincial Children's Hospital on the Republican Road, west to Hualian Supermarket on the Huayuan Street which is gathering area for the minorities in Xining City. Current width of the road is 18 meters and the road a life road.

Functionally, the plan maintains the current characteristics of life service, through demolition, changing, reataining, building and other means to change the current space environment, creating a space environment which is pleasant, comfortable and has some characteristics of street folk. In the image, using traditional ethnic architectural planning for the mother symbols to form a national architectural style that can appropriately reflect the characteristics of folk customs street.

For the formation of "folk customs street", while taking into account the many constraints of planning and implementation, the plan determines first, second phase and long-term implementation plan and gives clear planning objectives. Phase one determines a modified street section of Huayuan to South Avenue, a model for the formation of customs section of street. Phase two transforms Gonehe Road to South Avenue section to complete the overall environmental regulation of the street. In addition, on the basis of mid-term implementation, the plan transforms the mosque in the south block to improve the supporting functions and enhance the quality of the environment.

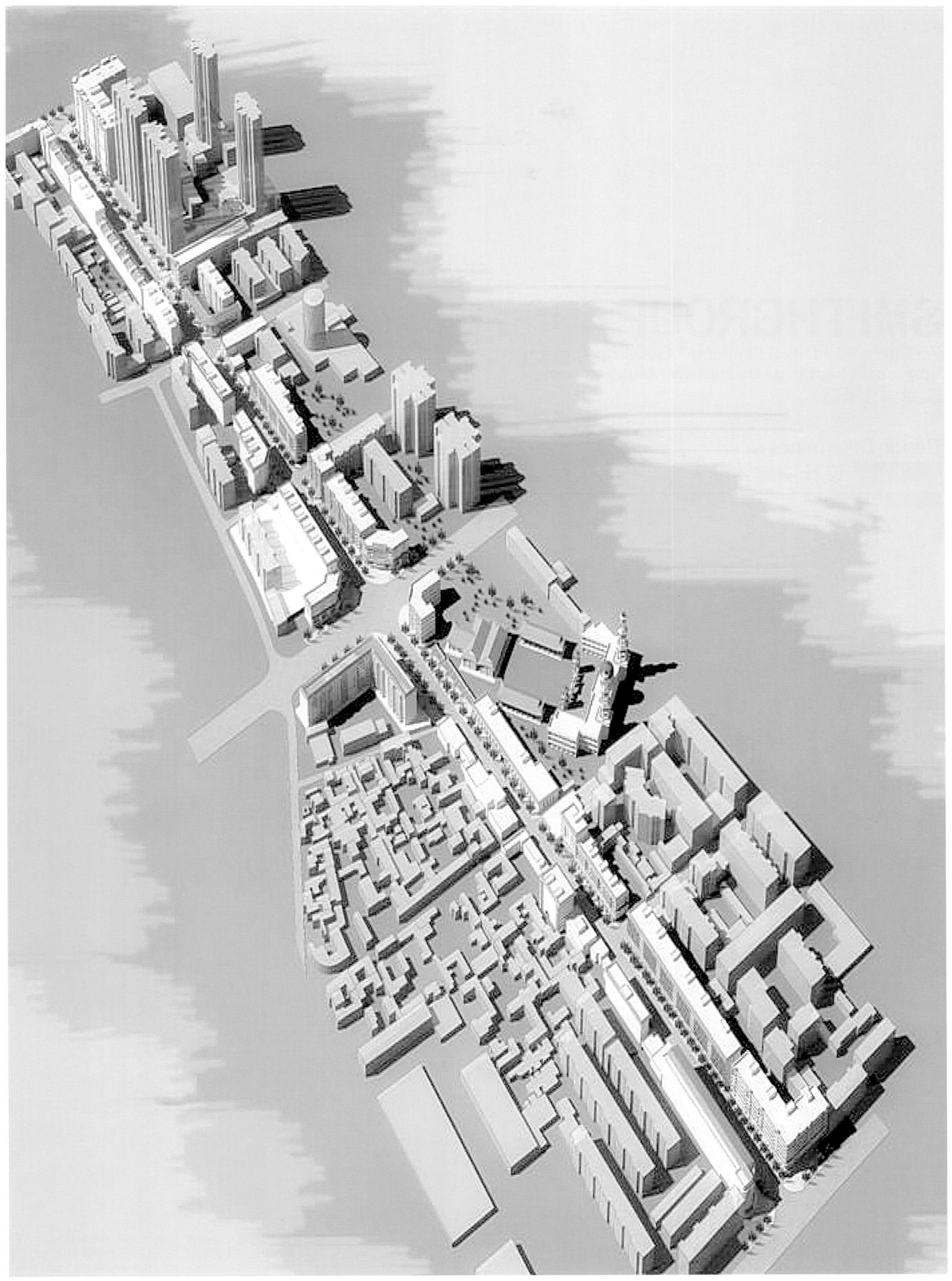

SMITHGROUP cna

Architecture · Urban Planning · Landscape Design
Interiors Design · Art Consultant · Lighting Design
Since 1853
www.cna-group.com

Design Determines . . .

设计决定······

1   2   3   4   5   6   7   8   9   10 ······

01 规划设计类项目
01 Urban Planning

滨水地区
Water Front
城市街区
City Block

城市空间
Urban Space

旅游休闲
Travel and Leisure

02 景观设计类项目
02 Landscape Design

无锡新区站前商务区控制性详细规划暨城市设计项目

Wuxi New Area Station Business District Regulatory
Planning & Urban Design

Urban Planning   2009·12

无锡新区站前商务区控制性详细规划暨城市设计
Wuxi New Area Station Business District Regulatory Planning & Urban Design

项目名称：无锡新区站前商务区控制性详细规划暨城市设计
项目地点：中国 / 江苏 / 无锡
建筑面积：327 万平方米
编制时间：2009 / 12
业主：无锡新区城际铁路站前商务区办公室 / 无锡新区城际铁路站前商务区投资开发有限公司
规划类型：控制性详细规划及城市设计

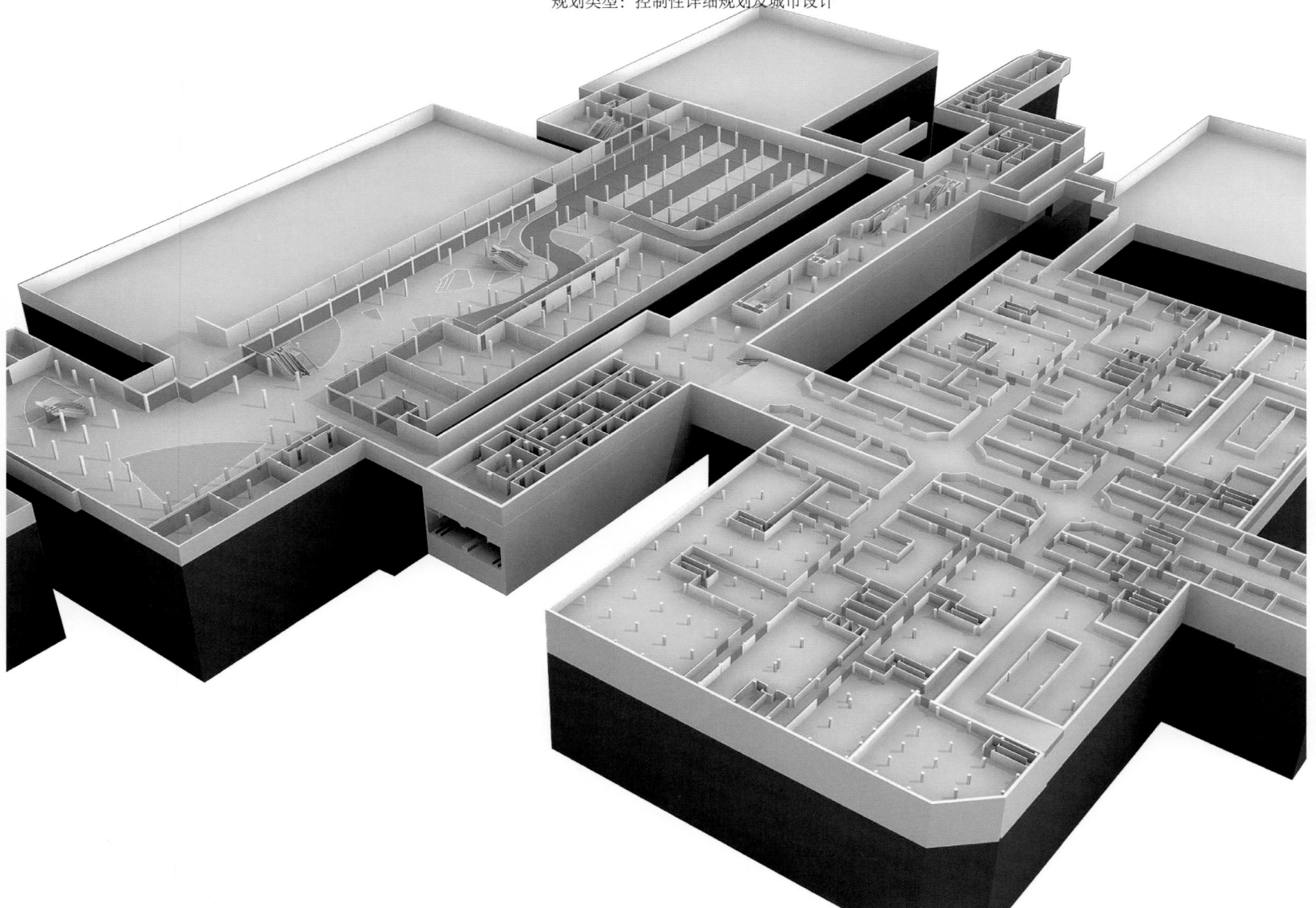

项目位于无锡市南部，沪宁城际铁路东北侧，西临太湖科技园和太湖新城，东靠无锡机场，距无锡机场约 4 公里。

该项目规划主要包含了三个规划阶段

阶段 1：站前地区详细设计，规划主要针对站前地区及其核心区，对如何紧密结合轨道交通站点的建设来促进地区发展作一个前瞻性的设想，并对站前核心区进行了详细设计。

阶段 2：站前商务区控制性详细规划，在《站前地区详细设计》基础上往南扩展形成更大范围的站前商务区，为促进无锡新区城际铁路站前商务区的有序发展，加强该区的规划管理和建设管理，特别是编制规划。

阶段 3：站前商务区城市设计，以控制性详细规划为指导，对整体城市空间作进一步的安排与组织，对核心区及重要节点的空间作深入研究，以期创造一个充满生机活力的火车站地区。

The project is located in the south of Wuxi City, northeastern part of Shanghai-Nanjing Intercity Railway, west of Taihu Lake and New Town Science and Technology Park, 4 km from Wuxi Airport to the east.

The project planning mainly consists of three planning stages:

Phase 1: detailing design of station areas, the plan focuses on the station area and its core area, proposes a vision of how to closely combine the construction of rail transit stations to promote regional development and carry out the detailed design of the core area.

Phase 2: station business district regulatory plan, on the basis of "detailed design of station areas" extended to the south, forms a wider range of station business district to promote inter-city railway station and orderly development of the business district, promoting the area's planning, management and construction management, especially the preparation of planning.

Phase 3: station business district urban design, under the guidance of regulatory plan, arranges the overall urban space, conducts in-depth research for the important node in the core areas to create an exuberant and vigourous station area .

① ②　① 地下一层建筑
　　　② 规划总平面图

系列分析图

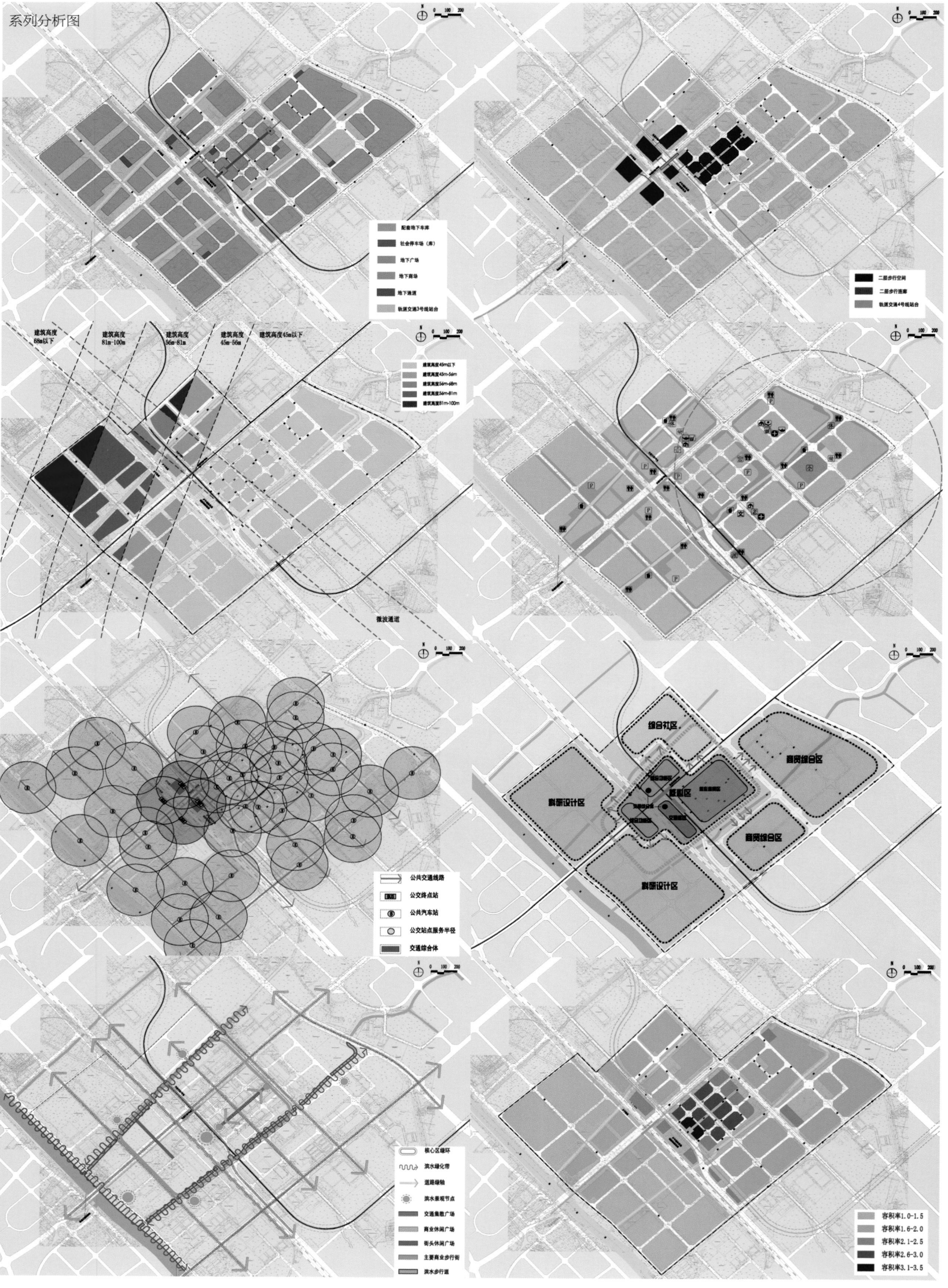

配套地下车库
社会停车场（库）
地下广场
地下商场
地下通道
轨道交通3号线站台

二层步行空网
二层步行连廊
轨道交通4号线站台

建筑高度
68m以下

建筑高度
81m-100m

建筑高度
56m-81m

建筑高度
45m-56m

建筑高度45m以下

建筑高度45m以下
建筑高度45m-56m
建筑高度56m-68m
建筑高度68m-81m
建筑高度81m-100m

微波通道

综合社区

创意设计区

庭社区

商贸综合区

创意设计区

商贸综合区

公共交通线路
公交终点站
公共汽车站
公交站点服务半径
交通综合体

核心区绿环
滨水绿化带
道路绿轴
滨水景观节点
交通集散广场
商业休闲广场
街头休闲广场
主要商业步行街
滨水步行道

容积率1.0-1.5
容积率1.6-2.0
容积率2.1-2.5
容积率2.6-3.0
容积率3.1-3.5

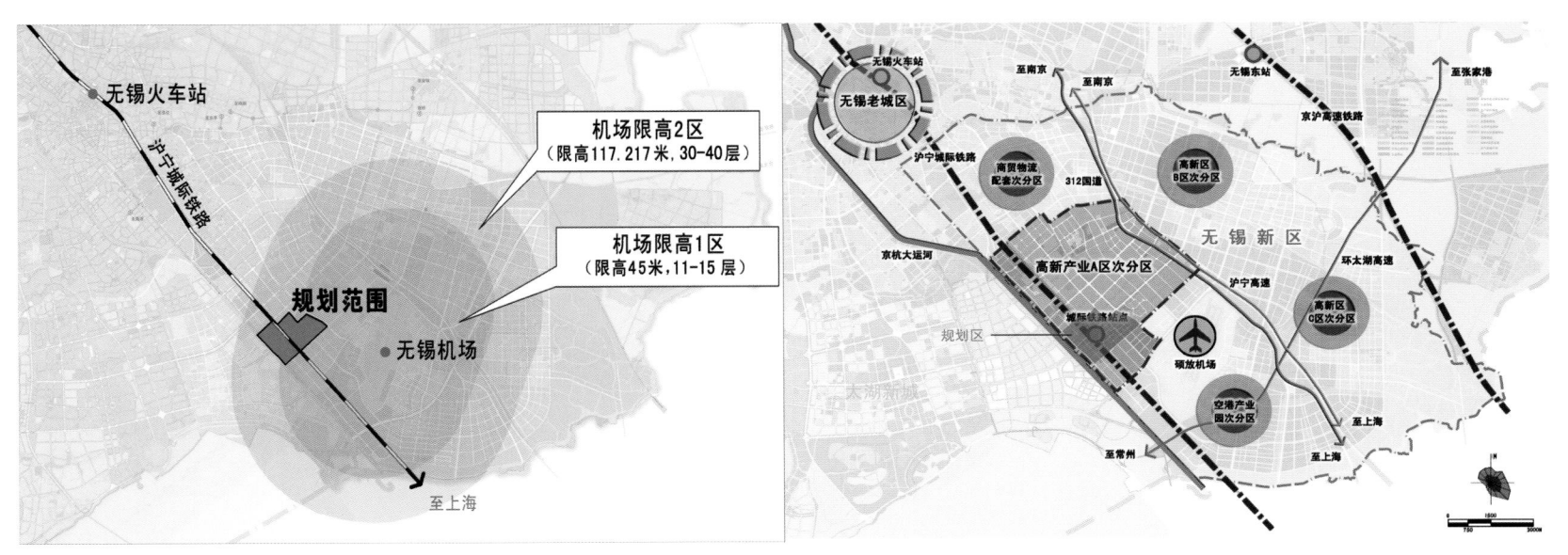

机场净空分析

区位分析图

规划总平面图

西宁火车站地区控制性详细规划及站前核心区详细规划

Xining Railway Station

Regulatory Plan and The Detailed Planning of Core Area

西宁火车站地区控制性详细规划及站前核心区详细规划
Xining Railway Station Regulatory Plan and the Detailed Planning of Core Area of Front Station Area

项目名称：西宁火车站地区控制性详细规划及站前核心区详细规划
项目地点：中国 / 青海 / 西宁
建筑面积：627 万平方米
编制时间：2010 / 09
业主：西宁市城乡规划建设局
规划类型：控制性详细规划 / 详细规划

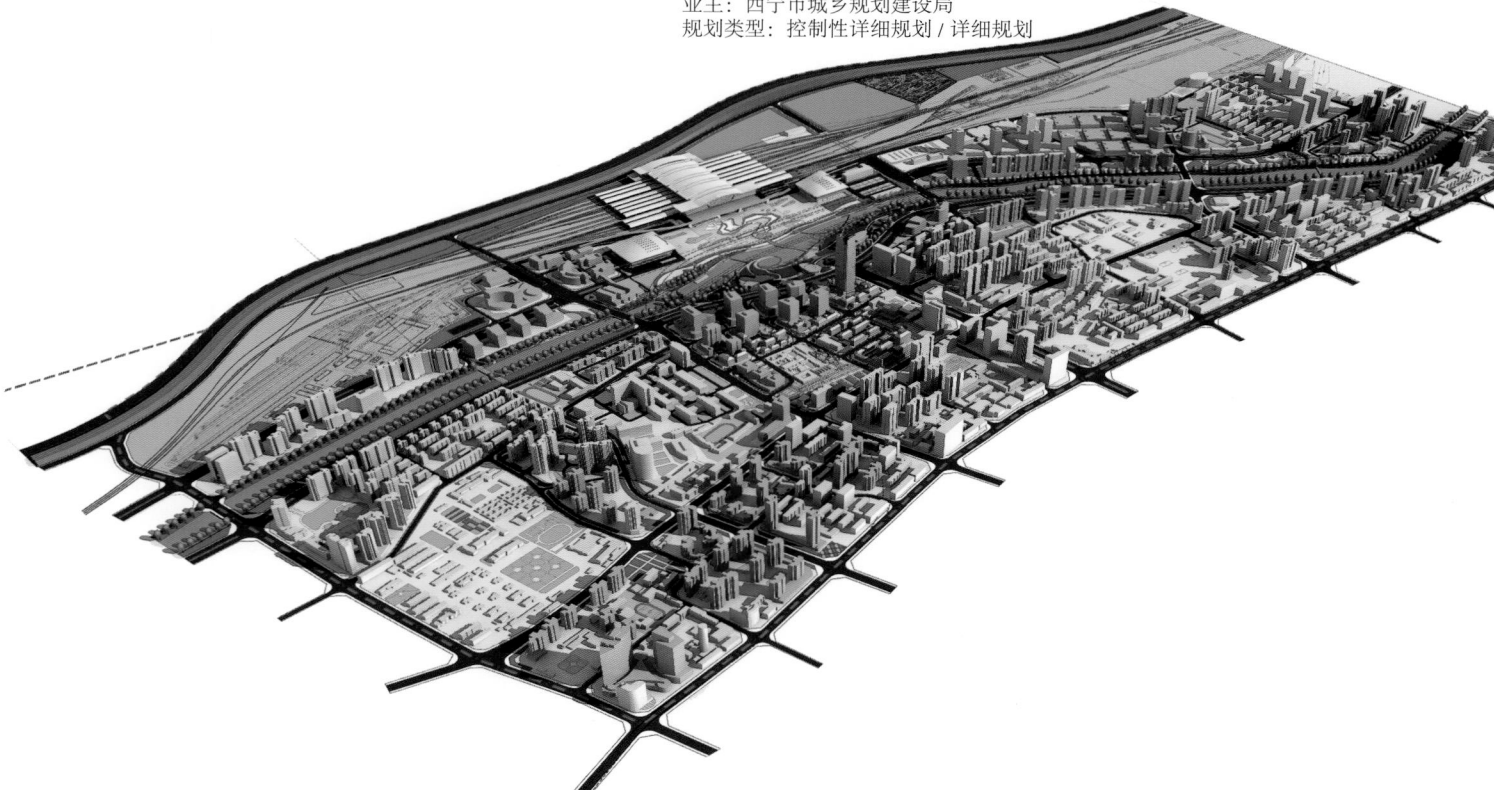

规划区北起兰西高速路，南至东关大街、大众街及八一路一线，东为湟中路，西临五一路及花园北路。

规划区位于西宁中心城区的十字型商业布局的东部，处于城市商业经济中心强辐射范围内，地段土地价值较高，又拥有得天独厚的环境优势。

规划区毗邻西宁市中心地区，邻近西宁市东西发展轴，城市南北发展轴从中穿过。在此区域内建设城市沿河景观绿化将为中心城区提供城市绿化公共空间，使市中心区的功能更加完善，成为西宁市展示现代化、人性化与自然景观完美结合的标志性窗口。

本次规划总体功能定位为——"青海省西宁市门户，外向经济服务中心"，包含两层主要涵义：

1. 城市中心十字商业金融中心向东拓展区。

2. 集多种交通集散、城市景观展示、综合商务开发为一体的综合性现代交通商务中心。

根据西宁总体的布局结构，同时结合湟水河两岸的自然、历史、交通等多种条件，本次规划形成富有特色的规划区布局结构。总体概括为：一带串联、井字构架、南北双核、两岸互通、三区互动。

项目规划总建筑面积约 627 万平方米，平均容积率为 1.1。

The planning district starts from Lanxi Highway in the north, south to Dongguan Street, Dazhong Street and Bayi Road, east to Huangzhong Road, west to Wuyi Road and Huayuan Road.

The planned area is located in the east of cross-shaped business layout of Xining city which is a strong radiation area of commercial and economic center of the city. It not only has a high land value, but also has unique environmental advantages.

The planned area is adjacent to the downtown area of Xining, near the Xining city east to west development axis, the development of the city north-south axis running through it. Construction of landscape planting in this region will provide urban green public space, so that the downtown area has a more perfect function,which makes it a perfect window showing modernity, humanity and the natural landscape of Xining city.

The overall function of the planning has been defined as "the portal of Xining, Qinghai Province, export-oriented economic service center", which contains two major implications:

First, the downtown cross-shaped business and financial center eastward expansion area.

Second, a modern transport business center with multi-traffic distribution, urban landscape display, integrated business development.

According to the overall layout of Xining, combined with natural, historical, traffic and other conditions along the two sides of Huangshui River, this plan forms special layout structure of the area. It is generalized as follows: connected with one ship, Tic-Tac-architecture, north-south dual-core, cross-bank exchange, interaction between the three areas.

The total construction area of the project is about 6,270,000 square meters, with average floor area ratio of 1.1.

区域规划总平面图

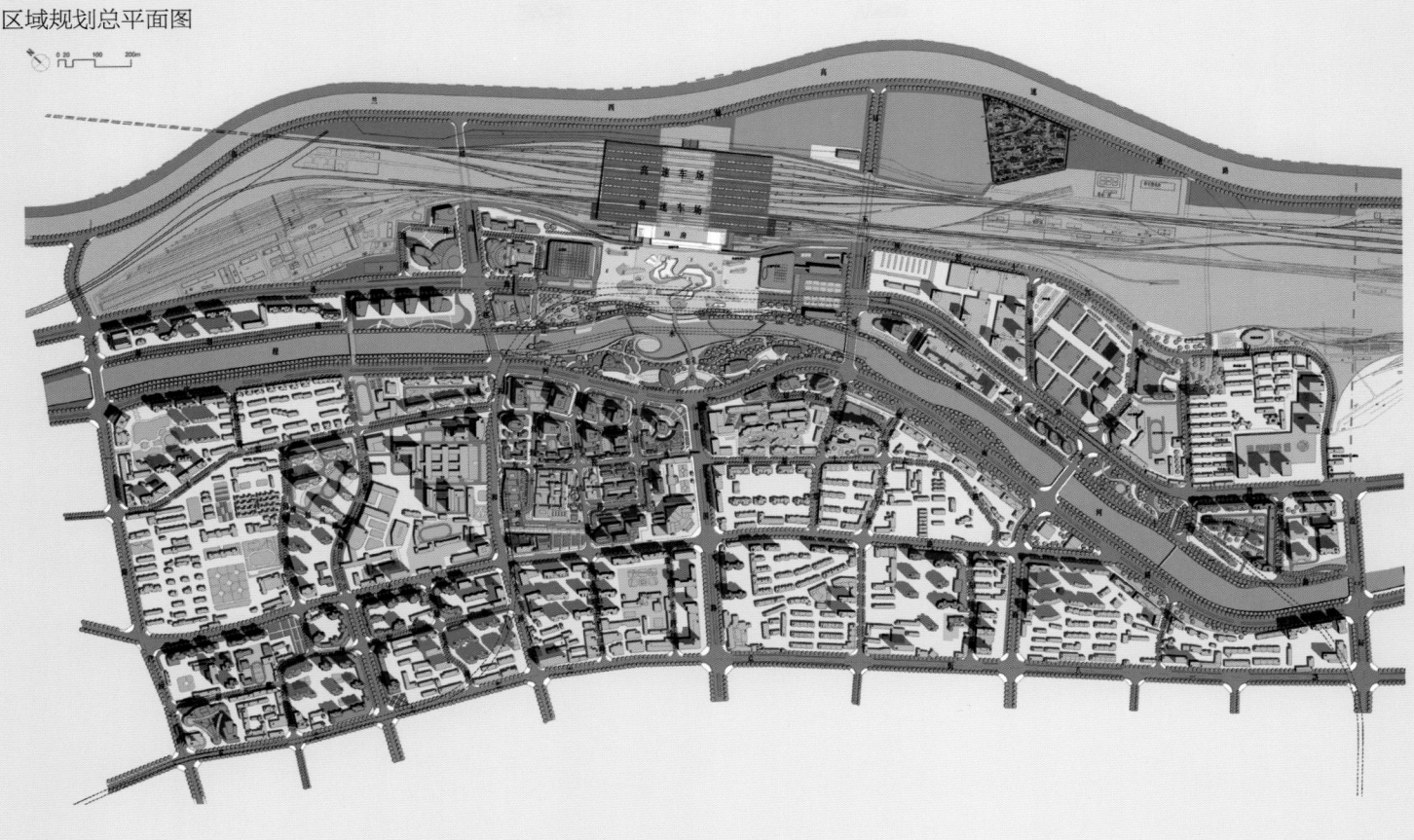

区域规划用地图

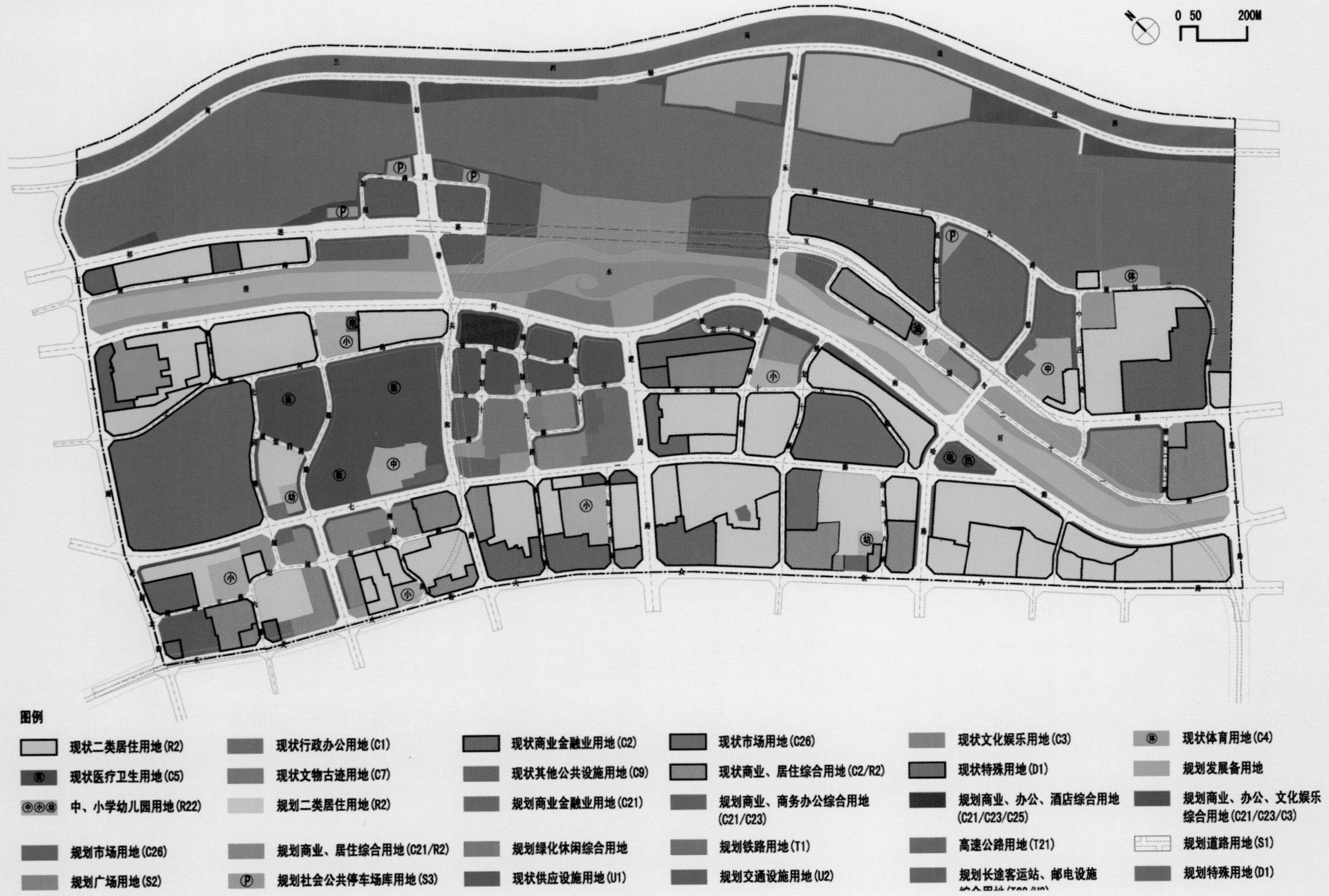

0 50   200M

图例

现状二类居住用地(R2)		现状行政办公用地(C1)		现状商业金融业用地(C2)		现状市场用地(C26)		现状文化娱乐用地(C3)		现状体育用地(C4)
现状医疗卫生用地(C5)		现状文物古迹用地(C7)		现状其他公共设施用地(C9)		现状商业、居住综合用地(C2/R2)		现状特殊用地(D1)		规划发展备用地
中、小学幼儿园用地(R22)		规划二类居住用地(R2)		规划商业金融业用地(C21)		规划商业、商务办公综合用地(C21/C23)		规划商业、办公、酒店综合用地(C21/C23/C25)		规划商业、办公、文化娱乐综合用地(C21/C23/C3)
规划市场用地(C26)		规划商业、居住综合用地(C21/R2)		规划绿化休闲综合用地		规划铁路用地(T1)		高速公路用地(T21)		规划道路用地(S1)
规划广场用地(S2)		规划社会公共停车库用地(S3)		现状供应设施用地(U1)		规划交通设施用地(U2)		规划长途客运站、邮电设施综合用地(T22/U3)		规划特殊用地(D1)

591

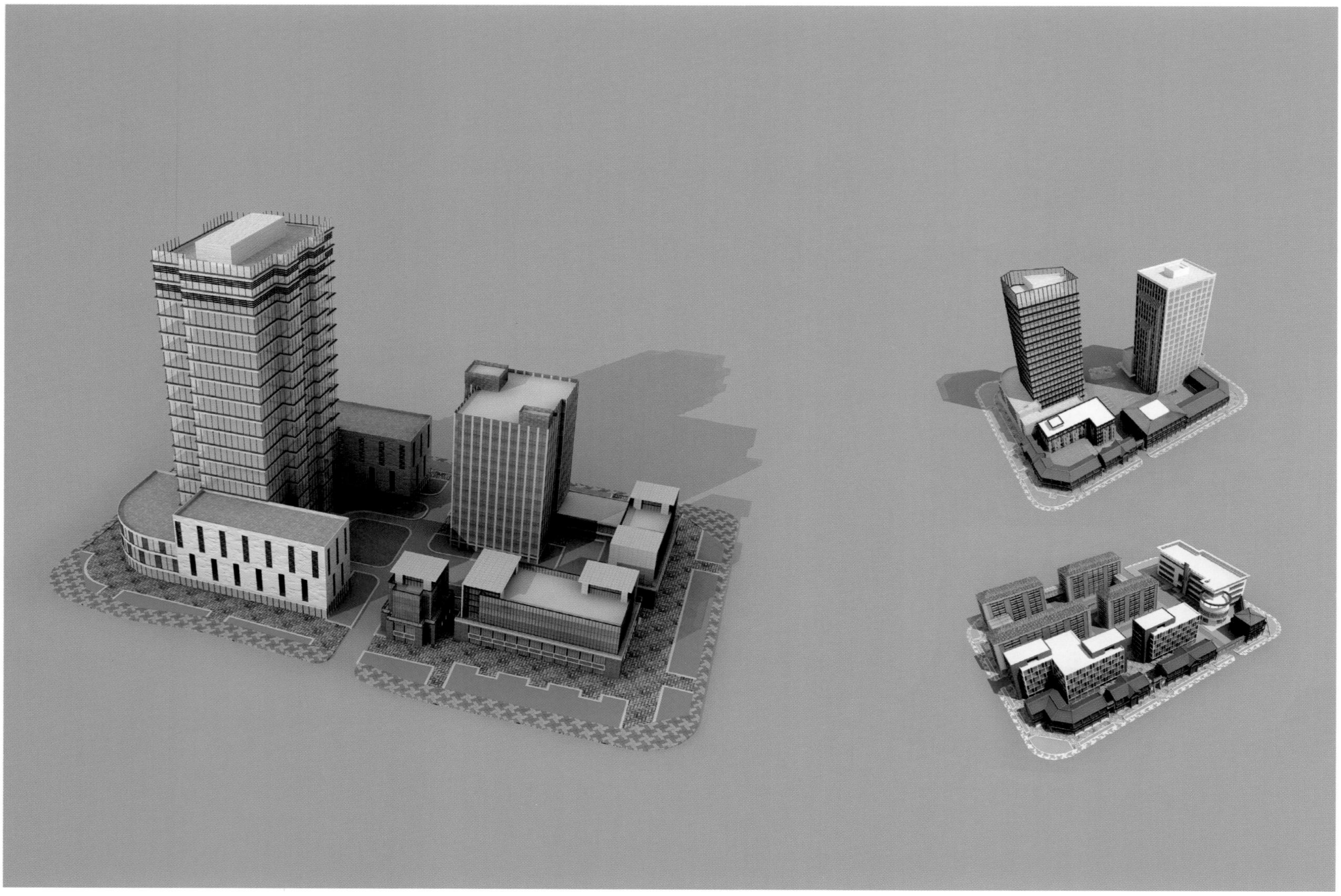

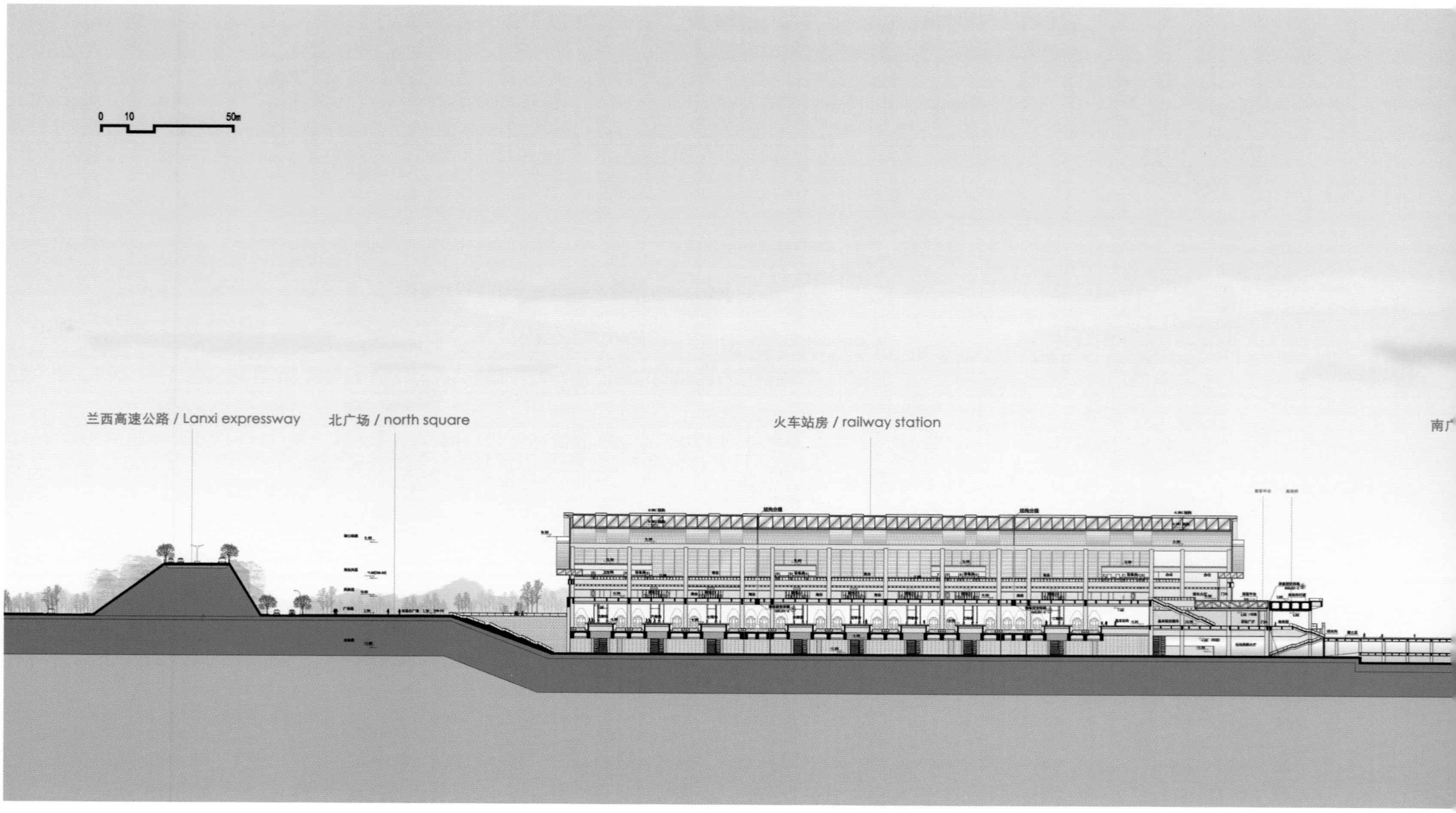

| 0 | 10 | 50m |

兰西高速公路 / Lanxi expressway　　北广场 / north square　　火车站房 / railway station　　南广

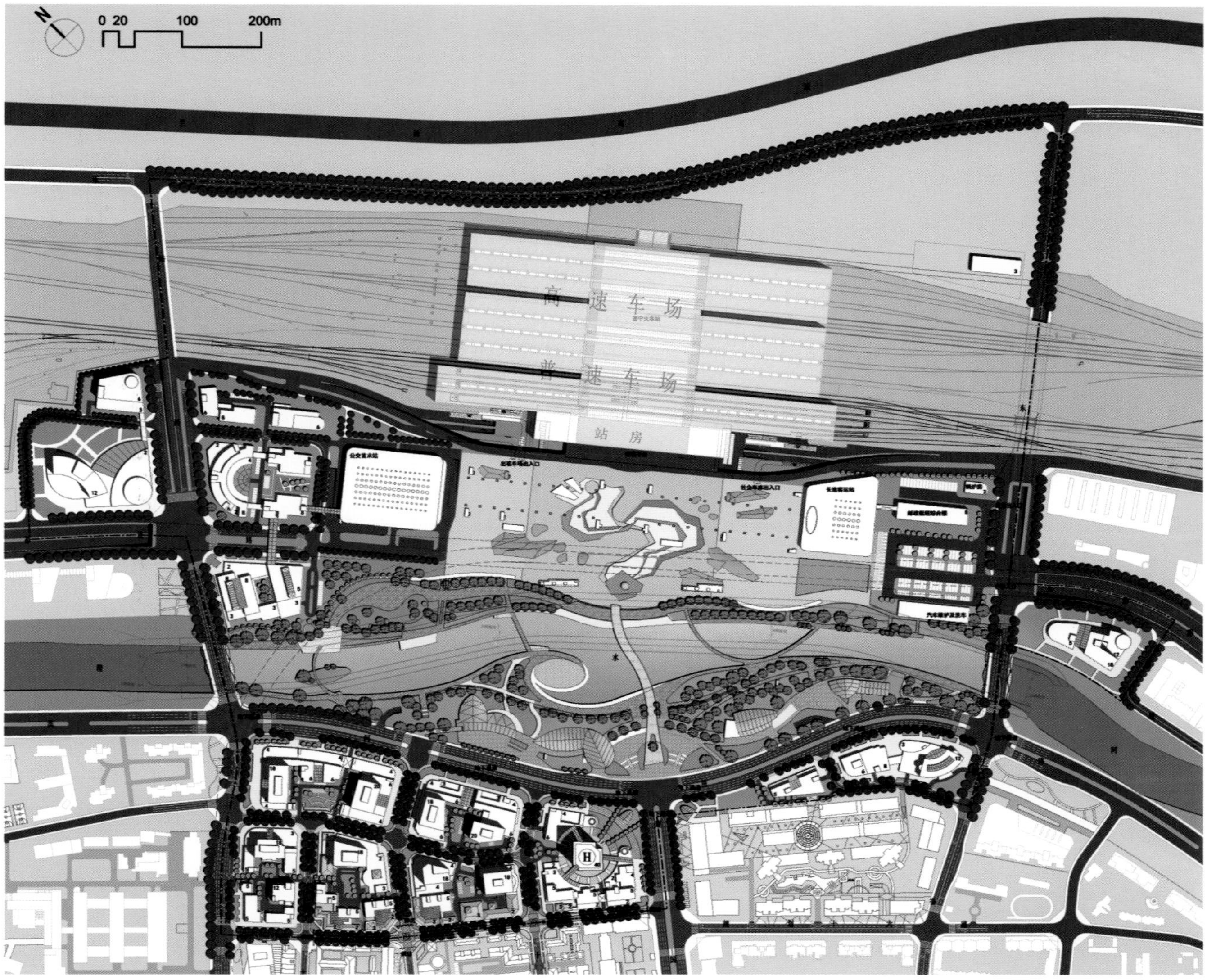

N
0 20 100 200m
高 速 车 场
普 速 车 场
站 房

核心区规划总平面图

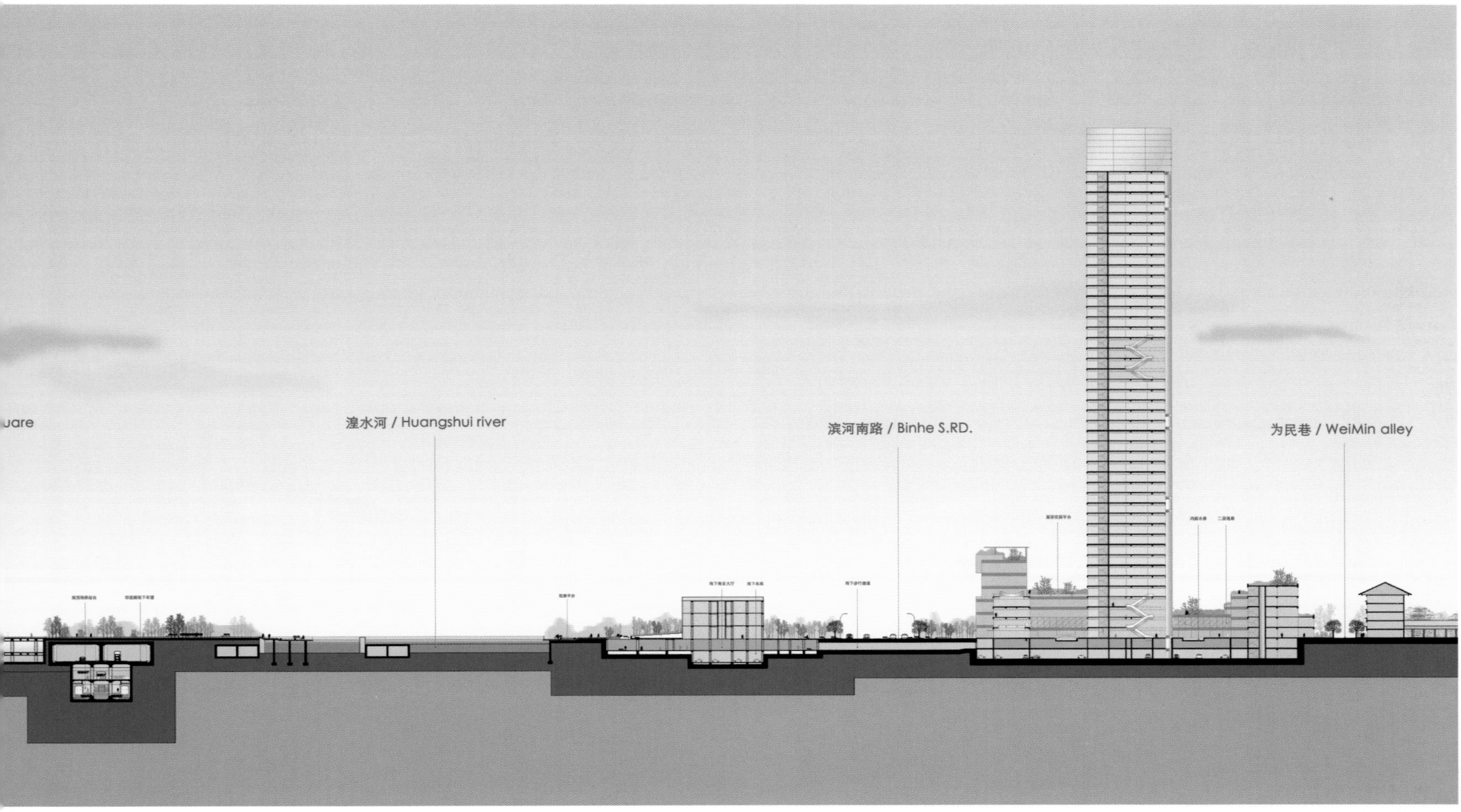

uare
湟水河 / Huangshui river
滨河南路 / Binhe S.RD.
为民巷 / WeiMin alley

南北剖面图

松江新城国际生态商务区城市风貌控制及城市设计方案

Songjiang New City International Business District Urban Ecological
Style Control and Urban Design

Urban Planning | 2010·08

松江新城国际生态商务区城市风貌控制及城市设计
Songjiang New City International Business District Urban Ecological Style Control and Urban Design

项目名称：松江新城国际生态商务区城市风貌控制及城市设计
项目地点：中国 / 上海
建筑面积：273.44 万平方米
编制时间：2010 / 08
业主：上海松江新城国际生态商务区管委会
规划类型：城市设计

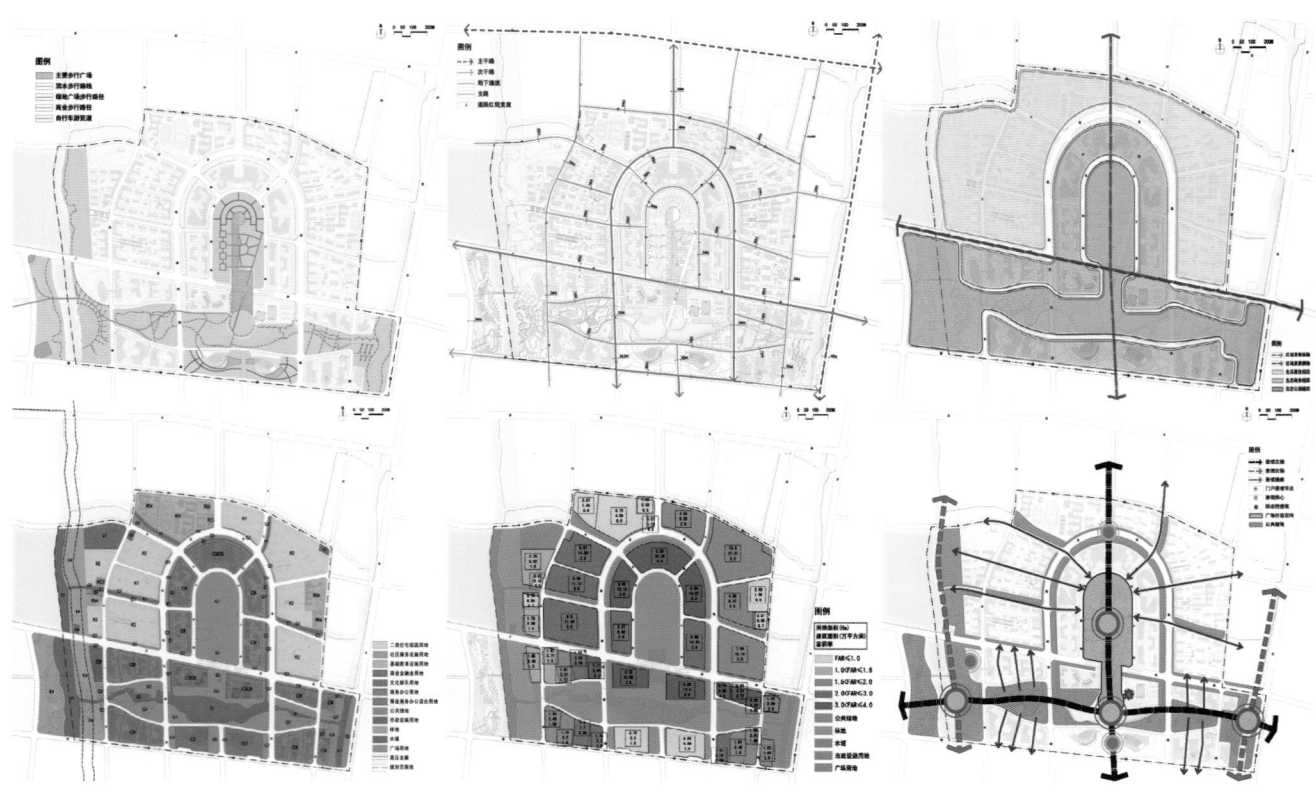

项目位于松江新城主城东北部，北临余山旅游度假区，西近松江大学城，东接松江工业区，南望松江老城，拥有良好的产业和环境优势。

项目定位：以现代高端商务为核心，打造成集文化休闲、商业服务、生活居住于一体的自然生态、品牌突出、品质卓越的现代化国际商务区，成为长三角地区一流的现代商务盆地，并将其打造成整体高效的商务之都、结构紧凑的核心之城、景观多样的生态乐园、丰富趣味的游览天堂。

规划为圈层式空间布局，以生态公园为核心圈层，内圈形成生态商务区，设置现代商务功能，并植入文化、商业设施，通过与水系、绿地的巧妙融合，打造成为松江新城主城区的新地标，外圈形成配套齐全的生态居住组团。

规划总建筑面积为 273.44 万平方米，其中居住建筑总面积约为 119.2 万平方米，公共服务设施建筑总面积约为 154.2 万平方米。地块平均容积率为 1.0。

The project is located in the northeast of Songjiang New City, north to Sheshan Resort, near the West Songjiang University City, east to Songjiang Industrial Zone, south to Songjiang Old City. It boasts a good industrial and environmental advantage.

Project position: centered on modern high-end business, creating a high-quality modern international business district with culture and leisure, business services, life living as modern business-class basin Yangtze River Delta region, making it core business city of overall efficiency, diverse ecological park, rich in interesting tour paradise.

The plan is the circle-type layout, using the ecological park as the core circle, Business District for the inner circle, setting the modern business function and implanting cultural, commercial facilities, the ingenious fusion with water, green space, creating a new landmark for Songjiang New City. The outer ring supports the formation of a full range of eco-living group.

The total planned capacity is 2,734,400 square meters, including residential buildings with a total area of about 1,192,000 square meters. A total construction area of public service facilities is approximately 1,542,000 square meters. The average floor area ratio is 1.0.

规划总平面图

大连市体育中心暨配套开发项目一期地块规划设计
Dalian Municipal Sports Center and Supporting Development Phase1 Planning and Design

项目名称：大连市体育中心暨配套开发项目一期地块规划设计
项目地点：中国 / 辽宁 / 大连
建筑面积：237.5 万平方米
编制时间：2010 / 05
业主：上海众稳房地产经纪有限公司
规划类型：规划方案

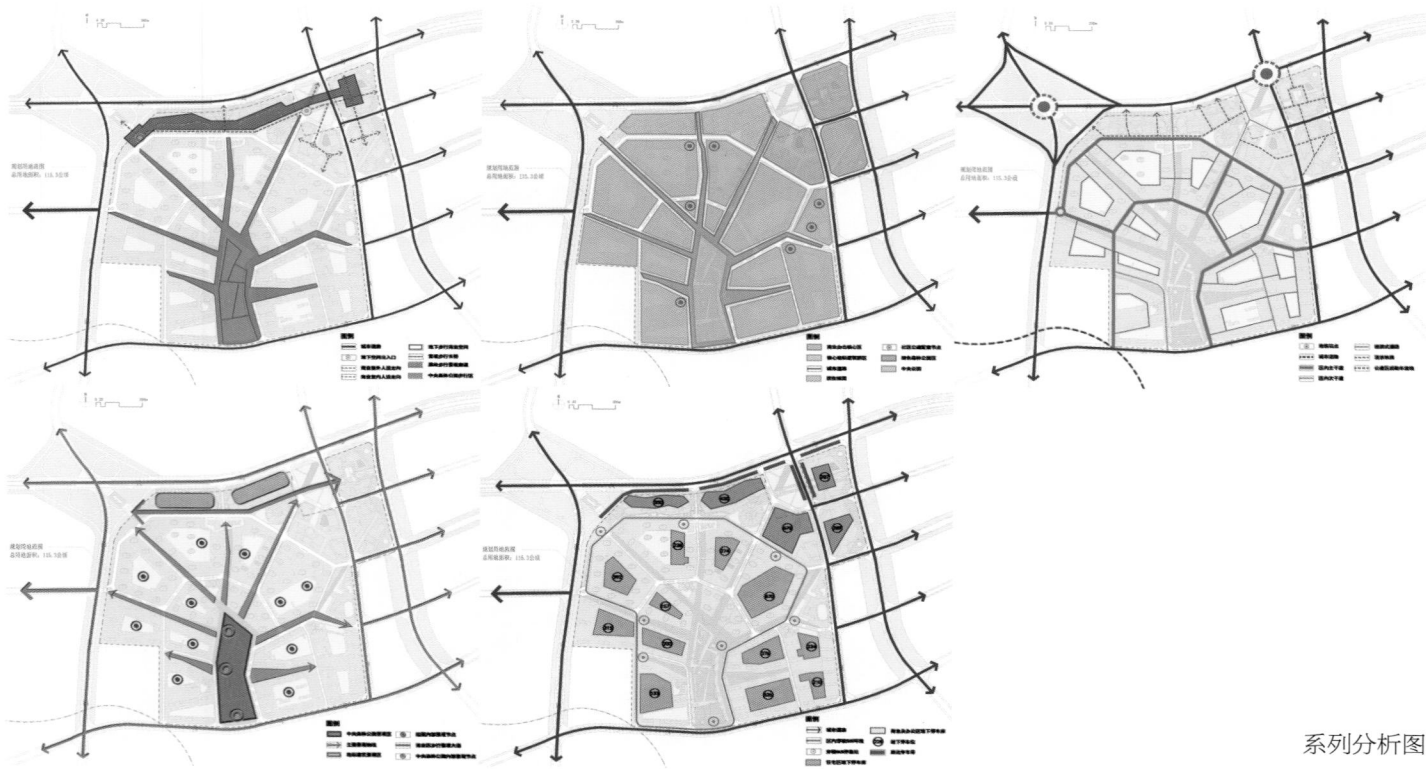

系列分析图

规划项目位于体育新城南侧，与体育新城一路之遥，但基地内部及周边区域生活配套设施较为匮乏，居住氛围不够浓厚。

规划目标在于将项目打造成中国生态健康城的标杆，并根据整体区域特点和优势区位，依托市级体育场馆和良好交通环境打造集商业办公、商务会展、金融贸易、文化娱乐、体育休闲、居住生活六大功能为一体的，体现大连风情、滨海特色的生态型、园林化、复合型、多元化的城市混合区域。

依据目标和定位侧重点的不同，规划提出两个方案进行比选。

"大连森林"方案以生态、和谐和可持续发展理念为指导，将森林、果园、花园、绿色等自然元素融入项目建设，以中央森林公园为核心向外辐射形成多轴渗透的指状空间结构，营造出自然、生态的环境特色。

该方案中规划可开发用地为 101.31 万平方米，总建筑面积为 237.5 万平方米，可容纳 5.06 万人。

The project is located in the south of Sports New City, a stone's throw from Sports New City , but the surrounding life facilities are inadequate and life atmospere is not strong enough.

Planning objectives of the project is to establish it as the model of Chinese ecological health city, according to regional characteristics and advantages of the overall location, relying on the municipal sports stadium and good traffic environment, building mixed urban areas integrating six major functions commercial office, commercial exhibition, finance and trade, culture and entertainment, sports and leisure, residentice as one of life, reflecting an ecological, gardenesque,complex and diverse urban area with Dalian style and coastal features.

Based on different objectives and positioning, the plan puts forward two options for comparison.

"Dalian Forest" uses ecological, harmonious and sustainable development concept as guidelines, mixing the forests, orchards, gardens, green and other natural elements into the project, using the Central Forest Park as the core to form multi-axis penetration of radiation finger space structure outward, creating a natural and ecological environment characteristic.

The planned land area is 1,013,100 m² with a total construction area of 2,375,000 square meters which can accommodate 50,600 people.

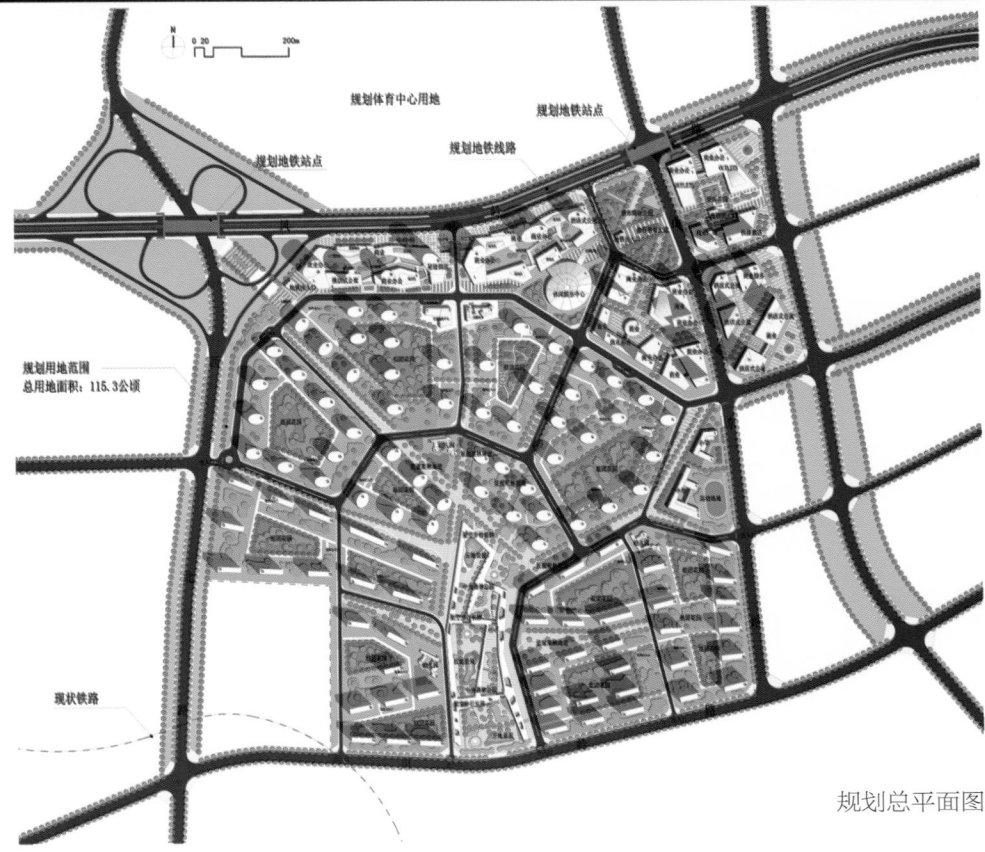

规划总平面图

无锡新区超大规模集成电路产业园概念规划
Wuxi New District Ultra-large Integrated Circuit Industrial Park Concept Planning

项目名称：无锡新区超大规模集成电路产业园概念规划
项目地点：中国 / 江苏 / 无锡
建筑面积：166.8 万平方米
编制时间：2010 / 07
业主：无锡新区管委会
规划类型：规划方案

项目位于无锡新区北部，距离无锡机场 15 公里。北起泰山路、西至锡仕路，东临沪宁高速公路，南至新二路。

规划以形成产业配套合理、市政配套完备、生活配套全面的综合性园区为目标，打造拥有完备便捷生活服务设施和高品质景观休闲空间的无锡新区超大规模集成电路产业园。在园区功能设置上主要以超大规模集成电路制造功能为核心，囊括产品研发、设计、孵化、培训、商务等功能。

规划以"生态"为核心理念，在规划设计中强调土地利用、绿色交通、生态建设、节能减排和资源利用板块间的有机联系，围绕五大板块构建物质空间各子系统，并从思想、技术和管控上进行生态应对，提出具体的解决思路。

规划区总建筑面积为 166.8 万平方米，其中现状建筑面积 75.10 万平方米，规划新增建筑面积 91.7 万平方米，地块平均容积率为 0.83。

The project is located in the north of Wuxi New District, 15 km away from Wuxi Airport. It ranges from Taishan Road in the north, west to Tin Shi Road, east to Shanghai-Nanjing Expressway, south to Xin'er Road.

The planning aims to form a comprehensive park integrated with reasonable supporting industries, complete municipal supporting, fully life support zone, building a comprehensive and convenient service facilities and high quality leisure space landscape VLSI Wuxi New District Industrial Park. Features of the park center on the ultra-large integrated circuit manufacturing capabilities, including functions like product development, design, incubation, training and business.

The planning uses "ecology" as the core concept, emphasizing organic links in planning land use, green transportation, ecological construction, energy conservation and resource use between the plates, building five major sections of the subsystem physical space and making specific solutions from the ideas, technologies and control on the ecological response.

Total the planning building capacity is 1.668 million square meters, including building area of 751,000 square meters, planning new construction area of 917,000 square meters, and average floor area ratio of 0.83.

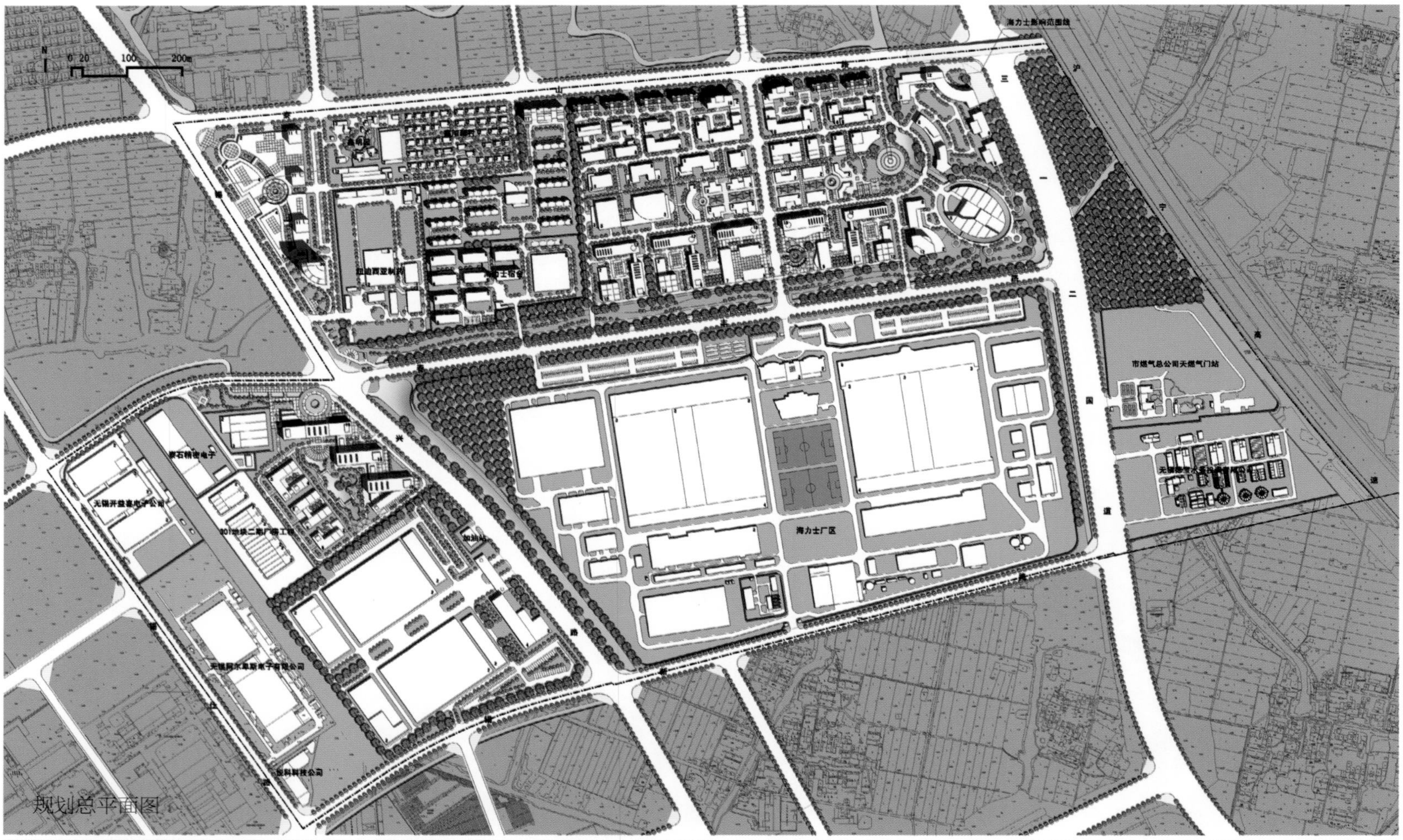

规划总平面图

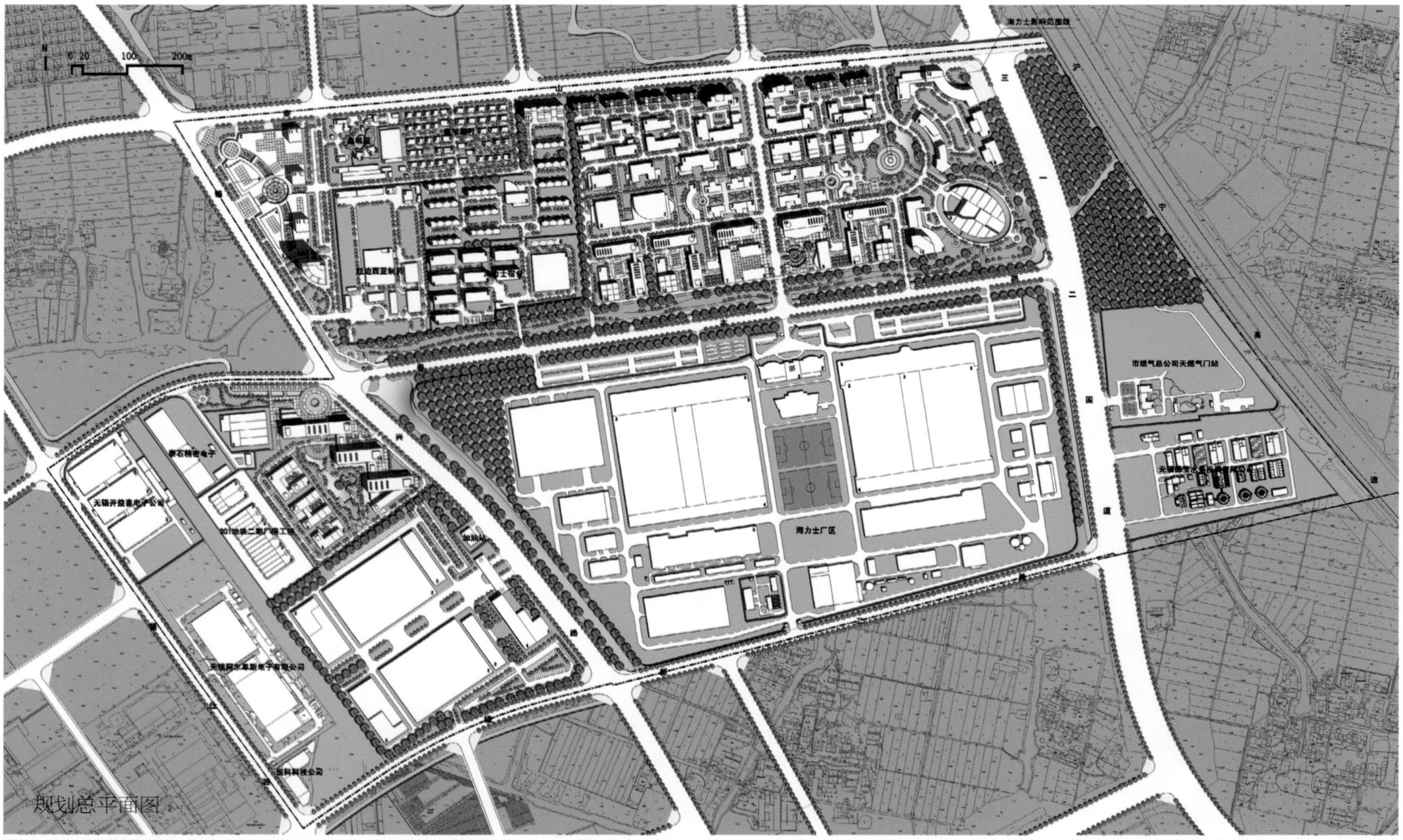

规划总平面图

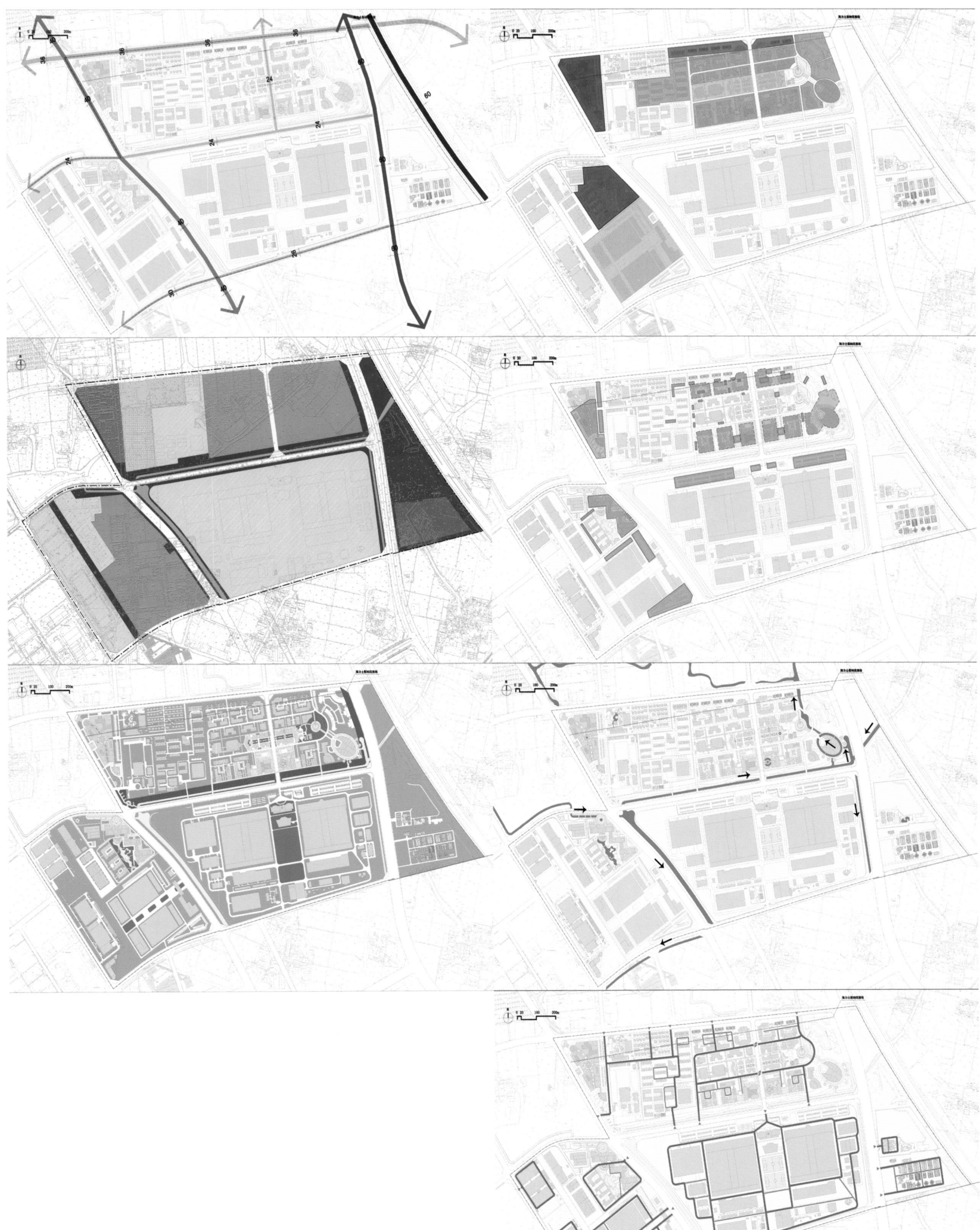

系列分析图

杭州铁路东站枢纽地区城市绿地景观系统概念规划
Hangzhou East Railway Station Urban Green Space Landscape System Conceptual Design

项目名称：杭州铁路东站枢纽地区城市绿地景观系统概念规划
项目地点：中国 / 浙江 / 杭州
建筑面积：1 000 万平方米
编制时间：2008 / 07
业主：杭州铁路及东站枢纽建设指挥部、杭州市规划局
规划类型：城市设计

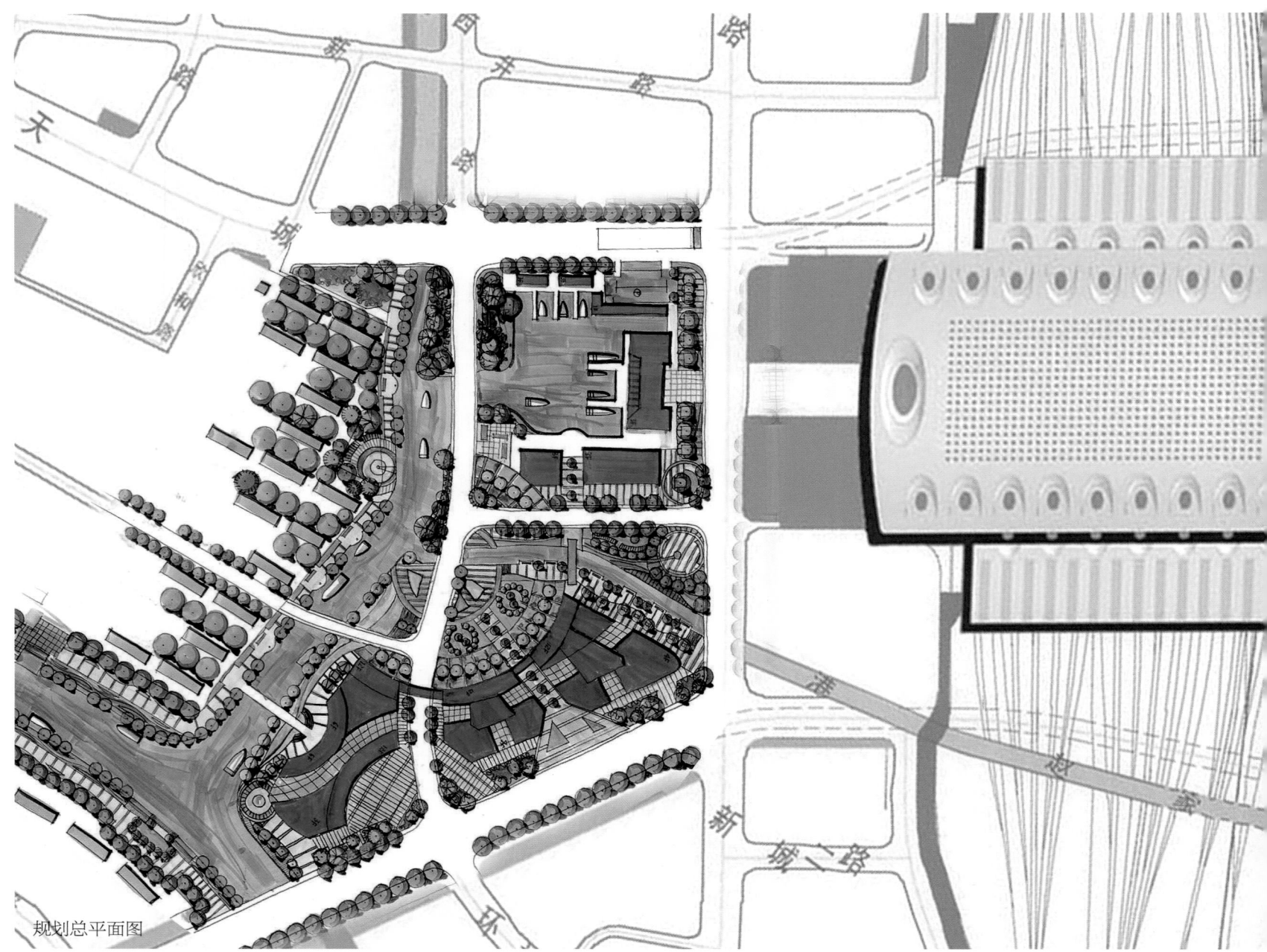

规划总平面图

杭州未来的新东站将整合各种地面长途交通系统，集客运专线、城际铁路、磁浮交通、干线铁路、地铁、客运、公交、运河码头等多种交通形式和配套服务设施于一体，成为浙江省规模最大的现代化综合交通枢纽中心，杭州未来陆上交通的重要门户。

规划基地不仅是杭州未来最重要的交通节点，形成多中心格局的又一城市中心，其更是杭州未来最为重要的景观门户、展示杭州的直接窗口。因而，如何能打造具有"城东新城"特色又能反映杭州城市性格的门户景观是本次规划思考的又一个重点。

The new Hangzhou Eastern Railway Station will integrate different ground long-distance transportation systems of passenger line, intercity rail, maglev transportation, trunk railway, subway, passenger transport, bus, canal terminals and other forms of transport and service facilities into one, becoming the largest modern integrated transport hub in Zhejiang Province, as well as an important gateway to the future land transportation in Hangzhou.

The planning site is not only an important transportation node to build another multi-center pattern of urban centers, but also the most important landscape portal to future Hangzhou. Thus, how to build a "Eastern New Town" with feature and portal landscape is an important thinking of our plan.

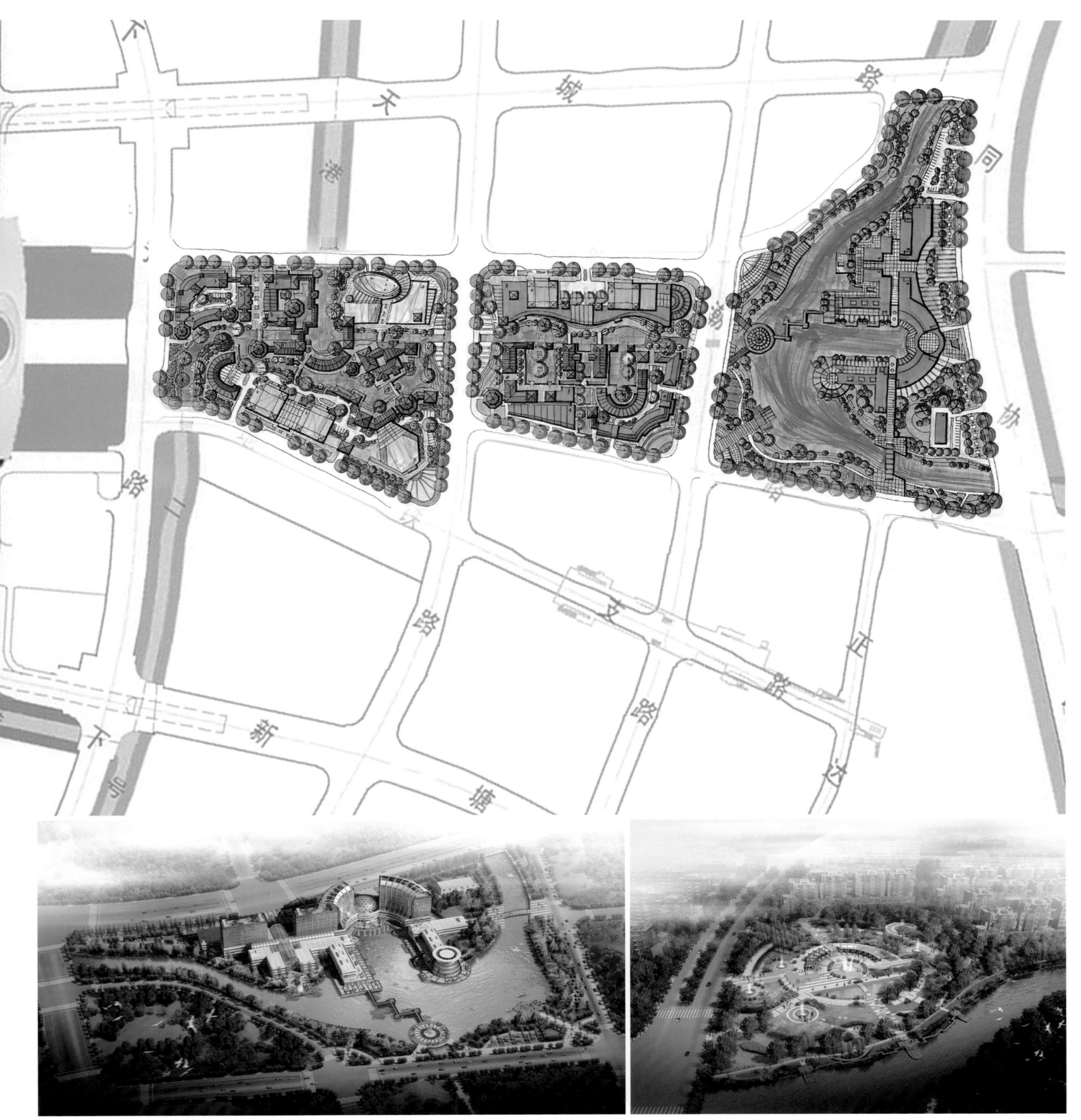

607

城际轨道交通"翠亨站"片区规划设计
Intercity Rail Transit "Tsuiheng Station" Planned Design

项目名称：城际轨道交通"翠亨站"片区规划设计
项目地点：中国 / 广东 / 中山
建筑面积：235 万平方米
编制时间：2008 / 01
业主：远洋地产（中山）开发有限公司
规划类型：概念规划

规划总平面图

南蓢镇区以规划翠亨站站点为核心形成 "翠亨站"片区。其用地范围包括：北至逸和路，东、西、南至规划路，总用地规模 2.35 平方公里。南蓢镇位于珠三角所确定的聚合区域核心功能的区域发展 "脊梁"上，是中山市东部新城的行政、商业、科研教育中心，以高新技术研发、制造职能为主，大力培育区域性居住职能。未来南蓢镇发展 "以旅游产业为主导，以文化产业为支柱，发展创意产业、高科技产业和综合服务业的滨海生态型文化新城"。至 2020 年南蓢镇城镇建设用地规模约 40 平方公里，人口规模约 30 万人。

"Tsuiheng Station" is north to Yihe Road, east, west, south to the planning road, with a total site area of 2.35 square kilometers. It is the administrative, commercial, research and education center of Zhongshan East New Town, mainly to cultivate the regional housing functions. The future development of the Nan Lang Town is taking the tourism industry as the lead, the cultural industry as the pillar, to develop creative industries, high-tech industry and comprehensive services in the coastal eco-cultural town. By 2020, the scale of urban construction land will be about 40 square kilometers with a population size of about 0.3 million people.

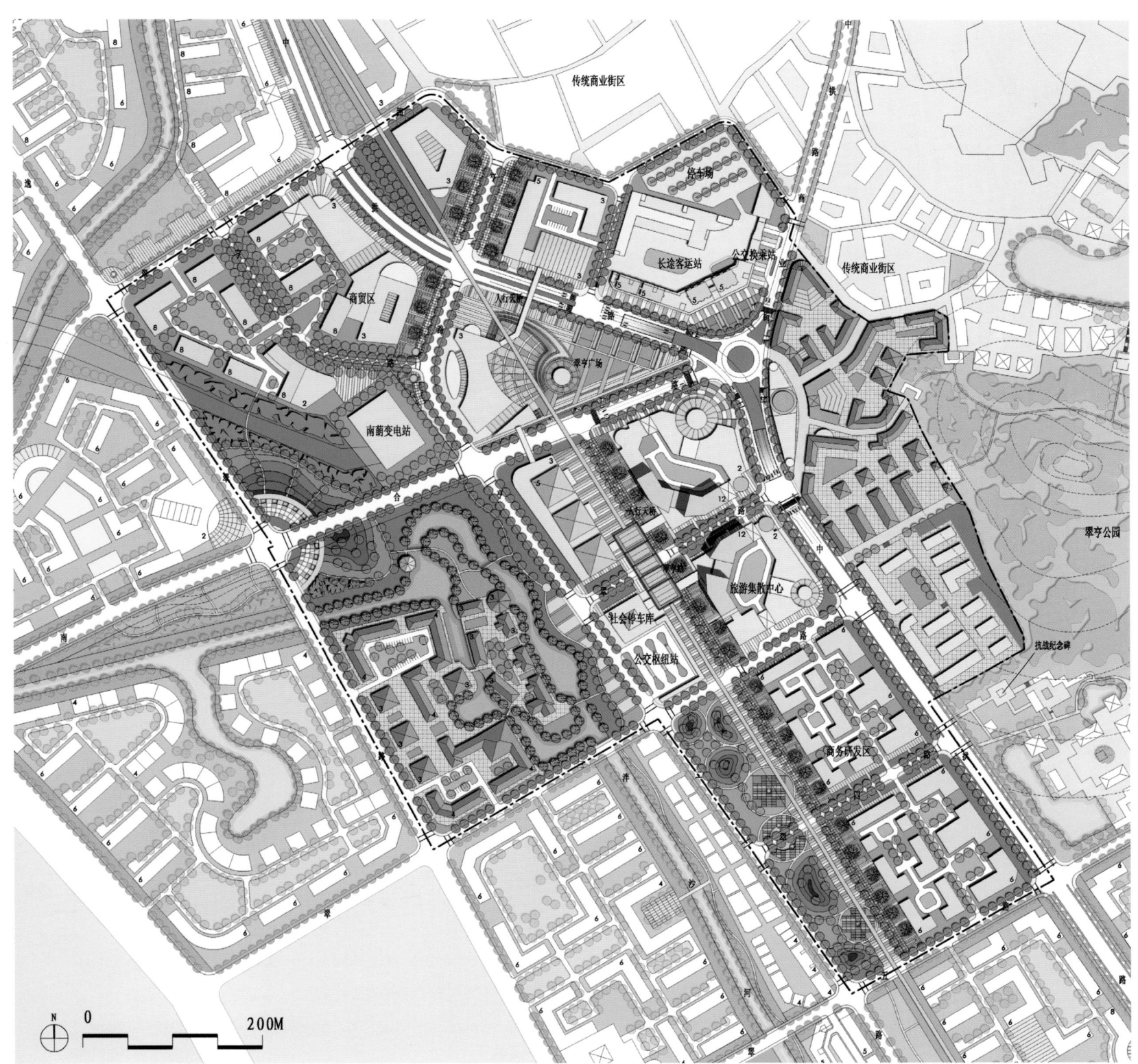

传统商业街区

停车场

长途客运站　公交换乘站

传统商业街区

商贸区

MTR站

翠亨广场

南葡变电站

翠亨公园

水行天桥

旅游集散中心

抗战纪念碑

社会停车库

公交枢纽站

商务研发区

N 0　　　　200M

站点核心区规划总平面图

芜湖东方蓝海大砻坊周边地块总体规划
Wuhu Oriental Blue Ocean Dalongfang Surrounding Plot Overall Plan

项目名称：芜湖东方蓝海大砻坊周边地块总体规划
项目地点：中国 / 安徽 / 芜湖
建筑面积：158.7 万平方米
编制时间：2009 / 03
业主：芜湖东方蓝海房地产开发有限公司
规划类型：城市设计

规划总平面图

规划基地位于芜湖城市东部，是芜湖城市东扩战略的重要战略点。

规划在 158 万平方米的项目用地内，布局近 50 万平方米的保障性住房、150 万平方米的中高档商品住房和近 30 万平方米的商业建筑，并针对商业和居住功能提出三个规划原则与目标。

商业功能——形成等级结构清晰的商业街道网络；创造不同类型商业形态，增强吸引力；化整为零、模数设计，灵活面对市场需求。

居住功能——形成产品类型划分清晰、组团完整、便于分期开发居住地块。

地区形象——化零为整，形成完整、标志性地区空间形象。

The planning base is located in the east of Wuhu city and it is an important strategic point of Wuhu city's eastward expansion strategy.

Inside 1,580,000 square meters of project area, the plan layouts of 500,000 square meters affordable housing, 1,500,000 square meters commercial residential building and 300,000 square meters commercial buildings and proposes three planning principles and objectives according to commercial and residential functions.

Business function-forming a clear hierarchy of commercial streets network; enhancing the attractiveness of different types of business forms; decentralizing, modular design, flexible to the market demand.

Housing function-forming a residential area with clear product type, group integrity, convenient for development in phases.

The regional image-making parts into a whole, forming a complete and landmark image of the region.

北部生活社区
东部生活社区
西部生活社区
中部生活社区
南部生活社区
N
0
200M

图例
城市主干道
城市次干道
区主路
区支路
限时商业步行街
公交总站
N
0
200M

简约的新古典风格
稳重的新古典风格
现代与传统相对比的风格
时尚的新古典风格
地区景观轴线
图例
次要商业街道
童装玩具特色商业街
(兼北部社区商业中心职能)
文化体育特色商业街区
(兼东部社区商业中心职能)
文化旅游特色商业街区
地区商业核心区
主要商业街道
中部社区商业中心
西部社区商业中心
规划建议打造形成景观联系道路
次要商业街道
南部社区商业中心
亲水休闲特色商业街区
N
0
200M

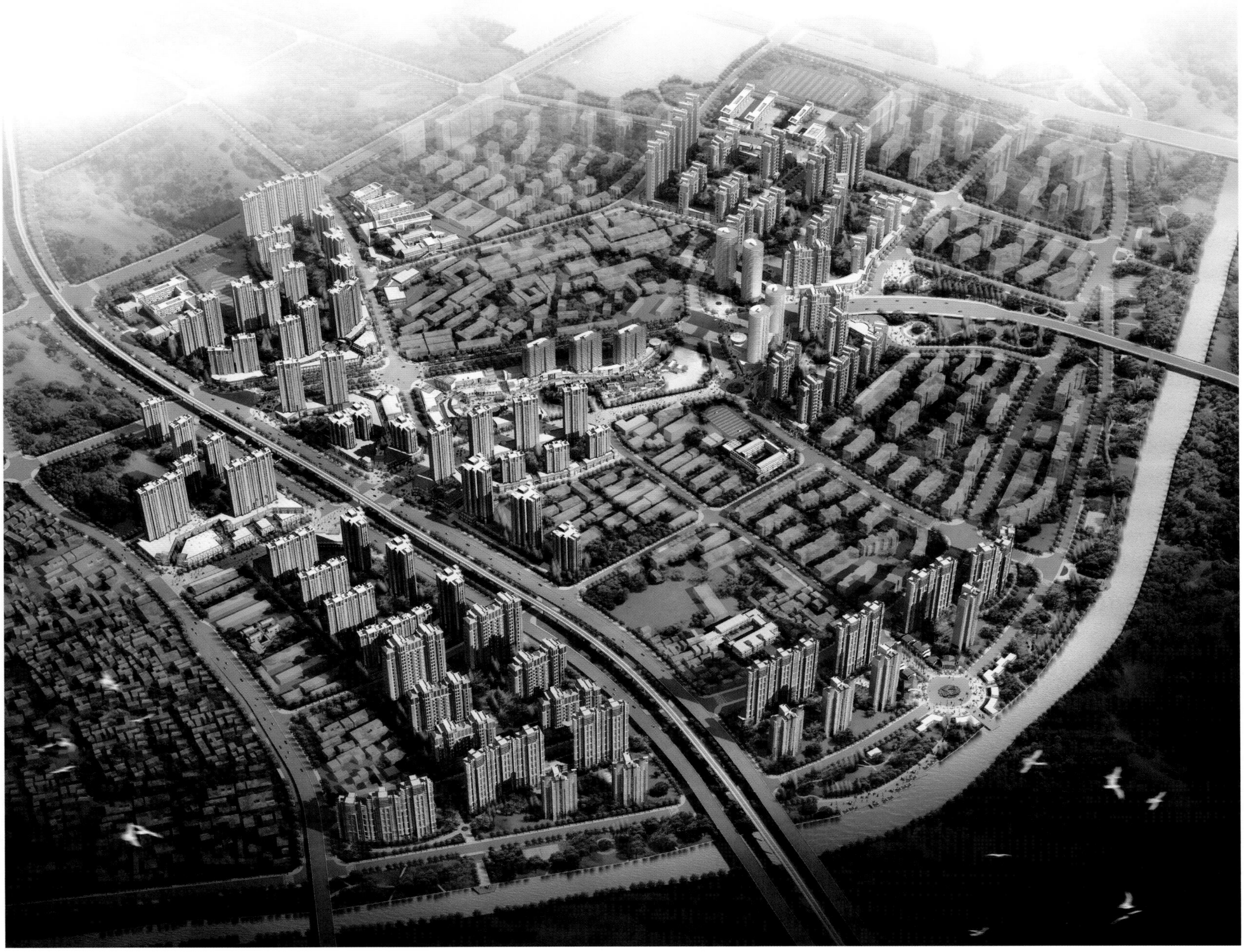

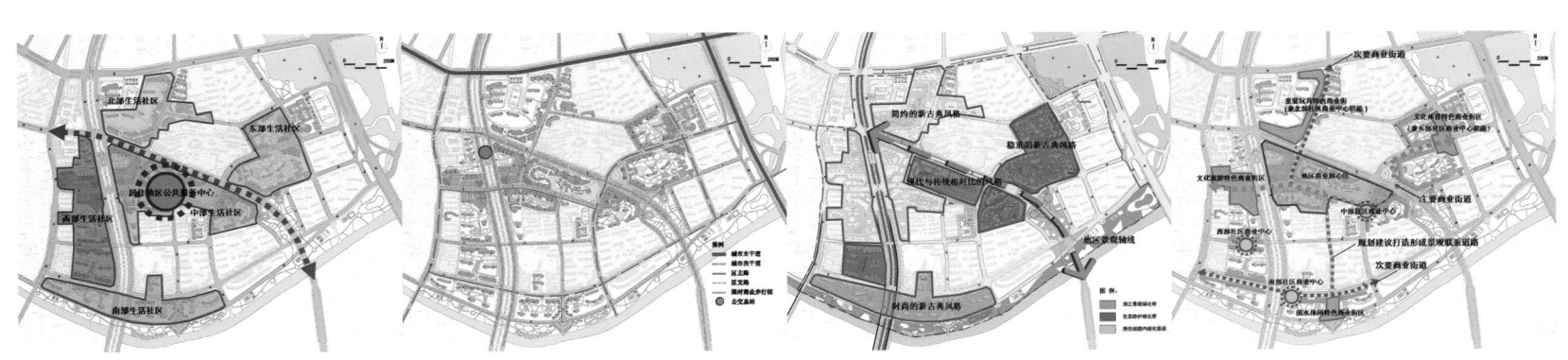

上海市轨道交通 5 号线环城西路站地区开发概念规划与城市设计
Shanghai Metro Line No.5 Western Station Regional Development Conceptual Plan and Urban Design

项目名称：上海市轨道交通 5 号线环城西路站地区开发概念规划与城市设计
项目地点：中国 / 上海
建筑面积：140 万平方米
编制时间：2007 / 02
业主：上海市奉贤区轨道交通指挥部
规划类型：城市设计

规划总平面图

项目位于上海奉贤南桥新城西北角，其上一站到奉浦站，下一站至南桥新城站。

本次规划分站点地区和核心地区两个层次进行设计。站点地区规划范围为：东到南桥路，西到环城西路，南至浦南运河，北至国顺路；核心区规划范围为：东到秀南路以东，西到环城西路，南至运河北路，北至国顺路。

环城西路站点地区功能定位为"上海市远郊新城辅助站"，即以环城西路为中心的站点地区，将逐步发展成为南桥新城北部地区多功能综合中心，服务整个新城西部和北部区域。规划在结合国内外站点地区开发经验的基础上，对整个南桥新城西北地区总体结构作了一定的调整，同时在站点核心区范围对城市空间布局、城市风貌作了进一步的规划控制。

This project is located in the northwestern corner of Nanqiao New City, Fengxian District, Shanghai, the last station to Fengpu station, next to the Nanqiao New City station.

The planning is divided into two levels: station area and the core area. The station area covers: east to the Nanqiao Road, west to West Huancheng Road, south to Punan Canal, north to Guoshun Road; the core area ranges from east to Xiunan Road, west to West Huancheng Road, south to Punan Canal, north to Guoshun Road.

The function of West Huancheng Road station area is a "suburb of Shanghai Metro auxiliary station", which is using Huancheng Road area as the center of the site to gradually develop a multi-functional integrated northern Nanqiao Village region, serving the west and north areas of the entire New City. Based on regional development experience in domestic and international sites, designers make some adjustments on the overall structure of Nanqiao New City northwest region, while further controlling and planning the site layout of urban space, and urban landscape of the core area.

N
0 25 50M

环城路
国 顺 西 路

社会停车场

北部商务办公中心
公交换乘站
出租车站点
公交换乘站
工业综合展览中心
北广场
人行廊道

轨道交通5号线
(通往南桥新城站)

轨道交通5号线
(通往奉浦站)

环城西路站

团 南 公 路

高架人行廊道

规
划
西
路

规
划
路

南广场

出租车站点

商业服务中心

商务
商业
商务办公

自行车专用道

南部商务办公中心

住宅组团

南
沿街商业
民旺苑小区(在建中)

商业中心
公交换乘站
入口广场
住宅
幼稚园
入口

站点核心区规划总平面图

国顺西路
团南公路
规划二路
规划一路
轨道站屋

规划一路
规划二路
团南公路
国顺西路
轨道站屋

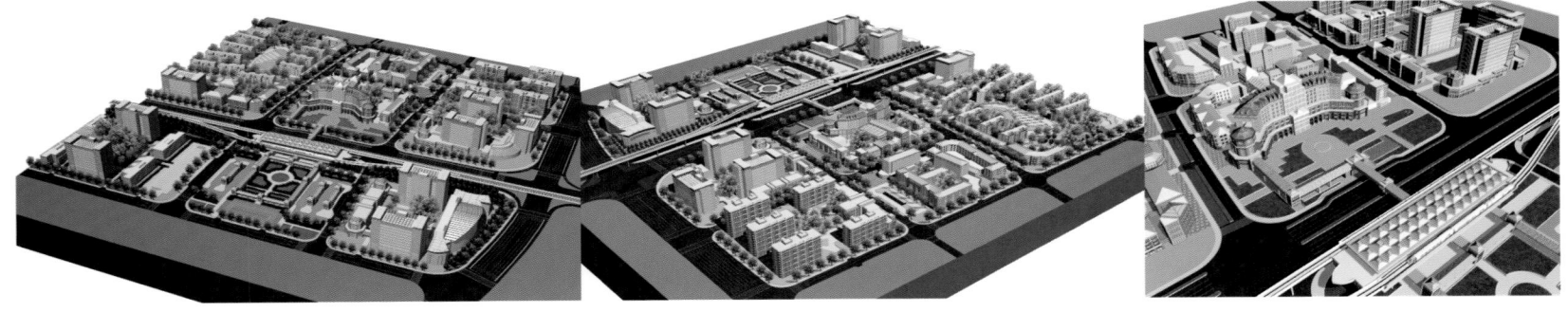

虹桥综合交通枢纽地区概念性详细规划及重要地区城市设计
Detailed Planning of Hongqiao Integrated Transport Hub and Major Areas Urban Design

项目名称：虹桥综合交通枢纽地区概念性详细规划及重要地区城市设计
项目地点：中国 / 上海
建筑面积：2 630 万平方米
编制时间：2007 / 04
业主：上海申虹投资发展有限公司
规划类型：城市设计

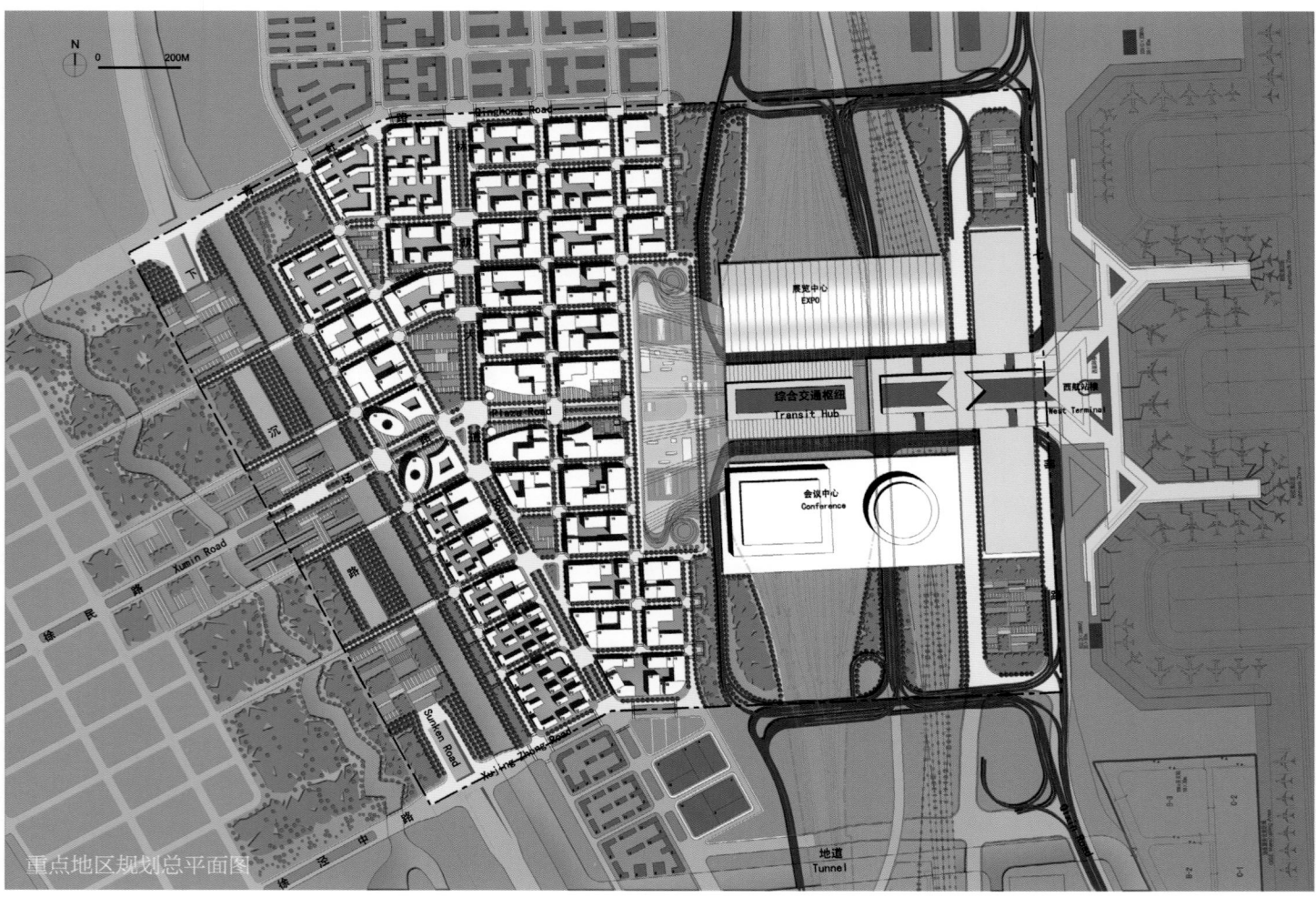

重点地区规划总平面图

为进一步促进长三角地区社会经济快速发展，增强上海城市的综合服务功能，上海市政府经与铁道部和民航总局共同研究，提出在虹桥机场西侧建设虹桥综合交通枢纽的战略构想。枢纽将包括虹桥机场新航站楼、京沪高速铁路车站，沪杭磁浮线路也将引入该枢纽并连接浦东机场，同时设置城市轨道交通、地面公交、服务长三角的城际轨道交通、长途高速大巴等，充分满足不同服务范围、不同交通方式之间的换乘需求。

本次规划包括：1. 区域结构研究以及土地使用布局，范围为 26.3 平方公里；2. 概念性详细规划，范围为 12.4 平方公里；3. 重点地区城市设计，约 2.7 平方公里。

本次规划与美国 SOM 公司密切合作，提出九条操作规则，贯穿于整个规划合作单位过程中，体现了先进的规划理念和对城市演变的深刻理解。

In order to further promote the rapid development of the Yangtze River Delta economic region and enhance Shanghai's comprehensive service function, Shanghai Municipal Government and the Ministry of Railways and Civil Aviation Administration, put forward strategic concept of establishing Hongqiao integrated transport hub at the western side of Hongqiao Airport. Hongqiao Airport hub includes a new terminal building for Hongqiao Airport, the Beijing-Shanghai Express Railway and Shanghai-Hangzhou maglev line will also be introduced into the hub and link the Pudong airport. It sets the urban rail transportation, ground transportation, services, intercity rail transport into the Yangtze River Delta, long-distance high-speed bus, etc., to fully meet the different ranges of services, transfer needs between different modes of transport demand.

The plan is divided into three levels: 1. land-use layout and regional structure research. The range is 26.3 square kilometers; 2. detailed conceptual planning. The range is 12.4 square kilometers; 3. key areas urban design, about 2.7 square kilometers.

We work closely with SOM and propose nine rules within the entire planning process, reflecting the advanced planning concepts and a deep understanding of the city's evolution.

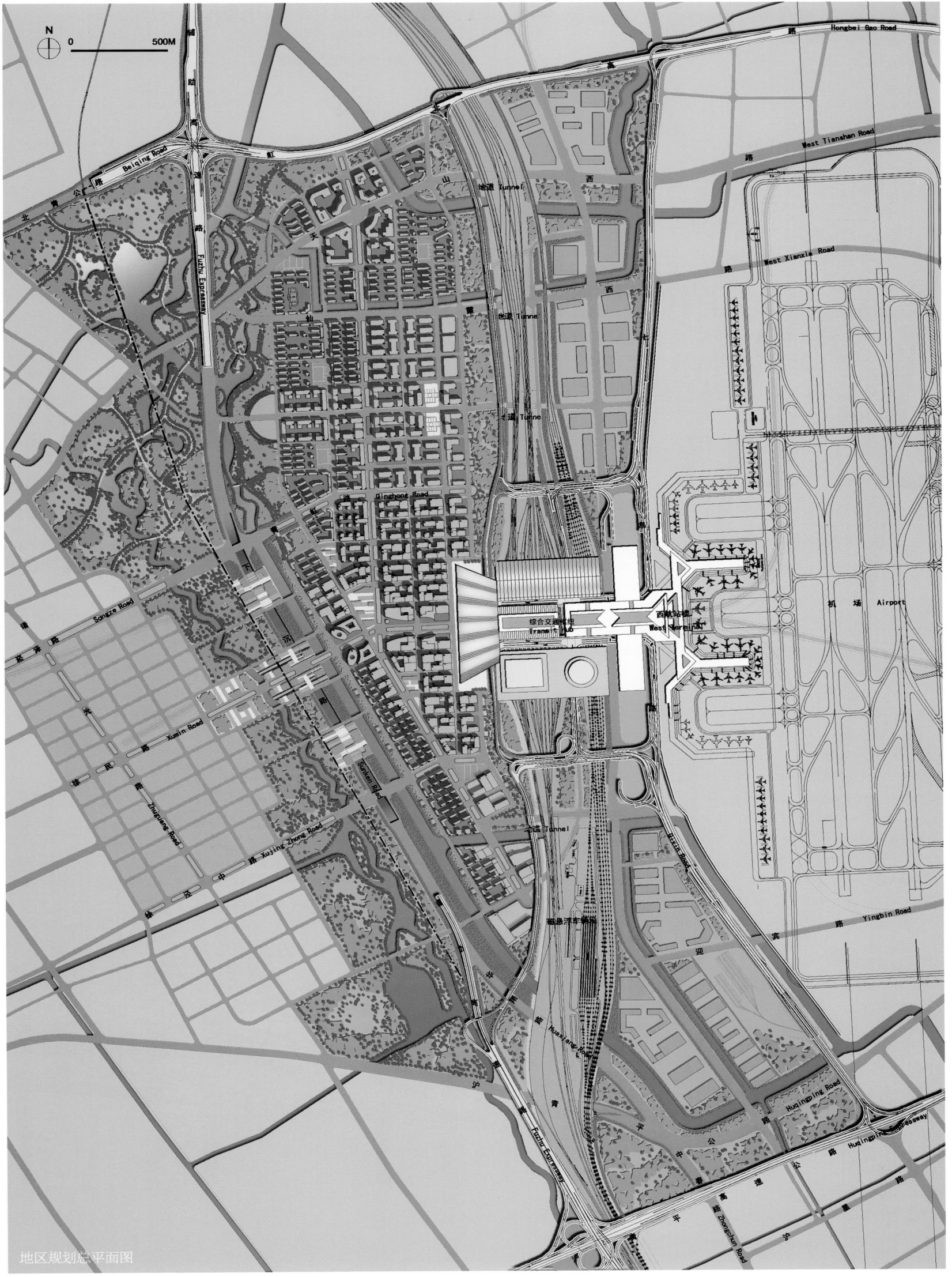

N
0   500M

Hongbei Gao Road

West Tianshan Road

West Xianxia Road

机场 Airport

地道 Tunnel

西藏站楼
West Terminal

综合交通枢纽
Hub

Beiqing Road

北青公路

虹桥辅道

Fuzhu Expressway

天山路

Dinghong Road

Songze Road

Xumin Road

Sunken Road

Zhuguang Road

Xujing Zhong Road

物流 Tunnel

迎宾高速车道桥

Yingbin Road

Huaxiang Road

Huqingping Road

Fuzhu Expressway

Huqingping Expressway

Zhongchun Road

地区规划总平面图

615

新疆·克拉玛依·乌尔禾区总体规划及整体城市设计
West Karamay New City Conceptual Master Plan

项目名称：克拉玛依市西部新城概念性总体规划
项目地点：中国 / 新疆维吾尔族自治区 / 克拉玛依
建筑面积：2 200 万平方米
编制时间：2007 / 10
业主：克拉玛依城市规划管理局
规划类型：概念规划

乌尔禾区位于"油城"克拉玛依市东北部，以世界魔鬼城著称。

2003 年，乌尔禾区委、区政府结合乌尔禾资源和地缘等优势，提出将乌尔禾区打造成"生态旅游新城"的发展目标，为乌尔禾今后的发展指明了方向。作为克拉玛依城市发展战略的一部分，规划提出在乌尔禾建设集生活、工作、休闲、旅游、度假于一体的"生态旅游新城"，将其打造成我国北疆地区的旅游度假胜地及旅游集散地、克拉玛依城市后花园和城市北部门户形象、体现社会和谐、生活富足，且兼具北疆独特地域风貌的绿洲度假休闲小镇。

在规划定位基础上，规划提出了相应的城镇发展战略，并分三个阶段实现。

阶段一：加强巩固城区中心建设，打破管理壁垒，在乌尔禾城区、风城油田作业区乌尔禾前线生活基地和 137 团之间形成统一的管理机制。

阶段二：向北发展旅游产业，在现有旅游资源基础上，提升乌尔禾乡农家乐档次，增设大型旅游项目，形成旅游集散中心，建设品牌旅游产品。

阶段三：向南沿白杨河发展生态休闲度假产业，建设滨河主题公园、高尔夫球场及生态住宅开发，提升环境品质，吸引外来资本，平衡政府资金。

规划乌尔禾区城镇总建筑面积为 2 200 万平方米，平均容积率为 0.62，可容纳城镇人口 5 万人。

Urho District is located in the northeast Karamay "Oil City" known as world-famous Ghost City.

In 2003, Urho district government, combined with Urho's resources and geographical advantages, proposed development goals of converting Urho area to an "eco-tourism Park", which pointed out the direction for future development. As part of the Karamay city development strategies, the plan proposed to construct "eco-tourism Park" integrated by living, work, leisure, travel, vacation. This will make Karamay city a resort and travel center of the northern area, back garden and northern gate of Karamay city image and a harmonious and prosperous northern oasis town of vacation and leisure with a unique regional style.

Based on the positioning in the planning, the plan put forward the corresponding development strategies and implementation in three stages.

Stage one: strengthen consolidation of city center, breaking the management barrier, to form a unified management system between the Urho City, Windy City oil field operation area and 137 Urho front groups.

Stage two: develop tourism industry to the north, based on existing tourism resources to enhance rural peasant grade, adding large-scale tourism projects to form tourism center, building brand tourism products.

Stage three: develop ecological and leisure industry along the BaiYang River in the south, construct theme parks, golf courses and eco-residential development to enhance the quality of the environment, attract foreign capital and balance government funds.

The total building area is 22,000,000 square meters, the average floor area ratio of 0.62, which can accommodate urban population of 50,000

规划总平面图

系列分析图

临海市杜桥镇城市中心区
Linhai City Duqiao Town Centre Urban Design

项目名称：临海市杜桥镇城市中心区
项目地点：中国 / 浙江 / 临海
建筑面积：200 万平方米
编制时间：2009 / 06
业主：临海市杜桥镇人民政府
规划类型：城市设计

规划总平面图

项目位于临海市东南部，紧邻台州市椒江区。临海滨海新城建设序幕的拉开，为杜桥镇进一步的跨越式发展提供了新的动力与契机。

通过对杜桥所在区域的宏观分析，规划确定在杜桥中心区内安排城镇公共服务中心，眼镜研、展、销一体区，高档山水特色住区和远景轨道站点发展区等功能。

通过对杜桥镇整体空间结构的研究，规划以金脉、银脉、地脉、文脉为线索进行城市设计，注重对城市建筑空间和城市绿化开放空间进行合理布局，同时对传统空间肌理、文脉进行研究，建议保留、改造大多数现状居民点，对重要的庙宇、祠堂进行保留，并采用传统街坊式的模式对基地内的新建建筑进行规划布局，以期形成具有当地特色、富有传统文化内涵的城市风貌。

The project is located in the southeast of Linhai City, near Taizhou City. Construction of Linhai City provided a new impetus and opportunity for the development of Duqiao town.

Through macro analysis of Duqiao region, the planning determines functions of Duqiao town as district public service center, glasses research, development, marketing integration area, high-end residential landscape features and long-term track site development area.

Through research of the overall spatial structure of Duqiao, the plan introduces gold vein, silver vein, land, culture as clues for urban design, focusing on rational distribution of open space like urban architecture and urban green space, investigating of traditional space and cultural pulse, proposing to retain important temples and ancestral halls and planning the layout with traditional texture neighborhood-style patterns to form a cityscape with local feature and rich in traditional culture.

青浦新城东区中心城市设计
Qingpu New City East District Centre Urban Design

项目名称：青浦新城东区中心城市设计
项目地点：中国 / 上海
建筑面积：185.9 万平方米
编制时间：2008 / 07
业主：上海市青浦区规划和土地管理局
规划类型：城市设计

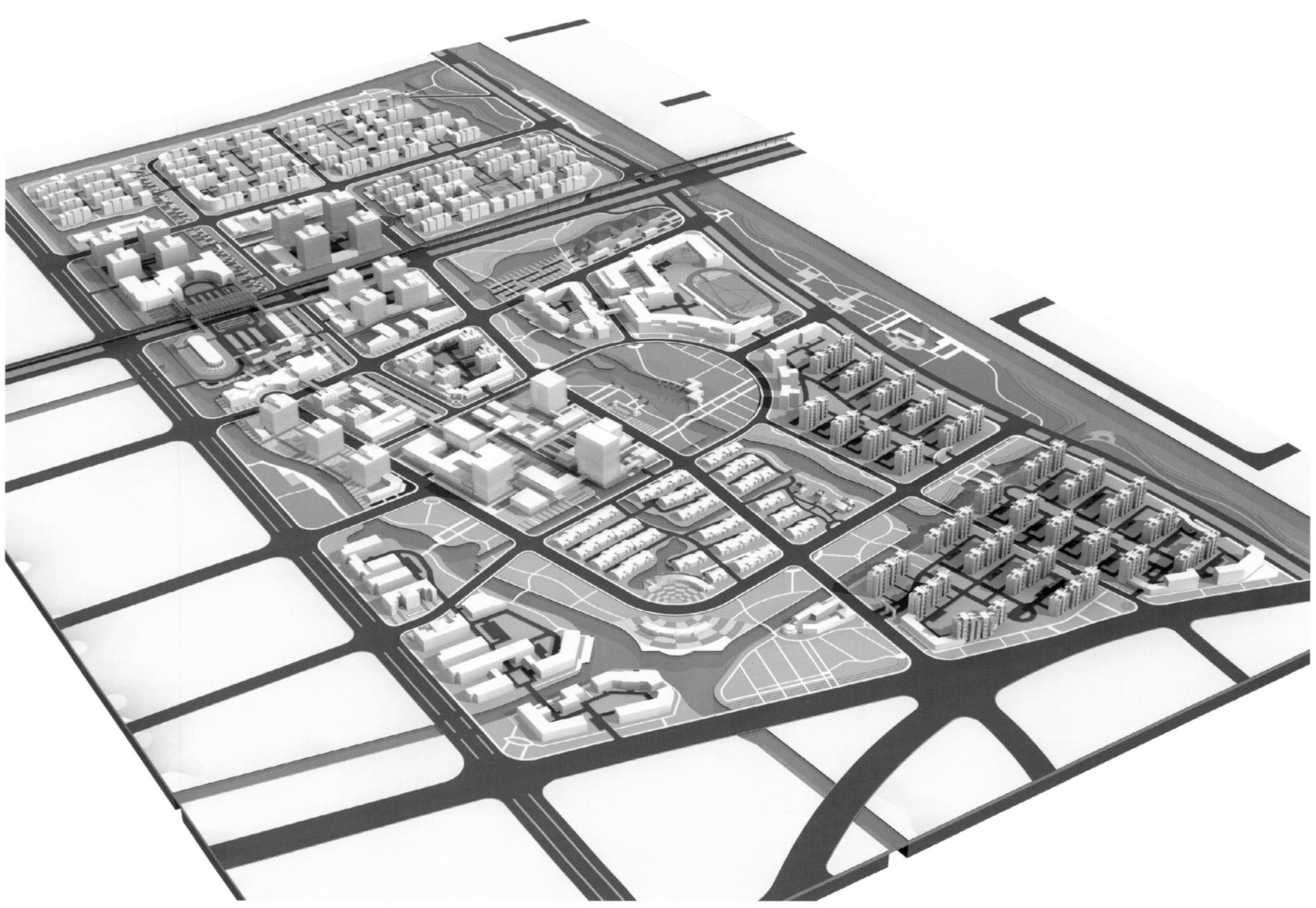

项目位于东片区东北角，具体范围：北至上达河、南至公园路、西至汇金路、东至油墩港，规划用地面积约 185.9 万平方米，占东片区总用地的 28%。

作为上海西大门，青浦位于城市东西向的沪青发展轴上，城区规模、开发强度将大大加强，发展模式和层次也会提升。其中轨道交通将作为实现这一飞越的重要促进手段和动力。

规划东片区中心的具体定位为：青浦新城轨道交通东端站点门户区、新城东片区的公共活动中心、具有"新水乡"气质的城镇中心区。在定位基础上，规划对土地使用、交通组织等方面做进一步深化与细化，对城市空间进一步研究推敲，尤其是加入了轨道交通站点对东片区发展的影响这一重要因素。

This project is located in the northeast corner of East Area. The specific range includes: north to the Shangda River, south to Gongyuan Road, west to Huijin Road, west to Youdun Port. The planned land area is about 1,859,000m² covering 28% of the total East Area.

As the west gate of Shanghai, Qingpu is located in east to west Shanghai Qingpu development axis. The urban scale and intensity of development will greatly enhance the development of models and levels, within which rail traffic will become the important means and motivation for achieving this leap.

East Area Planning can be specifically defined as: gate area of east Qingpu District Rail Transport, public activity center of the new East Area, with the "New Water" quality of the town center. Based on the positioning, the plan further deepens and refines aspects including land use, traffic organization. It further studies on the urban space, in particular, important factor that is the impact of inclusion of East Area Rail Transit Stations.

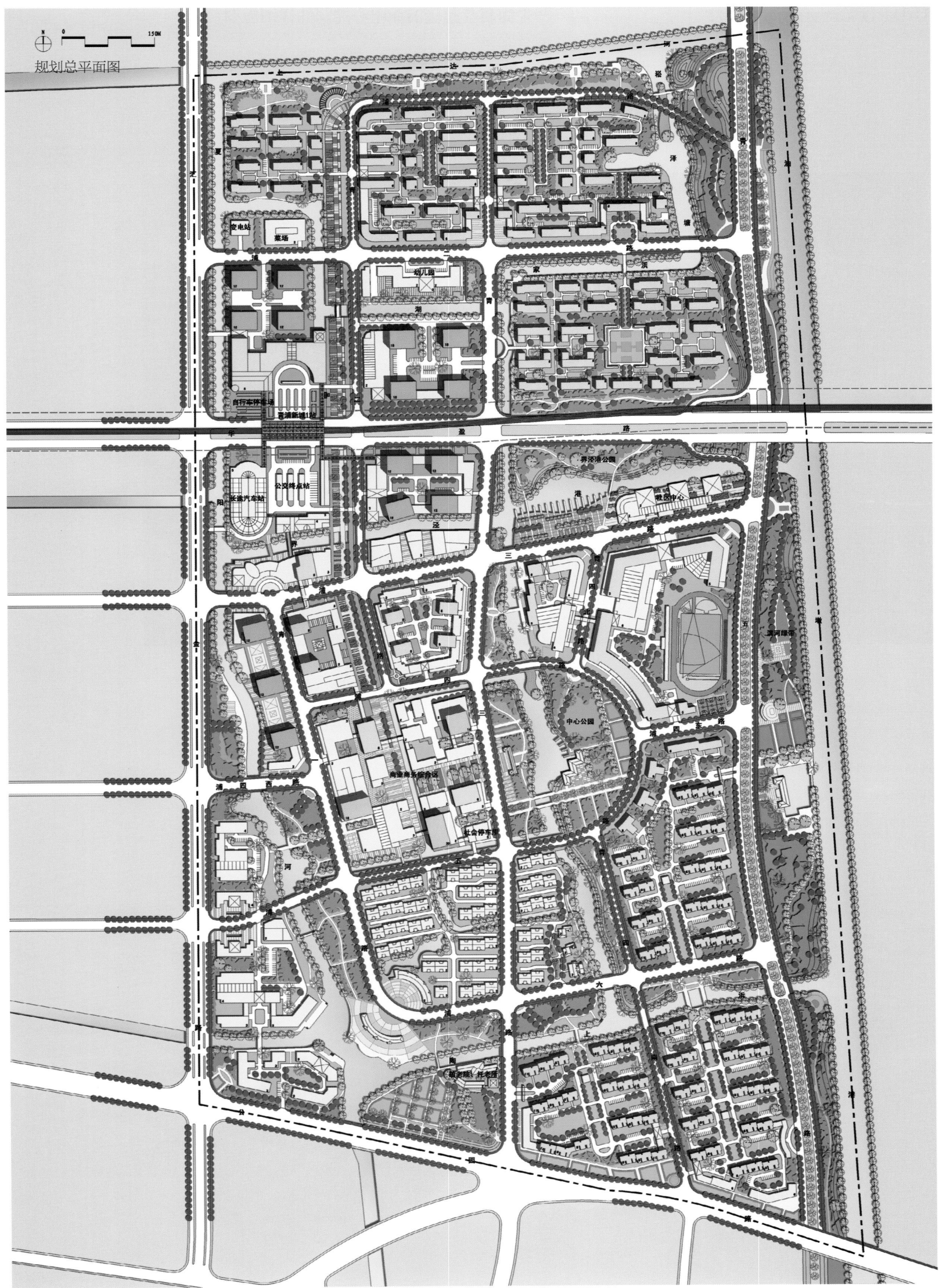

规划总平面图

上海赵巷超级商业中心控制性详细规划
Shanghai Zhaoxiang Super Shopping Centre Controlled Plan

项目名称：上海赵巷超级商业中心控制性详细规划
项目地点：中国 / 上海
建筑面积：266 万平方米
编制时间：2005 / 08
业主：上海崧泽商业投资发展有限公司
规划类型：控制性详细规划

上海赵巷超级商业中心选址于上海青浦区赵巷镇东部，318国道与方黄公路交汇处。该项目已被列入上海市 43 个现代服务业聚集区规划。

规划将其总体目标定位于：上海西部商业新形象、长三角商贸走廊、商业旅游的结合区、高尚生活的示范区、生活品质的优化区。此外，规划力求突出商业中心的文化性、多样性、综合性、便捷性、环境优美、整体结构分期实施等六大特点。

Shanghai Zhaoxiang Super Shopping Center is located at Zhaoxiang Town in Qingpu District, Shanghai, the joining point of 318 National Road and Fanghuang highway. The project has been listed among the 43 modern service industry gathering areas in Shanghai.

The overall objective of planning will be: new business image of the western Shanghai, the Yangtze River Delta trade corridor, a combination of business and tourism areas, demonstration areas of noble life, optimization area of life quality. In addition, the plan seeks to highlight the six major characteristics of the commercial center including cultural diversity, comprehensiveness, convenience, beautiful environment, the phased implementation of overall structure.

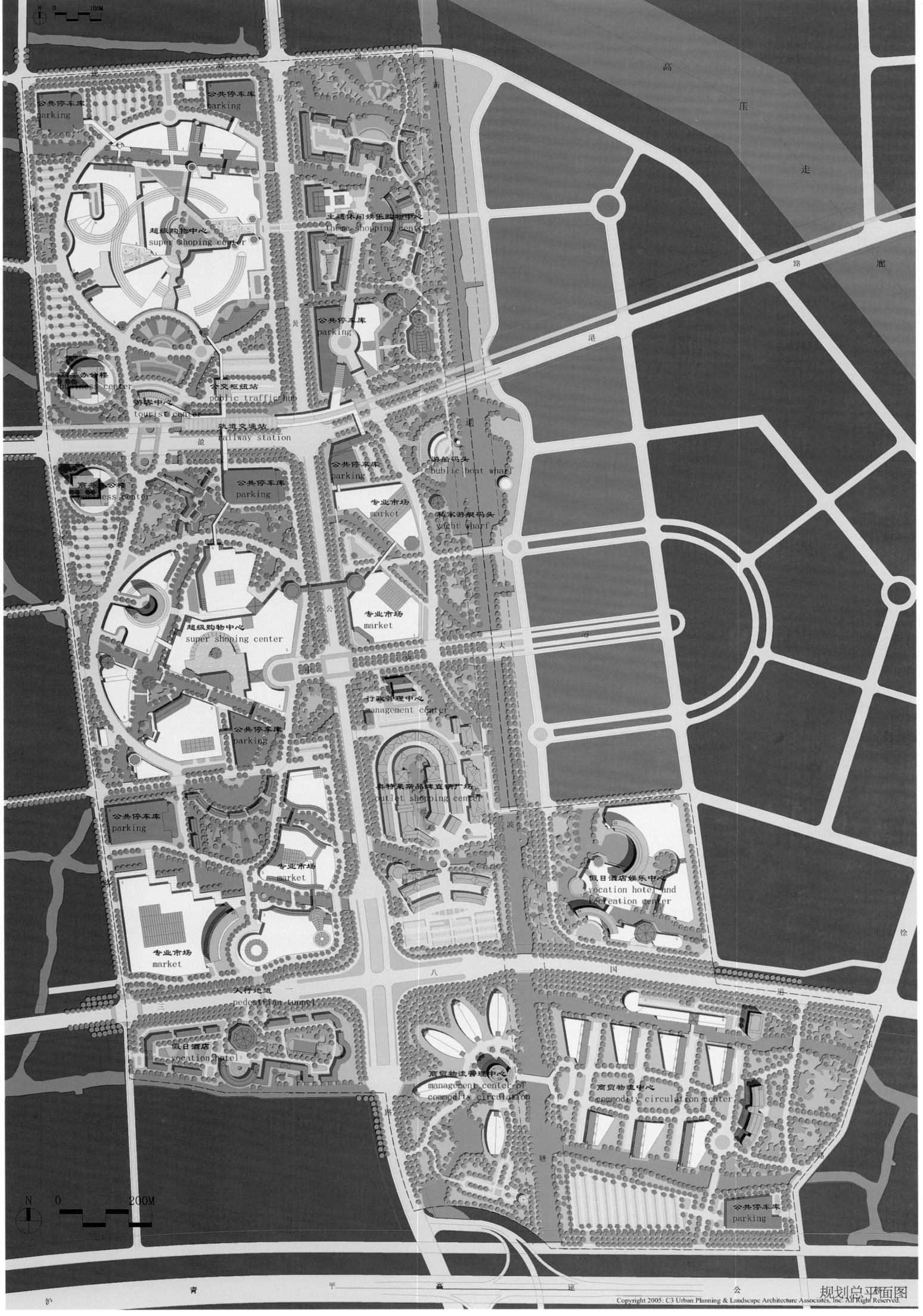

公共停车库
parking

公共停车库
parking

超级购物中心
super shoping center

主题休闲娱乐购物中心
theme shopping center

公共停车库
parking

商务办公楼
business center

公交枢纽站
public traffic hub

旅游中心
tourist center

轨道交通站
railway station

公共停车库
parking

游艇码头
public boat wharf

商务办公楼
business center

公共停车库
parking

专业市场
market

私家游艇码头
yacht wharf

超级购物中心
super shoping center

专业市场
market

行政管理中心
management center

公共停车库
parking

奥特莱斯品牌直销广场
outlet shopping center

公共停车库
parking

专业市场
market

假日酒店娱乐中心
vocation hotel and
recreation center

专业市场
market

人行地道
pedestrian tunnel

假日酒店
vocation hotel

商贸物流管理中心
management center of
commodity circulation

商贸物流中心
commodity circulation center

公共停车库
parking

规划总平面图

上海青浦区夏阳街道村落体系规划
Shanghai Qingpu District Xiayang Village Master Planning

项目名称：上海青浦区夏阳街道村落体系规划
项目地点：中国 / 上海
建筑面积：1 502 万平方米
编制时间：2006 / 02
业主：上海市青浦区夏阳街道办事处
规划类型：总体规划

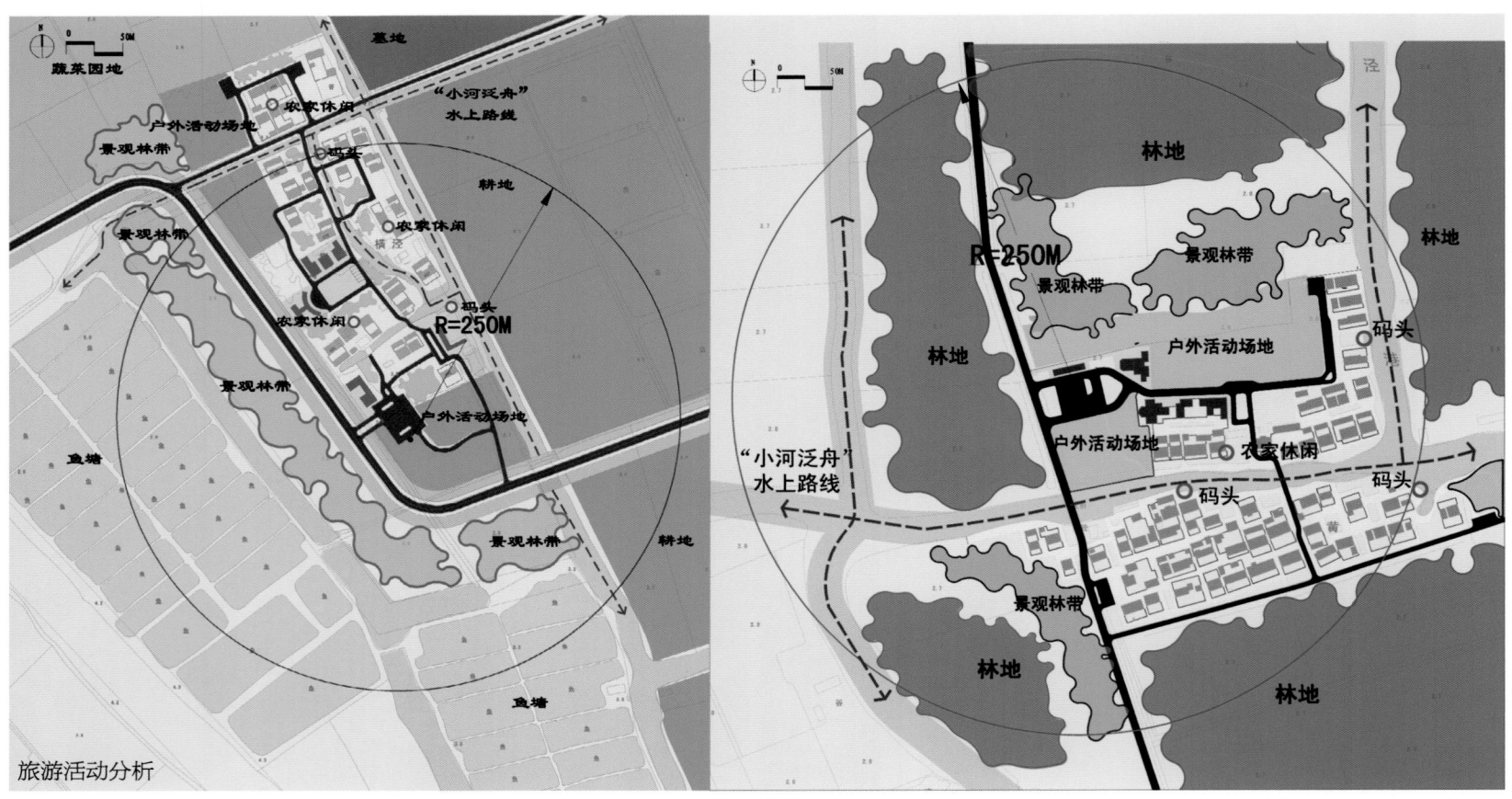

旅游活动分析

项目地处上海市青浦区的中南部，紧邻青浦新城，北至沪青平高速公路 (A50)、外青松公路一线，东南至油墩港及青浦区界，西至东大盈港一侧。

本次规划分两个层次

第一层次为夏阳街道农村地区即本规划区范围内村落体系规划。在现状调查和分析基础上，对村落体系中的土地使用、道路交通体系、产业空间布局、水系控制、公共设施布局、市政设施改造等方面做出相应的规划布局。同时对乡村整体风貌和"农家乐"旅游系统作规划控制。

第二层次为中心村和"农家乐"村的控制导则即结构性规划，包括对东西两大片区中的两个中心村和九个"农家乐"村在用地规模、人口分布、用地布局、公共设施布局、道路交通系统、风貌景观控制、开发强度控制等方面制定相应的规划控制条件。

This project is located in central and southern Qingpu area close to Qingpu New City, Huqingping Expressway (A50) to the north, external Qing Chung Road, Youdun Port and Qingpu District to the south east, west to the eastern side of Daying Port.

The planning can be divided into two levels.

The first level is the systematic analysis of Xiayang streets rural area, which is the planning area in the village. Based on investigation and analysis, the plan has arranged layout based on land use, road traffic system, industrial layout, water system control, public facilities layout, municipal facilities of the village system. Change of the layout has been made accordingly. At the same time planning and control has been carried out for the overall rural area.

The second level is the control guideline which is structural planning for village center and the "Happy Farm" village, including two patches in the East and West Village, two village centers and nine "Happy Farm" villages in the aspects of land size, population distribution, land distribution, the layout of public facilities, roads, transportation systems, landscape style control, intensity control, and so the development plan can develop appropriate control conditions.

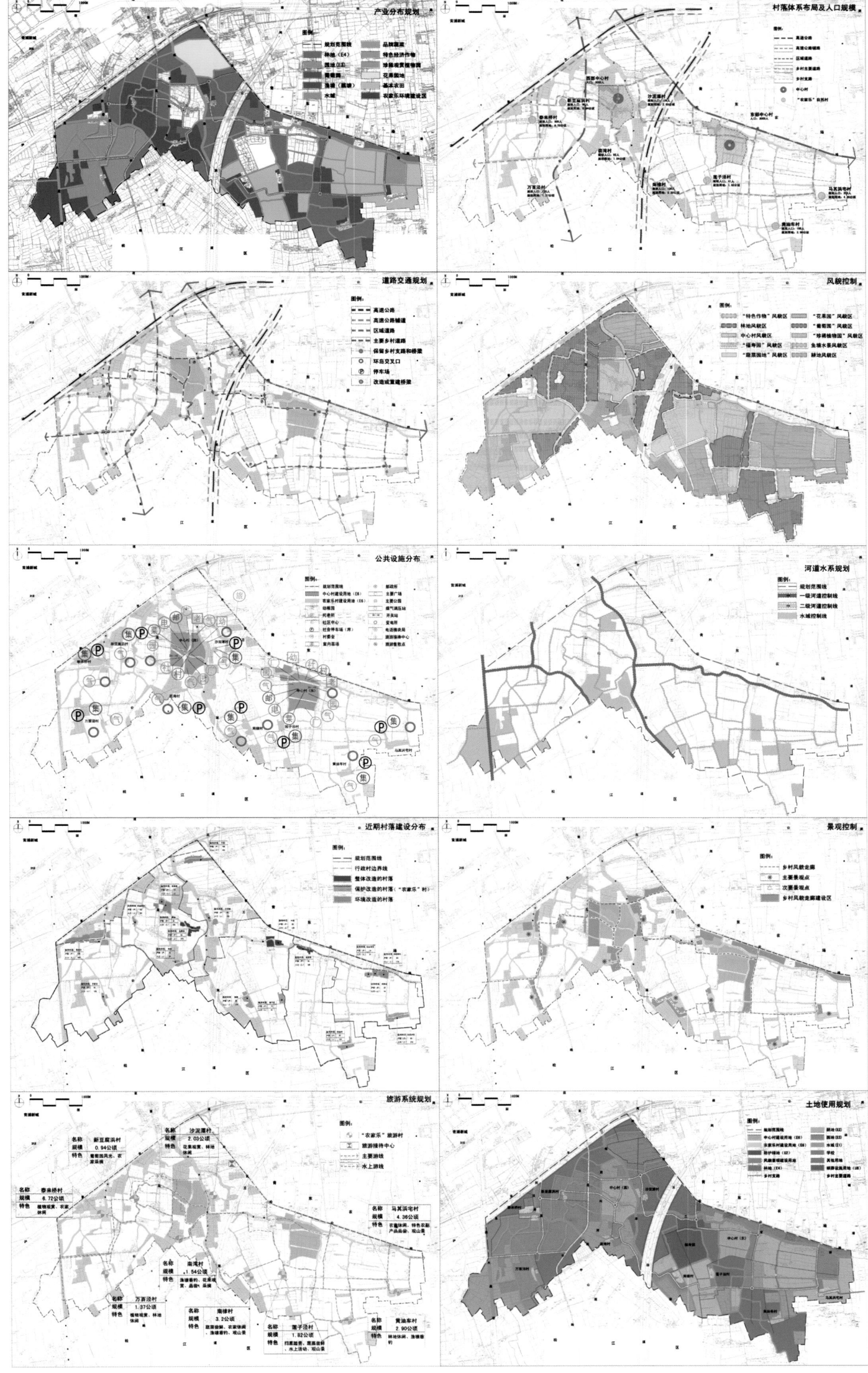

沈北新城总体发展概念规划及重点地段城市设计
Overall Development Concept Plan of Shenbei New City and Urban Design of Key Sections

项目名称：沈北新城总体发展概念规划及重点地段城市设计
项目地点：中国 / 辽宁 / 沈阳
建筑面积：19 100 万平方米
编制时间：2006 / 05
业主：沈阳市沈北新区开发委员会
规划类型：总体规划和城市设计

本次规划研究有四个任务范围：第一个范围为沈北地区总体发展研究；第二个范围为沈北新区约 890 平方公里沈北新城总体发展研究；第三个范围为沈北新城约 191 平方公里，包括新城中心区城市设计；第四个范围为重点区域，主要是蒲河两岸重要地段城市设计。

其中本项目的合作单位中国城市规划设计研究院承担前两个范围的研究工作。

There are four objectives for this research: the first range is the overall development of Shenbei area; the second is the overall development of about 890 square kilometers Shenbei New Area; the third range is about 191 square kilometers Shenbei New City, including design for urban Town Center; the fourth range is key urban design area including the main area along Pu River.

The partner of the project is the Chinese Academy of Urban Planning and Design, which takes the first two areas of work.

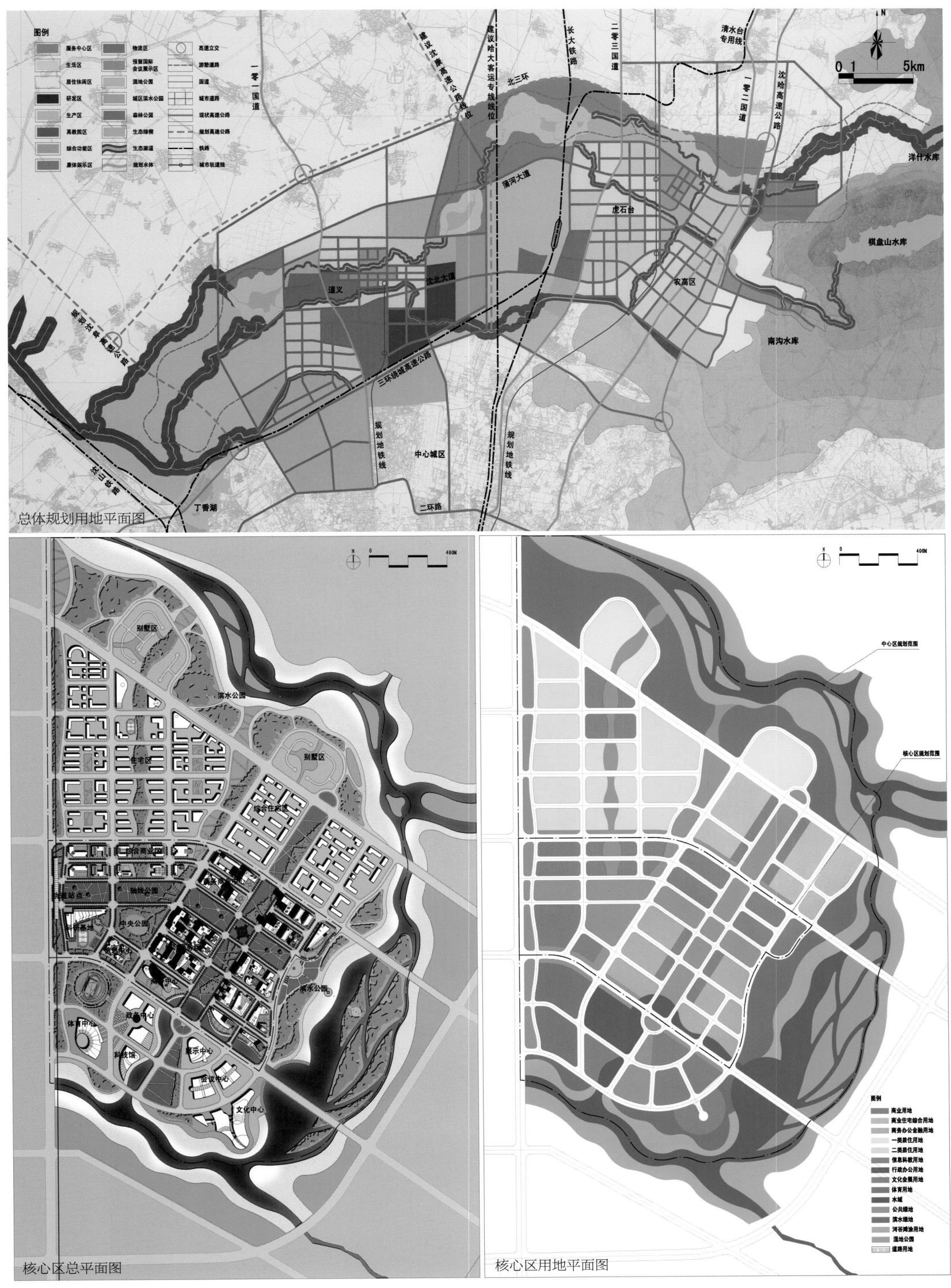

总体规划用地平面图

核心区总平面图

核心区用地平面图

无锡新区旺庄金融科技商务区及珠江路沿线地区城市设计
Urban Design of Wuxi New District Wangzhuang Financial Technology Business Districts and Along the Zhujiang Road

项目名称：无锡新区旺庄金融科技商务区及珠江路沿线地区城市设计
项目地点：中国 / 江苏 / 无锡
建筑面积：金融科技商务区 268.7 万平方米；珠江路沿线地区 11.34 万平方米
编制时间：2009 / 09
业主：无锡市人民政府新区管理委员会
规划类型：城市设计

项目位于无锡新区长江片区国际商贸商务功能板块内。规划范围包含金融科技商务区和珠江路沿线地区。

规划从资源整合、功能置换的角度去实现城市空间的转型，合理安排生产与生活的相互关系，通过居住、商务办公、商业金融、休闲娱乐等功能的不断植入，实现从现状的工业功能逐渐向都市性功能转换。秉承"资源整合"的开发模式，规划力求将金融科技商务区塑造成为品质卓越、环境优美、富有特色的都市金融科技商务综合区。

而珠江路沿线地区作为城市特色功能区，规划将其主导功能确定为集专业运动、体育健身、休闲娱乐于一体的综合性运动健身场所。以规划的手段来引导人们培养良好的生活习惯，提升生活品质，从而使得全体市民拥有健康的心态和良好的体态，引领城市健康生活，树立城市健康形象。

The project is located in international trade business function block in Yangtze plate, Wuxi New Area. The planning area contains the financial technology business district and areas along Zhujiang Road.

The planning achieves the transformation of urban space from the integration of resources and functional replacement, arranges the relationship between production and life reasonably, transfers the current industry function to the function of urbanity. By holding development model of "resources integration", the planning shapes the financial technology business district with high quality, beautiful environment, distinctive urban features.

The areas along Zhujiang Road feature as special urban area. The plan determines its dominant feature as a comprehensive sports and fitness place composed of professional sports, sports and fitness, leisure and entertainment. Leading people to develop good habits by means of planning, it improves quality of life so that the whole community has a healthy attitude and good posture, leading a healthy city life and establishing a healthy urban image.

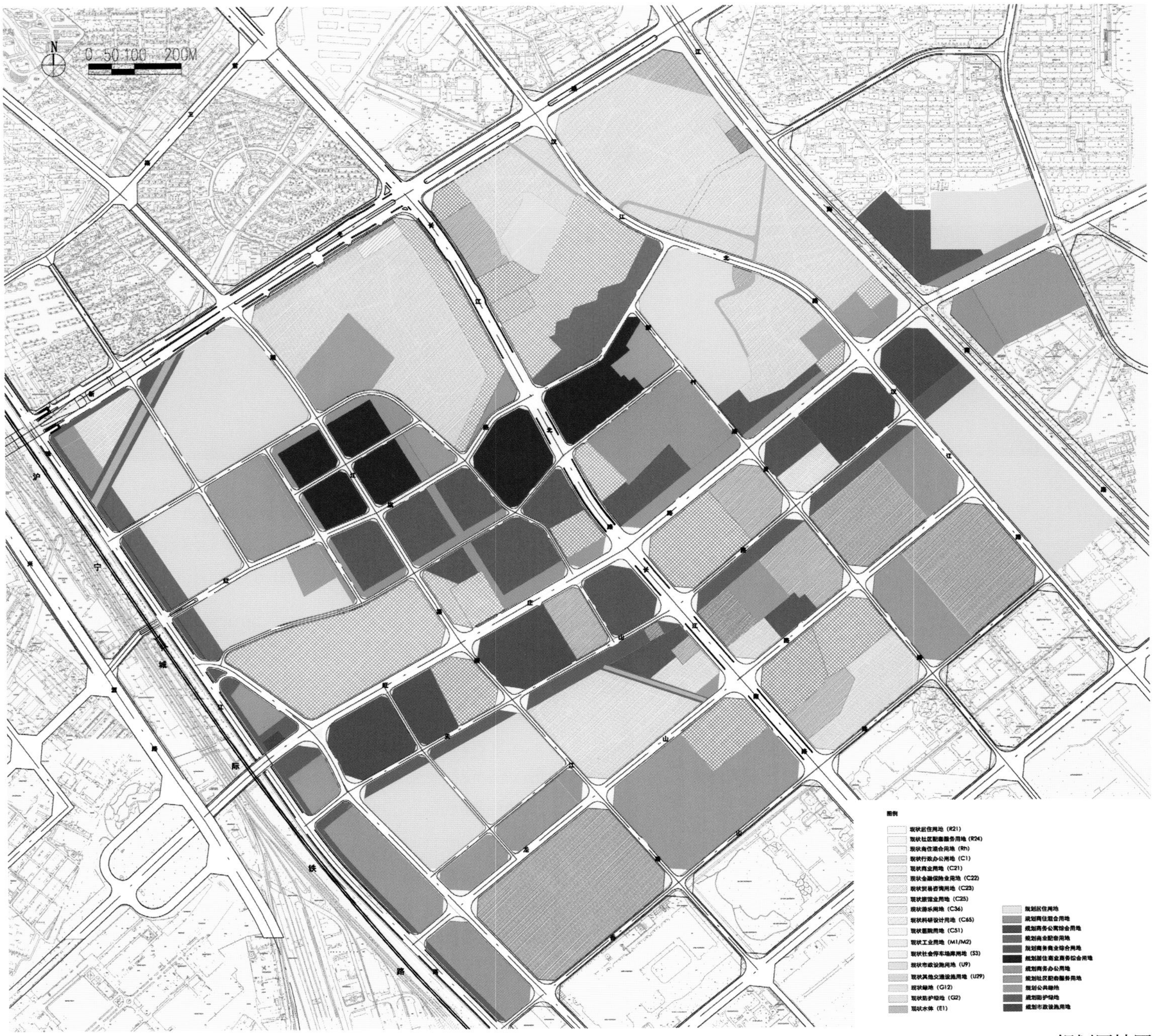
0 50 100 200M
图例
现状居住用地（R21）
现状社区配套服务用地（R24）
现状商住混合用地（Rh）
现状行政办公用地（C1）
现状商业用地（C21）
现状金融保险业用地（C22）
现状贸易咨询用地（C23）
现状旅馆业用地（C25）
现状游乐用地（C36）
现状科研设计用地（C65）
现状医院用地（C51）
现状工业用地（M1/M2）
现状社会停车场库用地（S3）
现状市政设施用地（U9）
现状其他交通设施用地（U29）
现状绿地（G12）
现状防护绿地（G2）
现状水体（E1）
规划居住用地
规划商住混合用地
规划商务公寓综合用地
规划商业配套用地
规划商务商业综合用地
规划居住商业商务综合用地
规划商务办公用地
规划社区配套服务用地
规划公共绿地
规划防护绿地
规划市政设施用地

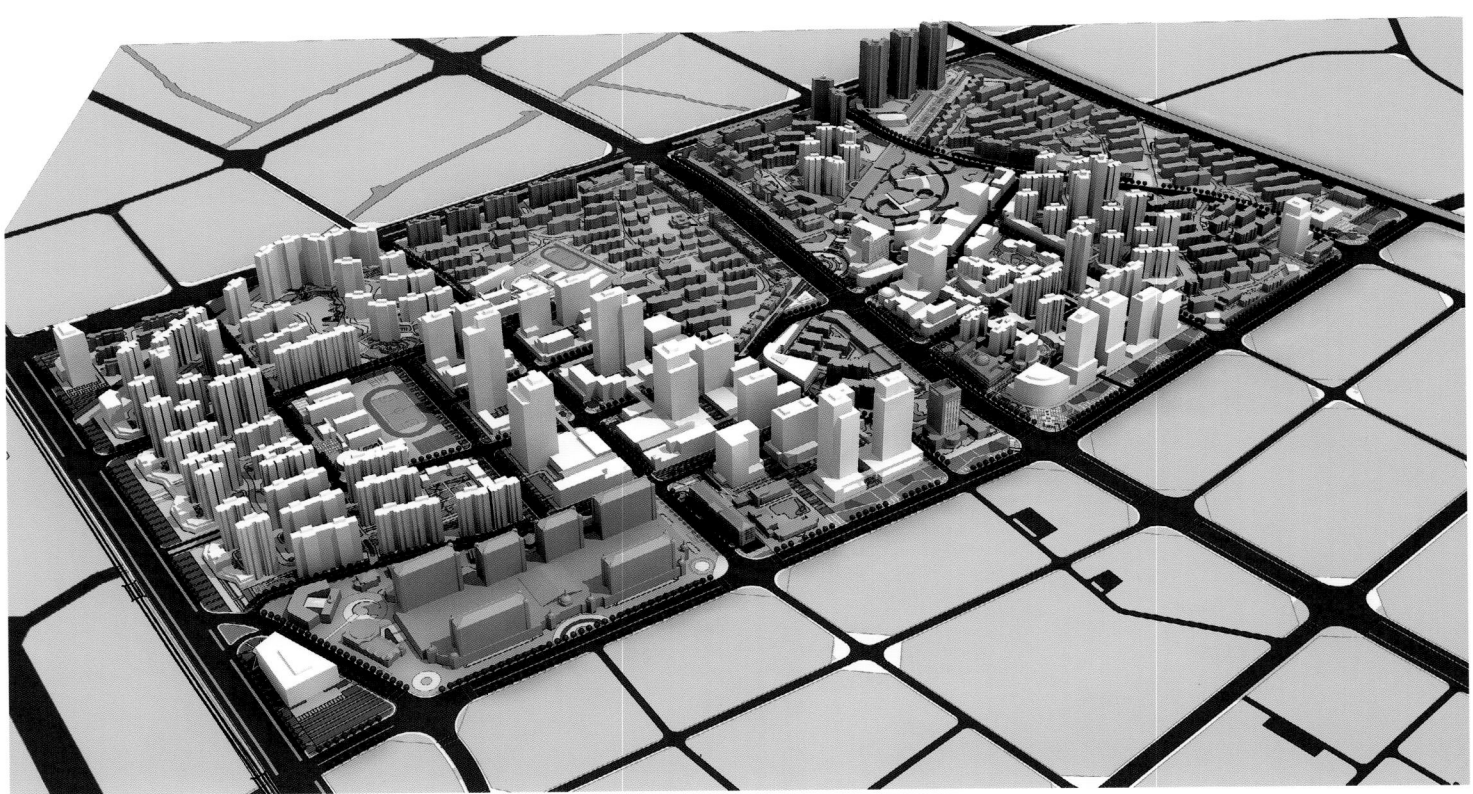

规划用地图

芜湖市城市夜景亮化总体规划
Wuhu City Night Lighting Master Plan

项目名称：芜湖市城市夜景亮化总体规划
项目地点：中国 / 安徽 / 芜湖
建筑面积：78 200 万平方米
编制时间：2009 / 05
业主：芜湖市房屋管理局
规划类型：专项规划

规划范围与芜湖市中心城区总体规划（2006~2020 年）的范围相一致，规划期限为 2009~2020 年，其中近期规划为 2009~2010 年，远期规划期限为 2020 年。

规划由总体框架与详细规划两个层次组成。总体框架可作为城市景观亮化工程建设的指导性、策略性和纲领性文件，主要内容为：针对城市五区、四轴的总体框架，明确形成各个景观风貌区以及与风貌区密切相关的线要素、点要素，对其进行规划定位，使其能指导城市夜景灯光景观详细规划设计的进行，并提出具体可行的实施措施与策略，合理安排近期建设项目；详细规划则在总体规划框架的指导下，对框架内所确定的重点要素进行详细的规划设计，指导灯光建设的具体实施。

The scope of the plan is consistent with the overall plan for downtown Wuhu (2006-2020), with planning period of 2009-2020, of which the recent planning period is from 2009 to 2010, while long-term planning period is from 2010 to 2020.

The plan is composed of two levels-overall framework and detailed planning. General framework can be used as guiding, strategic and programmatic documents for construction of the urban landscape lighting project. Main contents are as follows: in accordance with the overall framework according to five urban areas and four axis, specifies the formation of the line elements, points feature closely related with various landscape character areas and landscape areas, positions it so that it can guide the urban night landscape lighting design and implements concrete practical measures and strategies, carries out reasonable arrangements for short-term construction projects; under the guidance of overall planning framework, the plan conducts detailed planning for identified key elements, guiding design and construction of the light construction.

扬州蜀冈西峰 111 号地块规划设计
Yangzhou Shugang Xifeng Lot 111 Planning Design

项目名称：扬州蜀冈西峰 111 号地块规划设计
项目地点：中国 / 江苏 / 扬州
建筑面积：60.44 万平方米
编制时间：2006 / 06
业主：浙江北辰建设投资有限公司
规划类型：详细规划

扬州蜀冈西峰 111 号地块位于扬州市蜀冈瘦西湖风景区内。

本次规划分两个层次。第一层次是 60.44 万平方米范围内整个地区的概念性方案；第二层次是 18.4 万平方米范围内的一期地块详细规划。规划以"连南北贯东西"为总体构架，以"水"为造景要素，以现代本土建筑风格为基调，力图为扬州市构筑高起点的社区形象。

Yangzhou Shugang Xifeng Lot No. 111 is located in Yangzhou Shugang Slender West Lake Scenic Area.

The planning is divided into two levels. The first level is the conceptual plan throughout 604,400m²; the second level is the detailed Phase 1 plan within the 184,000m² plot. The plan has the overall framework of "consistent with north-south and go through west to east", and uses "water" as landscaping elements and tone of modern local architectural style, trying to build a high-starting point community image for Yangzhou City.

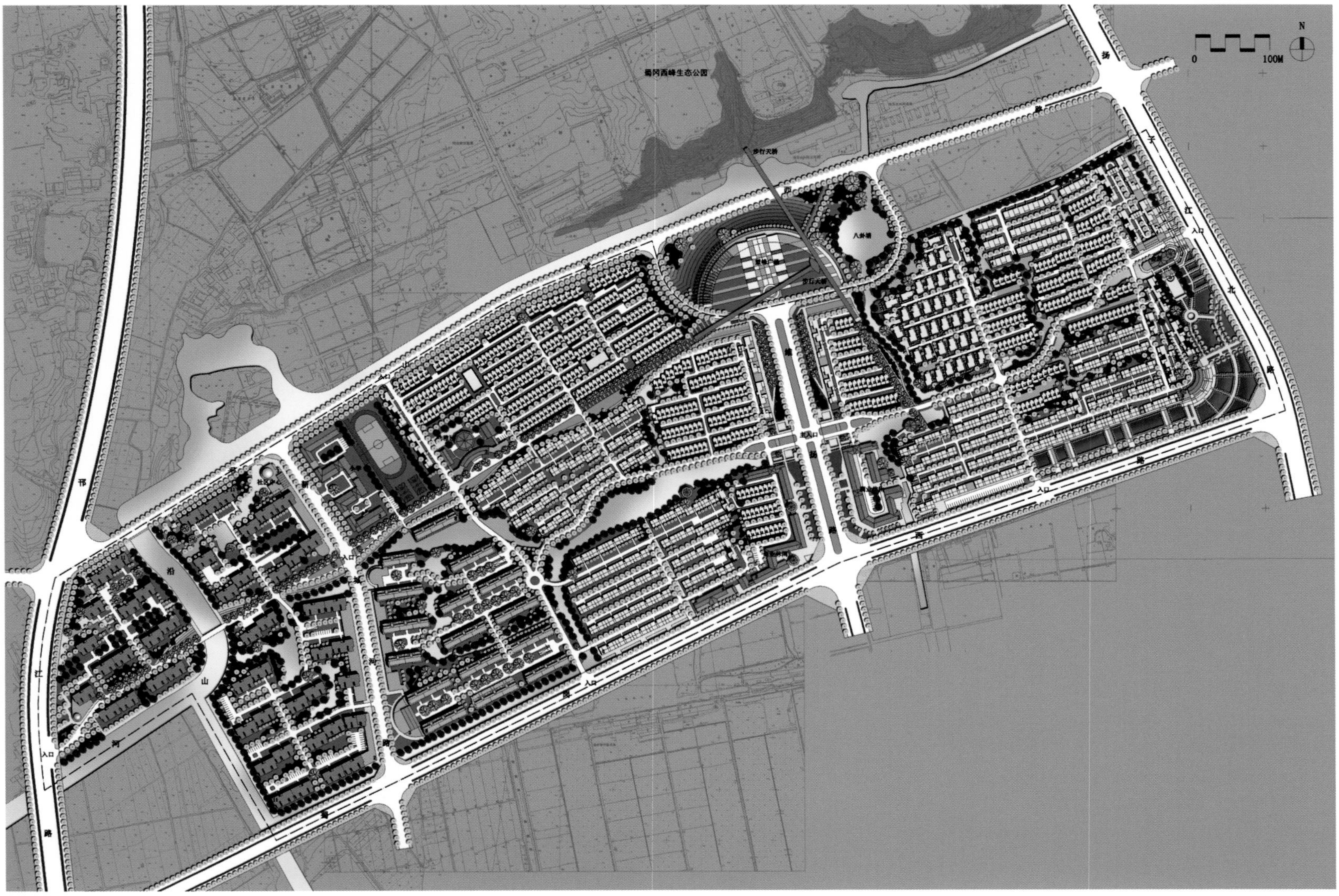

规划总平面图

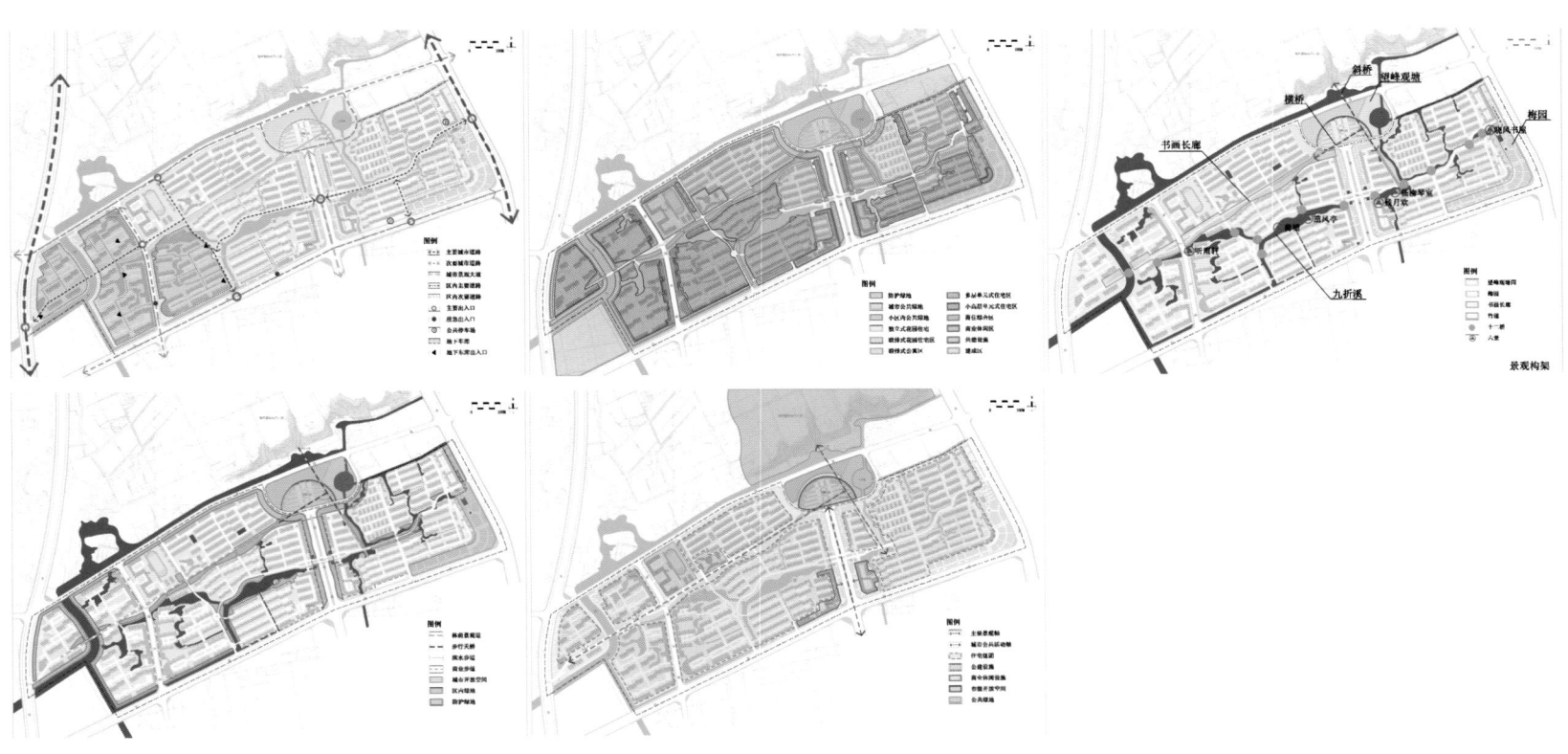

景观构架

上海朱家角中心镇总体规划
Shanghai Zhujiajiao Town Master Plan

项目名称：上海朱家角中心镇总体规划
项目地点：中国 / 上海
建筑面积：800 万平方米
编制时间：2001 / 10
业主：上海市青浦区城市规划管理局
规划类型：总体规划

朱家角镇位于青浦区行政区中西部，北接江苏省昆山市，东侧接青浦镇、东南与松江区接壤，西临淀山湖，西南与西岑镇、练塘镇相连。全镇陆域土地面积 92.56 平方公里，淀山湖水域面积 45.72 平方公里。

本次朱家角镇城镇总体规划的编制共分两个层次：一是朱家角中心镇镇域结构规划；二是朱家角中心镇镇区总体规划。规划充分吸收了"朱家角中心镇风貌景观规划国际方案征集"中三家设计机构所提交方案中的精华部分，加以归纳总结，并力争在土地利用层面上加以充分体现，使朱家角镇城镇建设向"布局合理，建筑风貌独特，具有 21 世纪时代特征的江南水乡新型城镇"目标迈进。

Zhujiajiao Town, is located in the midwest of Qingpu Administrative District, which faces Kunshan City, Jiangsu Province in the north, east to the Qingpu Town, bordering Songjiang District in the southeast, west to Dianshan Lake, southwest to Xi cen Town and Liantang Town. It covers a land area of 92.56 square kilometers. The water area of Dianshan Lake is 45.72 square kilometers.

The overall master planning of Zhujiajiao Town is divided into two levels: first, the structural plan of Zhujiajiao center town; second, the overall plan of Zhujiajiao central town. The plan fully absorbs the essence of "international program solicitation of Zhujiajiao town center landscape planning " program submitted by three design agencies, after summarization, striving to fully reflect in the respect of land use level so as to make construction of Zhujiajiao Town achieve the goal of "the 21st century new Yangtze River towns with reasonable layout and unique architectural style ".

SMITHGROUP cna

Architecture · Urban Planning · Landscape Design
Interiors Design · Art Consultant · Lighting Design
Since 1853

www.cna-group.com

Design Determines . . .

设计决定……

1   2   3   4   5   6   7   8   9   10 ······

01 规划设计类项目
01 Urban Planning

滨水地区
Water Front
城市街区
City Block
城市街区
City Block

旅游休闲
Travel and Leisure

02 景观设计类项目
02 Landscape Design

淀山湖国家级旅游度假区规划设计
Dianshan Lake Tourism Area Master Planning
Urban Planning | 2010·08

淀山湖国家级旅游度假区规划设计
Dianshan Lake Tourism Area Master Planning

项目名称：淀山湖国家级旅游度假区规划设计
项目地点：中国 / 上海
建筑面积：161 万平方米
编制时间：2010 / 08
业主：上海湖区建设开发有限公司
规划类型：概念规划

淀山湖位于上海青浦西翼，邻接江苏省。湖呈葫芦形，西接太湖，东连黄浦江，有上海的母亲湖之称。湖水碧澄如镜，沿岸烟树迷蒙，富有江南水乡的韵味。规划淀山湖旅游度假区位于淀山湖西面，元荡湖东面，是两湖交界的地段，拥有独特的区位优势、良好的生态环境和丰富的旅游资源。

在保护自然生态环境的前提下，规划多种旅游度假产品，提供层次多样、内容丰富的旅游度假活动，满足不同客户需求，并保持自然环境与项目开发之间的平衡；发展特色旅游项目，创造具有当地风情、独特性的活动场所，以调整身心、放松自我、娱乐生活为目的，以体验和享受为主要形式，以舒适环境、趣味活动、特色服务为吸引点，为旅游者提供食宿、游憩、娱乐、休息和其他度假休闲等多项服务的综合性旅游服务场所，实现人与自然生态的直接交流和沟通，营造江南水乡独特的度假氛围。

为满足以上项目需求，本次规划主要设计了星级度假酒店、独立别墅及会所、帆船俱乐部、生态农庄、旅游配套商业等一系列活动场所，总建设用地为 460.07 万平方米，总建筑面积为 161 万平方米。

Dianshan lake is located in the west of Shanghai Qingpu District, adjacent to Jiangsu Province. The lake is gourd-shaped, west to Taihu Lake, east of Huangpu River and it is the mother river of Shanghai. The lake is very clean like a mirror. Along the river there are many trees, full of southern Yangtze River scenery. The Planning Dianshan Lake Resort is located in the west of Dianshan River, east of Yuandang Lake which is the junction of the two lakes. It boasts a unique geographical, ecological environment and rich tourism resources.

Under the premise of protecting the natural environment, a variety of planning vacation products have been carried out to provide vacational activities of various levels and rich in content, meeting different needs of customers, and maintaining the natural environment, balance between project development; the development of tourism projects, creates activity with local customs, unique places to adjust the body and mind, relax themselves, for the purpose of entertainment, experiencing and enjoying the comfortable environment, fun activities, special services of the tourist attractions. Tourists are provided with accommodation, recreation, entertainment, rest and leisure and many other holiday services. Integrated tourism services achieves a direct and natural ecology of exchange and communication, creating a unique holiday atmosphere of southern Yangtze River.

To meet the requirements of these projects, this plan designed star resort hotels, independent villas and clubs, yacht club, eco-farm and a series of tourism activities. The total construction area is 4,600,700 m² with a total construction area of 1,610,000 square meters. To establish a brand image of the tourist resort, the plan creates areas attraction, landmarks and fosters regional energy to promote tourism and development, attracting capital and popularity, enhancing regional investment capacity to increase the value of the area and promote the region's rapid development.

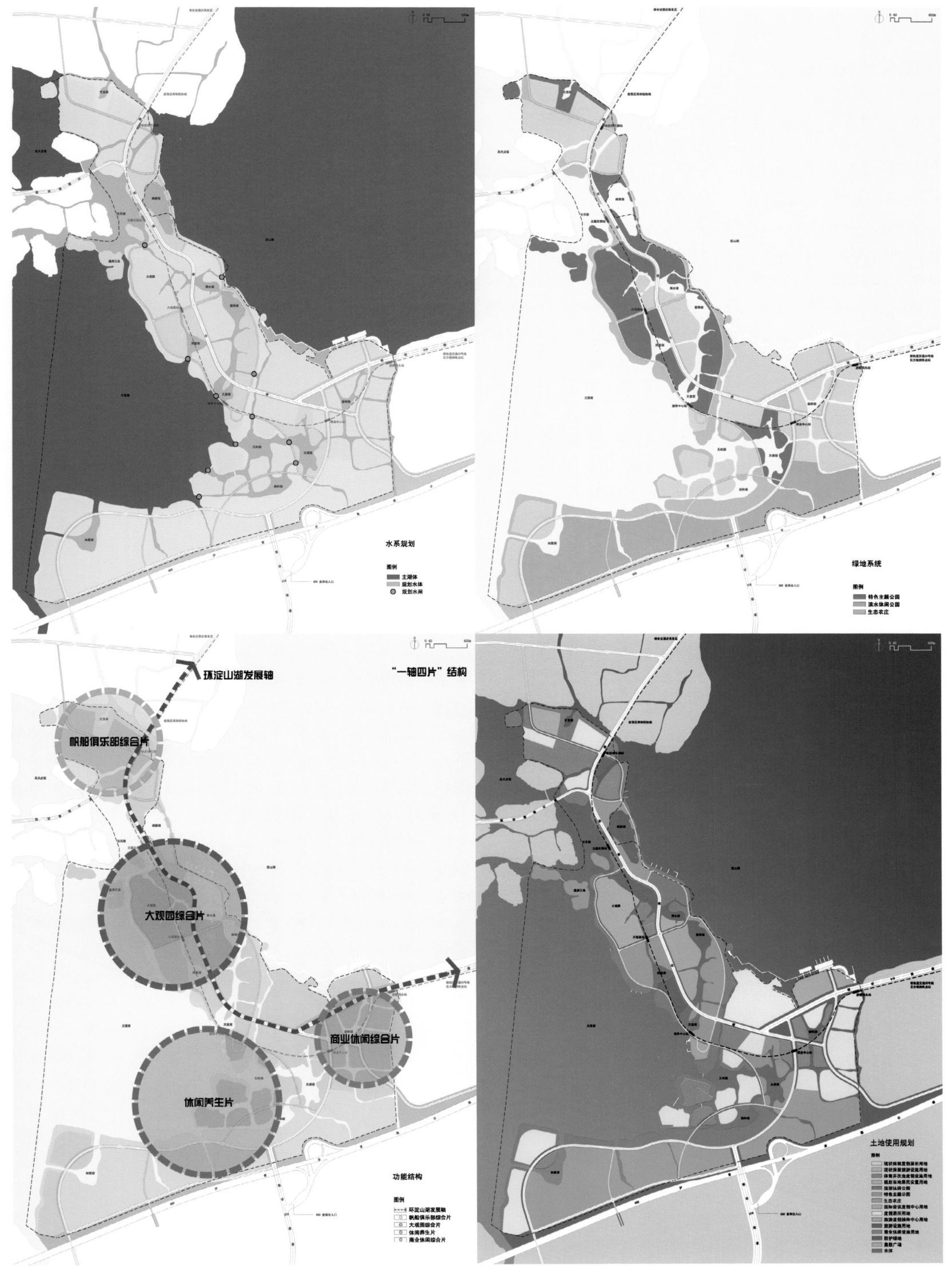

水系规划

图例
- 主湖体
- 规划水体
- 规划水闸

绿地系统

图例
- 特色主题公园
- 滨水休闲公园
- 生态农庄

"一轴四片"结构

环淀山湖发展轴

帆船俱乐部综合片

大观园综合片

商业休闲综合片

休闲养生片

功能结构

图例
- 环淀山湖发展轴
- 帆船俱乐部综合片
- 大观园综合片
- 休闲养生片
- 商业休闲综合片

土地使用规划

图例

N   0 40   400m

接会议酒店商务区

度假区高架轻轨线

元宝湖

昊天贞篱

帆船俱乐部站

桃源湖

云天湖

主题乐园站

蓬洲三岛

秀水湖

大坡湖站

翡翠湖

涡湖湖

接轨道交通20号线
东方绿洲终点站

元荡湖

天章湖

春树湖

楼特中心站

天屿湖

天清湖

润和湖

如墨湖

050   沪青平高速公路   318

050 金泽出入口

规划总平面图

正阳·南平旅游度假区概念规划设计

Zhengyang Nanping Resort Planning and Design Concept

Urban Planning | 2010·07

正阳·南平旅游度假区概念规划设计
Zhengyang Nanping Resort Planning and Design Concept

项目名称：正阳·南平旅游度假区概念规划设计
项目地点：中国 / 海南 / 陵水
建筑面积：179.41 万平方米
编制时间：2010 / 07
业主：上海正阳投资集团有限公司
规划类型：概念规划

国营南平农场位于海南省东南侧的陵水县，距陵水县城约18 公里，往北可通过保陵公路联系陵水、保亭、万宁及海南北面其他城市，往南可通过东线高速公路快速联系三亚。

三亚是海南省主要的旅游资源地，如今包括陵水在内的"大三亚旅游圈"已经成为海南岛旅游的新热点，将重点发展酒店业、文化娱乐、商业餐饮、体育休闲、康体保健等产业。与此同时，南平农场拥有海南省单泉眼流量最大的温泉资源，其温泉旅游价值的高效利用已被提上议程，南平农场温泉资源项目开发已列入海南省农垦总局和陵水县政府的重点开发项目。这些都为南平旅游提供了很好的发展契机。

依据区域宏观定位，结合规划基地的现状条件，规划给出以下功能与形象定位。

功能定位：以倡导传播慢生活文化为特色，植入保健康体、运动休闲、商务会议、婚庆旅游、度假居住等功能，立足国际旅游岛，辐射大陆及泛北部湾经济区，打造富有地域特点的海南省一流热带旅游度假目的地及集散地。

形象定位：麓林河谷、慢活小镇、诗意栖居、养生天堂。

根据南平地方特色及现状建设条件，规划为南平旅游度假区设置了若干个主题项目：直升机通航游览项目、温泉主题项目、高尔夫主题项目、健康养生主题项目、婚庆主题公园，另外还包括山地户外运动公园、商务会议及休闲商业等项目。为该项目提供层次多样、内容丰富的旅游度假活动，满足不同客户需求，平衡旅游淡旺季客流峰谷变化；创造具有当地风情、独特性、适合性的活动场所，树立旅游度假地的品牌形象；着力塑造地区的吸引点、标志点，培植地区活力，以旅游促发展，吸引资本和人气，提升地区招商能力，提升地区土地价值，推动地区快速发展。

规划总建设用地面积为 314.4 万平方米，总建筑面积为179.41 万平方米，平均容积率为 0.22。

State-run Nanping Farm is located in the southeast of Lingshui County, Hainan Province, about 18 km away from the Lingshui County, linking Lingshui in the north, Baoting,wanning and other cities in the north of Hainan through Baoling highway, and links to Sanya through eastern expressway in the south.

Sanya is the main tourist resource of Hainan Province. "Big Sanya tourism circle" including Lingshui has become the new hot spot for tourism in Hainan Island. The focus has been put on developing hotels, entertainment, business dining, sports and leisure, sports, health and other industries. At the same time, Nanping Farm boasts the largest single flow of hot spring resources in Hainan Province. The value of its efficient use of hot sping tourism has been put on the agenda; development of Nanping hot spring farm project has been included in Hainan Land Reclamation Bureau and Lingshui county government as key development projects. These have provided a good development opportunity for Nanping tourism.

Based on macro positioning of the region, combined with the status of conditions of planning base, the planning gives the following features and image positioning.

Functional orientation: to promote dissemination slow of life culture, uphold the healthy, harmonious, sustainable development concept, with various functions of health care, sports, business meetings, wedding travel, vacation housing and other functions, based on international tourism island, radiating the mainland and the Beibuwan economic zone, creating first-class tourist resort and one distribution center with geographical features of Hainan.

Image orientation: beautiful valley, non-rushing living, poetic dwelling, healthy heaven.

According to local characteristics and current construction conditions of Nanping, the planning sets a number of topics for Nanping tourist resort, including helicopter navigation tours, hot spring theme project, golf theme project, health theme project, wedding theme parks, as well as outdoor mountain sports park, business meetings, leisure business and other projects. The plan arranges a variety of vacation products for the project and provides various levels of content-rich vacation activities to meet different needs of customers to balance change of passenger flow in different tourist seasons. At the same time the development of tourism projects creates a venue of local unique, suitable style. It also establishes the brand image of tourist resorts. To shape the region's attractions and landmarks, it fosters regional energy ,promoting tourism and development, attracting capital and popularity, investment, enhancing regional capacity and the value of the land area, promoting the region's rapid development.

The total planned construction area is 3,144,000 square meters, with a total construction area of 1,794,100 square meters with the average floor area ratio of 0.22.

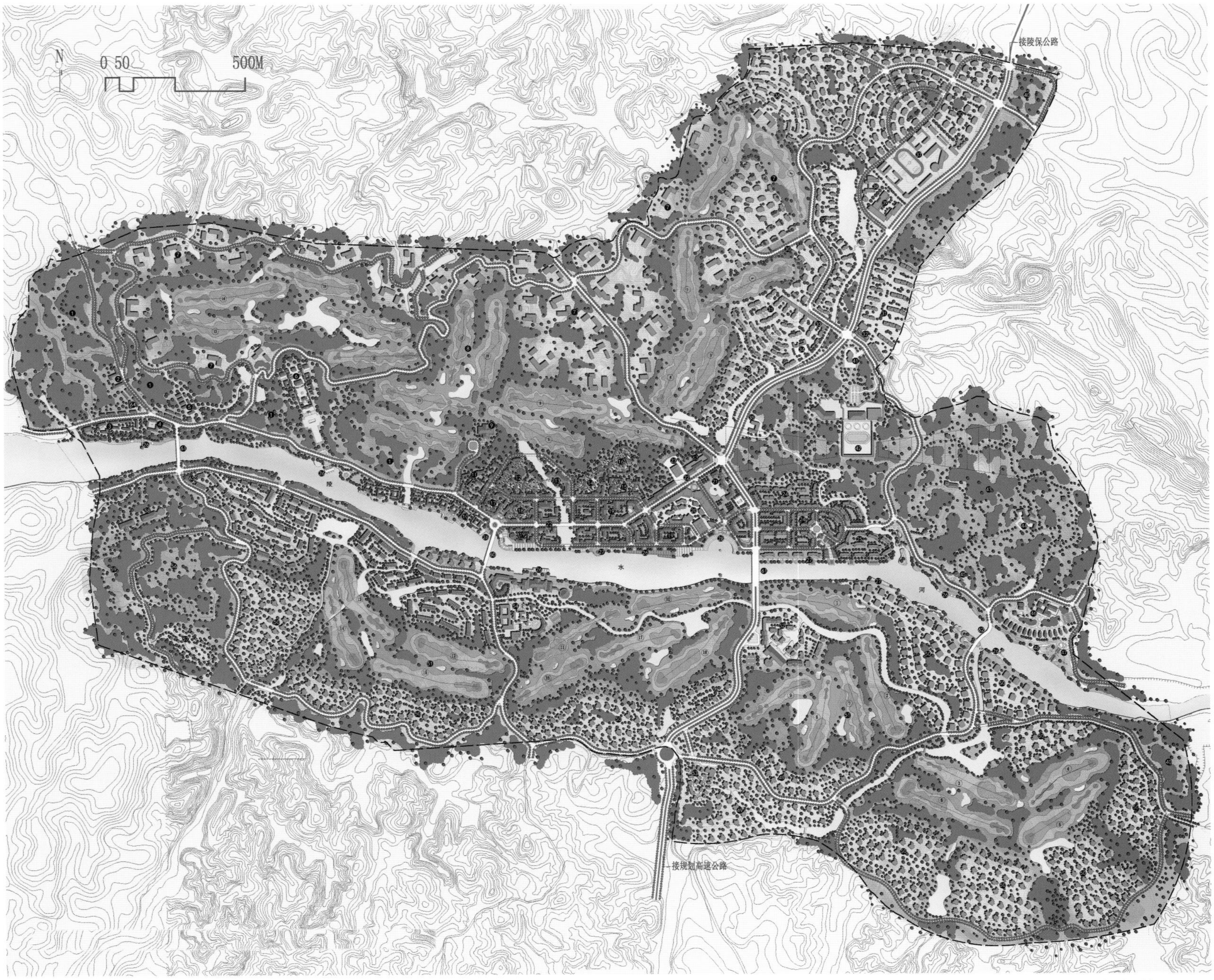

规划总平面图

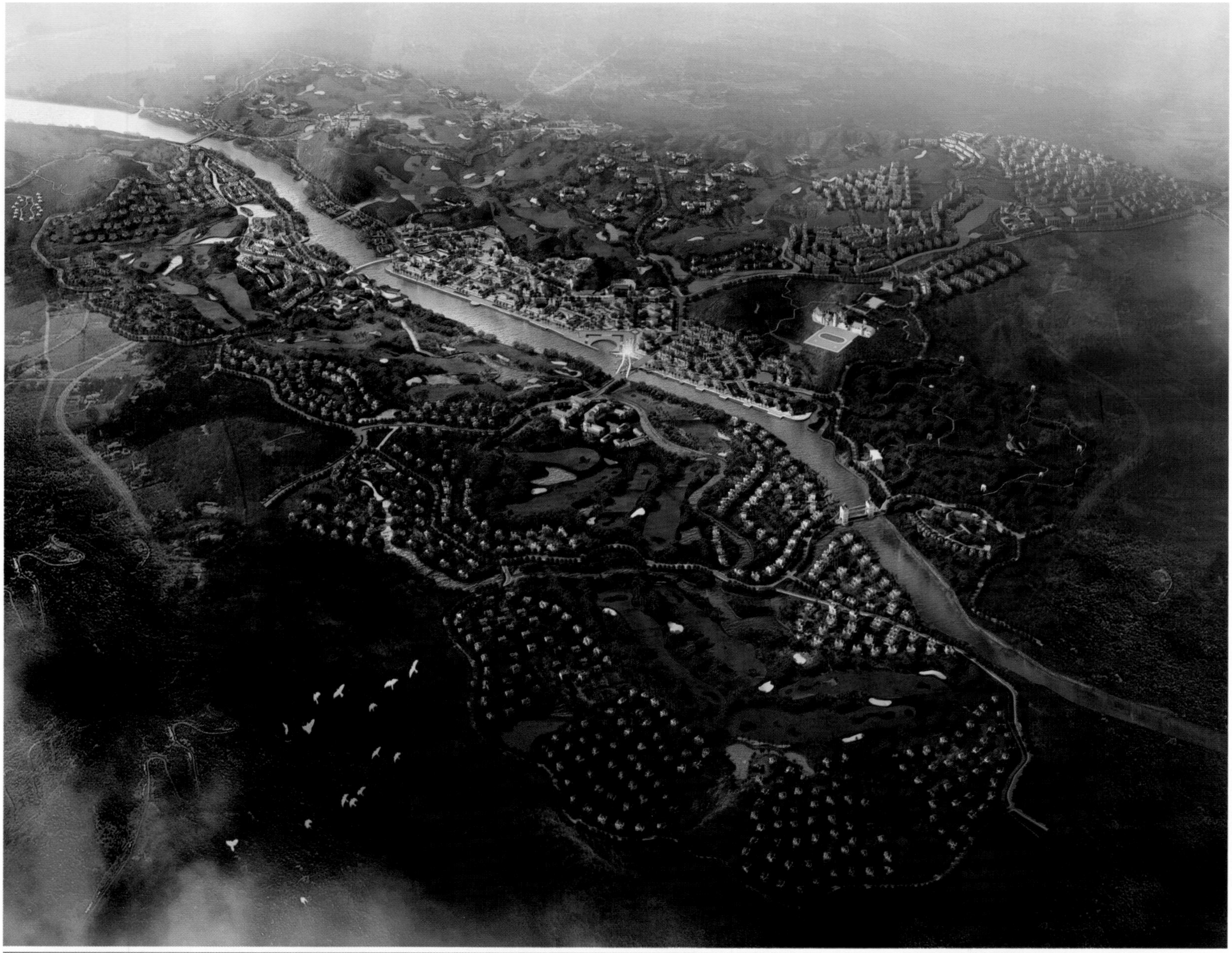

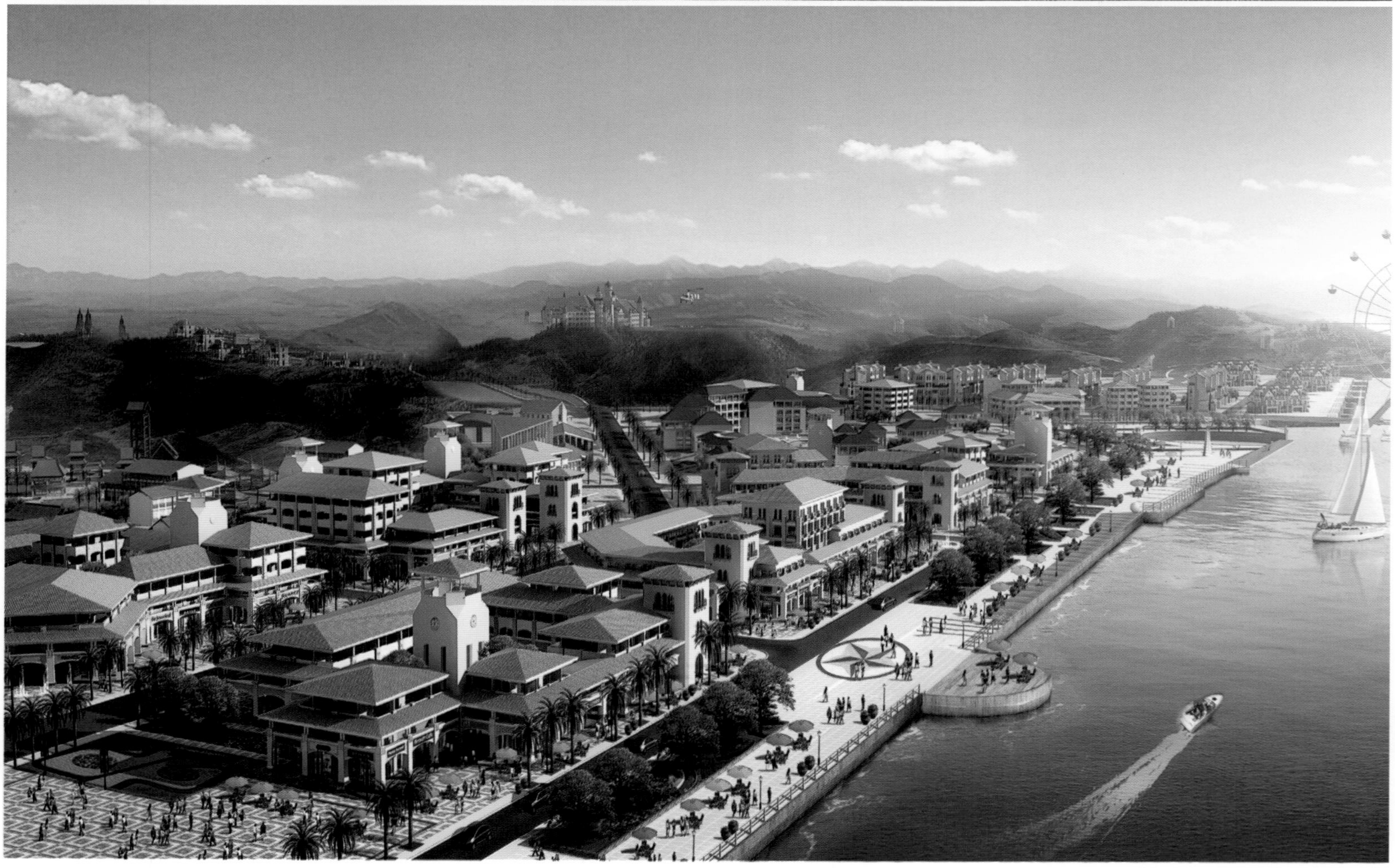

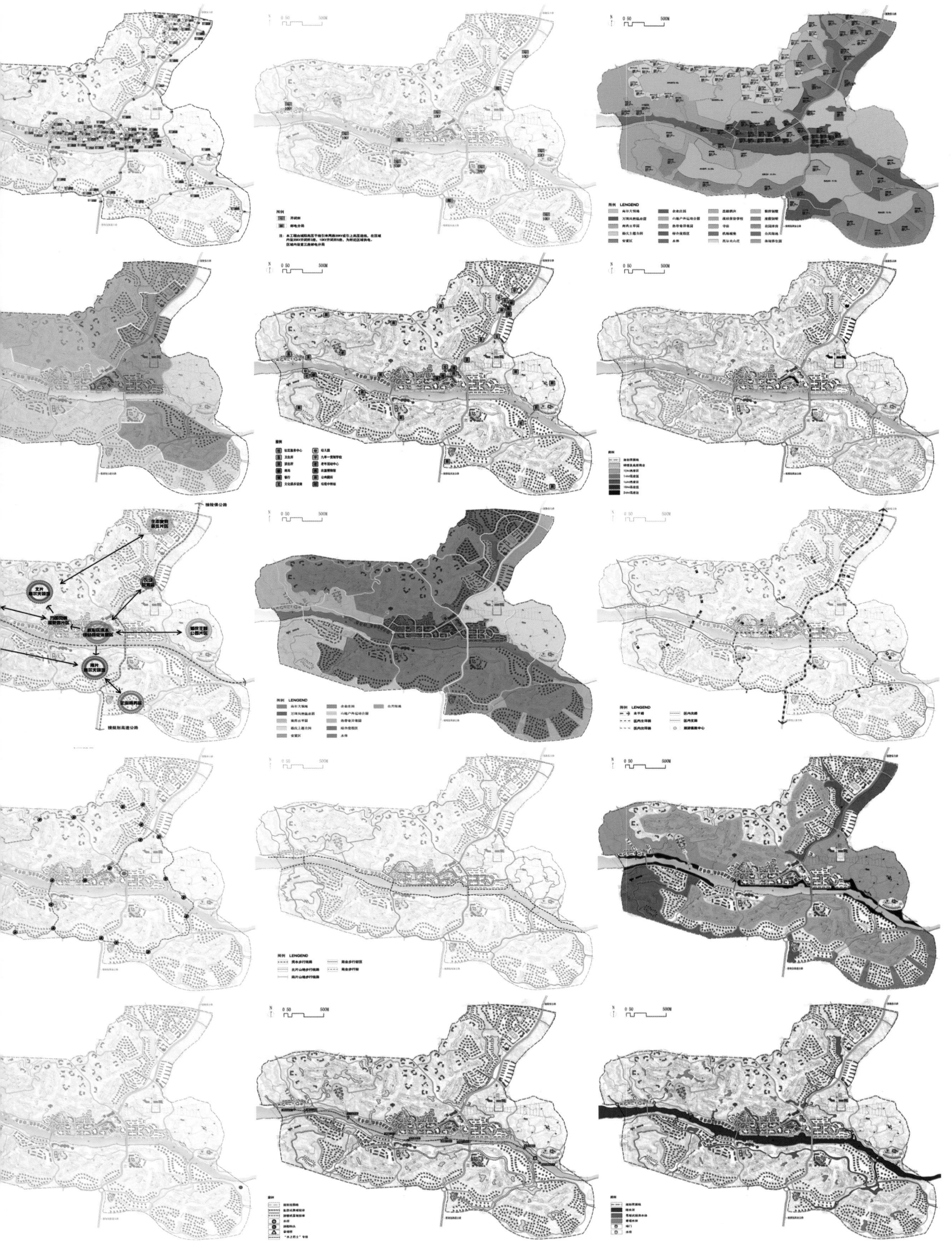

浦江仙华文景园控制性详细规划及城市设计

Pujiang Xianhua Wenjing Park Regulatory Plan and Urban Design

项目名称：浦江仙华文景园控制性详细规划及城市设计
项目地点：中国 / 浙江 / 金华
建筑面积：20.75 万平方米
编制时间：2010 / 08
业主：浦江县仙华文景建设工程指挥部
规划类型：控制性详细规划和城市设计

项目位于浦江县县城北部，距离著名 4A 旅游风景区—仙华山旅游景区 10 公里。

本次规划旨在打造高品质、高起点、高规格的浦江仙华文景书画产业园，满足浦江提升城市文化品位与内涵的要求，将园区定位为：以浦江书画产业的研究、展示、展览、销售、生产为核心，集合书画主题活动、群众文化活动、非物质遗产文化活动的场馆和场地，配套高品质休闲度假酒店、休闲娱乐商业设施、高端书画家村等功能的文化园区，并将其城市风貌定位为：新中式、水景园林式园区。

为了指导空间的有序发展，本次规划在控制性详细规划层面加强了城市设计的引导，重点对空间轴线、门户入口、重要路径、空间节点、空间界面和视线通廊等城市要素进行控制和设计，保证规划区形成高品质的城市空间和具有特色的空间特征。

规划园区总建筑面积 20.75 万平方米，其中公共设施总建筑面积 15.95 万平方米，住宅总建筑面积 4 万平方米，其他建筑面积 0.8 万平方米。园区内可容纳居住人口 378 人，就业岗位 6 345 个。

The project is located in the north of Pujiang County, 10 km from the famous 4A tourist scenic spot-Xian Hua Mountain Scenic resort.

The plan aims to create high-quality, high starting point, high standard of Pujiang Wenjing Industrial Park to satisfy Pujiang's requirements of upgrading urban culture and content. The park is defined as: venues of a collection of calligraphy and painting, mass cultural, intangible heritage and cultural activities using the River painting industry, research, exhibition, exhibitions, sales, production as the core theme, with high-quality resort hotel, leisure and entertainment business facilities, high-end features such as painting the cultural village park and its urban style has been defined as: the new Chinese style, water features garden- style park.

To guide the orderly development of space, this plan enhances the urban design at the controlled detailed level, to control and design focusing on urban elements like space axis, entrance portal, an important path, space nodes, the interface with the line of sight through gallery space to ensure the formation of high-quality characteristics of urban space and the space characteristics.

The total construction area is 207,500 square meters, total construction area of public facilities including 159,500 square meters of residential, gross floor area of 40,000 square meters and other construction area of 8,000 square meters. The park can accommodate 378 people and 6,345 jobs.

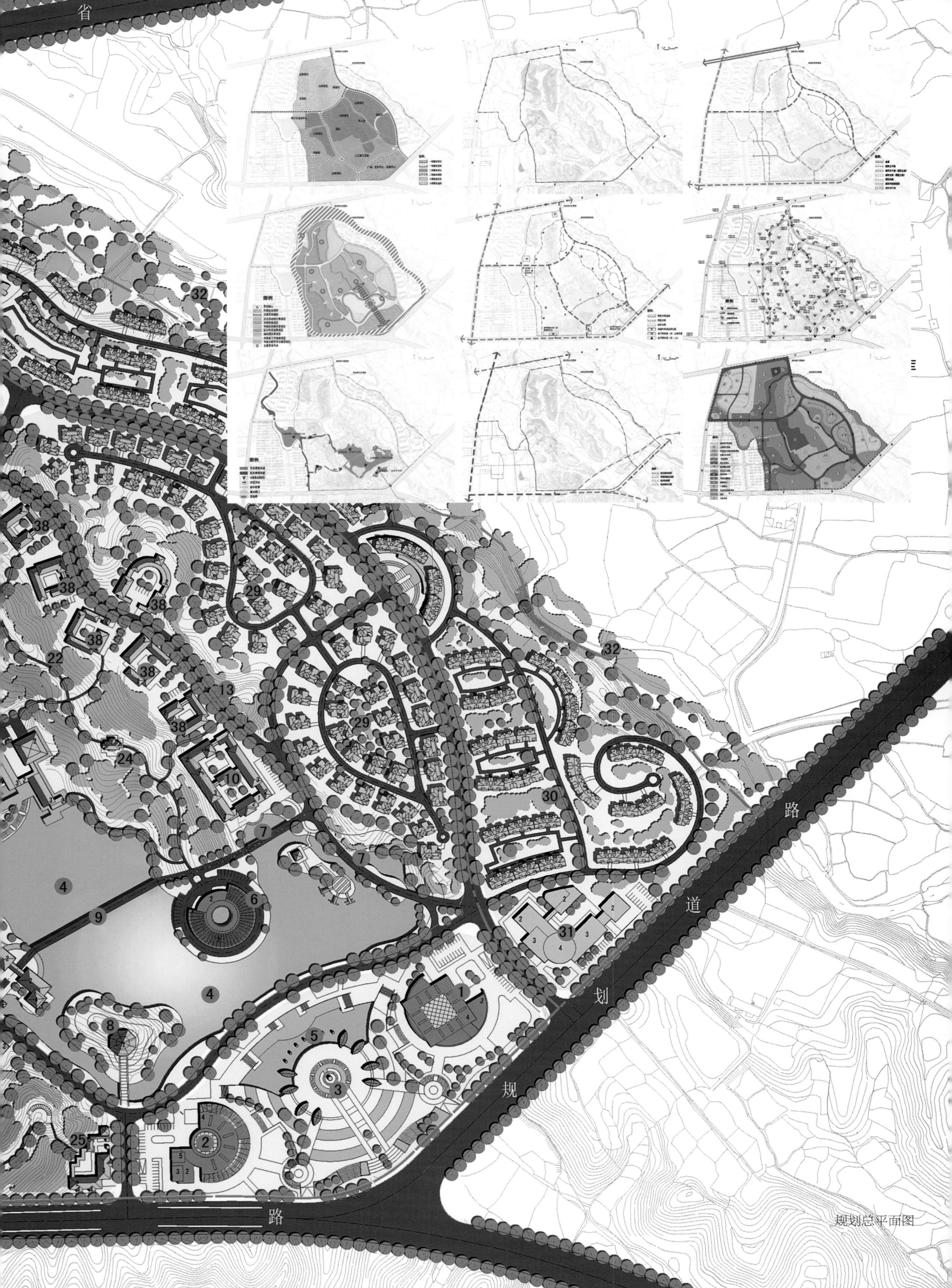

省

32

38

38

38

22

38

13

38

24

10

29

29

7

7

30

32

路

道

划

4

9

6

31

1

8

5

4

3

规

25

2

路

规划总平面图

汤泉东方修建性详细规划设计
Tangquan East Constructive Detailed Planning and Design

项目名称：汤泉东方修建性详细规划设计
项目地点：中国 / 江苏 / 南京
建筑面积：103.6 万平方米
编制时间：2010 / 08
业主：南京国泉投资开发有限公司
规划类型：修建性详细规划

规划基地位于南京城市北部、老山国家森林公园北麓的汤泉镇；拥有良好的地下热泉和冷泉资源，当地居民自古就有泡温泉的习俗。基地东侧为汤泉镇镇区，西侧为太吉度假村，南侧为老山国家森林公园，北侧为数条铁路线通过的轨道交通走廊。现今太吉温泉度假村已在南京及周边区域颇具知名度，但对汤泉旅游资源的利用尚显单薄，其规模和档次仍有待提升，而随着南京建设与经济的快速发展，汤泉的温泉旅游价值的高效利用被提上了议程。

依据区域景观定位，结合规划基地的现状条件，规划将汤泉东方定位为集温泉旅游、休闲度假、创意工坊、生态居住等复合功能于一体的长三角高档生态旅游胜地，并将其打造成泛东方风格的中国长三角绿色假日经济圈生态温泉主题度假胜地，具国际风范的全球生态宜居花园社区和高标准健康养生首选之地。

规划总建筑面积为 103.6 万平方米，地块平均容积率为 0.44。

The planning base is located in the north of Nanjing City, and at the northern foot of Old Mountain National Forest Park.

It boasts good underground hot springs and cold seeps resources. The local residents have the custom of bathing in hot springs since ancient times. The base has Tangquan town area in the east, Daji resort in the west, old Mountain National Forest Park in the south, rail transportation consisted of corridor several railway line in the north. Currently Daji Resort has been quite well known in Nanjing and its surrounding areas, but the use of hot spring tourism resources are still weak, and its size and grade need to be improved. With construction and rapid economic development of Nanjing, the effective use of hot spring spa tourism value has been put on the agenda.

Based on regional macro positioning, combined with the status of the planning base, create Tangquan Oriental spa high-end eco-tourist destination in Yangtze River Delta with tourism, leisure, creative workshop, eco-living and other functions, and make it a Green eco spa theme resort of pan - Oriental-style economic zone of China Yangtze River Delta, a garden world livable communities with international style and a high standard health regimen.

The total planned construction area is 1.036,000 square meters, the average floor area ratio is 0.44.

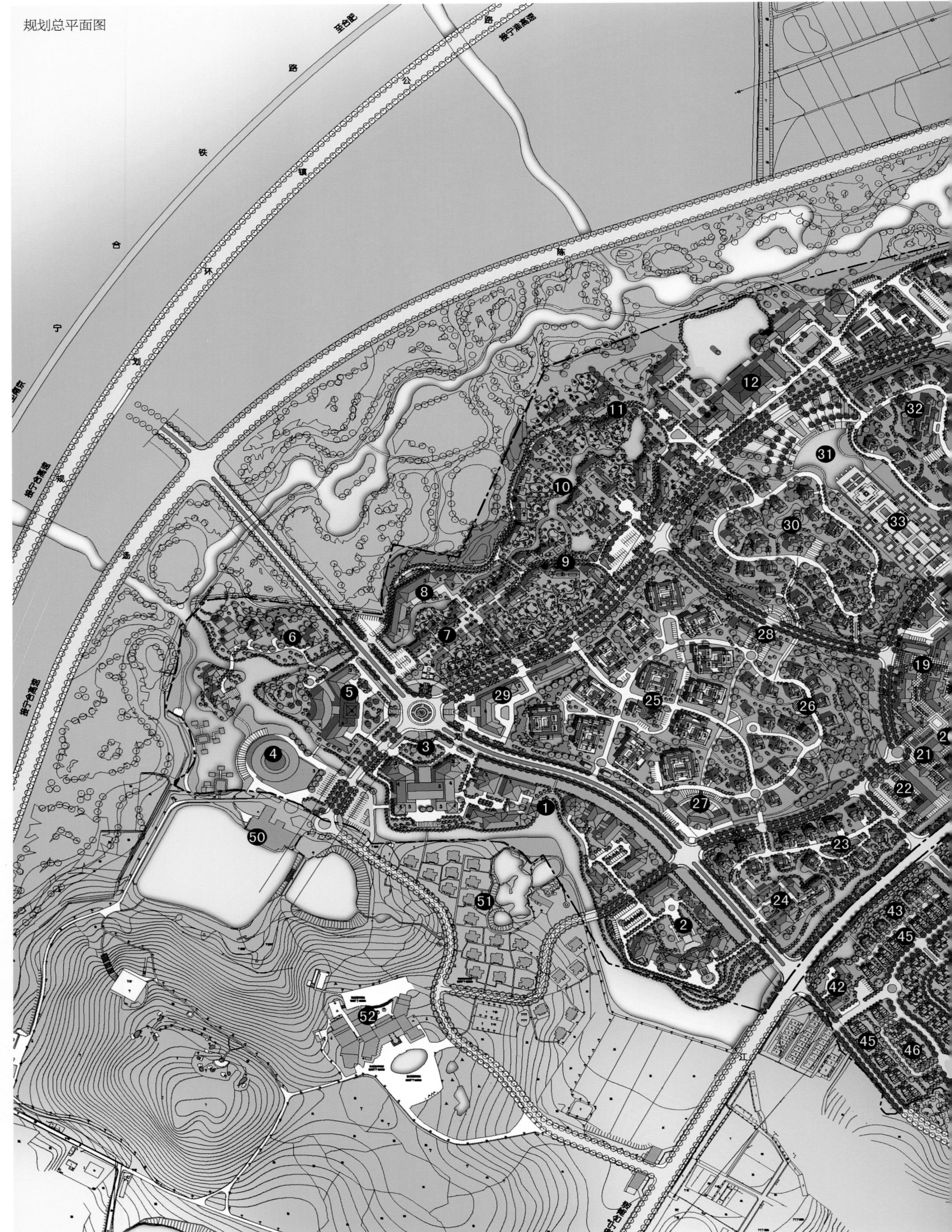

规划总平面图

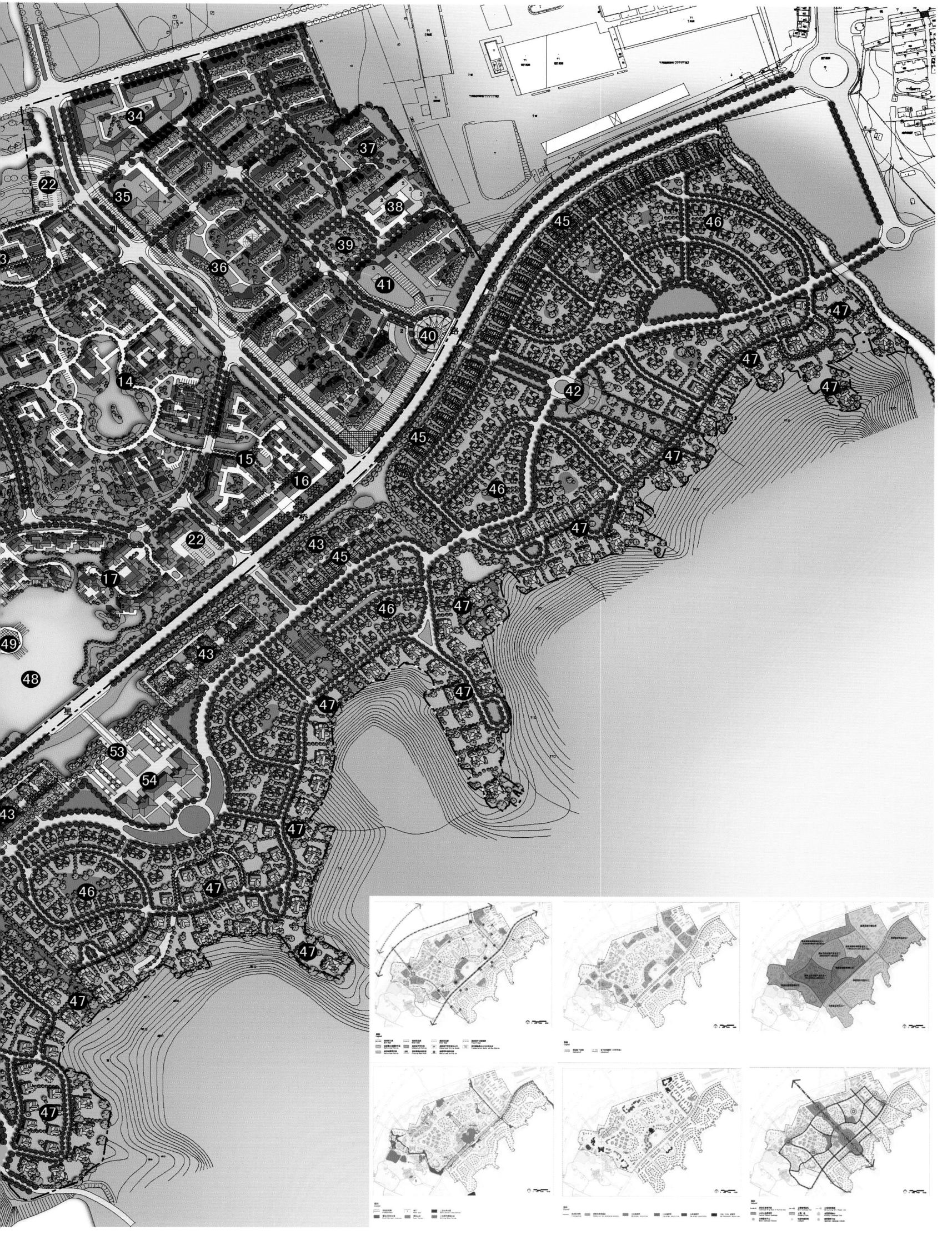

厦门厦金湾项目规划设计
Xiamen "Golden Bay" Plan

项目名称：厦门厦金湾项目规划设计
项目地点：中国 / 福建 / 厦门
建筑面积：56 万平方米
编制时间：2009 / 08
业主：上海正阳投资集团有限公司
规划类型：概念规划

项目地块南邻观音山海滨旅游休闲区（临时用地），西面以环岛路为界，北面现为路桥公司储备用地，南北长约 1.2 公里。

规划重点在于研究基地的土地利用、交通组织方式，项目布局、建筑群空间组合模式与形象，景观、绿化系统组织安排，以及技术经济指标的概算，为下一部详细规划提供一个框架性的草案。

规划提出"闽东南第一活力海滩"的构想，以"丰富的功能配套、市民公众的场所、热力的海湾形象、自然归真的环境"的功能定位，精心的景观设计，打造自然生态的园区环境，使人们陶醉其中，享乐其中。

规划建议采用新东南亚的建筑形态。建筑采用玻璃、金属等现代材料，配合木材、砖等传统材料创造一个暗喻传统东南亚风格的现代建筑，既能传承东南亚固有的传统建筑文脉，又能与时俱进不失现代之美感，创造一个能穿越时空对话历史的新东南亚园林建筑群。

The site is located to the south of Guanyin Mountain tourism area (temporary site), bounded by Huandao Road in the west, north of reserve land for Luqiao Company, which has a north and south length of about 1.2 km.

The plan focuses on research-based planning of land use, traffic organization mode, project layout, buildings combined mode and arrangement for image space, landscape, green system and estimated economic indicators, which provide a draft framework for detailed planning of the next step.

The plan proposes the concept of "the first dynamic southeastern Fujian Beach", using the functional orientation of "rich supporting function, public places, hot image of the beach, natural environment" and careful landscape design to create the natural ecological park environment in which people are intoxicated.

The plan uses new Southeast Asia architectural style. Architecture uses modern materials like glass and metal with traditional materials of wood, brick and others to create a modern architectural metaphor of traditional Southeast Asian style, which both inherents in the traditional architecture heritage context and the beauty of modern times to create a Southeast Asian architecture complex, which can communicate through time and space with the history .

规划总平面图

绍兴市会稽山度假休闲中心规划
Shaoxing Kuaijishan Mountain Resort & Recreation Center

项目名称：绍兴市会稽山度假休闲中心规划
项目地点：中国 / 浙江 / 绍兴
建筑面积：314 万平方米
编制时间：2003 / 10
业主：绍兴宝业集团股份有限公司
规划类型：详细规划

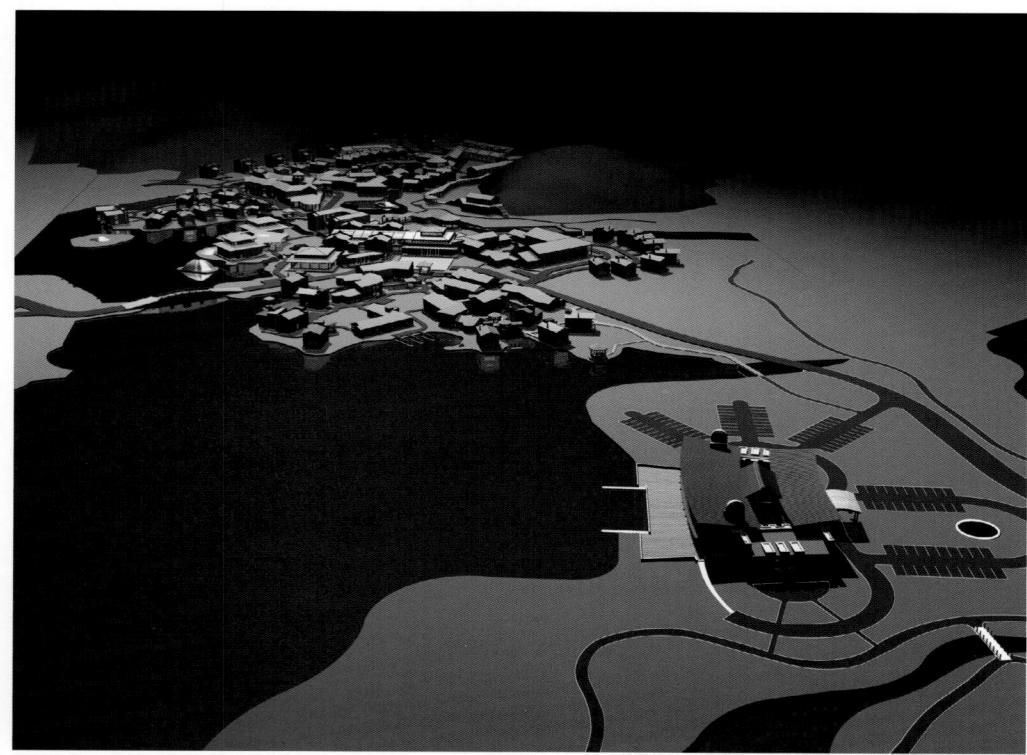

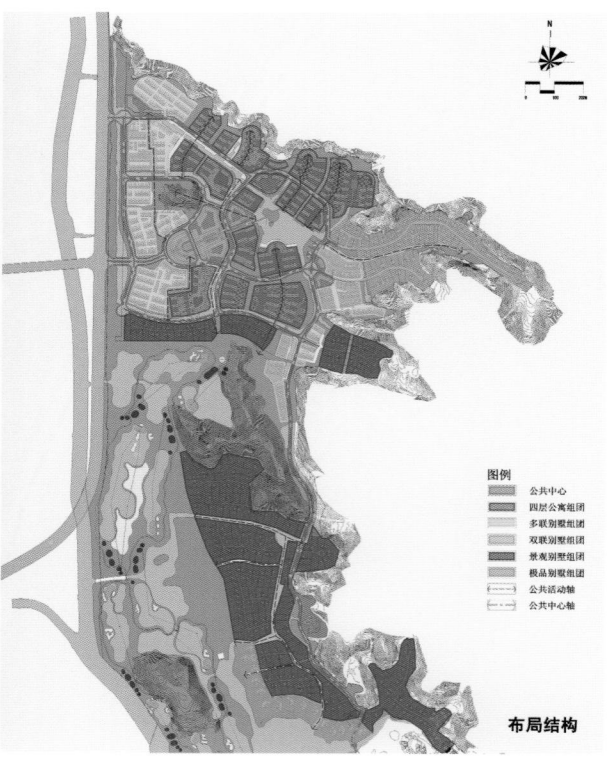

布局结构

项目位于浙江省绍兴市越城区、浙江省会稽山旅游度假区内。北起美女山南坡山脚，向东至岙底，并沿山脚向南至四丰，西至若耶溪、平水江和东湖路一线。

规划将会稽山度假休闲中心分三大块内容，北部为景观房产区，南部为高尔夫球场区，作为水体的洄涌湖穿插其中。设计充分体现出尊重自然、保护自然、利用自然的原则，同时体现地方性历史文化特色，具备超前性，以高尔夫球场、洄涌湖、会稽山大酒店为布局核心，形成山、水和高尔夫球场三大景观元素合一的格局。

The project is located in Yuecheng District, Shaoxing City, Zhejiang Province, inside Kuaijishan Mountain tourist resort centre. It faces the southern slope of Beauty Mountain foot, east to the beautiful mountains and along Sifeng in the south, west to the Ruoye Lake, Pingshui River and East Lake Road.

The planning divided Kuaijishan Mountain tourist resort centre into three parts: the northern part is the real estate landscape, and the southern area is a golf course with the Huiyong Lake inserted. The design is fully reflecting the principle of respect for nature, conservation, use of natural. It also reflects the characteristics of local history and culture, with advancement and golf courses, Huiyong Lake, Kuaijishan Grand hotel as the core, to form a pattern integrated with landscape elements like mountains, water and golf course.

规划总平面图

滨海火山国家地质公园旅游度假区
Coastal Volcano National Geological Park Resort

项目名称：滨海火山国家地质公园旅游度假区
项目地点：中国 / 福建 / 漳州
建筑面积：736.5 万平方米
编制时间：2008 / 10
业主：厦门明昇集团有限公司
规划类型：概念规划

规划漳州滨海旅游休闲度假区选址基地位于漳州市东南部，龙海市隆教畲族乡。距离漳州 60 公里，距离厦门 120 公里。项目占地约 736.5 万平方米，南北长约 4.2 公里，宽约 2 公里。用地范围内包括漳州滨海火山国家地质公园的一部分用地。

我们的规划理念是，海岸线资源的最大化及合理利用，营造沙滩、海水、蓝天、白云和人相互交融的特色滨海景观；火山公园的自然景观与人文活动的融合与互动；塑造地区的吸引点、公共中心、标志点，培植地区的活力；策划特色活动，创造具有当地风情、独特性、适合性的主题公园；将休闲度假、自然环境与激情运动和休闲养生相结合。

项目的总体定位为：龙海市隆教湾旅游经济开发区重要的组成部分，是以火山地质公园为核心，以中国大陆第一海岸为目标的海湾型多功能度假区即世界海洋火山奇观、中国大陆第一海岸。

项目主题为：海洋火山探险、火山康复度假、海洋文化体验、休闲体育。

Zhangzhou Tourist Resort is located in the southeast part of Zhangzhou City, which is 60 kilometers away from Zhangzhou, and 120 kilometers away from Xiamen. The site area is 7,365,000 square meters, with 4.2 kilometers long from south to north, and 2 kilometers wide. Part of the site will be used for Zhangzhou Coastal Volcano National Geological Park.

Our planning philosophy is to maximize the coastline resources and rational use, create characteristics of coastal landscape interacting with sand beach, blue sky, white clouds and people; achieve integration and interaction of volcano park's natural landscape and human activity; create area attractions, public centers, landmarks; plan special events, and create a local style, uniqueness, suitability of the theme park; integrate the leisure and natural environment with leisure health and passion sports.

Our project is orientated as a major component of Longjiao Bay Tourism Economic Development zone, taking volcanic park as the core, and the first China Coast as the goal, to be a multifunctional resort area.

Project themes are marine volcanic exploration, volcanic rehabilitation vacation, marine cultural experiences and leisure sports.

规划总平面图

长泰海西国际城规划设计
Chang tai Haixi International City Plan Design

项目名称： 长泰海西国际城规划设计
项目地点： 中国 / 福建 / 漳州
建筑面积： 484 万平方米
编制时间： 2009 / 12
业主： 中冶集团中冶（福建）房地产开发有限公司
规划类型： 概念规划

项目位于厦漳泉闽南"金三角"的核心腹地，是漳州长泰县马洋溪生态旅游区的核心项目之一，项目东距厦门市 30 公里，南去漳州市区 20 公里，北往泉州市 100 公里，交通优势突出。

本次规划分为三个层面：第一块层面用地范围 4 239 亩，第二层面为原范围以北新增约 2 000 亩，第三层面为原范围以东新增约 1 020 亩，三大范围共计约 7 200 亩，约合 484 万平方米。

鉴于项目规模较大，为了更好地与未来开发接轨，提高规划的实施弹性，本规划在 7 200 亩规划的层面做了多方案比选，力争取得规划设计与市场机制的平衡。

根据项目目前地处生态旅游区的特点，项目总体定位为：坐拥长泰县马洋溪生态版块，以山体公园为外环境，以高尔夫景观为内核的稀缺性休息度假社区。在建筑风格分布上本案分为：亚洲风格，欧洲风格，美洲风格三大版块，每个版块里面根据景观品质及特质又做了对应的下一级风格布置，并将世界各地经典建筑风格融入其中，力争打造一个国际化风格的大型楼盘。

The project is located in central hinterland of Xiamen Zhangzhou Quanzhou "Golden Triangle", which is one of the core areas of Zhangzhou Changtai county Mayang stream ecological tourism zone. It is 30 kilometers from Xiamen in the east, 20 kilometers from Zhangzhou city in the south, 100 kilometers from Quanzhou city in the north with a good transportation advantage.

The plan is divided into three levels: the first is a land of 4,239 acres; the second level extends about 2,000 acres in the north; the third level extends about 1,020 acres in the east. The total area is about 7,200 acres or about 4,840,000 squrea meters.

Because of the large scale project, in order to better integrate with the future development and improve the implementation flexibility of the plan, the plan conducts a multi-program selection in level of 7,200 acres and strives for a balance between plan and market mechanism.

According to the characteristics of eco-tourism area, the project has been defined as: boasting Changtai County Mayang stream ecological zone, using Mountain Park as the outside environment to build a resort community with golf landscapes. In the architectural style, the case is divided into three major sections of Asian style, European style and American style, each of which arranges next level of style based on landscape quality and character and combines with classical architectural styles around the world, striving to create a large-scale real estate of international style.

N 0 20 200m

至兴泰工业区

规划总用地范围
总用地面积：２８２公顷(4239亩)

11万伏高压走廊

新增用地面积：133公顷(约2000亩)

闽南台商子弟学校

新增用地面积：66.7公顷(1000亩)

马洋溪河道蓝线

发现之旅住宅区(二期)

至长泰

原一期B区用地范围

天藤桃园

发现之旅住宅区(一期)

连氏大酒店

十里村

至厦门

规划总平面图

SMITHGROUP cna

Architecture · Urban Planning · Landscape Design
Interiors Design · Art Consultant · Lighting Design

Since 1853

www.cna-group.com

Design Determines . . .

设计决定······

| | 1 | 2 | 3 | 4 | 5 | 6 | 7 | 8 | 9 | 10 | ······ |

01 规划设计类项目
01   Urban Planning

滨水地区
Water Front
城市街区
City Block
城市街区
City Block
旅游休闲
Travel and Leisure

02 景观设计类项目
02 Landscape Design

玉树扎曲河景观设计
Yushu Zhaqu River

Landscape Design | 2010·08

青海省玉树州结古镇扎曲河和巴曲河滨水景观规划设计
Waterfront Landscape Planning and Design of the Zhaqu River and the Baqu River of Jiegu Town, Yushu Prefecture, Qinghai Province

项目名称：青海省玉树州结古镇扎曲河和巴曲河滨水景观规划设计
项目地点：中国 / 青海 / 玉树
建筑面积：约 1 218.8 万平方米
编制时间：2010 / 05~2010 / 08
业主：青海省规划局
服务范围：景观规划设计 / 方案

在地震灾区条件非常艰苦的前提下，试图营造高原滨水小镇宁静、独特的气氛和魅力，为灾后重建提供一条个性发展之路，同时重新规划的景观设计在解决小镇交通的同时将地区的景观特色、文化特色高度浓缩起来，形成中国三江源独特的风景。

The design tries to create a tranquil and unique atmosphere and charm in the waterfront small town on the plateau under very tough conditions of the earthquake-stricken area, to provide an individualistic development road for the post-disaster reconstruction. The new landscaping design solves the traffic problem, as well as highly concentrates localized landscape features and cultural features, to form a unique landscape view.

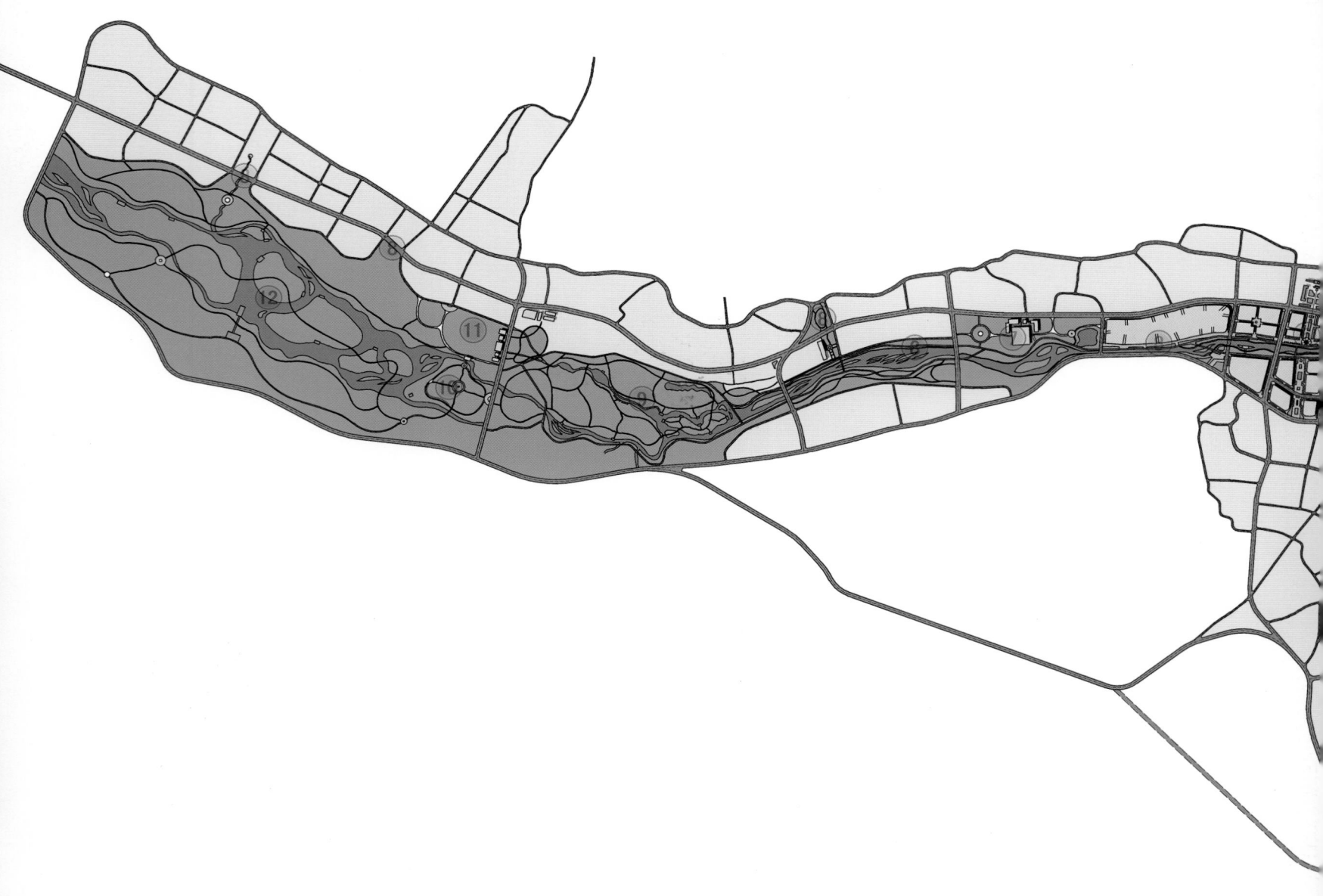

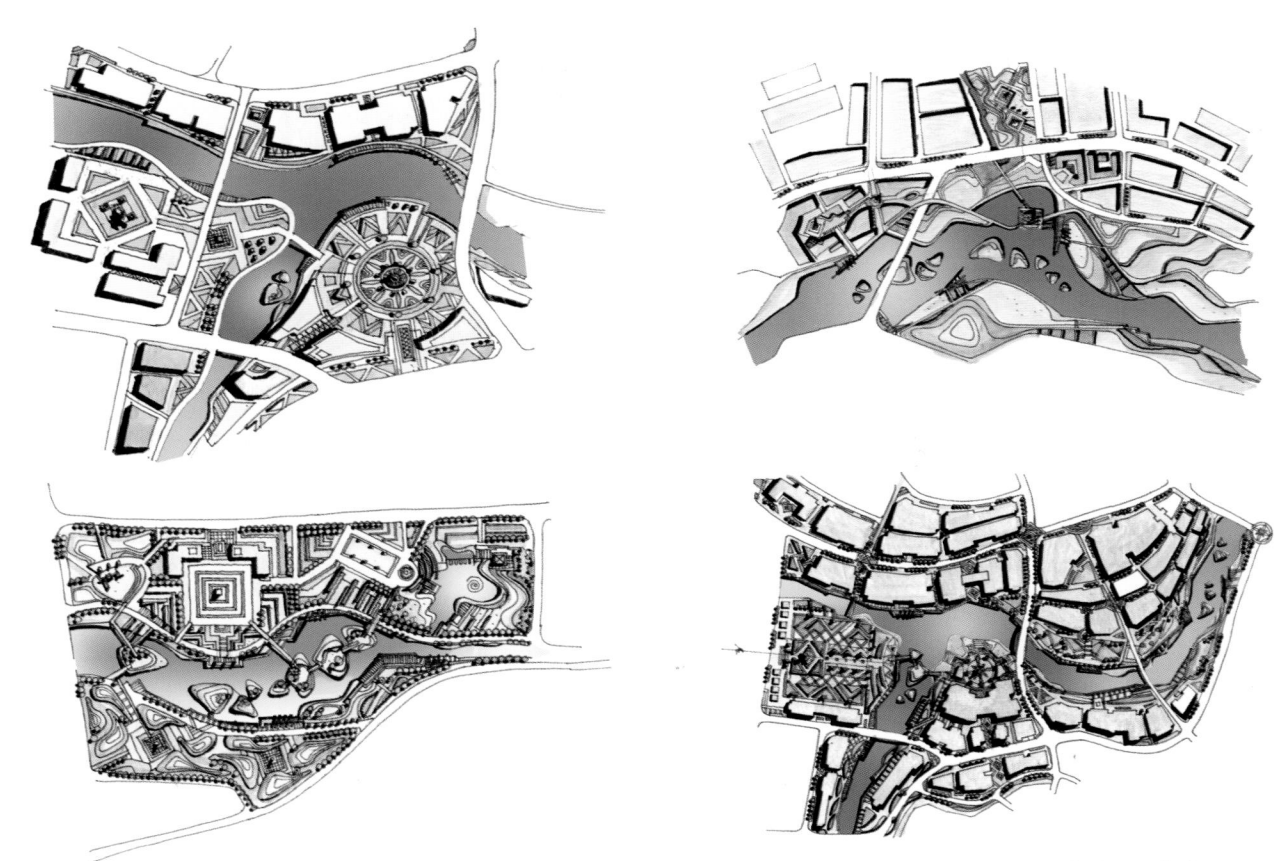

重要景观节点平面图

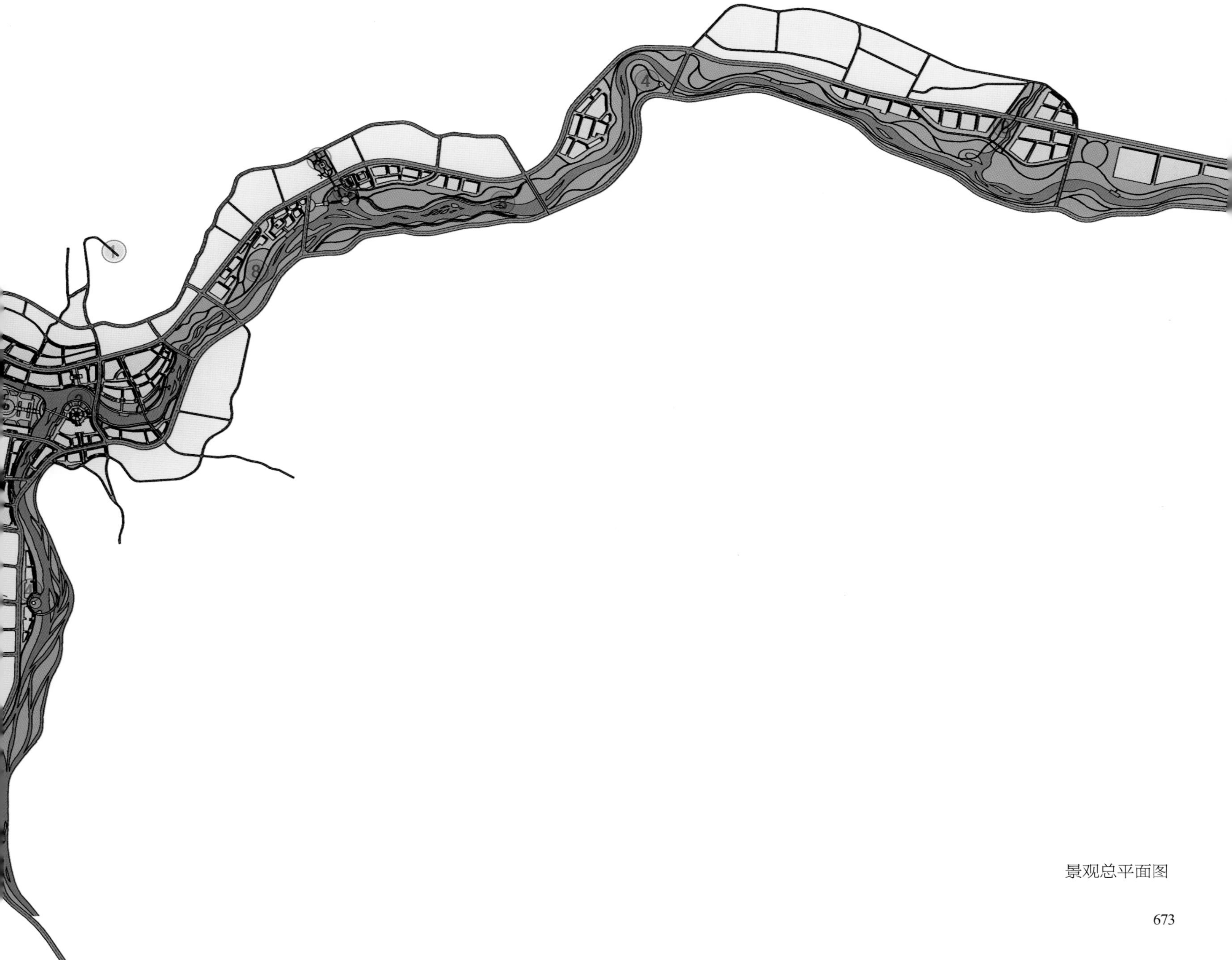

景观总平面图

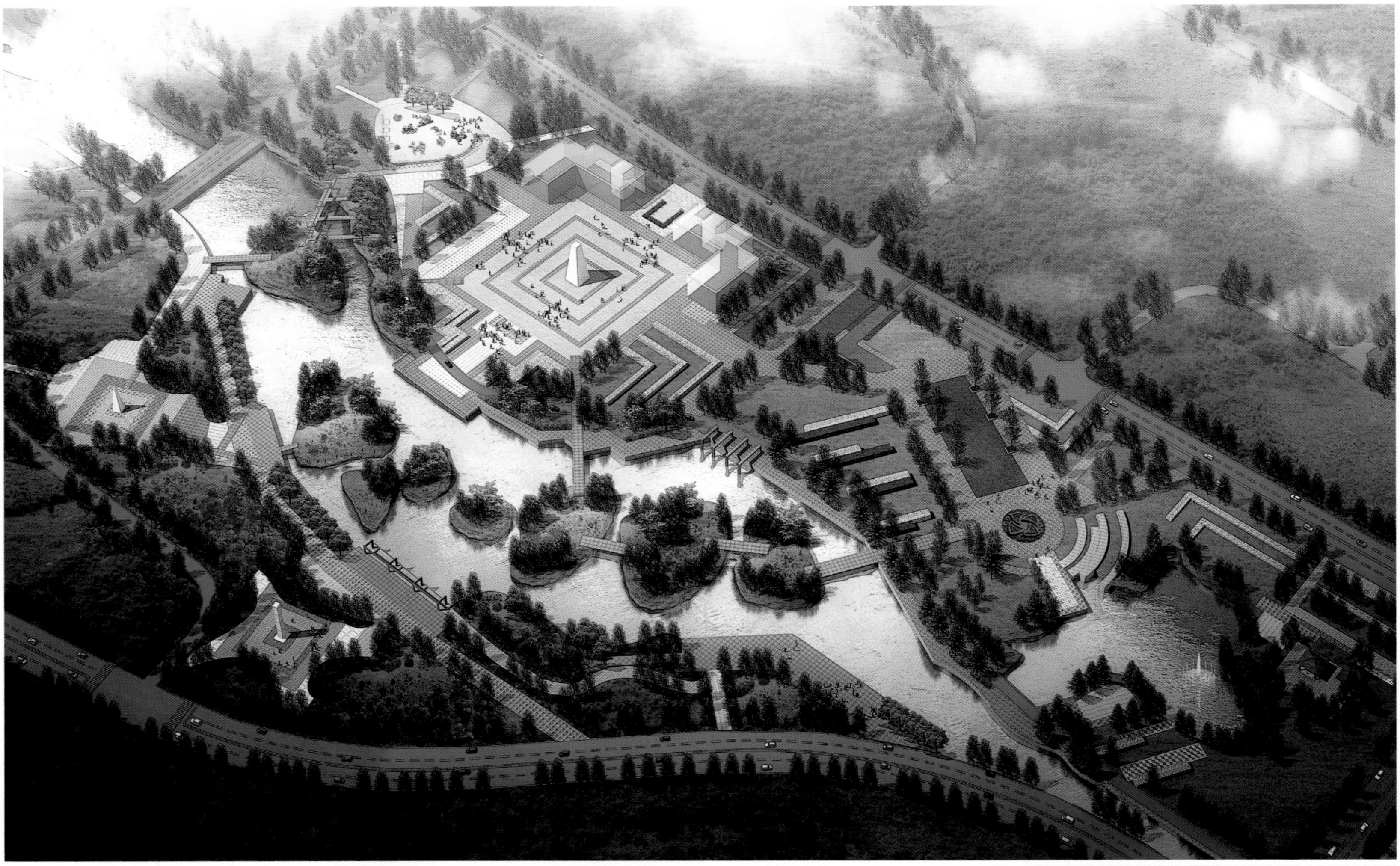

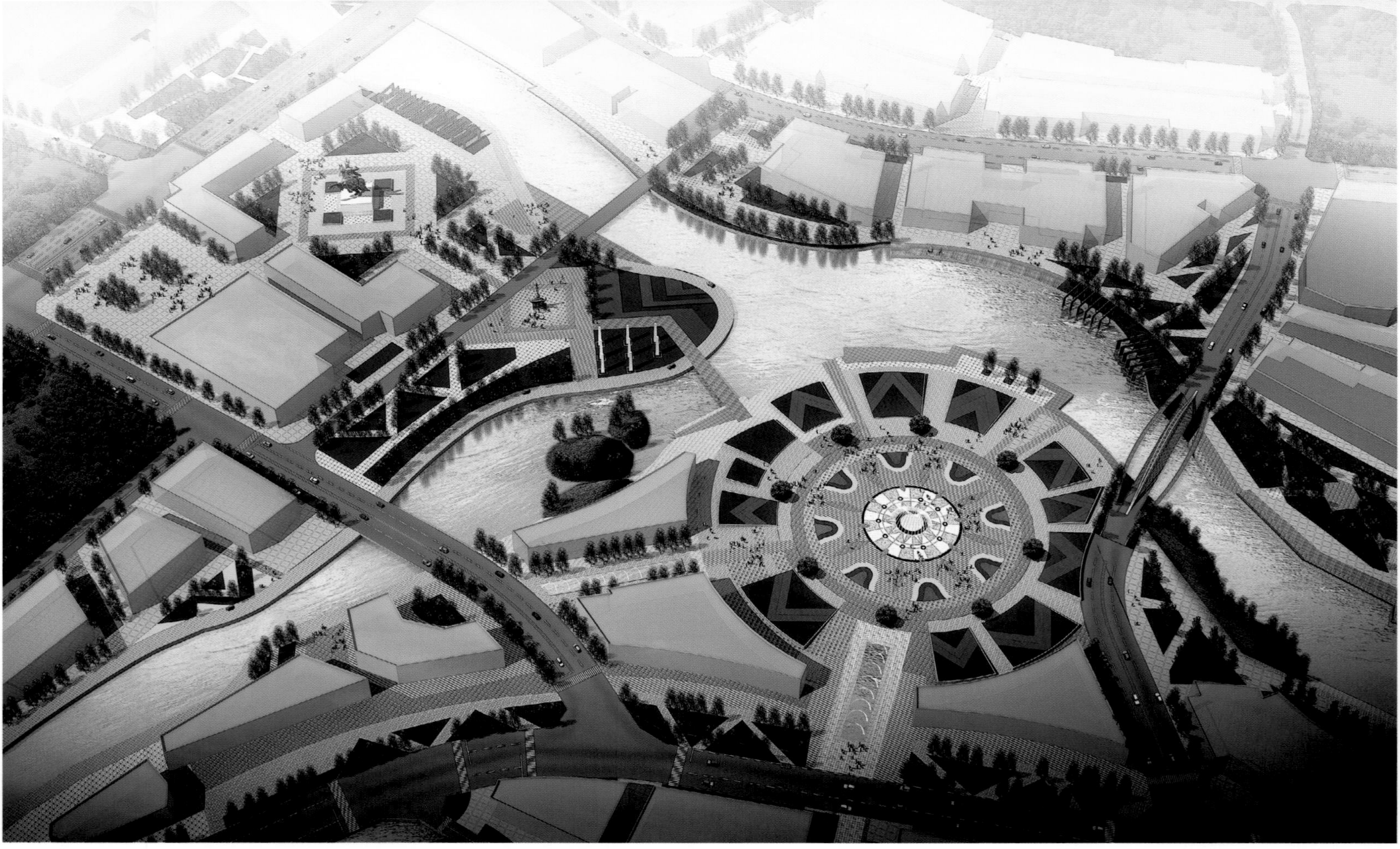

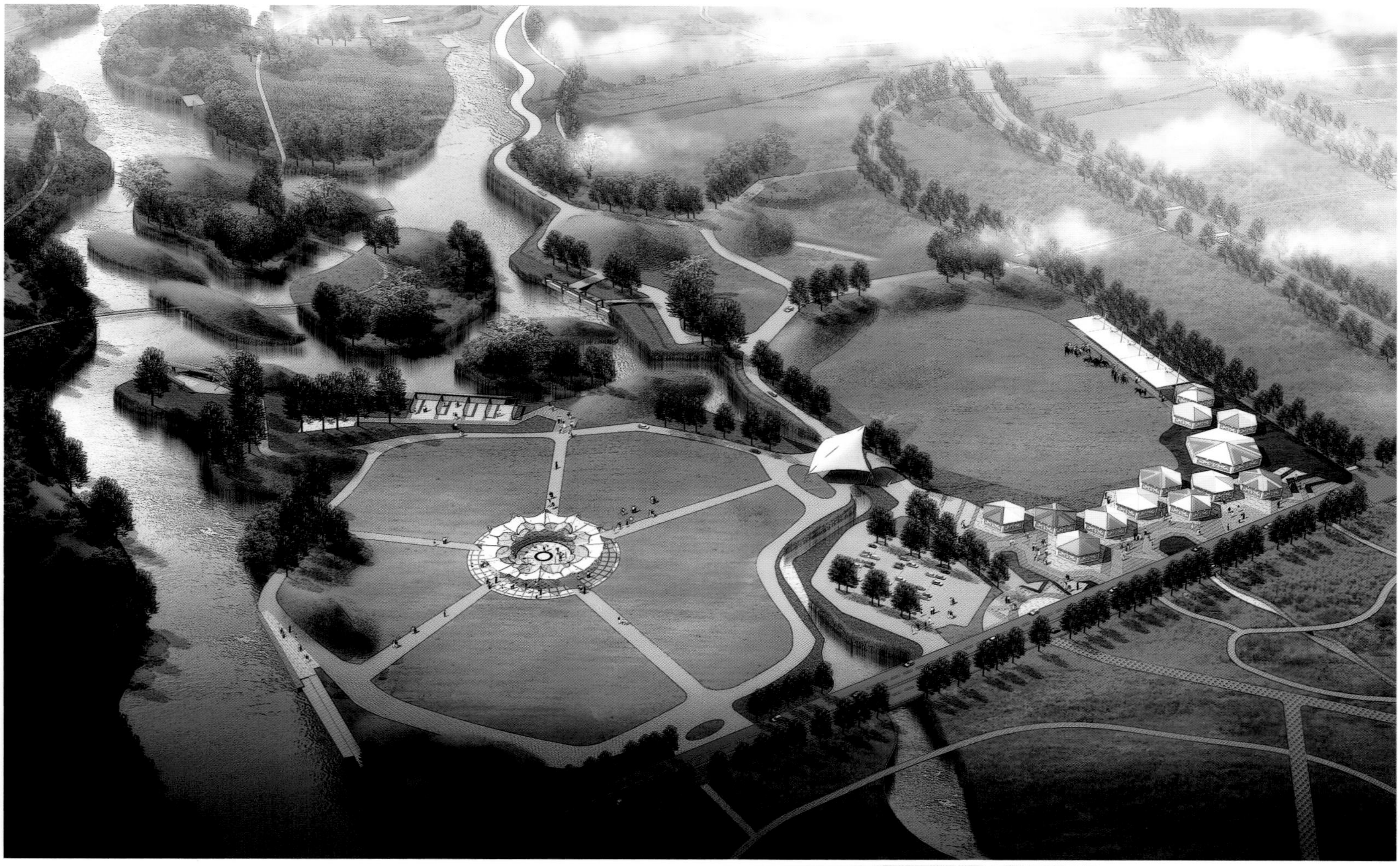

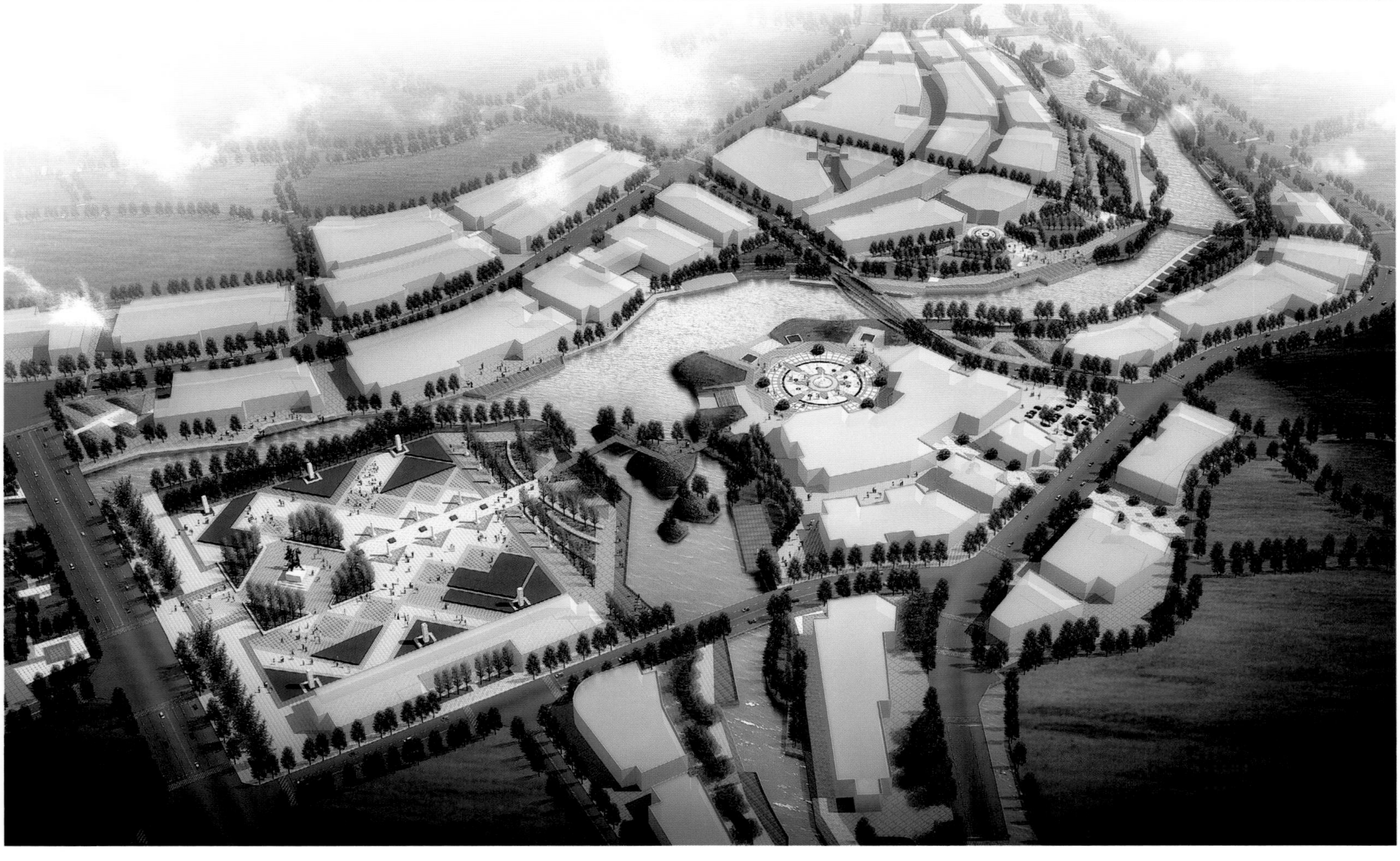

苏宁环球璞邸景观设计
Landscape Design of Suning Universal Pudi Landscape Plan

项目名称：苏宁环球璞邸景观设计
项目地点：中国 / 江苏 / 南京
建筑面积：9 600 平方米
编制时间：2009 / 12
业主：苏宁环球集团
服务范围：方案 / 施工图

设计希望在嘈杂的城市环境中，通过景观设计手法，创造出一个安静、典雅、具有古典气质并适合人居住和活动的场所。

By way of the landscape design method, the design aims to create a tranquil, elegant, classic residence and sports place in the noisy urban environment.

节点细部平面图

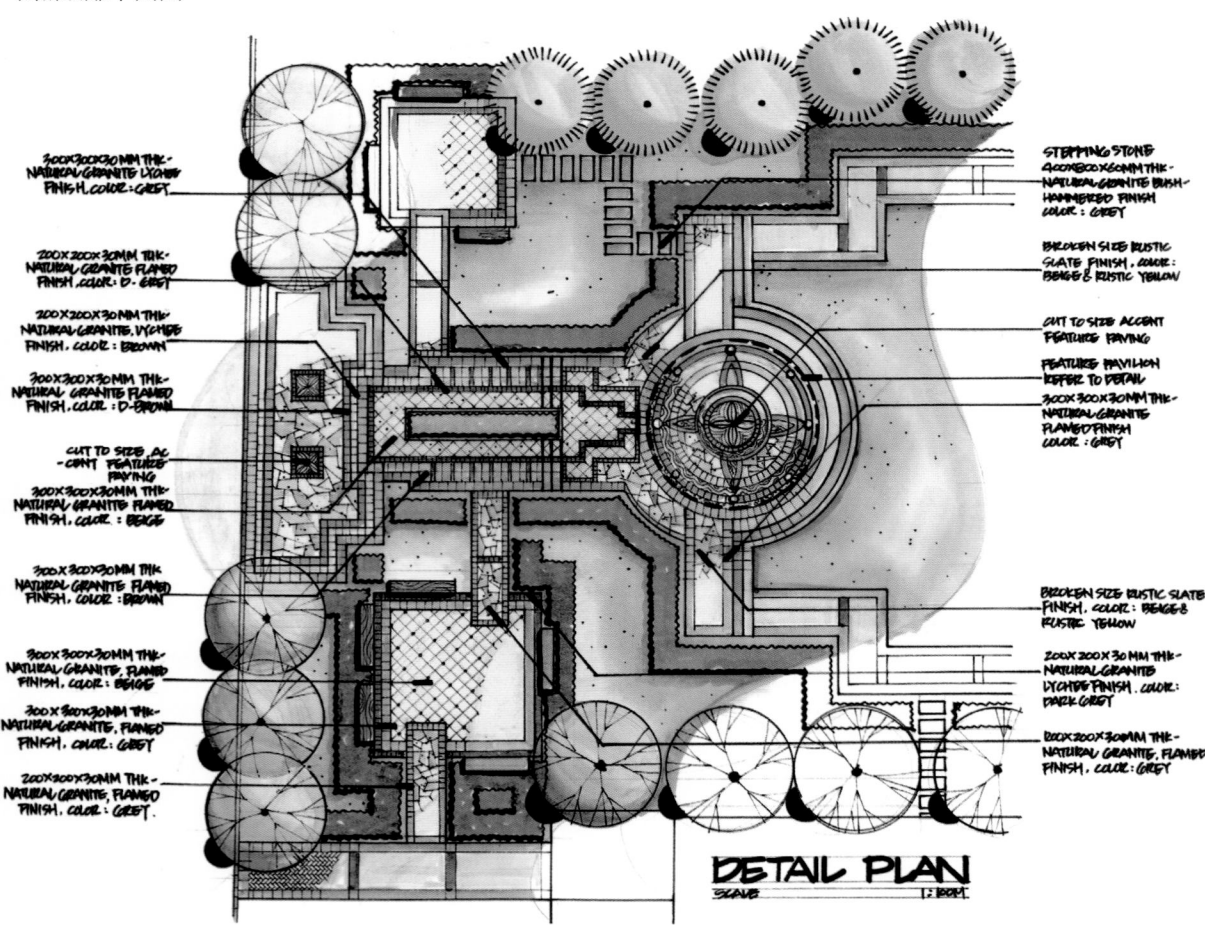

DETAIL PLAN
SCALE 1:100M

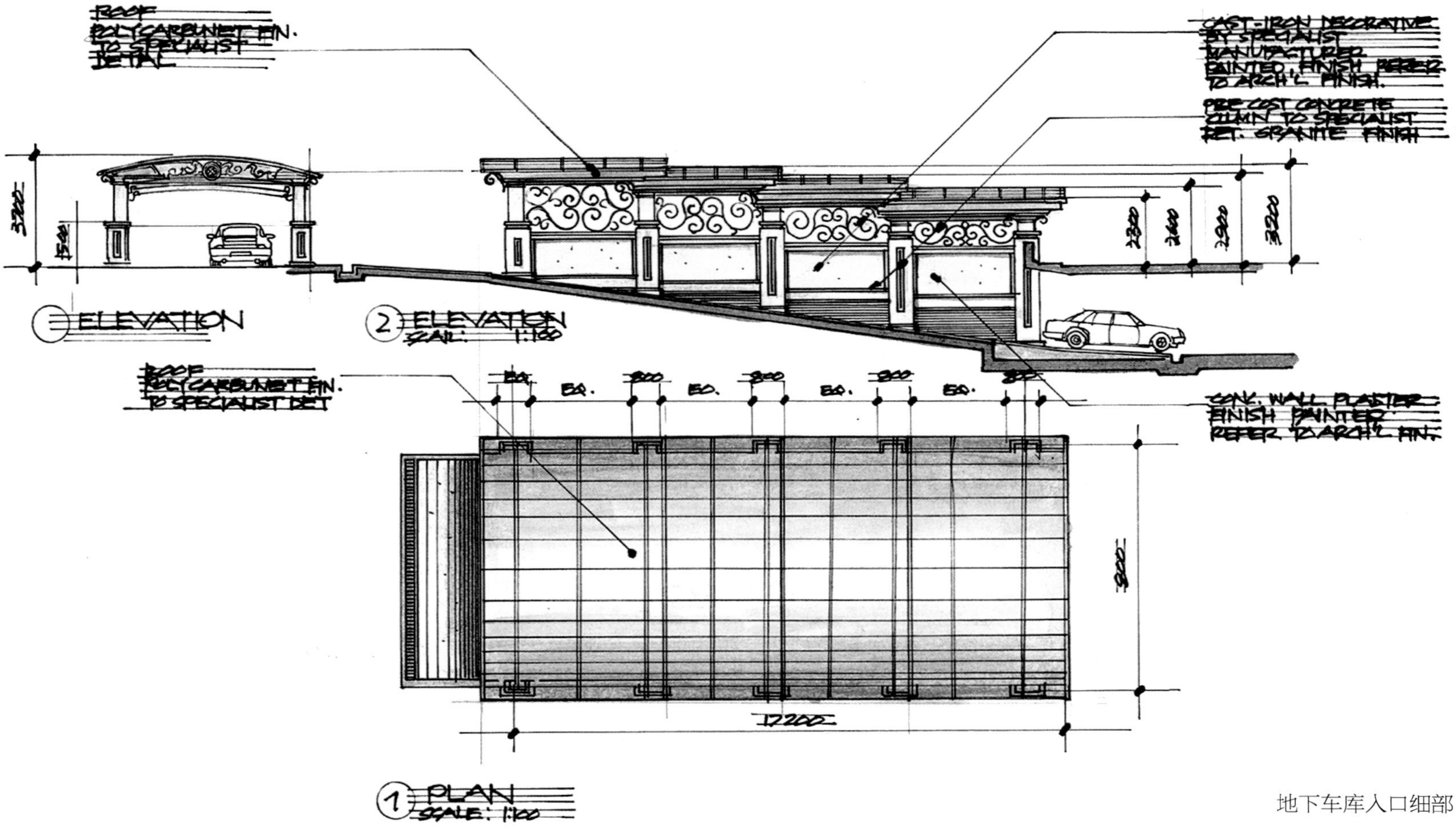

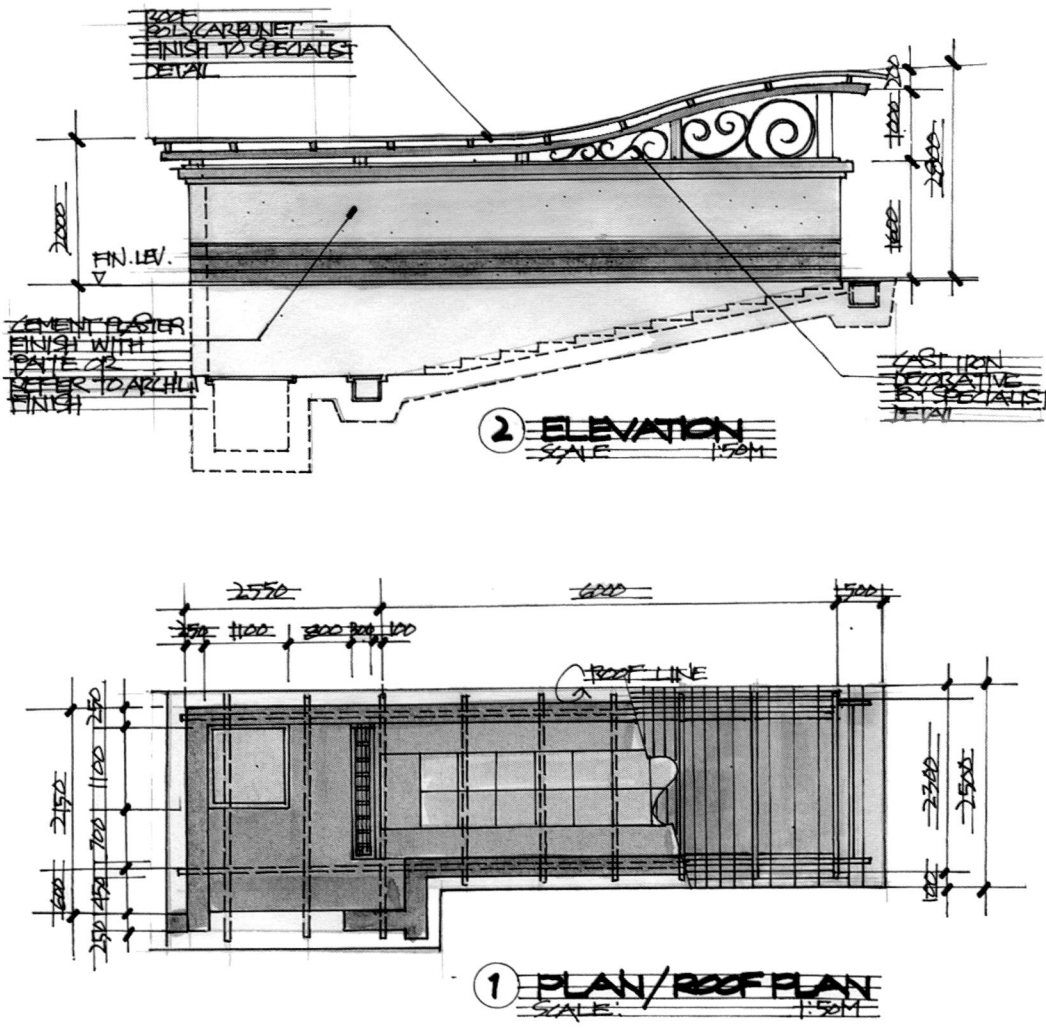

ROOF
POLYCARBONET FIN.
TO SPECIALIST
DETAIL

CAST-IRON DECORATIVE
BY SPECIALIST
MANUFACTURER
PAINTED FINISH REFER
TO ARCH'L FINISH.

PRE-CAST CONCRETE
COLUMN TO SPECIALIST
DET. GRANITE FINISH

ELEVATION

2 ELEVATION
SCALE 1:100

ROOF
POLYCARBONET FIN.
TO SPECIALIST DET.

CONC. WALL PLASTER
FINISH PAINTED
REFER TO ARCH'L FIN.

1 PLAN
SCALE: 1:100

地下车库入口细部

ROOF
POLYCARBONET
FINISH TO SPECIALIST
DETAIL

FIN. LEV.

CEMENT PLASTER
FINISH WITH
PATTE OR
REFER TO ARCH'L
FINISH

2 ELEVATION
SCALE 1:50M

CAST IRON
DECORATIVE
BY SPECIALIST
DETAIL

ROOF LINE

1 PLAN/ROOF PLAN
SCALE: 1:50M

地下人行入口细部

679

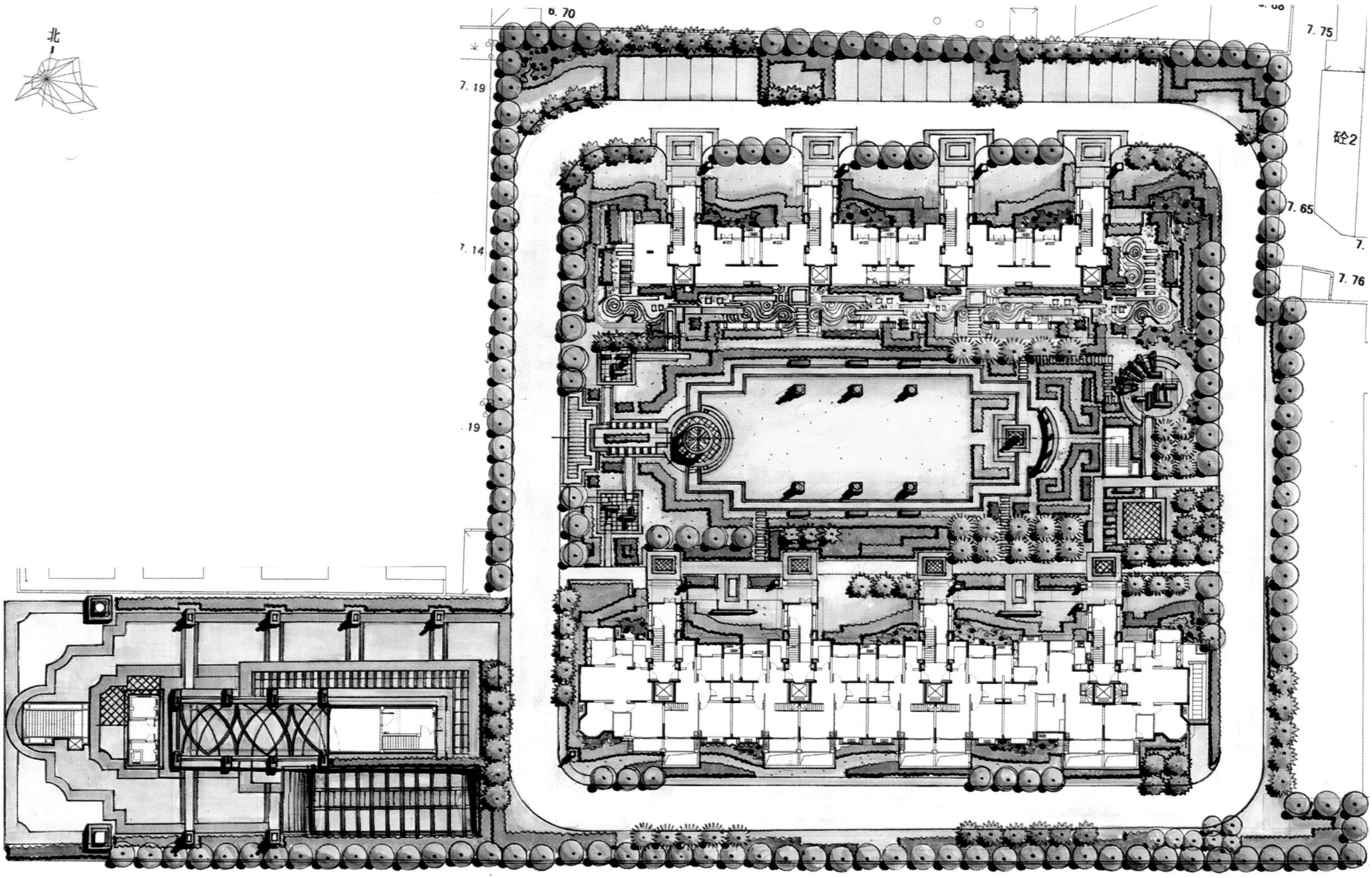

第一轮方案总平面图

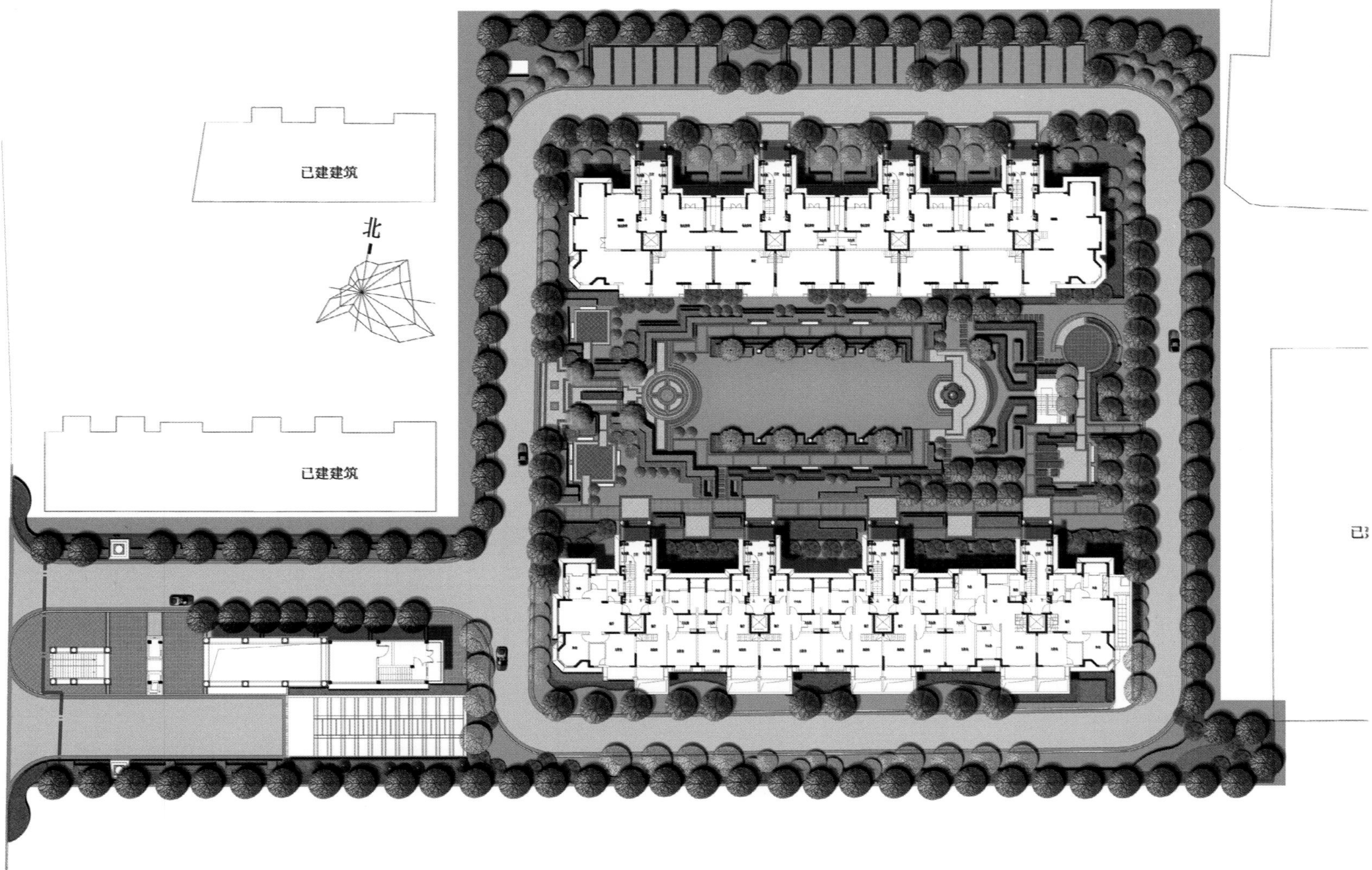

北

已建建筑

北

已建建筑

实施方案总平面图

浦江仙华文景园文化公园
Pujiang Xianhua Wenjing Cultural Park

项目名称：浦江仙华文景园文化公园
项目地点：中国 / 浙江 / 金华
建筑面积：28 000 平方米
编制时间：2010 / 05
建造时间：2010~2011
业主：浦江县仙华文景建设工程指挥部
设计深度：方案 / 扩初 / 施工图

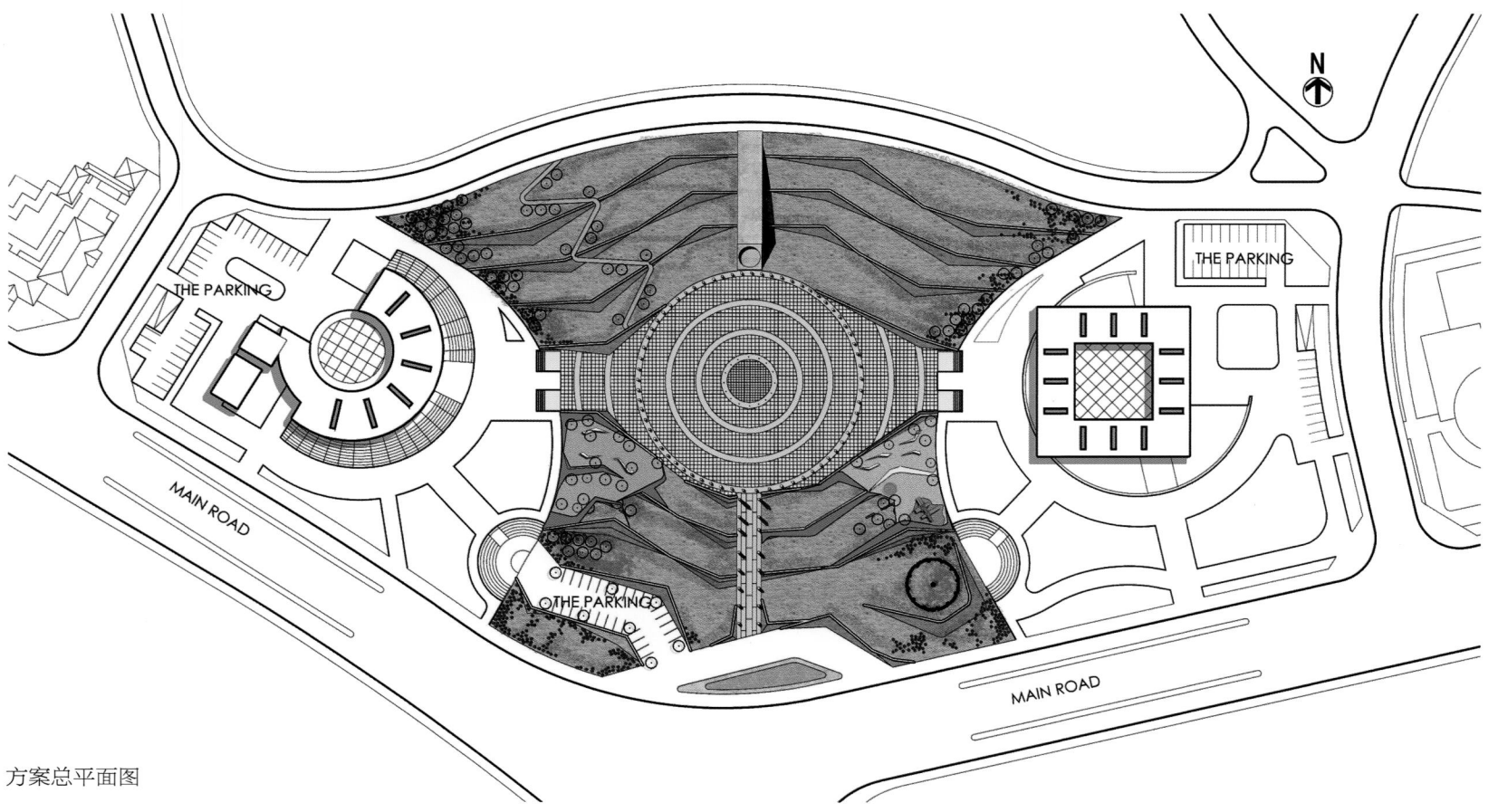

方案总平面图

设计师意图创造简洁、整体、流畅的视觉空间，同时集文化、人的活动以及城市的功能于一体。

The designers try to create a clear, complete and fluent visual space, combined with cultural, human activities and city functions.

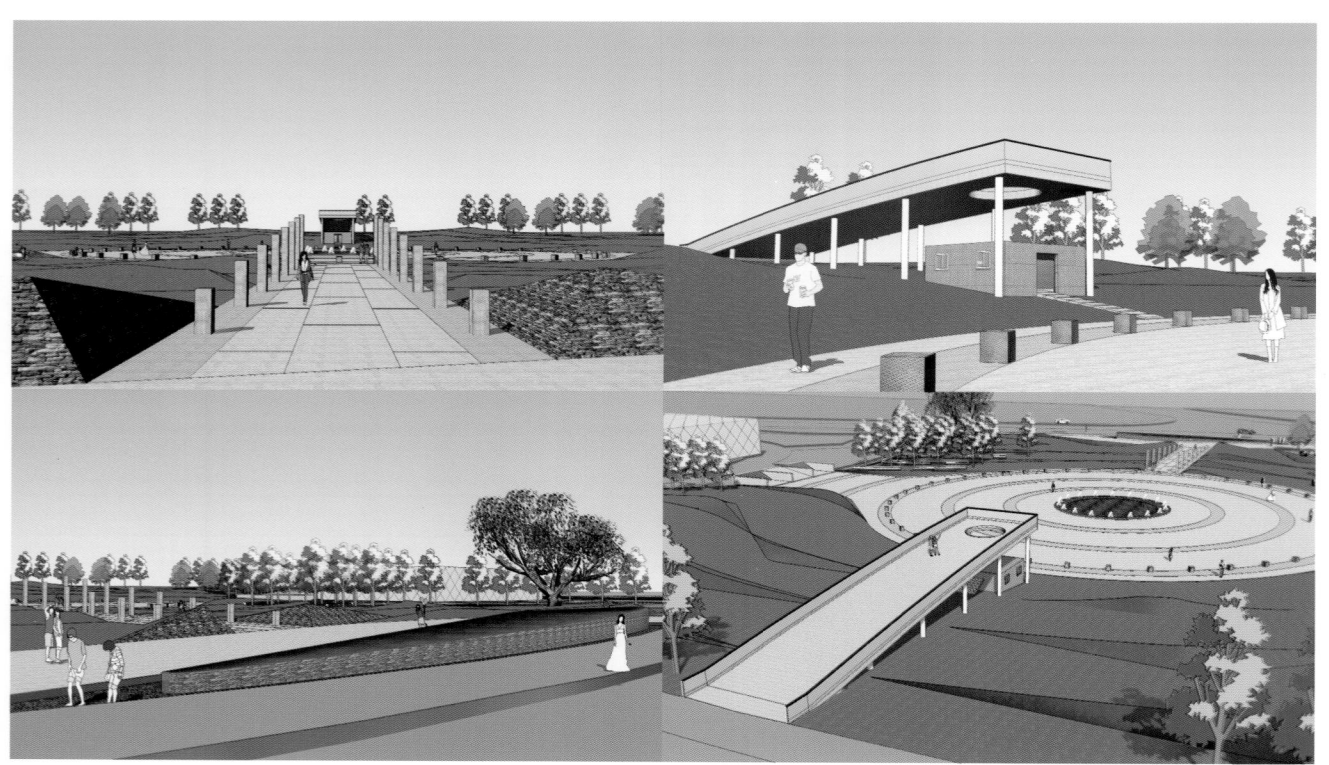

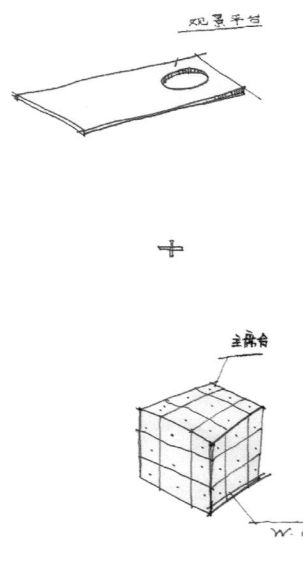

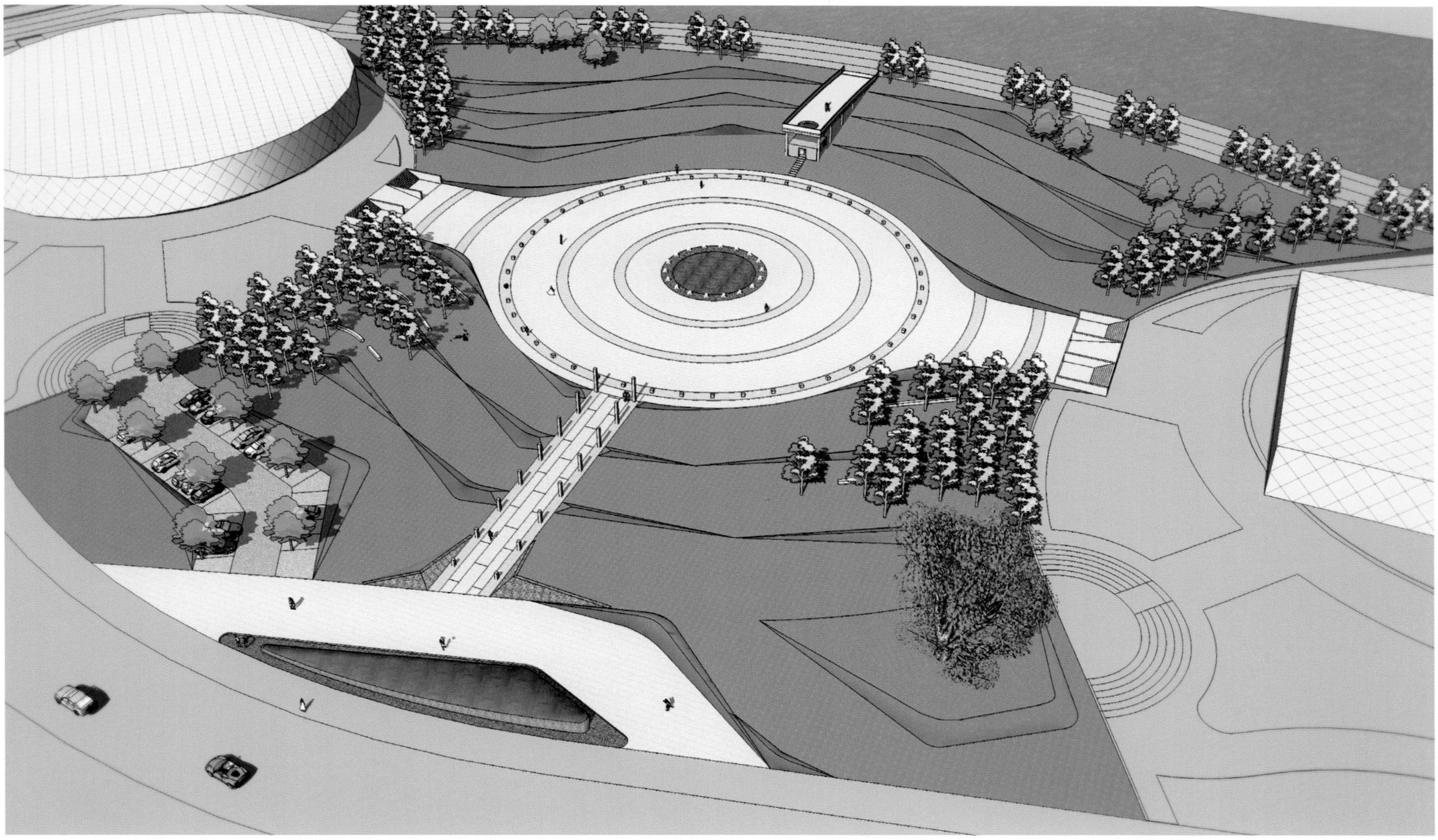

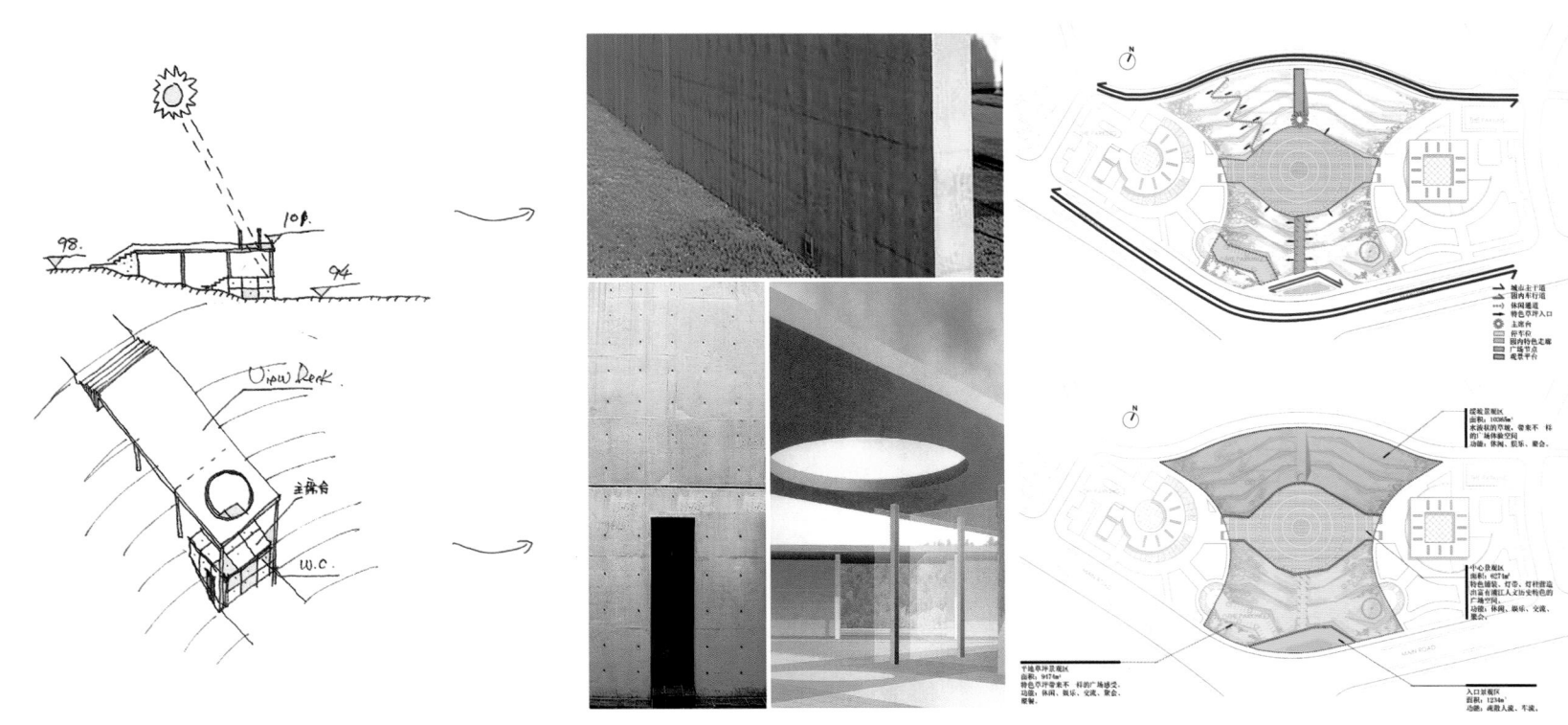

无锡新区长江路
Wu Xi New Area Changjiang Road

项目名称：无锡新区长江路
项目地点：中国 / 江苏 / 无锡
长度：2.29 公里
编制时间：2010 / 05
业主：无锡市新区规划建设环保局
服务范围：方案

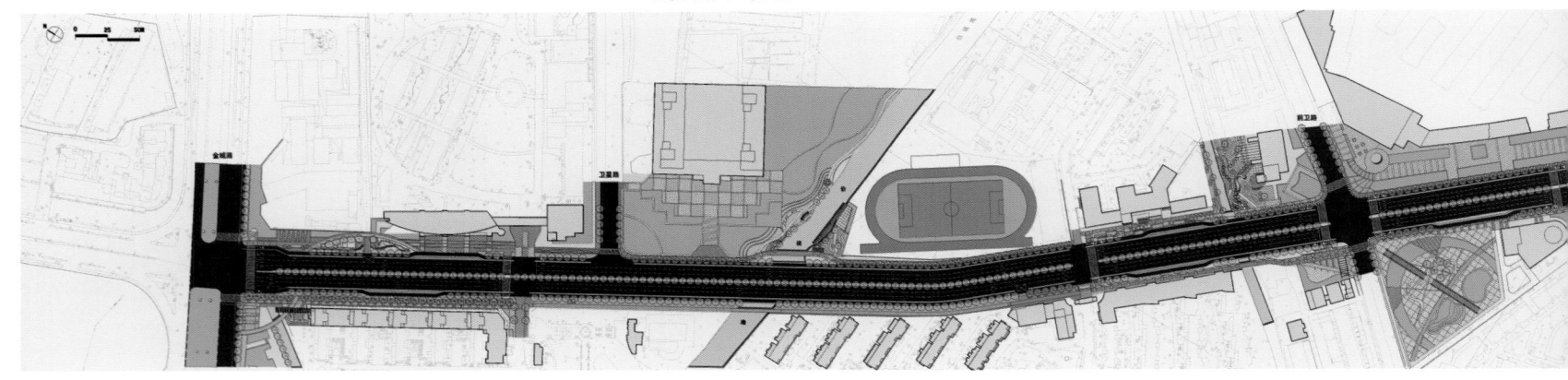

通过景观设计和城市设计手法，改变原有新区道路尺度巨大、缺乏个性、不适合行人活动等现在城市新区建设中普遍存在的通病。

By means of landscape and urban design methods, the original common faults in the construction of newly-developed area in modern city will be solved, e.g. large-scale roads in the original new district, lack of individuality, unsuitability for pedestrian activities, etc.

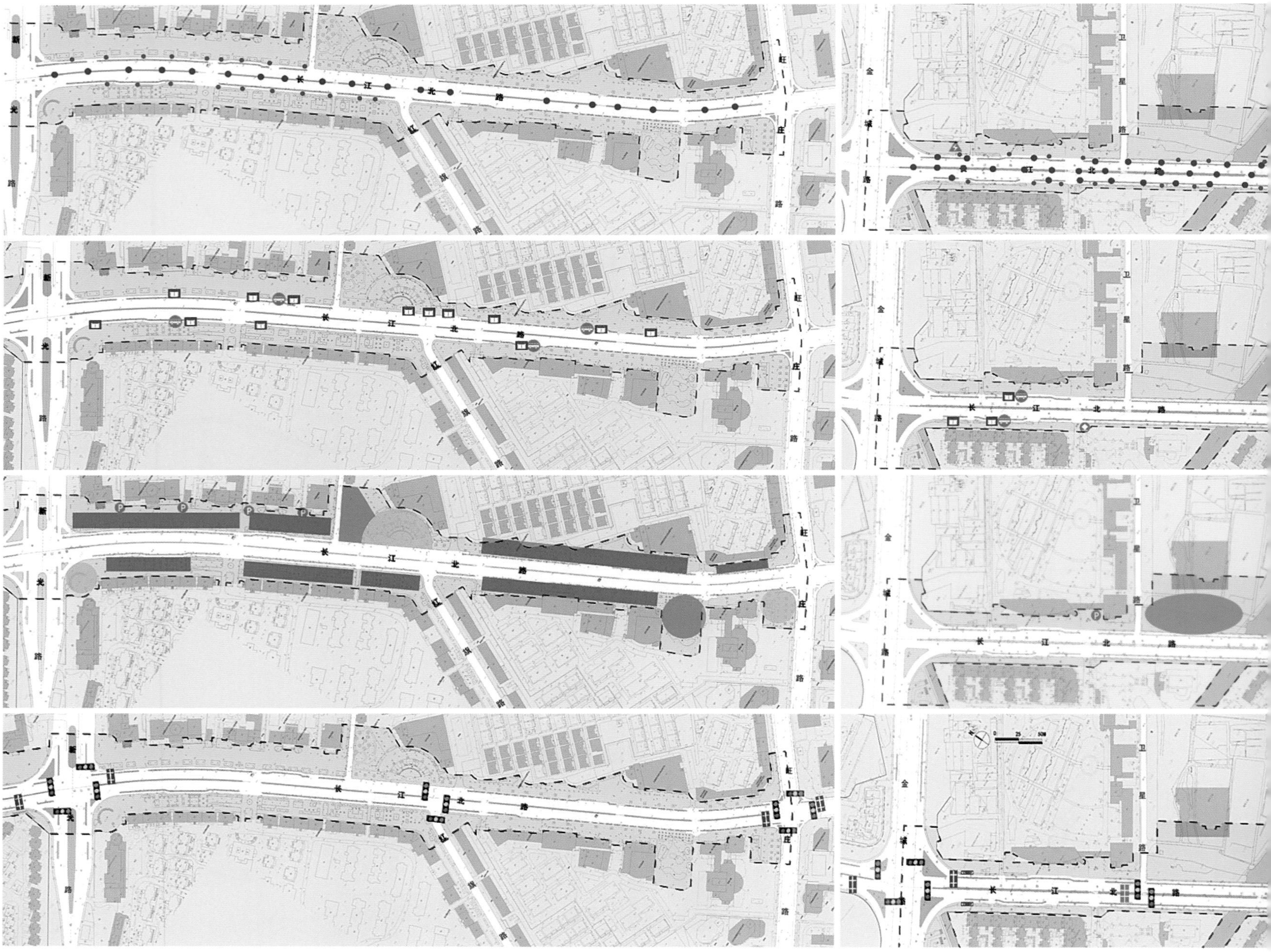

景观总平面图

系列分析图

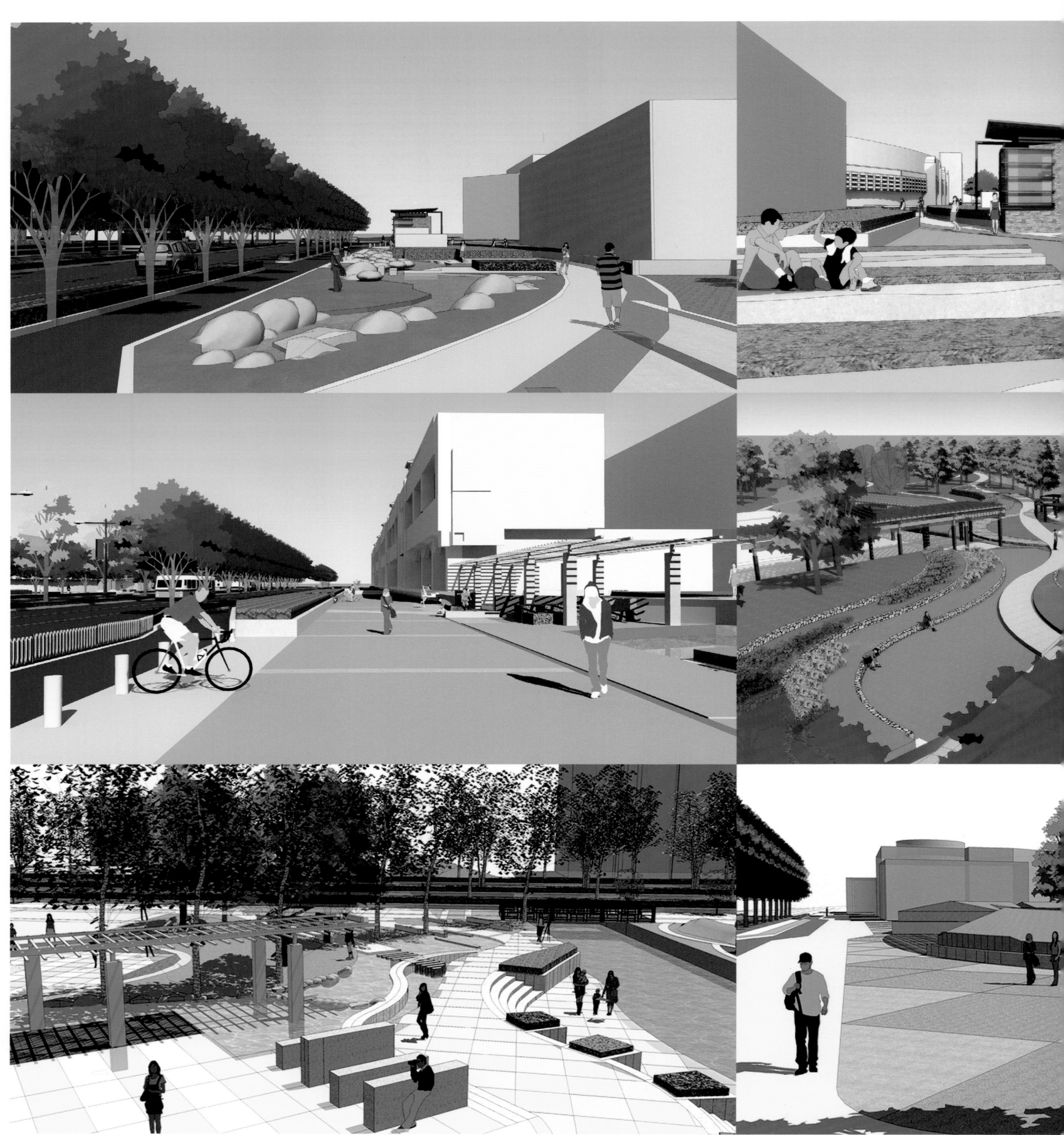

哈尔滨金色莱因国际商业广场
Harbin Golden Rhine International Commercial Square

项目名称：哈尔滨金色莱因国际商业广场
项目地点：中国 / 黑龙江 / 哈尔滨
建筑面积：29 000 平方米
编制时间：2010 / 09
业主：哈尔滨华鸿房地产有限公司
服务范围：方案 / 扩初 / 施工图

设计特点：在狭小的空间内，通过景观设计手法，创造一种高贵、奢华、有品味的气氛。

Design characteristics: a dignified, luxurious and tasty atmosphere is created in a narrow space, through landscape design techniques.

屋顶景观总平面图

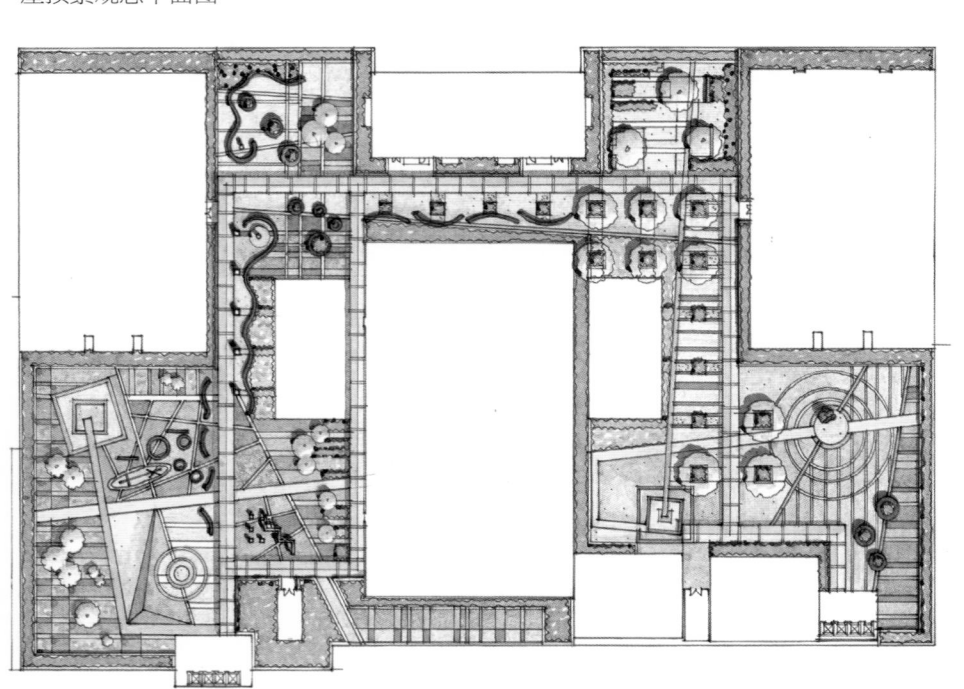

地面景观总平面图

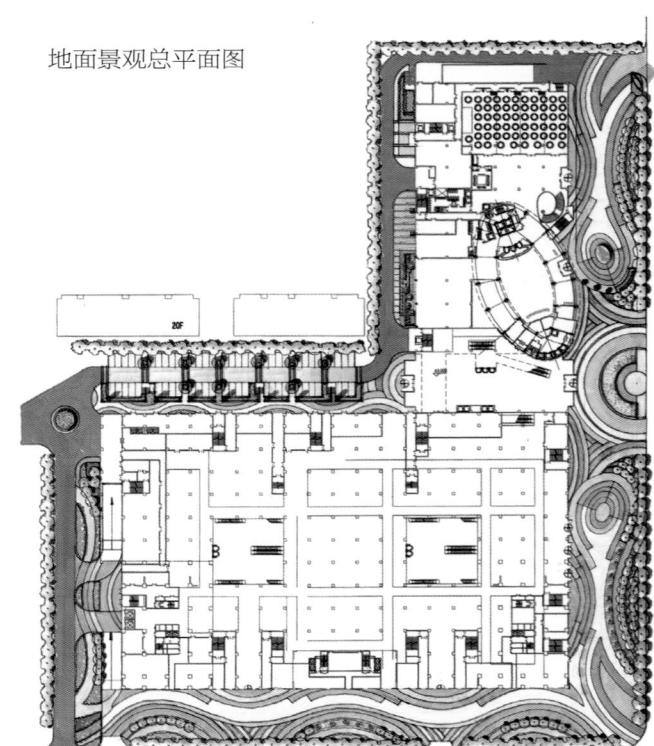

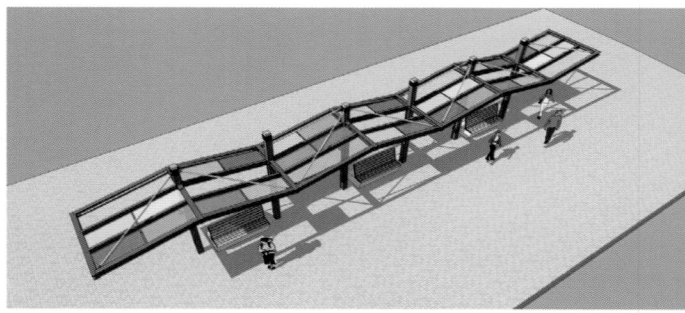

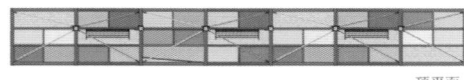

顶平面

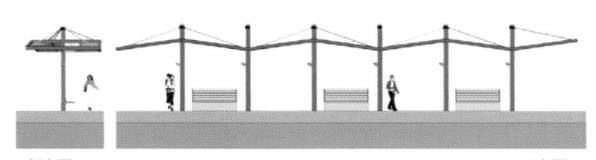

侧立面　　　　　　　　　　　　　正立面

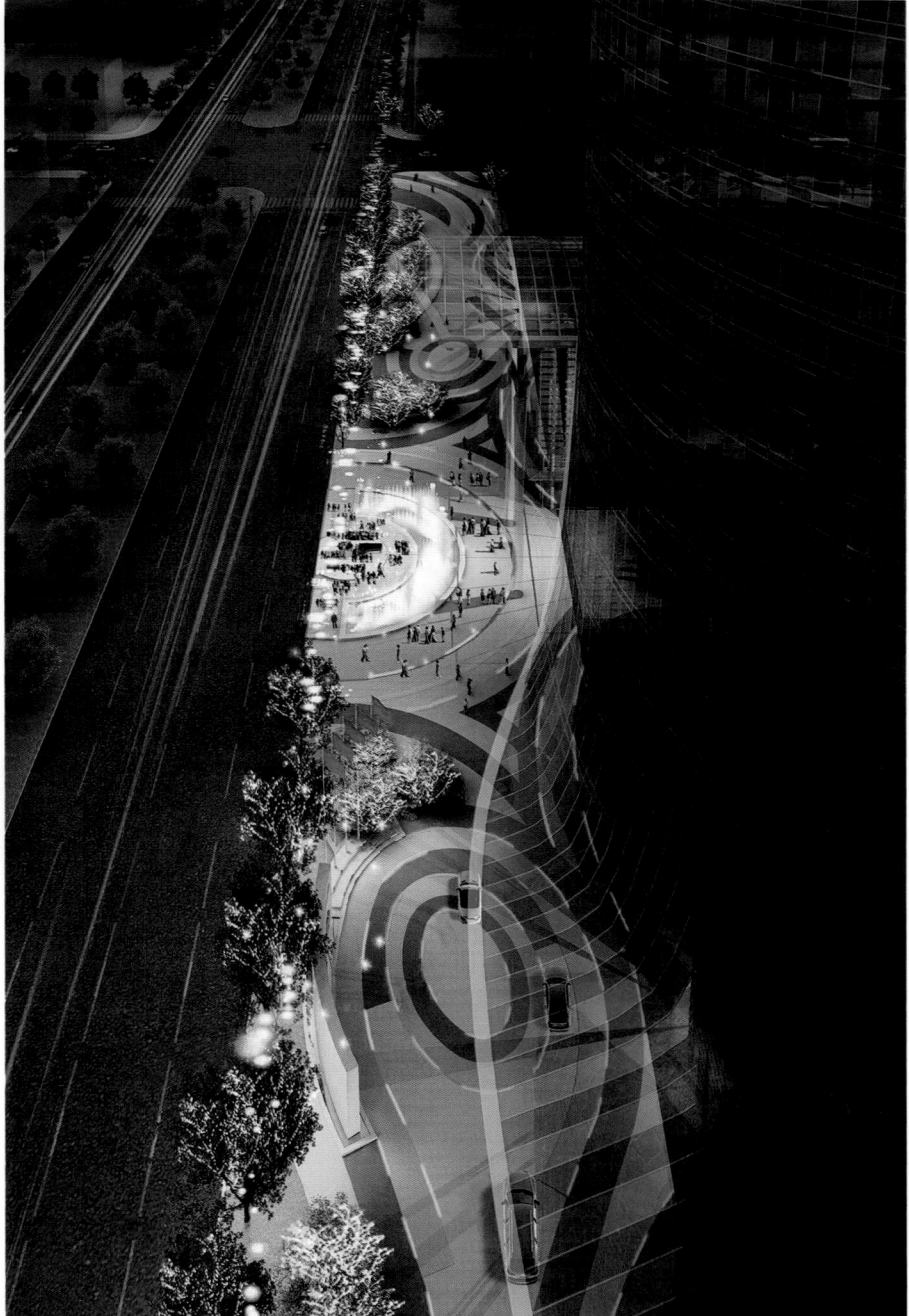

静安寺地区城市景观设计
Jing'an Temple Neighborhood Urban Landscape Design

项目名称：静安寺地区城市景观设计
项目地点：中国／上海
建筑面积：8 000 平方米
编制时间：2009／03
业主：上海市静安区城市规划管理局
服务范围：方案／扩初设计

设计理念

成长崛起中的魅力静安、夜色无限的活力静安。

设计特点

景观总体定位上以植物与水为主题。

采用"水体元素"，体现出静安的生生不息，同时采用的"植物元素"又相对静止，体现出静安寺地区宜动宜静的特点，给人以可停留可休憩的有庇护感的休闲感受。

在此次设计中，我们致力于创造出一个独特的、开放的、交通便捷的个性空间，同时能与周边地区相融合。其设计应具备可应变性，能够根据情况的变化提供不同的选择。

Design Concept

Charming Jing'an on the rise, energetic Jing'an with boundless night scenes.

Design Characteristics

The landscape overall orientation is towards the theme of plant and water.

The "water body element "is adopted, to embody the perpetual life and growth of Jing'an, while the "plant element "adopted at the same time is relatively motionless, embodying both dynamic and static characteristics of the Jing'an Temple district, and creating a nestle-down leisure feeling of stay and rest for people.

In the course of this design, we devote ourselves to creating a unique, open and traffic convenience space, meanwhile capable of mingling with other functional facilities in the neighborhood. Its design should be adaptable, capable of providing different options according to circumstantial changes.

下沉广场细部改造图

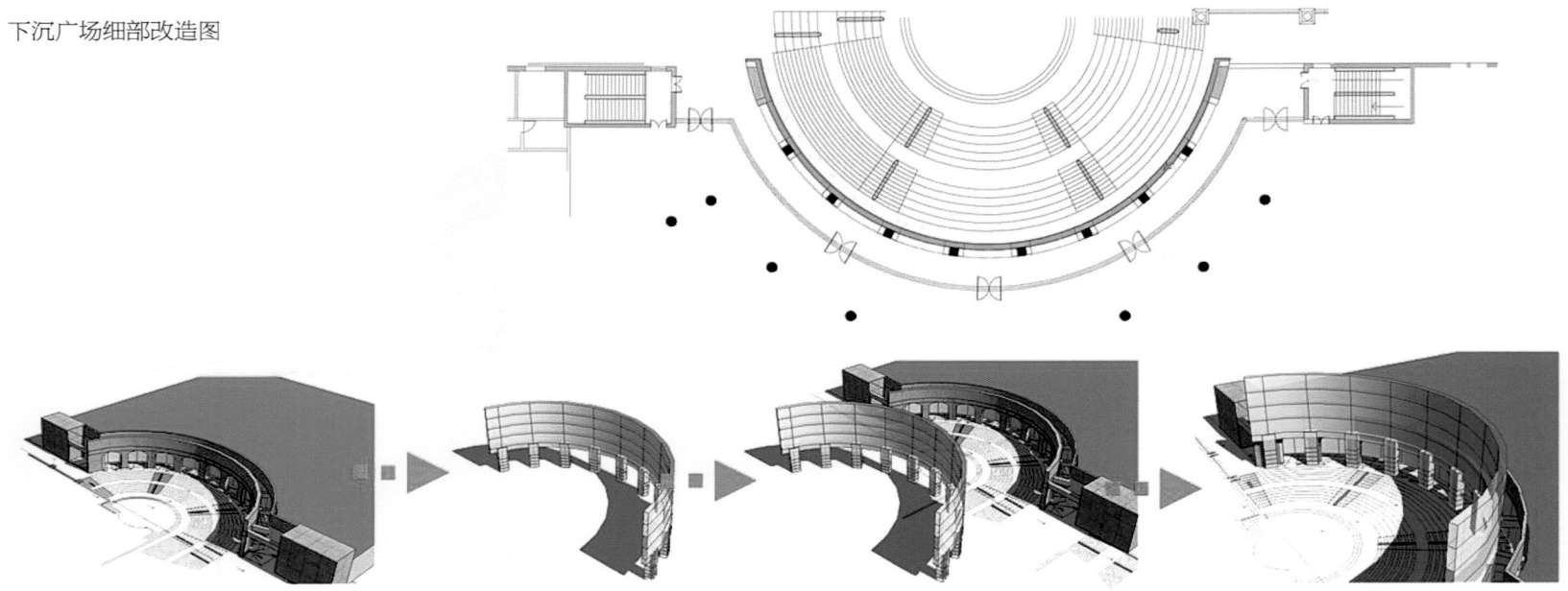

巢湖西坝口
Chaohu West Dam

项目名称：巢湖西坝口
项目地点：中国 / 安徽 / 巢湖
建筑面积：景观总用地为 7.3 万平方米
编制时间：2008 / 07
业主：巢湖市绿化园林局
服务范围：方案 / 扩初 / 施工图

设计特点

把景观设计与水利工程结合在一起，在艺术与工程的实施之间找到了平衡。

Design Characteristics

The landscape design and irrigation works are combined, and a balance is found between art and engineering practice.

景观总平面图

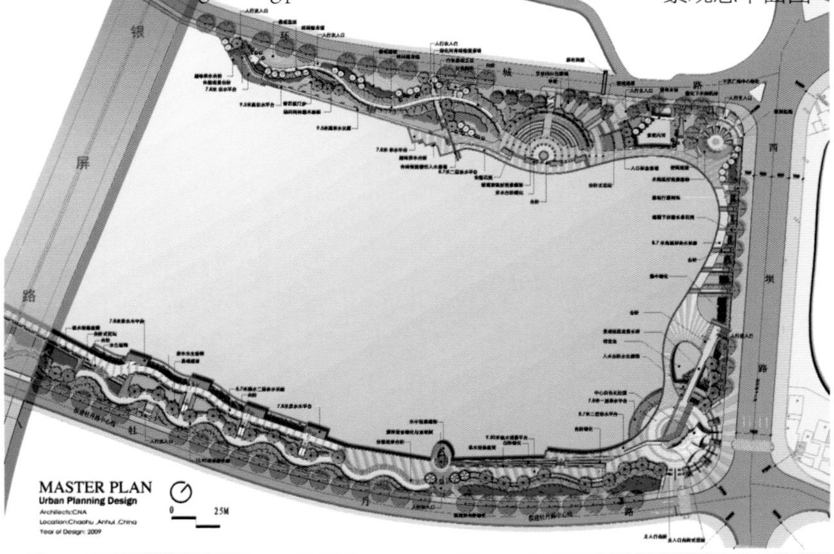

MASTER PLAN
Urban Planning Design

驳岸细部设计图

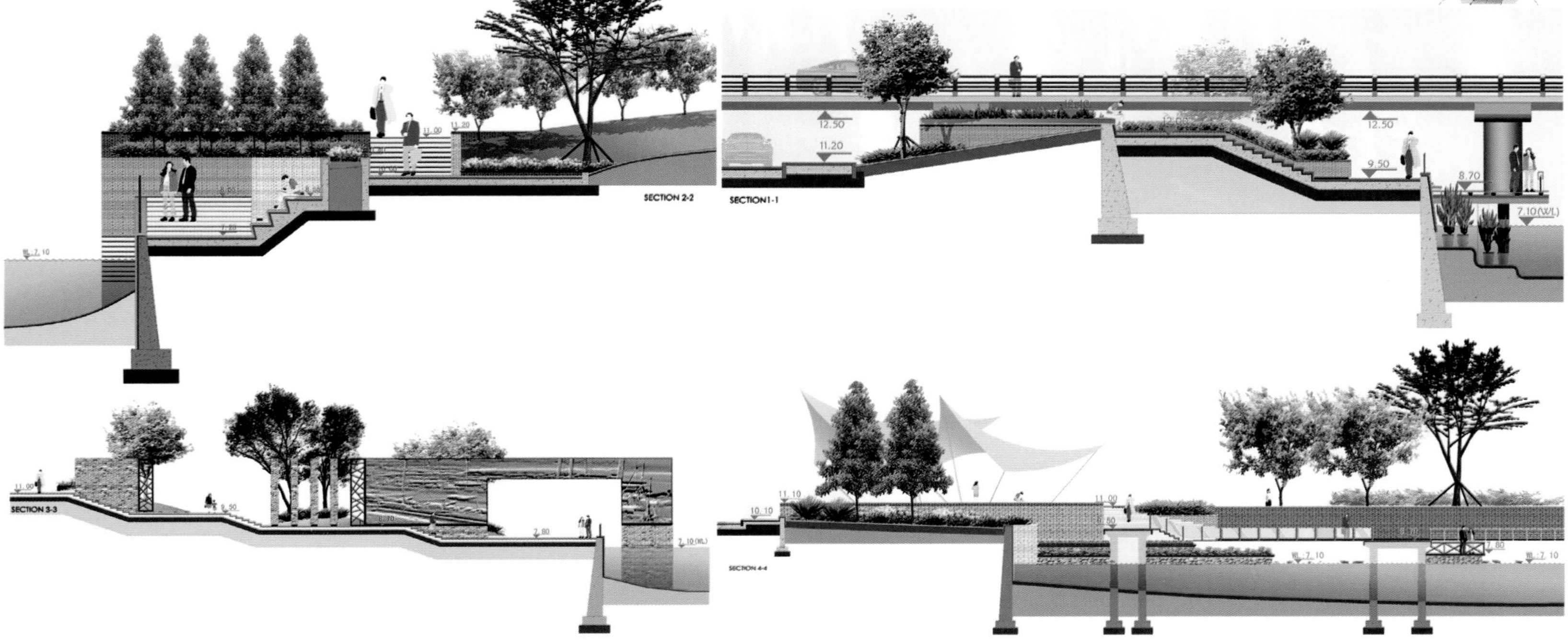

上海越洋广场
Shanghai Yueyang Square

项目名称：上海越洋广场
项目地点：中国 / 上海
建筑面积：2 000 平方米
编制时间：2010 / 04
施工时间：2010 / 09
业主：上海越洋房地产开发有限公司
服务范围：方案 / 扩初 / 施工图

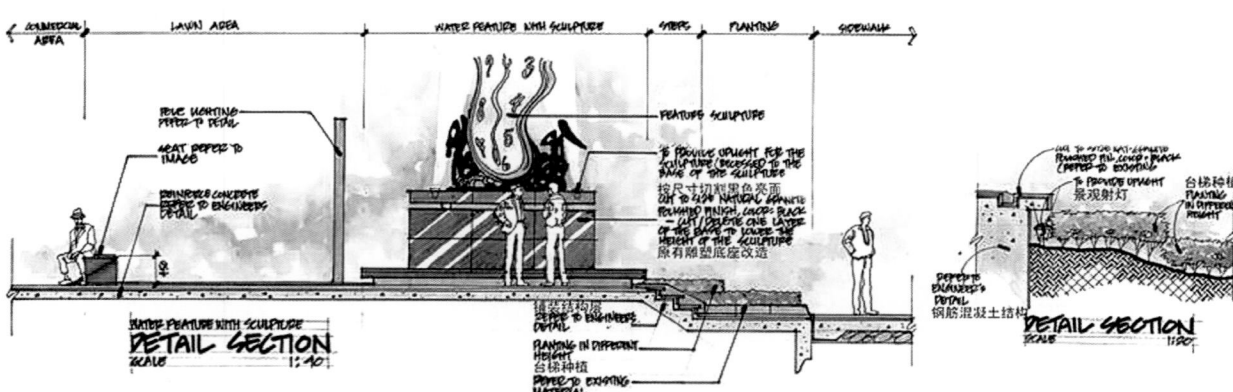

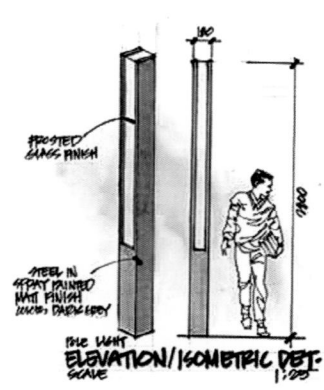

方案以简洁、现代为主基调，以实际功能为大前提，以求达到最直观的视觉效果，及通达的交通流线。小型的休闲木平台，既提供了亲水空间，也为后期的商业活动预留了充足的扩展空间，雕塑感强烈的休闲座凳，为狭小的空间增添了别样风情。

作为本案的最大亮点"达利雕塑"，通过种植池的设计，在一定程度上起到隔离人群、保护雕塑的作用，也充分预留给市民、游客更为广阔的互动空间。

The plan has the keynote of concision and modernity, and the major premise of actual function, so as to achieve the most direct visual effect, and unobstructed traffic flow lines. The small-sized recreational leisure wooden platforms provide a water-friendly space, and also allow enough expansion space for later business activities, and the recreational leisure seats of a strong sculptural sense add a distinctive flavor to the narrow space.

"Dali Sculpture", as the biggest highlight of this project, through the planting bed design, to some extent plays the role of isolating crowds and protecting the sculpture, also adequately allows a broader interaction space for citizens and tourists.

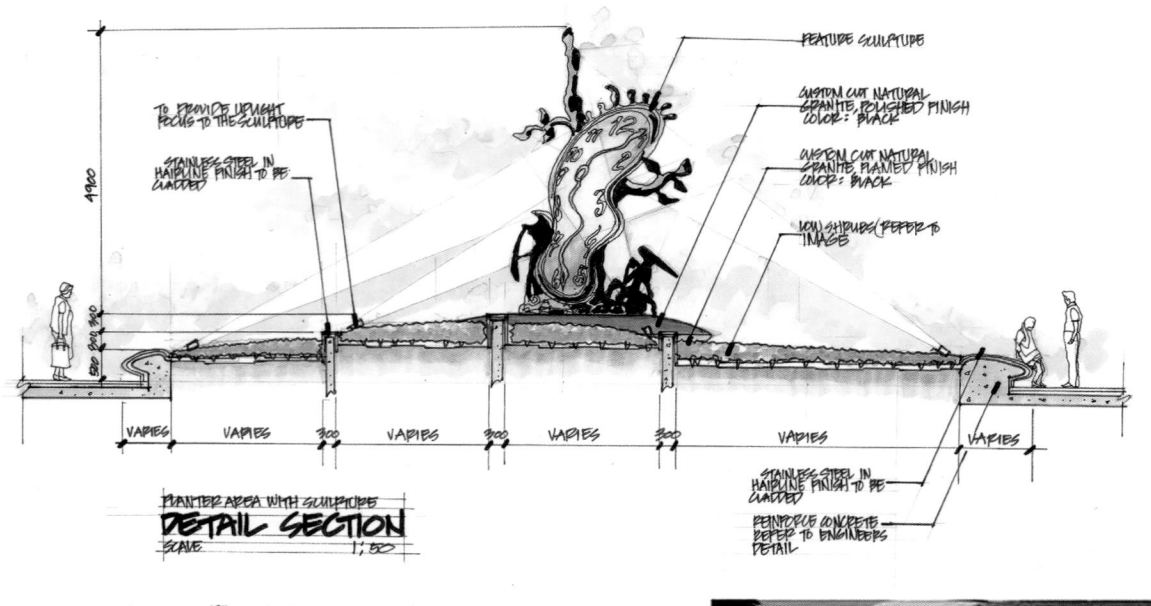

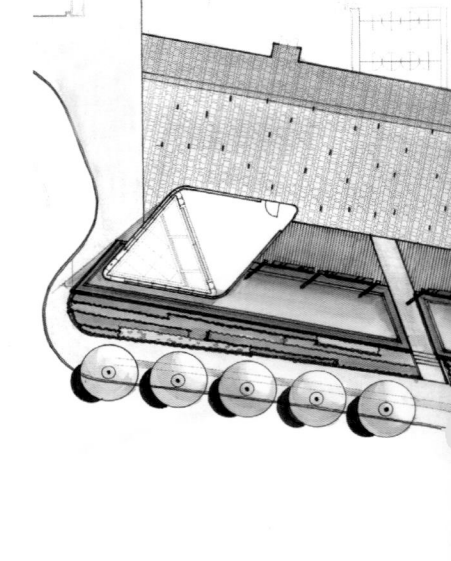

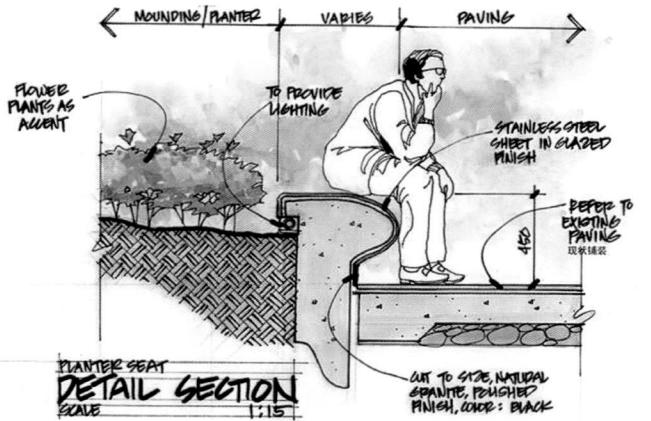

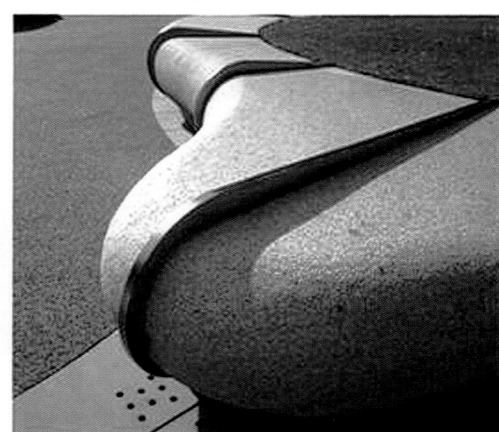

Yue Yang Piazza Landscape Design

Architects: CNA
Location: Shang Hai, China
Year of completion: 2010

LOCATIAN

Yue Yang Piazza Landscape Design

Architects: CNA
Location: Shang Hai, China
Year of completion: 2010

LOCATIAN

Yue Yang Piazza Landscape Design

Architects: CNA
Location: Shang Hai, China
Year of completion: 2010

LOCATIAN

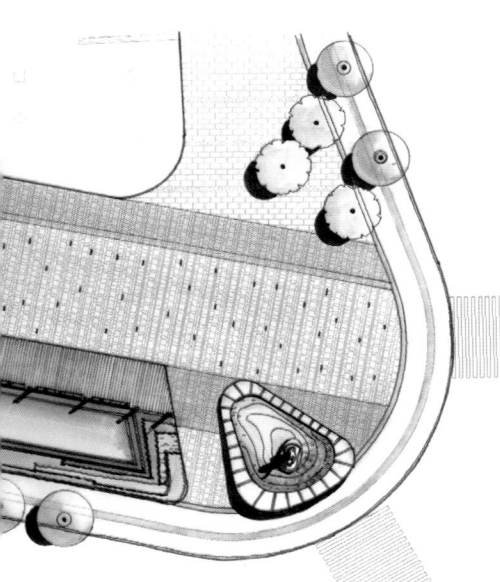

景观总平面图

淄博创业颐丰花园
Zibo Chuangye Yifeng Garden

项目名称：淄博创业颐丰花园
项目地点：中国 / 山东 / 淄博
建筑面积：规划总用地为 21.21 万平方米
编制时间：2009 / 12
业主：山东创业房地产开发有限公司
服务范围：方案 / 扩初 / 施工图

MASTER PLAN
Urban Planning Design

景观总平面图

创业颐丰花园位于淄博市新城区，城市核心区北部，是淄博新城区的标志性地段，四周被三条城市主要干道和一条区内支路围绕，绿地率达到 40%。小区规划采取 2 个街坊布局，结构清晰，每个街坊有各自的中心绿地和主次空间轴线组织，使小区形成统一的有机整体。

景观设计师在设计过程中，意在延续建筑的特点，体现新古典主义的景观设计风格，以"健康，和谐，生态，乐观"为设计主线，采用简约而不简单的设计手法与选材，打造宁静、古典、大气优雅的绿色生态的社区环境。景观元素以高低错落的植物配置、优雅的水体曲线、神秘精致的镂空铁艺、图案化的地面铺装、细腻的景观小品，一气呵成地营造出花园式地标式的高档社区环境。该项目在 09 年被评为"淄博市国家康居示范工程"。

Chuangye Yifeng Garden, located in the new urban district of Zibo city, to the north of the city heart, is the marker sector of the Zibo new urban district, surrounded by three arteries and one minor road inside the district circumferentially, with the greening rate up to 40%. The community planning adopts the layout of 2 neighborhoods. The structure is clear. Each neighborhood has its own central green space and the organization of primary and secondary spatial axes, so that the community is formed into a unified organic whole.

The landscape architect, in the design process, intends to extend the characteristics of the architecture, embody the neo-classical landscape design style, with the design principle of "health, harmony, ecology, optimism", and the terse yet non-simple design technique and material selection, to forge a serene, classic, grand, and elegant green ecological community. The landscape elements of plant configuration, elegant water-body curves, mysterious delicate hollow iron crafts, patterned ground pavement, and exquisite landscape ornaments, smoothly and harmoniously create a high-grade community environment of garden type and landmark type.

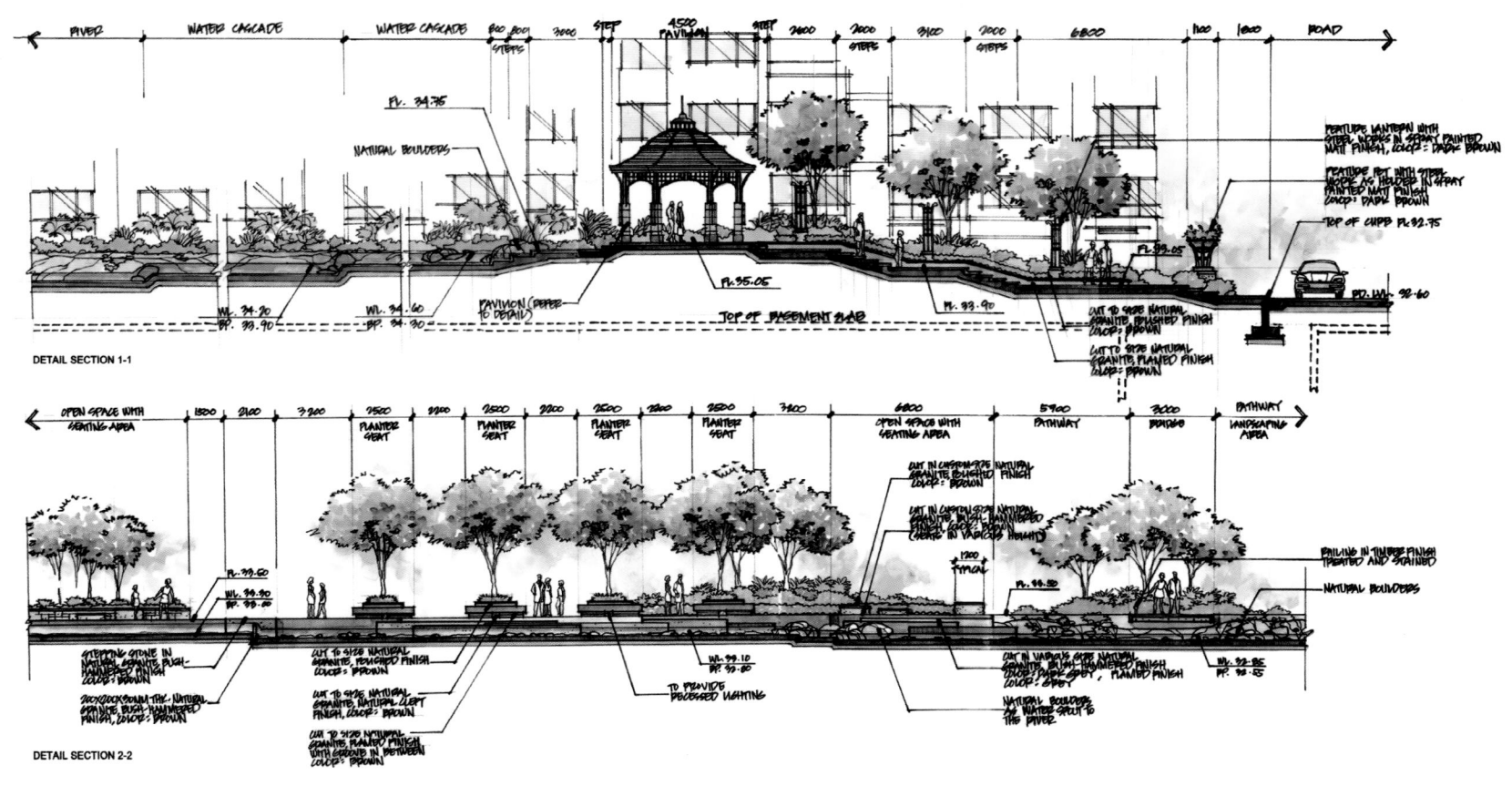

DETAIL SECTION 1-1

RIVER | WATER CASCADE | WATER CASCADE | STEP | PAVILLION | STEP | STEPS | ROAD

FEATURE LANTERN WITH STEEL WORKS IN SPRAY PAINTED MATT FINISH, COLOR: DARK BROWN
FEATURE POT WITH STEEL WORKS AS ABOVE IN SPRAY PAINTED MATT FINISH, COLOR: DARK BROWN
TOP OF CURB FL.92.75
CUT TO SIZE NATURAL GRANITE, POLISHED FINISH, COLOR: BROWN
CUT TO SIZE NATURAL GRANITE, FLAMED FINISH, COLOR: BROWN

NATURAL BOULDERS
PAVILION REFER TO DETAIL
TOP OF BASEMENT SLAB

DETAIL SECTION 2-2

OPEN SPACE WITH SEATING AREA | PLANTER SEAT | PLANTER SEAT | PLANTER SEAT | PLANTER SEAT | OPEN SPACE WITH SEATING AREA | PATHWAY | BRIDGE | PATHWAY LANDSCAPING AREA

CUT IN CUSTOM SIZE NATURAL GRANITE BUSH FINISH, COLOR: BROWN
CUT IN CUSTOM SIZE NATURAL GRANITE, BUSH HAMMERED, COLOR: IN VARIOUS HEIGHTS
RAILING IN TIMBER FINISH TREATED AND STAINED
NATURAL BOULDERS
STEPPING STONE IN NATURAL GRANITE, BUSH HAMMERED, COLOR: BROWN
300X300X50MM THK. NATURAL GRANITE, BUSH HAMMERED FINISH, COLOR: BROWN
CUT TO SIZE NATURAL GRANITE, NATURAL CLEFT FINISH, COLOR: BROWN
CUT TO SIZE NATURAL GRANITE, RADIUS FINISH WITH GROOVE IN BETWEEN COLOR: BROWN
TO PROVIDE RECESSED LIGHTING
CUT IN VARIOUS SIZE NATURAL GRANITE, BUSH HAMMERED FINISH, FLAMED FINISH COLOR: GREY
NATURAL BOULDER AS WATER SPOUT TO THE PAVER

CUT IN RANDOM SIZE NATURAL STONE FINISH, COLOR: DIFFERENT TONE OF BROWN
50X100MM TIMBER DECK TREATED AND STAINED
PLANTING AREA IN BETWEEN PLANTING, 50MM THK. NATURAL GRANITE, BUSH-HAMMERED NATURAL CLEFT COLOR: BROWN
500X500X50MM STONE STEPPING STONE NATURAL TREATED AND STAINED WITH 25MM GAP IN BETWEEN

PAVILLION REFER TO DETAIL
100X100X30MM THK. NAT. GRANITE POLISHED FINISH, COLOR: BROWN
NATURAL BOULDERS
CUT ACCORDING TO RADIUS NATURAL STONE FINISH, COLOR: DIFFERENT TONE OF BROWN

300X300X30MM THK. NATURAL GRANITE FLAMED FINISH COLOR: BROWN
300X300X50MM THK. NATURAL GRANITE, NATURAL CLEFT FINISH COLOR: DARK BROWN
ARCHWAY REFER TO DETAIL
300X300X50MM THK. NATURAL STONE FIN. COLOR: DIFFERENT TONE OF BROWN
100X100X30MM THK. COBBLE STONE FINISH COLOR: BEIGE
(2 ROWS) 150X150X50MM THK. NATURAL GRANITE FLAMED FINISH COLOR: DARK BROWN
300X300X30MM THK. NATURAL GRANITE BUSH-HAMMERED FINISH, COLOR: BROWN
FEATURE POT REFER TO DETAIL
FEATURE LIGHT REFER TO DETAIL
300X300X30MM THK. NATURAL GRANITE FLAMED FINISH COLOR: BEIGE
400X400X30MM THK. NATURAL GRANITE, BUSH-HAMMERED FINISH, COLOR: BROWN
TYP. BASEMENT ENTRANCE REFER TO DETAILS

TREE... DETAIL
STEPPING STONE CUT IN RANDOM SIZE NATURAL GRANITE, BUSH-HAMMERED FINISH, COLOR: BROWN
5-15MM PEBBLE FINISH COLOR: BEIGE
PEBBLEWASH FINISH COLOR: BROWN
GROOVE LINE
BUBBLER JET
BROKEN SIZE SLATE FINISH, COLOR: BROWN

PART PLAN 1A
DETAIL PLAN
SCALE 1:300

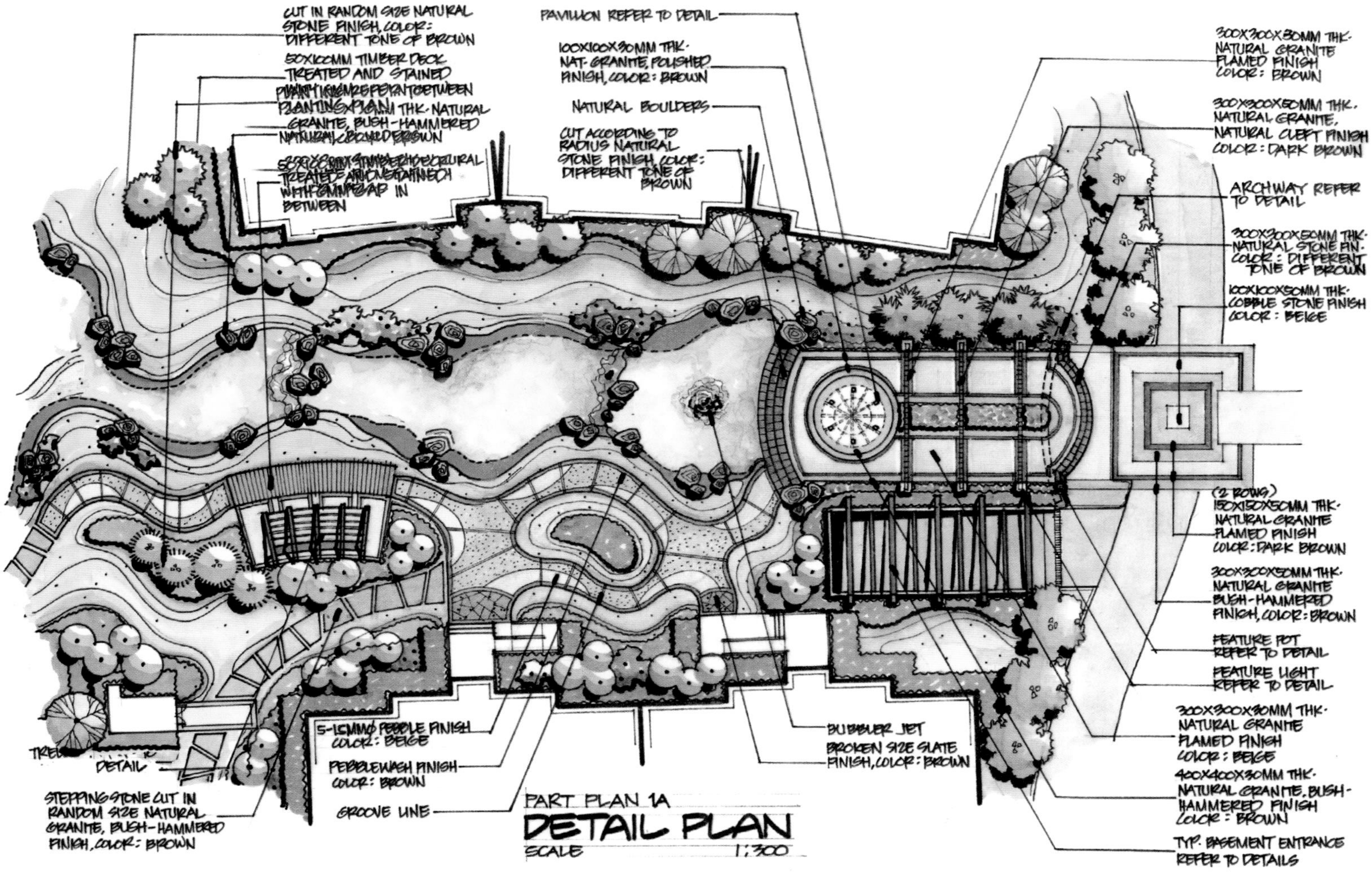

中央水景局部细节图

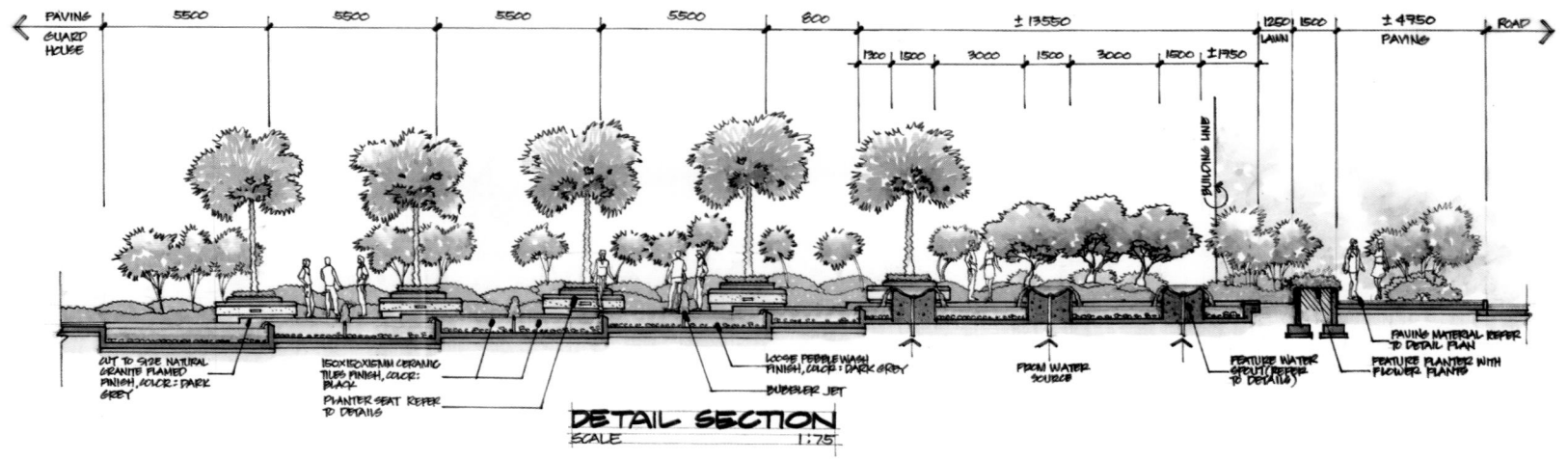

Top section dimensions: 5500 | 5500 | 5500 | 5500 | 800 | ±13550 | 1250 1500 | ±4750

PAVING GUARD HOUSE

1300 1500 3000 1500 3000 1500 ±1750 LAWN PAVING ROAD

BUILDING LINE

DETAIL SECTION
SCALE 1:75

Labels (left to right):
- CUT TO SIZE NATURAL GRANITE FLAMED FINISH, COLOR: DARK GREY
- 150X150X10MM CERAMIC TILES FINISH, COLOR: BLACK
- PLANTER SEAT REFER TO DETAILS
- LOOSE PEBBLEWASH FINISH, COLOR: DARK GREY
- BUBBLER JET
- FROM WATER SOURCE
- FEATURE WATER SPOUT (REFER TO DETAILS)
- PAVING MATERIAL REFER TO DETAIL PLAN
- FEATURE PLANTER WITH FLOWER PLANTS

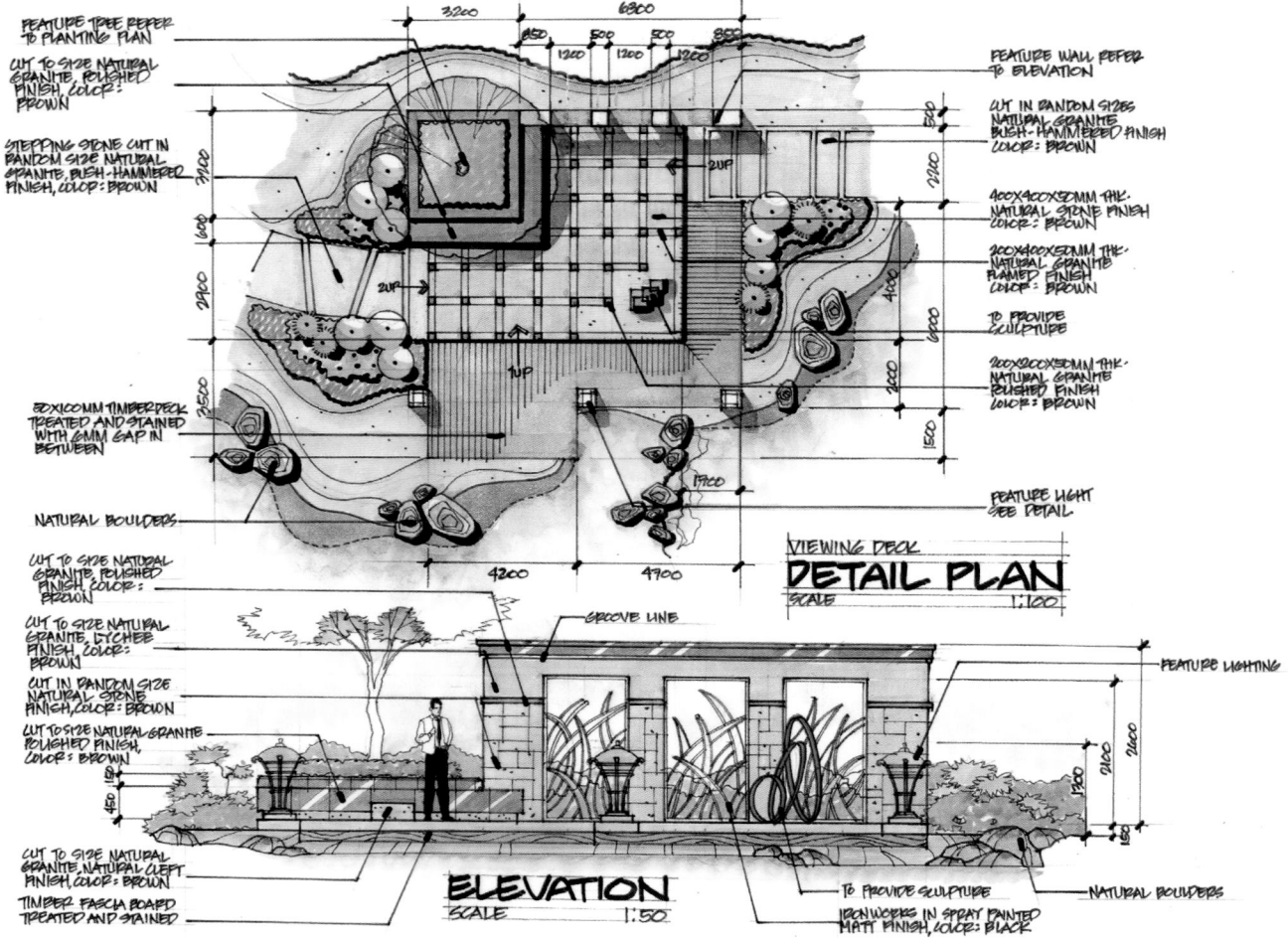

Top dimensions: 3200 | 6800 | 850 500 500 800 | 1200 1200 1200

Left labels:
- FEATURE TREE REFER TO PLANTING PLAN
- CUT TO SIZE NATURAL GRANITE, POLISHED FINISH, COLOR: BROWN
- STEPPING STONE CUT IN RANDOM SIZE NATURAL GRANITE, BUSH-HAMMERED FINISH, COLOR: BROWN
- 50X100MM TIMBER DECK TREATED AND STAINED WITH 5MM GAP IN BETWEEN
- NATURAL BOULDERS

Right labels:
- FEATURE WALL REFER TO ELEVATION
- CUT IN RANDOM SIZES NATURAL GRANITE BUSH-HAMMERED FINISH COLOR: BROWN
- 400X400X50MM THK. NATURAL STONE FINISH COLOR: BROWN
- 200X400X50MM THK. NATURAL GRANITE FLAMED FINISH COLOR: BROWN
- TO PROVIDE SCULPTURE
- 200X200X50MM THK. NATURAL GRANITE POLISHED FINISH COLOR: BROWN
- FEATURE LIGHT SEE DETAIL

2UP / 1UP markings

VIEWING DECK
DETAIL PLAN
SCALE 1:100

Lower left labels:
- CUT TO SIZE NATURAL GRANITE, POLISHED FINISH, COLOR: BROWN
- CUT TO SIZE NATURAL GRANITE, LYCHEE FINISH, COLOR: BROWN
- CUT IN RANDOM SIZE NATURAL STONE FINISH, COLOR: BROWN
- CUT TO SIZE NATURAL GRANITE POLISHED FINISH, COLOR: BROWN
- CUT TO SIZE NATURAL GRANITE, NATURAL CLEFT FINISH, COLOR: BROWN
- TIMBER FASCIA BOARD TREATED AND STAINED

- GROOVE LINE
- FEATURE LIGHTING

ELEVATION
SCALE 1:50

- TO PROVIDE SCULPTURE
- IRON WORKS IN SPRAY PAINTED MATT FINISH, COLOR: BLACK
- NATURAL BOULDERS

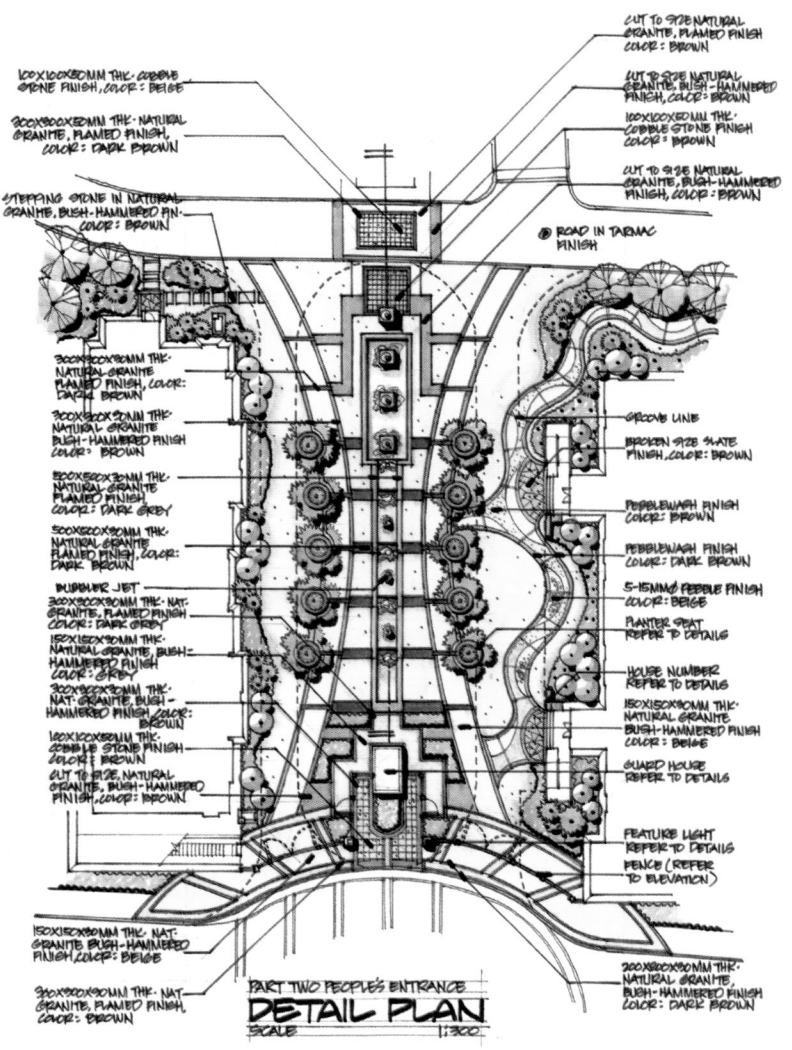

100x100x50MM THK. COBBLE STONE FINISH, COLOR: BEIGE
300x300x50MM THK. NATURAL GRANITE, FLAMED FINISH, COLOR: DARK BROWN
CUT TO SIZE NATURAL GRANITE, FLAMED FINISH
CUT TO SIZE NATURAL GRANITE, BUSH-HAMMERED FINISH, COLOR: BROWN
100x100x50MM THK. COBBLE STONE FINISH, COLOR: BROWN
CUT TO SIZE NATURAL GRANITE, BUSH-HAMMERED FINISH, COLOR: BROWN

STEPPING STONE IN NATURAL GRANITE, BUSH-HAMMERED FIN. COLOR: BROWN
ROAD IN TARMAC FINISH

300x300x50MM THK. NATURAL GRANITE FLAMED FINISH, COLOR: DARK BROWN
300x300x50MM THK. NATURAL GRANITE BUSH FLAMED FINISH, COLOR: BROWN
300x300x50MM THK. NATURAL GRANITE FLAMED FINISH, COLOR: DARK GREY
300x300x50MM THK. NATURAL GRANITE FLAMED FINISH, COLOR: DARK BROWN
BUBBLER JET
300x300x50MM THK. NAT. GRANITE, FLAMED FINISH, COLOR: DARK BROWN
150x150x50MM THK. NATURAL GRANITE, BUSH-HAMMERED FINISH, COLOR: GREY
300x300x50MM THK. NAT. GRANITE, BUSH-HAMMERED FINISH, COLOR: BROWN
100x100x50MM THK. COBBLE STONE FINISH, COLOR: BROWN
CUT TO SIZE NATURAL GRANITE, BUSH-HAMMERED FINISH, COLOR: BROWN

150x150x50MM THK. NAT. GRANITE BUSH-HAMMERED FINISH, COLOR: BEIGE
300x300x50MM THK. NAT. GRANITE, FLAMED FINISH, COLOR: BROWN

GROOVE LINE
BROKEN SIZE SLATE FINISH, COLOR: BROWN
PEBBLEWASH FINISH COLOR: BROWN
PEBBLEWASH FINISH COLOR: BROWN
5-15MMØ PEBBLE FINISH COLOR: BEIGE
PLANTER SEAT REFER TO DETAILS
HOUSE NUMBER REFER TO DETAILS
150x150x50MM THK. NATURAL GRANITE BUSH-HAMMERED FINISH COLOR: BEIGE
GUARD HOUSE REFER TO DETAILS
FEATURE LIGHT REFER TO DETAILS
FENCE (REFER TO ELEVATION)

PART TWO PEOPLE'S ENTRANCE
DETAIL PLAN
SCALE 1:300

南入口细部平面图

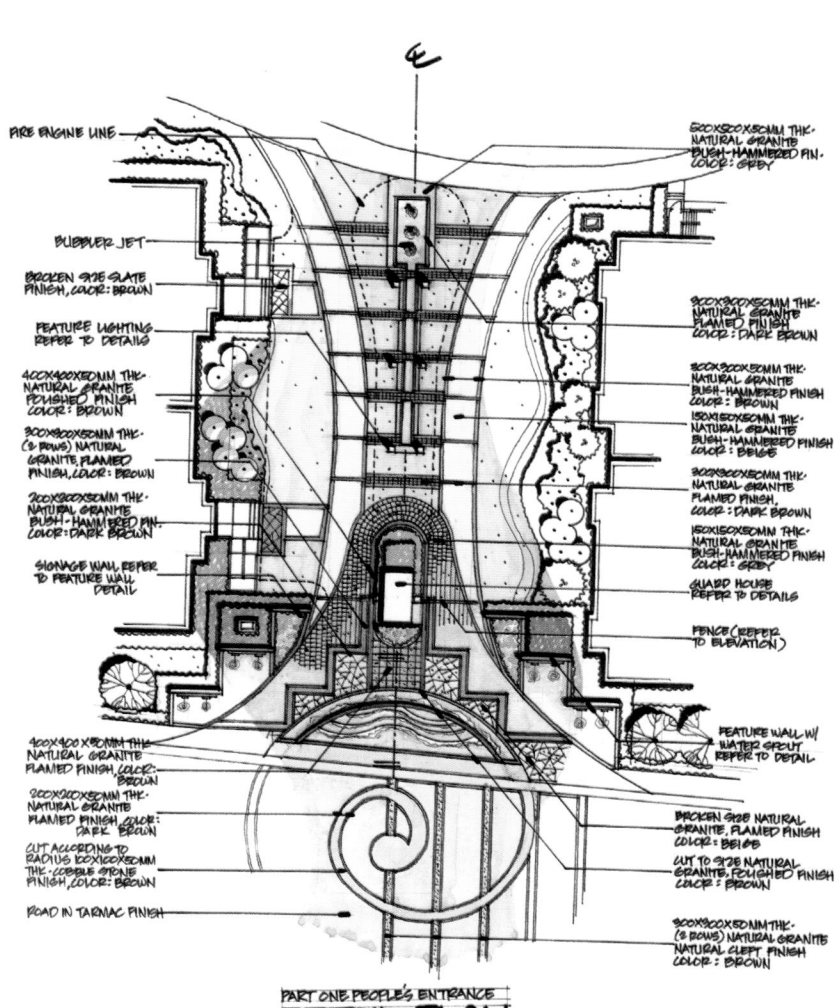

FIRE ENGINE LINE
300x300x50MM THK. NATURAL GRANITE BUSH-HAMMERED FIN. COLOR: GREY

BUBBLER JET
BROKEN SIZE SLATE FINISH, COLOR: BROWN
FEATURE LIGHTING REFER TO DETAILS
400x400x50MM THK. NATURAL GRANITE POLISHED FINISH, COLOR: BROWN
300x300x50MM THK. (3 ROWS) NATURAL GRANITE, FLAMED FINISH, COLOR: BROWN
300x300x50MM THK. NATURAL GRANITE BUSH-HAMMERED FIN., COLOR: DARK BROWN
SIGNAGE WALL, REFER TO FEATURE WALL DETAIL

400x400x50MM THK. NATURAL GRANITE FLAMED FINISH, COLOR: BROWN
300x300x50MM THK. NATURAL GRANITE FLAMED FINISH, COLOR: DARK BROWN
CUT ACCORDING TO RADIUS 100x100x50MM THK. COBBLE STONE FINISH, COLOR: BROWN
ROAD IN TARMAC FINISH

300x300x50MM THK. NATURAL GRANITE FLAMED FINISH, COLOR: DARK BROWN
300x300x50MM THK. NATURAL GRANITE BUSH-HAMMERED FINISH, COLOR: BROWN
150x150x50MM THK. NATURAL GRANITE BUSH-HAMMERED FINISH, COLOR: BEIGE
300x300x50MM THK. NATURAL GRANITE FLAMED FINISH, COLOR: DARK BROWN
150x150x50MM THK. NATURAL GRANITE BUSH-HAMMERED FINISH, COLOR: GREY
GUARD HOUSE REFER TO DETAILS
FENCE (REFER TO ELEVATION)

FEATURE WALL W/ WATER SPOUT REFER TO DETAIL
BROKEN SIZE NATURAL GRANITE, FLAMED FINISH, COLOR: BEIGE
CUT TO SIZE NATURAL GRANITE, POLISHED FINISH, COLOR: BROWN
300x300x50MM THK. (3 ROWS) NATURAL GRANITE NATURAL CLEFT FINISH, COLOR: BROWN

PART ONE PEOPLE'S ENTRANCE
DETAIL PLAN
SCALE 1:300

东入口西部平面图

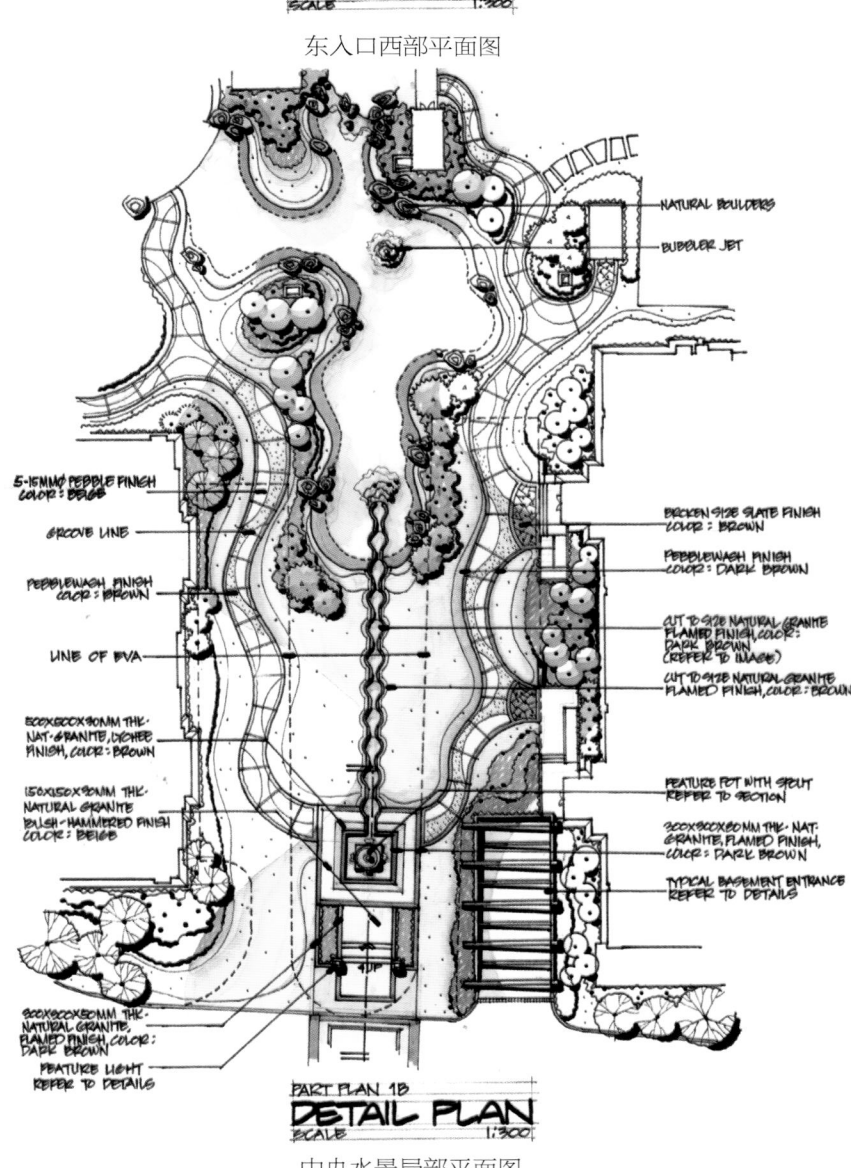

NATURAL BOULDERS
BUBBLER JET

5-15MMØ PEBBLE FINISH COLOR: BEIGE
GROOVE LINE
PEBBLEWASH FINISH COLOR: BROWN
LINE OF EVA
300x300x50MM THK. NAT. GRANITE, LYCHEE FINISH, COLOR: BROWN
150x150x50MM THK. NATURAL GRANITE BUSH-HAMMERED FINISH, COLOR: BEIGE
FEATURE LIGHT REFER TO DETAILS

BROKEN SIZE SLATE FINISH COLOR: BROWN
PEBBLEWASH FINISH COLOR: DARK BROWN
CUT TO SIZE NATURAL GRANITE FLAMED FINISH, COLOR: DARK BROWN (REFER TO IMAGE)
CUT TO SIZE NATURAL GRANITE FLAMED FINISH, COLOR: BROWN
FEATURE POT WITH SPOUT REFER TO SECTION
300x300x50MM THK. NAT. GRANITE, FLAMED FINISH, COLOR: DARK BROWN
TYPICAL BASEMENT ENTRANCE REFER TO DETAILS

PART PLAN 1B
DETAIL PLAN
SCALE 1:300

中央水景局部平面图

697

广西北海国宾馆
Guangxi Beihai State Guest house
项目名称：广西北海国宾馆
项目地点：中国 / 广西 / 北海
建筑面积：规划总用地为 36 万平方米
编制时间：2010 / 03
业　主：广西旅游投资集团有限公司
服务范围：方案

北海地处广西南端，距南宁 206 公里，市区南北西三面环海，与海南省隔海相望。邻近东南亚诸国，地理位置优越。

整个规划用地呈半岛状，顺应地势，东高西低，依山傍海，景色优美，东部背靠国家森林公园，植物葱郁，视野开拓，有良好的景观基础和生态优势。西面和西南面临大海，活动区域直达海滩。

项目建设包括了 10 幢高档别墅、接待用房、会议中心、休闲活动中心等，景观设计目标是依托良好的生态基础，最大限度地利用地形优势，造山理水，力求打造一个国际一流的集会晤、接待、度假为一体的国际顶级国宾馆。

景观设计从结合地形与地域特点的角度出发，使用现代的设计手法，融合蕴含东南亚风情的设计风格，配以高端的室外酒店设施，如自然露天的 SPA 馆、户外的按摩池、水疗吧、跨越在山崖之上的无边界水景等等。山有棱，水无痕，树有形，而景无垠，在这得天独厚的山水之间，享受 海水，阳光，沙滩，泳池，密林的旖旎风光。

Beihai lies in the southern end of Guangxi, 206 kilometers from Nanning. It is surrounded by the sea on the three sides of south, north and west, facing Hainan Province across the sea. Adjacent to Southeast Asian countries, its geographic location is excellent.

The whole planned land to be used is in the shape of a peninsula, adapts to the topography of the high east and the low west, has beautiful scenes for the neighboring hills and the adjacent sea, backs on the national forest park in the east. The vegetation is lush, the horizon broad, and there is an excellent landscape basis and ecological advantage. It faces the open sea in the west and the southwest, and the activity region reaches as far as the beach.

The project construction includes 10 high-grade villas, the reception house, conference center, leisure activity center, etc. The objective of the landscape design is, relying on the excellent ecological basis, to optimize the geographical advantage to build hills and adjust water, and to elaborate an international first-class state guest house providing meeting, reception and holiday services integrated.

The landscape design, starting from the angle of combining geographical and regional characteristics, employing modern design techniques, incorporates the design style full of Southeast Asian romance, complete with high-end outdoor hotel facilities, such as the natural open-air spa pavilion, outdoor massage pool, spa bar, the boundless waterscape striding over the cliff and so on. The hills are edged, the streams traceless, and the trees shaped, while the scenery is boundless. In between these hills and streams, you can enjoy the sea water, sunshine, sand beaches, swimming pool, and the charming and gentle scenery of dense woods.

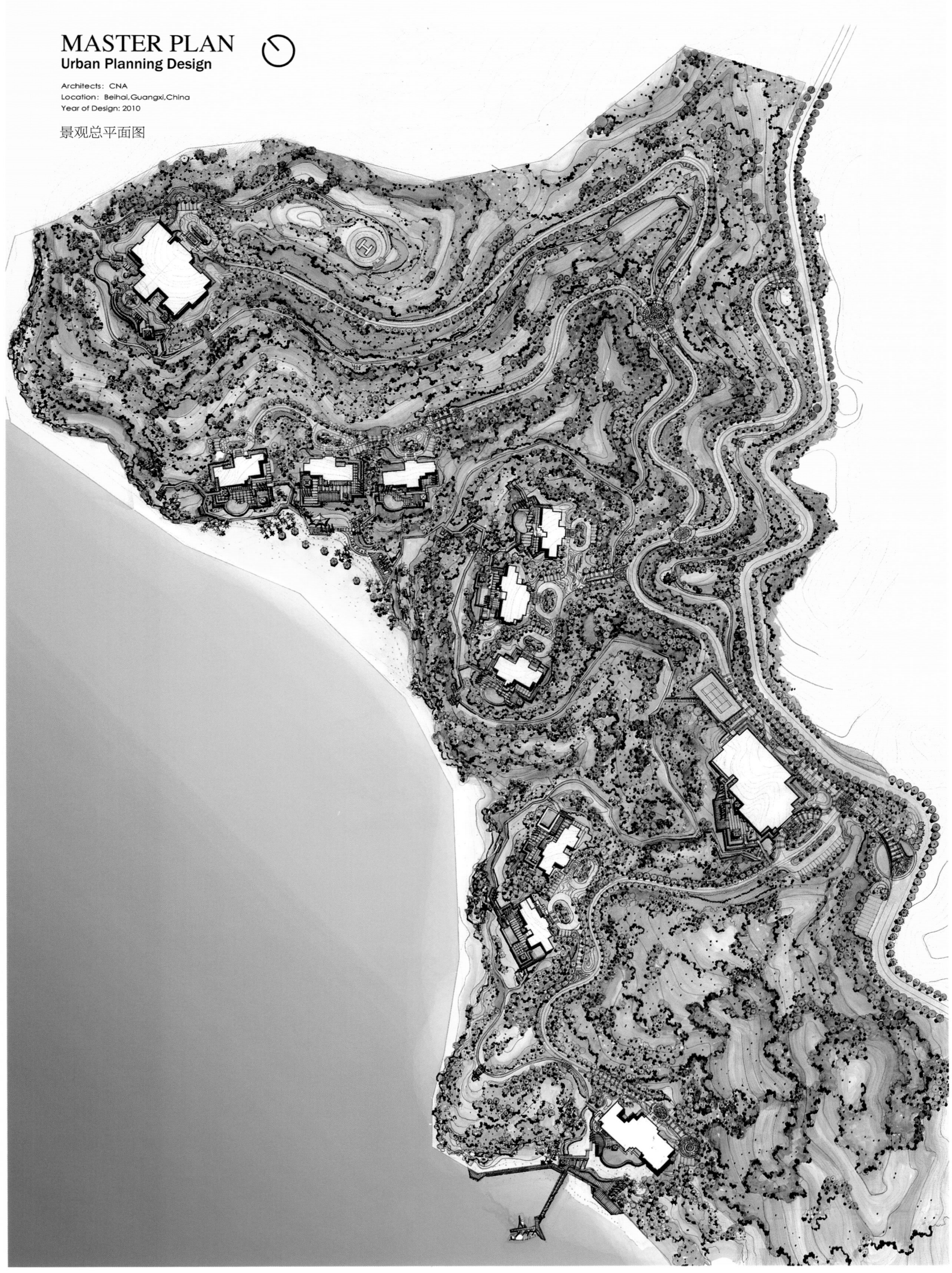

MASTER PLAN
Urban Planning Design

Architects: CNA
Location: Beihai,Guangxi,China
Year of Design: 2010

景观总平面图

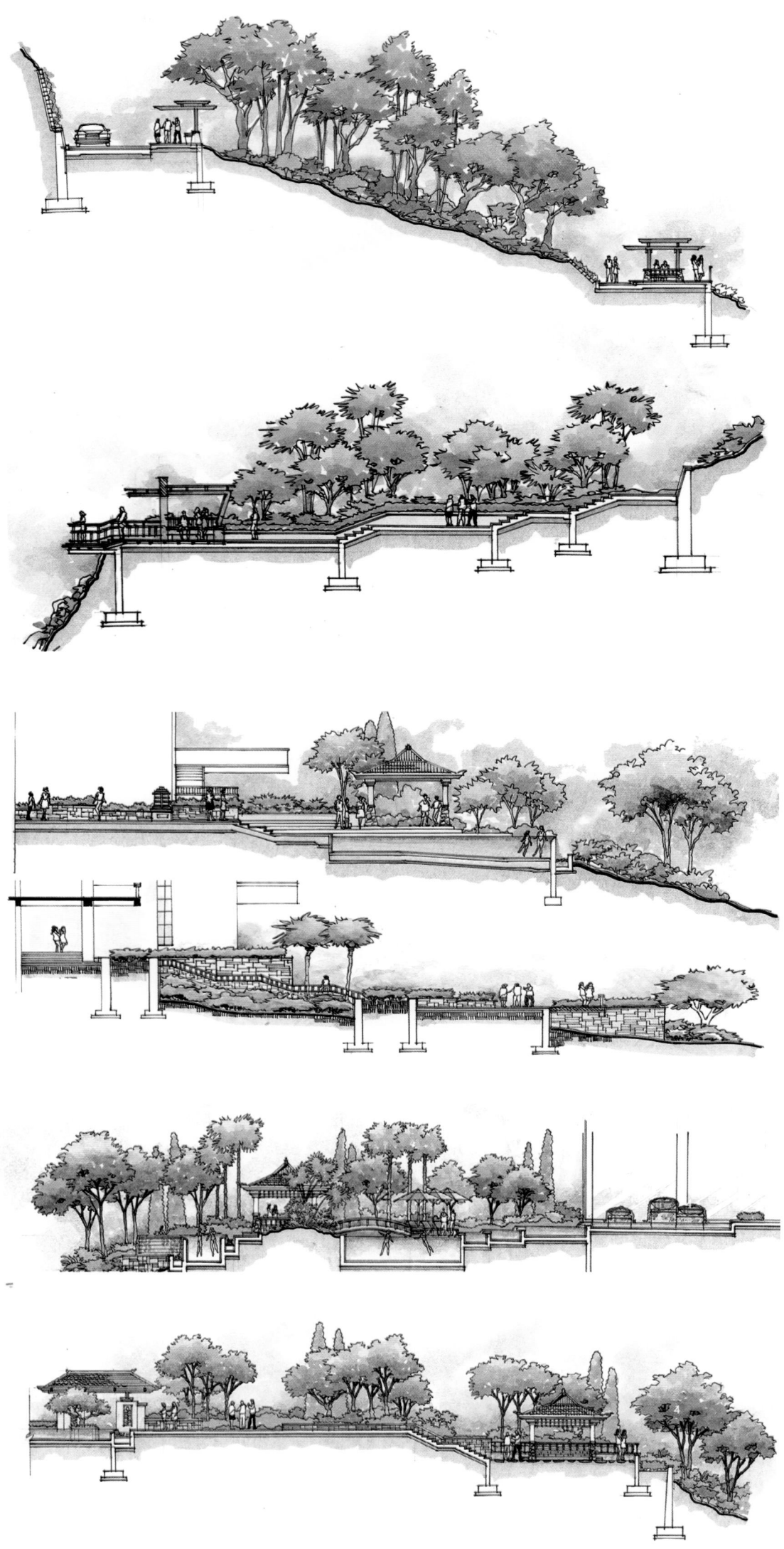

不同坡度的断面设计图

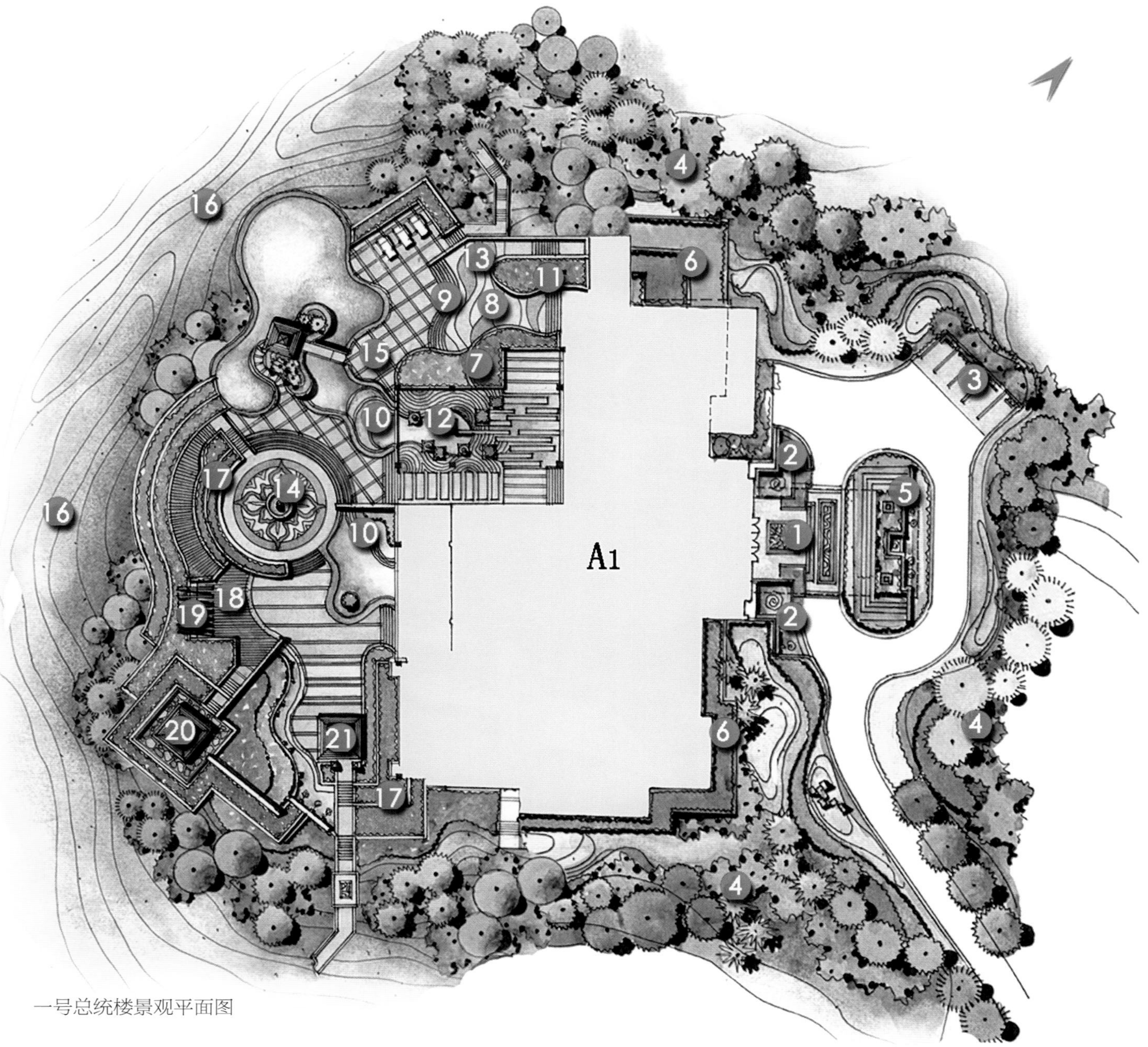

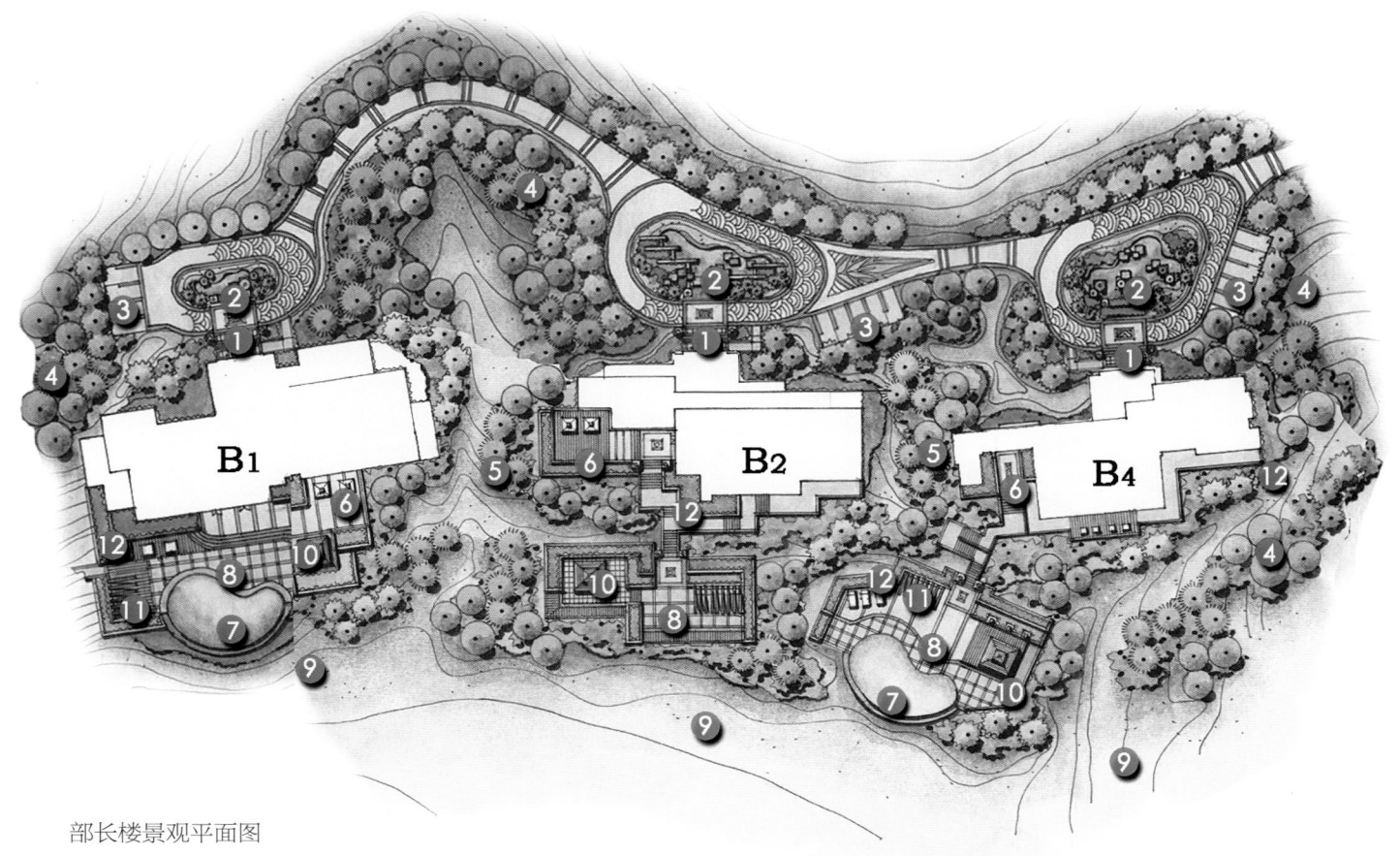

一号总统楼景观平面图

部长楼景观平面图

二号总统楼泳池细部图

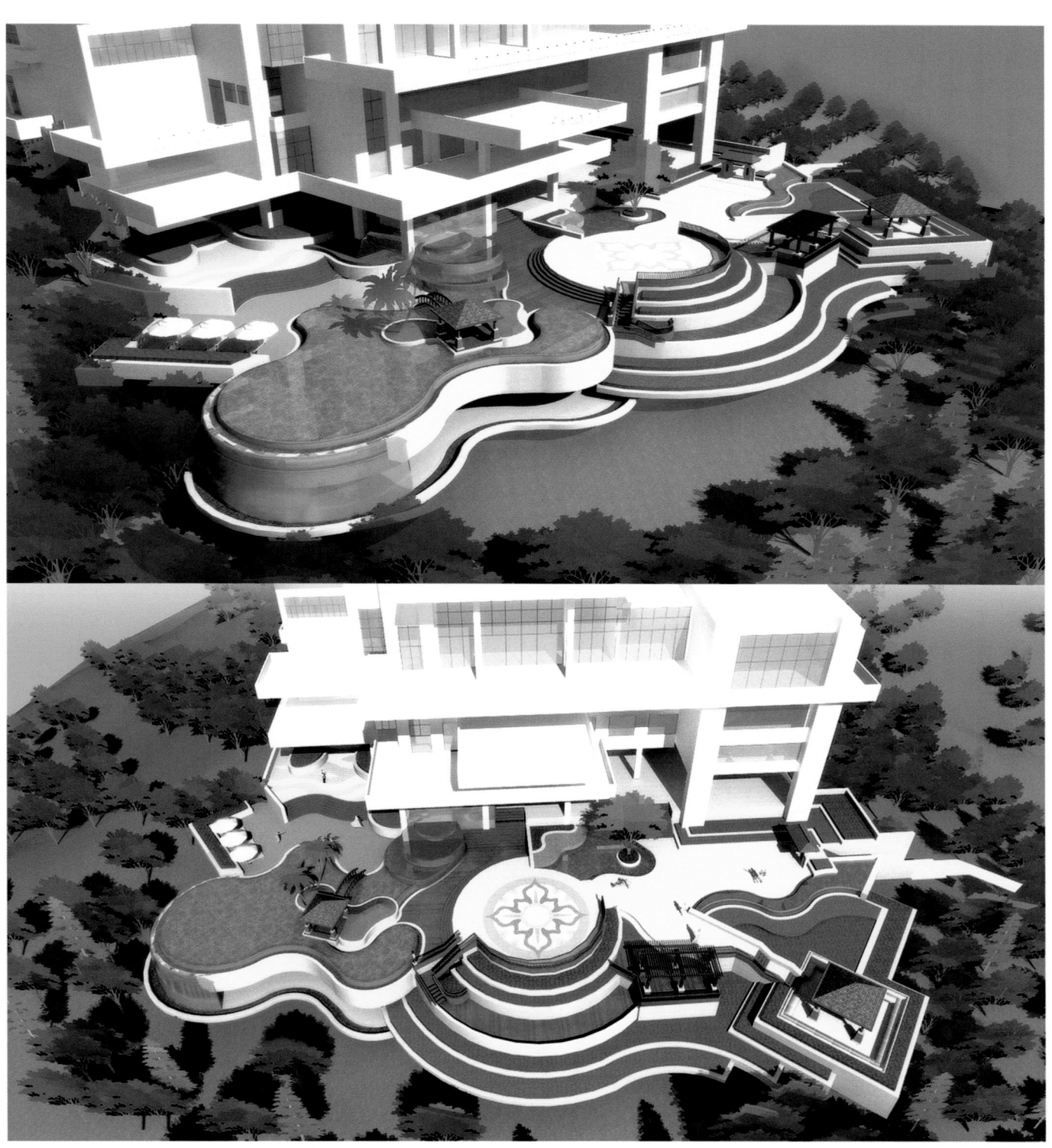

一号总统楼水池细部图

金山嘴海鲜文化园
Jinshanzui Seafood Culture Park

项目名称：金山嘴海鲜文化园
项目地点：中国 / 上海
建筑面积：规划总用地为 3.36 万平方米
编制时间：2009 / 07
业主：上海金吉置业有限公司
服务范围：方案 / 施工图

设计特点：在平淡的自然环境中，通过设计创造，形成吸引人的视觉焦点，从而形成独特的景观环境和商业效果。

Design characteristics: making the attractive visual focuses by means of design creation in the prosaic environment so that a unique landscape environment and commercial effect can be achieved.

景观总平面图

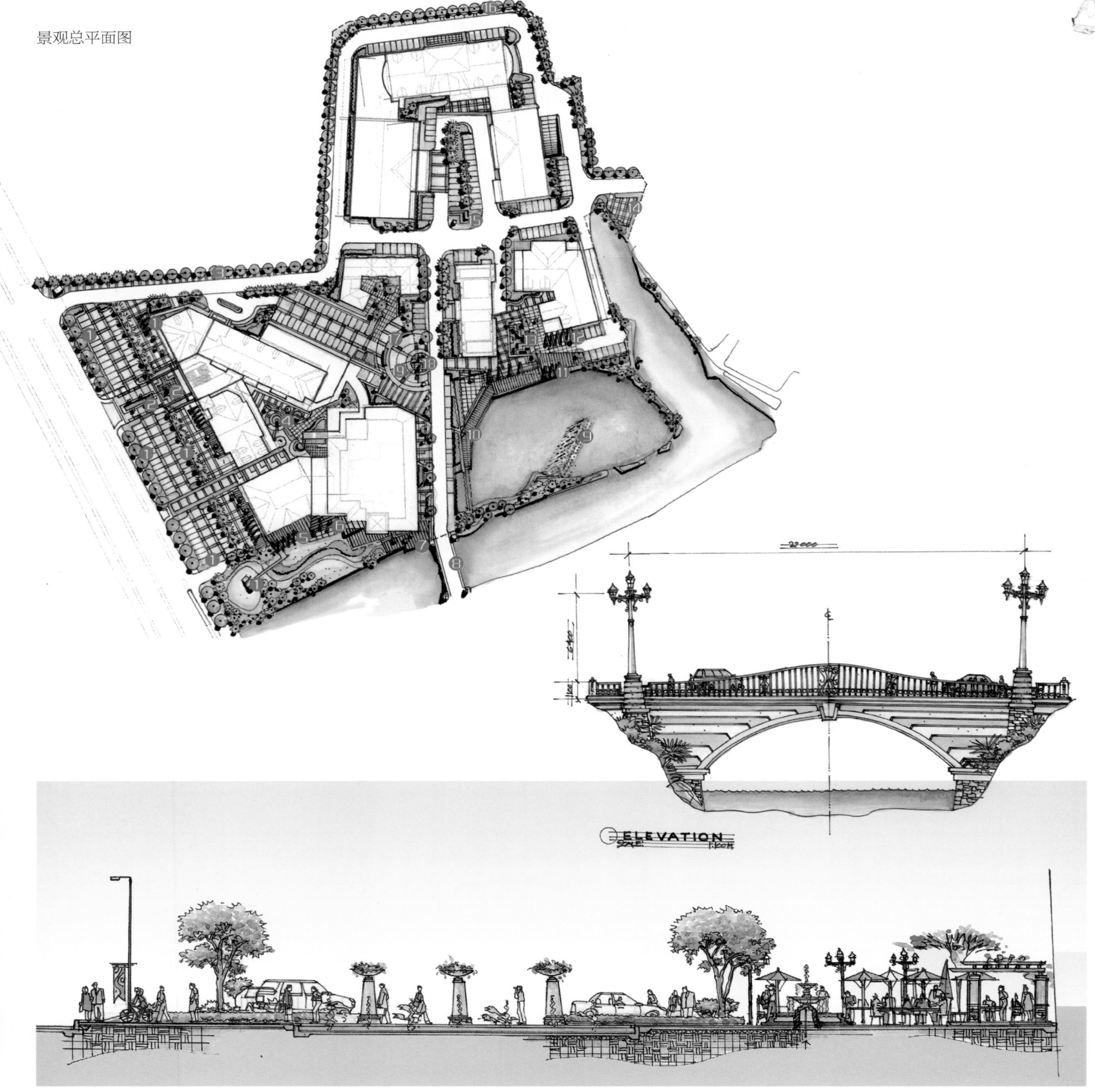

ELEVATION

天津南开区老城厢景观项目

Tianjin Nankai District Old Town Landscaping Project

项目名称：天津南开区老城厢景观项目
项目地点：中国 / 天津
建筑面积：规划总用地为 6.56 万平方米
编制时间：2008 / 06
业主：天津中新名都房地产开发有限公司
服务范围：方案 / 施工图

景观设计以古朴、简洁、吉祥、宁静为主要基调，围合半封闭的空间院落形式，形成主街区宽、胡同窄、内庭花园又豁然开朗的空间序列。

Design characteristics: in an insipid natural environment, attractive visual focuses are formed through design creation, so that a unique landscape environment and commercial effect are achieved.

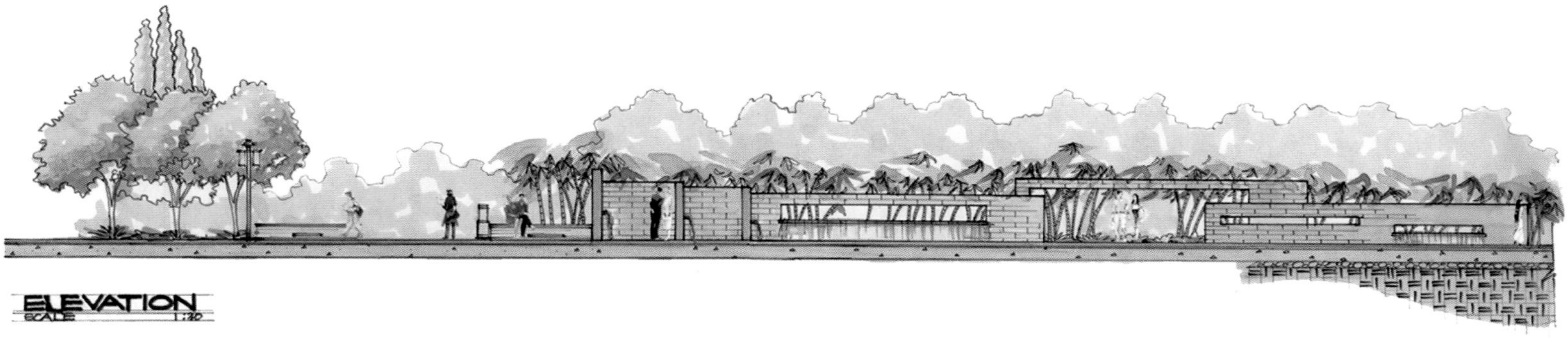

ELEVATION
SCALE 1:30

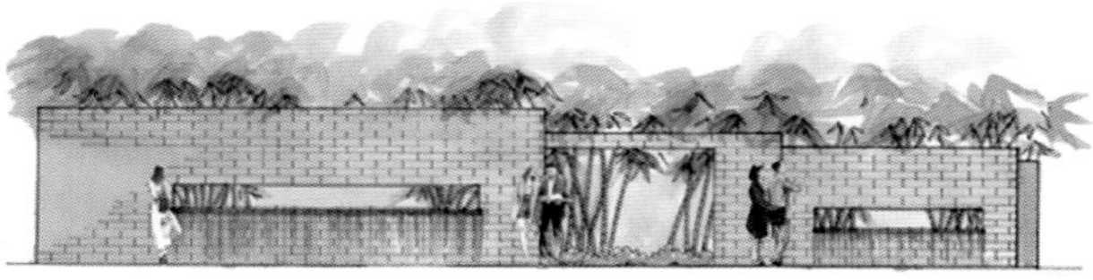

ELEVATION
SCALE 1:30

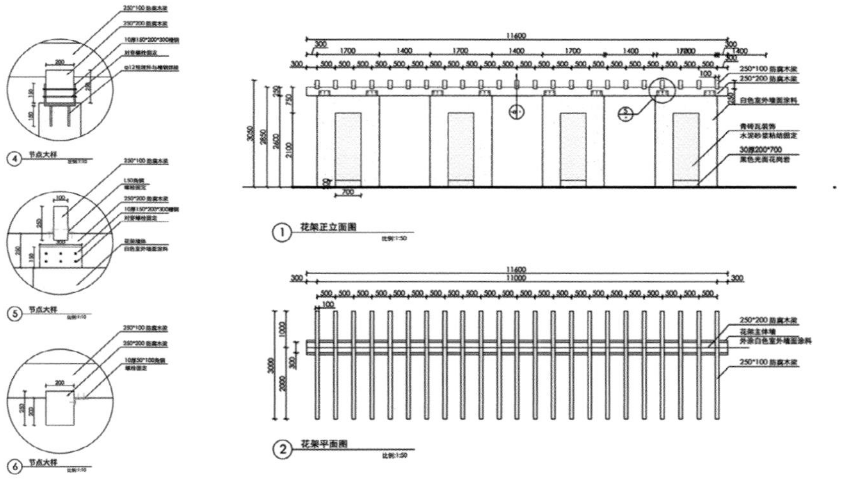

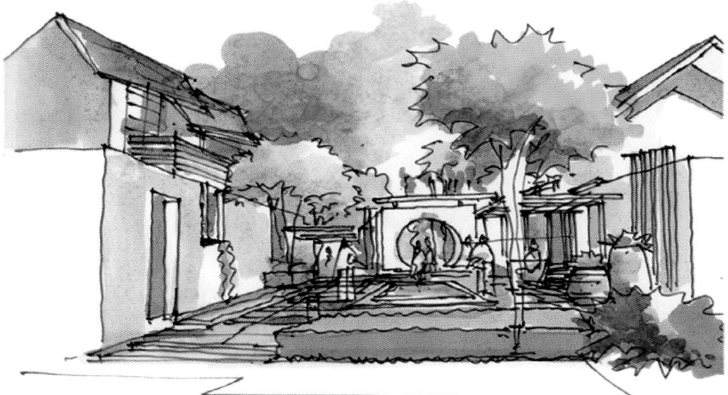

①NEIGBORHOOD SEMI-COMMON SPACE PERSPECTIVE

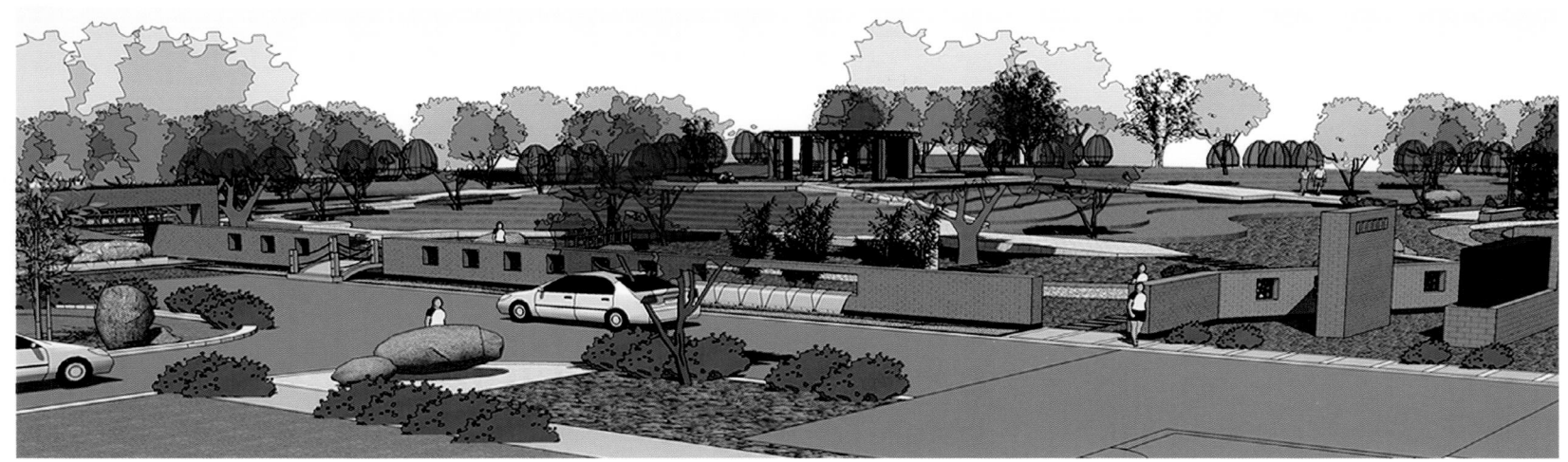

上虞国际时代广场
Shangyu International Times Square

项目名称：上虞国际时代广场
项目地点：中国 / 浙江 / 上虞
建筑面积：规划总用地为 4.12 万平方米
编制时间：2008 / 01
业主：上虞国际时代广场置业有限公司
服务范围：方案 / 扩初 / 施工图

实施方案总平面图

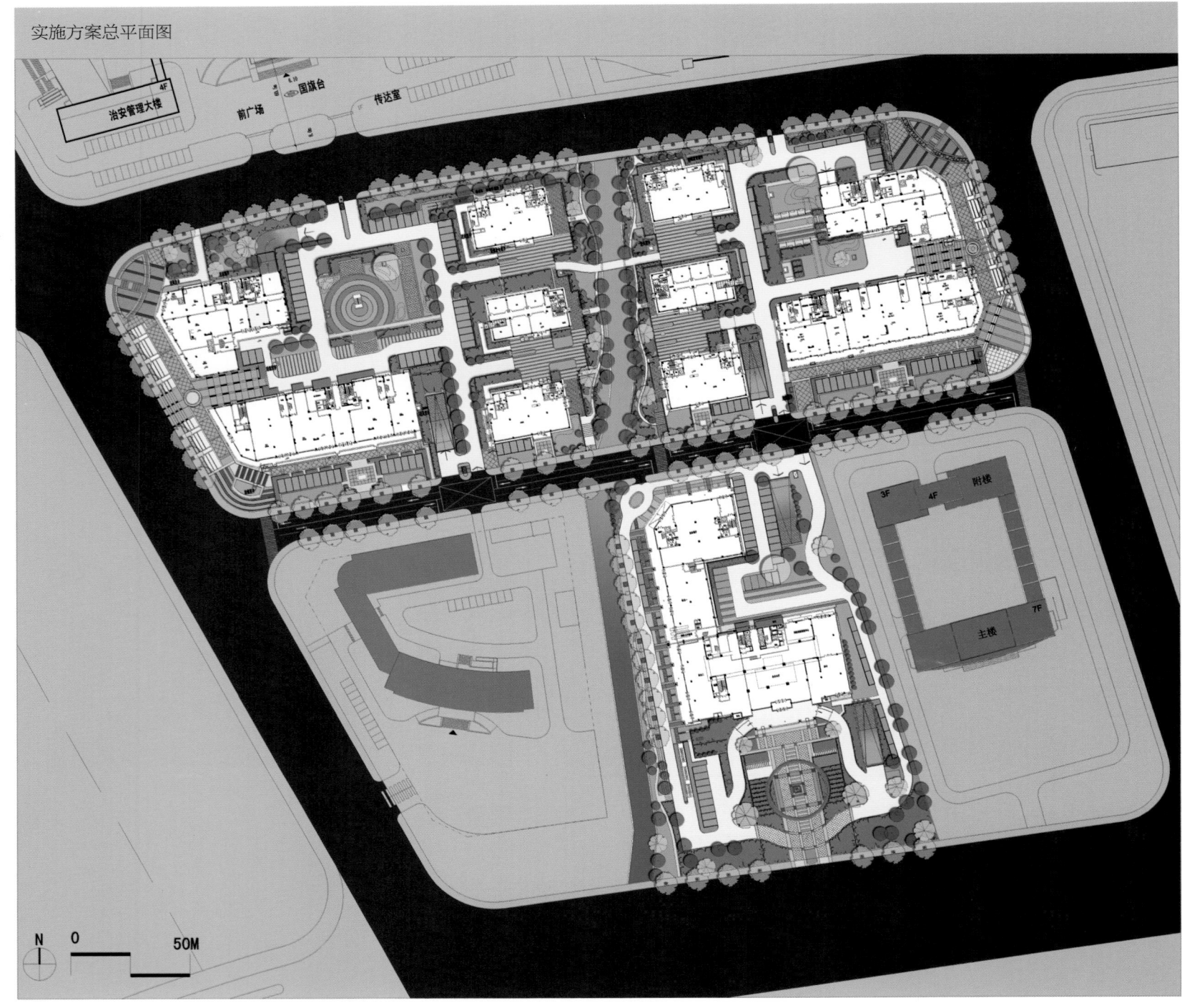

N　0　　　50M

设计特点：从设计到后期施工，全过程控制，最大程度地实现最初方案中的设计理念。

Design characteristics: from the design to the later construction the entire process is controlled, to implement the design ideas in the original plan to the greatest extent.

西宁七一路、为民巷景观设计
Xining Qiyi Road, Weimin Lane Landscaping Design

项目名称：西宁七一路、为民巷景观设计
项目地点：中国 / 青海 / 西宁
长度：500 米
编制时间：2009 / 06
业主：西宁市城辉建设投资有限公司
服务范围：方案

设计特点：在保留原有的建筑和街道尺度下，通过改造立面和街道环境，以较低的费用，创造具有当地文化和特点的特色街景。

Design characteristics: with the original building and street dimensions to be preserved, a characteristic streetscape with the local culture and features is created by reforming the facades and street environment at a relatively low cost.

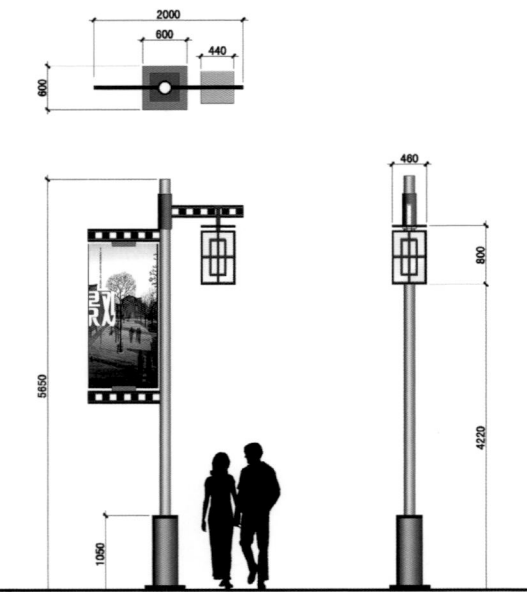

入口牌楼　黑色铁艺树池　图案式地面铺装　花坛种植池　影壁　黑色铁艺树池　景观广告灯柱　深色花岗岩地面铺装　入口牌楼
道路盲道　花坛种植池　暖色地砖　新增围墙　青砖铺砌　花坛种植池　仿古石材　道路盲道
图案式地面铺装　景观灯柱　新增围墙

景观总平面图

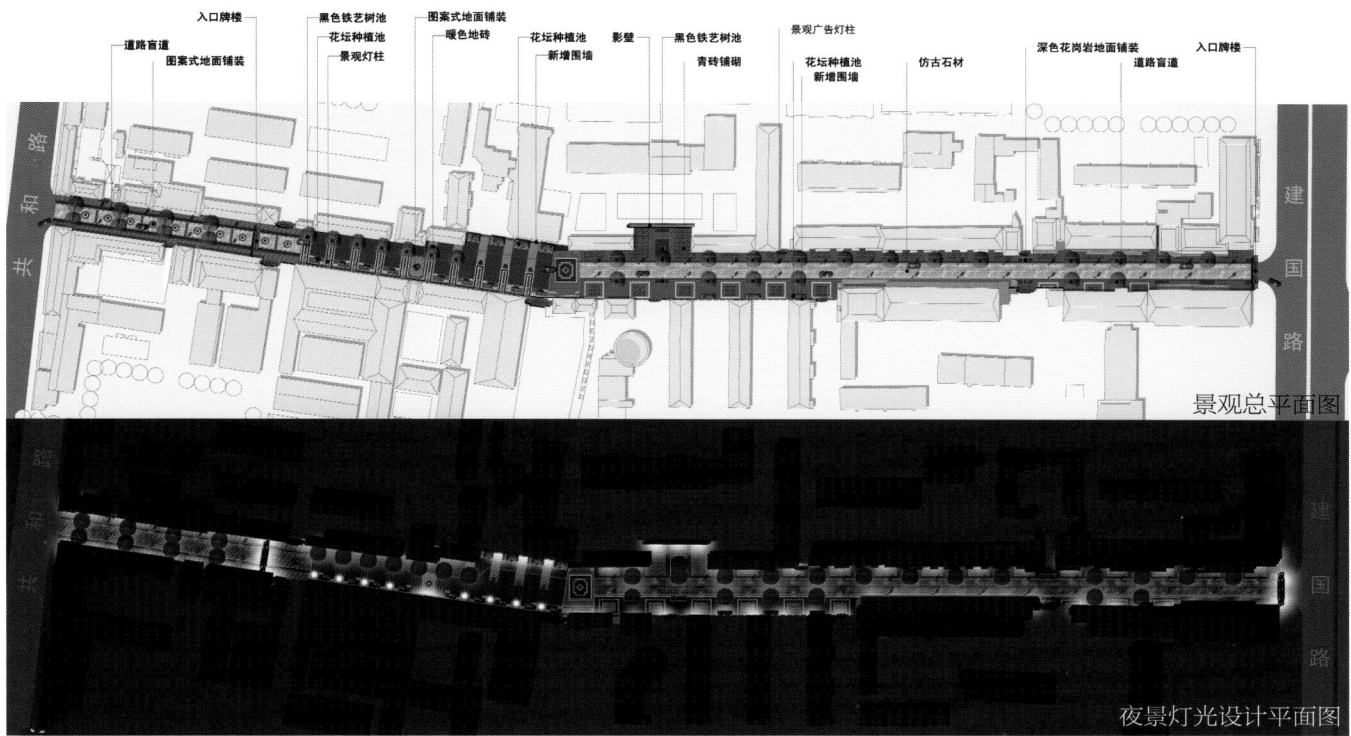

夜景灯光设计平面图

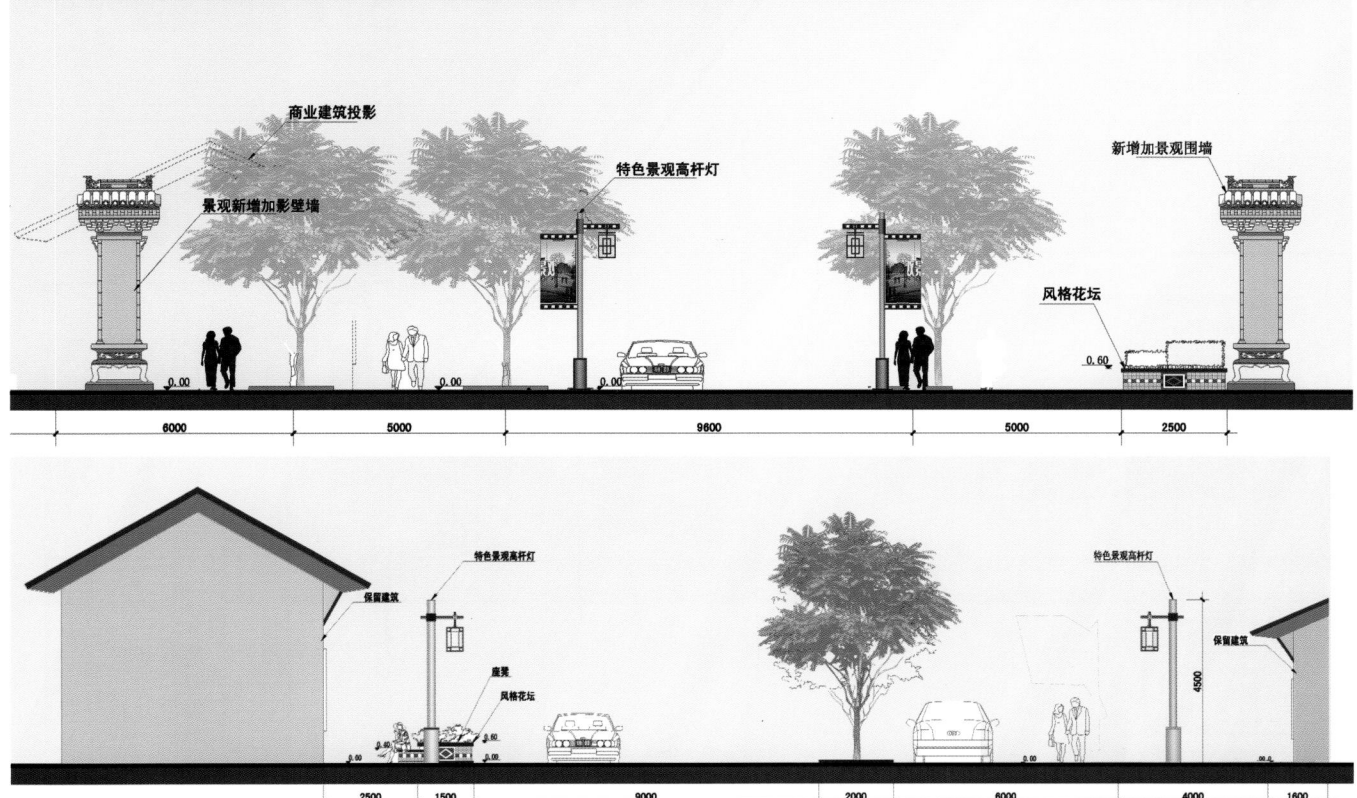

商业建筑投影
景观新增加影壁墙
特色景观高杆灯
新增加景观围墙
风格花坛

6000　5000　9600　5000　2500

保留建筑
特色景观高杆灯
座凳
风格花坛
特色景观高杆灯
保留建筑

2500　1500　9000　2000　6000　4000　1600

沪宁城际铁路无锡新区站站前广场景观设计
Station Front Square Design of Inter-City Railway Wuxi New Area Station

项目名称：沪宁城际铁路无锡新区站站前广场景观设计
项目地点：中国 / 江苏 / 无锡
建筑面积：47 890 平方米
编制时间：2009 / 07
业主：无锡新区城际铁路站前商务区投资开发有限公司
服务范围：方案 / 扩初 / 施工图

无锡新区城际铁路站是一个现代化、集约型的地区综合交通枢纽，是沪宁城际铁路大动脉上的一颗璀璨明珠。

广场采用下沉式设计，地面层主要满足进站候车和公交、社会车辆等停车功能；地下层主要服务于出站，同时满足快速连接地铁、出租车等公共交通功能，地下层四周还设有餐饮、零售、公厕、寄存、问询等服务性场所，以满足旅客的各种需求。站前广场的景观设计使得该工程从交通功能广场转换成可识别性强的地区中心广场，给人带来强烈的、浓厚的吴文化主题广场印象的同时，也是一个具有节能环保功能和可持续发展的现代高科技广场。

Wuxi New District Intercity Railway Station is a modern and intensive regional comprehensive traffic hub, and a bright pearl on the Shanghai-Nanjing Intercity Railway artery.

For the plaza the sunken-type design is adopted, the ground level is mainly to meet the needs of entering and waiting at the station as well as of the parking of public vehidcl; the underground level mainly serves the departure from the station, meanwhile fulfills the functions of a quick connection with the metro, taxi and other public traffic facilities, and around the underground level, catering, retail, public convenience, depositing, inquiry and other service facilities are also provided, so that passengers' various needs are satisfied. The landscape design of the station front plaza transforms this project into a highly distinguishable district centre plaza from a traffic function plaza, and while a strong and rich Wu Culture theme plaza image is brought to people, it is also a hi-tech plaza of energy-saving eco-protection function and sustainable development.

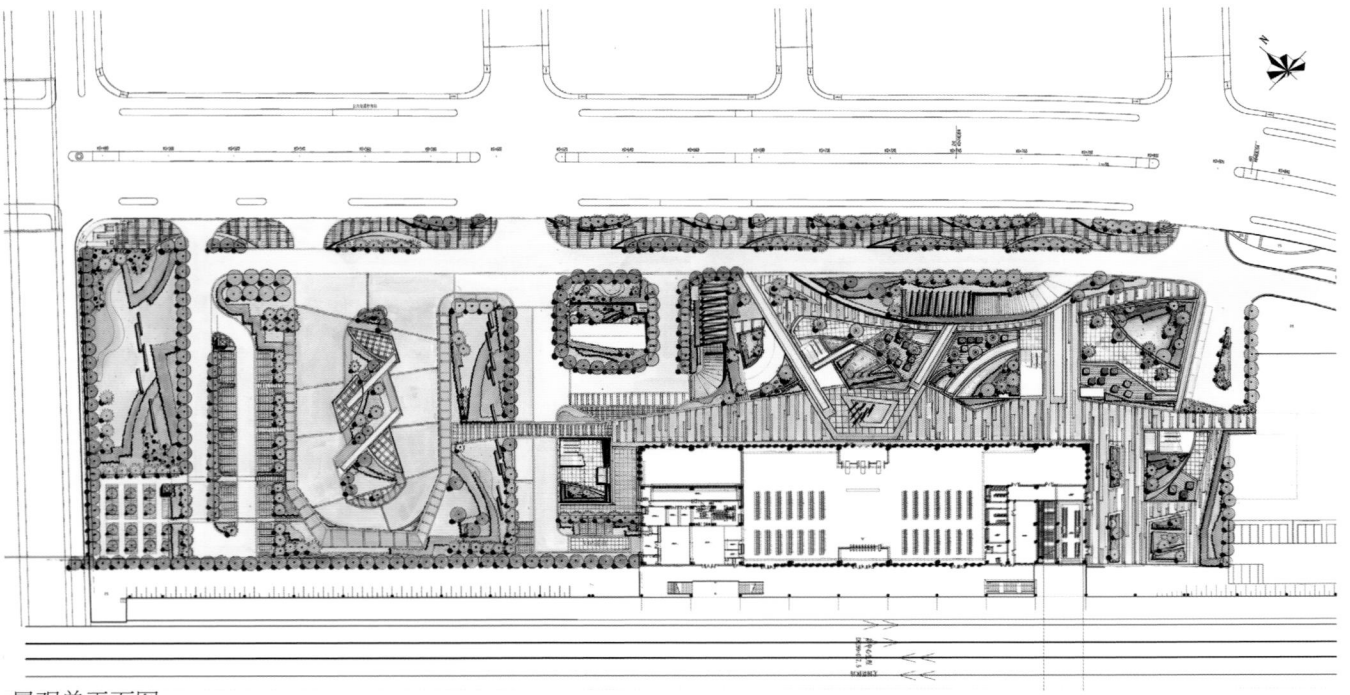

景观总平面图

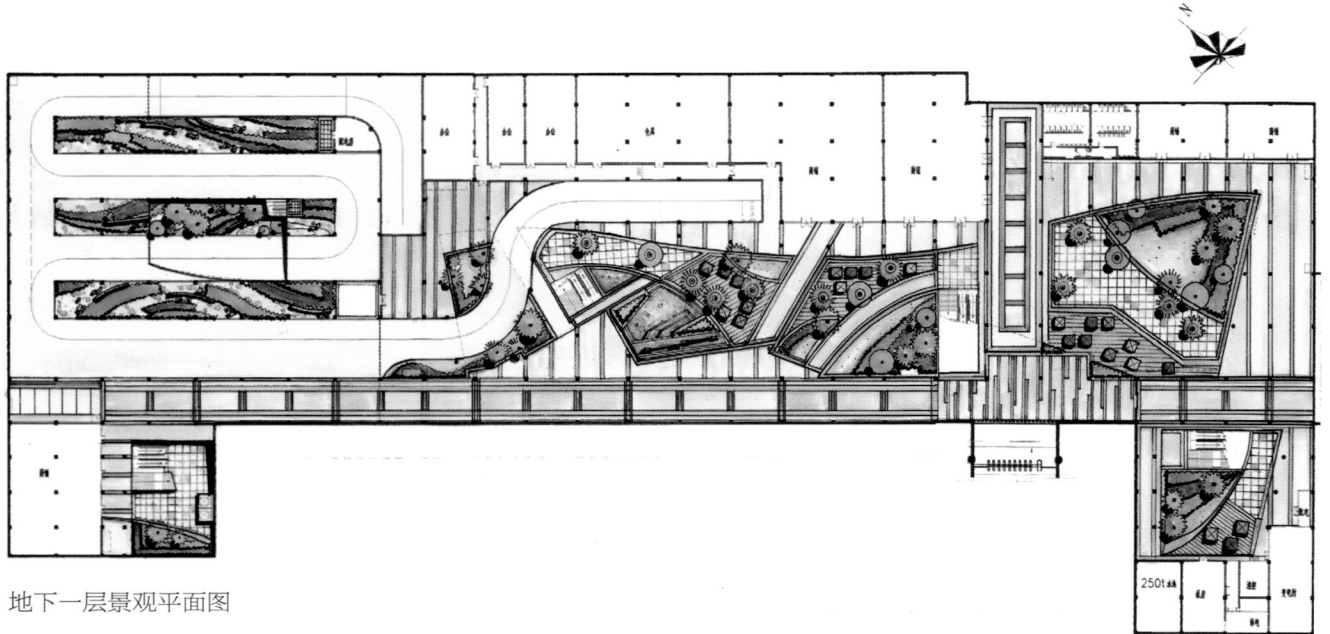

地下一层景观平面图

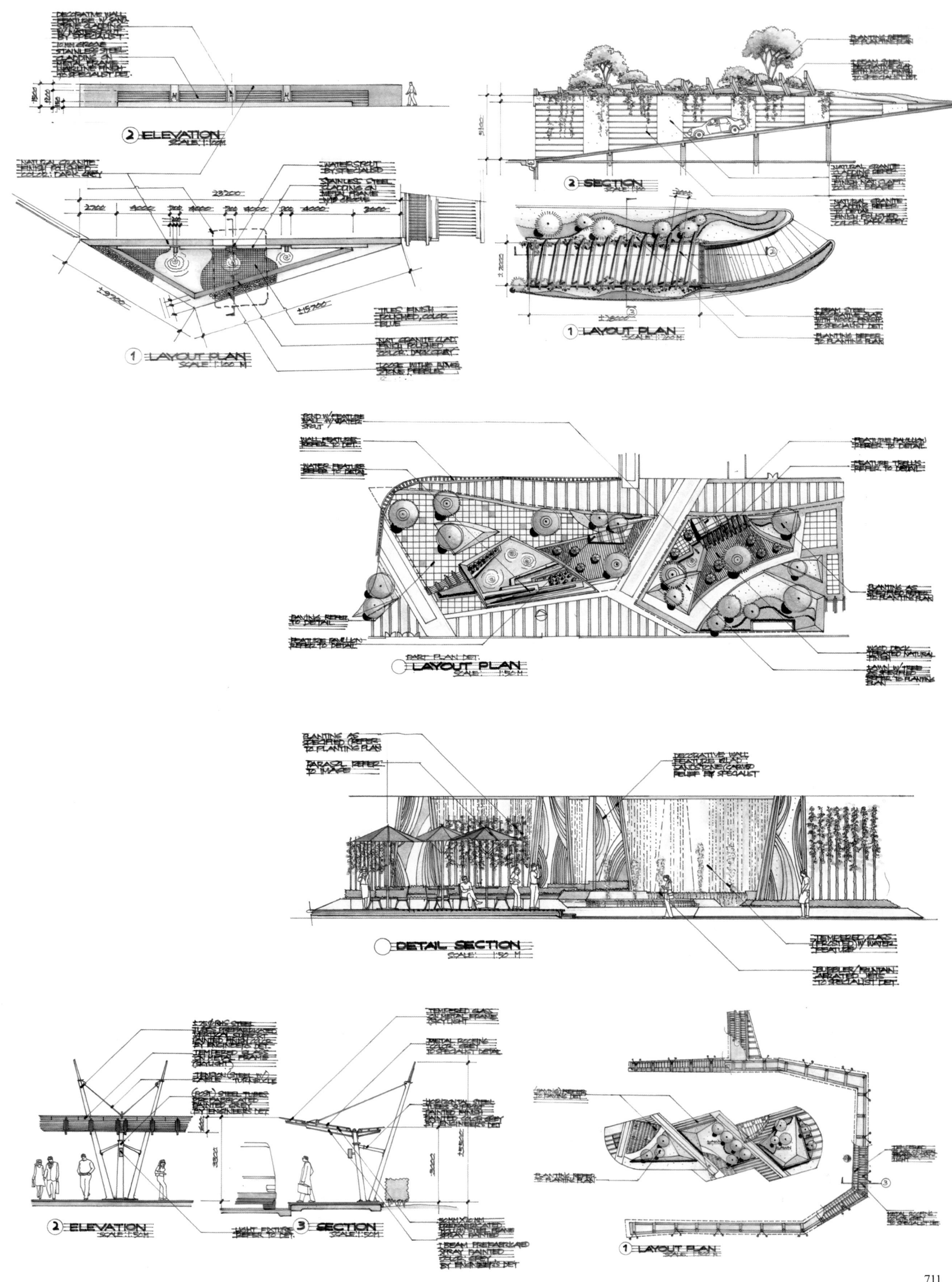

建成后的细部照片图

金华市区市民广场改造工程
Civil Plaza Reconstruction Project in Jinhua Downtown Area

项目名称：金华市区市民广场改造工程
项目地点：中国 / 浙江 / 金华
建筑面积：74 650 平方米
编制时间：2009 / 08
业主：金华市规划局
服务范围：景观方案设计
2

设计特点：设计师试图以一种新颖的设计手法，以流畅的景观空间，创造出一个独特的市民广场。它不是传统的、单调乏味、缺乏人气的硬铺大广场，而是一个从视觉和活动线路出发，完全考虑市民使用和活动的、同时具有当地文化韵味的市民广场。

Design characteristics: the designer tries to create a unique civil plaza by a novel design technique, and with a fluent landscape space. Different from the traditional, tedious and unfeeling hard-paved large plaza, it is a civil plaza which, on the basis of vision and activity routes, is completely intended for civil use and activity, meanwhile has the local cultural charm.

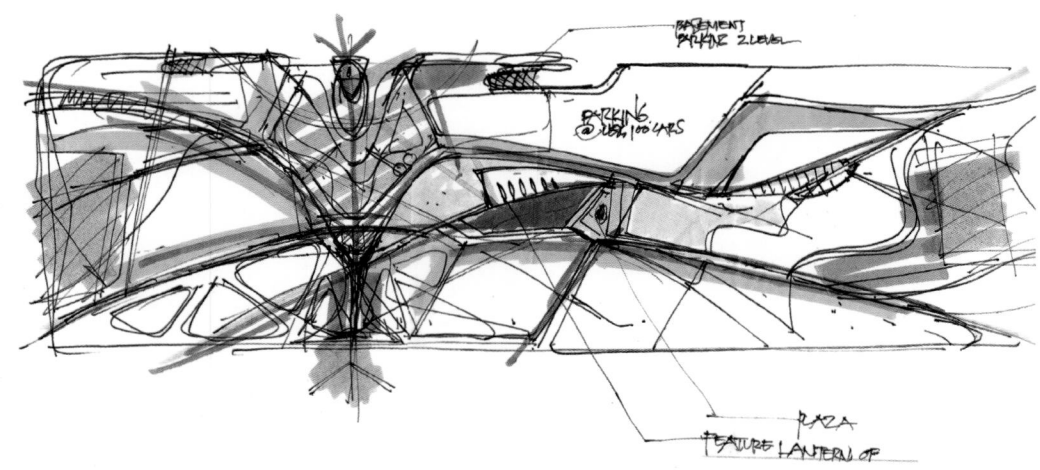

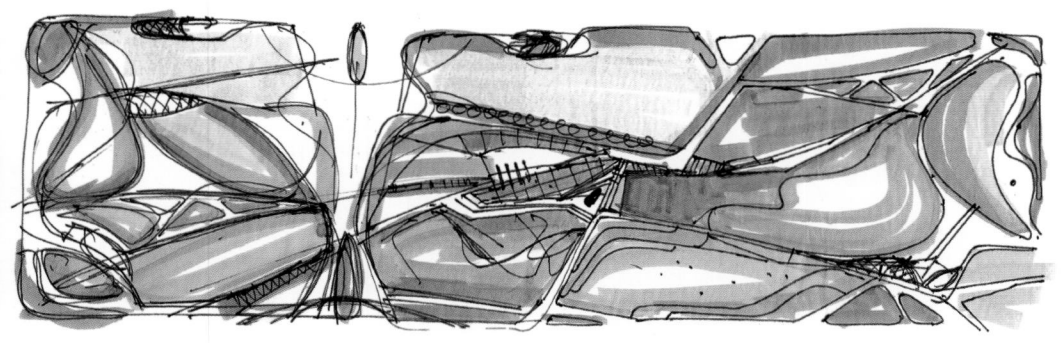

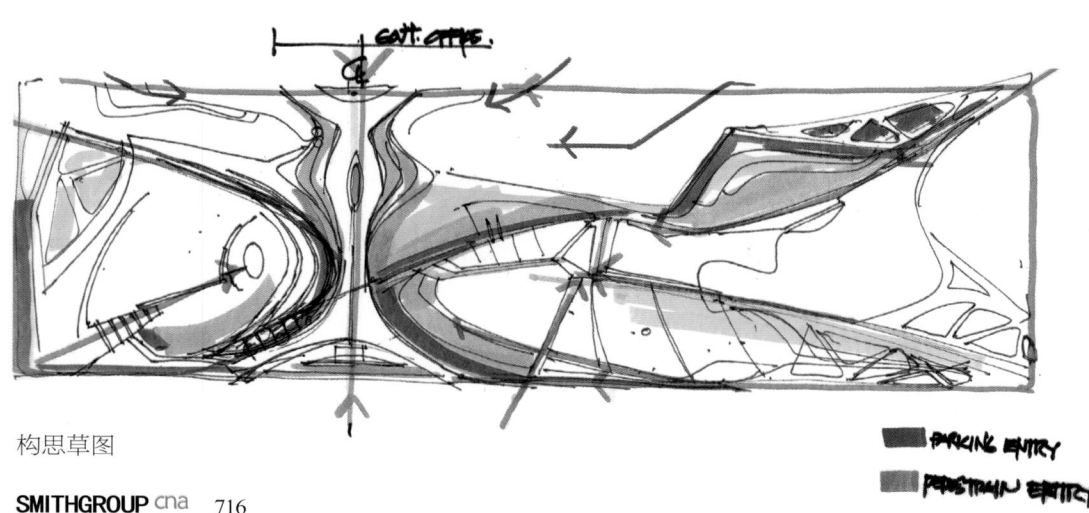

构思草图

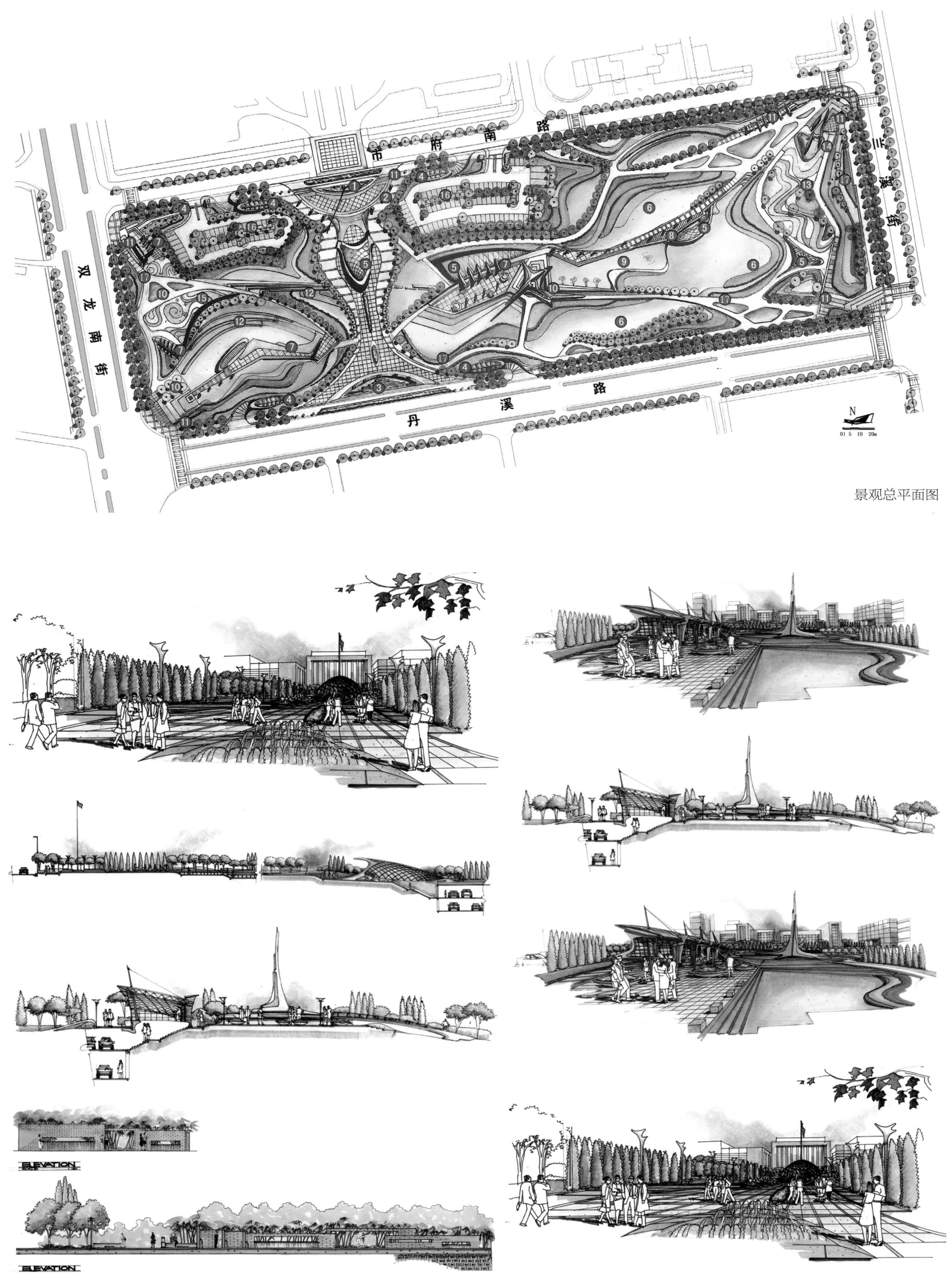

景观总平面图

ELEVATION

ELEVATION

淄博圣亚颐丰花园
Zibo Shengya Yifeng Garden

项目名称：淄博圣亚颐丰花园
项目地点：中国 / 山东 / 淄博
建筑面积：84 530 平方米
编制时间：2009 / 12
业主：山东淄博圣亚房地产开发有限公司

SMITHGROUP cna

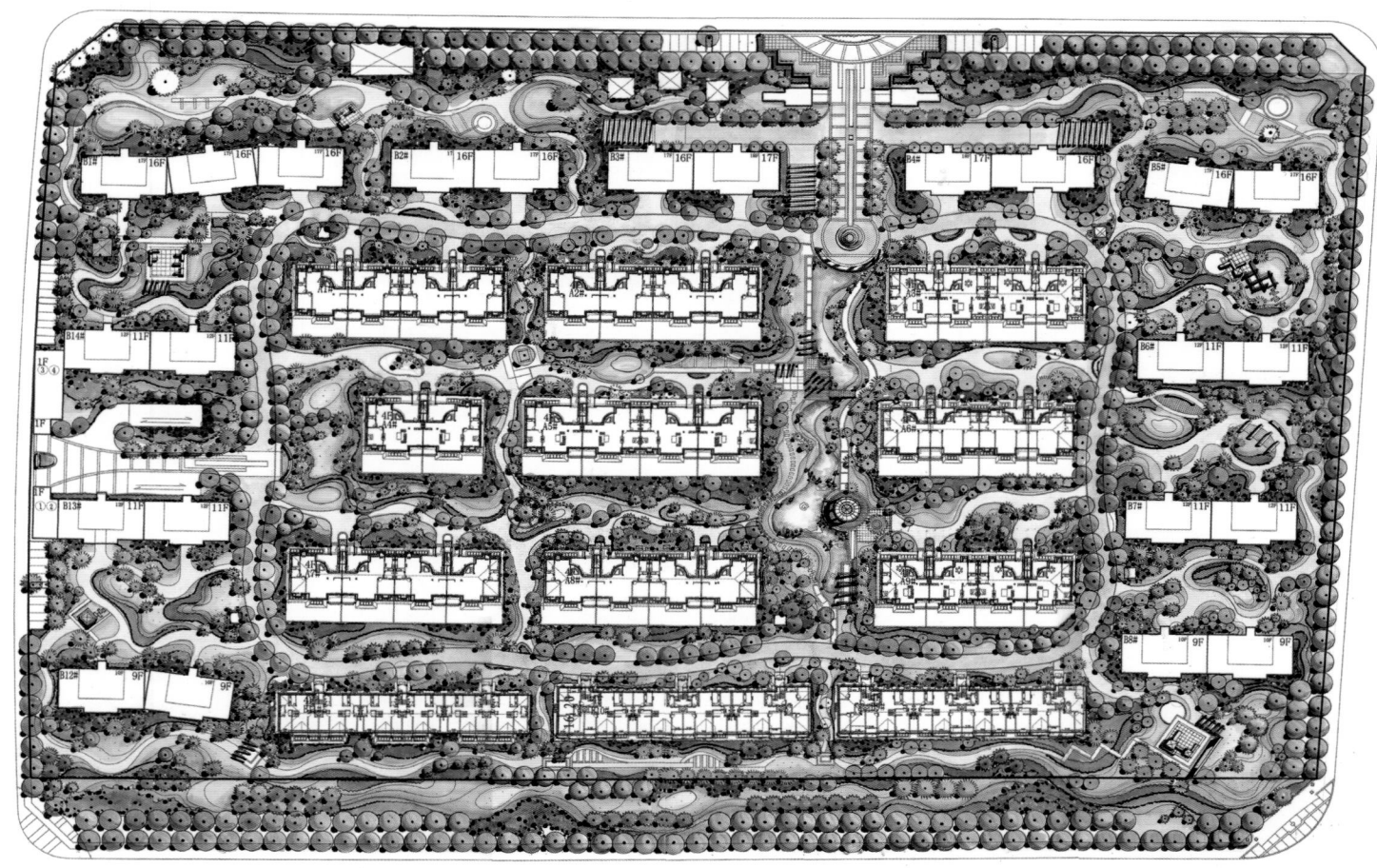

圣亚颐丰花园位于淄博市新城区，西邻淄博新城最大的休闲生态公园，有着良好的外部生态环境。 小区由 14 幢新古典主义风格的小高层和 9 幢花园洋房组成，容积率低，空间丰富。

景观设计布局整体大气，北入口与中央水景构成整个小区的主轴线，流畅的步行系统勾勒出生动灵活的主题空间；环道内外高程相差约一米，再加上局部的微地形设计，给人带来丰富的视觉感受；将浪漫情怀与现代人对生活的需要相结合，展现一种华贵、典雅、时尚、现代的新都市空间——"出则繁华，入则宁静"的都市住宅区。

Shengya Yifeng Garden is located in the new urban area of Zibo city, neighboring the biggest recreational leisure eco-park of the new town of Zibo in the west, and has a good external ecological environment. The community consists of 14 low-rise buildings of neoclassical style and 9 garden houses, with a low plot ratio and rich space.

The landscape design has a grand overall arrangement, with the northern entry and the central waterscape constituting the main axis of the whole community, and the fluent system outlining a lively and agile thematic space; the ring road has its inside and outside elevations differed by one meter or so, in addition to the partial micro-topographic design, provides people with rich visual experience; the landscape space is organized by combining romantic sentiments and modern needs for life, to exhibit a splendid, elegant, fashionable and modern new metropolitan space-the metropolitan residential area of "the bustling prosperity outside and the peaceful serenity inside".

景观总平面图

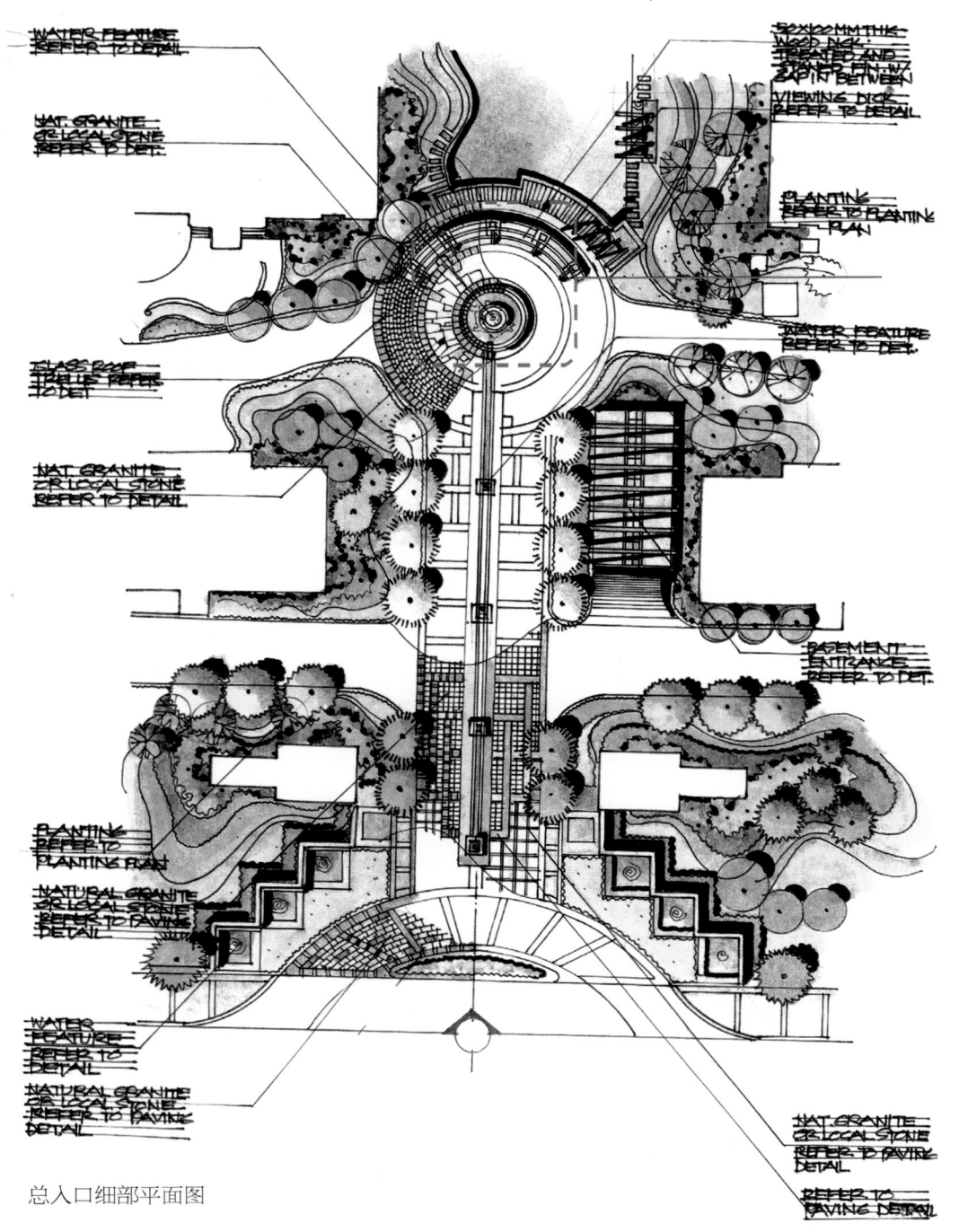

总入口细部平面图

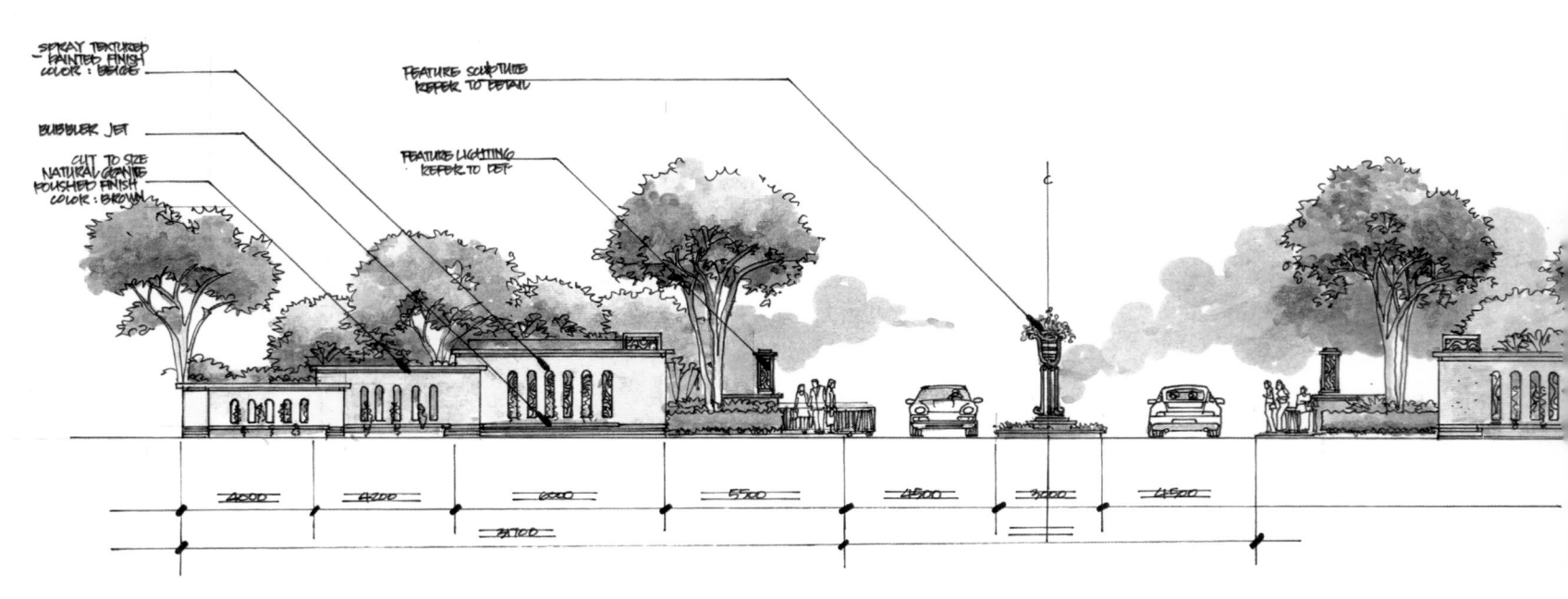

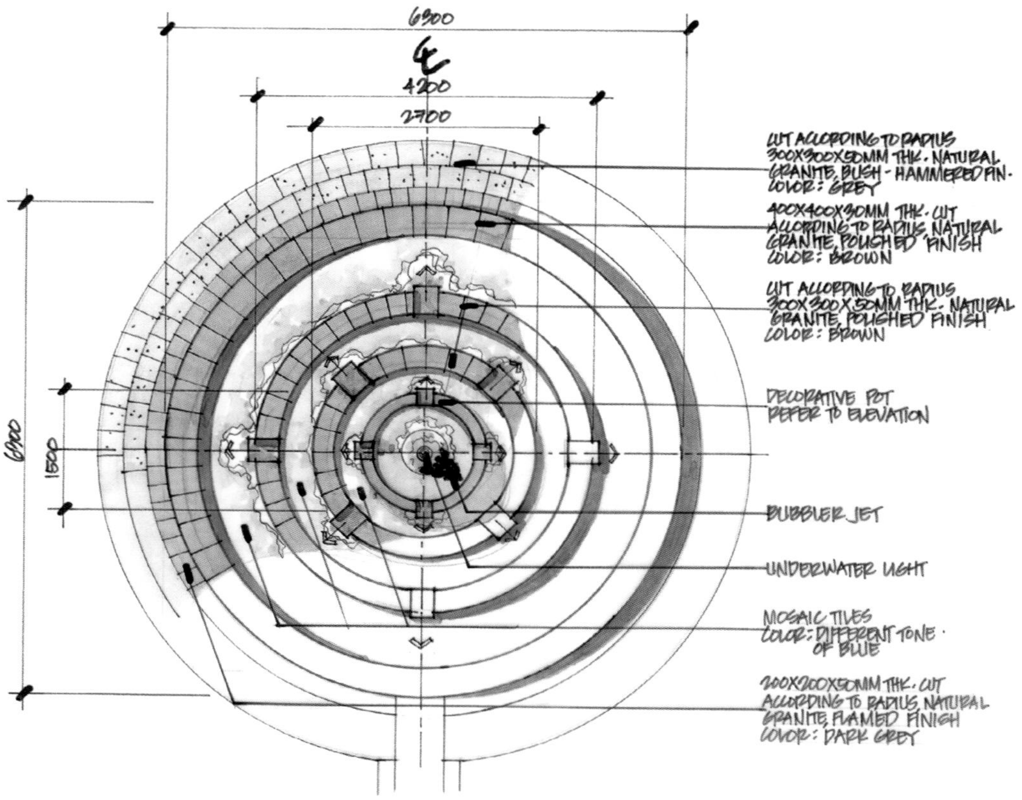

6300

4200

2700

6300

1500

CUT ACCORDING TO RADIUS
300X300X50MM THK. NATURAL
GRANITE BUSH - HAMMERED FIN.
COLOR: GREY

400X400X50MM THK. CUT
ACCORDING TO RADIUS NATURAL
GRANITE POLISHED FINISH
COLOR: BROWN

CUT ACCORDING TO RADIUS
300X300X50MM THK. NATURAL
GRANITE POLISHED FINISH
COLOR: BROWN

DECORATIVE POT
REFER TO ELEVATION

BUBBLER JET

UNDERWATER LIGHT

MOSAIC TILES
COLOR: DIFFERENT TONE
OF BLUE

200X200X50MM THK. CUT
ACCORDING TO RADIUS, NATURAL
GRANITE FLAMED FINISH
COLOR: DARK GREY

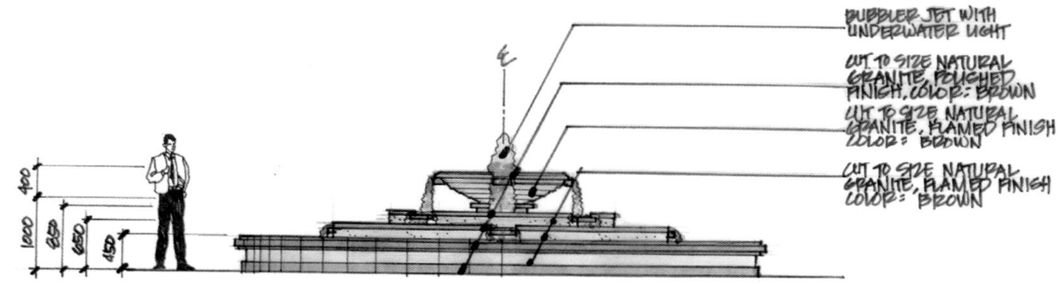

BUBBLER JET WITH
UNDERWATER LIGHT

CUT TO SIZE NATURAL
GRANITE POLISHED
FINISH. COLOR: BROWN

CUT TO SIZE NATURAL
GRANITE FLAMED FINISH
COLOR: BROWN

CUT TO SIZE NATURAL
GRANITE FLAMED FINISH
COLOR: BROWN

400

1000

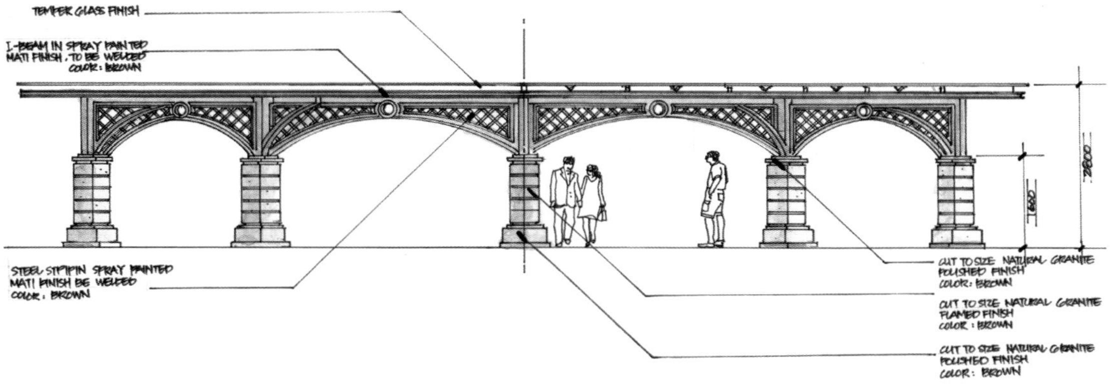

TEMPER GLASS FINISH

I-BEAM IN SPRAY PAINTED
MATT FINISH. TO BE WELDED
COLOR: BROWN

STEEL STRIP IN SPRAY PAINTED
MATT FINISH BE WELDED
COLOR: BROWN

2800

1600

CUT TO SIZE NATURAL GRANITE
POLISHED FINISH
COLOR: BROWN

CUT TO SIZE NATURAL GRANITE
FLAMED FINISH
COLOR: BROWN

CUT TO SIZE NATURAL GRANITE
POLISHED FINISH
COLOR: BROWN

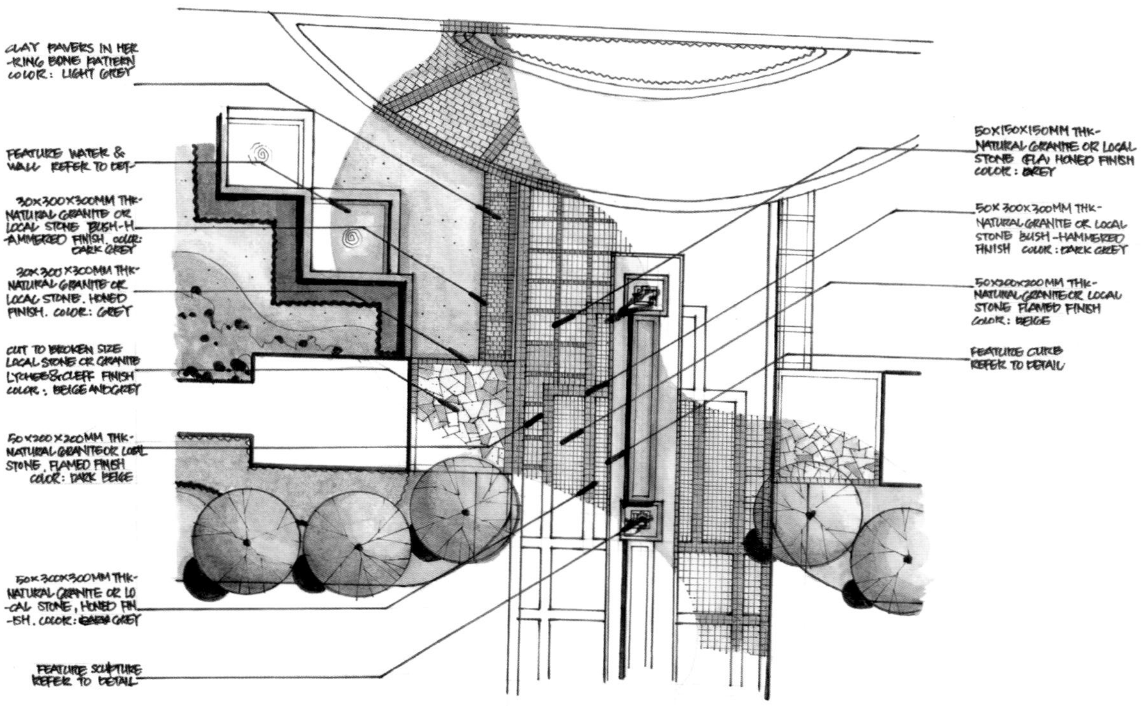

CLAY PAVERS IN HER
-RING BONE PATTERN
COLOR: LIGHT GREY

FEATURE WATER &
WALL REFER TO DET.

30X300X300 THK
NATURAL GRANITE OR
LOCAL STONE BUSH-H
-AMMERED FINISH. COLOR:
DARK GREY

30X300X300MM THK
NATURAL GRANITE OR
LOCAL STONE. HONED
FINISH. COLOR: GREY

CUT TO BROKEN SIZE
LOCAL STONE OR GRANITE
LYCHEE&CLEFT FINISH
COLOR: BEIGE AND GREY

50X300X300MM THK
NATURAL GRANITE OR LOCAL
STONE. FLAMED FINISH
COLOR: DARK BEIGE

50X300X300MM THK
NATURAL GRANITE OR LO
-CAL STONE, HONED FIN
-ISH. COLOR: DARK GREY

FEATURE SCULPTURE
REFER TO DETAIL

50X150X150MM THK
NATURAL GRANITE OR LOCAL
STONE (FLA HONED FINISH
COLOR: GREY

50X300X300MM THK
NATURAL GRANITE OR LOCAL
STONE BUSH-HAMMERED
FINISH. COLOR: DARK GREY

50X300X300MM THK
NATURAL GRANITE OR LOCAL
STONE FLAMED FINISH
COLOR: BEIGE

FEATURE CURB
REFER TO DETAIL

721

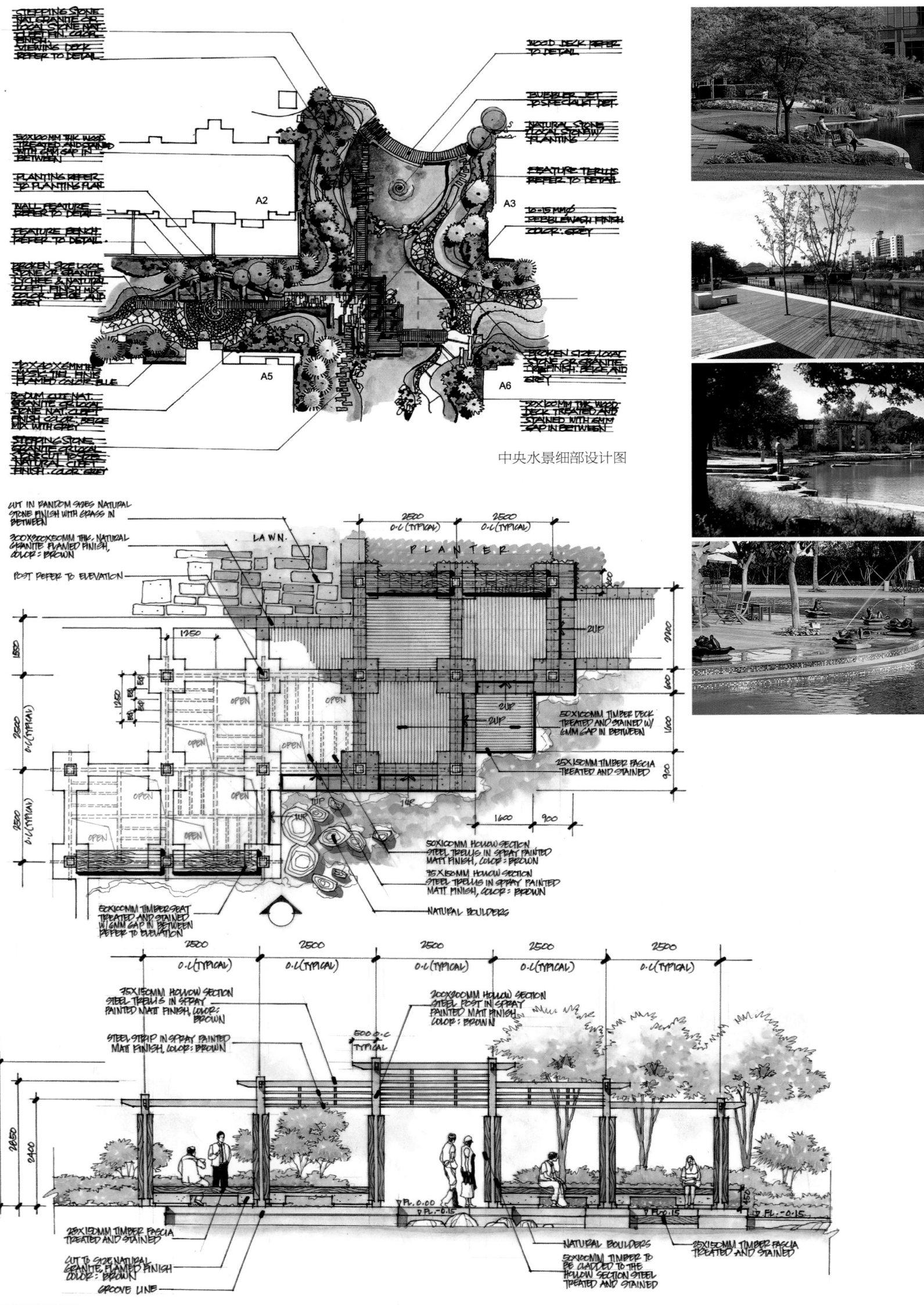

中央水景细部设计图

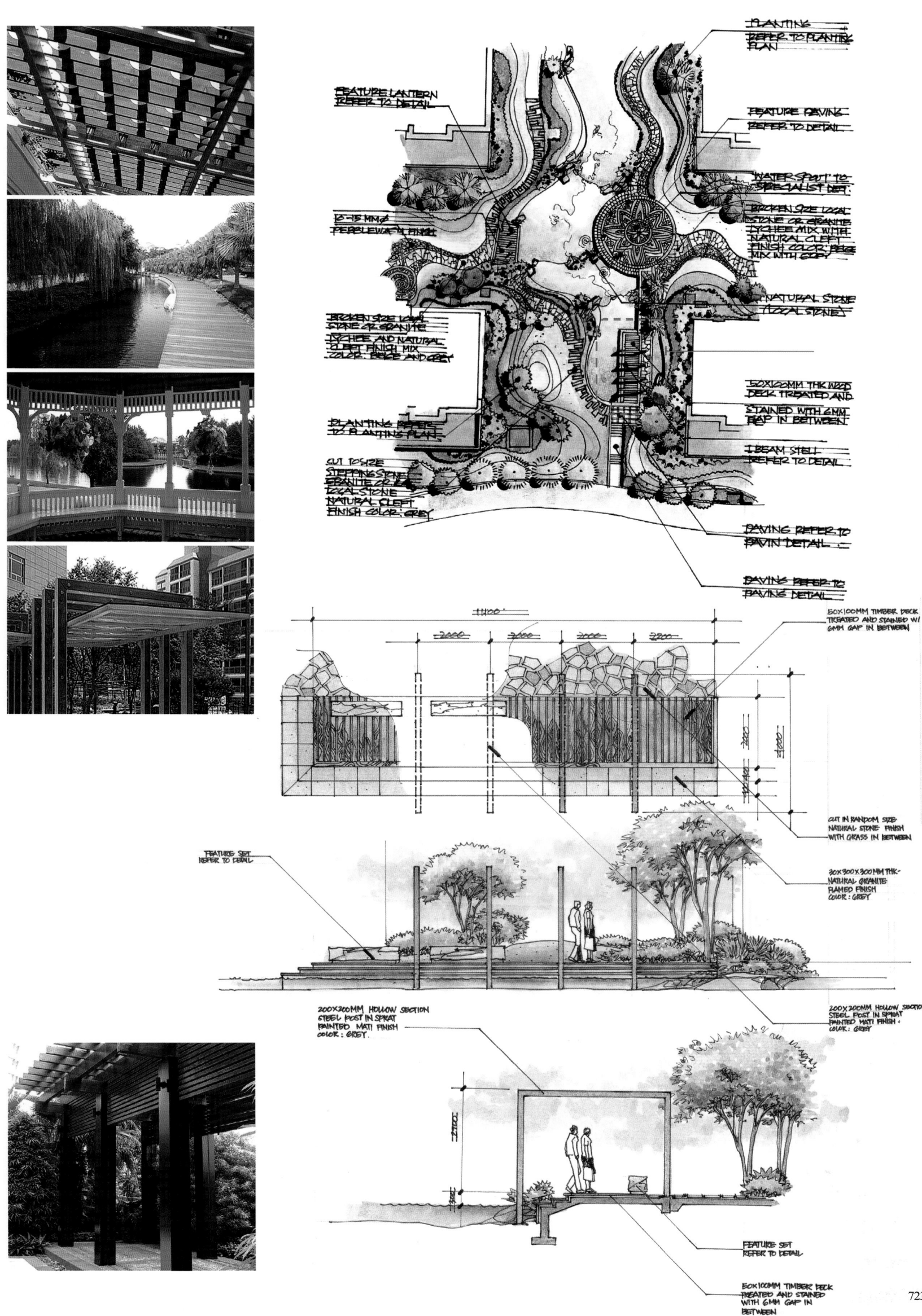

PLANTING
REFER TO PLANTING
PLAN

FEATURE LANTERN
REFER TO DETAIL

FEATURE PAVING
REFER TO DETAIL

WATER SPOUT TO
SPECIALIST DET.

10-15 MMØ
PEBBLEWASH FINISH

BROKEN SIZE LOCAL
STONE OR GRANITE
LYCHEE MIX WITH
NATURAL CLEFT
FINISH COLOR: BEIGE
MIX WITH GREY

NATURAL STONE
LOCAL STONE

BROKEN SIZE LOCAL
STONE OR GRANITE
LYCHEE AND NATURAL
CLEFT FINISH MIX
COLOR: BEIGE AND GREY

50X100MM THK WOOD
DECK TREATED AND
STAINED WITH 6MM
GAP IN BETWEEN

I BEAM STEEL
REFER TO DETAIL

PLANTING REFER
TO PLANTING PLAN

CUT TO SIZE
STEPPING STONE
GRANITE OR
LOCAL STONE
NATURAL CLEFT
FINISH COLOR: GREY

PAVING REFER TO
PAVING DETAIL

PAVING REFER TO
PAVING DETAIL

11100

2000 2000 2000 2200

50X100MM TIMBER DECK
TREATED AND STAINED W/
6MM GAP IN BETWEEN

CUT IN RANDOM SIZE
NATURAL STONE FINISH
WITH GRASS IN BETWEEN

30X300X300 MM THK
NATURAL GRANITE
FLAMED FINISH
COLOR: GREY

FEATURE SET
REFER TO DETAIL

200X200MM HOLLOW SECTION
STEEL POST IN SPRAY
PAINTED MAT'L FINISH
COLOR: GREY.

200X200MM HOLLOW SECTION
STEEL POST IN SPRAY
PAINTED MAT'L FINISH
COLOR: GREY

FEATURE SET
REFER TO DETAIL

50X100MM TIMBER DECK
TREATED AND STAINED
WITH 6MM GAP IN
BETWEEN

723

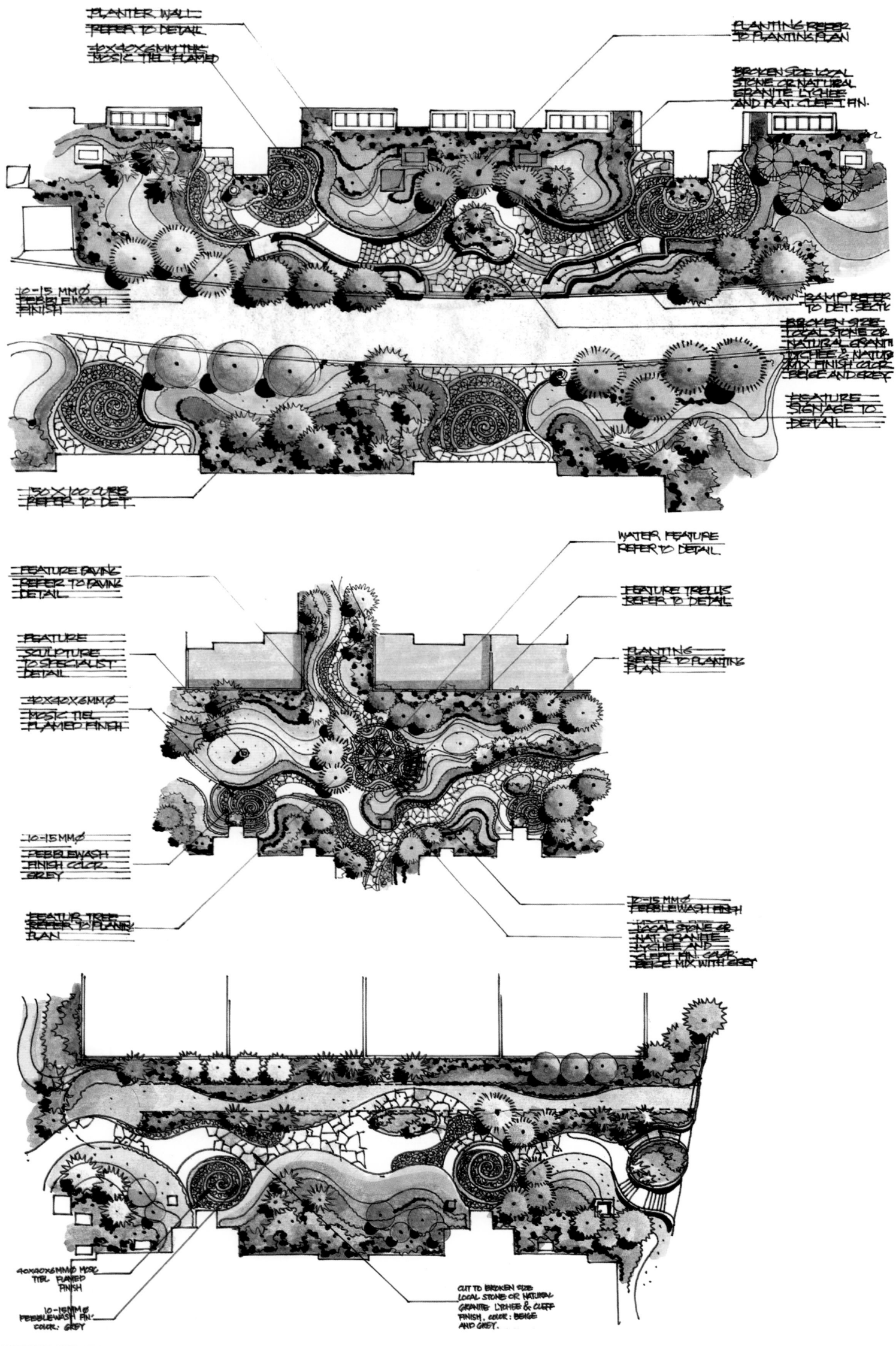

PLANTER WALL
REFER TO DETAIL

40X40X5MM THK
MOSIC TIEL FLAMED

PLANTING REFER
TO PLANTING PLAN

BROKEN SIZE LOCAL
STONE OR NATURAL
GRANITE LYCHEE
AND NAT. CLEFT FIN.

10-15 MM∅
PEBBLEWASH
FINISH

RAMP REFER
TO DET. SECTN

BROKEN SIZE
LOCAL STONE OR
NATURAL GRANITE
LYCHEE & NATURAL
MIX FINISH COLOR
BEIGE AND GREY

FEATURE
SIGNAGE TO
DETAIL

150 X 100 CURB
REFER TO DET.

WATER FEATURE
REFER TO DETAIL

FEATURE PAVING
REFER TO PAVING
DETAIL

FEATURE TRELLIS
REFER TO DETAIL

FEATURE
SCULPTURE
TO SPECIALIST
DETAIL

PLANTING
REFER TO PLANTING
PLAN

40X40X5MM∅
MOSIC TIEL
FLAMED FINISH

10-15 MM∅
PEBBLEWASH
FINISH COLOR
GREY

10-15 MM∅
PEBBLEWASH FINISH

FEATURE TREE
REFER TO PLANTING
PLAN

LOCAL STONE OR
NAT. GRANITE
LYCHEE AND
CLEFT FIN. COLOR
BEIGE MIX WITH GREY

40X40X5MM∅ MOSC
TIEL FLAMED
FINISH

10-15MM∅
PEBBLEWASH FIN.
COLOR: GREY

CUT TO BROKEN SIZE
LOCAL STONE OR NATURAL
GRANITE LYCHEE & CLEFF
FINISH. COLOR: BEIGE
AND GREY.

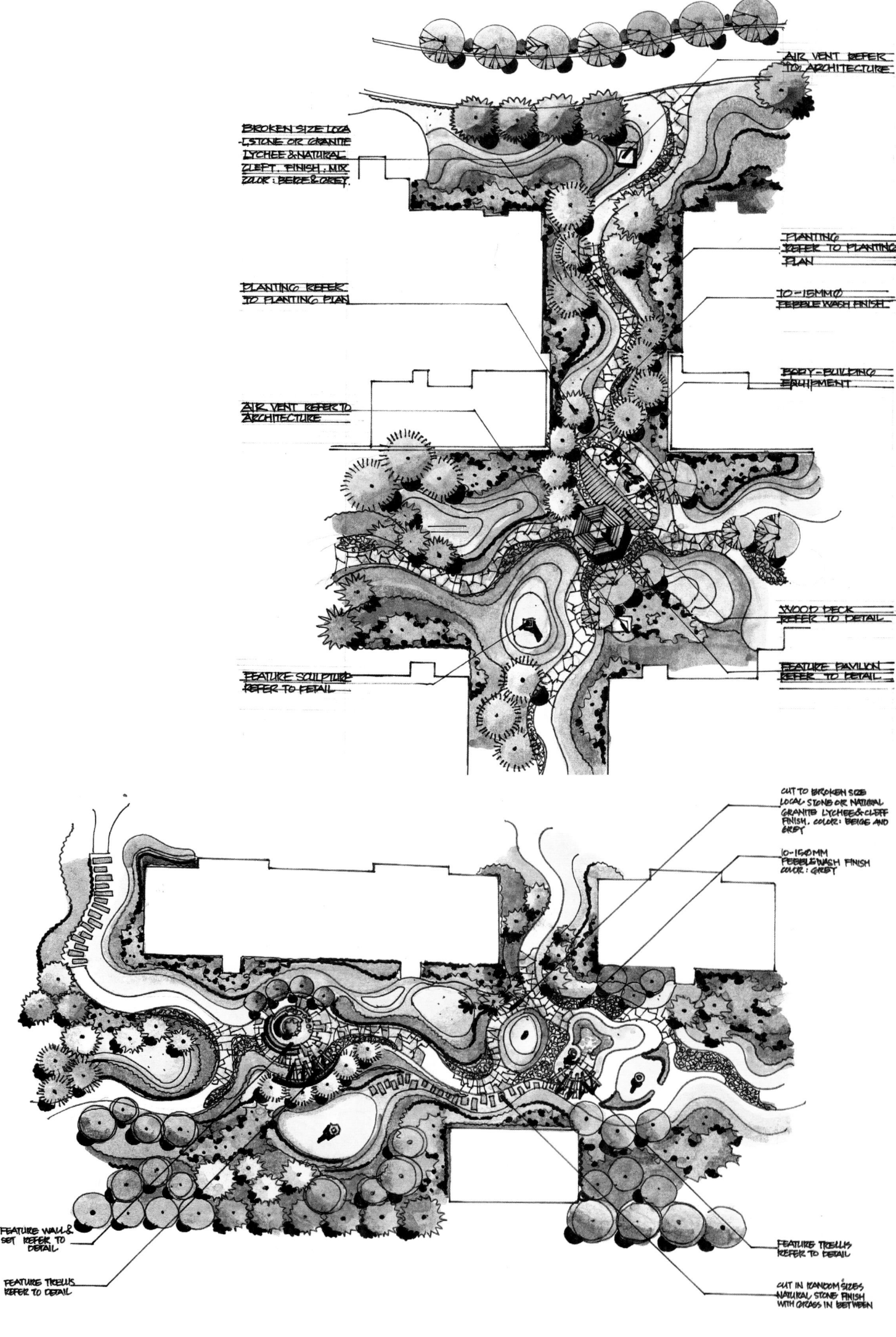

BROKEN SIZE LOZA
-STONE OR GRANITE
LYCHEE & NATURAL
2LEFT. FINISH. MIX
COLOR: BEIGE & GREY.

AIR VENT REFER
TO ARCHITECTURE

PLANTING
REFER TO PLANTING
PLAN

PLANTING REFER
TO PLANTING PLAN

10 - 15MMØ
PEBBLE WASH FINISH

BODY - BUILDING
EQUIPMENT

AIR VENT REFER TO
ARCHITECTURE

WOOD DECK
REFER TO DETAIL

FEATURE PAVILION
REFER TO DETAIL

FEATURE SCULPTURE
REFER TO DETAIL

CUT TO BROKEN SIZE
LOCAL STONE OR NATURAL
GRANITE LYCHEE & CLEFT
FINISH. COLOR: BEIGE AND
GREY

10 - 15Ø MM
PEBBLE WASH FINISH
COLOR: GREY

FEATURE WALL &
SET REFER TO
DETAIL

FEATURE TRELLIS
REFER TO DETAIL

FEATURE TRELLIS
REFER TO DETAIL

CUT IN RANDOM SIZES
NATURAL STONE FINISH
WITH GRASS IN BETWEEN

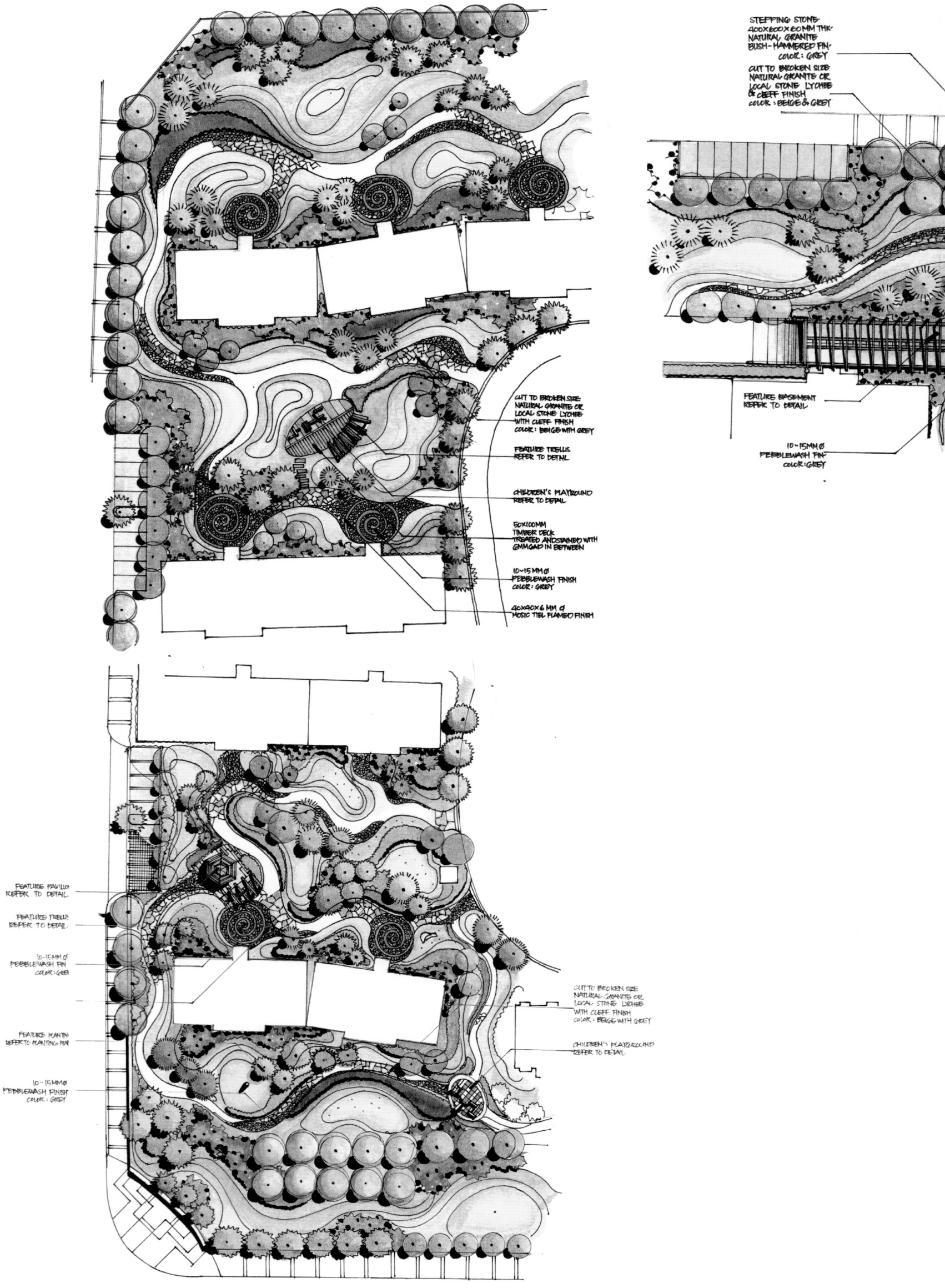

STEPPING STONE
400×600×60MM THK·
NATURAL GRANITE
BUSH-HAMMERED FIN·
COLOR: GREY

CUT TO BROKEN SIZE
NATURAL GRANITE OR
LOCAL STONE LYCHEE
& CLEFF FINISH
COLOR: BEIGE & GREY

FEATURE BASEMENT
REFER TO DETAIL

10~15MM Ø
PEBBLEWASH FIN·
COLOR: GREY

CUT TO BROKEN SIZE
NATURAL GRANITE OR
LOCAL STONE LYCHEE
WITH CLEFF FINISH
COLOR: BEIGE WITH GREY

FEATURE TRELLIS
REFER TO DETAIL

CHILDREN'S PLAYGROUND
REFER TO DETAIL

50×100MM
TIMBER DECK
TREATED AND STAINED WITH
6MM GAP IN BETWEEN

10~15 MM Ø
PEBBLEWASH FINISH
COLOR: GREY

40×40×6 MM Ø
MOSIC TIEL FLAMED FINISH

FEATURE PAVILION
REFER TO DETAIL

FEATURE TRELLIS
REFER TO DETAIL

10~15MM Ø
PEBBLEWASH FIN·
COLOR: GREY

FEATURE PLANTING
REFER TO PLANTING PLAN

10~15MM Ø
PEBBLEWASH FINISH
COLOR: GREY

CUT TO BROKEN SIZE
NATURAL GRANITE OR
LOCAL STONE LYCHEE
WITH CLEFF FINISH
COLOR: BEIGE WITH GREY

CHILDREN'S PLAYGROUND
REFER TO DETAIL

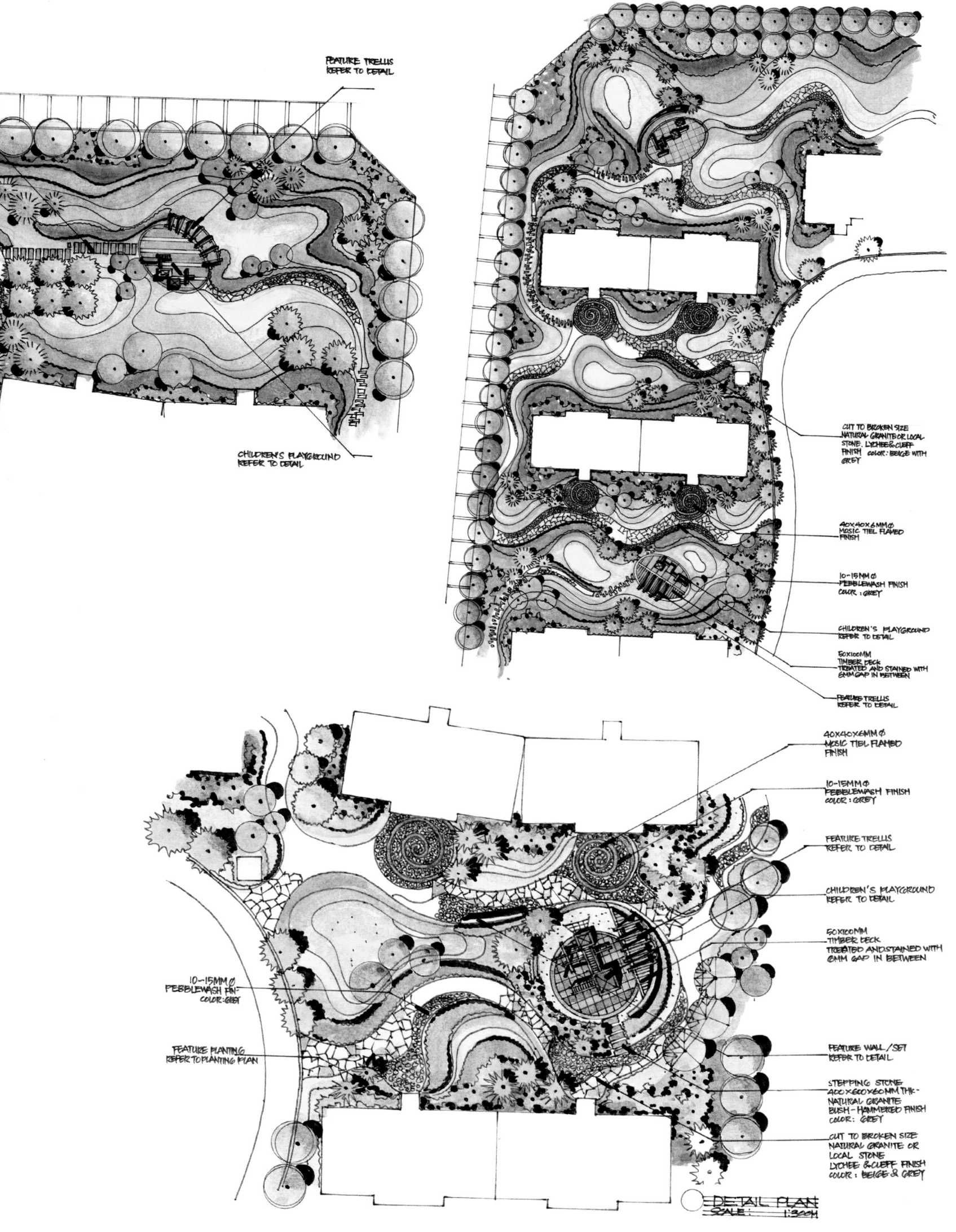

FEATURE TRELLIS
REFER TO DETAIL

CHILDREN'S PLAYGROUND
REFER TO DETAIL

CUT TO BROKEN SIZE
NATURAL GRANITE OR LOCAL
STONE. LYCHEE & CLEFT
FINISH COLOR: BEIGE WITH
GREY

40X40X6MM Ø
MOSIC TIEL FLAMED
FINISH

10-15MM Ø
PEBBLEWASH FINISH
COLOR : GREY

CHILDREN'S PLAYGROUND
REFER TO DETAIL

50X100MM
TIMBER DECK
TREATED AND STAINED WITH
6MM GAP IN BETWEEN

FEATURE TRELLIS
REFER TO DETAIL

40X40X6MM Ø
MOSIC TIEL FLAMED
FINISH

10-15MM Ø
PEBBLEWASH FINISH
COLOR : GREY

FEATURE TRELLIS
REFER TO DETAIL

CHILDREN'S PLAYGROUND
REFER TO DETAIL

50X100MM
TIMBER DECK
TREATED AND STAINED WITH
6MM GAP IN BETWEEN

FEATURE WALL / SET
REFER TO DETAIL

STEPPING STONE
400X600X60MM THK-
NATURAL GRANITE
BUSH - HAMMERED FINISH
COLOR: GREY

CUT TO BROKEN SIZE
NATURAL GRANITE OR
LOCAL STONE
LYCHEE & CLEFT FINISH
COLOR : BEIGE & GREY

10-15MM Ø
PEBBLEWASH FIN-
COLOR : GREY

FEATURE PLANTING
REFER TO PLANTING PLAN

DETAIL PLAN
SCALE :   1:200H

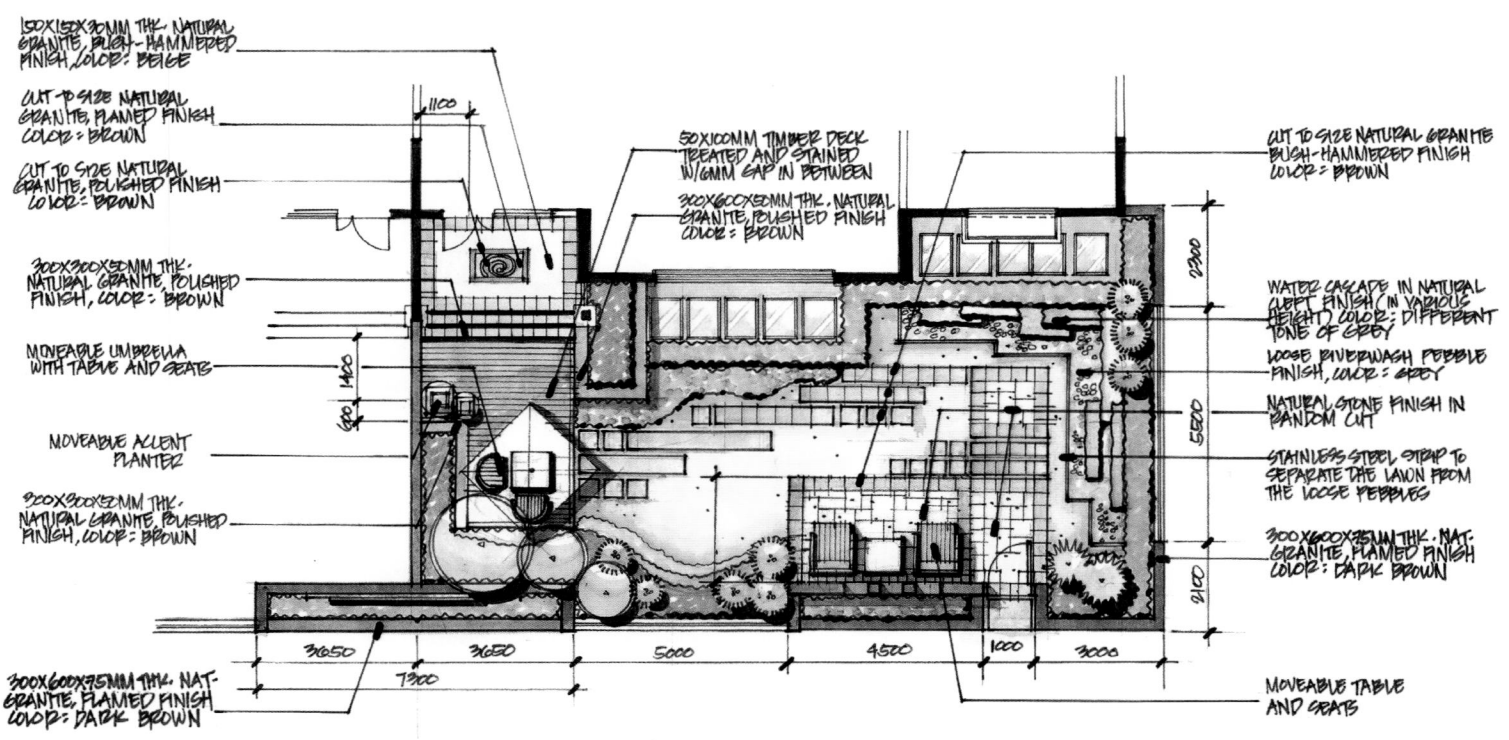

150X150X30MM THK. NATURAL
GRANITE, BUSH-HAMMERED
FINISH, COLOR: BEIGE

CUT TO SIZE NATURAL
GRANITE, FLAMED FINISH
COLOR: BROWN

CUT TO SIZE NATURAL
GRANITE, POLISHED FINISH
COLOR: BROWN

300X300X50MM THK.
NATURAL GRANITE, POLISHED
FINISH, COLOR: BROWN

MOVEABLE UMBRELLA
WITH TABLE AND SEATS

MOVEABLE ACCENT
PLANTER

900X300X52MM THK.
NATURAL GRANITE, POLISHED
FINISH, COLOR: BROWN

300X600X75MM THK. NAT.
GRANITE, FLAMED FINISH
COLOR: DARK BROWN

50X100MM TIMBER DECK
TREATED AND STAINED
W/6MM GAP IN BETWEEN

300X600X50MM THK. NATURAL
GRANITE, POLISHED FINISH
COLOR: BROWN

CUT TO SIZE NATURAL GRANITE
BUSH-HAMMERED FINISH
COLOR: BROWN

WATER CASCADE IN NATURAL
CLEFT FINISH, IN VARIOUS
HEIGHT, COLOR: DIFFERENT
TONE OF PEBBLE

LOOSE RIVERWASH PEBBLE
FINISH, COLOR: GREY

NATURAL STONE FINISH IN
RANDOM CUT

STAINLESS STEEL STRIP TO
SEPARATE THE LAWN FROM
THE LOOSE PEBBLES

300X600X75MM THK. NAT.
GRANITE, FLAMED FINISH
COLOR: DARK BROWN

MOVEABLE TABLE
AND SEATS

MOVEABLE UMBRELLA
WITH TABLE AND SEATS

HOLLOW SECTION STEEL IN
SPRAY PAINTED MATT FINISH
COLOR: DARK BROWN

STEEL STRIP IN SPRAY PAINTED
MATT FINISH, COLOR:
DARK BROWN

300X600X75MM THK. NAT.
GRANITE, FLAMED FINISH
COLOR: DARK BROWN

FLOWER PLANTS REFER
TO PLANTING PLAN

SPRAY TEXTURED PAINTED
FINISH, COLOR: LIGHT
BROWN

STEELWORKS IN SPRAY
PAINTED MATT FIN. COLOR: BLACK

BROKEN SIZE NATURAL
STONE FINISH, COLOR: DIFFERENT
TONE OF BROWN

300X600X30 MM THK. NATURAL
GRANITE, NATURAL CLEFT FINISH
COLOR: BROWN

30-50MM Ø RIVERWASH
PEBBLE FINISH, COLOR:
BROWN

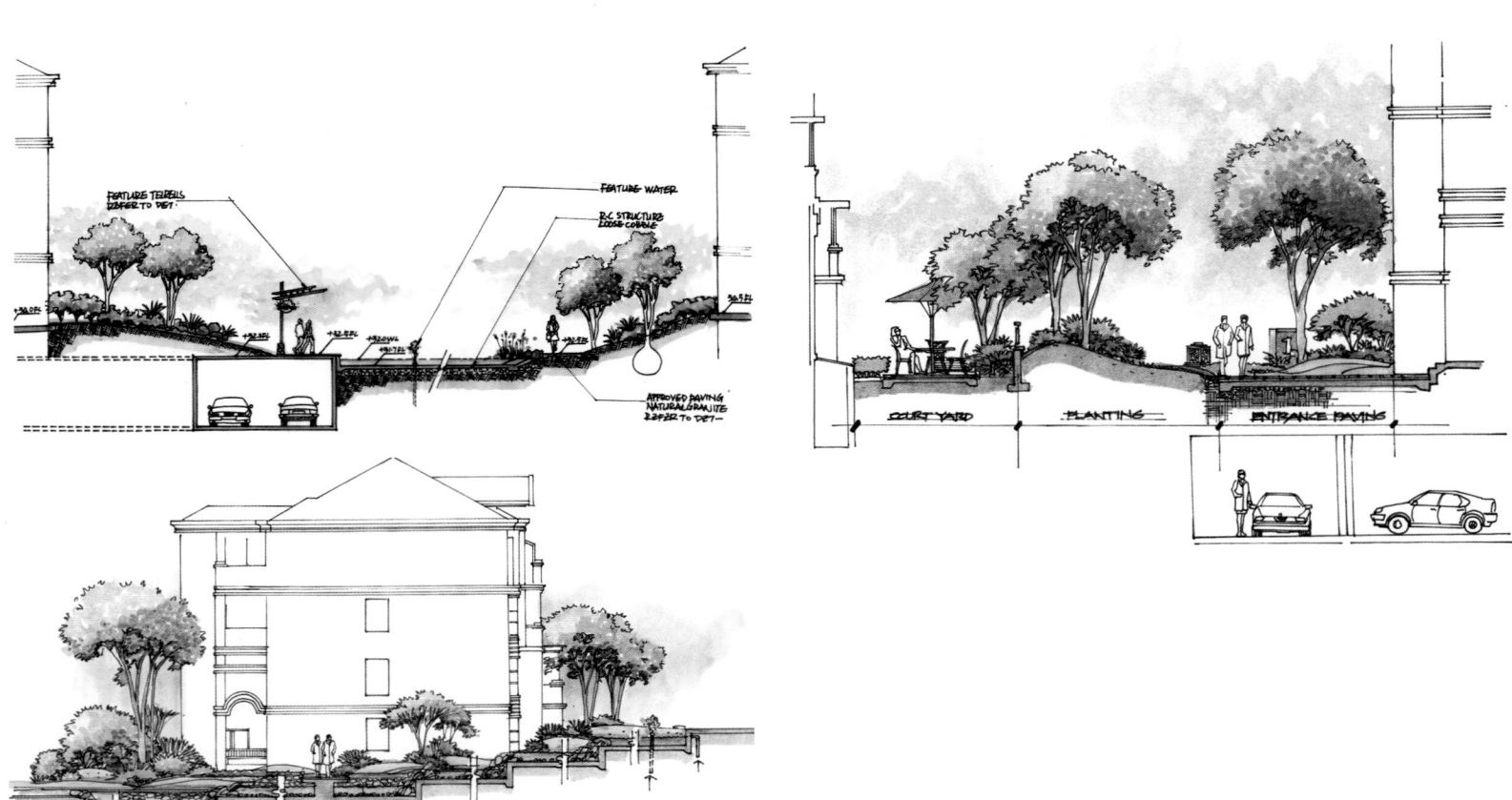

FEATURE TRELLIS
REFER TO DET.

FEATURE WATER

R.C. STRUCTURE
LOOSE COBBLE

APPROVED PAVING
NATURAL GRANITE
REFER TO DET.

COURT YARD   PLANTING   ENTRANCE PAVING

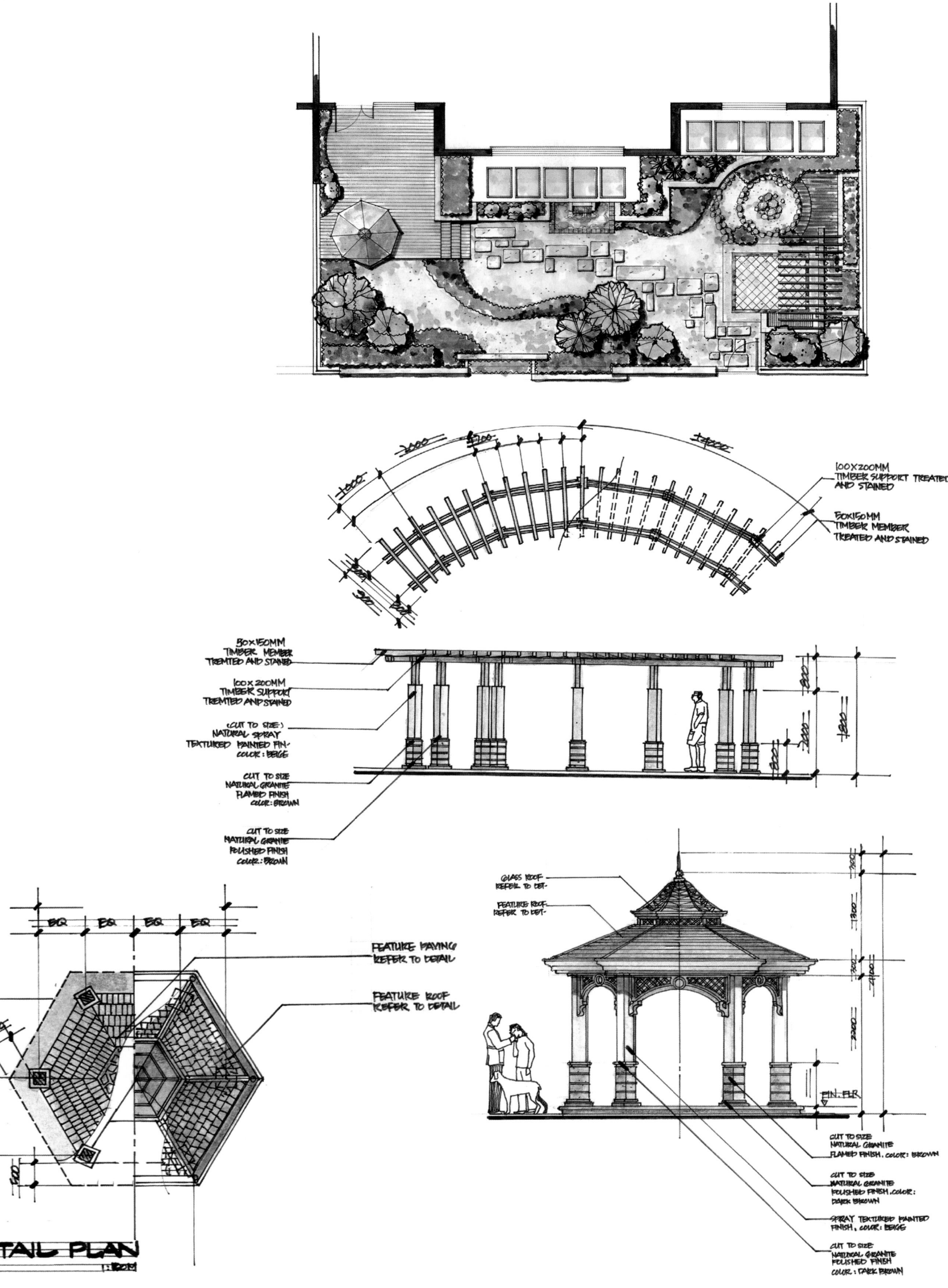

100x200MM
TIMBER SUPPORT TREATED
AND STAINED

50x150MM
TIMBER MEMBER
TREATED AND STAINED

50x150MM
TIMBER MEMBER
TREATED AND STAINED

100x200MM
TIMBER SUPPORT
TREATED AND STAINED

(CUT TO SIZE)
NATURAL SPRAY
TEXTURED PAINTED FIN.
COLOR: BEIGE

CUT TO SIZE
NATURAL GRANITE
FLAMED FINISH
COLOR: BROWN

CUT TO SIZE
NATURAL GRANITE
POLISHED FINISH
COLOR: BROWN

FEATURE PAVING
REFER TO DETAIL

FEATURE ROOF
REFER TO DETAIL

GLASS ROOF
REFER TO DET.

FEATURE ROOF
REFER TO DET.

CUT TO SIZE
NATURAL GRANITE
FLAMED FINISH, COLOR: BROWN

CUT TO SIZE
NATURAL GRANITE
POLISHED FINISH ,COLOR:
DARK BROWN

SPRAY TEXTURED PAINTED
FINISH, COLOR: BEIGE

CUT TO SIZE
NATURAL GRANITE
POLISHED FINISH
COLOR : DARK BROWN

FIN. FLR.

EQ   EQ   EQ   EQ

DETAIL PLAN
SCALE   1:150M

729

西宁火车站站前景观设计
Xining Railway Station Front Landscape Design

项目名称：西宁火车站站前景观设计
项目地点：中国 / 青海 / 西宁
建筑面积：239 214 平方米
编制时间：2010 / 05
业主：西宁市政府
服务范围：方案 / 扩初 / 施工图

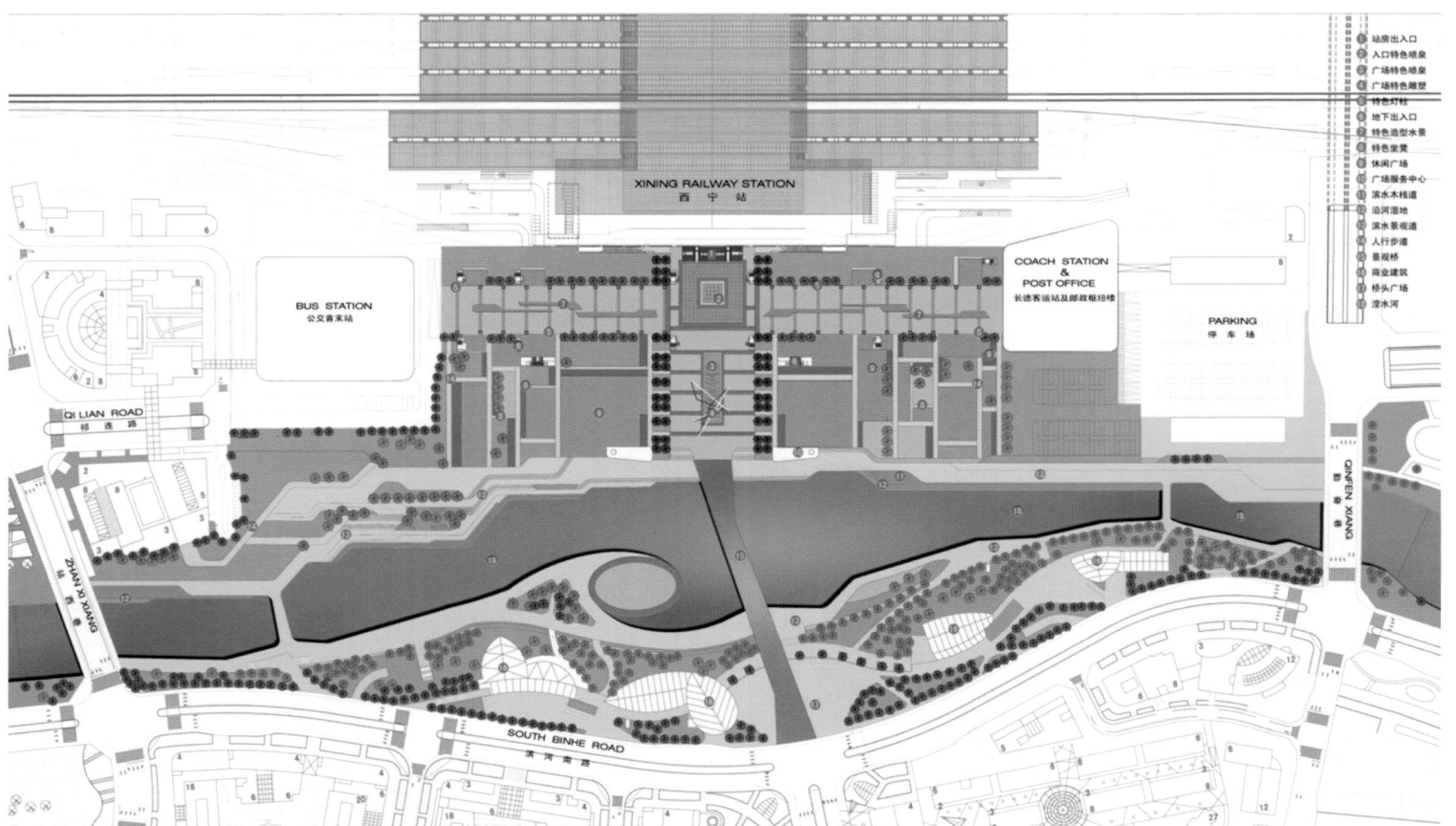

西宁火车站是青海省乃至整个西部最重要的区域性交通枢纽，同时也是西宁城市建设发展的一个里程碑。本方案在多个规划和建筑方案的基础之上，整合美化城市交通，合理划分规划用地，处理好集散广场及周边环境，满足旅客游人流动、停留、聚集、购物、休息、观赏等需求；是各种具有创新意义和示范价值的设计的最佳实践区，同时也是为提高城市生活质量所做的公认的、创新的和有价值的各种实践方案最佳实践区；另通过宽阔的主题广场，供游人参与探讨交流城市管理及发展的话题。

Xining Railway Station is the most important regional traffic hub in Qinghai Province and even in the whole west of China. It is also a milestone in the urban construction development of Xining. This project will, on the basis of various plans and buildings, integrate and beautify the city traffic, reasonably demarcate planned lands to be used, and deal well with the distribution square and neighboring circumstances, to meet passengers and tourists' needs for flowing, staying, congregating, shopping, resting and sightseeing, etc; meanwhile it is also the optimal practice area for various innovative meanings and exemplary values, and the various recognized, innovative and valuable practice plans made to improve the quality of urban life; in addition, the broad thematic square can be the topic for tourists participating in the discussion and exchange on city management and development.

景观总平面图

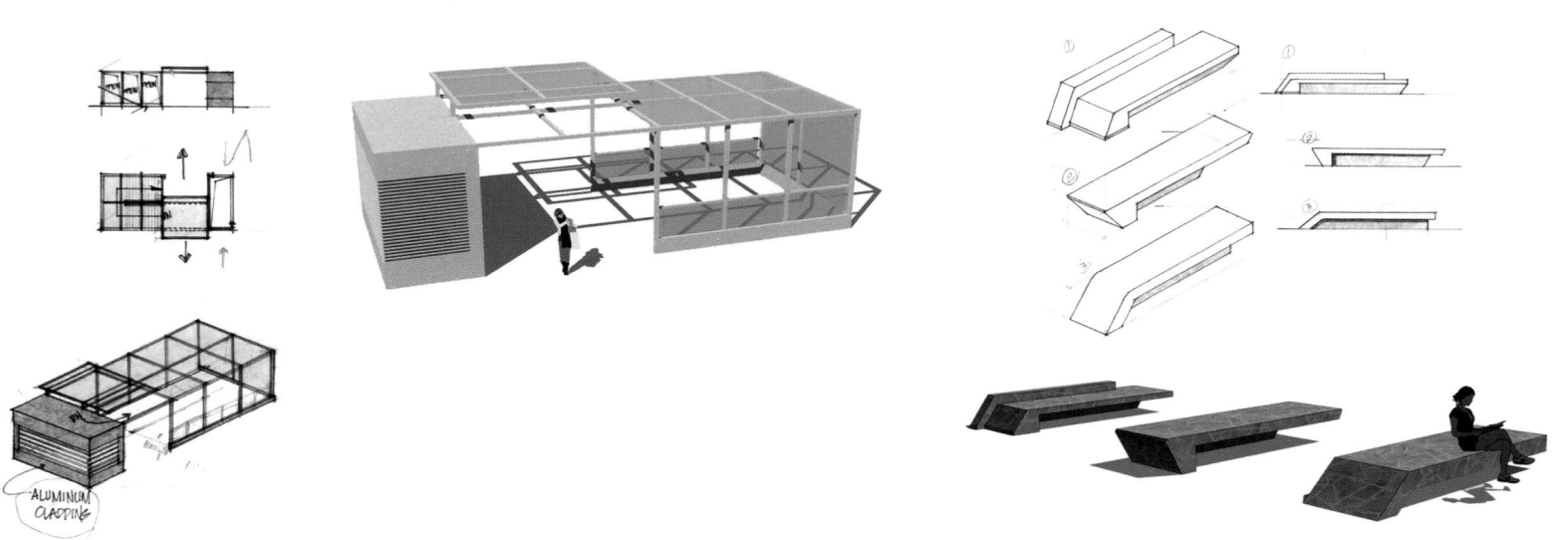

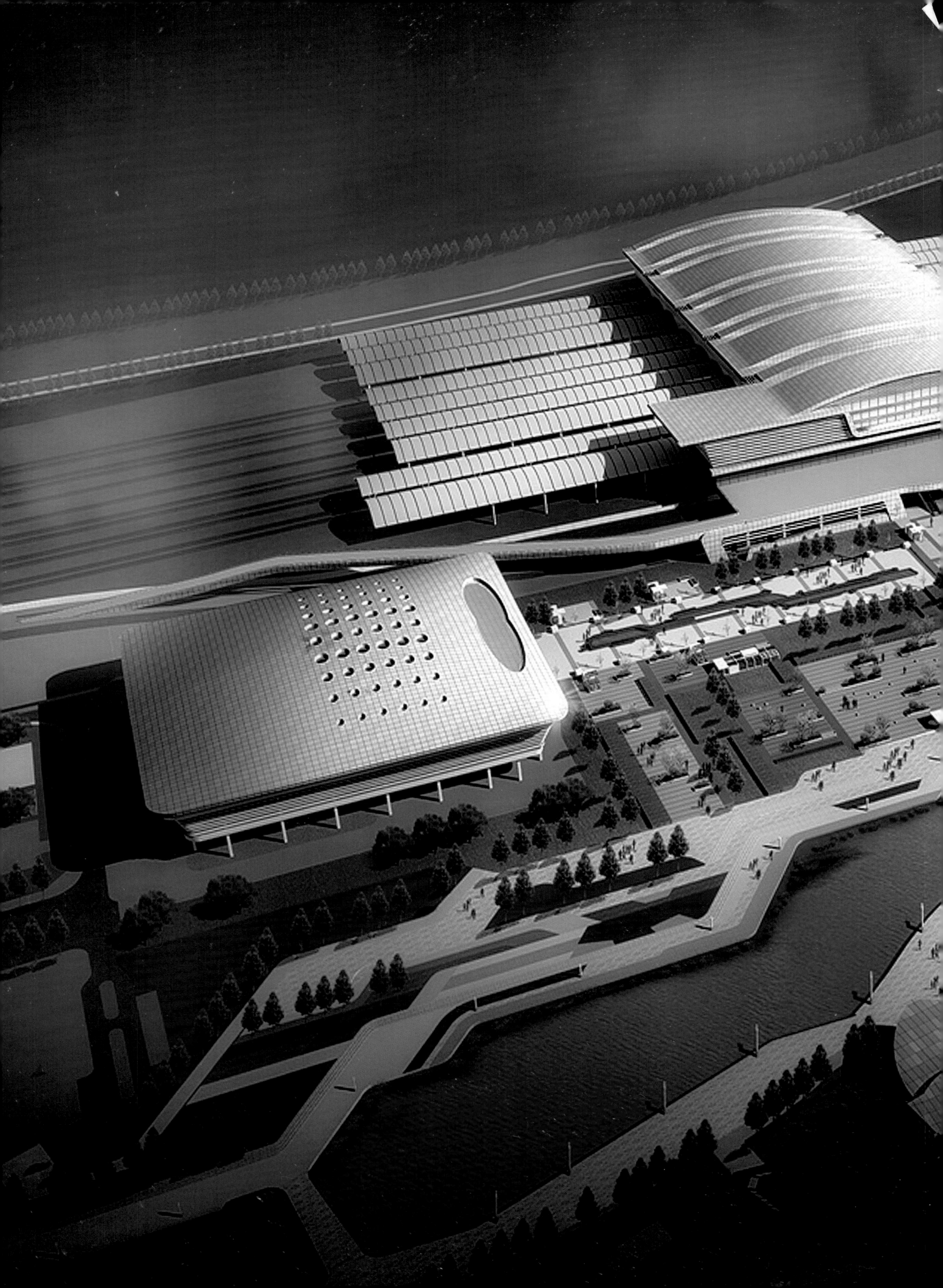

青城山泰达研发社区
Qingcheng Mountain Taida R&D Community

项目名称：青城山泰达研发社区
建筑地点：中国 / 四川 / 成都
建筑面积：规划总用地约为 34.7 万平方米
编制时间：2008 / 11
业主：成都泰达房地产开发有限公司
服务范围：方案 / 扩初 / 施工图

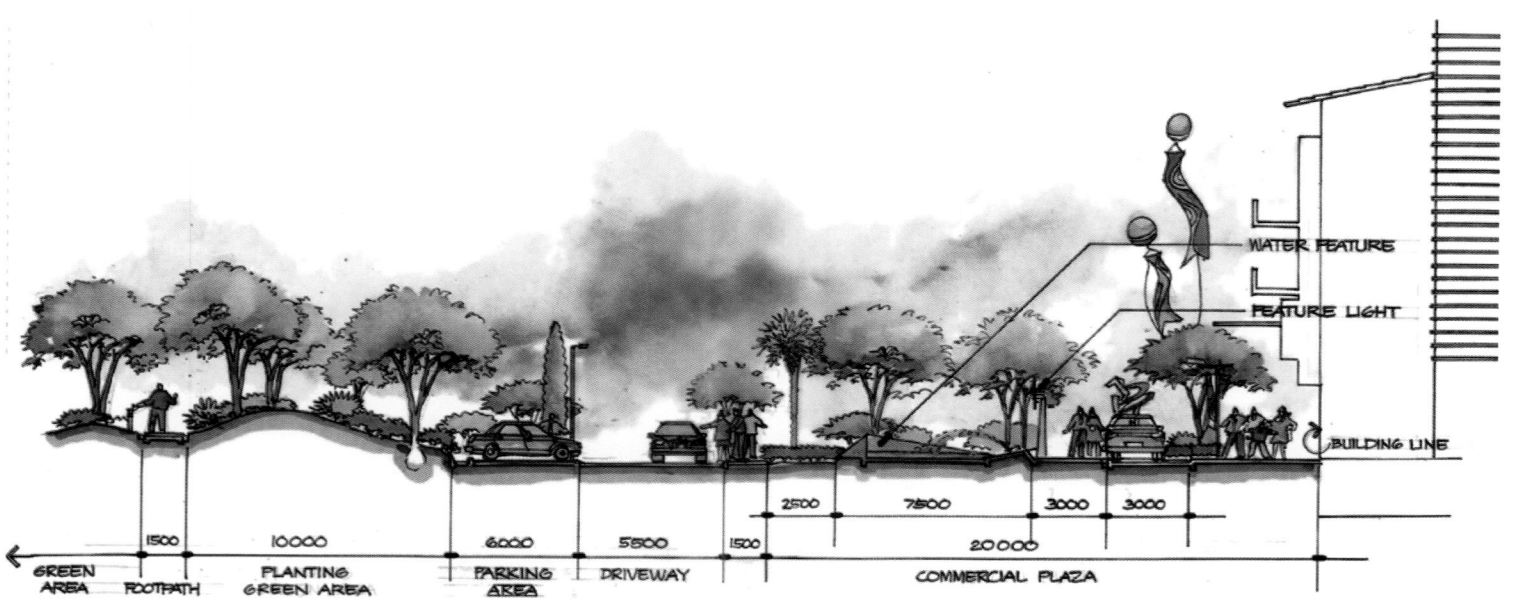

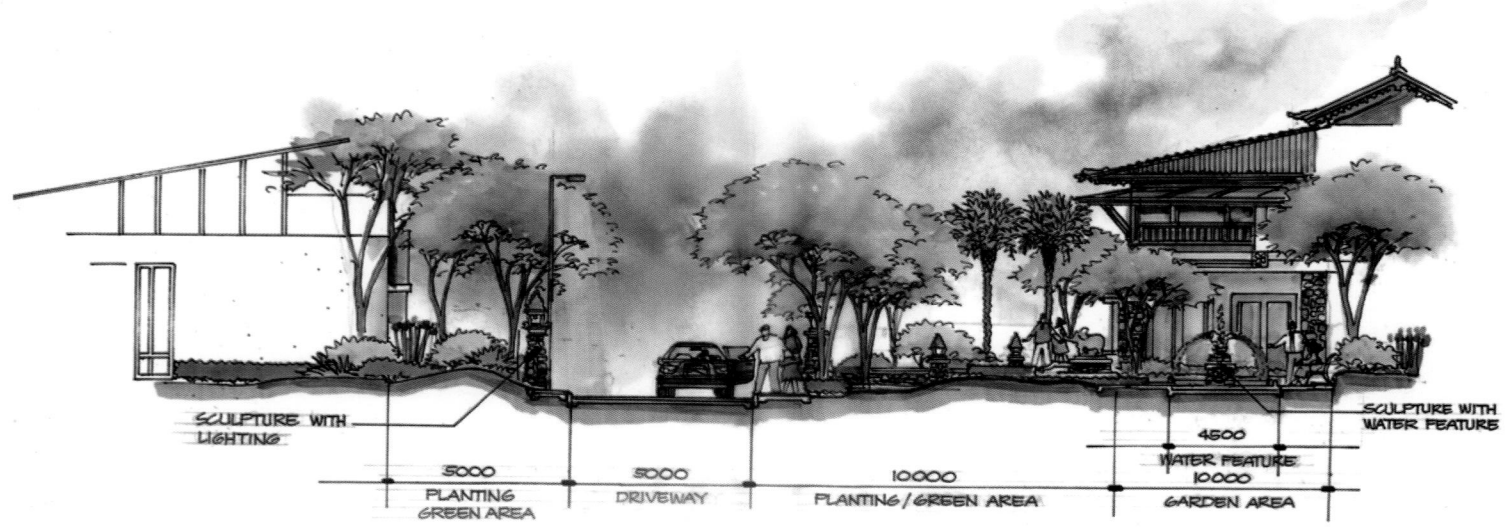

青城山位于成都西北都江堰市境内，距成都市区 75 公里，因山上树木茂盛、四季常青，故历来享有"青城天下幽"的美誉。本项目地块正位于青城山脚下，西临绕镇道，南靠东软大道，其中景观面积约 16.5 万平方米。是集商业休闲广场、文化产业办公、城市度假酒店、居住养身别墅，四部分为一体的综合性时尚社区。

景观设计紧跟建筑设计的步伐，体现有机、生长、叠合的总图规划设计理念，将西蜀建筑原有的地域特点，既与现有环境有机结合，融入既有时代气息，又具有充满东南亚韵味与情趣的景观特点，形成青城山脚下特有的泛亚洲风格综合体。景观设计师致力于创造一处细腻而丰富的人文景象，通过大量的泛东南亚元素的运用，展现出一个集商业、产业、办公、餐饮娱乐、度假、酒店、多功能为一体的城市新景观空间。

Qingcheng Mountain, located in Dujiangyan City, northwest of Chengdu, is 75km away from Chengdu area. The mountain, which is surrounded by numerous peaks and is shaped like a city, is dubbed "the most peaceful and secluded mountain under heaven" and combines perfectly with its evergreen scenery. This project is at the foot of the mountain, to the west of Rouzhen Road, to the south of Dongruan Avenue, of which the landscaping area is about 165,000 m². It includes the commercial recreation square, cultural industry office work, city holiday hotels, and residential health-preserving villas. They are integrated into a whole to form a comprehensive fashionable community.

Keeping up the step of architectural design, the landscape design reflects the design concept of organicity, growth, and superposition, which forms the unique Asian style complex at the foot of Qingcheng Mountain by combining the original regional features of the west Sichuan architecture and the present circumstances into an organic unity, incorporating the existing spirit of the times and abounding in the landscaping characteristics of Southeast Asian flavor and exoticism so as to offer a new urban landscape space which integrates the multiple functions of commerce, industry, office work, catering, entertainment, vacation and hotel.

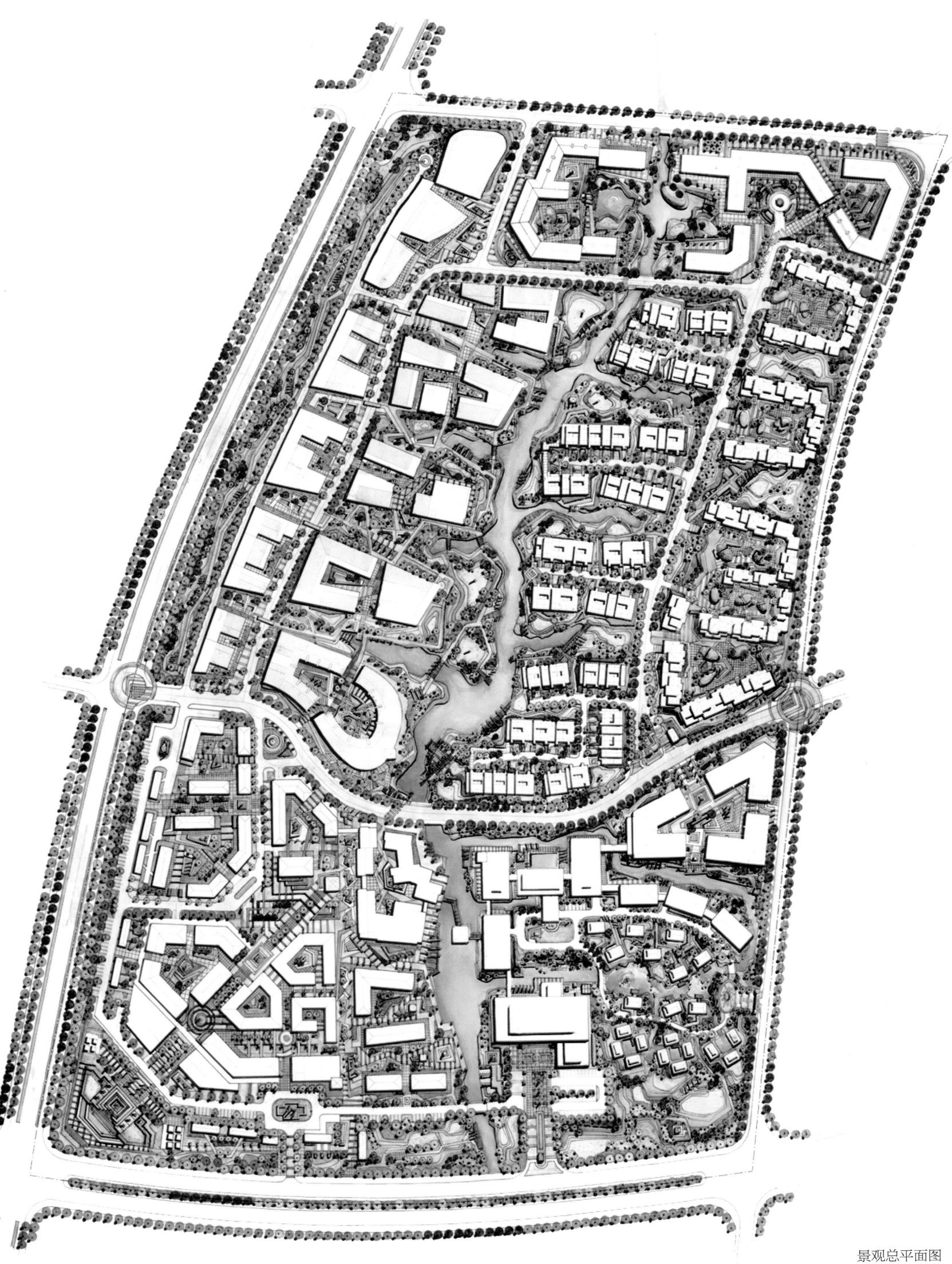

景观总平面图

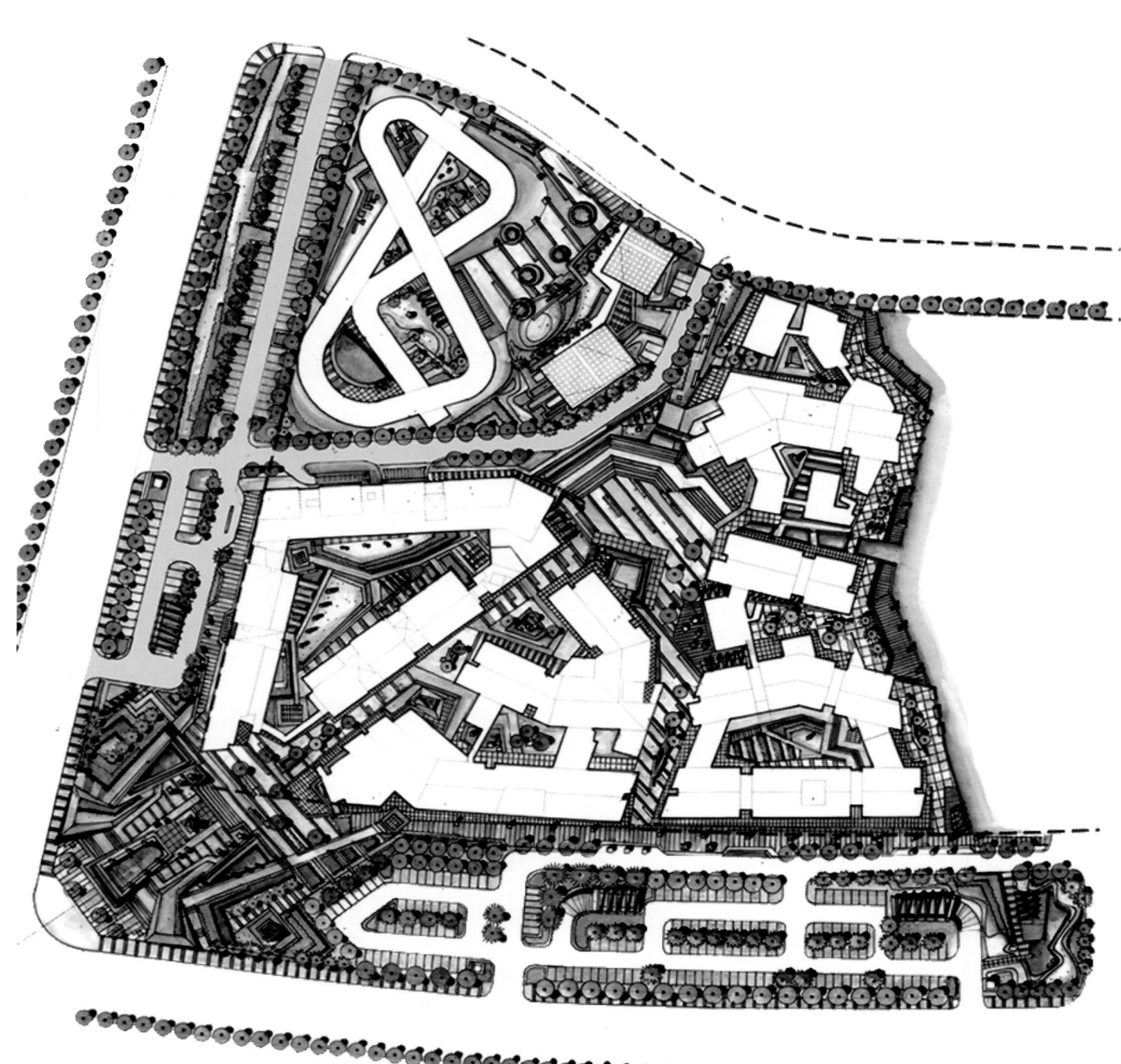

景观总平面图

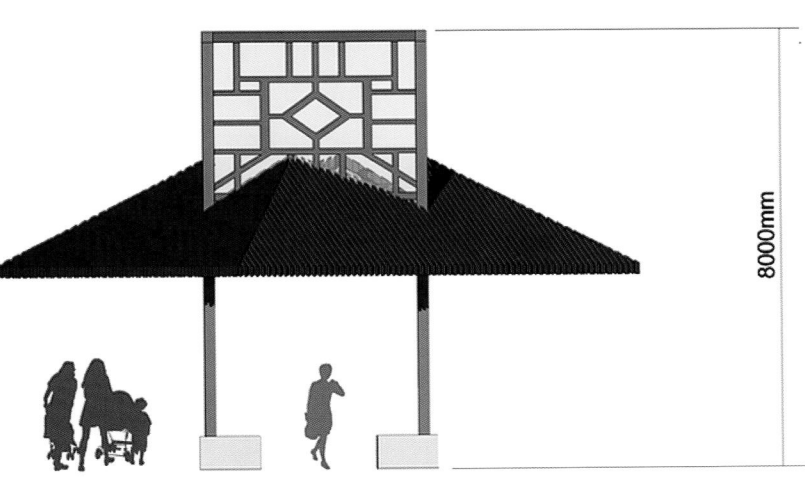

戏台亭窗花元素意向象

戏台亭立面

8000mm

戏台亭剖面

戏台亭透视

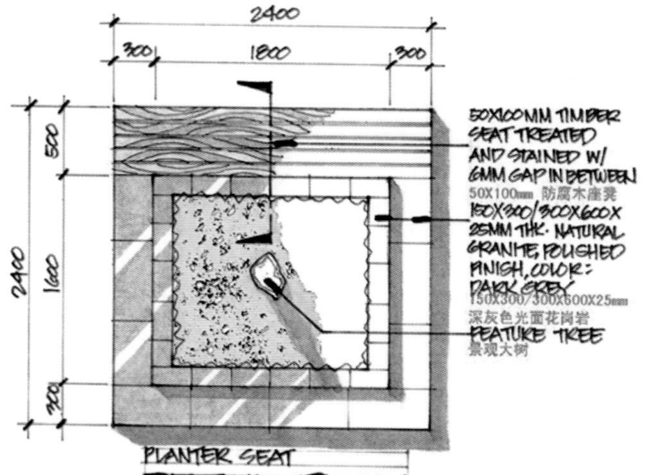

PLANTER SEAT
DETAIL PLAN
SCALE 1:30

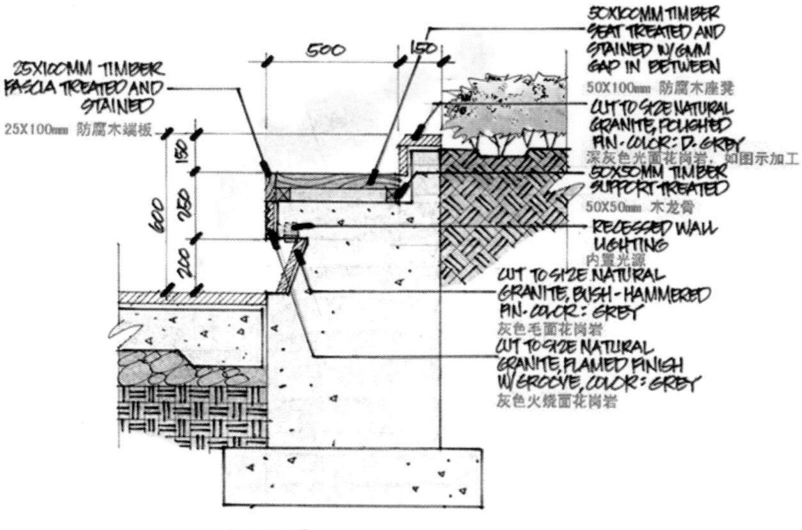

2400
300 1800 300
500 1600 300

50X100MM TIMBER
SEAT TREATED
AND STAINED W/
6MM GAP IN BETWEEN
50X100mm 防腐木座凳
150X300/300X600X
25MM THK. NATURAL
GRANITE, POLISHED
FINISH, COLOR:
DARK GREY
150X300/300X600X25mm
深灰色光面花岗岩
FEATURE TREE
景观大树

DETAIL SECTION
SCALE 1:15

500 150

50X100MM TIMBER
SEAT TREATED AND
STAINED W/6MM
GAP IN BETWEEN
50X100mm 防腐木座凳
CUT TO SIZE NATURAL
GRANITE, POLISHED
FIN. COLOR: D. GREY
深灰色光面花岗岩, 如图示加工
50X50MM TIMBER
SUPPORT TREATED
50X50mm 木龙骨
RECESSED WALL
LIGHTING
内置光源
CUT TO SIZE NATURAL
GRANITE, BUSH-HAMMERED
FIN. COLOR: GREY
灰色毛面花岗岩
CUT TO SIZE NATURAL
GRANITE, FLAMED FINISH
W/ GROOVE, COLOR: GREY
灰色火烧面花岗岩

25X100MM TIMBER
FASCIA TREATED AND
STAINED
25X100mm 防腐木端板

150 250
600 250 200

ELEVATION
SCALE 1:30

600 100 350 150

FEATURE TREE
景观大树
25X100MM TIMBER
FASCIA TREATED AND
STAINED
25X100mm 防腐木端板
CUT TO SIZE NATURAL
GRANITE, POLISHED
FINISH, COLOR: DARK GREY
CUT TO SIZE NATURAL
GRANITE, FLAMED
FINISH W/ GROOVE
COLOR: GREY
灰色开槽火烧面花岗岩
CUT TO SIZE NATURAL
GRANITE, BUSH-
HAMMERED FINISH
COLOR: GREY
灰色蓟毛面花岗岩

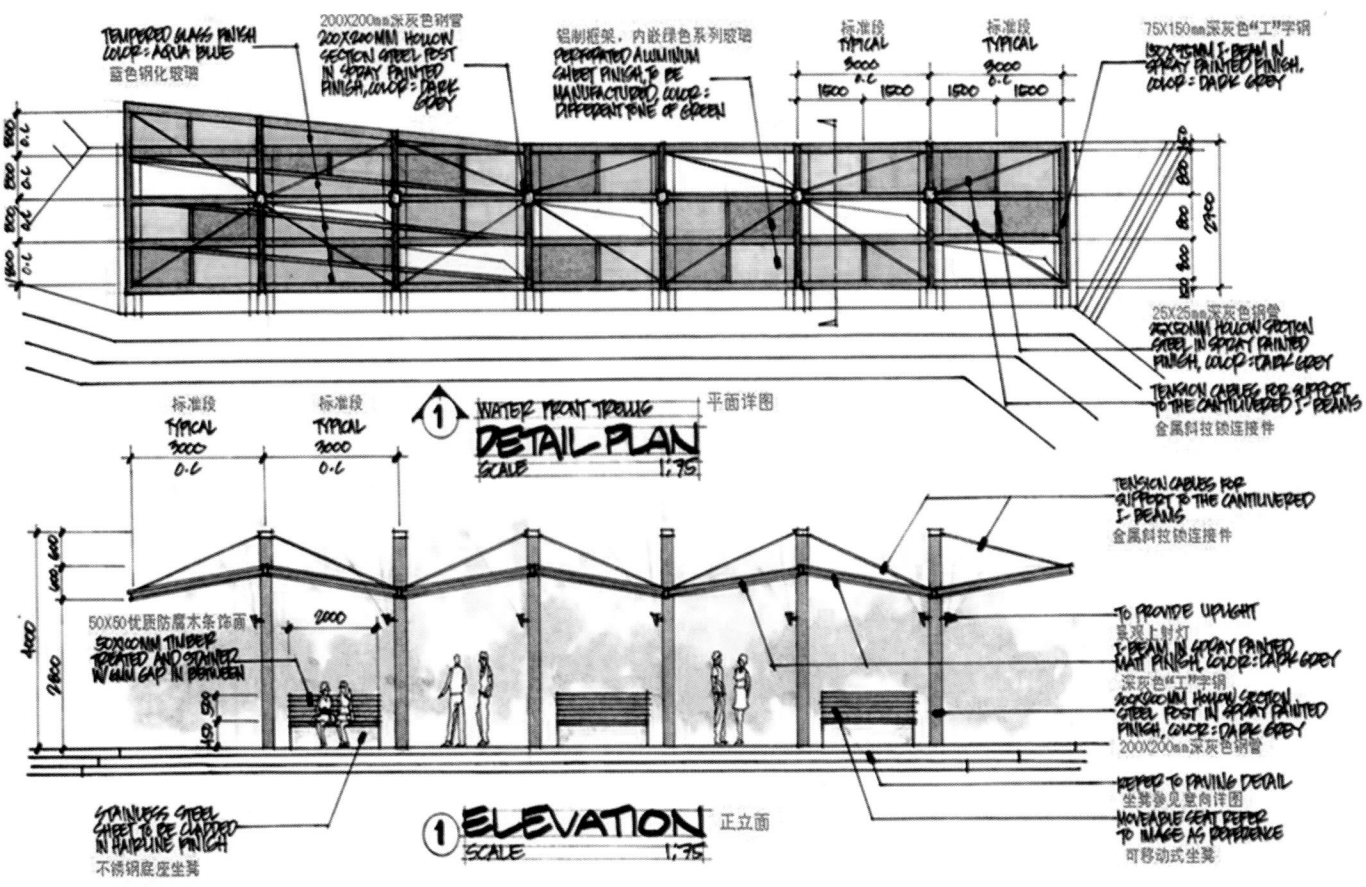

TEMPERED GLASS FINISH
COLOR: AQUA BLUE
蓝色钢化玻璃

200X200MM深灰色钢管
200X200MM HOLLOW
SECTION STEEL POST
IN SPRAY PAINTED
FINISH, COLOR: DARK
GREY

铝制框架, 内嵌绿色系列玻璃
PERFORATED ALUMINUM
SHEET FINISH, TO BE
MANUFACTURED, COLOR:
DIFFERENT TONE OF GREEN

标准段 TYPICAL 3000
1500 1500 标准段 TYPICAL 3000
1500 1500 O.C.

75X150mm深灰色"工"字钢
150X75MM I-BEAM IN
SPRAY PAINTED FINISH,
COLOR: DARK GREY

25X25mm深灰色钢管
25X25MM HOLLOW SECTION
STEEL IN SPRAY PAINTED
FINISH, COLOR: DARK GREY

TENSION CABLES FOR SUPPORT
TO THE CANTILEVERED I-BEAMS
金属斜拉锁连接件

① WATER FRONT TRELLIS 平面详图
DETAIL PLAN
SCALE 1:75

标准段 TYPICAL 3000 O.C.
标准段 TYPICAL 3000 O.C.

50X50优质防腐木条饰面
50X50MM TIMBER
TREATED AND STAINED
W/6MM GAP IN BETWEEN

2000

STAINLESS STEEL
SHEET TO BE CLADDED
IN HAIRLINE FINISH
不锈钢底座凳

① **ELEVATION** 正立面
SCALE 1:75

TENSION CABLES FOR
SUPPORT TO THE CANTILEVERED
I-BEAMS
金属斜拉锁连接件

TO PROVIDE UPLIGHT
草坪上射灯
I-BEAM IN SPRAY PAINTED
MATT FINISH, COLOR: DARK GREY
深灰色"工"字钢
200X200MM HOLLOW SECTION
STEEL POST IN SPRAY PAINTED
FINISH, COLOR: DARK GREY
200X200mm深灰色钢管

REFER TO PAVING DETAIL
坐凳参见意向详图
MOVEABLE SEAT REFER
TO IMAGE AS REFERENCE
可移动式坐凳

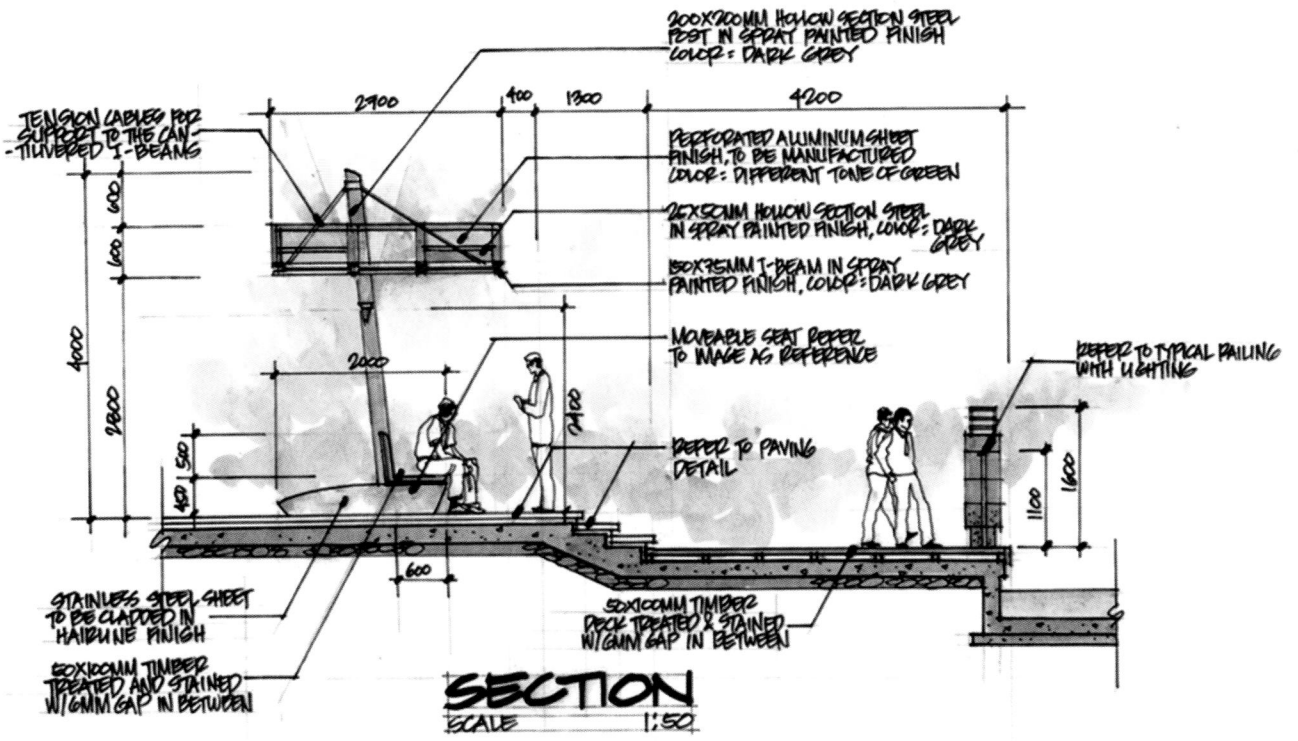

200X200MM HOLLOW SECTION STEEL
POST IN SPRAY PAINTED FINISH
COLOR: DARK GREY

2900 400 1300 4200

TENSION CABLES FOR
SUPPORT TO THE CAN-
-TILVERED I-BEAMS

PERFORATED ALUMINUM SHEET
FINISH, TO BE MANUFACTURED
COLOR: DIFFERENT TONE OF GREEN

25X25MM HOLLOW SECTION STEEL
IN SPRAY PAINTED FINISH, COLOR: DARK
GREY

150X75MM I-BEAM IN SPRAY
PAINTED FINISH, COLOR: DARK GREY

2000

MOVEABLE SEAT REFER
TO IMAGE AS REFERENCE

REFER TO PAVING
DETAIL

REFER TO TYPICAL RAILING
WITH LIGHTING

STAINLESS STEEL SHEET
TO BE CLADDED IN
HAIRLINE FINISH

50X100MM TIMBER
TREATED AND STAINED
W/6MM GAP IN BETWEEN

50X100MM TIMBER
DECK TREATED & STAINED
W/6MM GAP IN BETWEEN

SECTION
SCALE 1:50

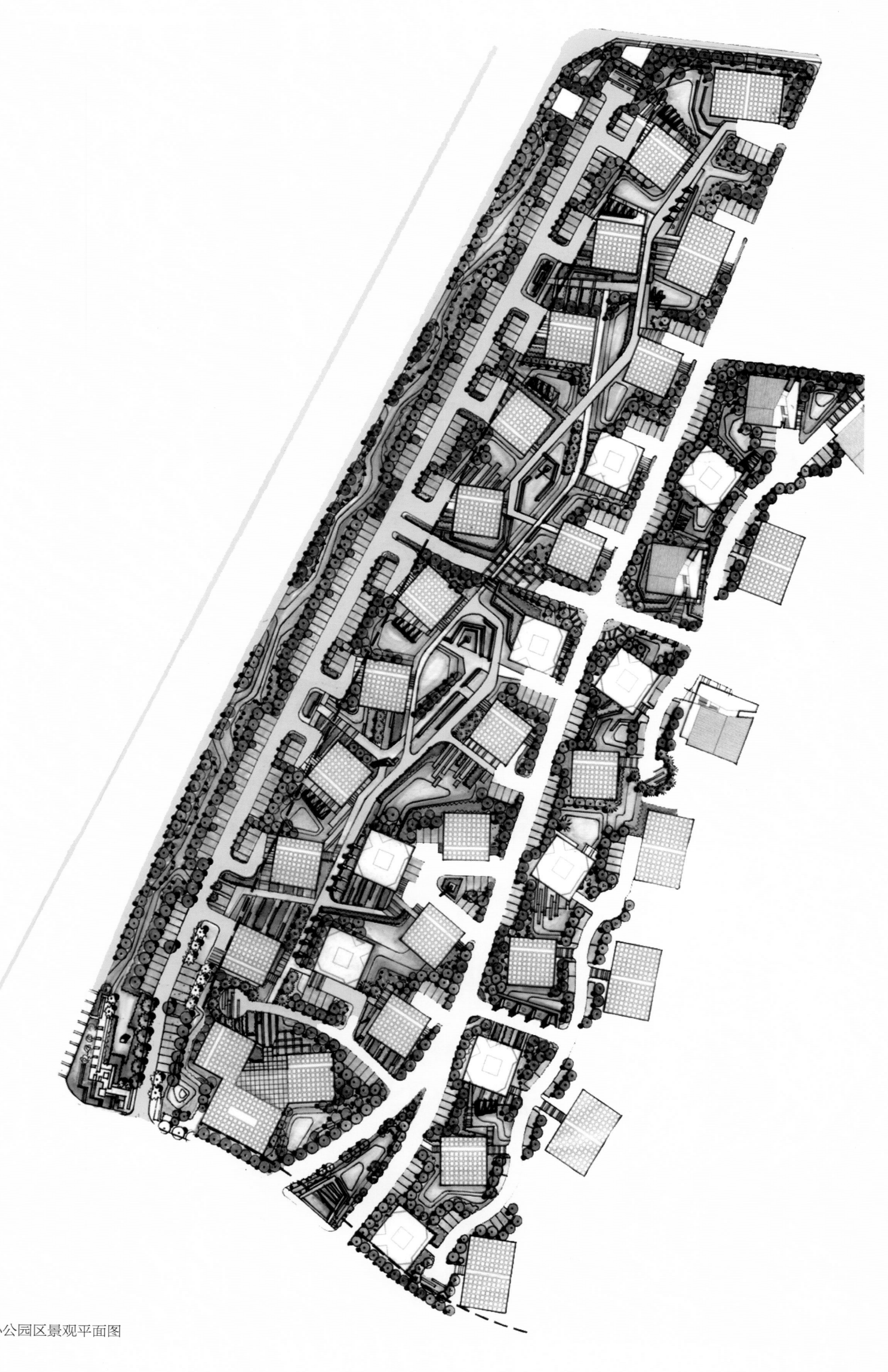

企业办公园区景观平面图

西侧入口景观设计图

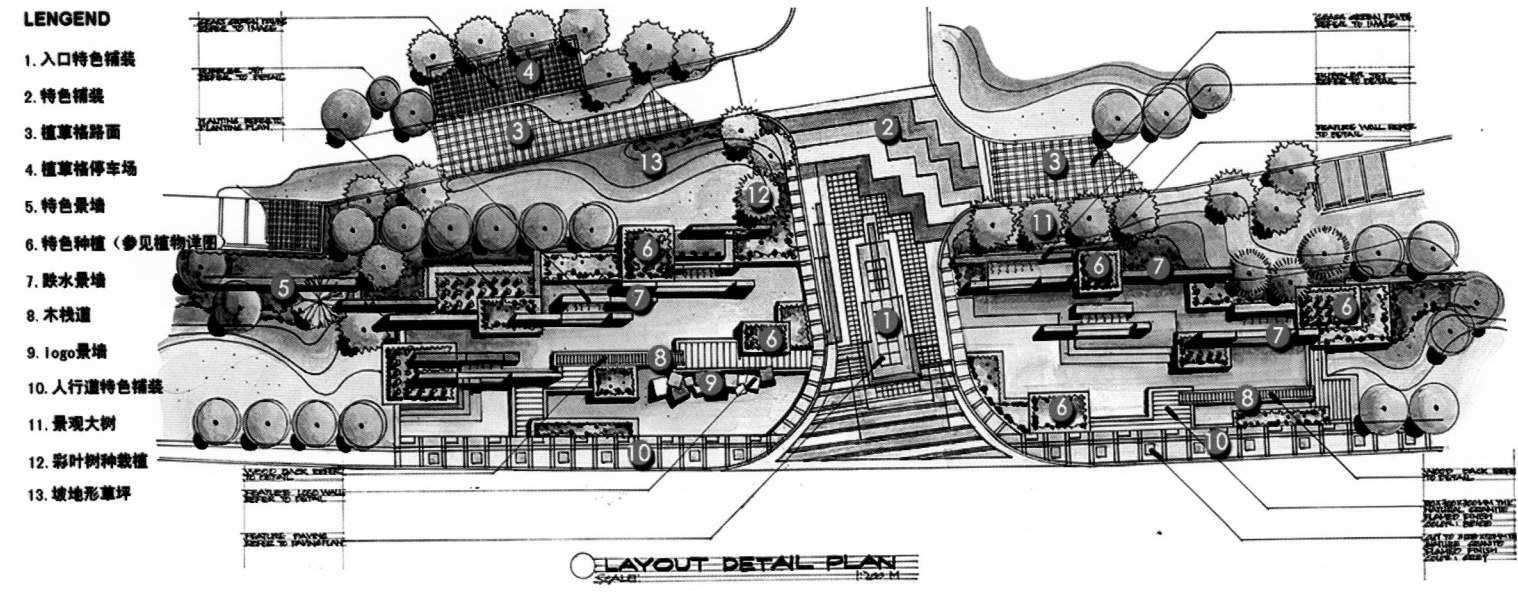

LENGEND

1. 入口特色铺装
2. 特色铺装
3. 植草格路面
4. 植草格停车场
5. 特色景墙
6. 特色种植（参见植物详图）
7. 跌水景墙
8. 木栈道
9. logo景墙
10. 人行道特色铺装
11. 景观大树
12. 彩叶树种栽植
13. 坡地形草坪

LAYOUT DETAIL PLAN

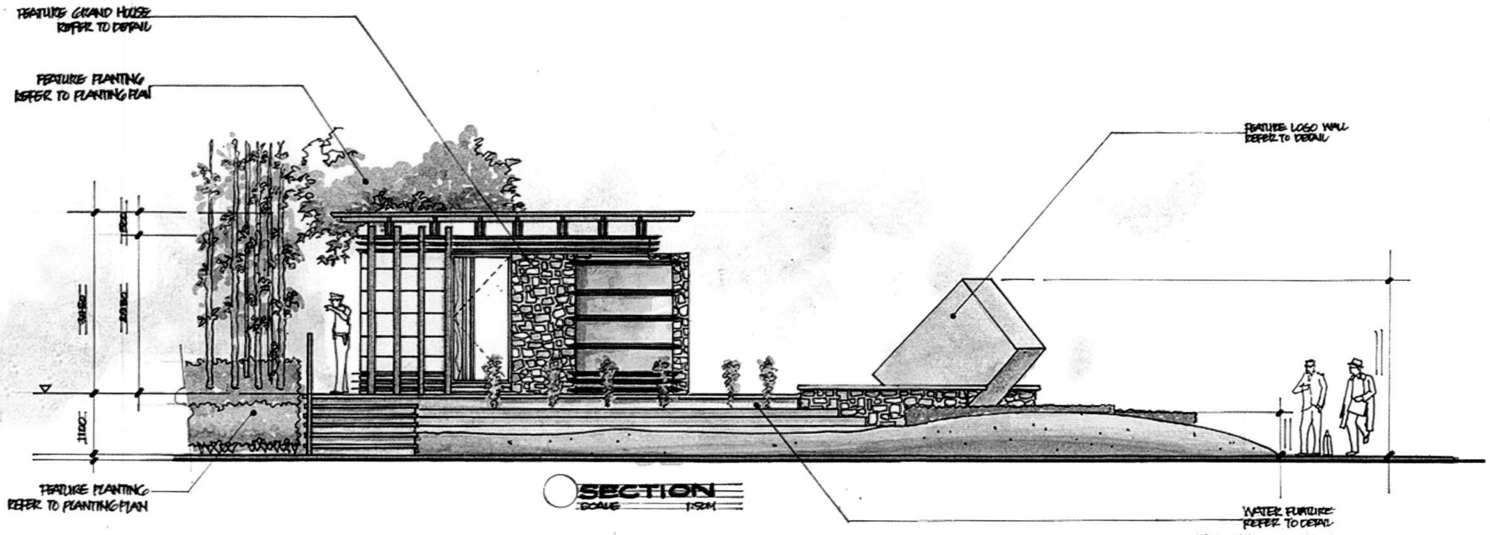

企业办公园区入口景观设计图

商业区设计酒店景观设计图

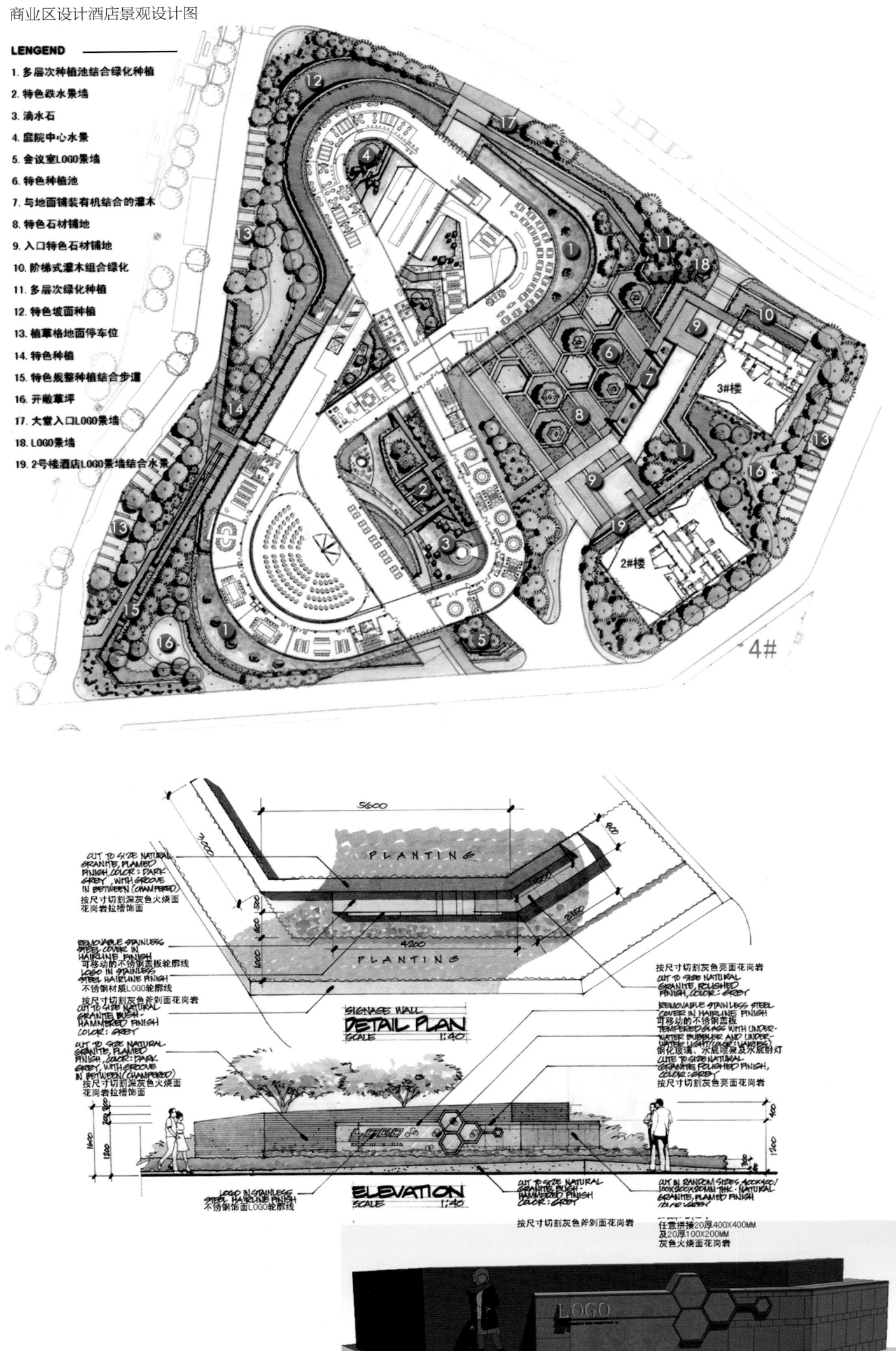

LENGEND

1. 多层次种植池结合绿化种植
2. 特色跌水景墙
3. 滴水石
4. 庭院中心水景
5. 会议室LOGO景墙
6. 特色种植池
7. 与地面铺装有机结合的灌木
8. 特色石材铺地
9. 入口特色石材铺地
10. 阶梯式灌木组合绿化
11. 多层次绿化种植
12. 特色坡面种植
13. 植草格地面停车位
14. 特色种植
15. 特色规整种植结合步道
16. 开敞草坪
17. 大堂入口LOGO景墙
18. LOGO景墙
19. 2号楼酒店LOGO景墙结合水景

4 FLOORS

OPTION 1
@ ELEVATION

FEATURE
WALL (SCULPTURE)

SIGNAGE W/
WATER FEATURE

FEATURE
LANTERN

FEATURE
WALL

SIGNAGE
W/ WATER
FEATURE

FEATURE
SCULPTURE/
WATER FEATURE

@ PROPOSED
LAYOUT PLAN

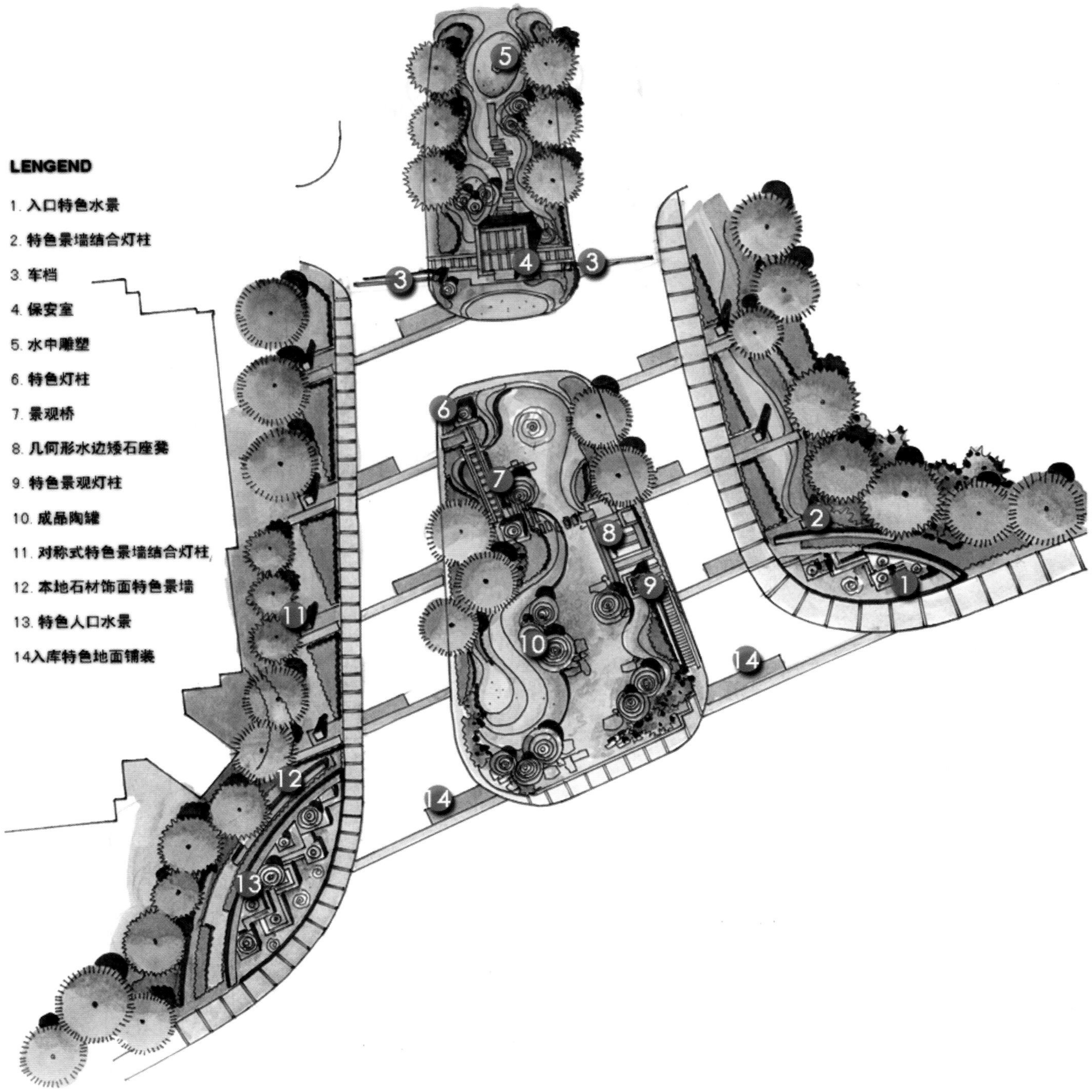
LENGEND
1. 入口特色水景
2. 特色景墙结合灯柱
3. 车档
4. 保安室
5. 水中雕塑
6. 特色灯柱
7. 景观桥
8. 几何形水边矮石座凳
9. 特色景观灯柱
10. 成品陶罐
11. 对称式特色景墙结合灯柱
12. 本地石材饰面特色景墙
13. 特色人口水景
14入库特色地面铺装

ROAD
PA
POND
WOOD DECK
SECTION
SCALE 1:50 M
STEPPING STONE
FEATURE POT
FEATURE BENCH
FEATURE POT W/LIGHTING

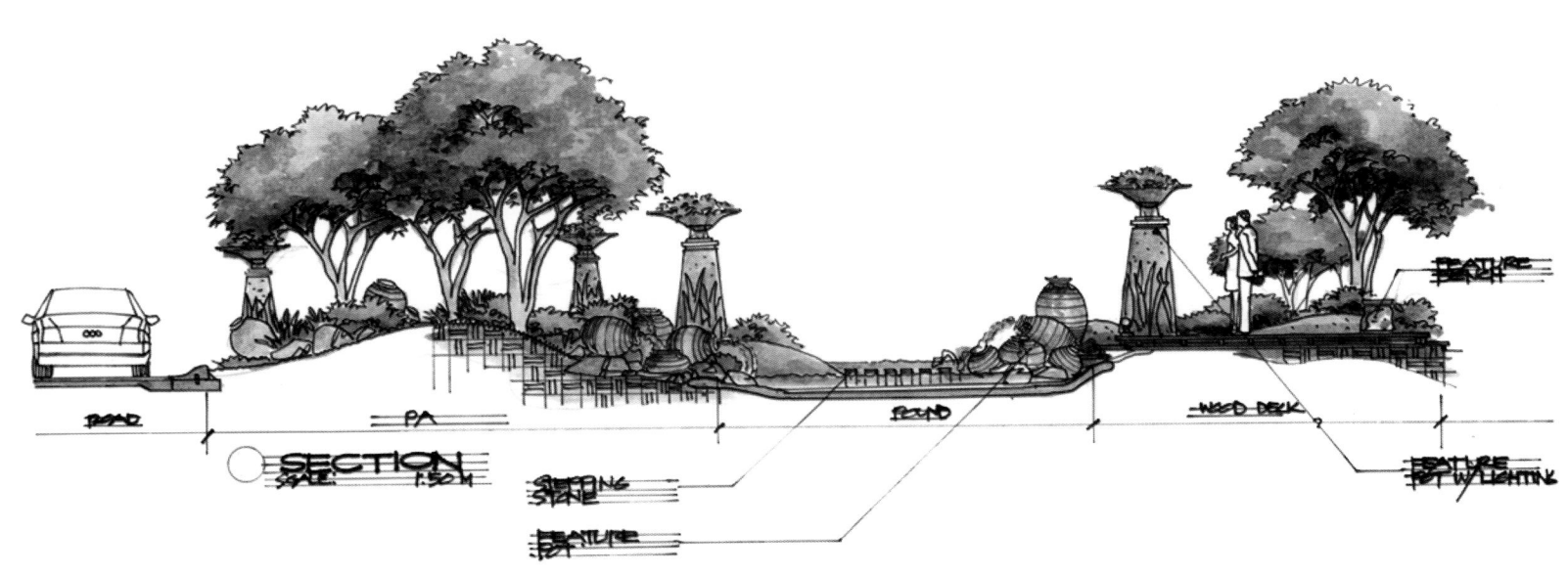

图书在版编目（CIP）数据

十年：设计决定一切 /思纳史密斯集团编. — 天津：
天津大学出版社，2012.7
ISBN 978-7-5618-4246-1

Ⅰ.①十… Ⅱ.①思… Ⅲ.①建筑设计—作品集—
世界—现代 Ⅳ.①TU206

中国版本图书馆CIP数据核字（2011）第282629号

十年 设计决定······
TenYears　Design Determines…

主编：朱轶俊
编委会：张振涛、安庆东、李超、曹洁、王玫、傅威、杨冬冬

出 版 人：	杨欢
责任编辑：	油俊伟
装帧设计：	C_Lucida Art
翻　　译：	李 超、秦 琳、郑先皎
摄　　影：	朱轶俊、曹 洁、谭维琦等

出版发行： 天津大学出版社
天津市卫津路 92 # 天津大学内 (300072)
t. +86-22-27403647（发行部）
　+86-22-27402742（邮购部）
publish.tju.edu.cn

成品尺寸：	240 mm × 330 mm
印　张：	46.75
字　数：	739 千
制　版：	上海瑞时印刷有限公司
印　刷：	上海瑞时印刷有限公司
版　次：	2012年7月第1版
印　次：	2012年7月第1次印刷
书　号：	ISBN 978-7-5618-4246-1
定　价：	698.00元